Ram Charitra Maurya
Inorganic Chemistry

Also of Interest

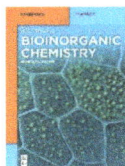

Bioinorganic Chemistry.
Physiological Facets
Ram Charitra Maurya, 2021
ISBN 978-3-11-072729-6, e-ISBN 978-3-11-072730-2

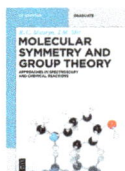

Molecular Symmetry and Group Theory.
Approaches in Spectroscopy and Chemical Reactions
R. C. Maurya, J. M. Mir, 2019
ISBN 978-3-11-063496-9, e-ISBN 978-3-11-063503-4

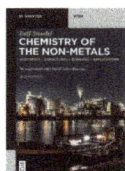

Chemistry of the Non-Metals
Ralf Steudel, 2020
ISBN 978-3-11-057805-8, e-ISBN 978-3-11-057806-5

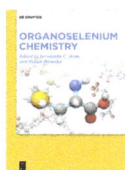

Organoselenium Chemistry
Brindaban C. Ranu, Bubun Banerjee (Eds.), 2020
ISBN 978-3-11-062224-9, e-ISBN 978-3-11-062511-0

Ram Charitra Maurya

Inorganic Chemistry

———

Some New Facets

DE GRUYTER

Author
Prof. Dr. Ram Charitra Maurya
Ph.D., D.Sc, CChem FIC (India), CChem FRSC (UK)
Professor of Inorganic Chemistry
Former Head, Department of Chemistry and Pharmacy, and
Dean, Faculty of Science
Rani Durgavati University
Jabalpur 482001 (M. P.), India
rcmaurya1@gmail.com

ISBN 978-3-11-072725-8
e-ISBN (PDF) 978-3-11-072728-9
e-ISBN (EPUB) 978-3-11-072740-1

Library of Congress Control Number: 2020952731

Bibliographic information published by the Deutsche Nationalbibliothek
The Deutsche Nationalbibliothek lists this publication in the Deutsche Nationalbibliografie;
detailed bibliographic data are available on the Internet at http://dnb.dnb.de.

© 2021 Walter de Gruyter GmbH, Berlin/Boston
Cover image: Piranka/Gettyimages
Typesetting: Integra Software Services Pvt. Ltd.
Printing and binding: CPI books GmbH, Leck

www.degruyter.com

Dedicated to
My wife
Mrs. Usha Rani Maurya,
who has always been a source of inspiration for me
throughout my growth
&
My sons, Ashutosh and Animesh; daughter, Abhilasha; and son-in-law, Adarsh
for
encouragement

Preface

During my long journey of teaching span of more than 40 years in three different universities at B.Tech., M.Sc., M.Phil. and Ph.D. levels, I had always realized the lack of textbooks in Inorganic Chemistry that covers most of the topics for in-depth teaching of the subject matters. Moreover, I was also in quest of reading materials in the form of research papers, reviews and books, so that curriculum at Postgraduate, M.Phil. and Ph.D. coursework levels in any university of India and abroad too can be revised for updating manpowers in Chemistry for manifold applications. The present book, entitled *Inorganic Chemistry: Some New Facets*, incorporating 10 chapters, is designed to meet out the objectives mentioned above. Although many more topics are still not covered in this book only because of looking over the too much bulk of the book, they will be given attention in its next volume.

The book also aims to assist students in preparing for competitive examinations, viz. NET, GATE, SLET and Doctoral Entrance Test (DET) particularly in India. Each chapter ends with different multiple choice, short answer and long answer questions, covering the topics discussed in the chapter to allow an opportunity to the students for their self-evaluation.

In completing a book of this nature, one accumulates gratefulness to the previous authors and editors of books, research papers, reviews and monographs on the relevant topics. I have consulted these sources freely and borrowed their ideas and views with no hesitation in preparing the present manuscript. These sources are acknowledged and listed in bibliography, and I am highly thankful to these authors.

Moreover, the present book is the outcome of my teaching of the subject for more than 40 years to several batches of Masters, M.Phil. and Ph.D. coursework students at the Department of Chemistry, Atarra P.G. College, Atarra (Bundelkhand University, Jhansi, U.P., India) and Rani Durgavati University, Jabalpur (M.P.), India. I have benefited enormously from the response, questions and criticisms of my students.

Looking over the problems of students in learning the subject, I have tried my best to present the subject with a student-friendly approach, that is, expressing it in an interactive manner and in simple language with many illustrative examples. Moreover, the mathematical parts wherever required are given in details to make the subject easily understandable. Therefore, I hope that the book will serve as a text for M.Sc., M.Phil. and Ph.D. coursework students of Chemistry.

My endeavour will be amply rewarded if the book is found helpful to the students and teachers. Despite serious attempts to keep the text free of errors, it would be presumptuous to hope that no error has crept in. I shall be grateful to all those who may care to send their criticism and suggestions for the improvement of the book on my e-mail ID (rcmaurya1@gmail.com).

https://doi.org/10.1515/9783110727289-202

The writing of this book was initiated in September 2018 at 4515 Wavertree Drive, Missouri City, Texas 77459, USA, during our stay with my daughter, Dr. Abhilasha, and son-in-law, Dr. Adarsh, for which they deserve thanks.

Last but not the least, I am thankful to my wife, Mrs. Usha Rani Maurya, for her patient understanding of the ordeal which she had to undergo due to my almost one-sided attention during the completion of this challenging task. I am also indebted to my students for their encouragement and cooperation.

October 2, 2020   Ram Charitra Maurya
B-95, Priyadarshini Colony, Dumna Airport Road,   (Author)
Jabalpur (M.P.), India

Contents

Chapter III
Chemistry of borane and related compounds: structure, bonding and topology —— 71

Chapter X
Some aspects of safe and economical inorganic experiments at UG and PG levels —— 451

Chapter I
Valence shell electron pair repulsion (VSEPR) theory: principles and applications

1.1 Introduction

This theory was first formulated by Sidgwick and Powell (1940) based on the repulsions between electron pairs, known as *valence shell electron pair repulsion* (VSEPR) *theory to* explain molecular shapes and bond angles of molecules of **non-transition elements**. Later on Gillespie and Nyholm (1957) developed an extensive rationale (basis/underlying principle) called VSEPR model of molecular geometry.

According to this theory, the shape of a given species (molecule or ion) depends on the number and nature of electron pairs surrounding the central atom of the species.

1.2 Postulates of VSEPR theory: Sidgwick and Powell

The various postulates of this theory are as follows:
(i)   The unpaired electrons in the valence shell of central atom form bond pairs (bps) with surrounding atoms while paired electrons remain as lone pairs (lps).
(ii) The electron pairs surrounding the central atom repel each other. Consequently, they stay as far apart as possible in space to attain stability.
(iii) The geometry and shape of the molecule depend upon the number of electron pairs (bond pair as well as lone pair) around the central atom.
(iv) The geometrical arrangements of electron pairs with different number of electron pairs around central atom are given in Table 1.1.

1.3 Rules proposed by Gillespie and Nyholm

The following rules have been proposed by **Ronald Gillespie** and **Ronald Sydney Nyholm of University College of London** to explain the shape of a number of polyatomic molecules or ions.

1.3.1 Rule 1. Spatial arrangement of electron pairs around the central atom of a given molecule/ion

The electron already present in the valence shell of the central atom of the given species plus the electron acquired by the central atom as a result of bonding with

https://doi.org/10.1515/9783110727289-001

Table 1.1: Shapes of the various molecules depending upon the number of shared electrons around the central metal atom.

Number of shared electron pairs around central atom	Geometry of the molecule	Shape of the molecule	Examples
2		Linear	BeF_2, $BeCl_2$, MgF_2, $MgCl_2$
3		Triangular planar	BF_3, BCl_3, AlF_3, $AlCl_3$
4		Tetrahedral	CH_4, CCl_4, SiH_4, $SiCl_4$
5		Trigonal bipyramidal	PF_5, PCl_5, SbF_5
6		Octahedral	SF_6, TeF_6

other atoms are called the valence shell electrons. Half of this gives the number of electron pairspresent in the valence shell of the central atom of the given species.

Electron pairs present in the valence shell of the central atom occupy localized orbitals which arrange themselves in space in such a way that they keep apart from one another as for as possible. This gives minimum energy and maximum stability to the species. As there can be only one definite orientation of orbitals corresponding to minimum energy, a molecule or ion of a given substance has a definite shape, that is, a definite geometry.

When the central atom in a molecule is surrounded by bonded electron pair only, the molecule will have a regular geometry or shape. The geometry depends on the number of bonded electron pairs as given in Table 1.1.

1.3.2 Rule 2. Regular and irregular geometry: presence of hybrid orbitals containing bond pairs and lone pairs

If the central atom is surrounded only by orbitals containing shared pair of electrons, that is, bps), and there are no hybrid orbitals containing lone pairs (lps) of electrons in the valence shells, the molecule has a regular geometry.

If, however, the central atom is surrounded by one or more hybrid orbitals containing lone pairs of electrons in the valence shells, the bond angle gets distorted from the value expected for a particular geometry of molecule or ions. Hence, with the change in the magnitude of the bond angles, the shape of the molecule or ions gets distorted.

Cause of distortion in geometry/change in bond angles

The strength of the repulsions between the electron pairs in a valence shell decrease in the order: lone pair (lp)–lone pair (lp) > lone pair (lp)–bond pair (bp) > bond pair (bp)–bond pair (bp). In other words, lone pairs and bond pairs are not equivalent but lone pairs exert a greater effect on bond angles than bond pairs.

This rule can be easily understood because a bonding electron pair is under the influence of two nuclei (two +ve centres) whereas a lone pair is under the influence of only one nucleus. Thus, a lone pair is expected to occupy a broader orbital with a greater electron density radially distributed closer to the central atom than the bond pair electrons, which are drawn out between two positive centres. In other words, the lone pair occupies more space on the surface of the central atom than a bonding pair. Hence, it will repel the electron pair in the neighbouring orbitals more strongly. This is shown diagrammatically in Figure 1.1.

Consequently, the presence of one or more orbitals with lone pairs has the effect of altering the bond angles to a significant extent. The molecule will not retain any regular geometry now.

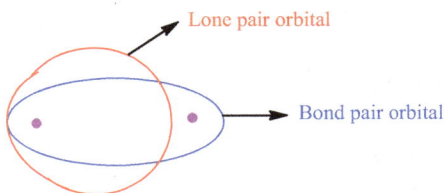

Figure 1.1: Spatial difference between a lone pair and a bond pair.

This rule may be used to explain:
(i) The well-known bond angle decrease in the series CH_4, 109.5° (bp–bp) > NH_3, 107.3° (lp–bp) > H_2O, 104.45° (lp–lp) (Figure 1.2).

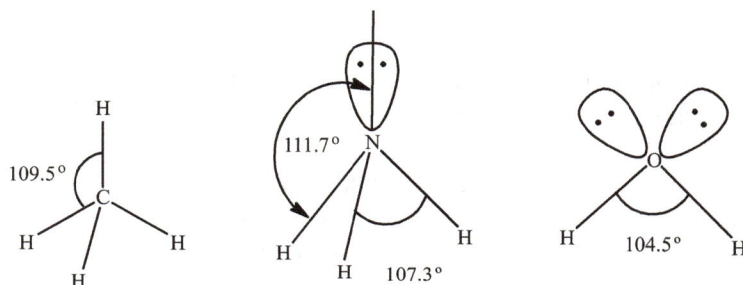

Figure 1.2: Effect of bp–bp, bp–lp and lp–lp on bond angles.

The presence of one lone pair over the central atom nitrogen in ammonia reduces the bond angle from 109.5° to 107.3° while with the presence of two lone pairs over oxygen in water molecule, reduction in bond angle is more compared to methane and ammonia because of the more lp–lp repulsion.
(ii) It is expected that the lone pair will exert a greater effect in contraction of bond angle than a lone electron. The bond angles in the following series (Figure 1.3) substantiate this.

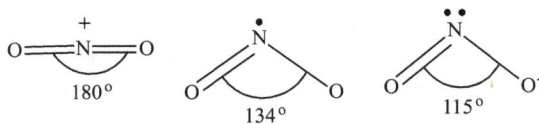

Figure 1.3: Effect of a lone pair, a lone electron and no electron over central atom on bond angles.

The NO_2^+ (2 bps, counting only σ-electrons = 2e pair) ion is isoelectronic with CO_2 (2 bps, counting only σ-electrons) and similarly, it will adopt a linear structure with two π bonds.

The nitrite ion, NO_2^-, contains one π bond, two σ bonds and one lone pair (2 bps + 1 lp, counting only σ-electrons = 3e pair). The resulting structure is,

therefore, expected to be trigonal, with bond angle 120° to a first approxima-
tion. The lone pair should be expected to expand at the expense of the bonding pairs, and the bond angle is found to be 115°.

NO_2 contains an unpaired electron. It may be considered to be a NO_2^- from
which one electron has been removed from the least electronegative atom, N
atom. Instead of having a lone pair on N, it has a single electron in an approximate trigonal orbital. Since a single electron would be expected to repel less
than two electrons, the bonding electron pairs move so as to open up bond
angle (134°) and reduce the repulsion between them.

(iii) This rule offers a reasonable explanation of why the lone pairs occupy equatorial
positions rather than axial in the trigonal bipyramidal (TBR) electron arrangement of AX_4E, AX_3E_2, AX_2E_2 molecules (E = lp), and occupy *trans*-positions rather
than *cis*-positions in the octahedral electron (O_h) arrangement of AX_4E_2 molecules (Figure 1.4). The answers are as follows:

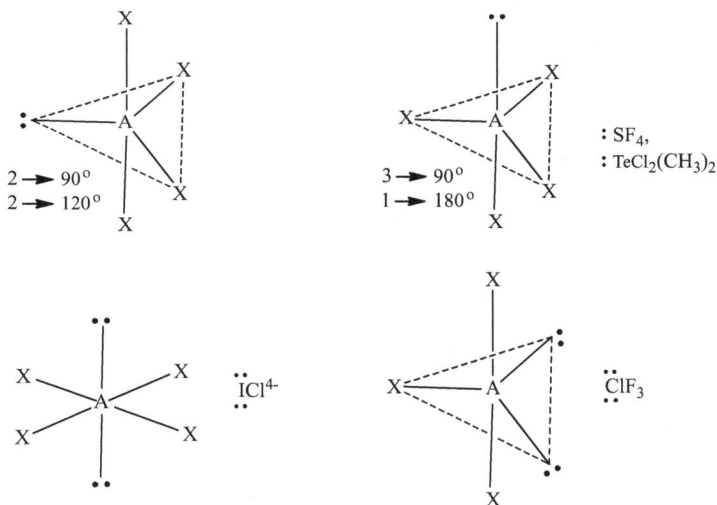

Figure 1.4: Structures of some molecules having 5 and 6 valence shell electron pairs.

(a) In the TBP, a lone pair in an equatorial position has two nearest neighbours
at 90° and two next nearest neighbours at 120°, whereas an axial lone pair
has three nearest neighbours at 90° and one next nearest neighbours at 180°.

(b) If we accept the postulates that the repulsion due to Pauli forces decreases
rapidly with distance, or with angle between electron pairs, then repulsion
will be minimum when lone pairs occupy equatorial positions rather than
axial one.

(c) In O_h case, lone pairs at axial positions will be at 180 (i.e. maximum far
apart), and thus feel minimum repulsion.

1.3.3 Rule 3. Effect of electronegativity: repulsions exerted by bond pairs decrease as the electronegativity of the bonded atom increases

The high electronegativity of the bonded atom (with the central atom) pulls the σ-bonded electron pair away from the central atom nucleus, thereby contracting and thinning out the orbital.

Hence, more the electronegative of the bonded atom, more the electron density is displaced towards it in the σ-bond. This reduces repulsion between the bond pairs, and hence allows the lone pair(s) to expand more.

Both the factors causes decrease in the bond angles. The reduction in bond angle with increase in electronegativity may be viewed in the following examples:

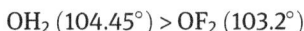

$$NH_3\ (107.3°) > NF_3\ (102°)$$

$$OH_2\ (104.45°) > OF_2\ (103.2°)$$

$$PI_3\ (102°) > PBr_3\ (101.5°) > PCl_3\ (100°)$$

$$AsI_3\ (101°) > AsBr_3\ (100.5°) > AsCl_3\ (98.4°)$$

1.3.4 Rule 4. Multiple bonds exert a stronger repulsion

Multiple bonds are composed of two or three electron pair domains (space of influence) so that together they take up more angular space than a bonding pair domain. In other words, bond angles decrease in the order multiple bond–multiple bond > multiple bond–single bond > single bond–single bond.

As the multiple bonds exert a stronger repulsion than single bonds, the molecule adjusts its geometry to reflect it. The following molecules (Figure 1.5) illustrate specific application of this rule.

1.3.5 Rule 5. Repulsions between electron pairs in filled shells are larger than those between electron pairs in incomplete shells

In a filled shell, the orbitals effectively fill all of the available space around an atom. Because of the maximum repulsion, anything that attempts to reduce an angle between such filled orbitals will be strongly resisted by Pauli repulsion forces, preventing appreciable orbital overlap.

The application of this rule is made in the following examples:

(i)   The valence shell (2s 2p, with capacity of $2 + 6 = 8e$) of the first row atoms, **Li** to Ne, are completely filled by four electron pairs in their binary compounds. Therefore, for molecules, AX_4, AX_3E, AX_2E_2 $(E = lp)$, where both A and X are

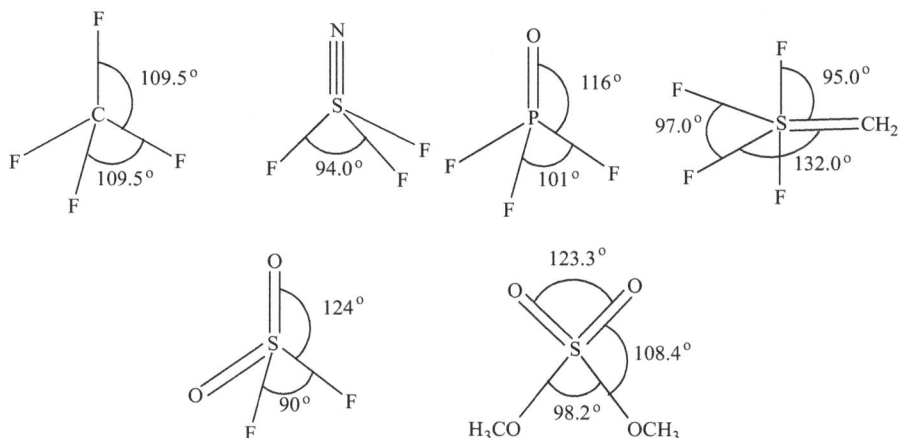

Figure 1.5: Bond angles in some multiple bonded molecules showing that bond angle decreases in the order: multiple–multiple bond > multiple–single bond > single–single bond.

first row atoms, the bond angles will not deviate very greatly from 109.5°. In fact, they are always within several degrees of this value.

The largest deviation is found to occur in NF_3 which has an angle of 102.1°, a deviation of 7.4°. The high electronegativity of F is responsible for this (**Rule 3**).

(ii) The large difference in bond angles of H_2O and H_2S: OH_2 (104.4°) $\gg$ SH_2 (92.2°) may be explained using this rule. The central atoms (O, S) in these molecules have eight electrons in their valence shells (6e of O/S + 2e of 2H's). Thus, the valence shell of O atom in H_2O which contains four orbitals (one 2s and three 2p orbitals, with maximum capacity of 8e) is completely filled while that of S in H_2S, which contains nine orbitals (one 3S, three 3p and five 3d with maximum capacity of 18e) is incompletely filled.

So, due to decreased repulsion between electron pairs in incomplete shell of S, bond angles are smaller in H_2S compared to bond angles in H_2O wherein filled orbitals have greater repulsion.

A similar explanation may be given to justify the following observed bond angle decreases:

(a) OH_2 (104.5°) > SH_2 (92.2°) > SeH_2 (91°) > TeH_2 (89.5°)
(b) $OH (CH_3)$ (109°) $\gg$ $SH (CH_3)(100°)$
(c) NH_3 (107.3°) $\gg$ PH_3 (93.3°) > AsH_3 (91.8°) > SbH_3 (91.3°)
(d) $N (CH_3)_3$ (109°) $\gg$ $P (CH_3)_3$ (102.5°) > $As (CH_3)_3$ (96°)

1.4 Exceptions to the VSEPR model, ligand–ligand repulsion and the ligand close packing model

Over the years, some failures of the VSEPR model to give a correct qualitative explanation of the deviations of bond angles from the polyhedral angles of 90°, 109.5° and 120° have been used as a basis for claims that the model does not have a sound physical basis, but later on, it was shown by Gillespie that the model is, indeed, soundly based on the Pauli principle. Nevertheless, there are a significant number of exceptions to the VSEPR model.

It has been realized in recent year that exceptions to the VSEPR model are primarily due to the neglect of ligand–ligand repulsions by the VSEPR model. The importance of neighbouring ligand interactions in determining geometry and reaction rates has been recognized for many years but it was not generally appreciated that the interactions between geminal ligands can also be of importance.

The importance of ligand repulsions was finally realized by Gillespie (2008) when his research group noticed that ligand–ligand distances in a large number of molecular fluorides of the elements of second period are essentially constant for a given central atom, independent of the coordination number and the presence of other ligands. These results suggested strongly that the ligands in these molecules may be regarded as close packed. In other words, ligand–ligand repulsions must be of considerable importance in determining the geometry of these molecules and probably other molecules too. Moreover, his research group also studied ligand–ligand distances in molecules with other ligands such as Cl, Br and CH_3 and found that inter ligand distances for these ligands bonded to a second period central atom are also essentially constant.

The recognition of the importance of ligand–ligand repulsions enables some of the failures of the VSEPR model for the main group elements to be easily explained.

Examples

(i) Though chlorine is more electronegative than hydrogen, the ClOCl bond angle in Cl_2O (111.2°) is larger than the HOH angle (104.5°) in H_2O (Figure 1.6) because the large size of the Cl ligand prevents the Cl ligands from being pushed any closer together by the lone pair. Indeed, the ClOCl bond angle is larger than the tetrahedral angle which cannot be explained by the VSEPR model alone.

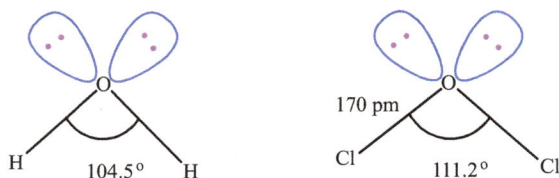

Figure 1.6: Large variation in bond angle of H_2O and Cl_2O.

(ii) The bond angle in PH_3 (93.8°) is significantly smaller than the bond angles in both PF_3 (97.8°) and in PCl_3 (100.3°) (Figure 1.7) despite the fact that F and Cl are both more electronegative than H. The reason for these deviations from the VSEPR rules is now obvious. The hydrogen is much smaller than any other ligand with a ligand radius decreasing from 102 pm on boron to 82 pm on nitrogen so that $H \cdots H$ repulsions only become important at smaller ligand–ligand distances and smaller bond angles than in comparable molecules with other ligands. Because of the large size of the Cl ligands the lone pairs in Cl_2O are unable to push the ligands closer together than 111.2° and similarly the lone pair in PCl_3 (100°) and PF_3 (98°) cannot push these ligands as close together as in PH_3 (94°).

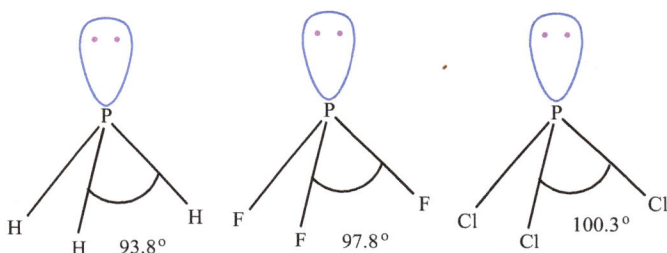

Figure 1.7: Variation in bond angle of PH_3, PF_3 and PCl_3 in spite of single lone pair on central atom in each.

The early success of the electronegativity rule in fact obscured the importance of ligand–ligand repulsions in many molecules. For example, the bond angle increasing in the series PF_3 (98°), PCl_3 (100°), PBr_3 (101°) (Figure 1.8) is consistent with the decreasing electronegativity of the ligand, but it is also consistent with the increasing size of the ligand.

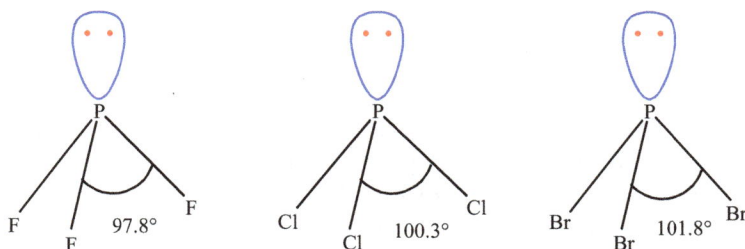

Figure 1.8: Effect of increasing size of ligand on bond angle in PH_3, PCl_3 and PBr_3.

Overall, ligand size explains those bond angles that are not consistent with the electronegativity rule, but also those that are consistent with the rule. So, it is reasonable to replace the electronegativity rule of the original VSEPR model with the rule that bond angles increase with ligand size. Since bond angles can be predicted from the ligand radii in close-packed AX_2E_2 and AX_3E molecules where the bonds are all the same length, and can be calculated if the bond lengths in $AXYE_2$ and $AXYZE$ molecules are known, the LCP model adds a semi-quantitative aspect to the VSEPR model.

1.5 Applications of VSEPR theory

The rules of VSEPR theory discussed earlier can be applied to know the preferred/actual structure of various molecules and ions as illustrated as follows:

(i) Structure of sulphur tetrafluoride, SF₄

The molecule contains 10e [6e (S) + 4e (F)] in the valence shell of S, four **bps** and one **lp**. In order to let each pair of electrons as much room (space) as possible, the appropriate geometry will be a TBP. In this geometry, two possible arrangements of the **bps** and **lp** are as shown in the structures (I) and (II) (Figure 1.9). In structure (I), the **lp** occupies one of the equatorial positions while in Structure (II) one of the axial positions is occupied by **lp**.

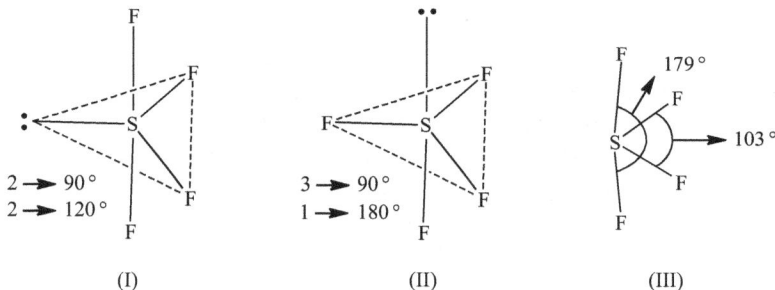

Figure 1.9: Possible and actual structures of SF₄.

In structure (I), the lp is facing two bps at 90° and two bps at 120° while in structure (II), the lp is facing three bps at 90° and one bps at 180°. In this situation, structure (I) feels minimum repulsion compared to structure (II), and hence, structure (II) is preferred. Consequently, the lp in structure (I) affects the ∠F-S-F (axial) to reduce to 179° and the ∠F-S-F (equatorial) to 103°. Structure (III) (Figure 1.9) is the experimentally determined structure of SF₄.

(ii) Structure of bromine trifluoride, BrF$_3$

The molecule contains 10e [7e(Br) + 3e (F)] in the valence shell of Br, three **bps** and two **lps**. Again, the approximate structure is TBP with the two lone pairs occupying either at equatorial positions (I) or at axial positions (II) (Figure 1.10).

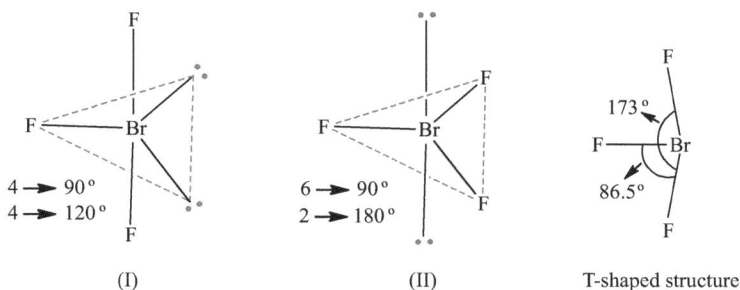

Figure 1.10: Possible and preferred structures of BrF$_3$.

Obviously, the structure (I) feels minimum repulsion compared to (II). The distortion from lone pair repulsion causes the axial F atoms to be bent away from a linear arrangement so that the molecule is a slightly 'bent T-shaped' with bond angle of 85.5°.

Similarly, the T-shaped structure of ClF$_3$ with bond angle 87.5° can be explained.

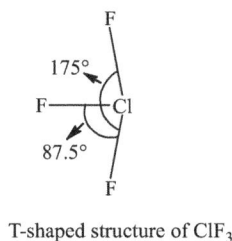

T-shaped structure of ClF$_3$

(iii) Structure of dichloroiodate(I) anion, ICl$_2^-$

The ICl$_2^-$ contains 10e [7e (I) + 2e (Cl) + 1 negative charge] in the valence shell of iodine (I), two bps and three lps. Again, the approximate geometry of the anion by the VSEPR theory is TBP. So, the two possible arrangements of bps and lps are as shown in structures (I) and (II) (Figure 1.11), and the structure (I) will have minimum repulsion. The effect of three lps at the equatorial positions in structure (I) on the **Cl–I–Cl** (axial) bond angle will be zero. Hence, the anion will adopt a linear structure as shown in structure (III) (Figure 1.11).

(iv) Structure of tetrachoroiodate(III) anion, ICl$_4^-$

The ICl$_4^-$ contains 12e [7e (I) + 4e (Cl) + 1 negative charge] in the valence shell of I, 4bps and 2lps. The approximate geometry of the anion should be octahedral. The

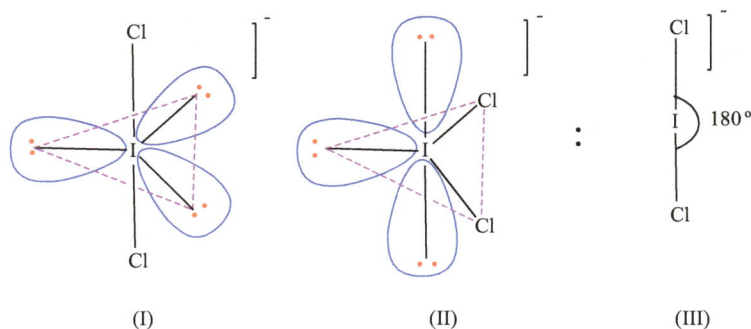

Figure 1.11: Possible and actual structures of ICl_2.

two possible structures (I) and (II) (Figure 1.12) may be taken. The structure (II) is acceptable because the anion will feel minimum repulsion when the lps are trans to each other. The presence of two lone pairs opposite to each other at axial positions, however, causes no change in Cl–I–Cl bond angles. Hence, the anion will have a square planar structure (III) (Figure 1.12).

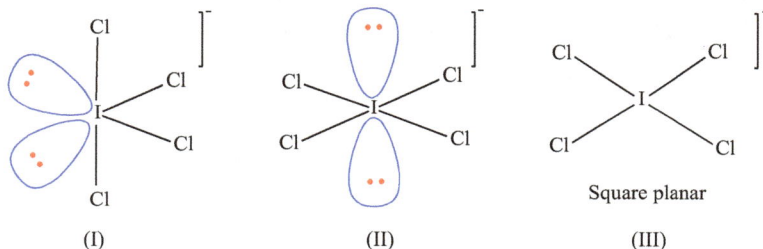

Figure 1.12: Possible and actual structures of ICl_4.

(v) Structure of pentafluorotellurate(IV) anion⁻, TeF_5^-

Inthe TeF_5^-, the tellurium atom has 12e [6e (Te) + 5e (Cl) + 1 negative charge] in the valence shell, five bond pairs and one lone pair. The most stable arrangement for six pairs of electrons is the octahedral as shown in the structure (I) (Figure 1.13). It is notable here that the lone pair is occupying one of the axial positions in order to have minimum repulsion. The presence of the single lone pair of electrons at one of the axial positions causes adjacent F atoms (4 F atoms of the equatorial plane) to move upward somewhat. The resulting structure is, therefore, a square pyramid with the Te atom below the plane of four F atoms (II) (Figure 1.13).

(vi) Structures/shapes of noble gas compounds

The VSEPR theory adequately predicts the shapes of noble gas compounds.

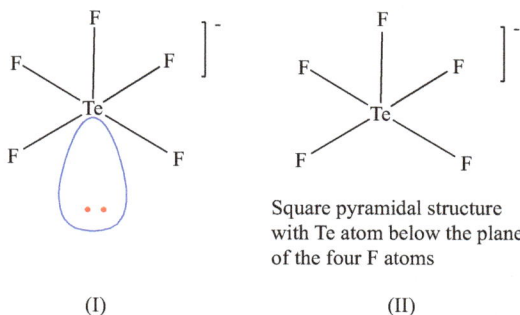

Square pyramidal structure
with Te atom below the plane
of the four F atoms

(I)   (II)

Figure 1.13: Possible octahedral and actual square pyramidal structures of TeF$_5$.

(a) XeF$_2$

In xenon difluoride (XeF$_2$), Xe has (8 + 2), that is, 10 electrons (2bps + 3lps) in the valence shell. These are accommodated in five sp^{3}d hybrid TBP orbitals. The three lps occupy equatorial positions to minimize lone pair repulsion giving linear F–Xe–F bonds (Figure 1.14).

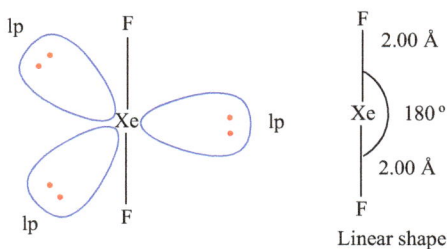

Figure 1.14: Possible octahedral and actual linear structure of XeF$_2$.

(b) XeF$_4$

In XeF$_4$, the 12 valence shell electrons (4 bps + 2 lps) around Xe are in six sp^{3}d^2 hybrids octahedral orbitals. The two lps occupy two axial (trans) positions giving a square planar structure (Figure 1.15) for XeF$_4$.

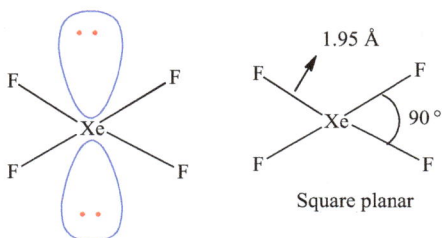

Figure 1.15: Possible octahedral and actual square planar structure of XeF$_4$.

(c) XeF$_6$

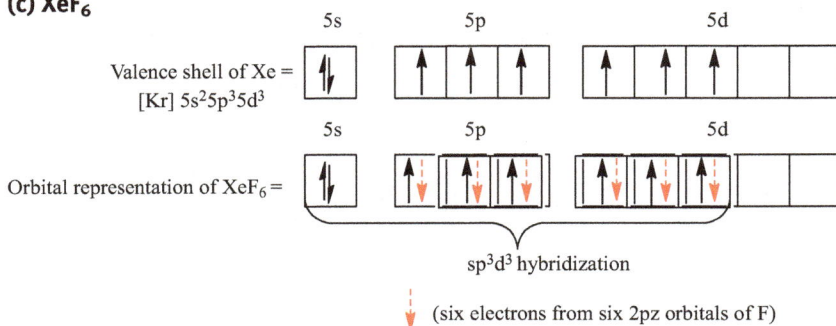

Valence shell of Xe = [Kr] $5s^2 5p^3 5d^3$

Orbital representation of XeF$_6$ =

sp^{3}d^3 hybridization

↓ (six electrons from six 2pz orbitals of F)

The shape of XeF$_6$, however, can not be obtained from VSEPR theory. The molecule is octahedral with a distorted triangular face to accommodate the lp at the centre (Figure 1.16).

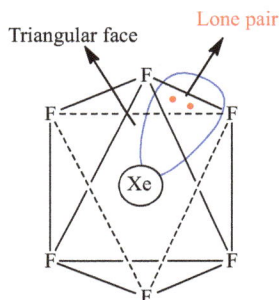

Figure 1.16: Octahedral structure of XeF$_6$ with distorted triangular face to accommodate the lone pair.

(d) IF$_2^-$, IF$_4^-$ and IF$_6^-$

IF$_2^-$ (10e), IF$_4^-$ (12e) and IF$_6^-$ (14e) are isoelectronic with XeF$_2$ (10e), XeF$_4$ (12e) and XeF$_6$ (14e), respectively, and have the same shape.

(e) XeO$_3$

In XeO$_3$, considering only σ-bonds, Xe has eight electrons (3 bps + 1 lp) in the valence shell. These are accommodated in four sp^3 hybrid tetrahedral orbitals (hybridization scheme given later) with one position occupied by a lp giving pyramidal shape to the molecule as shown in Figure 1.17.

(f) XeO$_4$

In XeO$_4$, Xe has eight electrons (4 bps) in the valence shell taking into account only σ-bonds. These are accommodated in four sp^3 hybrid orbitals. With four bps and four π-bonds, the shape of the molecule is tetrahedral as shown in Figure 1.18.

5s   5p   5d

Valence state of Xe ⟶
[Kr] $5s^2 5p^3 5d^3$

sp^3 hybridization

lp   bp   bp   bp

Orbital representation of XeO_3 ⟶

σ   σ   π   π   π

$O_8 = 1s^2\, 2s^2\, 2p_x^2\, 2p_y^1\, 2p_z^1$ ⟶

$2p_y$ $2p_z$ $2p_y$ $2p_z$ $2p_y$ $2p_y$

O   O   O

Xe   or   Xe

O ‖ O   O ‖ O
   O   O

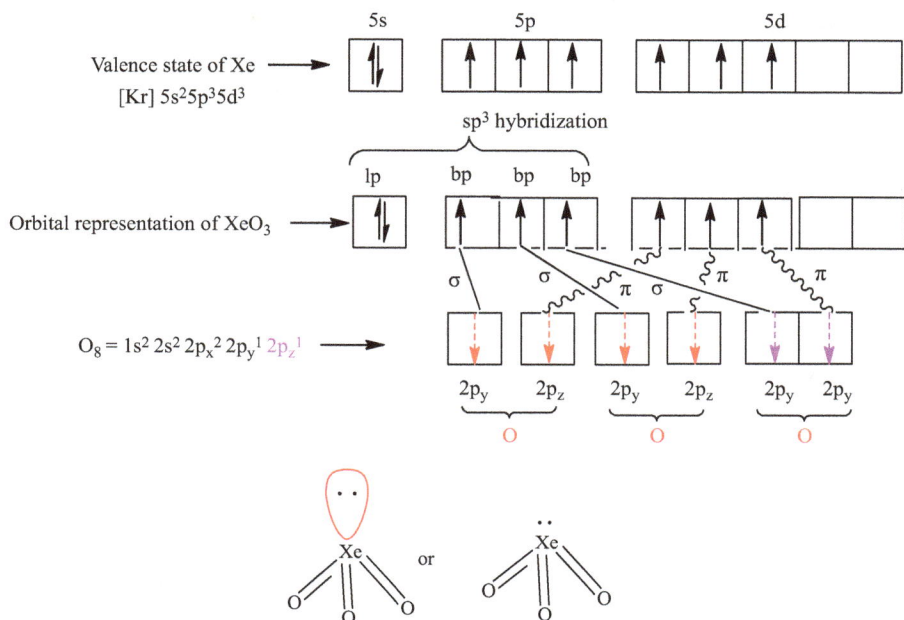

Figure 1.17: Pyramidal structure of XeO_3.

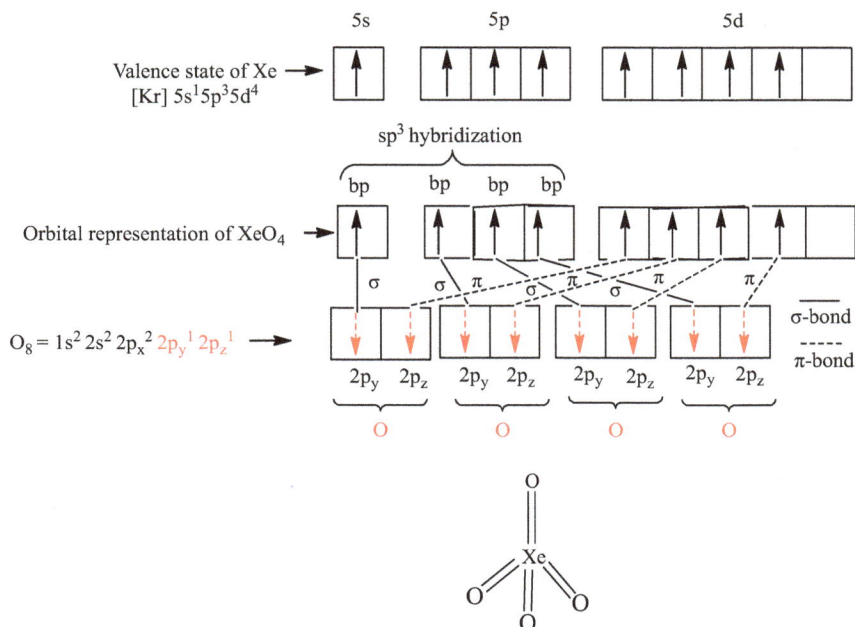

5s   5p   5d

Valence state of Xe ⟶
[Kr] $5s^1 5p^3 5d^4$

sp^3 hybridization

bp   bp   bp   bp

Orbital representation of XeO_4 ⟶

σ   σ   π   σ   π   σ   π   π

$O_8 = 1s^2\, 2s^2\, 2p_x^2\, 2p_y^1\, 2p_z^1$ ⟶

σ-bond
---- π-bond

$2p_y$ $2p_z$ $2p_y$ $2p_z$ $2p_y$ $2p_z$ $2p_y$ $2p_z$

O   O   O   O

O
‖
Xe
O ⁄   ⁄ O
   O

Figure 1.18: Tetrahedral structure of XeO_4.

(g) $[XeO_6]^{4-}$

In the present case, Xe has eight electrons (4 bps) in the valence shell taking into account only σ-bonds. With six bps and two π-bonds, the shape of the molecule is octahedral as shown in Figure 1.19. The four unused electrons in the four p_y orbitals justify four negativechargesover the anion (**see the sp^3d^2 hybridization scheme shown later**).

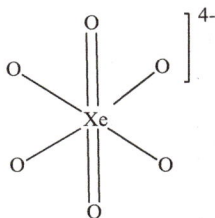

Figure 1.19: Octahedral structure of $[XeO_6]^{4-}$ with six σ and two π-bonds.

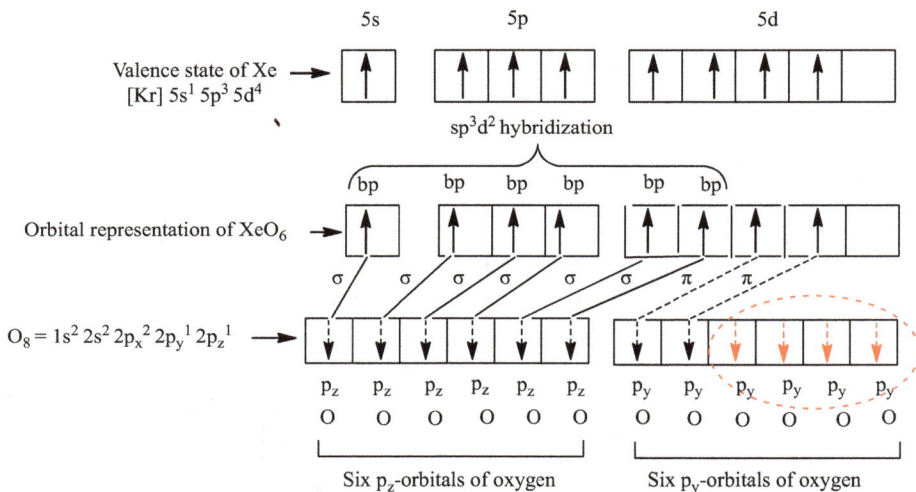

sp^3d^2 hybridization scheme in $[XeO_6]^{4-}$

(h) $XeOF_4$

In $XeOF_4$, Xe has 12 electrons (5 bps + 1 lp) in the valence shell considering only σ-bonds. These are accommodated in six sp^3d^2 hybrid octahedral orbitals with one of the axial positions occupied by a lp. With five bps, a lp and a π-bonds, the shape of the molecule is octahedral as shown in Figure 1.20 leading to square pyramidal shape of the molecule.

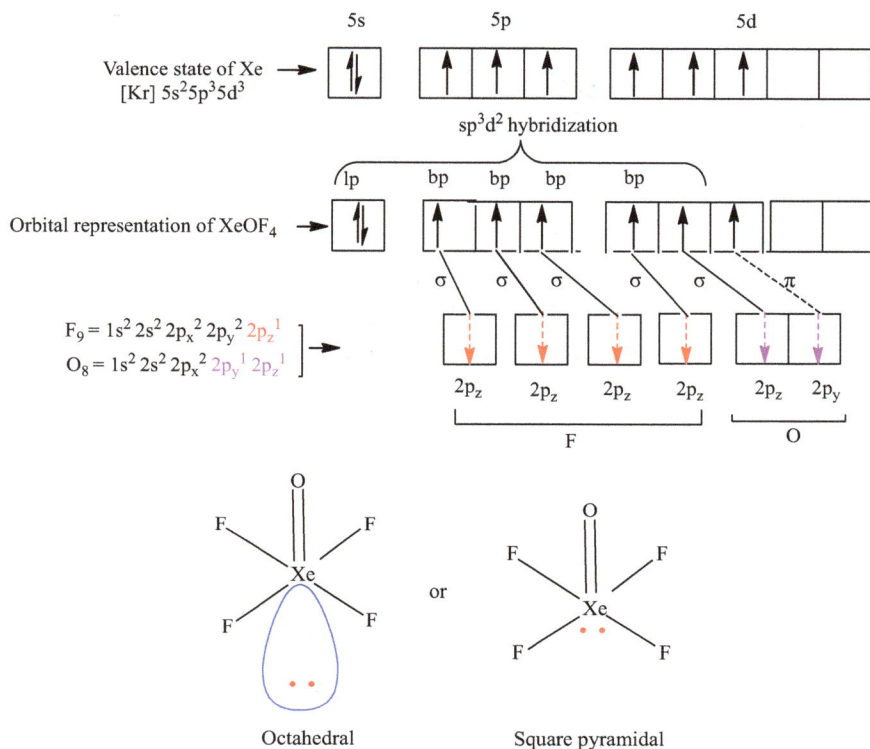

Figure 1.20: Square pyramidal structure of $XeOF_4$.

(i) XeO_2F_2

In XeO_2F_2, X has 10 electrons (4 bps + 1 lp) in the valence shell taking into account only sigma bonds. These are accommodated in five sp^3d hybrid TBP orbitals with one of the equatorial positions occupied by a lp, giving 'see-saw' shape to XeO_2F_2 (Figure 1.21). As evidenced from the orbital diagram, two π-bonds are also present in the equatorial position.

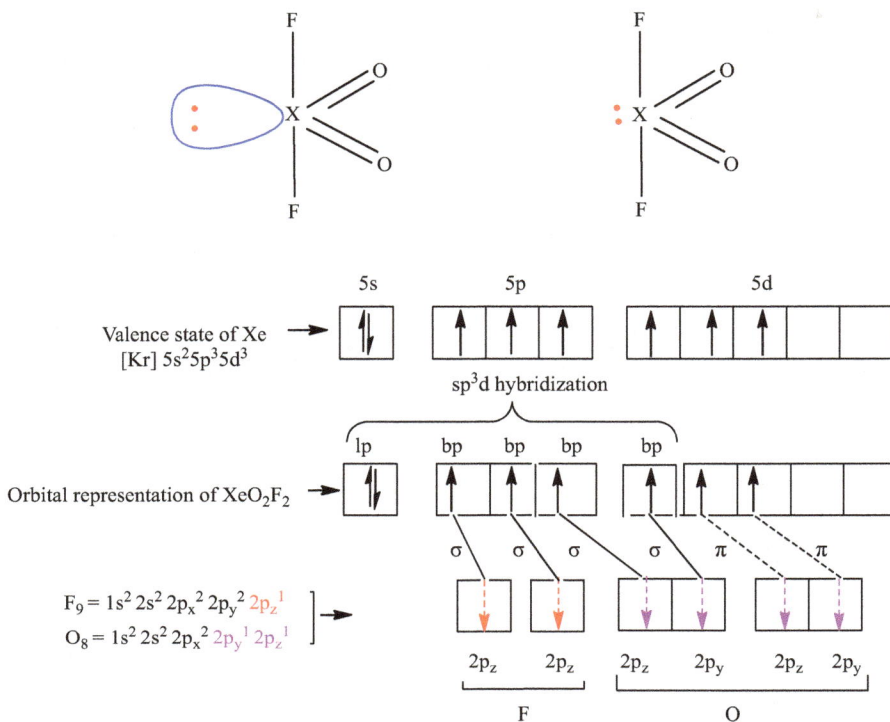

Figure 1.21: See-saw shape to XeO_2F_2.

(j) XeO_3F_2

In XeO_3F_2, Xe has 10 electrons (5 bps) in the valence shell. These are accommodated in five sp^3d hybrid orbitals. With 5bps and three π-bonds, the shape of the molecule is TBP as shown in Figure 1.22.

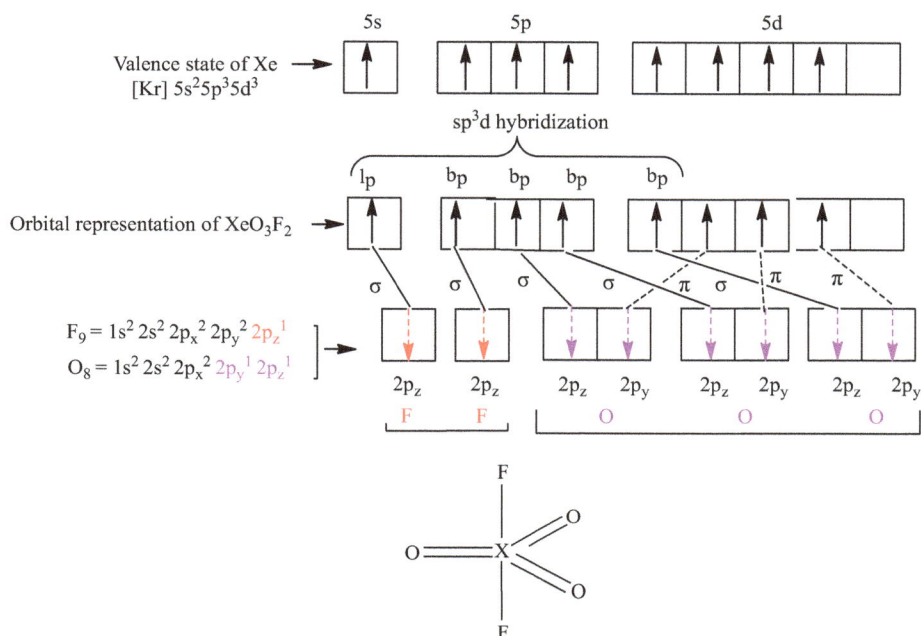

Figure 1.22: Trigonal bipyramidal structure of XeO_3F_2.

The following table summarizes the geometry of molecules containing bond pair and lone pair of electrons:

Total no. of electron pairs in valence Shell of central atom	State of hybridization of central atom	No. of bond pairs	No. of lone pairs	Arrange-ment of lone pairs and bond pairs	Shape of molecules	Examples
2	sp	2	0	Linear	Linear	BeF_2, $BeCl_2$, CO_2, CS_2, NO_2^+
3	sp^2	3	0	Triangular planar	Triangular planar	$AlCl_3$, BF_3, SO_3, CO_3^{2-}
			1	Triangular planar	V-shape	SO_2, $PbCl_2$, $SnCl_2$, NO_2^-
4	sp^3	4	0	Terahedral	Terahedral	CH_4, CCl_4, SiH_4^+, XeO_4
		3	1	Terahedral	Pyramidal	NH_3, PH_3, PCl_3, H_3O^+, XeO_3
		2	2	Terahedral	Bent	H_2O, NH_2^-

(continued)

Total no. of electron pairs in valence Shell of central atom	State of hybridization of central atom	No. of bond pairs	No. of lone pairs	Arrange-ment of lone pairs and bond pairs	Shape of molecules	Examples
5	sp^3d	5	0	Trigonal bipyramidal	Trigonal bipyramidal	PF_5, PCl_5, $SbCl_5$, XeO_3F_2
		4	1	Trigonal bipyramidal	See-saw	SF_4, TeF_4, XeO_2F_2
		3	2	Trigonal bipyramidal	T-shaped	ClF_3, BrF_3
		2	3	Trigonal bipyramidal	Linear	XeF_2, ICl_2^-
6	sp^3d^2	6	0	Octahedral	Octahedral	SF_6, SeF_6
		5	1	Octahedral	Square pyramidal	IF_5, BrF_5, $XeOF_4$
		4	2	Octahedral	Square planar	XeF_4

1.6 Shortcomings of VSEPR theory

Though satisfactory for many species, the VSEPR theory has many shortcomings, namely,

(i) This theory does not explain the shapes of molecules having very polar bonds. For example, Li_2O should have the same structure as H_2O but it is linear.

(ii) This theory is unable to explain the shapes of molecules having extensive π-electron systems.

(iii) Though ionic in solid state, the alkaline earth halides are covalent in vapour phase, where some of them have a bent V-shape. This can not be explained as the alkaline earth ions, after the formation of dihalides do not have any electron pair (**lp**) on them.

(iv) It fails for most of the 14-electron systems. Though the expected pentagonal bipyramidal structure for sp^3d^3 hybridization is observed for IF_7 and ReF_7 where nolp exists, it fails for the species with lps. Thus it does not give a correct structure for XeF_6 and SbF_6^{3-} (distorted octahedron in which the lp is trying to emerge out of the triangular face) (Figure. 1.23).

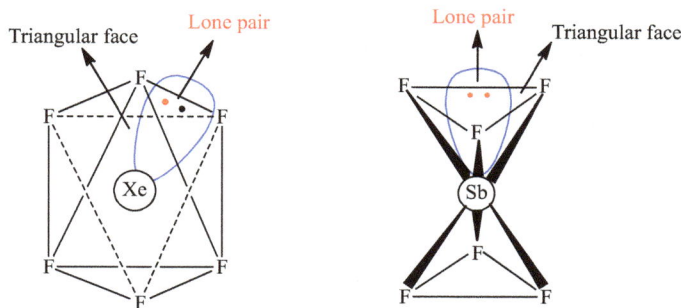

Figure 1.23: Structures of XeF_6 and SbF_6^{3-} in which the lone pair of electrons is emerging out of the triangular face.

1.7 Bent rule

In general both s- and p-orbitals provide insufficient overlap compared to that of hybrid orbitals. The relative overlap of hybrid orbitals decreases in order $sp > sp^2 > sp^3 \gg p$. The differences in bonding resulting from hybridization effects on overlap can be seen from the following table:

Table 1.2: % s-character, bond length and bond angles of some molecules.

Molecule	Hybridization	s-character	C-H bond energy (kJ/mole)	Bond angle	C-H bond length (pm)
$HC \equiv CH$	sp	50%	~506	~180°	106
$H_2C = CH_2$	sp^2	33%	~444	~120°	107
CH_4	sp^3	25%	410	109.5°	109
$CH^{\cdot}$	~p	0%	~335	90°	112

(i)   From the above table, it is clear that the C-H bond in acetylene is shorter and stronger than in hydrocarbons with less s-character. The hybridization in hydrocarbons is dictated by the stoichiometry and stereochemistry.

(ii) For the involvement of s-character in hybridization to achieve better bond strength in different types of molecules, an empirical rule was given by Bent known as bent's rule. This rule states that
   – 'More electronegative substituents with a central atom "prefer" hybrid orbitals having less-character and more electropositive substituents "prefer" hybrid orbitals having mores-character.'
   – This empirical rule observed in certain compounds is substantiated my molecular orbital calculations.

The Bent's rule can be observed in the following examples:
(i) A good example of Bent's rule is provided by the fluoromethanes.

Molecule	H–C–H bond angle	F–C–F bond angle
CH_3F	110–112°	–
CH_2F_2	111.9° ± 0.4°	108.3° ± 0.1°
CHF_3	–	108.8° ± 0.75°

In difluoromethane, CH_2F_2 (C atom having electronegative fluoro substituents), the F–C–F bond angle is less than 109.5°(108.3°), indicating less than 25% s-character, but the H-C-H (C atom having electropositive substituents H atoms) bond angle in this fluoromethane is larger (111.9°) and hence the C–H bond has more s-character. The bond angles in fluoromethanes yields similar results.

(ii) A second example of Bent's rule is provided by chlorofluorides of phosphorous PCl_xF_{5-x} and the alkylphosphorous fluorides R_xPF_{5-x}.

Before coming to these two cases, let us have a look over the effect of differences in hybrid bond strength in PCl_5 molecule involving sp^3d hybridization. Contrary to the equivalent and symmetric four sp^3 hybrid orbitals directed to the four corners of a regular tetrahedron and six d^2sp^3 hybrid orbitals directed to the six corners of an octahedron, sp^3d hybrid orbitals are not equivalent; instead, they may be considered as a combination of a $p_zd_z^2$ hybrid and sp_xp_y hybrid.

The former makes two linear hybrid orbitals bonding axially and the latter forms the trigonal equivalent bonds. The sp^2 hybrid orbitals are capable of forming stronger bonds (33% s-character), and it is found that they are 15 pm shorter than the longer weak axial bonds (0% s-character, pdz^2 hybridization).

When the electronegativities of the substituents on P atom differ, as in the mixed chlorofluorides, PCl_xF_{5-x} and the alkylphosphorous fluorides, R_xPF_{5-x}, it is experimentally observed that the more electronegative substituents *occupies the axial position* (0% s-character, pdz^2 hybridization) and the less electronegative substituents is equatorial bonded (more s-character).

The tendency of the more electronegative substituents to seek out (search for) the low electronegative $p_zd_z^2$ apical orbital in TBP structure is often termed as apicophilicity. It is well illustrated in a series of oxosufuranes of the type shown:

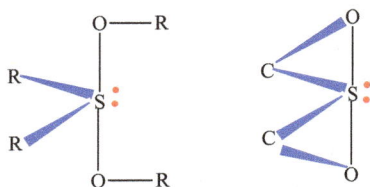

1.7.1 Consistency of Bent's rule with Gillespie's VSEPR model

Bent's rule is also consistent with, and may provide alternative rationalization for, Gillespie's VSEPR model. Thus, the Bent's rule prediction that highly electronegative substituents will attract p-character and reduce bond angles is compatible with the reduction in angular molecule of the bonding pair when held tightly by an electronegative substituent.

Strong, s-character covalent bonds require larger volume in which to **bond**. Thus, doubly bonded oxygen, despite the highly electronegativity of oxygen, seeks s-rich orbital because of the shortness and better overlap of the double bond. Again, the explanation whether in purely s-character terms (**Bent's rule**) or in larger angular volume for a double bond (VSEPR), predicts the correct structure.

Exercises

Multiple choice questions/fill in the blanks

1. VSEPR theory is applicable for molecular shapes and bond angles of molecules or ions of
    (a) Transition elements
    (b) Non-transition elements
    (c) Inner transition elements
    (d) All of these

2. Repulsion is minimum in
    (a) Bond pair–bond pair
    (b) Bond pair–lone pair
    (c) Lone pair–lone pair
    (d) None of these

3. The number of electrons in the valence shell of the central atom in PCl_5 is
    (a) 8   (b) 10   (c) 12   (d) None of these

4. The number of electron pair in the valence shell of the central atom in SF_6 is
    (a) 4   (b) 5   (c) 7   (d) None of these

5. The number of lone pair of electrons in XeF_2 molecule is
    (a) 1   (b) 2   (c) 3   (d) 4

6. Sort out the linear species from the following
    (a) SO_2 (b) NO_2^+ (c) NO_2 (d) NO_2^-

7. The trend in variation of bond angles in NH_3, PH_3, AsH_3 and SbH_3 is
   (a) $NH_3 = PH_3 < AsH_3 < SbH_3$
   (b) $NH_3 < PH_3 < AsH_3 < SbH_3$
   (c) $NH_3 \gg PH_3 > AsH_3 > SbH_3$
   (d) None of the above

8. The geometry of NO_2^+ is
   (a) Bent   (b) Linear   (c) Tetrahedral   (d) None of these

9. Among the following which molecule contains a central atom that has incompletely filled shell?
   (a) NF_3 (b) $N(CH_3)_3$ (c) $OH(CH_3)$ (d) SbH_3

10. The number of electron pairs in valence shell of the central atom of HF molecule is
    (a) 2   (b) 3   (c) 4   (d) 5

11. Which one of the following molecule has distorted geometry?
    (a) $BeCl_2$ (b) SF_6 (c) H_2Se (d) TeF_6

12. The shape of ammonia molecule is
    (a) Linear
    (b) Triangular planar
    (c) Tetrahedral
    (d) Triangular pyramidal

13. The geometry of PCl_5 is
    (a) Pentagonal
    (b) Square pyramidal
    (c) Tetrahedral
    (d) None of these

14. The pair having similar geometry is
    (a) BF_3, NH_3 (b) IF_5, PF_5 (c) $BeCl_2$, H_2O (d) None of these

15. The bond angle in H_2S is
    (a) Greater than in NH_3
    (b) Same as in $BeCl_2$
    (c) Same as in CCl_4
    (d) Larger than in H_2Se and smaller than in H_2O

16. Which of the following molecule is not tetrahedral?
    (a) CF_4 (b) SiF_4 (c) SF_4 (d) CH_4

17. In OF_2, the bond angle is
    (a) 180°   (b) 109.5°   (c) >109.5°   (d) <109.5°

18. The structure of XeF_4 is
    (a) Square planar
    (b) TBP with one position occupied by lone pair
    (c) Octahedral with *trans* positions occupied by two lone pairs

19. Which of the following molecule has regular geometry?
    (a) H_2O (b) PF_3 (c) XeF_4 (d) SF_6

20. The molecule of ClF_3 adopts
    (a) Tbp   (b) Triangular planar   (c) T-shaped   (d) None of these

21. Bond angle . . . . in the order: multiple-multiple bond > multiple-single bond > single-single bond.

22. Bond angle in NF_3 is . . . . than NH_3.

23. A molecule having one or two lone pair(s) of electrons will occupy . . . . position(s) in TBP structure for minimum repulsion.

24. An octahedral molecule with two lone pairs of electrons will occupy . . . . positions for minimum repulsion.

25. The bond angle in H_2O molecule is . . . .

Short answer type questions

1. Why does a lone pair of electrons repel more than a bond pair of electrons?
2. Why is bond angle in water smaller than that in ammonia?
3. Explain why the structure of ICl_2^- is linear.
4. Explain the structure of ClF_3 on the basis of VSEPR theory.
5. Explain why the structure of tetrachoroiodate(III) anion is square planar.
6. Explain why NO_2^+ is linear while NO_2^- is bent.
7. Explain with reason the difference in bond angles of OH_2 and SH_2.
8. Account for the geometry of IF_4^-.
9. Which molecule CH_4 or NH_3 has bigger bond angle and why?
10. Explain how the structure of XeO_4 is tetrahedral.
11. Explain why the $XeOF_4$ issquare pyramidal in shape.
12. Justify why the structure of XeO_2F_2 is 'see-saw' like.
13. Justify that the structure of SF_4 is 'see-saw' like.
14. Justify the large variation in bond angle of H_2O and Cl_2O.
15. Explain the significant variation in bond angle of PH_3, PF_3 and PCl_3 in spite of single lone pair on central atom P in each.

Long answer type questions

1. What is VSEPR theory? How is it useful in understanding the geometry of molecules?
2. Based on VSEPR theory, explain the geometry of NH_3 and H_2O. Why CH_4, NH_3 and H_2O have different shapes?
3. Explain why the geometries of CH_4, NH_3 and H_2O are different from one another.
4. Explain lp–lp > lp–bp > bp–bp repulsion. Describe the geometry of NH_3, H_2O and PF_5 on its basis.
5. Having the same number of electrons in the valence shell of the central atom of molecules CH_4, NH_3 and H_2O have different shapes.
6. Using the concept of VSEPR theory, discuss the structure of ICl_2^- and SF_6.
7. Describe the salient features of valance shell electron pair repulsion theory.
8. Explain the geometries of IF_7 and SF_4 on the basis of VSEPR theory.
9. Explain the structures f ICl_4^- and TeF_5^- using the concept of VSEPR theory.
10. How is VSEPR theory helpful in actual shapes of XeF_2 and XeF_4 molecules?
11. Justify the pyramidal shape of XeO_3 and tetrahedral structure of XeO_4 on the basis of valance shell electron pair repulsion theory.
12. The structure of $[XeO_6]^{4-}$ is octahedral with six sigma- and two pi-bonds using the concept of VSEPR theory and hybridization. Also justify the presence of four negative charges over the molecule.
13. Justify that XeO_2F_2 is 'see-saw'-shaped with two π-bonds and a lone pair on the central atom.
14. Explain the TBP structure of XeO_3F_2 with three π-bonds using the concept of VSEPR theory and hybridization.
15. Explain how Bent's rule is consistent with Ronald Gillespie's VSEPR model.
16. Explain Bent's rule with suitable examples. Explain how this rule provides alternative rationalization for Gillespie's VSEPR model.
17. Explain the wide variations in bond angles of NO_2^+, NO_2 and NO_2^- on the basis of valance shell electron pair repulsion theory.
18. Explain with reasonable explanation that why lone pairs are placed in equatorial and axial positions in TBP and octahedral compounds, respectively, for minimum repulsion.
19. With suitable examples, explain the effect of multiple bonds on bond angles in multiple bonded molecules.
20. Present an account of exceptions to the VSEPR model, ligand–ligand repulsion and the LCP model with suitable examples.
21. Justify why the lone pairs occupy equatorial positions rather than axial in AX_4E type molecules and occupy *trans-* rather than *cis*-positions in octahedral electronic arrangement of AX_4E_2 type molecules (E = lone pair of electrons).

Chapter II
Delocalized π-bonding in polyatomic molecules: molecular orbital approach

2.1 Localized and delocalized bonds

2.1.1 Localized σ- and π-bonds

A number of molecules contain σ-bonds only as in case of NH_3 and CH_4. For such molecules, the valence bond theory involves the appropriate number of localized two electron bonds. Molecular orbital (MO) approach for such molecules describes the behaviour of electrons by means of orbitals that are localized. When more than two electrons hold two nuclei together, then the orbitals may be divided into localized σ- and π-types depending upon their symmetry. For example, in the N_2 molecules, bonding involves one σ- and two π-orbitals $[(\sigma 1s)^2 (\sigma\star 1s)^2(\sigma 2s)(\sigma\star 2s)^2(\pi 2py^2 = \pi 2pz^2)(\sigma 2px)^2]$. Similarly, the σ- and π-orbitals for $CH_2=CH_2$ and $CH=CH$ are localized between two carbon atoms. Such molecules involving localized σ- and π-bonds are not of any interest under the purview of this topic.

2.1.2 Delocalized π-bond(s)

The formation of polyatomic molecules through delocalized π-bond(s)/bonding means, we can no longer distinguish π-electron pairs and bonding pairs of atoms. Instead, we consider electron pairs as belonging to a group of atoms and holding the molecule together.

2.2 Basic principles of molecular orbital theory: polyatomic molecules or ions involving delocalized π-bonding

MO treatment of polyatomic molecules involving delocalized π-bonding is somewhat more complicated because the atomic orbitals overlapping may results in the formation of large number of MOs since the number of atoms involved in the mutual overlap of their orbitals may itself be large.

Essentials of the theory in the present case remains the same as that for the simple diatomic molecules, however, the following generalizations are notable
(i) Delocalized π-bonding is best described by multi-centred bonds, which involve π-MOs (π-MOs).

https://doi.org/10.1515/9783110727289-002

(ii)   The total number of π-MOs formed in any overlap is always equal to the number of atomic orbitals involved.

(iii) The energy of various π-MOs formed in an overlap is worked out with the help of linear combination of atomic orbitals (LCAOs) method.

(iv)   Among these MOs, increasing number of nodes signify the MO of increasing energy.

(v)   A nodal plane containing the nuclei does not produce any anti-bonding effect, but a node between two nuclei implies an anti-bonding interaction.

(vi)   As the MO wave function is obtained by the LCAOs, its formation can be represented by the equation:

$$\psi_{M(bonding)} = C_1\psi_1 + C_2\psi_2 + C_3\psi_3 + C_4\psi_4 + \ldots C_n\psi_n$$

where ψ_1, ψ_2, ψ_3 and ψ_n are wave functions of atomic orbitals and C_1, C_2, C_3 and C_n are contributions of the respective atomic orbitals towards the resultant MOs.

2.2.1 Various steps involved in working out delocalized π-MOs

(i)   Find the basic shape of the molecule or ion either experimentally or from the VSEPR theory using the number of bps and lps on the central atom.

(ii)   Add up the total number of electrons in the valence shell of all the atoms involved, and add or subtract electrons as appropriate to form ion.

(iii) Calculate the number of electrons used in σ-bonds and lone pairs, and by subtracting this from the total number of electrons obtained as above, determine the number of electrons which can participate in π-bonding.

(iv)   Count the number of AOs which can take part in π-bonding. Combine these AOs to give the same number of MOs, which will be delocalized over all of the atoms. Decide whether MOs are bonding, non-bonding or anti-bonding, and feed the appropriate number of π-electrons into the MOs.

(v)   The number of π-bonds formed can easily be determined from the π-MOs which has been filled.

The overall bond order of the molecule or ion is equal to:

$$\frac{\text{Number of Bonding electrons } (\sigma \text{ and } \psi \text{ both}) - \text{Number of Antibonding electrons}}{2 \times \text{number of bond centers}}$$

2.3 Delocalized π-bonding in non-cyclic polyatomic molecules or ions

2.3.1 Ozone molecule (O_3)

There are six valence cell electrons over central 'O' considering only σ-bond formation (excluding π electrons). So, the structure of ozone molecule is triangular planar structure with one lp over central 'O' (Figure 2.1).

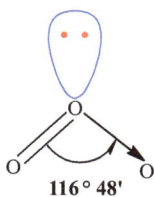

Figure 2.1: Triangular planar structure of ozone molecule with one lp over central 'O'.

O_3 forms a V-shaped molecule. Both the bond lengths are 1.278 Å, and the bond angle is 116° 48'. It may be assumed that the central O atom (O_B) uses roughly sp^2 hybrid orbitals for σ bonding. The same may be considered for the other two oxygens, O_A and O_C (Figure 2.2).

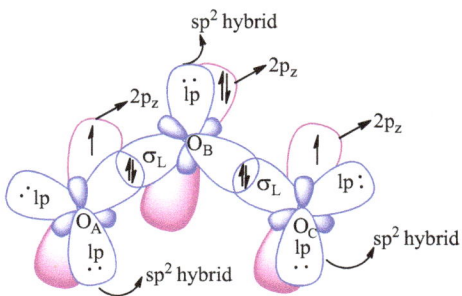

Figure 2.2: Positions of O atoms (O_A, O_B and O_C) and σ-overlaps of sp^2 hybrid orbitals in O_3 molecule.

The presence of delocalized π-bonding in the O_3 molecule is best explained by delocalized three-centred π-bonding as follows:

(i)   There is a total of 18e in the valence shell considering six electrons from each of the three O-atoms ($2s^2 2p^4$).

(ii) The central oxygen atom (O_B) forms a σ-bond with two end oxygens, O_A and O_C, which accounts for **4e**. As shown in the hybridization given in Figures 2.3 and 2.2, the central O atom (O_B) uses two of its singly filled sp^2 hybrid orbitals, and the remaining one is having a **lp**.

$O_8 = 1s^2\, 2s^2\, 2p^4$; Valence shell electronic configuration = $2s^2 2p^4$

Valence shell electronic configuration of O_A (ground state) $\longrightarrow$ $2s^2\ \underbrace{2p_x^2\ 2p_y^1}_{lp\quad lp}\ \overset{\diagup \sigma-\text{bond}}{2p_z^1}$ $\Big\{$ sp^2 hybridization

Valence shell electronic configuration of O_B (ground state) $\longrightarrow$ $2s^2\ \underbrace{2p_x^1\ 2p_y^1}_{lp}\ \overset{\diagdown \sigma-\text{bond}}{2p_z^2}$ $\Big\{$ sp^2 hybridization

Valence shell electronic configuration of O_A (ground state) $\longrightarrow$ $2s^2\ \underbrace{2p_x^2\ 2p_y^1}_{lp\quad lp}\ 2p_z^1$ $\Big\{$ sp^2 hybridization

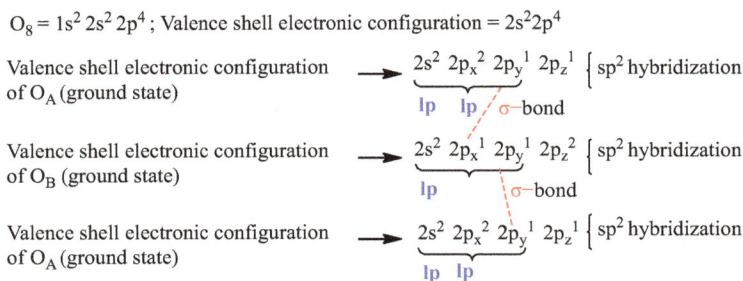

Figure 2.3: Sp^2 hybridization scheme in O_3 molecule.

(iii) The end O atoms (O_A and O_C) after σ-bonds with O_B using their sp^2 hybrid orbitals, are now left with **2lp's** on each. Hence, **5lp's** (O_B: 1lp + O_A: 2lp's + O_C: 2lp's = 5lp's) and **2σ-bonds** together account for **14e**, thus leaving **4e** (18e−14e = 4e) for π-bonding.

(iv) The unused **2pz** orbitals on each of the three O atoms involve themselves in mutual π overlaps, giving a total of three delocalized π_D MOs (π_D MOs). The three modes of combinations resulting in these π_D MOs may be represented as shown in Figure 2.4.

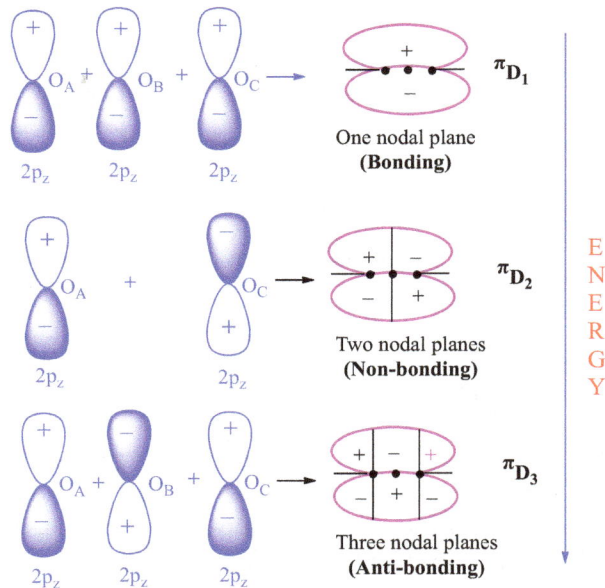

Figure 2.4: Formation of three π_D MOs in O_3 molecule.

(v)   The three π_D MOs, in terms of wave functions, can also be represented as:

$$\pi_{D1} = C_1 \Psi O_A(p_z) + C_2 \Psi O_B(p_z) + C_3 \Psi O_C(p_z)$$

$$\pi_{D2} = C_4 \Psi O_A(p_z) + C_6 \Psi O_C(p_z)$$

$$\pi_{D3} = C_7 \Psi O_A(p_z) + C_8 \Psi O_B(p_z) + C_9 \Psi O_C(p_z)$$

Here, in case for π_{D2}, the value of C_5 for $\psi O_B(p_z)$ is obviously zero, and hence it does not take part in this interaction. In fact, a nodal plane passes through O_B nucleus. It is also obvious from Figure 2.4 that π_{D1} is most bonding as it possesses only one nodal plane that contains nuclei.

(vi) The π_{D2} containing two nodal planes is of higher energy compared to π_{D1}. The highest energy MO is π_{D3} because of having three nodal planes.

(vii) It is notable that π_{D2} is of non-bonding character as the adjacent atoms O_A and O_C do not take part in overlap. On the other hand, π_{D1} and π_{D3} are clearly bonding and anti-bonding MOs.

(viii) A simplified MO energy level diagram of O_3 molecule involving both localized-σ and delocalized-π overlaps is given in Figure 2.5.

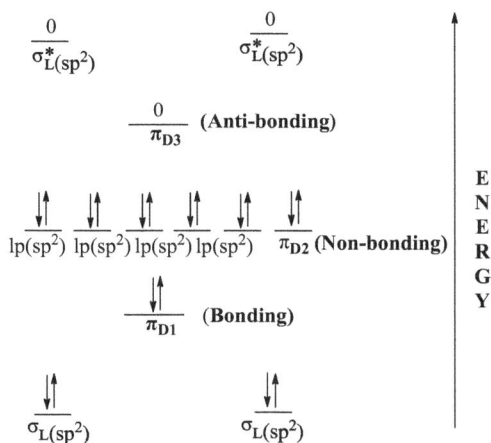

Figure 2.5: Simplified MO energy level diagram of O_3 molecule.

(ix) Of the 4πes available, two are filled in the lowest energy bonding π_{D1} and the remaining two is accommodated in the non-bonding π_{D2}. The highest anti-bonding π_{D3} remains vacant.

(x)   The **5lps** of O_A, O_B and O_C atoms remain accommodated in their respective sp^2 hybrid orbitals. As these orbitals are not involved in bonding (overlapping), they will be equivalent in energy to that of non-bonding π_{D2}.

(xi) It is notable that σ* orbitals formed in the σ overlap will be of still higher energy compared to even π_{D3}, and hence they too will remain vacant.

Bond order of O–O bond in ozone (O$_3$)

Bond order of O–O bond

$$= \frac{\text{No. of bonding electrons } (\sigma + \pi) - \text{No. of antibonding electrons}}{2 \times \text{No. of bonding centres}}$$

$$= \frac{\{4(\sigma) + 2(\pi)\}}{2 \times 2} = \frac{6}{4}$$

$$= 1.5$$

Thus, in O$_3$ molecule, the O–O bond is neither one nor two. In fact, in O$_3$, there are two σ-bonds (O$_A$–O$_B$ and O$_A$–O$_C$) and one π-bond, which is delocalized over O$_A$–O$_B$ and O$_A$–O$_C$.

2.3.2 Nitrogen dioxide (NO$_2$) molecule

This is an odd electron molecule in terms of VBT, which can be represented as shown in Figure 2.6. The number of valence shell electrons surrounding the central N considering only σ bonding is five (2.5 electron pairs, that is, 2 bp + 1 lone electron). Based on the VSEPR theory, the geometry of the molecule is triangular planar. The O–N–O bond angle is 134°15′ because of smaller effect in contraction of bond angle due to a lone electron than the lone pair as in NO$_2^-$ (3 electron pairs; 2 bp + 1 lp).

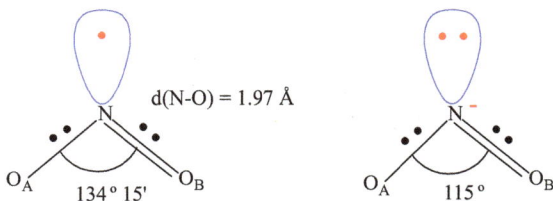

Figure 2.6: Triangular planar structure of NO$_2$ and NO$_2^-$.

The bond angle suggests that the N atom of the NO$_2$ molecule utilizes sp^2 hybrid orbitals for σ bonding (Figure 2.7) leaving 2p$_z$ orbitals free to involve in π-overlap. A similar hybridization may be considered for the other two oxygen atoms O$_A$ and O$_B$. Positions of the atoms and σ-overlaps have been shown in Figure 2.7. The hybridization scheme (Figure 2.8) also makes it clear.

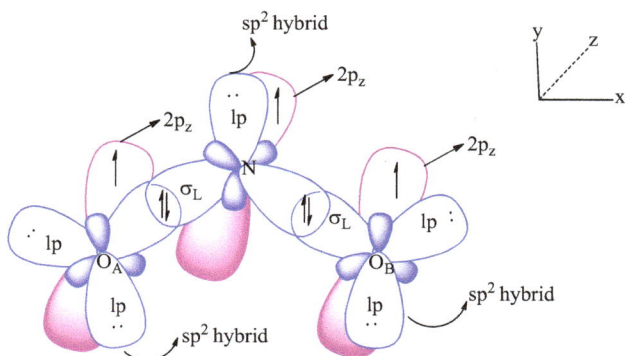

Figure 2.7: Positions of N and O (O_A and O_B) atoms and σ-overlaps of sp^2 hybrid orbitals in NO_2.

$O_8 = 1s^2\,2s^2\,2p^4$; Valence shell electronic configuration $= 2s^2 2p^4$

$N_7 = 1s^2\,2s^2\,2p^3$; Valence shell electronic configuration $= 2s^2 2p^3$

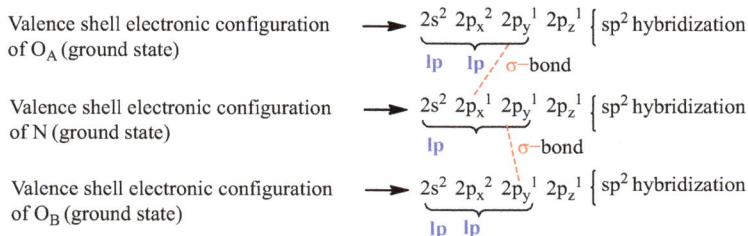

Valence shell electronic configuration of O_A (ground state) $\longrightarrow$ $2s^2\ 2p_x{}^2\ 2p_y{}^1\ 2p_z{}^1$ $\{ sp^2$ hybridization
lp   lp σ−bond

Valence shell electronic configuration of N (ground state) $\longrightarrow$ $2s^2\ 2p_x{}^1\ 2p_y{}^1\ 2p_z{}^1$ $\{ sp^2$ hybridization
lp   σ−bond

Valence shell electronic configuration of O_B (ground state) $\longrightarrow$ $2s^2\ 2p_x{}^2\ 2p_y{}^1\ 2p_z{}^1$ $\{ sp^2$ hybridization
lp lp

Figure 2.8: Sp^2 hybridization scheme in NO_2 molecule.

The presence of delocalized π-bonding in the NO_2 molecule is best explained by delocalized three-centred π-bonding as follows:

(i)   The molecule has a total of 17 valence electrons [N(5) + O_A (6) + O_B (6) = 17e], out of which 10e are in the form of lps. The central nitrogen atom forms a σ-bond with two end oxygens, O_A and O_B which accounts for **4e**. After σ overlaps N is left with one, and each O is left with two sp^2 hybrids orbitals. These five sp^2 hybrid orbitals accommodate 5lps.

(ii)   After taking into account of 5lps (**10e**) and 2σ bonds (**4e**) (10 + 4 = **14e**), only **3e** (17−14 = 3) remain to be accommodated. These **3e** are ultimately accommodated in delocalized $π_D$ orbitals resulting from the mutual overlap of three unused $2p_z$ atomic orbitals (Figure 2.9). The π overlap is of the same type as in the case of O_3 molecule.

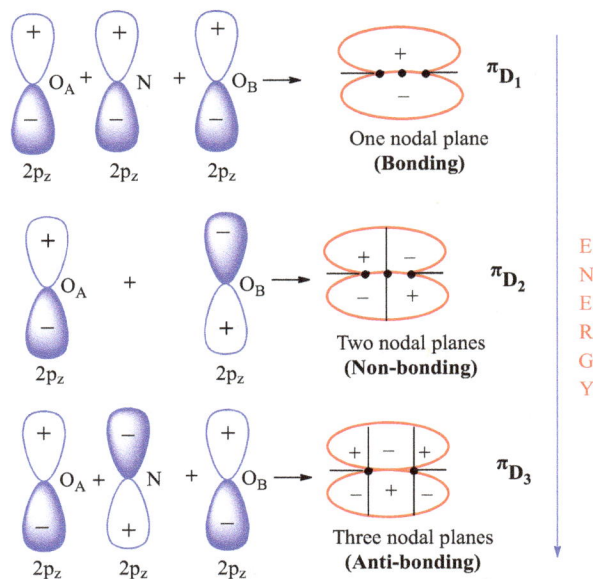

Figure 2.9: Formation of three π_D MOs in NO_2 molecule.

(iii)   A simplified MO energy level diagram of NO_2 molecule involving both local-ized-σ and delocalized-π overlaps is given in Figure 2.10.

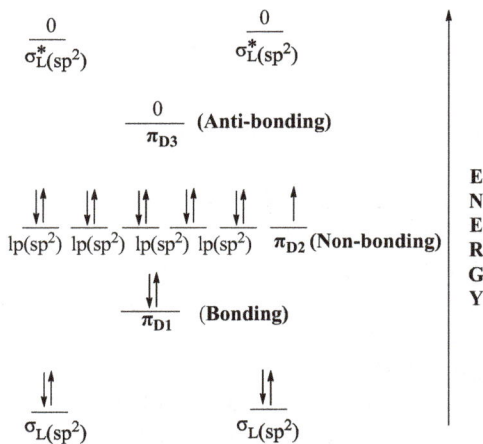

Figure 2.10: Simplified MO energy level diagram of NO_2 molecule.

(iv)   Of the 3πes available, two are filled in the lowest energy bonding π_{D1} and the remaining one is accommodated in the non-bonding π_{D2}. The highest anti-bonding π_{D3} remains vacant.

(v)   The **5lps** of N, O_A and O_B atoms remain accommodated in their respective sp^2 hybrid orbitals. As these orbitals are not involved in bonding (overlapping), they will be equivalent in energy to that of non-bonding π_{D2}.

(vi) It is notable that σ* orbitals formed in the σ overlap will be of still higher energy compared to even $\pi_{D3,}$ and hence they too will remain vacant.

Bond order of N–O bond in ozone (NO₂)

Bond order of N–O bond

$$= \frac{\text{No. of bonding electrons } (\sigma + \pi) - \text{No. of antibonding electrons}}{2 \times \text{No. of bonding centres}}$$

$$= \frac{\{4(\sigma) + 2(\pi)\}}{2 \times 2} = \frac{6}{4}$$

$$= 1.5$$

Thus, in NO_2 molecule, the N–O bond is neither one nor two. In fact, in NO_2, there are two σ-bonds (N–O_A and N–O_B) and one π-bond, which is delocalized over N–O_A and N–O_B.

The MO treatment explains very efficiently the properties of this molecule as evident from the following:

(i)   It does not have tendency to form NO_2^+ ion because of the fact that no electron is present in any high energy anti-bonding orbital. The highest energy lone (single) electron of πD_2 does not have much tendency to ionize due to its non-bonding character.

(ii)   It easily forms a dimmer (N_2O_4) due to the tendency of the lone electron present in the non-bonding πD_2 orbital to pair up.

(iii) It can be easily converted into NO_2^- as the incoming electron is to be accommodated in a non-bonding πD_2 which is not of high energy.

2.3.3 Nitrite ion (NO₂⁻)

The number of valence shell electrons surrounding the central N considering only σ bonding is six (3 electron pairs, i.e. 2 bp + 1 lp). Based on the VSEPR theory, the geometry of the molecule is triangular planar (Figure 2.11). The ONO bond angle is 115° because of greater effect in contraction of bond angle due to a lone pair of electron in NO_2^- compared to the lone electron in NO_2.

The bond angle suggests that the N atom of the NO_2^- molecule utilizes sp^2 hybrid orbitals for σ bonding (Figure 2.12) leaving $2p_z$ orbitals free to involve in π-overlap. A similar hybridization may be considered for the other two oxygen atoms O_A and O_B. Positions of the atoms and σ-overlaps have been shown in Figure 2.12. The hybridization scheme (Figure 2.13) also makes it clear.

Figure 2.11: Triangular planar structure of NO_2^-.

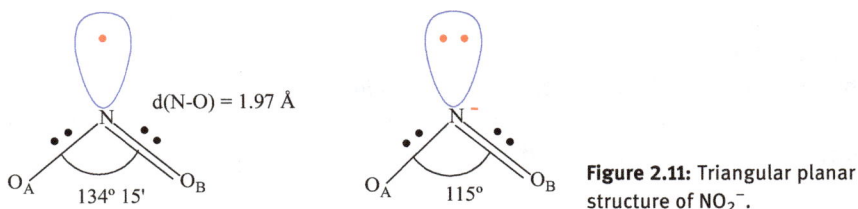

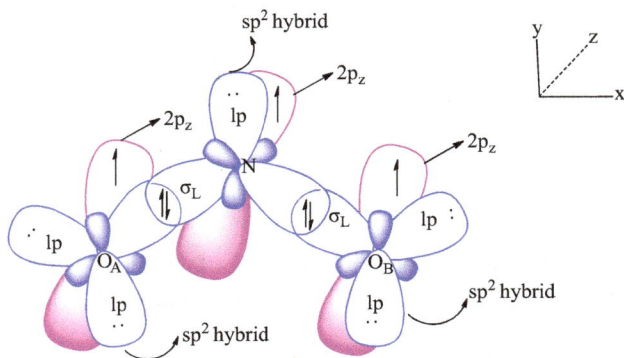

Figure 2.12: Positions of N and O (O_A and O_B) atoms and σ-overlaps of sp^2 hybrid orbitals in NO_2^-.

$O_8 = 1s^2\,2s^2\,2p^4$; Valence shell electronic configuration = $2s^2 2p^4$

$N_7 = 1s^2\,2s^2\,2p^3$; Valence shell electronic configuration = $2s^2 2p^3$

Figure 2.13: sp^2 Hybridization scheme in NO_2^- molecule.

The presence of delocalized π-bonding in the NO_2^- molecule is best explained by delocalized three-centred π-bonding as follows:

(i)   The molecule has a total of 18 valence electrons [N⁻(6) + O_A (6) + O_B (6) = 18e], out of which 10e are in the form of lps. The central nitrogen atom forms a σ-bond with two end oxygens, O_A and O_B which accounts for **4e**. After σ overlaps N is left with one, and each O is left with two sp^2 hybrids orbitals. These five sp^2 hybrid orbitals accommodate **5lps**.

(ii)   After taking into account of 5lps (**10e**) and 2σ bonds (**4e**) (10 + 4 = **14e**), only **4e** (18 − 14 = 4) remain to be accommodated. These **4e** are ultimately accommodated in delocalized π_D orbitals resulting from the mutual overlap of three unused $2p_z$ atomic orbitals (**Figure 2.14**). The π overlap is of the same type as in the case of NO_2 molecule.

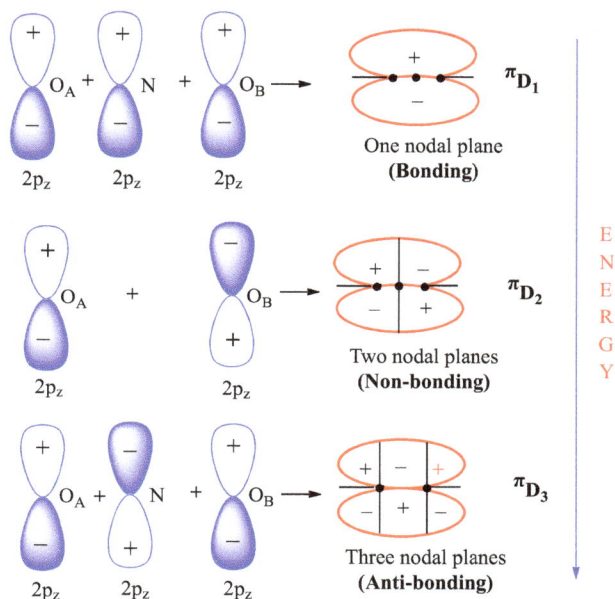

Figure 2.14: Formation of three π_D MOs in NO_2^- molecule.

(iii) A simplified MO energy level diagram of NO_2 molecule involving both localized-σ and delocalized π-overlaps is given in Figure 2.15.

(iv)   Of the $4\pi es$ available, two are filled in the lowest energy bonding π_{D1} and the remaining two are accommodated in the non-bonding π_{D2}. The highest antibonding π_{D3} remains vacant.

(v)   The **5lps** of N, O_A and O_B atoms remain accommodated in their respective sp^2 hybrid orbitals. As these orbitals are not involved in bonding (overlapping), they will be equivalent in energy to that of non-bonding π_{D2}.

(vi)   It is notable that σ* orbitals formed in the σ overlap will be of still higher energy compared to even $\pi_{D3,}$ and hence they too will remain vacant.

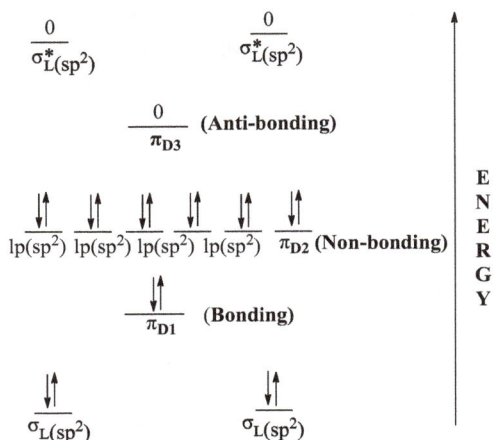

Figure 2.15: Simplified MO energy level diagram of NO_2^- molecule.

Bond order of N–O bond in nitrite ion (NO_2^-)

Bond order of N–O bond

$$= \frac{\text{No. of bonding electrons } (\sigma + \pi) - \text{No. of antibonding electrons}}{2 \times \text{No. of bonding centres}}$$

$$= \frac{\{4(\sigma) + 2(\pi)\}}{2 \times 2} = \frac{6}{4}$$

$$= 1.5$$

Thus, in NO_2^- ion, the N–O bond is neither one nor two. In fact, in NO_2^-, there are two σ bonds (N–O_A and N–O_B) and one π-bond, which is delocalized over N–O_A and N–O_B.

2.3.4 Delocalized π-bonding in hydrazoic acid

In hydrazoic acid, the three N atoms are collinear, and the molecule has the following structure (Figure 2.16):

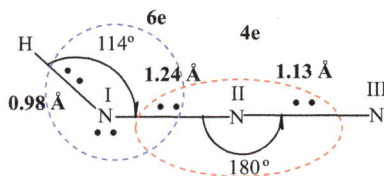

Figure 2.16: Structure of hydrazoic acid.

The bond angles HN^IN^{III} (114°) and $N^IN^{II}N^{III}$ (180°) are due to the presence of 3e pairs (2bp + one lp) on N^I and 2e pairs (2bp) on N^{II}, respectively, based on VSEPR theory.

The reduction in bond length of $N^{II}–N^{III}$ (1.13 Å) compared $N^I–N^{II}$ (1.24 Å) can be very well explained on account of delocalized π-bonding in this molecule, the details of which are as follows:

(i)   The hydrazoic acid molecule contains a total of **16** valence electrons: $[N^I (5) + N^{II} (5) + N^{III} (5) + H (1)] = $ **16e.**

(ii) As the bond angle HN^IN^{II} is 114° and the grouping $N^IN^{II}N^{III}$ is linear, N^I utilizes sp^2 hybrid orbitals for $2\sigma_Ls$ with H and N^{II}. The 3rd sp^2 hybrid orbital of N^I will then accommodate its lp. On the other hand, N^{II} utilizes sp hybrid orbitals for a σ_L with N^I and N^{III}. N^{III} may also employ sp hybrid orbitals, one for σ bonding with N^{II} and the other for accommodating its lp. The 2lps and $3\sigma Ls$ account for 10e. The said hybridization is given in Figure 2.17.

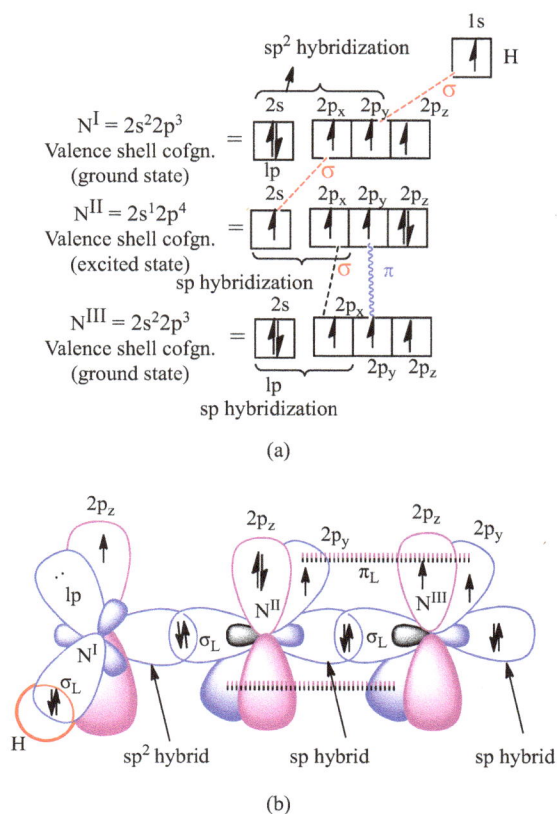

(a)

(b)

Figure 2.17: Hybridization scheme in hydrazoic acid.

(i)   Since N^{II} and N^{III} use sp-hybrid orbitals for σ bonding, both of them retain $2p_y$ and $2p_z$ orbitals for π interactions, while N^{I} is now left with one p-orbital ($2p_z$) for the purpose. Now, a localized π interaction of $2p_y$ orbitals ensures a π_L bond between N^{II} and N^{III}. This accounts for two more electrons, leaving only 4e [16 − (10 + 2) = 4e)] to be accommodated.

(ii)   A π interaction involving $2p_z$ orbitals of all three nitrogens yields three delocalized MOs, a bonding (πD_1), a non-bonding (πD_2) and an anti-bonding (πD_3), as a result of three types of combinations given in Figure 2.18 similar to O_3. Two electrons are accommodated in πD_1 (bonding) and the two electrons in πD_2 (non-bonding). This leaves all the σ* and π* orbitals in the molecule totally vacant.

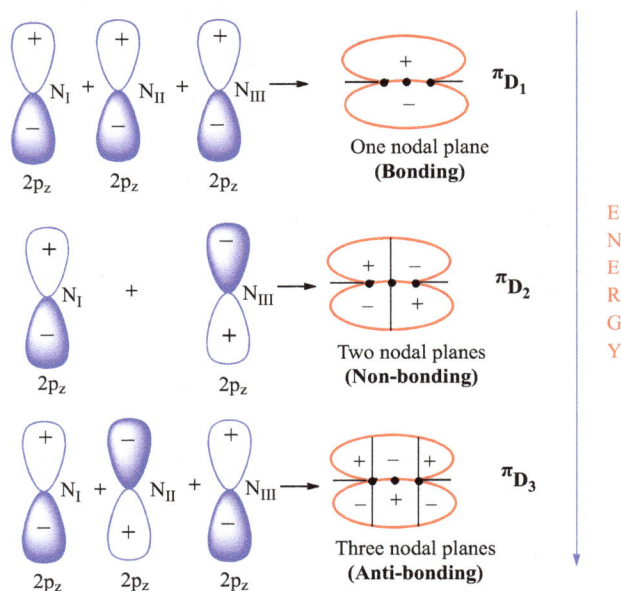

Figure 2.18: Formation of three delocalized π-MOs in hydrazoic acid.

(iii)   A qualitative MO diagram of the molecule is given in Figure 2.19 and the bonding details in N_3H and other related molecules are represented in Figure 2.20.

$$\overline{\sigma_L^*(sp^2\text{-}s)} \quad \overline{\sigma_L^*(sp^2\text{-}sp)} \quad \overline{\sigma_L^*(sp\text{-}sp)}$$

$$\overline{\pi_L^*}$$

$$\overline{\pi_{D_3}(\textbf{Anti-bonding})}$$

$$\overline{\uparrow\downarrow}\atop{lp(sp^2)} \quad \overline{\uparrow\downarrow}\atop{lp(sp)} \quad \overline{\uparrow\downarrow}\atop{\pi_{D_2}\,(\textbf{N.B.})}$$

$$\overline{\uparrow\downarrow}\atop{\pi_{D_1}(\textbf{Bonding})}$$

$$\overline{\uparrow\downarrow}\atop{\pi_L}$$

$$\overline{\uparrow\downarrow}\atop{\sigma_L(sp^2\text{-}s)\atop(Ni^I\text{-}H)} \quad \overline{\uparrow\downarrow}\atop{\sigma_L(sp^2\text{-}sp)\atop(Ni^I\text{-}Ni^{II})} \quad \overline{\uparrow\downarrow}\atop{\sigma_L(sp\text{-}sp)\atop(Ni^{II}\text{-}Ni^{III})}$$

E N E R G Y

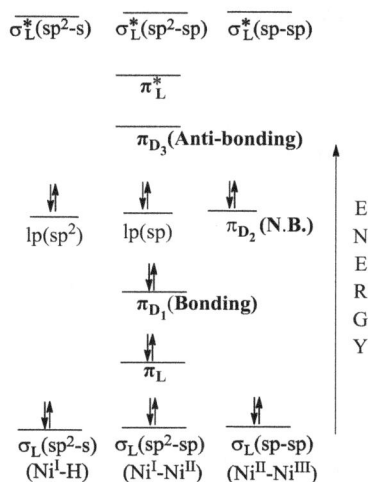

Figure 2.19: MO diagram of N_3H.

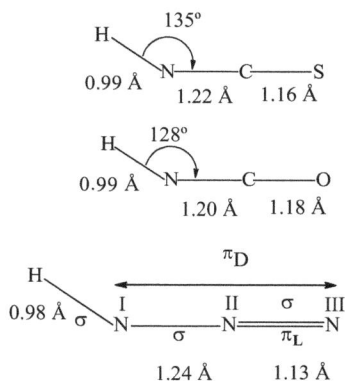

Figure 2.20: Bonding details in N₃H and other related molecules.

2.3.5 Delocalized π-bonding in hydrazoic ion (N_3^-)

Azide ion differs from N_3H only in the disappearance of N–H bond, and as a consequence, acquisition of a negative charge. X-ray studies have been made of several ionic azides. They contain linear symmetrical ions with N–N bond length close to 1.18 Å (Figure 2.21).

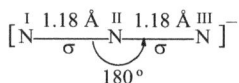

Figure 2.21: Linear symmetrical structure of azide ion.

The presence of two delocalized π-bonding in the ion may be worked out as follows:

(i)   The azide ion contains a total of 16 valence electrons [N^I (5) + N^{II} (5) + N^{III} (5) + one anionic charge = 16e].

(ii) As the $N^I N^{II} N^{III}$ grouping is linear in the anion, sp hybridization can be envisaged in all the three nitrogens. For convenience of further discussion, the sp hybridization scheme (Figure 2.22) involved in the anion is given below.

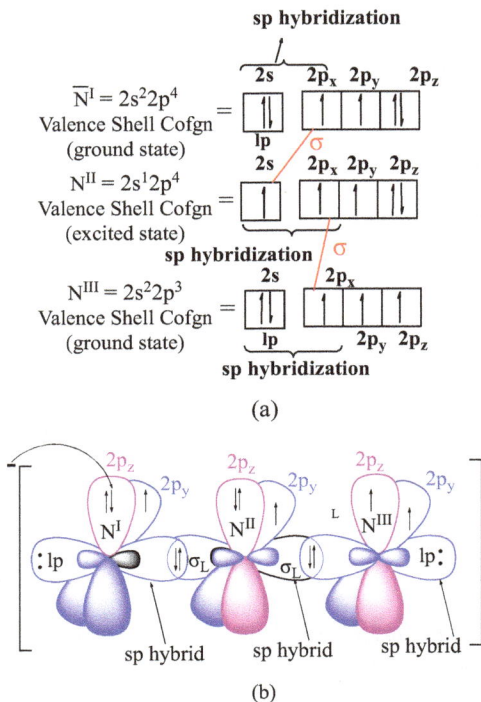

(a)

(b)

Figure 2.22: Hybridization scheme in hydrazoic acid anion.

(iii) As shown in the scheme, two σ_L are formed due to overlapping of two sp-hybrid orbitals of N^{II} with two neighbouring sp-hybrid orbitals, one from N^I and other from N^{III}. One unused hybrid orbital each on N^I and N^{III} will accommodate lone pairs. These are responsible for the excellent coordinating properties of this ion. The $2\sigma_L$ and 2lps account for 8e, leaving 8e to be accommodated.

(iv) The unused pure 2py and 2pz orbitals on all three nitrogens undergo π-interactions ($2p_y$-$2p_y$-$2p_y$ and $2p_z$-$2p_z$-$2p_z$ of similar symmetry) yielding doubly degenerate sets of localized bonding, non-bonding and antibonding π-MOs (Figure 2.23). In each case, combination remains the same as in case of O_3. Two electrons go to each bonding and non-bonding π-MOs, leaving π* and σ* orbitals vacant.

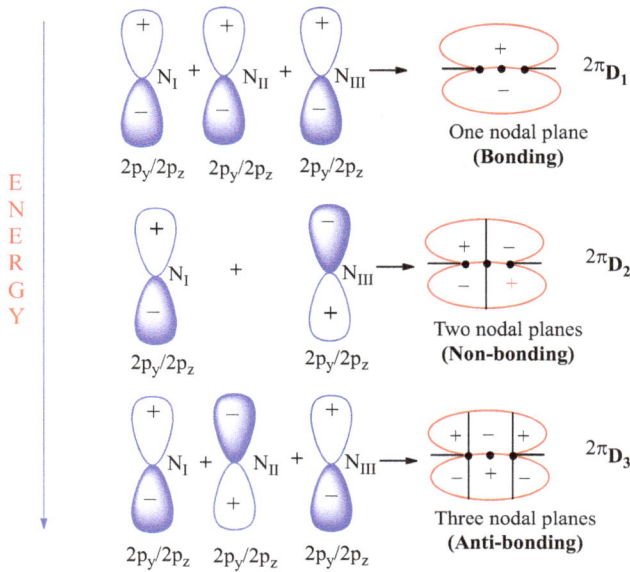

Figure 2.23: Formation of six delocalized π-MOs in hydrazoic acid anion.

(v) A qualitative MO diagram of the molecule is given in Figure 2.24(a). The total bonding in the molecule may be represented as shown in Figure 2.24(b).

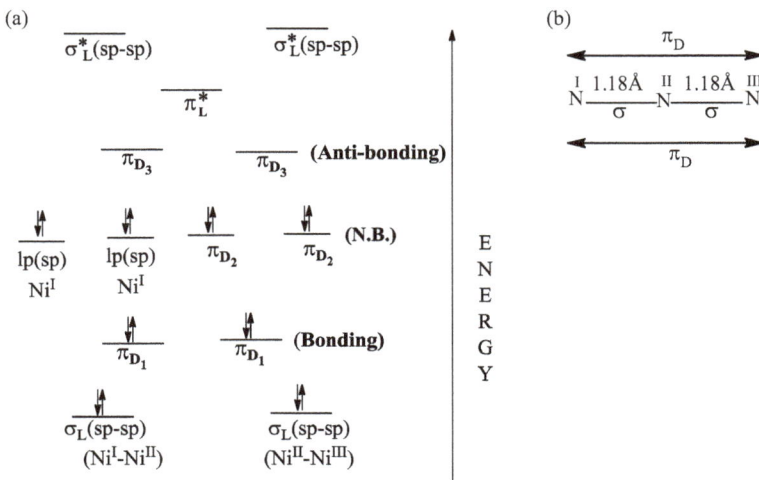

Figure 2.24: (a) MO diagram of N_3^-, (b) Bonding details in N_3^-.

Bond order of N—N bond in azide ion (N₃⁻)

Bond order of N—N bond

$$= \frac{\text{No. of bonding electrons } (\sigma + \pi, \text{ both}) - \text{No. of antibonding electrons}}{2 \times \text{No. of bonding centres}}$$

$$= \frac{\{4(\sigma) + 4(\pi)\}}{2 \times 2} = \frac{8}{4}$$

$$= 2$$

The same bond order (2) for both the N—N bonds, N^I–N^{II} and N^{II}–N^{III}, is responsible for the same and reduced bond lengths (1.18 Å) for both N^I–N^{II} and N^{II}–N^{III} bonds compared to N^I–N^{II} bond length (1.24 Å) of hydrazoic acid.

2.3.6 Delocalized π-bonding in nitrate ion (NO₃⁻)

X-ray diffraction studies have shown that the nitrate ion is a planar symmetrical, the three O atoms occupy the corners of an equilateral triangle. Each O–N–O bond angle is ~120°. The N–O bond length, 1.22 Å, is shorter than single bond length 1.36 Å, which suggests resonating structures in NO_3^- (Figure 2.25).

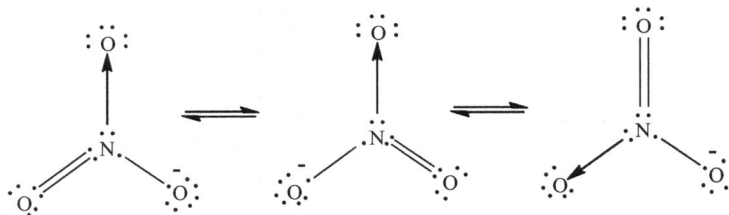

Figure 2.25: Resonating structure of NO_3^-.

Based on VSEPR theory, the triangular planar structure of NO_3^- with bond angles O–N–O (120°) is due to the presence of 3e pairs (3bp) on the central N of NO_3^-.

The delocalized π-bonding in the anion may be worked out as given below.

(i)   The ion contains a total of 24 [5 (N) + 3 × 6 (3O) + 1 negative charge = 24] valence electrons.

(ii) The planar triangular structure of the anion with O–N–O bond angle of ~120° strongly suggests that the central N atom undergoes sp^2 hybridization. The three oxygens may also be considered to involve sp^2 hybridization (Figure 2.26). This gives the formation of $3\sigma_L$. The two unused sp^2 hybrid orbitals on each of the oxygen accommodate 6lps. Thus, a total of 18e are accounted for and 6e remain to be accommodated.

$N_7 = 1s^2 2s^2 2p^3$; Valence shell electr. confgn. $= 2s^2 2p^3$
(ground state)

; Valence shell electr. confgn. $= 2s^1 2p^4$
(excited state)

$O_8 = 1s^2 2s^2 2p^4$; Valence shell electr. confgn.
(ground state)

Valence shell electr. confgn.
of O_A^- (ground state) + 1-ve charge $\longrightarrow$ $2s^2\ 2p_x^2 2p_y^1 2p_z^2$ $\left\{ \begin{array}{l} sp^2\ \text{hybri-} \\ \text{dization} \end{array} \right.$
$(2s^2\ 2p^5)$
lp   lp   lp
σ bond

Valence shell electr. confgn.
of N (excited state) $2s^1\ 2p^4$ $\longrightarrow$ $2s^1\ 2p_x^1 2p_y^1 2p_z^2$ $\left\{ \begin{array}{l} sp^2\ \text{hybri-} \\ \text{dization} \end{array} \right.$
lp

σ bond

Valence shell electr. confgn.
of O_B (ground state) $2s^2\ 2p^4$ $\longrightarrow$ $2s^2\ 2p_x^2 2p_y^1 2p_z^1$ $\left\{ \begin{array}{l} sp^2\ \text{hybri-} \\ \text{dization} \end{array} \right.$
lp   lp   σ bond

Valence shell electr. confgn.
of O_C (ground state) $2s^2\ 2p^4$ $\longrightarrow$ $2s^2\ 2p_x^2 2p_y^1 2p_z^1$ $\left\{ \begin{array}{l} sp^2\ \text{hybri-} \\ \text{dization} \end{array} \right.$
lp   lp

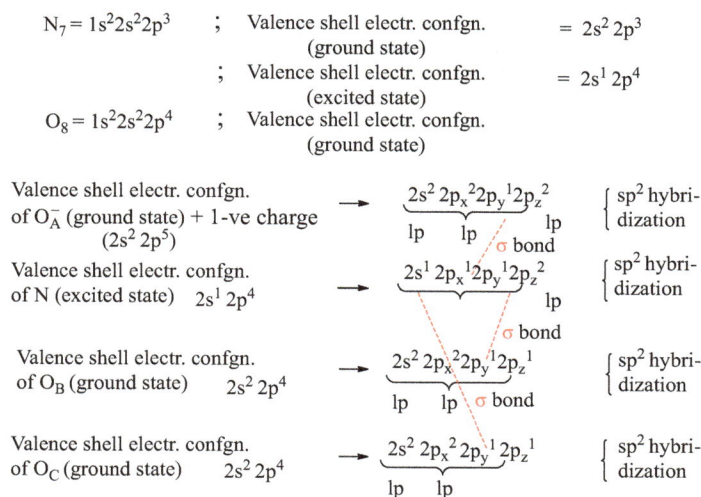

Figure 2.26: Hybridization scheme in NO_3^- ion.

(iii)   The four unused $2p_z$ orbitals, of which two are full filled (O_A and N) and two
are half filled (O_B and O_C), on π-interactions will give rise to 4π-MOs, one
bonding, two non-bonding to accommodate 2lps and one anti-bonding. The
four unused $2p_z$ orbitals along with σ-overlaps are shown in Figure 2.27.

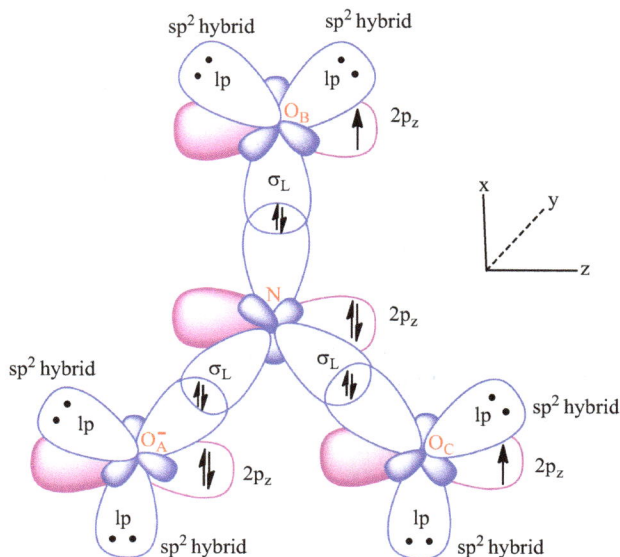

Figure 2.27: Orbital overlap in NO_3^-.

(iv) The four different modes of combinations of four $2p_z$ AOs in the formation of four delocalized πMOs (Figure 2.28) said above can be understood as follows:

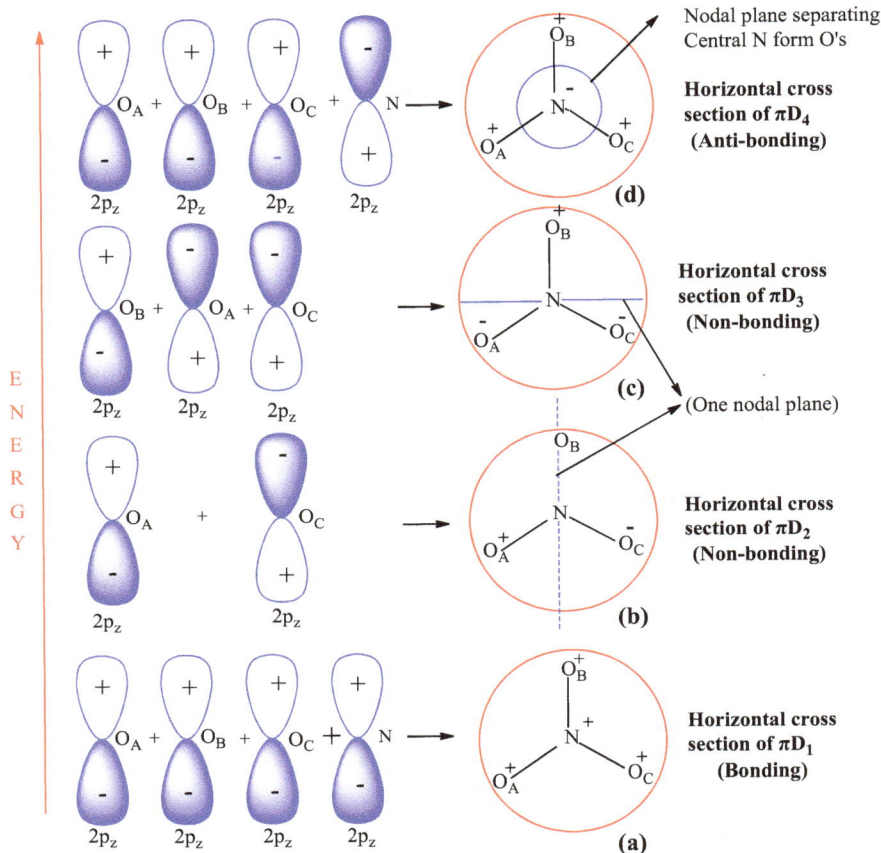

Figure 2.28: Formation of four delocalized πMOs in NO_3^-.

(a) Delocalized bonding MO (π_{D1})

It is formed because of the combination: $\pi_{D1} = C_1\psi O_A(p_z) + C_2\psi O_B(p_z) + C_3\psi O_C(p_z) + C_4\psi N(p_z)$. In this mode of combination, the positive lobes of all the four atoms are on one side while negative lobes are on the other side of the central nodal plane. A horizontal cross section of the ion can then be shown as given in Figure 2.28(a).

(b) Delocalized nonbonding MO (π_{D2} and π_{D3})

These will be formed from the following combinations:

$$\pi_{D2} = C_5\psi O_A(p_z) - C_6\psi O_C(p_z) \quad \text{and}$$
$$\pi_{D3} = C_7\psi O_B(p_z) - C_8\psi O_A(p_z) - C_9\psi O_C(p_z)$$

Here, π_{D2} and π_{D3} are formed when the central nitrogen atom does not take part in overlap (i.e. coefficient C in the wave function for N is reduced to zero). In these a nodal plane is passed through the N nucleus. In fact, in one of the non-bonding MOs, π_{D2}, value of C for one of the oxygen O_B is also reduced to zero. These two non-bonding MOs are degenerate as they possess only one effective nodal plane (i.e. a nodal plane separating the nuclei). The nature of these MOs has been made clear with the help of cross section (Figure 2.28(b) and (c)).

(c) Delocalized anti-bonding MO (π_{D4})

It is formed as a result of the combination: $\pi_{D4} = C_{10}\psi O_A(p_z) + C_{11}\psi O_B(p_z) + C_{12}\psi O_C(p_z) - C_{11}\psi N(p_z)$. In this mode of combination, the positive lobes of all the oxygens are on the same side of the central nodal plane while that of the central N is on the other side. Thus, the $2P_z$ orbitals of oxygens are favourably disposed for a bonding overlaps, though it does not materialize since the $2P_z$ orbital of central N strongly repels all of them making the whole thing extremely antibonding. It means that the formation of a nodal plane completely separating the central N from all the peripheral oxygens. This situation can be visualized as given in Figure 2.28(d).

A qualitative MO diagram of the molecule is given in Figure 2.29. Out of **6e** waiting to be accommodated after the formation of σ-skeleton, **2e** go to π_{D1} (bonding) and

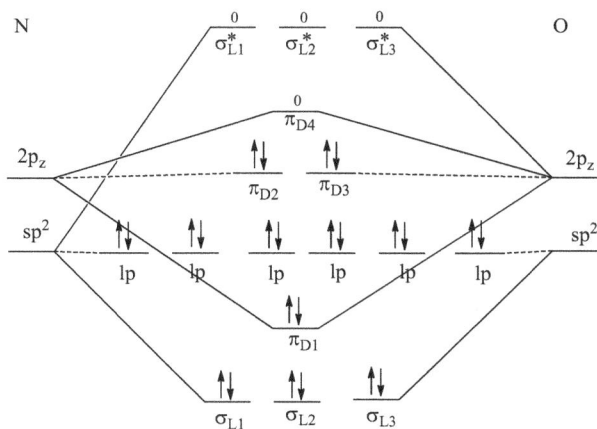

Figure 2.29: Qualitative energy level diagram of MOs formed due to interaction of sp^2 hybrid and $2p_z$ orbitals of N and O in NO_3^-.

4e to two non-bonding orbitals π_{D2} and π_{D3}. The anti-bonding π_{D4} remains vacant and so do σ_L^* orbitals.

Bond order of N–O bond in nitrite ion (NO₃⁻)

Bond order of N–O bond

$$= \frac{\text{No. of bonding electrons } (\sigma + \pi, \text{ both}) - \text{No. of antibonding electrons}}{2 \times \text{No. of bonding centres}}$$

$$= \frac{\{6(\sigma) + 2(\pi)\}}{2 \times 3} = \frac{8}{6}$$

$$= 1.33$$

Thus, in **NO₃⁻** ion, the N–O bond is neither one nor two. In fact, in **NO₃⁻**, there are three σ-bonds (N–O$_A$, N–O$_B$ and N–O$_C$) and one π-bond, which is delocalized over N–O$_A$, N–O$_B$ and N–O$_C$. This is why the bond order of N–O is 1.33.

2.3.7 Delocalized π-bonding in BF₃

The B–F bond distance in the only σ-bonded molecule should be approximately 1.51 Å (calculated value) while the experimentally determined value is 1.31 Å. This wide difference in the calculated and experimental value can be explained as due to additional delocalized π-bonding involving the unused (vacant) $2p_z$ orbital (after sp^2 hybridization) of **boron atom** and one filled p orbital of right symmetry ($2p_z$) **on each F atom** (see hybridization Figure 2.30).

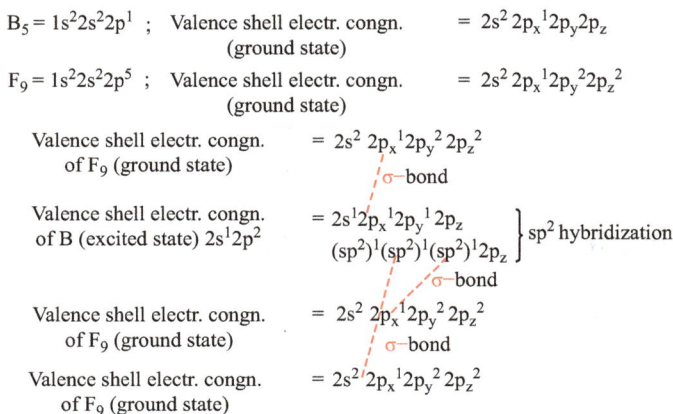

$B_5 = 1s^2 2s^2 2p^1$; Valence shell electr. congn. $= 2s^2\, 2p_x^1 2p_y 2p_z$
(ground state)

$F_9 = 1s^2 2s^2 2p^5$; Valence shell electr. congn. $= 2s^2\, 2p_x^1 2p_y^2 2p_z^2$
(ground state)

Valence shell electr. congn. $= 2s^2\, 2p_x^1 2p_y^2\, 2p_z^2$
of F_9 (ground state) $/\sigma$–bond

Valence shell electr. congn. $= 2s^1 2p_x^1 2p_y^1\, 2p_z$
of B (excited state) $2s^1 2p^2$ $(sp^2)^1 (sp^2)^1 (sp^2)^1 2p_z$ $\Big\}$ sp^2 hybridization
                                                    $/\sigma$–bond

Valence shell electr. congn. $= 2s^2\, 2p_x^1 2p_y^2\, 2p_z^2$
of F_9 (ground state) $/\sigma$–bond

Valence shell electr. congn. $= 2s^2 2p_x^1 2p_y^2\, 2p_z^2$
of F_9 (ground state)

Figure 2.30: Hybridization scheme in BF₃ molecule.

The mutual overlap of these four $2p_z$ orbitals gives one bonding, two non-bonding (degenerate) and one antibonding delocalized π-MOs as shown in Figure 2.31.

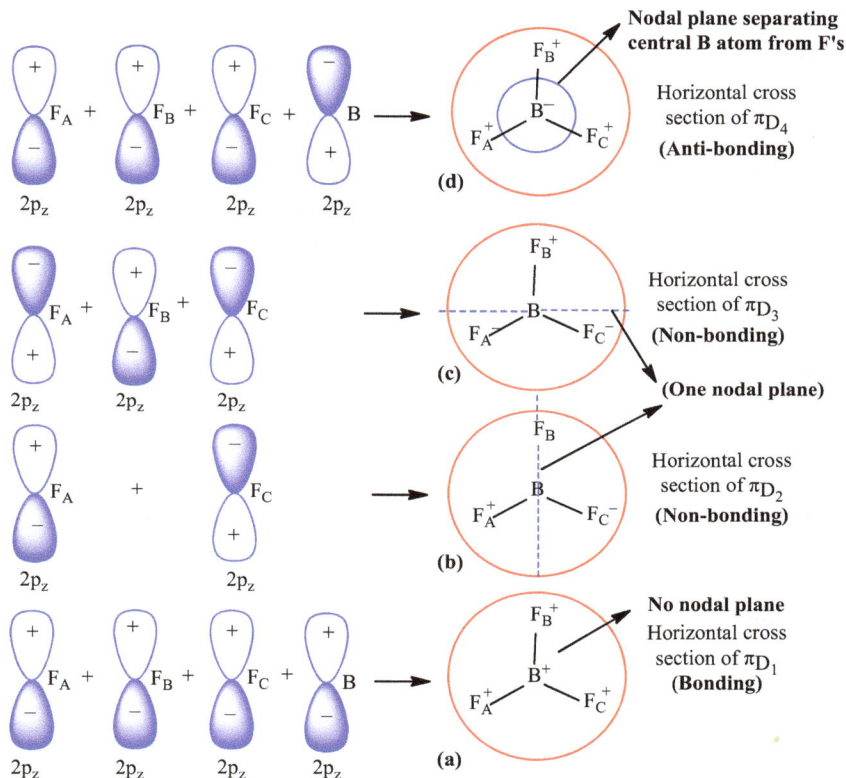

Figure 2.31: Formation of delocalized π-MOs by mutual overlapping of four $2p_z$-orbitals.

A simplified MO energy level diagram of BF_3 molecule involving both localized-σ and delocalized-π overlaps is given in Figure 2.32. Of the **6e** available for π-bonding, **2e are** accommodated in the bonding and **4e** in non-bonding orbitals. This accounts for one delocalized π-bond and consequent shortening of B–F bond distance.

$\overset{*}{\sigma}_L(sp^2-2p_x)$ $\overset{*}{\sigma}_L(sp^2-2p_x)$ $\overset{*}{\sigma}_L(sp^2-2p_x)$
   (B-F)   (B-F)   (B-F)

π_{D_3} (Anti-bonding)

π_{D_2} π_{D_2} (Non-bonding)

π_{D_1} (Bonding)

E
N
E
R
G
Y

$\sigma_L(sp^2-2p_x)$ $\sigma_L(sp^2-2p_x)$ $\sigma_L(sp^2-2p_x)$
  (B-F)   (B-F)   (B-F)

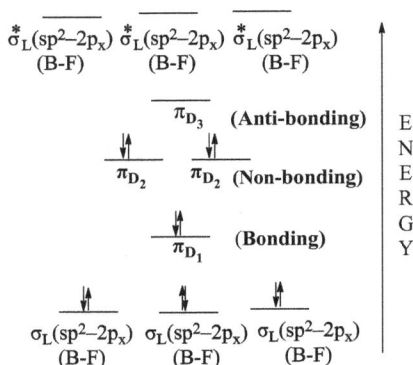

Figure 2.32: M.O. diagram of BF_3 molecule involving both localized σ– and delocalized π–bonds.

B–F bond order in BF_3

Bond order of B–F bond

$$= \frac{\text{No. of bonding electrons } (\sigma + \pi, \text{ both}) - \text{No. of antibonding electrons}}{2 \times \text{No. of bonding centres}}$$

$$= \frac{\{6(\sigma) + 2(\pi)\}}{2 \times 3} = \frac{8}{6}$$

$$= 1.33$$

Thus, in BF_3, the B–F bond is neither one nor two. In fact, in BF_3, there are three B–F σ–bonds and one B–F π-bond, which is delocalized over all the three B–F σ-bonds. This is why the bond order of B–F is 1.33.

2.3.8 Delocalized π-bonding in Cu_2O molecule

In Cu_2O (the cuprite mineral of Cu), each O atom is surrounded tetrahedrally by four Cu atoms, in exactly the same way, as each Si atom is surrounded tetrahedrally by four O atoms in mineral Cristobalite of Si (a form of SiO_2) (Figure 2.33).
In Cu_2O, the Cu^+ ion has the d^{10} electronic configuration, and the oxygen is in the form of dianion (O^{2-}). Compounds of Cu^+ are normally colourless but the Cu_2O is red (Figure 2.34). The red **colour** of this compound is due to charge transfer transitions within delocalized π–bonding MOs.

The formation of delocalized π–bonding in the compound can be understood as follows:

(i)   Because of the presence of 2e pair (2bp) on the central O [Cu:O:Cu], the shape of the **Cu_2O** molecule should be linear on the basis of valance shell electron pair repulsion theory.

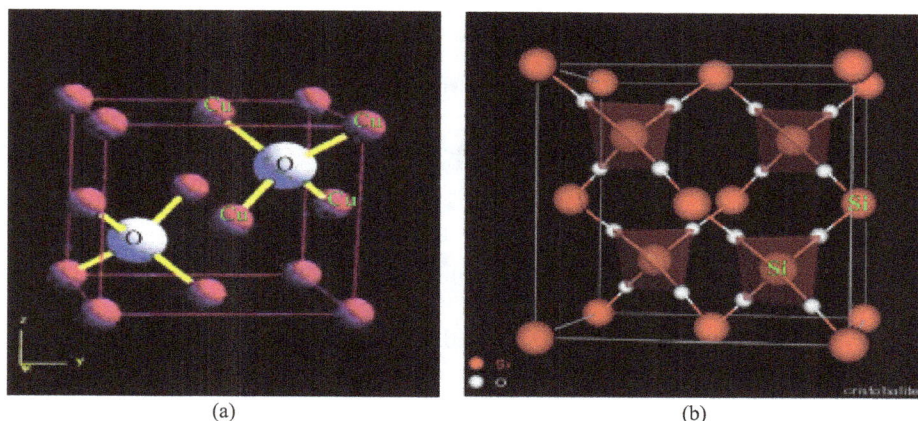

(a)   (b)

Figure 2.33: (a) Crystal structure of cuprite (Cu_2O). (b) Crystal structure of cristobalite (SiO_2).

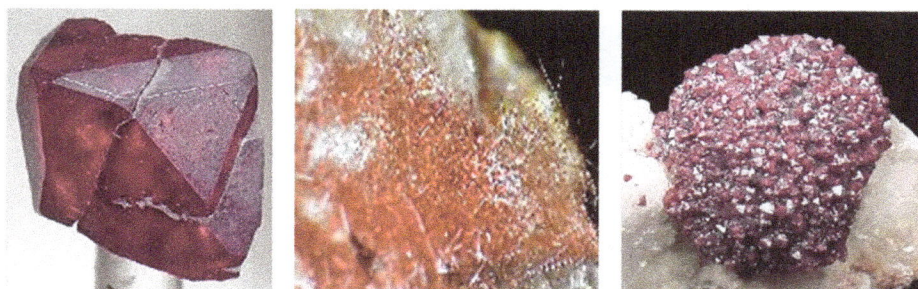

Figure 2.34: Colours of cuprite mineral in different mines.

(ii)   The molecule contains a total of 8e [$2 \times 1(2$ Cu, $4s^1) + 6e$ (O, $2s^2 2p^4$) = 8e] in the valence shell of oxygen.

(iii) The linear Cu–O–Cu structure suggests that the central atom [O] in Cu_2O involves sp hybridization as shown in the hybridization scheme.

(iv)   It is notable here that in considering sp hybridization (Figure 2.35) in the formation of Cu_2O molecule, only ground-state configuration of Cu ($3d^{10}4s^1$) not Cu^+ ($3d^{10}$) is taken into account.

(v)   The two sp-hybrid orbitals of the central oxygen overlap with half–filled 4s orbital of two Cu atoms to give two σ-bonds, while $2p_y$ and $2p_z$ orbitals of oxygen each containing a pair of electrons remain unused.

(vi)   These $2p_y$ and $2p_z$ orbitals of O atom interact with the respective 4p orbitals of the Cu atoms in π fashion to result in the formation of a doubly degenerate set of delocalized π_{D1} (bonding), π_{D2} (non-bonding) and π_{D3} (anti-bonding) MOs.

$$\overset{\sigma}{Cu}\text{—}O\overset{\sigma}{\text{—}}Cu$$

$O_8 = 1s^2 2s^2 2p^4;$ Valence shell electr. confgn. $= 2s^2\, 2px^1 2py^1 2pz^2$
(ground state)

Valence shell electr. confgn. ⟶ $3d^{10}4s^1 4p_x\ 4p_y\ 4p_z$
of Cu (ground state)
                                           σ bond

Valence shell electr. confgn. ⟶ $2s^1 2p_x^1\ 2p_y^2\ 2p_z^2$ ⟶ sp hybridization
of O (excited state) $2s^1\, 2p^5$
                                           $(sp)^1\ (sp)^1\ 2p_y^2\ 2p_z^2$
                                           σ bond

Valence shell electr. confgn. ⟶ $3d^{10}4s^1\ 4p_x\ 4p_y\ 4p_z$
of Cu (ground state)

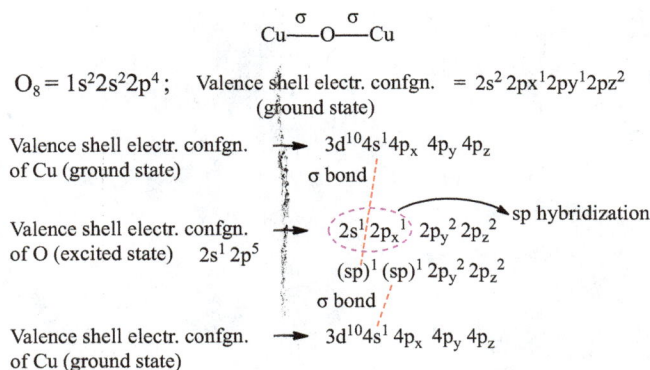

Figure 2.35: Hybridization scheme in Cu_2O.

(vii) The three different possible modes of combinations of 3AOs ($4p_y$, $2p_y$, $2p_y$ or $4p_z$, $2p_z$, $2p_z$) in the formation of 3MOs (Figure 2.36) said above can be understood as follows:

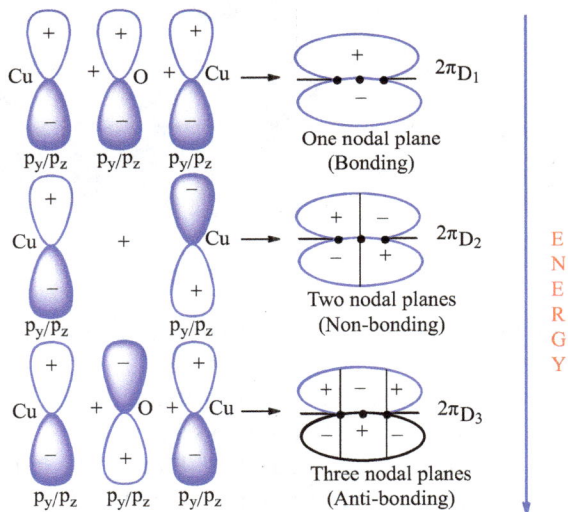

Figure 2.36: Formation of three doubly degenerate set of π_D MOs in Cu_2O.

(viii) It is notable that the π overlap is quite weak due to large difference in the energy of combining orbitals of Cu and O. Also the energy gap between π_{D1} (bonding) and π_{D2} (nonbonding) is quite small.

A simplified MO energy level diagram of Cu_2O molecule involving both localized-σ and delocalized-π overlaps is given in Figure 2.37. The available **4e** for π-bonding **are** accommodated inthe **two** bonding orbitals π_{D1}.

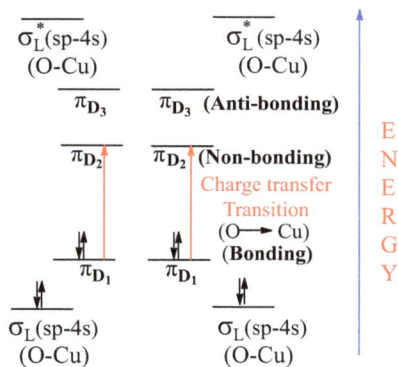

Figure 2.37: MO diagram of Cu_2O showing charge transfer transitions.

The excitation of an electron from π_{D1} to π_{D2} amounts/accounts to charge transfer (O→Cu) since the former is largely concentrated on O and the latter on Cu atoms. This apparent shift of electron density when the compound is exposed to light, causes it to appear red coloured (see MO diagram).

2.4 Delocalized π-bonding in cyclic molecules

This category comprises of organic molecules of the composition C_nH_n. They possess a ring of carbon atoms linked through σ-bonding.

The following points are notable in case of cyclic molecules of the composition C_nH_n.

(i)   Carbon atoms use sp^2 hybrid orbitals for σ bonding.

(ii) The total number of σ-bonds formed is 2 n [n (C–C) and n(C–H) bonds].

(iii) The total number of p_z-orbitals available for delocalized π-interaction is n, and each of these orbitals contains one electron. Consequently, $n\pi$ MOs are formed ultimately.

(iv) Beyond the first energy level, π MOs occur in degenerate pairs when n is odd, while for n even, beyond the first energy level, degenerate pairs occur except for highest energy level.

(v)   The number of bonding and anti-bonding orbitals may not be equal

2.4.1 Benzene (C₆H₆)

Benzene is an example of cyclic molecule. It is a regular hexagon (Figure 2.38) with bond angle of 120°.

Figure 2.38: A regular hexagon benzene molecule.

The hybridization scheme (Figure 2.39) shows that each carbon undergoes sp^2 hybridization during the formation of molecule. The three sp^2 hybrid orbitals on each carbon form 3σ bonds, bond angle being 120°.

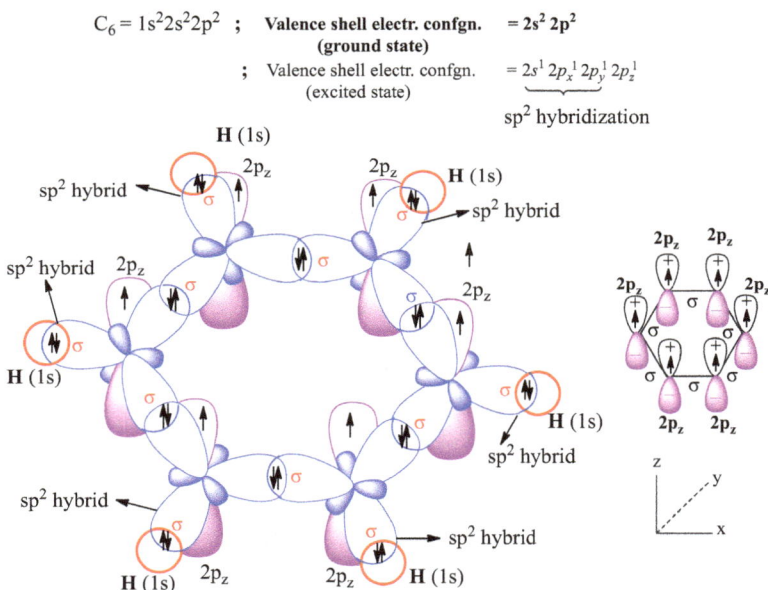

Figure 2.39: Hybridization scheme showing sp^2 orbital overlaps and formation of six (C–C) σ and (C–H) σ bonds in C_6H_6.

Each C atom is still left with a p_z-orbital $\perp r$ to the plane of the benzene C–C σ-skeleton. All the 12 (6C and 6H) atoms are co-planar, with the same C–C bond distance of 1.39 Å which is intermediate between ethylene (1.33 Å) and ethane (1.54 Å).

The formation of delocalized π-bonding in benzene can be understood as follows:

(i)   Total skeleton electrons in benzene involved in σ- and π-bonding apart from C–H σ-bonds = 6 × 4 – 6 = 18e.

(ii) The six $2p_z$ orbitals (one on each C) interact to give $6\pi_D$ MOs, three bonding and three anti-bonding. The six different modes of combination of $2p_z$ orbitals giving $6\pi_D$ MOs are shown in Figure 2.40.

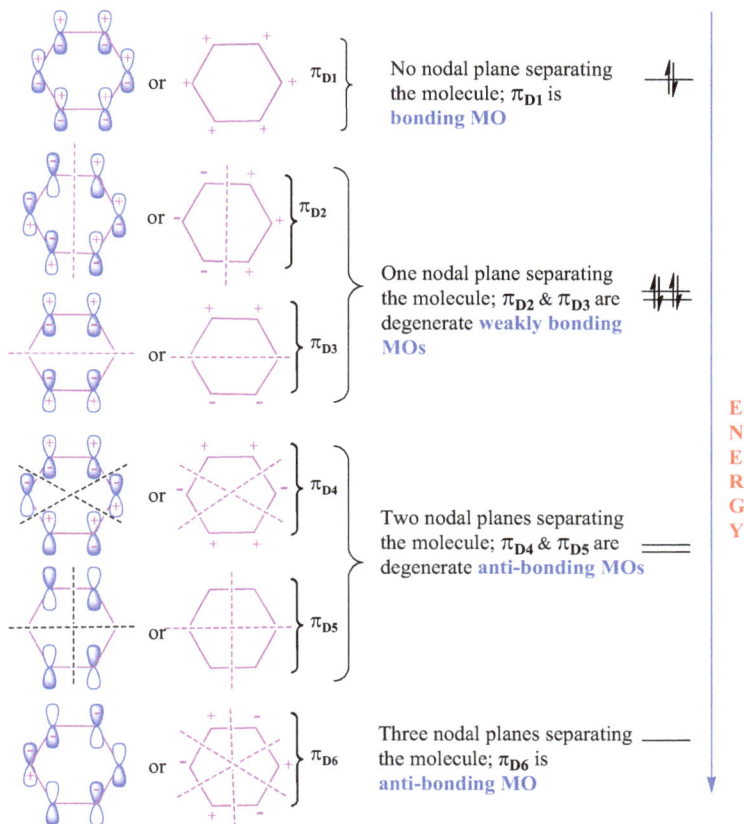

Figure 2.40: Formation of six π_D MOs in C_6H_6 molecule.

(iii) It is notable that the + and − sign in Figure 2.40 denote the signs of the top lobes of various $2p_z$ orbitals involved.

(iv) The six π-electrons of the molecules go to occupy three bonding orbitals π_DMOs ($\boldsymbol{\pi_{D1}}$, $\boldsymbol{\pi_{D2}}$ and $\boldsymbol{\pi_{D3}}$) while antibonding MOs $\boldsymbol{\pi_{D4}}$, $\boldsymbol{\pi_{D5}}$ and $\boldsymbol{\pi_{D6}}$ remain vacant. σL* also remain vacant. A simplified MO diagram **(including only C–C σ and π MOs)** is shown in Figure 2.41.

$\overline{\sigma^*_{LC-C}}$ $\overline{\sigma^*_{LC-C}}$ $\overline{\sigma^*_{LC-C}}$ $\overline{\sigma^*_{LC-C}}$ $\overline{\sigma^*_{LC-C}}$ $\overline{\sigma^*_{LC-C}}$ (6)

$\overline{\hspace{2cm}}$ (Anti-bonding) (2)
π_{D6}

$\overline{\pi_{D4}}$ $\overline{\pi_{D5}}$ (Anti-bonding) (1)

$\overline{\uparrow\downarrow}\;\pi_{D2}$ $\overline{\uparrow\downarrow}\;\pi_{D3}$ (Bonding) (2)

$\overline{\uparrow\downarrow}\;\pi_{D1}$ (Bonding) (1)

$\overline{\uparrow\downarrow}$ $\overline{\uparrow\downarrow}$ $\overline{\uparrow\downarrow}$ $\overline{\uparrow\downarrow}$ $\overline{\uparrow\downarrow}$ $\overline{\uparrow\downarrow}$ (6)
σ_{LC-C} σ_{LC-C} σ_{LC-C} σ_{LC-C} σ_{LC-C} σ_{LC-C}

ENERGY →

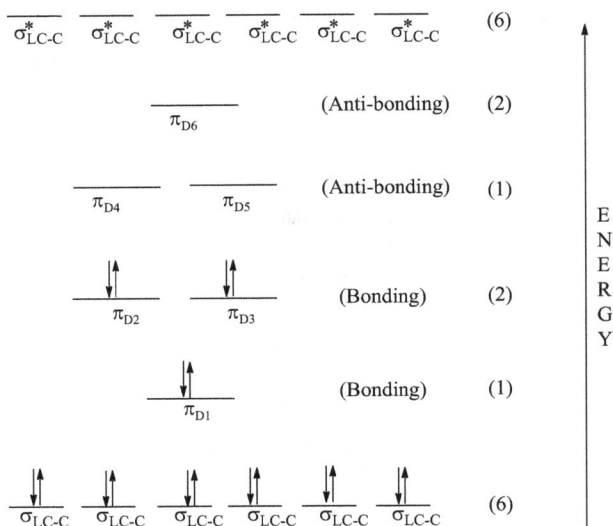

Figure 2.41: MO diagram of benzene molecule including only (C–C)σ and (C–C)π MOs.

C–C bond order in C_6H_6

Bond order of C–C bond

$$= \frac{\text{No. of bonding electrons } (\sigma + \pi,\ \text{both}) - \text{No. of antibonding electrons}}{2 \times \text{No. of bonding centres}}$$

$$= \frac{\{12(\sigma) + 6(\pi)\}}{2 \times 6} = \frac{18}{12} = \frac{3}{2}$$

$$= 1.5$$

Thus, in benzene, the C–C bond is neither one nor two. In fact, there are six σ-bonds and three π-bonds, which are delocalized over all the six σ-bonds. This is why the bond order of C–C is 1.5.

2.4.2 Cyclopentadienyl radical ($C_5H_5\cdot$)

It is a regular pentagon with bond angle of 120° (Figure 2.42). Similar to benzene, the MO treatment of this radical envisages/considers that each carbon undergoes sp^2 hybridization during its formation. The five sp^2 hybrid orbitals on each carbon form 5σ bonds, bond angle being 120°. However, each C atom is still left with a $2p_z$-orbital $\perp$r to the plane of C–C σ-skeleton (Figure 2.43). All the 10 (5C and 5H) atoms are co-planar, with the same C–C bond distance of 1.39 Å which is intermediate between ethylene (1.33 Å) and ethane (1.54 Å).

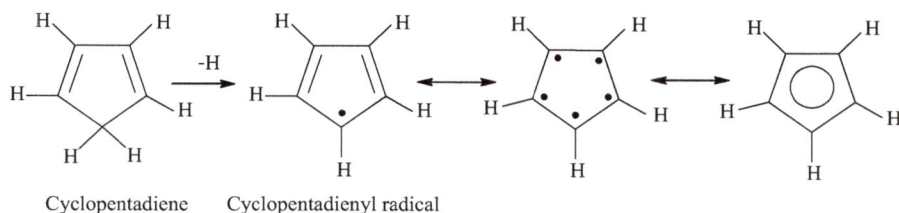

Figure 2.42: Regular pentagon structure of cyclopentadienyl radical.

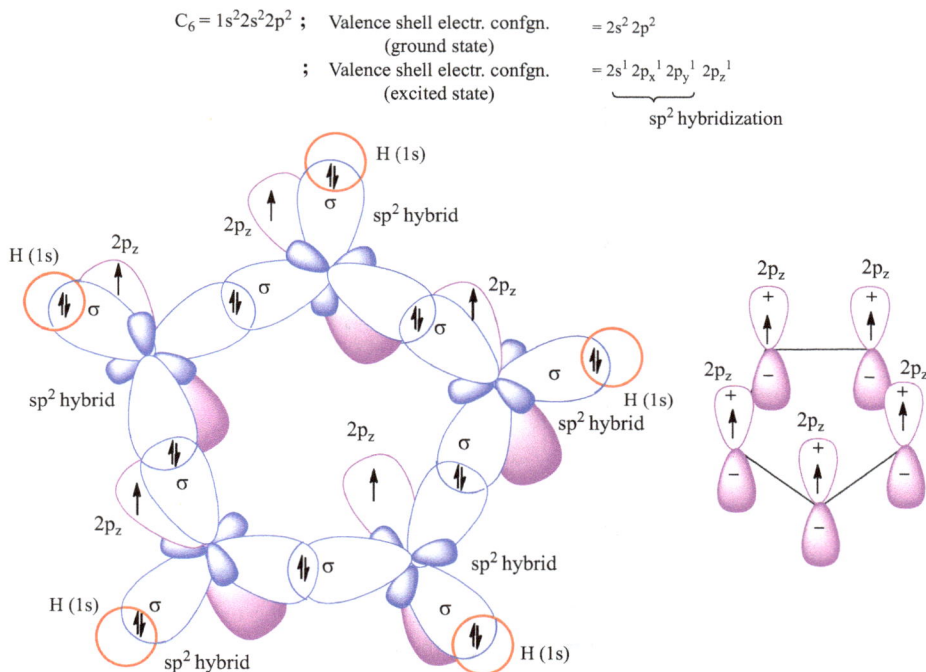

Figure 2.43: sp^2 Orbital overlaps and formation of five σ_L bonds in C$_5$H$_5$ radical.

The formation of delocalized π–bonding in cyclopentadienyl radical can be understood as follows:

(v) Total skeleton electrons in C$_5$H$_5$ radical involved in σ- and π-bonding apart from C–H σ-bonds = $5 \times 4 - 5 = 15e$.

(vi) The five 2p$_z$ orbitals (one on each C) interact to give 5π$_D$ MOs., three bonding and 2 anti-bonding. The five different modes of combination of 2p$_z$ orbitals giving 5πD MOs can be viewed as shown in Figure 2.44.

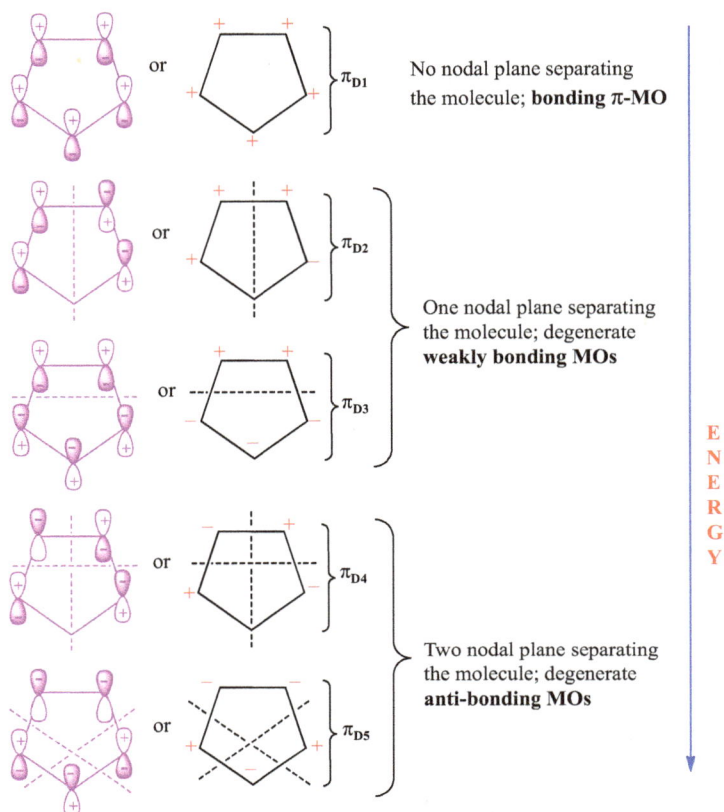

Figure 2.44: Formation of five π–MOs in C_5H_5 radical.

The 5π-electrons of the C_5H_5 radical occupy the three bonding orbitals (π_{D1}, π_{D2} and π_{D3}) while antibonding orbitals π_{D4} and π_{D5} remain vacant. $\sigma_L s^\star$ also remain vacant. A simplified MO diagram (including only C–C σ_L (σ_{LCC}) and π MOs) is shown in Figure 2.45.

$$\overline{\sigma^*_{LCC}}\ \overline{\sigma^*_{LCC}}\ \overline{\sigma^*_{LCC}}\ \overline{\sigma^*_{LCC}}\ \overline{\sigma^*_{LCC}}\qquad (5)$$

$$\overline{\quad}\ \underset{\pi_{D4}}{\quad}\ \underset{\pi_{D5}}{\overline{\quad}}\qquad \text{(Anti-bonding)}\quad(2)$$

$$\underset{\pi_{D2}}{\overline{\uparrow\!\downarrow}}\ \underset{\pi_{D3}}{\overline{\uparrow}}\qquad \text{(Weak bonding)}\quad(2)$$

$$\underset{\pi_{D1}}{\overline{\uparrow\!\downarrow}}\qquad \text{(Bonding)}\qquad(1)$$

$$\underset{\sigma_{LCC}}{\overline{\uparrow\!\downarrow}}\ \underset{\sigma_{LCC}}{\overline{\uparrow\!\downarrow}}\ \underset{\sigma_{LCC}}{\overline{\uparrow\!\downarrow}}\ \underset{\sigma_{LCC}}{\overline{\uparrow\!\downarrow}}\ \underset{\sigma_{LCC}}{\overline{\uparrow\!\downarrow}}\qquad(5)$$

E N E R G Y

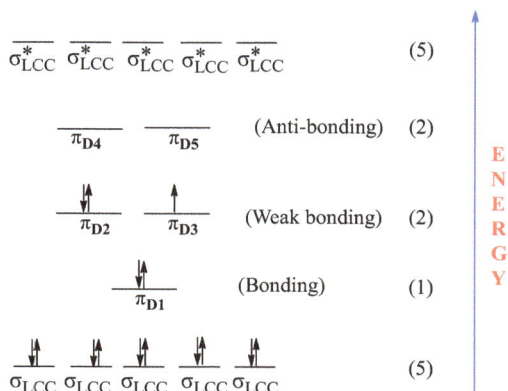

Figure 2.45: MO diagram of C_5H_5 radical including only (C–C) σ and (C–C)π MOs.

C–C bond order in C_5H_5 radical

Bond order of C–C bond

$$= \frac{\text{No. of bonding electrons }(\sigma+\pi,\text{ both}) - \text{No. of antibonding electrons}}{2\times \text{No. of bonding centres}}$$

$$= \frac{\{10(\sigma)+5(\pi)\}}{2\times5} = \frac{15}{10}$$

$$=1.5$$

Thus, in cyclopentadienyl radical, the C–C bond is neither one nor two. In fact, there are five σ-bonds and five π-electrons (2.5 π–bonds), which are delocalized over all the five σ-bonds. Hence, the bond order of C–C is 1.5.

2.5 Delocalized π-bonding in cyclic ions

This category comprises of a set of cyclic anions of the general composition $C_nO_n^{2-}$. They possess a ring of carbon atoms linked through σ-bonding.

The following points are notable in case of cyclic ions of the composition $C_nO_n^{2-}$.
(i)   Carbon as well as oxygen atoms use sp^2 hybrid orbitals for bonding. Two sp^2 orbitals on each of the oxygen contain lone pairs which are available for coordination.
(ii) The total number of σ_L bonds formed is 2n [n(C–C) and n(C–O) bonds].
(iii) Every C as well as O atom involves itself through the remaining p–orbitals (all in the same plane) into a mutual π–interaction. The total number of π_DMOs formed is thus 2n, unlike C_nH_n.

(iv)   Bonding and anti–bonding **π-MOs** are not equal in numbers. In fact, in $C_nO_n^{2-}$, a total of $2n\pi$-**MOs** (n bonding, two non-bonding and (n–2) anti-bonding) are formed.

Three ions, $C_4O_4^{2-}$ (present in the potassium salt of 3,4–diketocyclobutene diol, $K_2C_4O_4$), the croconate ion, $C_5O_5^{2-}$ and $C_6O_6^{2-}$ will be taken as examples of this category.

2.5.1 3,4–Diketocyclobutene dianion ($C_4O_4^{2-}$)

It is a square planar ion (Figure 2.46). The C–C and C=O bond lengths are 1.46 Å and 1.26 Å, respectively.

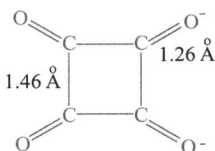

Figure 2.46: Square planar ion of $C_4O_4^{2-}$.

The formation of delocalized π-bonding in 3,4–diketocyclobutene dianioncan be understood as follows:

(i)   The anion contains a total of 42e [4 × 4 (4C) + 4 × 6 (4O) + two negative charge = 42**e**] valence electrons.

(ii)   The formation of $8\sigma_L$ bonds [4σ(C–C) and 4σ(C–O)] involving sp^2 hybridization on each carbon and oxygen in $C_4O_4^{2-}$ is shown in the hybridization scheme (Figure 2.47)

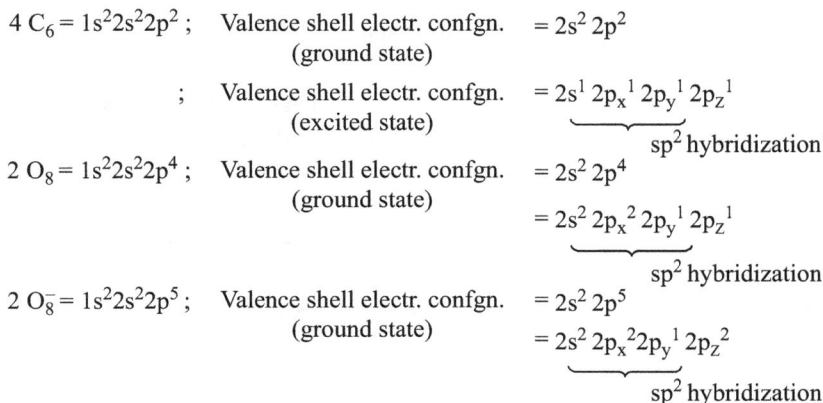

$4\,C_6 = 1s^2 2s^2 2p^2$;   Valence shell electr. confgn. $= 2s^2\,2p^2$
                                        (ground state)

                             ;   Valence shell electr. confgn. $= 2s^1\,2p_x{}^1\,2p_y{}^1\,2p_z{}^1$
                                        (excited state)
                                                                   $\underbrace{\phantom{2s^1\,2p_x{}^1\,2p_y{}^1}}$
                                                                   sp^2 hybridization

$2\,O_8 = 1s^2 2s^2 2p^4$;   Valence shell electr. confgn. $= 2s^2\,2p^4$
                                        (ground state)
                                                                   $= 2s^2\,2p_x{}^2\,2p_y{}^1\,2p_z{}^1$
                                                                   $\underbrace{\phantom{2s^2\,2p_x{}^2\,2p_y{}^1}}$
                                                                   sp^2 hybridization

$2\,O_8^- = 1s^2 2s^2 2p^5$;   Valence shell electr. confgn. $= 2s^2\,2p^5$
                                        (ground state)
                                                                   $= 2s^2\,2p_x{}^2 2p_y{}^1\,2p_z{}^2$
                                                                   $\underbrace{\phantom{2s^2\,2p_x{}^2 2p_y{}^1}}$
                                                                   sp^2 hybridization

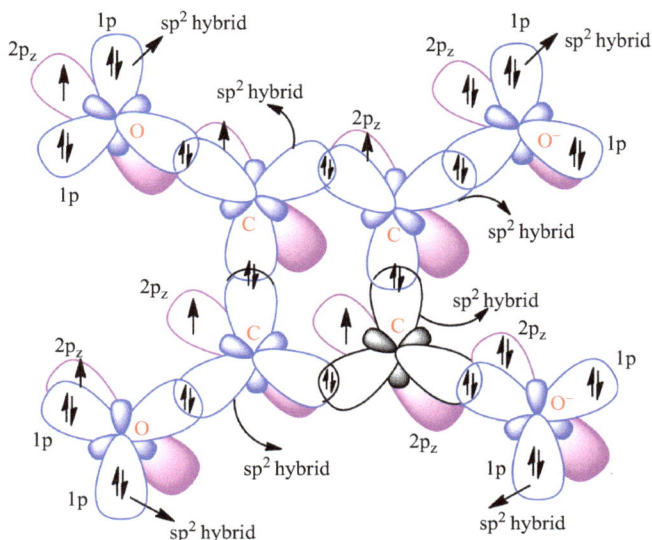

Figure 2.47: sp^2 Orbital overlaps and formation of 8σ$_L$ bonds in C$_4$H$_4^{2-}$.

(iii) As clear from the hybridization scheme, the formation of 8σ$_L$ [4(C–C) and 4(C–O)] and 8lps [2lps in the two sp^2 hybrid orbitals of each of the four oxygens] accounts for (16e + 16e) = 32e, leaving (42e − 32e = 10e) to be accommodated.

(iv) sp^2 hybridizations over four carbons and four oxygens leave one p$_z$ orbital on each of the four carbons (all singly filled) as well as each of the four oxygens, two singly filled and the two bearing a negative charge are fulfilled. Thus, a total of 10e {4e[(4p$_z$ − 4C) + 2e [2p$_z$ − 2O)] + 4e [2p$_z$ − 2O)] are to be accounted for.

(v) The mutual overlapping of the 8p$_z$ orbitals give four bonding (n), two non-bonding and two anti-bonding (n − 2 = 2) (total 8) π$_D$MOs. Out of the 10e, 8e go to four bonding MOs and the remaining 2e in the two non-bonding orbitals as unpaired spin. The two unpaired electrons in the two non-bonding MOs are due to two − ve charges over the anion. A simplified MO diagram for the anion is shown in below (Figure 2.48).

The delocalization of four π-bond over four C–C single bonds in C$_4$O$_4^{2-}$ is responsible for the C–C bond length of 1.46 Å, which is intermediate between ethylene (1.33 Å) (C=C) and ethane (1.54 Å) (C–C). Similarly, C=O bond length (1.26 Å) in this anion is lower than the C=O bond length (1.23 Å) in carbonyl compounds.

$$\overline{\sigma^*_{LCC}}\ \overline{\sigma^*_{LCC}}\ \overline{\sigma^*_{LCC}}\ \overline{\sigma^*_{LCC}}\ \overline{\sigma^*_{LCC}}\ \overline{\sigma^*_{LCC}}\ \overline{\sigma^*_{LCC}}\ \overline{\sigma^*_{LCC}} \quad (8)$$

$$\overline{\pi_{D7}}\qquad \overline{\pi_{D8}}\qquad \text{(Anti-bonding)}\ (2)$$

(18e) $\underline{\uparrow\downarrow}\ \underline{\uparrow\downarrow}\ \underline{\uparrow\downarrow}\ \underline{\uparrow\downarrow}\ \underline{\uparrow\downarrow}\ \underline{\uparrow\downarrow}\ \underline{\uparrow\downarrow}\ \underline{\uparrow\downarrow}$ (8) $\underline{\uparrow}\ \underline{\uparrow}$ (2)

lp(O) lp(O) lp(O) lp(O) lp(O) lp(O) lp(O) lp(O) $\pi_{D5}\ \pi_{D6}$

(Non-bonding)

(8e) $\underline{\uparrow\downarrow}\ \underline{\uparrow\downarrow}\ \underline{\uparrow\downarrow}\ \underline{\uparrow\downarrow}$ (Bonding) (4)

$\pi_{D1}\ \pi_{D2}\ \pi_{D3}\ \pi_{D4}$

(16e) $\underline{\uparrow\downarrow}\ \underline{\uparrow\downarrow}\ \underline{\uparrow\downarrow}\ \underline{\uparrow\downarrow}\ \underline{\uparrow\downarrow}\ \underline{\uparrow\downarrow}\ \underline{\uparrow\downarrow}\ \underline{\uparrow\downarrow}$ (8)

$\sigma_{LCC}\ \sigma_{LCC}\ \sigma_{LCC}\ \sigma_{LCC}\ \sigma_{LCC}\ \sigma_{LCC}\ \sigma_{LCC}\ \sigma_{LCC}$

E N E R G Y

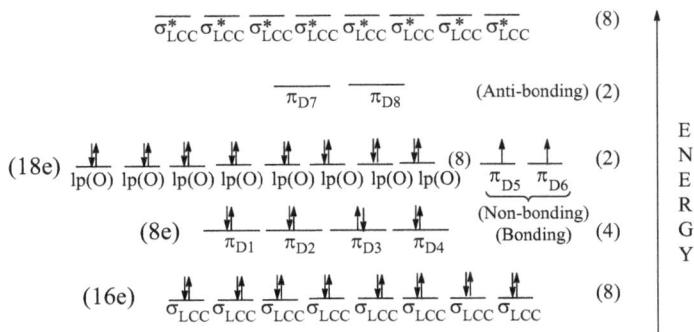

Figure 2.48: MO diagram of $C_4H_4{}^{2-}$.

2.5.2 Delocalized π-bonding in $C_5O_5{}^{2-}$

Croconate ion, $\mathbf{C_5O_5{}^{2-}}$, has almost perfect pentagonal symmetry (Figure 2.49). In this ion, the C–C and C=O bond lengths are 1.46 Å and 1.26 Å, respectively.

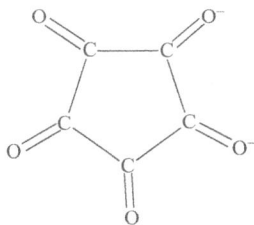

Figure 2.49: Pentagonal symmetrical structure of croconate ion, $C_5O_5{}^{2-}$.

The formation of delocalized π-bonding in croconate ion, $C_5O_5{}^{2-}$, can be understood as follows:

(i)   The anion contains a total of **52e** [5×4 (5C) + 5×6 (5O) + two −ve charge = **52e**] valence electrons.

(ii)   The formation of $10\sigma_L$ bonds [5σ(C–C) and 5σ(C–O)] involving sp^2 hybridization on each carbon and oxygen in $C_5O_5{}^{2-}$ is shown in Figure 2.50.

$5\ C_6 = 1s^2 2s^2 2p^2\ ;$ Valence shell electr. confgn. $= 2s^2\ 2p^2$
(ground state)

$;$ Valence shell electr. confgn. $= 2s^1\ 2p_x{}^1\ 2p_y{}^1\ 2p_z{}^1$
(excited state)

$\underbrace{\qquad\qquad}$
sp^2 hybridization

$3\ O_8 = 1s^2 2s^2 2p^4\ ;$ Valence shell electr. confgn. $= 2s^2\ 2p^4$
(ground state)

$= 2s^2\ 2p_x{}^2\ 2p_y{}^1\ 2p_z{}^1$

$\underbrace{\qquad\qquad}$
sp^2 hybridization

$2\ O_8^- = 1s^2 2s^2 2p^5\ ;$ Valence shell electr. confgn. $= 2s^2\ 2p^5$
(ground state)

$= 2s^2\ 2p_x{}^2 2p_y{}^1\ 2p_z{}^2$

$\underbrace{\qquad\qquad}$
sp^2 hybridization

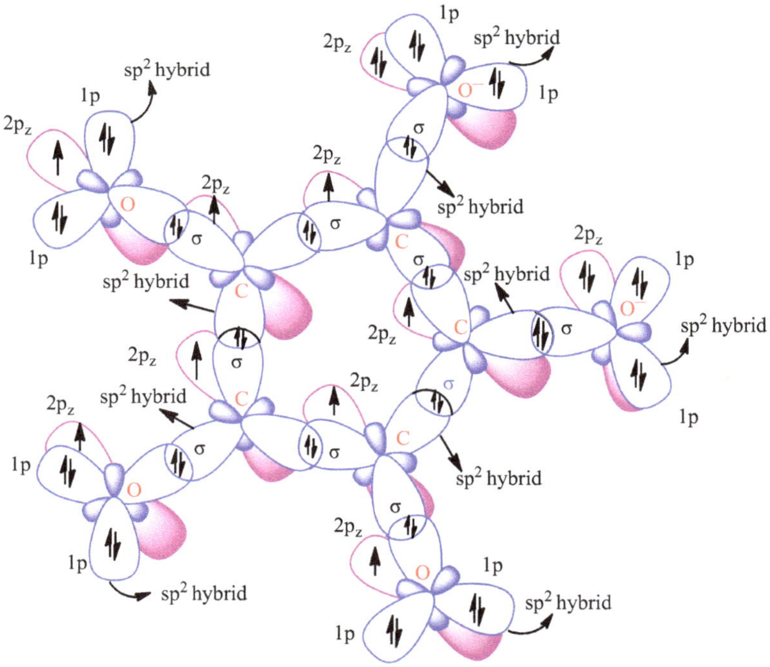

Figure 2.50: sp^2 Orbital overlaps and formation of $10\sigma_L$ bonds in $C_5H_5{}^{2-}$.

(iii) As clear from the hybridization scheme, the formation of **$10\sigma_L$ [5(C–C) and 5(C–O)]** and **10lps** [2lps in the two sp^2 hybrid orbitals of each of the five oxygens] accounts for **(20e + 20e) = 40e**, leaving 12e (52e – 40e = 12e) to be accommodated.

(iv) sp^2 hybridizations over five carbons and five oxygens leave one p_z orbital on each of the five carbons (all singly filled) as well as each of the five oxygens, three singly filled and the two bearing a negative charge are fulfilled. Thus, a total of 12e $\{5e[(4p_z - 5\,C) + 3e\,[2p_z - 3O)] + 4e\,[2p_z - 2O)]$ are to be accounted for.

(v) A simplified MO diagram of the anion comprising of $10\sigma_L$, $5\pi_D$ (n = 5) MOs (bonding), 10 lps, $2\pi_D$ MOs (non-bonding) and $3\pi D$ (n − 2 = 5 − 2 = 3) MOs (anti-bonding) are given in Figure 2.51. Again the presence of two unpaired electrons in two non-bonding orbitals is due to two −ve charges over the anion.

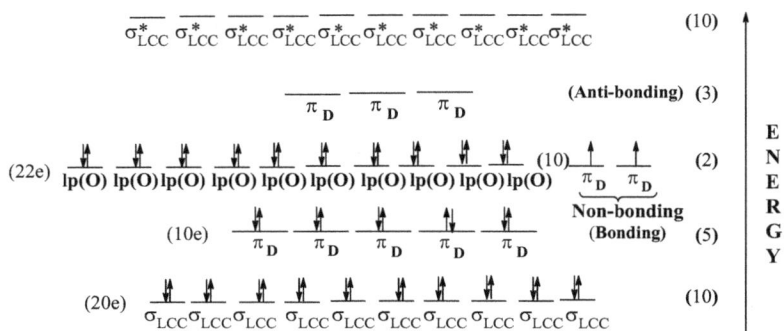

Figure 2.51: MO diagram of $C_5O_5{}^{2-}$ ion.

The delocalization of five π-bond over five C−C single bonds in $C_5O_5{}^{2-}$ is responsible for the C−C bond length of 1.46 Å, which is intermediate between ethylene (1.33 Å) (C=C) and ethane (1.54 Å) (C−C). Similarly, C=O bond length (1.26 Å) in this anion is lower than the C=O bond length (1.23 Å) in carbonyl compounds.

2.5.3 Delocalized π-bonding in $C_6O_6{}^{2-}$

This ion has almost perfect hexagonal symmetry (Figure 2.52). In this ion, the C−C and C=O bond lengths are 1.46 Å and 1.26 Å, respectively.

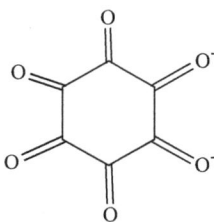

Figure 2.52: Hexagonal symmetrical structure of $C_6O_6{}^{2-}$.

The formation of delocalized π-bonding in $C_6O_6^{2-}$ can be understood as follows:

(i)   The anion contains a total of **62e** [6×4 (6 C) $+ 6 \times 6$ (6O) $+$ two $-ve$ charge $=$ **62e**] valence electrons.

(ii)   The formation of $12\sigma_L$ bonds [$6\sigma(C-C)$ and $6\sigma(C-O)$] involving sp^2 hybridization on each carbon and oxygen in $C_6O_6^{2-}$ is shown in Figure 2.53

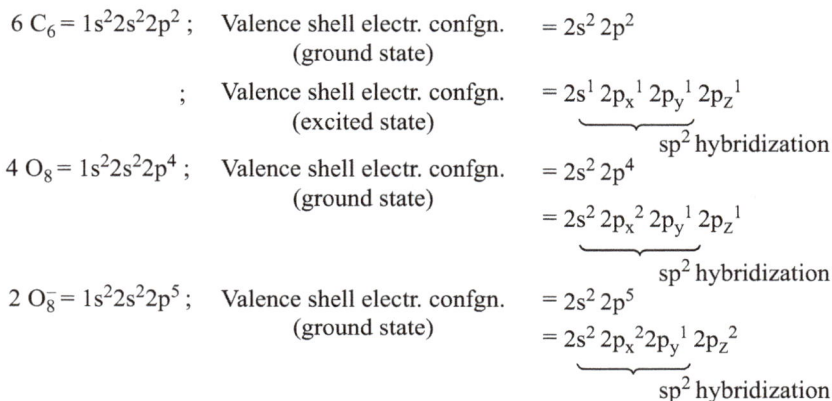

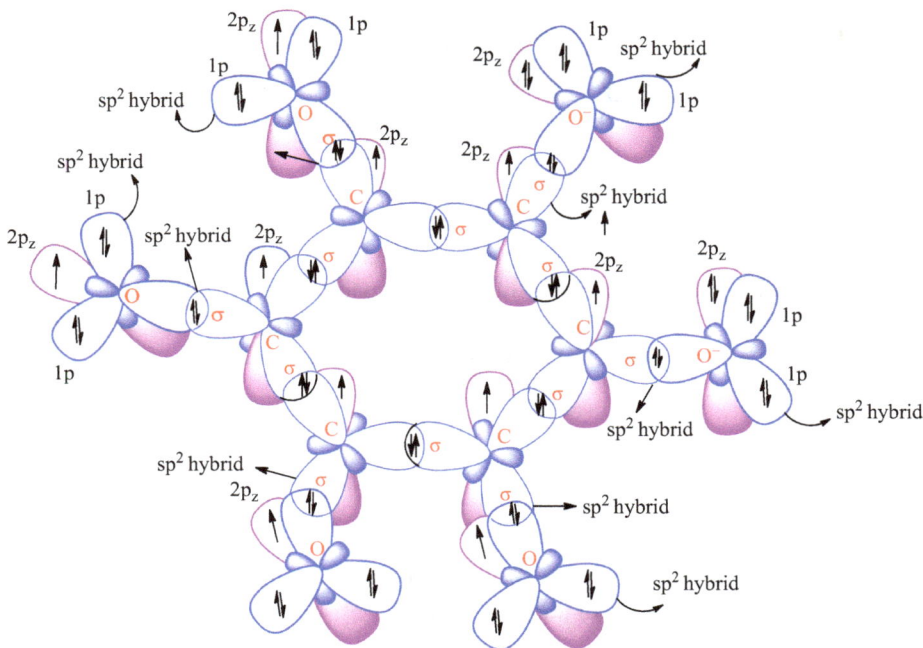

$6\ C_6 = 1s^2 2s^2 2p^2$;   Valence shell electr. confgn. $= 2s^2\ 2p^2$
(ground state)

;   Valence shell electr. confgn. $= 2s^1\ 2p_x{}^1\ 2p_y{}^1\ 2p_z{}^1$
(excited state)
$\underbrace{\qquad\qquad}_{sp^2 \text{ hybridization}}$

$4\ O_8 = 1s^2 2s^2 2p^4$;   Valence shell electr. confgn. $= 2s^2\ 2p^4$
(ground state)

$= 2s^2\ 2p_x{}^2\ 2p_y{}^1\ 2p_z{}^1$
$\underbrace{\qquad\qquad}_{sp^2 \text{ hybridization}}$

$2\ O_8^- = 1s^2 2s^2 2p^5$;   Valence shell electr. confgn. $= 2s^2\ 2p^5$
(ground state)
$= 2s^2\ 2p_x{}^2 2p_y{}^1\ 2p_z{}^2$
$\underbrace{\qquad\qquad}_{sp^2 \text{ hybridization}}$

Figure 2.53: sp^2 Orbital overlaps and formation of $12\sigma_L$ bonds in $C_6H_6^{2-}$.

(iii) From the hybridization scheme, it is well clear that the formation of **12σ_L** **[6(C–C) and 6(C–O)]** and **12lps** [2lps in the two sp^2 hybrid orbitals of each of the six oxygens] accounts for **(24e + 24e) = 48e**, leaving 14e (62e–48e = 14e) to be accommodated.

(iv) sp^2 Hybridizations over six carbons and six oxygens leave one p$_z$ orbital on each of the six carbons (all singly filled) as well as each of the six oxygens, four singly filled and the two bearing a negative charge are fulfilled. Thus, a total of 14e {6e[(4p$_z$–6C) + 4e [2p$_z$– 4O)] + 4e [2p$_z$– 2O)] are to be accounted for.

(v) A simplified MO diagram of the anion comprising of 12σ_L, 6π_D (n = 6) MOs (bonding), 12 lps, 2π_D MOs (non-bonding), and 4πD (n – 2 = 6 – 2 = 4) MOs (anti-bonding) are given in Figure 2.54. Again the presence of two unpaired electrons in two non-bonding orbitals is due to two –ve charges over the anion

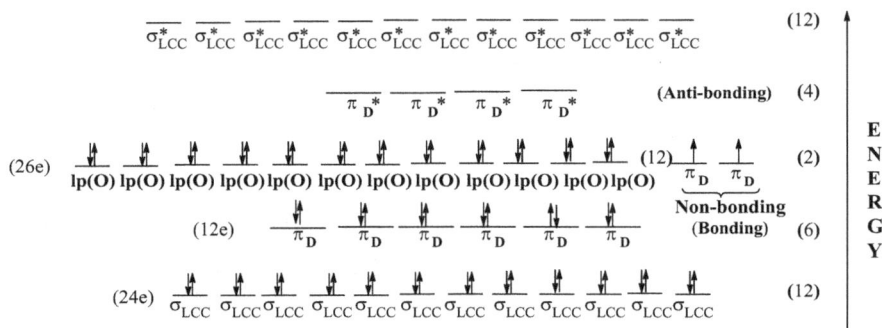

Figure 2.54: MO diagram of C$_6$H$_6{}^{2-}$ ion.

The delocalization of six π-bond over six C–C single bonds in C$_6$O$_6{}^{2-}$ is responsible for the C–C bond length of 1.46 Å, which is intermediate between ethylene (1.33 Å) (C=C) and ethane (1.54 Å) (C–C). Similarly, C=O bond length (1.26 Å) in this anion is lower than the C=O bond length (1.23 Å) in carbonyl compounds.

Exercises

Multiple choice questions/fill in the blanks

1. Sort out a molecule involving delocalized π-bonding from the following:
    (a) N$_2$ (b) H$_2$C=CH$_2$ (c) CO   (d) NO$_2$

2. The N–O bond order in NO$_2$ is:
    (a) 1.33   (b) 1.5   (c) 2   (d) None of these

3. In which one of the following species N–O bond order is 1.33?
   (a) NO_2 (b) NO_2^- (c) NO_3^- (d) None of these

4. The number of delocalized bonding π–MOs in benzene is:
   (a) 2   (b) 3   (c) 4   (d) 6

5. In cyclic molecule of the composition C_nH_n, the number of p_z-orbitals available for delocalized π-interaction is:
   (a) n–2   (b) n   (c) n + 2   (d) n + 1

6. The number of antibonding delocalized πMOs in cyclopentadienyl radical is:
   (a) 4   (b) 3   (c) 2   (d) None of these

7. The number of electrons available for π-interaction in cyclopentadienyl radical is:
   (a) 3   (b) 4   (c) 5   (d) None of the above

8. The geometry of croconate ion is:
   (a) Square planar   (b) Pentagonal symmetrical
   (c) Tetrahedral   (d) Hexagon

9. The maximum number of nodal planes in delocalized π–MOs of benzene is:
   (a) 1   (b) 2   (c) 3   (d) None of these

10. The number of nodal planes in antibonding delocalized π–MOs of benzene is:
    (a) 1   (b) 2   (c) 3   (d) 4

11. Which one of the following statements is correct for cyclic anions of the general composition $C_nO_n^{2-}$ with regard to the number of delocalized π–MOs?
    (a) two bonding, n non-bonding and $(n-2)$ anti-bonding
    (b) $(n-2)$ bonding, two non-bonding and n anti-bonding
    (c) n bonding, two non-bonding and $(n-2)$–anti-bonding
    (d) None of these

12. The bond order of C–C bond in benzene is:
    (a) 1   (b) 2   (c) 1.5   (d) None of these

13. In Cu_2O (the cuprite mineral of Cu) each O atom is surrounded tetrahedrally with Cu atoms in number:
    (a) 2   (b) 3   (c) 5   (d) None of these

14. Cuprite mineral have copper in + 1 oxidation state is:
    (a) Colourless   (b) Green   (c) Red   (d) Blue

15. Red colour of cuprite mineral is due to excitation of an electron from a delocalized bonding π–MO largely concentrated on O to a higher energy non-bonding π–MO largely concentrated on Cu called:
    (a) d–d transition   (b) π–π transition
    (c) σ–π transition   (d) Charge transfer transition

16. The bond length of B–F bond in BF_3 molecule is:
    (a) 1.51 Å   (b) 1.31 Å   (c) 1.25 Å   (d) None of these

17. In BF_3 molecule, the actual number of bonding non–bonding and antibonding delocalized π–MOs is:
    (a) Bonding (1), Non-bonding (2), Anti-bonding (1)
    (b) Bonding (2), Non-bonding (1), Anti-bonding (1)
    (c) Bonding (1), Non-bonding (1), Anti-bonding (2)
    (d) None of these

18. The bond length of N–O in NO_2^- is:
    (a) 1.36 Å   (b) 1.15 Å   (c) 1.22 Å   (d) None of these

19. The azide ion (N_3^-) contains structure with bond angle:
    (a) Bent structure with bond angle 115°
    (b) Linear symmetrical structure with bond angle 180°
    (c) Bent structure with bond angle 120°
    (d) None of these

20. Sort out the paramagnetic species form the following:
    (a) BF_3 (b) NO_2 (c) NO_2^- (d) NO_3^-

21. The bond angle in ozone molecule is . . . .

22. Magnetically, O_3 molecule is . . . . in nature.

23. Bonding π_D MOs in NO_2 molecule is . . . . in number.

24. The C–C bond length in cyclic $C_6O_6^{2-}$ ion is . . . .

25. The number of lone pairs in O_3 molecule is . . . .

26. The number of localized π-bond in hydrazoic acid is . . . .

Short answer type questions

1. Work out the formation of three delocalized π–MOs in O_3.
2. Briefly highlight the formation of six delocalized π–MOs in hydrazoic acid anion, N_3^-.
3. With regard to π–MOs, point out the salient features of cyclic molecules of the composition C_nH_n.
4. Explain why the experimentally determined B–F bond distance in BF_3 molecule is so short (1.31 Å) while the B–F bond distance in the only σ-bonded molecule should be approximately 1.51 Å.
5. Just draw the MO diagram of NO_3^- involving localized σ-bonding and delocalized π–bonding and calculate the N–O bond order in this anion.

6. Explain why the bond lengths (1.18 Å) for both N^I–N^{II} and N^{II}–N^{III} bonds in azide ion is so reduced compared to N^I–N^{II} bond length (1.24 Å) of hydrazoic acid.

7. Highlighting the formation of four delocalized πMOs in NO_3^-, give the number of bonding, non-bonding and anti-bonding πMOs.

8. Draw the MO diagram of BF_3 molecule involving both localized-σ and delocalized-π bonds, and justify that the molecule is diamagnetic with B–F bond order of 1.33.

9. Presenting the formation of three doubly degenerate set of π_D MOs in Cu_2O, show the charge transfer transitions responsible for its red colour.

10. Draw the hybridization scheme showing sp^2 orbital overlaps and formation of six (C–C)σ and (C–H)σ bonds in C_6H_6. Also calculate the C–C bond order in this molecule.

11. Draw the sp^2 orbital overlaps for the formation of five localized σ bonds in C_5H_5 radical and thus give the number and nature of unused p-orbitals for delocalized π-bonding.

12. Draw the MO diagram of $C_4H_4^{2-}$ and show that there are two unpaired electrons present in two non–bonding orbitals responsible of two negative charges on this cyclic anion.

13. Draw the orbital overlaps for the formation of $10\sigma_L$ bonds in $C_5H_5^{2-}$ and show that it contains 10 lone pairs and 10 unused p_z orbitals with 12 electrons.

14. Make a sketch of MO diagram of $C_6H_6^{2-}$ ion involving localized σ bonds and delocalized π-bonds, and show that this anion is paramagnetic with respect to two unpaired electrons.

15. Taking a suitable example of a molecule/ion involving delocalized π–bonding your own choice, draw its MO diagram and hence highlights on its magnetic behaviour.

Long answer type questions

1. Draw and discuss the MO diagram of NO_2^-.
2. Draw and discuss the MO diagram of a cyclic molecule.
3. Using the concept of delocalized π-bonding, explain the shorter B–F bond distance (1.31 Å) in BF_3 compared to B–F bond distance ~1.51 Å in the only σ-bonded molecule.
4. Present a detailed account of MO diagram of hydrazoic acid involving delocalized π-bonding.
5. How does the concept of delocalized π-bonding, explain the shorter C–C bond length and longer C=O bond length in $C_5O_5^{2-}$ ion?
6. Give an account of delocalized π-bonding in $C_6O_6^{2-}$ ion of hexagonal symmetry.

7. Describe in detail the delocalized π-bonding in 3,4-diketocyclobutene diol ($C_4O_4^{2-}$). How does this bonding explain the shorter C–C bond length (1.46 Å) and longer C=O bond length (1.26 Å) in $C_4O_4^{2-}$ ion?

8. Discuss the formation of MO diagram of benzene molecule including only (C–C)σ and (C–C)π MOs.

9. Explain how the concept of delocalized π-bonding in cuprite mineral (Cu_2O) mineral explains its appearance as red in colour.

10. Based on the concept of delocalized π-bonding, explain the N–O bond order (1.33) in nitrate ion (NO_3^-).

Chapter III
Chemistry of borane and related compounds: structure, bonding and topology

3.1 Introduction

Boranes are hydrides of boron. They are covalent compounds and are called boranes on analogy with alkanes. Borane chemistry began in 1912 with A. Stock's classic investigations and the numerous compounds prepared by his group during the following 20 years. During the past 50 years, the chemistry of boranes and the related carbaboranes and metalloboranes have been the major growth areas in inorganic chemistry, and interest continues intensified.

3.2 Importance of boranes

The importance of boranes stems from the following three factors:
- The completely unsuspected structural principle involved.
- The growing need to extend covalent molecular orbital (MO) bond theory considerably to cope with the unusual stoichiometries.
- The emergence of a versatile and extremely extensive reaction chemistry which parallels but is quite distinct from that of organic and organometallic chemistry.

This growing activity of boranes resulted (in the centenary year of Stock's birth) in the award of the 1976 Nobel Prize in Chemistry to W. N. Lipscomb (Harvard) for his studies of boranes which have illuminated the problems of chemical bonding.

3.3 Classification of boranes

Several neutral boranes of the general composition B_nH_m and even larger number of borane anions, $B_nH_m^{x-}$, have been characterized so far. For convenience of studies, these can be classified into five series. Though, examples of neutral boranes are not known for all the five classes.

3.3.1 *Closo*-boranes

The name is derived from the Greek word '*clovo*' which means a cage. These boranes have completely closed polyhedral clusters of *n-boron* atoms. The neutral

https://doi.org/10.1515/9783110727289-003

closo-boranes will be of the general composition B_nH_{n+2} but are not known, while certain anion of the composition $B_nH_n^{2-}$ are known. The structure of $B_6H_6^{2-}$ is shown in Figure 3.1.

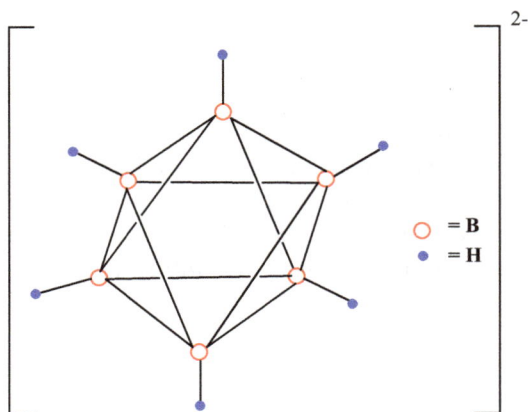

$O = B$
$\bullet = H$

Figure 3.1: Structure of *closo*-borane anion $B_6H_6^{2-}$.

3.3.2 *Nido*-boranes

The name is derived from the **Latin** word **'*Nidus*'** which means a nest. These boranes have non-closed structures in which the B_n cluster occupies *n*-corners of an $(n+1)$-cornered polyhedron. Neutral ***nido*-boranes** are of the general formula, B_nH_{n+4}. For example, B_2H_6, B_5H_9, B_6H_{10} and $B_{10}H_{14}$. The non-closed structures of B_2H_6 and B_5H_9 are shown in Figure 3.2.

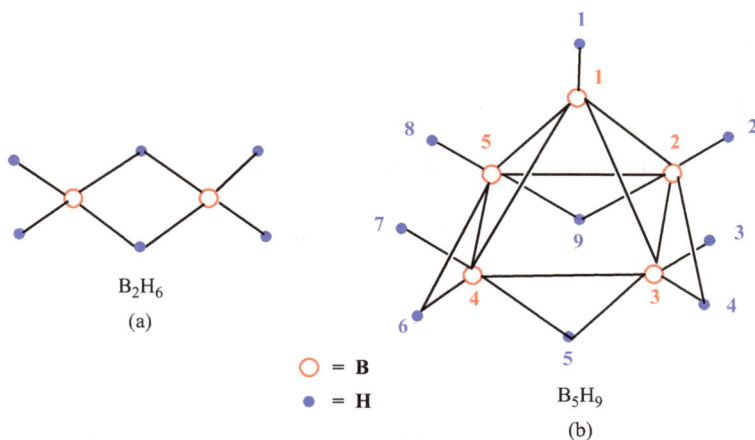

B_2H_6
(a)

$O = B$
$\bullet = H$

B_5H_9
(b)

Figure 3.2: The non-closed structure of (a) B_2H_6 and (b) B_5H_9.

The mono-anions $B_nH_{n+3}^-$ are formed by removal of one bridging hydrogen from the neutral B_nH_{n+4}; for example, $B_5H_8^-$, $B_{10}H_{13}^-$ and so on. Other anions in this series such as $B_4H_7^-$ and $B_9H_{12}^-$ are known, though the parent boranes have proved too difficult to isolate. The di-anions of the composition $B_nH_{n+2}^{2-}$; for example, $B_{10}H_{12}^{2-}$ and $B_{11}H_{13}^{2-}$ are also known.

3.3.3 *Arachno*-boranes

The name is derived from the **Greek** word '*Arachine*' which means a spiders web. These boranes have even more open clusters in which the **B atoms** occupy n adjoining corners of an $(n + 2)$-cornered polyhedron.

Arachno boranes known are of the following types in terms of neutral and anionic boranes:

Neural: B_nH_{n+6}; for example, B_4H_{10}, B_5H_{11} and B_6H_{12}.
Mono-anionic: $B_nH_{n+5}^-$; for example, $B_2H_7^-$, $B_3H_8^-$, $B_5H_{10}^-$, $B_9H_{14}^-$ and $B_{10}H_{15}^-$.
Di-anionic: $B_nH_{n+4}^{2-}$; for example, $B_{10}H_{14}^{2-}$

3.3.4 *Hypho*-boranes

The name is derived from the **Greek** word '*hyphe*' which means a net. Such boranes have the most open clusters in which the **B atoms** occupy n corners of an $(n + 3)$-cornered polyhedron, often visualizable as polyhedral fragments formed by removal of three contiguous/adjoining vertices of a complete polyhedron.

These boranes have the general formula B_nH_{n+8} for neutral boranes. However, no neutral boranes have yet been definitely established in this series but the known compounds B_8H_{16} and $B_{10}H_{18}$ may prove to be *hypho*-boranes.

3.3.5 *Conjuncto*-boranes

The name is derived from the **Latin** word '*conjuncto*' which means join together. Such boranes have structures formed by linking two or more of the preceding types of clusters together. They have general formula B_nH_m. At least five different structure types of interconnected borane clusters have been identified. They have the following features:
(a) Fusion by sharing single common B atom: for example, $B_{15}H_{23}$ (Figure 3.3).
(b) Formation of a direct two-centre B–B σ-bond between two clusters: for example, B_8H_{18}, that is, $(B_4H_9)_2$ (Figure 3.4); $B_{10}H_{16}$, that is, $(B_5H_8)_2$; $B_{20}H_{26}$, that is, $(B_{10}H_{13})_2$ and so on.

(c) Fusion of two clusters via 2B atoms at a common edge: for example, $B_{13}H_{19}$, $B_{14}H_{18}$, $B_{14}H_{20}$, $B_{16}H_{20}$, n-$B_{18}H_{22}$ (centrosymmetric), i-$B_{18}H_{22}$ (non-centrosymmetric) (Figure 3.5(a)) and so on.

(d) Fusion of two clusters via 3B atoms at a common face. No neutral borane or borane anion is yet known with this conformation, but the solvated complex $(MeCN)_2B_{20}H_{16} \cdot MeCN$ has this structure (Figure 3.5(b)).

(e) More extensive fusion involving four boron atoms.

Structures of some *conjuncto*-boranes are shown below:

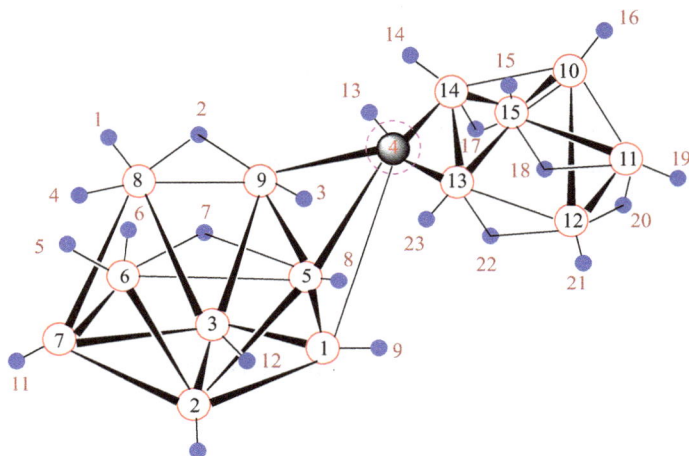

Figure 3.3: Structure of $B_{15}H_{23}$ sharing a common B atom.

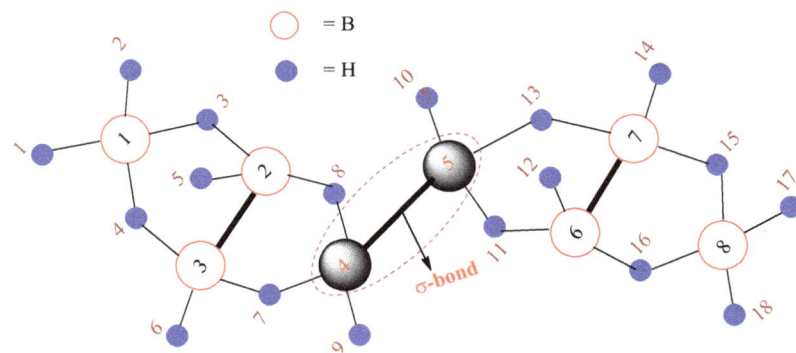

Figure 3.4: Structure of *conjuncto*-B_8H_{18} comprising of two B_4H_9 units linked by a direct B–B σ-bond.

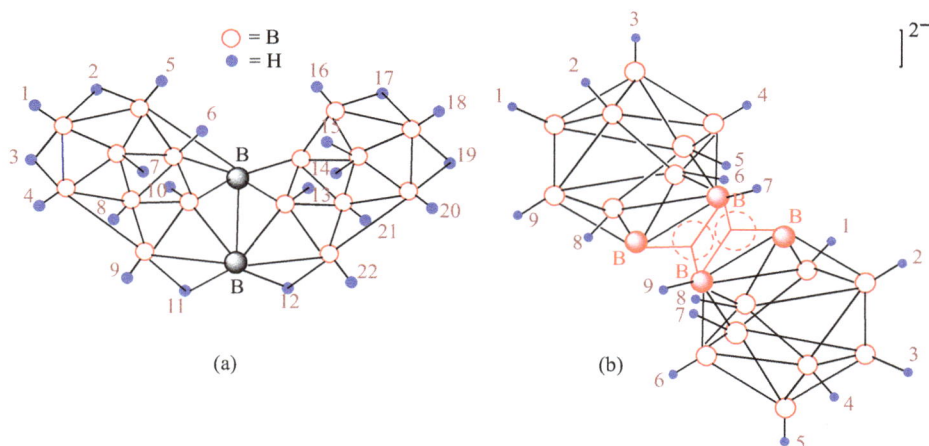

Figure 3.5: (a) Structure of i-$B_{18}H_{22}$ fused via 2B atoms at a common edge (b) Structure of $B_{20}H_{18}^{2-}$ ion. The two three-centre B–B–B bonds joining the two $B_{10}H_9^-$ units by red lines.

3.4 Nomenclature of boranes

The following three rules are followed in the nomenclature of boranes:

(i) Boranes are usually named by indicating the number of B atoms in a **Latin prefix** and the number of H atoms by an **Arabic number** in parentheses. For example:

B_2H_6:   Diborane (6)
B_4H_{10}:   Tetraborane (10)
B_5H_9:   Pentaborane (9)
B_5H_{11}:   Pentaborane (11)
B_6H_{10}:   Heaxborane (10)
B_8H_{12}:   Octaborane (12)
B_9H_{15}:   Nonaborane (15)
$B_{10}H_{14}$:   Decaborane (14)
$B_{20}H_{16}$:   Icosaborane (16)

(ii) The names of anions end in **'ate'** rather than **'ane'** and specify the number of H atom first and then B atom and the charge. For example:

$B_5H_8^-$:   Octahydropentaborate (1-)
$B_{10}H_{10}^{2-}$:   Decahydrodecaborate (2-)
$B_{10}H_{14}^{2-}$:   Tetradecahydrodecaborate (2-)

(iii) Further information can be provided by the optional inclusion of classes of boranes: for example, *closo-*, *nido-*, *arachno-*, *hypho-* or *conjuncto-* in italic form. For example:

$B_{10}H_{14}$:   *nido*-Decaborane (14)
$B_{10}H_{10}^{2-}$:   Decahydro-*closo*-decaborate (2-)
$B_{10}H_{14}^{2-}$:   Tetradecahydro-*arachno*-decaborate (-2)
$B_{10}H_{16}$:   *conjuncto*-Decaborane (16)

3.5 Properties of boranes

(i)   Boranes are colourless, diamagnetic, molecular compounds of moderate to low thermal stability.
(ii) The lower members are gases at room temperature but with increasing molecular weight they become volatile liquids or solids. Boiling points of boranes are approximately the same as those of hydrocarbons of similar molecular weights.
(iii) The boranes are all endothermic and their free energy of formation $\Delta G°$ is also positive. Their thermodynamic instability results from the exceptionally strong interatomic bonds in both elemental B and H_2 rather than the inherent weakness of the B–H bond. In this respect, boranes resemble the hydrocarbons.
(iv) It has been estimated that typical bond energies in boranes are: B–H_t (380), B–H–B (440), B–B (330) and B–B–B (380) kJ mol^{-1}, compared with bond energy of 436 kJ mol^{-1} for H_2.
(v)   Boranes are extremely reactive and several are spontaneously flammable in air. *Arachno*-boranes tend to be more reactive (also less stable to thermal decomposition) than *nido*-boranes and reactivity also diminishes with increasing molecular weight.
(vi) *Closo*-boranes anions are exceptionally stable and their general chemical behaviour has suggested the term 'three dimensional aromaticity'.

3.6 Metallaboranes/metalloboranes or borane complexes

Metallaboranes are borane cages containing one or more metal atoms, but no carbon atoms, in the skeletal framework. The 'metalla/metallo' usage is derived from the 'oxo/aza' convention of organic chemistry and denotes a metal replacing a skeletal boron atom. Most of the authors use the term metalloboranes and some one metallaboranes. But we will use the term metalloboranes for borane cages involving one or more metal atoms.

Reactions between boron hydrides and neutral metal reagents or more commonly between boron hydride anions and cationic metals or metal ligand units have been employed to generate a large number of metalloboranes ranging from very small four- and five-vertex species to multi-cage 'macropolyhedral' systems. With a few exceptions, these syntheses are uncontrolled, often affording products of novel structure but in wide varying yields. Consequently, synthetic routes having broad applicability are almost non-existent. For example:

(i) Co-pyrolysis of B_5H_9 and $[Fe(CO)_5]$ in a hot or cold reactor at 220/20 °C for three days gives an orange liquids (m. p. 5 °C) of formula $[1-\{Fe(CO)_3\}B_4H_8]$ having the structure shown in Figure 3.6.

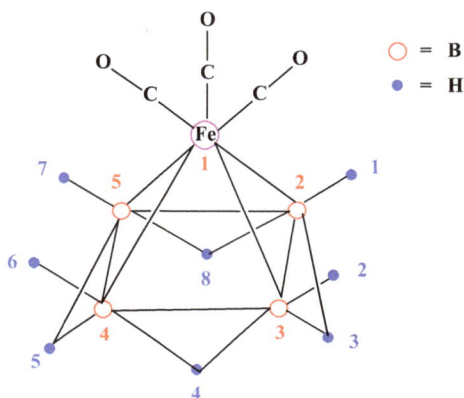

Figure 3.6: Structure of $[1-\{Fe(CO)_3\}B_4H_8]$.

(ii) The isoelectronic with complex $[1-\{Fe(CO)_3\}B_4H_8]$, the complex $[1-\{Co(\eta^5-C_5H_5)\}B_4H_8]$ (Figure 3.7) can be obtained as yellow crystals by pyrolysis at 200 °C of the corresponding basal derivative $[2-\{Co(\eta^5-C_5H_5)\}B_4H_8]$ (Figure 3.8) which is obtained as red crystals from the reaction of NaB_5H_8 and $CoCl_2$ with NaC_5H_5 in THF at −20 °C.

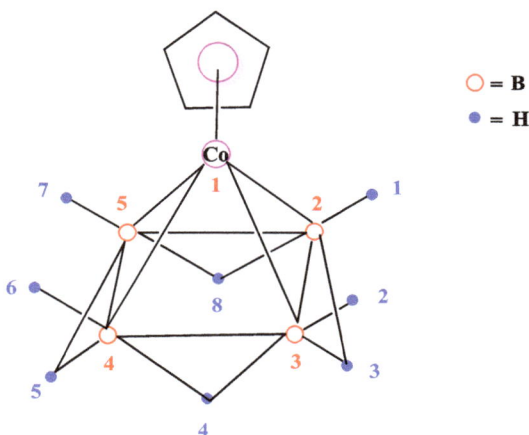

Figure 3.7: Structure of $[1-\{Co(\eta^5-C_5H_5)\}B_4H_8]$.

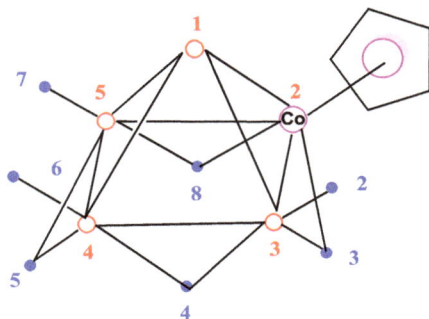

Figure 3.8: Structure of $[2\text{-}\{Co(\eta^5\text{-}C_5H_5)\}B_4H_8]$.

The course of both preparative reactions (i) and (ii) is obscure/unclear and other products are also obtained.

Since borane substrates, particularly the smaller ones, are quite reactive towards metal reagents, it is frequently the case that attack of the metal can occur at more than one sight. Moreover, the incorporation of the first metal atom into the cage may promote further reactions leading to di- or even tri-metallic products, as well as isomers. As a result, metalloboranes are often formed as complex mixtures of isolable species. For example:

(i) The first *closo*-metalloborane clusters were obtained from the reaction of the *nido*-$B_5H_8^-$ ion with $CoCl_2$ and $NaCp$ ($Cp = C_5H_5$), which generated both the *nido* complex $2\text{-}CpCoB_4H_8$ (Figure 3.9) and a family of Co_2, Co_3 Co_4 *closo*-metalloboranes.

(ii) Reactions of metal clusters with monoboron reagents such as BH_4^- or $BH_3 \cdot THF$ have been used to generate metal rich complexes, for example $Cp_4Co_4B_2H_4$, and metal bridge clusters that contain encapsulated boron atoms: for example (i) octahedral $H(CO)_{17}Ru_6B$ (Hong et al. 1989) and (ii) trigonal prismatic $[PPN][(H)_2(CO)_{18}Ru_6B]$ (Housecroft et al. 1992).

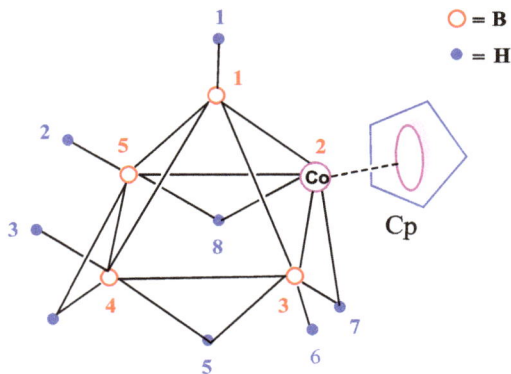

○ = **B**
● = **H**

Cp

Figure 3.9: Structure of $2\text{-}CpCoB_4H_8$.

3.7 Bonding and topology of boranes

Studies of boranes indicate the presence of the following types of bonds (Figure 3.10) in boranes:

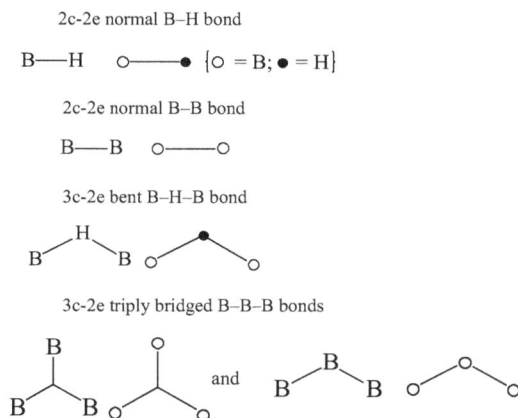

2c-2e normal B–H bond

B—H   ○———● {○ = B; ● = H}

2c-2e normal B–B bond

B—B   ○———○

3c-2e bent B–H–B bond

3c-2e triply bridged B–B–B bonds

and

Figure 3.10: Types of bonds in higher boranes.

The 2c-2e B–H and B–B bonds are usual covalent bonds. But the 3c-2e B–H–B and B–B–B bonds are unusual bonds that will be discussed in following section.

3.7.1 Bonding in boranes and higher boranes/boron clusters: molecular orbital approach

The bonding in Boranes may be explained on the basis of an idea of 3c-2e B–H–B and B–B–B types of bond proposed by H. C. Longuet-Higgins in 1943. As an undergraduate, he proposed the correct structure of the chemical compound diborane (B_2H_6), which was then unknown because it turned out to be different from structures in contemporary chemical structures in contemporary chemical valence theory. This was published in 1943 with his tutor, R. P. Bell (Longuet-Higgins and Bell 1943). This idea was later on refined by W. N. Lipscomb.

3.7.2 Molecular orbital approach

In simple MO theory of covalent bonding, two AOs of the same symmetry and almost similar energy linearly combine with each other to give rise two MOs (bonding and anti-bonding) by the linear combination of atomic orbitals method. Similarly, three

AOs of same symmetry and similar energy combine linearly and would give three MOs (number of MOs is always equal to the number of atomic orbitals combine).

3.7.3 Formation of B–H–B bond

As the attachment with each of the boron atom in borane is 4 (Figure 3.11(a)), sp^3 hybridization may be assumed in the formation of B_2H_6. Of the four sp^3 hybrid orbitals on each boron atom, two are used in the formation of two terminal $B–H_t$ bonds. The rest two sp^3 hybrid orbitals on each boron atom along with 1s atomic orbital of two hydrogen atoms, will now be utilized in the formation of two sets of B–H–B bonds. It is notable here that each B–H–B bond has only two electrons [one electron from one sp^3 hybrid orbital of boron (2nd sp^3 hybrid orbital is empty) and one electron from 1s atomic orbital of bridging hydrogen, H_b] (Figure 3.11(b)).

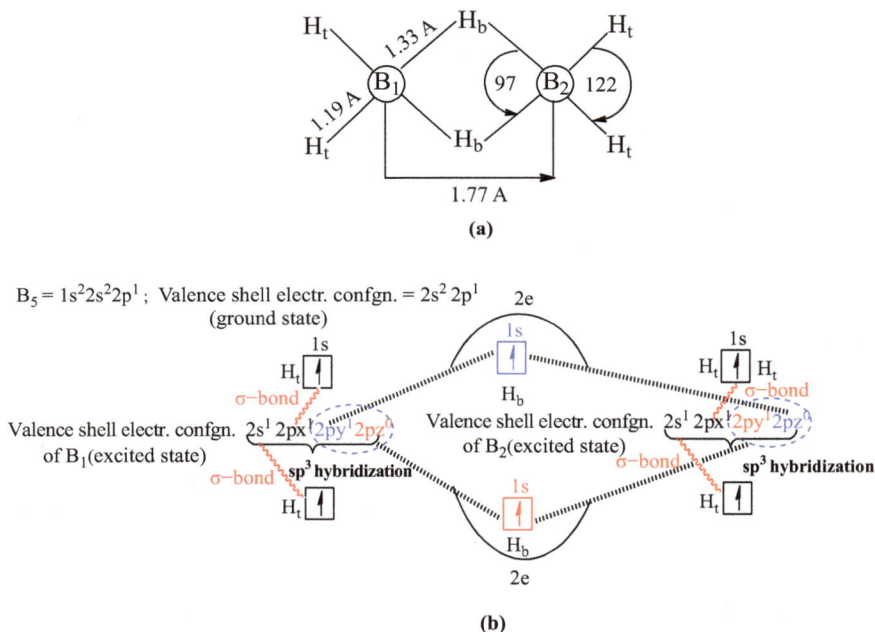

Figure 3.11: (a) Structure of biborane and (b) hybridization scheme in diborane.

The idea of linear combination of three AOs in borane can be visualized in the following manner:

Let ΨB_1 be one of the four sp^3 hybrid orbitals of one boron atom combine linearly with an sp^3 hybrid orbital ΨB_2 of another boron atom. This results in the formation of two MOs,

$$1/\sqrt{2}\ (\Psi B_1 + \Psi B_2) \quad (\text{I}) \text{ and} \tag{3.1}$$

$$1/\sqrt{2}\ (\Psi B_1 - \Psi B_2) \quad (\text{II}) \tag{3.2}$$

These two linear combinations (I) and (II) of ΨB_1 and ΨB_2 are pictorially shown in Figure 3.12.

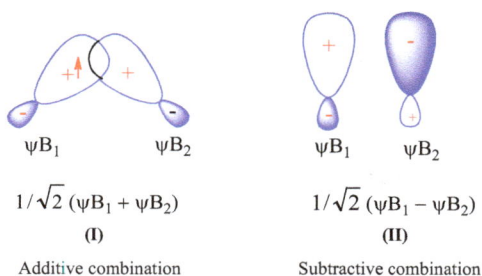

$$1/\sqrt{2}\ (\psi B_1 + \psi B_2)$$

(I)

Additive combination

$$1/\sqrt{2}\ (\psi B_1 - \psi B_2)$$

(II)

Subtractive combination

Figure 3.12: Additive and subtractive combination of ΨB_1 and ΨB_2.

The additive combination brings the positive lobe of ΨB_1 near the positive lobe of ΨB_2, making positive overlap possible between ΨB_1 and ΨB_2. This results in a decrease of energy of the MO formed by this combination. Contrary to this, the subtractive combination brings positive lobe of ΨB_1 near the negative lobe of ΨB_2 making only negative overlap possible between ΨB_1 and ΨB_2. This results in an increase of energy of the MO formed by this combination.

The relative energies of the two AOs (here AO mean sp^3 hybrid orbitals) ΨB_1 and ΨB_2 and resulting MOs are shown in Figure 3.13 along with their normalized wave functions.

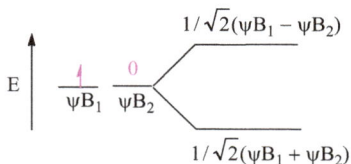

Figure 3.13: Relative energies of two AOs and resulting two MOs.

Let us now combine wave function of AO of H atom, ΨH with $1/\sqrt{2}(\Psi B_1 + \Psi B_2)$ and $1/\sqrt{2}(\Psi B_1 - \Psi B_2)$.

As shown in Figure 3.14, the MO, $1/\sqrt{2}(\Psi B_1 - \Psi B_2)$ cannot combine with ΨH either by addition or by subtraction.

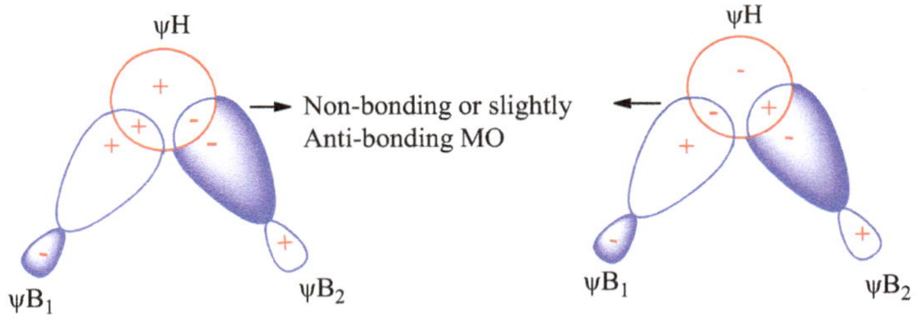

Additive combination of $1/\sqrt{2}(\psi B_1 - \psi B_2)$ with ψH. Net overlap is evidently zero. Hence, such a combination does not occur.

(III)

Subtractive combination of $1/\sqrt{2}(\psi B_1 - \psi B_2)$ with ψH. Net overlap is evidently zero. Hence, such a combination does not occur.

(IV)

Figure 3.14: Additive and subtractive combination of $1/\sqrt{2}(\Psi B_1 - \Psi B_2)$ with ΨH.

However, the MO, $1/\sqrt{2}(\Psi B_1 + \Psi B_2)$ would easily combine with ΨH both by additive and subtractive combinations as illustrated in Figure 3.15:

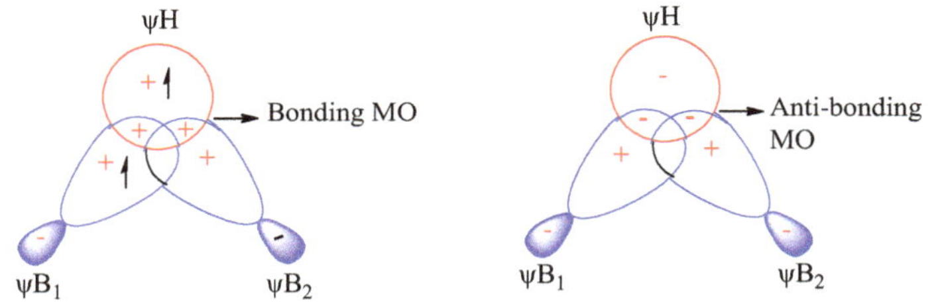

Additive combination of $1/\sqrt{2}(\psi B_1 + \psi B_2)$ with ψH. Net overlap is non-zero and +ve. Hence, such a combination is effective and leads to lowering of energy.

(V)

Subtractive combination of $1/\sqrt{2}(\psi B_1 + \psi B_2)$ with ψH. Net overlap is non-zero and -ve. Hence, such a combination is effective in the -ve sense and leads to increase in energy.

(VI)

Figure 3.15: Additive and subtractive combination of $1/\sqrt{2}(\Psi B_1 + \Psi B_2)$ with ΨH.

Thus, a linear combination of ΨB_1, ΨB_2 and ΨH results in three MOs, one is bonding (V), the other is non-bonding (III or IV) as ΨH does not affect the energy of $1/\sqrt{2}(\Psi B_1 - \Psi B_2)$ and the last one (VI) is strongly anti-bonding.

Hence, the above pictorial representations conclude that the linear combination of two-centre MO, $1/\sqrt{2}(\Psi B_1 \pm \Psi B_2)$ with ΨH giving rise to bonding, non-bonding and anti-bonding MOs along with their normalized wave functions, is shown in Figure 3.16.

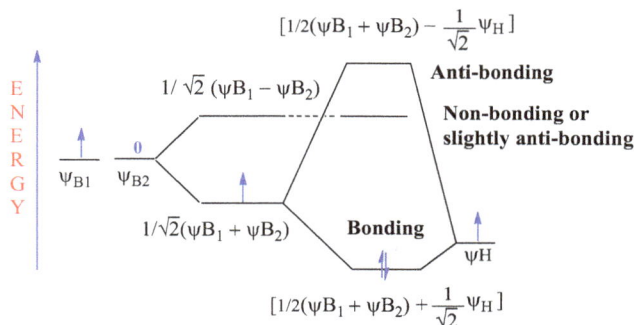

Figure 3.16: MO diagram for the formation of B–H–B bond in diborane.

It is notable here that the square of the coefficient of any atomic orbital wave function give the contribution of that wave function to the MO. In the normalized MO wave function, the sum of squares of the coefficients in it equals to unity.

The presence of two electrons in the lowest energy bonding MO is responsible for the formation and stability of 3c–2e B–H–B bond.

Bond order of B–H bond in B–H–B bond

Bond order of B–H bond

$$= \frac{\text{No. of electron in BO} - \text{No. of electron in AB orbital}}{2 \times \text{No. of bonding centre}}$$

$$= \frac{2-0}{2 \times 2} = \frac{2}{4}$$

$$= 0.5$$

Again, the bond order 0.5 of B–H bond in bridging B–H–B bond is self-explanatory for the stability of diborane (B_2H_6) molecule.

3.7.4 Formation of B–B–B bond

A similar 3c-2e B–B–B bond can be evoked by a linear combination of the three suitable orbitals with wave functions ΨB_1, ΨB_2 and ΨB_3 of the three boron atoms of higher borane clusters. Various combinations with respect to closed and open B–B–B bonds are pictorially shown below.

(a) Formation of closed B–B–B bond:

In the formation of closed B–B–B bond (Figure 3.17), wave functions of atomic orbitals to be combined are the sp^3-hybrid orbitals of each of the boron. These are ΨB_1, ΨB_2 and ΨB_3. Again, sp^3 hybrid orbitals corresponding to the wave functions ΨB_1 and ΨB_3 are having one electron and ΨB_2 is empty.

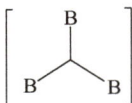

Figure 3.17: Closed B–B–B bond.

Again, the idea of linear combination of three AOs in higher boranes can be visualized in the following section.

Let ΨB_1 be one of the four sp^3 hybrid orbitals of one boron atom combined linearly with an sp^3 hybrid orbital ΨB_2 of another boron atom. This results in the formation of two MOs,

$$1/\sqrt{2}\ (\Psi B_1 + \Psi B_2) \quad \text{(I) and}$$

$$1/\sqrt{2}\ (\Psi B_1 - \Psi B_2) \quad \text{(II)}$$

The two linear combinations (I) and (II) of ΨB_1 and ΨB_2 are pictorially shown in Figure 3.18.

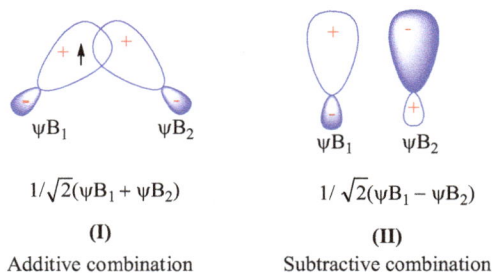

Figure 3.18: Additive and subtractive combination of ΨB_1 and ΨB_2.

Again, the additive combination brings the positive lobe of ΨB_1 near the positive lobe of ΨB_2, making positive overlap possible between ΨB_1 and ΨB_2. This results in a decrease of energy of the MO formed by this combination. Contrary to this, the subtractive combination brings positive lobe of ΨB_1 near the negative lobe of ΨB_2 making only negative overlap possible between ΨB_1 and ΨB_2. This results in an increase of energy of the MO formed by this combination.

Let us now combine wave function ΨB_3 of third boron atom, with $1/\sqrt{2}(\Psi B_1 + \Psi B_2)$ and $1/\sqrt{2}(\Psi B_1 - \Psi B_2)$. As shown below, the MO, $1/\sqrt{2}(\Psi B_1 - \Psi B_2)$ cannot combine with ΨB_3 either by addition or by subtraction as shown in Figure 3.19.

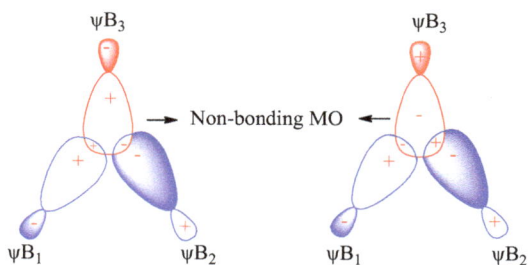

Additive combination of $1/\sqrt{2}\,(\psi B_1 - \psi B_2)$ with ψ_{B3}. Net overlap is evidently zero. Hence, such a combination does not occur.

(III)

Subtractive combination of $1/\sqrt{2}\,(\psi B_1 - \psi B_2)$ with ψ_{B3}. Net overlap is evidently zero. Hence, such a combination does not occur.

(IV)

Figure 3.19: Additive and subtractive combinations of ΨB_3 with $1/\sqrt{2}(\Psi B_1 - \Psi B_2)$. Net overlap is zero.

But, the MO, $1/\sqrt{2}(\Psi_{B1} + \Psi_{B2})$ would easily combine with ΨB_3 both by additive and by subtractive combinations as shown in Figure 3.20 given below:

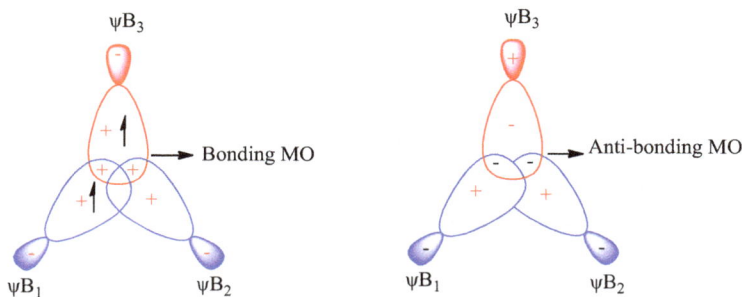

Additive combination of $1/\sqrt{2}\,(\psi B_1 + \psi B_2)$ with ψ_{B3}. Net overlap is non-zero and +ve. Hence, such a combination is effective and leads to lowering of energy.

(V)

Subtractive combination of $1/\sqrt{2}\,(\psi B_1 + \psi B_2)$ with ψ_{B3}. Net overlap is non-zero and -ve. and leads to increase in energy. Hence, such a combination is effective in the -ve sence.

(VI)

Figure 3.20: Additive and subtractive combinations of ΨB_3 with $1/\sqrt{2}(\Psi B_1 + \Psi B_2)$. Net overlap is non-zero.

(b) Formation of open B–B–B bond

In the formation of open B–B–B bond (Figure 3.21), the atomic orbitals involved in combination are one of the sp^3-hybrid orbitals of B_1 and B_2 and p-orbital of B_3. Again, B_1 and B_3 are half filled and B_2 is empty.

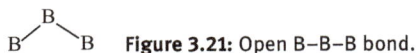

Figure 3.21: Open B–B–B bond.

The following three possible combinations giving the formation of bonding, non-bonding and anti-bonding MOs are shown in Figure 3.22.

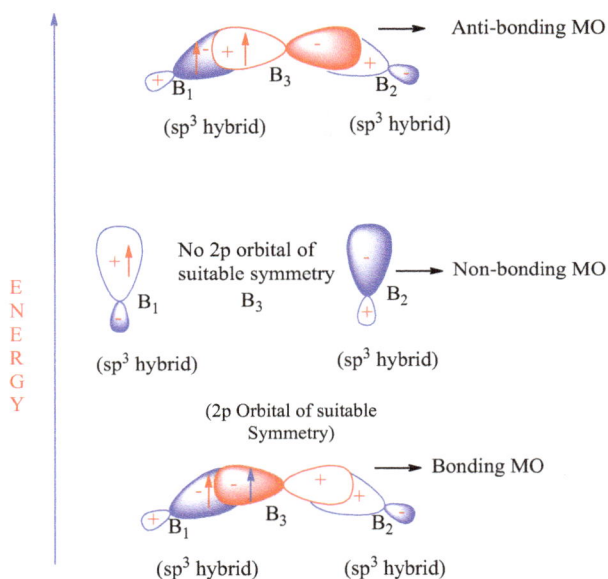

Figure 3.22: Overlapping scheme of orbitals of three boron orbitals resulting in the formation of bonding, non-bonding and anti-bonding MOs.

Based on above combinations, the resulting three-centred MO diagram for closed B–B–B and open B–B–B bonds are shown in Figure 3.23.

The presence of two electrons in the lowest energy bonding MO is responsible for the formation and stability of 3c-2e B–B–B bond in higher boranes.

$$\frac{1}{\sqrt{6}}[(\psi B_1 + \psi B_2) + 2\psi B_3]$$

Anti-bonding MO

$1/\sqrt{2}\,(\psi B_1 - \psi B_2)$

Non-bonding or slightly anti-bonding MO

E N E R G Y

ψB_1 0 ψB_2

$1/\sqrt{2}\,(\psi B_1 + \psi B_2)$ **Bonding MO**

ψB_3

$$\frac{1}{\sqrt{3}}(\psi B_1 + \psi B_2 + \psi B_3)$$

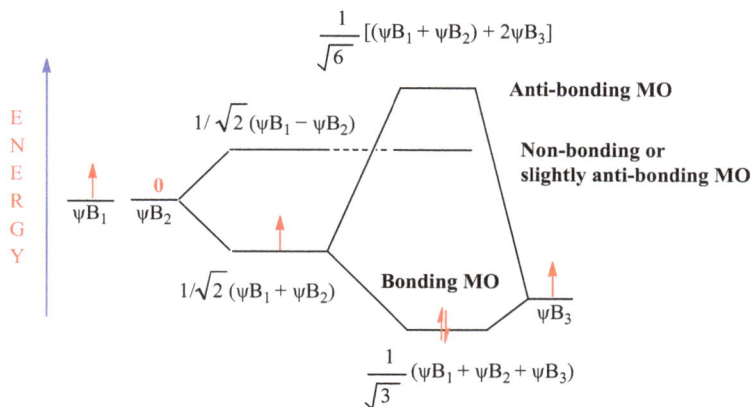

Figure 3.23: MO diagram for the formation of B–B–B bond in higher boranes.

Bond order of B–B bond in B–B–B bond

Bond order of B–B bond

$$= \frac{\text{No. of electron in BO} - \text{No. of electron in AB orbital}}{2 \times \text{No. of bonding centre}}$$

$$= \frac{2-0}{2\times 2} = \frac{2}{4}$$

$$= 0.5$$

Again, the bond order 0.5 of B–B bond in bridging B–B–B bond is self-evident for the stability of higher boranes.

3.8 Topology of boranes (Lipscomb's rule): s t y x four-digit coding of bonding in boranes

The overall bonding in boranes/higher boranes can be codified in a four-digit number, the so called styx number given by W. N. Lipscomb in 1963 (Lipscomb 1963). The significance of s, t, y and x are given below:

> s = No. of 3c–2e B–H–B bonds,
> t = No. of 3c–2e B–B–B bonds
> y = No. of 2c–2e B–B bonds and
> x = No. of BH_2 groups.

This four-digit coding system of bonding in boranes is also termed as ***topology of boranes.***

3.8.1 Utility of Lipscomb's rule

The number of atoms in a neutral boranes molecule is equal to twice the sum of s, t, y and x, that is,

$$\text{No. of atoms in neutral boranes} = 2(s + t + y + x)$$

The number of electron pairs involved in the bonding of the borane molecule is **n** plus the sum of the individual s t y x number, that is,

$$\text{No. of electron pairs in a borane molecule} = n + (s + t + y + x)$$

$$\text{Here, } n = \text{number of B-atoms in boron clusters}$$

3.8.2 Validity of Lipscomb's rule

Various types of bonding in boranes can be understood by considering plane projections of their respective structures. The validity of Lipscomb's rule can be seen in all the boranes.

Examples:
(i) Biborane (B_2H_6)

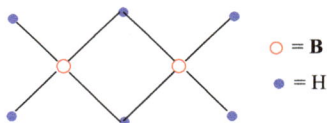

○ = **B**

● = **H**

(2002)

Total number of atoms $= 2 + 6 = 8$

Total number of valence electrons (VEs) $= 3 \times 2 + 6 = 12$

$$s = \text{No. of 3c-2e B–H–B bonds} = 2$$
$$t = \text{No. of 3c-2e B–B–B bonds} = 0$$
$$y = \text{No. of 2c-2e B–B bonds} = 0$$
$$x = \text{No. of BH}_2 \text{ groups} = 2$$

$$
\begin{aligned}
\text{Number of atoms in diborane} &= 2(s + t + y + x) \\
&= 2(2 + 0 + 0 + 2) \\
&= 2 \times 4 \\
&= 8 \ (\text{in } B_2H_6, \text{ total number of atoms is 8})
\end{aligned}
$$

No. of bond pairs in diborane molecule $= n + (s + t + y + x)$
$$= 2 + (2 + 0 + 0 + 2)$$
$$= 2 + 4$$
$= 6$ (total valence electrons in diborane is 12, that is, 6 electron pair)

Thus, the four-digit coding system of bonding in B_2H_6 is (2002). The **eight number of atoms** and **six electron pairs** calculated by Lipscomb's rule in diborane validate this rule.

(ii) Tetraborane (10): B_4H_{10}

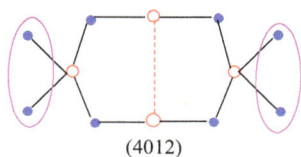

(4012)

Total number of atoms $= 4 + 10 = 14$
Total number of valence electrons (VEs) $= 3 \times 4 + 10 = 22$

s = No. of 3c-2e B–H–B bonds = 4
t = No. of 3c-2e B–B–B bonds = 0
y = No. of 2c-2e B–B bonds   = 1
x = No. of BH_2 groups   = 2

Number of atoms in tetraborane (10) $= 2(s + t + y + x)$
$$= 2(4 + 0 + 1 + 2)$$
$$= 2 \times 7$$
$= 14$ (in B_4H_{10}, total number of atoms is 14)

No. of electron pairs in tetraborane (10) molecule $= n + (s + t + y + x)$
$$= 4 + (4 + 0 + 1 + 2)$$
$$= 4 + 7$$
$= 11$ (total valence electrons in this boron cluster is 22, that is, 11e pairs)

Thus, the four-digit coding system of bonding in B_4H_{10} is (4012). The **14 number of atoms** and **11 electron pairs** calculated by Lipscomb's rule in **tetraborane (10)** validate this rule.

(iii) Pentaborane (9): B_5H_9

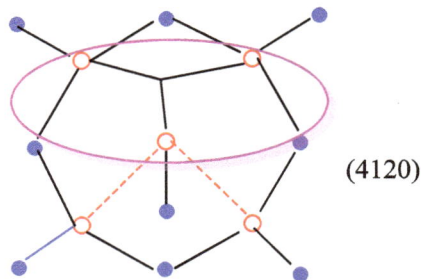

(4120)

Total number of atoms = 5 + 9 = 14
Total number of VEs = 3 × 5 + 9 = 24

s = No. of 3c-2e B–H–B bonds = 4
t = No. of 3c-2e B–B–B bonds = 1
y = No. of 2c-2e B–B bonds = 2
x = No. of BH_2 groups = 0

Number of atoms in **pentaborane (9)** = 2(s + t + y + x)
= 2(4 + 1 + 2 + 0)
= 2 × 7
= 14 (in B_5H_9, total number of atoms is 14)

No. of electron pairs in this molecule = n + (s + t + y + x)
= 5 + (4 + 1 + 2 + 0)
= 5 + 7
= 12 (total VEs in this boron cluster is 24, that is, 12 electron pairs)

Thus, the four-digit coding system of bonding in B_5H_9 is (4120). The **14 number of atoms** and **12 electron pairs** calculated by Lipscomb's rule in **pentaborane (9)** validate this rule.

(iv) Pentaborane (11): B_5H_{11}

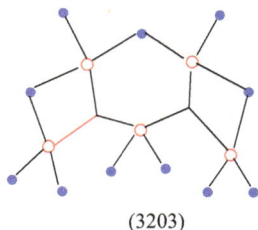

(3203)

Total number of atoms = 5 + 11 = 16
Total number of VEs   = 3 × 5 + 11 = 26

$\quad$ s = No. of 3c-2e B–H–B bonds = 3
$\quad$ t = No. of 3c-2e B–B–B bonds = 2
$\quad$ y = No. of 2c-2e B–B bonds   = 0
$\quad$ x = No. of BH_2 groups   = 3

Number of atoms in **pentaborane (11)** = 2(s + t + y + x)
$\qquad\qquad$ = 2(3 + 2 + 0 + 3)
$\qquad\qquad$ = 2 × 8
$\qquad\qquad$ = 16 (in B_5H_{11}, total number of atoms is 16)

No. of electron pairs in this molecule = n + (s + t + y + x)
$\qquad\qquad$ = 5 + (3 + 2 + 0 + 3)
$\qquad\qquad$ = 5 + 8
$\qquad\qquad$ = 13 (total VEs in this boron cluster is 26, that
$\qquad\qquad$ is, 13 electron pairs)

Thus, the four-digit coding system of bonding in B_5H_9 is (3203). The **16 number of atoms** and **13 electron pairs** calculated by Lipscomb's rule in **pentaborane (11)** show the validity of this rule.

(v) Hexaborane (10): B_6H_{10}

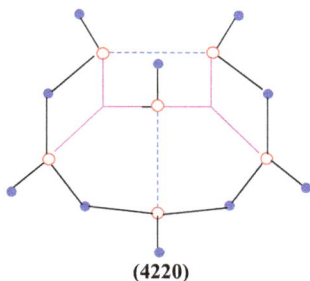

(4220)

Total number of atoms = 6 + 10 = 16
Total number of VEs   = 3 × 6 + 10 = 28

$\quad$ s = No. of 3c-2e B–H–B bonds = 4
$\quad$ t = No. of 3c-2e B–B–B bonds = 2
$\quad$ y = No. of 2c-2e B–B bonds   = 2
$\quad$ x = No. of BH_2 groups   = 0

Number of atoms in **hexaborane (10)** $= 2(s + t + y + x)$
$$= 2(4 + 2 + 2 + 0)$$
$$= 2 \times 8$$
$$= 16 \text{ (in } B_6H_{10}, \text{ total number of atoms is 16)}$$

No. of electron pairs in this molecule $= n + (s + t + y + x)$
$$= 6 + (4 + 2 + 2 + 0)$$
$$= 6 + 8$$
$$= 14 \text{ (total VEs in this boron cluster is 28, that}$$
$$\text{is, 14 electron pairs)}$$

Thus, the four-digit coding system of bonding in B_6H_{10} is (4220). The **16 number of atoms** and **14 electron pairs** calculated by Lipscomb's rule in **hexaborane (10)** indicate the validity of this rule.

(vi) Decaborane (14): $B_{10}H_{14}$

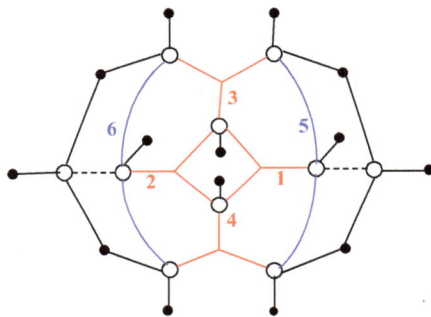

(4620)

Total number of atoms $= 10 + 14 = 24$
Total number of VEs $= 3 \times 10 + 14 = 44$

s = No. of 3c-2e B–H–B bonds = 4
t = No. of 3c-2e B–B–B bonds = 6
y = No. of 2c-2e B–B bonds = 2
x = No. of BH_2 groups = 0

Number of atoms in **decaborane (14)** $= 2(s + t + y + x)$
$$= 2(4 + 6 + 2 + 0)$$
$$= 2 \times 12$$
$$= 24 \text{ (in } B_{10}H_{14}, \text{ total number of atoms is 24)}$$

No. of electron pairs in this molecule $= n + (s + t + y + x)$

$$= 10 + (4 + 6 + 2 + 0)$$

$$= 10 + 12$$

$$= 22 \text{ (total VEs in this boron cluster is 44, that}$$
$$\text{is, 22 electron pairs)}$$

Thus, the four-digit coding system of bonding in $B_{10}H_{14}$ is (4620). The **24 number of atoms** and **22 electron pairs** calculated by Lipscomb's rule in **decaborane (14)** show the validity of this rule.

3.8.3 Calculation of total number of VEs and number of bonds in boranes without making use of Lipscomb's rule

Electron counting and orbital bookkeeping can easily be checked in above diagrams of higher boranes taken as examples in describing the validity of Lipscomb's rule. As each boron (B) has four valence orbitals (s + 3p), there should be four lines coming from each open circle (O) of boron (B). Likewise, as each B atom contributes three electrons and each H atom contributes one electron, the total number of valence electrons (VEs) for a borane of formula B_nH_m will be:

$$VEs = (3n + m)$$

And the number of bonds shown in the structure should be just half of this, that is,

$$\text{Number of bonds} = (3n + m)/2.$$

An approximate number of additional electrons should be added for anionic species. This can be checked taking examples of some boranes as follows:

(i) Hexaborane (10): B_6H_{10}

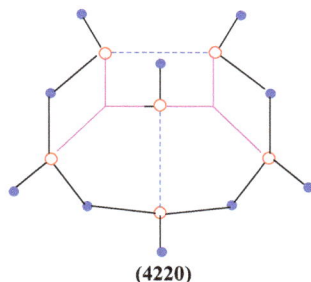

(4220)

Total number of VEs in $B_6H_{10} = 6 \times 3 + 10 \times 1 = 28 \ [VEs = (3n + m)]$

Hence,

Total number of bonds = 28/2 = 14 [Number of bonds = (3n + m)/2]

If we look over the structure of B_6H_{10}, we have

$$No. \text{ of 3c-2e B–H–B bonds} = 4$$

$$No. \text{ of 3c-2e B–B–B bonds} = 2$$

$$No. \text{ of 2c-2e B–B bonds} = 2$$

$$No. \text{ of 2c-2e B–H bonds} = 6$$

Thus, the total number of bonds in $B_6H_{10} = 4 + 2 + 2 + 6 = 14$. Hence, number of bonds in the present case is equal to one we have already calculated using the formula (3n + m)/2.

(ii) Decaborane (14): $B_{10}H_{14}$

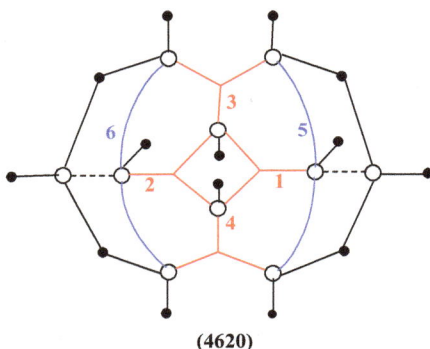

(4620)

Total number of VEs in $B_{10}H_{14} = 10 \times 3 + 14 \times 1 = 44$ [VEs = (3n + m)]

Total number of bonds = 44/2 = 22 [Number of bonds = (3n + m)/2]

An observation of the structure of $B_{10}H_{14}$, we have

$$No. \text{ of 3c-2e B–H–B bonds} = 4$$

$$No. \text{ of 3c-2e B–B–B bonds} = 6$$

$$No. \text{ of 2c-2e B–B bonds} = 2$$

$$No. \text{ of 2c-2e B–H bonds} = 10$$

Hence,

$$\text{Total number of bonds} = 22$$

Thus, the number of bonds (22) in the present case is equal to one we have already calculated using the formula (3n + m)/2.

(iii) Pentaborane (11): B_5H_{11}

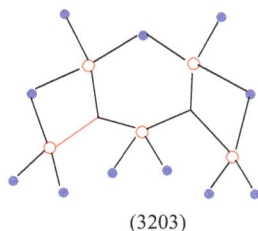

(3203)

Total number of VEs in $B_5H_{11} = 5 \times 3 + 11 \times 1 = 26$ [VEs = (3n + m)]

Total number of bonds $\quad = 26/2 = 12$ [Number of bonds = (3n + m)/2]

An observation of the structure of $B_{10}H_{14}$, we have

No. of 3c-2e B–H–B bonds = 3

No. of 3c-2e B–B–B bonds = 2

No. of 2c-2e B–B bonds $\quad = 0$

No. of 2c-2e B–H bonds $\quad = 8$

Hence,

Total number of bonds $\quad = 13$

Thus, the number of bonds (13) in in B_5H_{11} is equal to one we have already calculated using the formula (3n + m)/2.

3.8.4 Limitations of Lipscomb's topological scheme: Wade's rules relating the structures of boranes with their composition

The topological scheme has its limitations. For the symmetrical *closo*-boranes and even for the large open cluster boranes, it becomes difficult if not impossible to write simple, satisfactory structures of this sort. For B_5H_9, shown in Figure 3.24, the one structure shown is

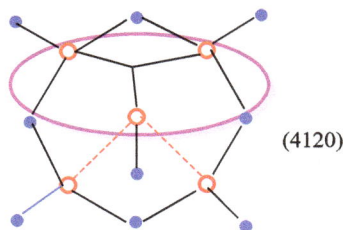

(4120)

Figure 3.24: Structure of B_5H_9.

incompatible with the high symmetry (C_{4v}) of the molecule. In this case, which is relatively simple, the difficulty is easily overcome by treating the structure shown as only one of four equivalent ones that together form a resonance hybrids. With increasing size and symmetry, however, this sort of approach becomes very cumbersome (because all canonical structures will not be equivalent) and it becomes desirable to employ an MO description, consistent with the full molecular symmetry and naturally incorporating delocalization.

Wade's rules

By combining empirical facts with the results of MO calculations, Wade (1976) derived a set of guidelines/rules for relating the structures of boranes with their compositions, which is also known as *polyhedral skeleton electron pair theory*. These rules pertain to boranes of formula $B_nH_m^{x-}$, where $m \geq n$ and $x \geq 0$ (that is, only neutral or anions are considered).

MO calculations have shown that independent of the value of n, if $m = n$ and the structure is *closo*, x will have to be 2 in order that all bonding orbitals be filled and all anti-bonding orbitals be empty. This means that there is (n + 1) pair of so called framework electrons, that is, electrons in the central B_n polyhedron. The $B_6H_6^{2-}$ (Figure 3.25) case may be taken to illustrate it.

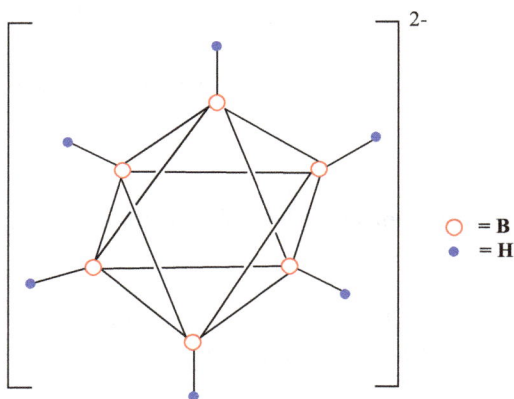

○ = B
● = H

Figure 3.25: Structure of *closo*-$B_6H_6^{2-}$.

Each B–H group supplies three atomic orbitals (one having been used to form the B–H bond) and two electrons to the octahedral B_6 frame work, that is, a total of **12e** by six B–H groups in $B_6H_6^{2-}$. The atomic orbitals combine to generate *seven bonding* **MOs** (not all of different energies since there are degeneracies) and *eleven anti-bonding* **MOs**. To fill all seven of the bonding MOs one more pair of electrons (**2e**) is required, hence the charge of –2. The MO diagram of *closo*-$B_6H_6^{2-}$ framework having [**12e** + **2e**] = **14e** along with electron filling is shown in Figure 3.26.

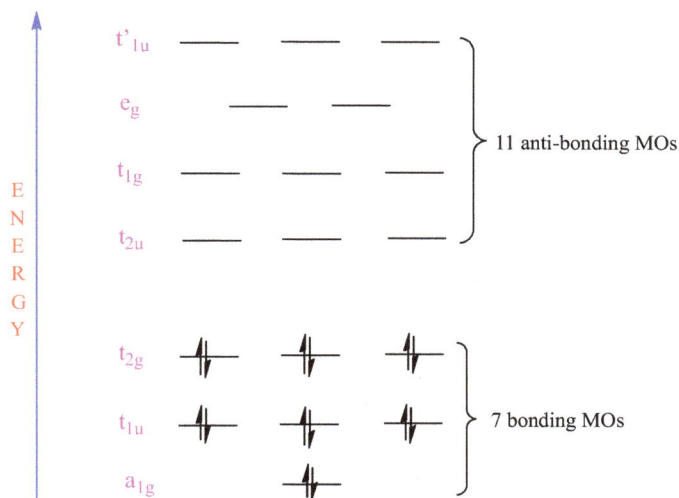

Figure 3.26: MO diagram *closo*-$B_6H_6^{2-}$ framework having 7 bonding and 11 anti-bonding MOs. Adopted from Greenwood and Earnshaws (1984).

Since, the same general result is obtained for all *closo*-boranes, we have the first of Wade's rule:

Rule I:

For a $B_nH_n^{x-}$ species, the preferred structure is a *closo* one (an n vertex polyhedron) with a preferred charge of 2–. There are then $(n + 1)$ pairs of framework electrons.

Next we consider $B_nH_m^{x-}$ species where $m > n$. There are a number of species where $m = n + 4$ and $x = 0$, such as B_5H_9 and B_6H_{10}. If we remove m-n protons from such a structure, we shall have $B_nH_n^{4-}$. MO calculations suggest that such a species, which has $(n + 2)$ pair of framework (that is, non-B–H) electrons, will have as its most stable structure the incomplete polyhedron obtained by removing one vertex from a polyhedron with $(n + 1)$ vertices. Thus, we have a *nido* structure. The second of Wade's rules states:

Rule II:

For a B_nH_{n+4} borane, where there are $(n + 2)$ pairs of framework electrons in the corresponding $B_nH_n^{4-}$ ion, a *nido* structure is preferred.

In a similar way, rules III and IV are derived, which state:

Rule III:

For a B_nH_m species, based on a $B_nH_n^{6-}$ anion, an *arachno* structure derived from an $(n + 2)$ vertex polyhedron is preferred. This requires $(n + 3)$ pairs of framework bonding electrons.

Rule IV:

For a B_nH_m species that derives from a $B_nH_n^{8-}$ anion, a *hypho* structure derived from an $(n+3)$ vertex polyhedron is preferred, requiring $(n+4)$ pairs of framework bonding electrons.

It is to be noted that these rules can be extended to isoelectronic anions, carbaboranes ($BH=B^-=C$) and also to metalloboranes, metallocarbaboranes and even to metal clusters, though they become less reliable the further one moves away from boron in atomic size, ionization energy, electronegativity and so on

More sophisticated and refined calculations lead to orbital populations and electron charge distribution within the boron molecules and to predictions concerning the sites of electrophilic and nucleophilic attack. In general, the highest electron charge density (and the preferred site of electrophilic attack) occurs at apical B atoms which are furthest removed from open faces. Conversely, the lowest electron charge density (and the preferred site of nucleophilic attack) occurs on B atoms involved in B–H–B bonding. The consistency of this correlation implies that the electron distribution in the activated complex formed during reaction must follow a similar sequence to that in the ground state. Bridged H atoms tend to be more acidic than terminal H atoms and are the ones first lost during the formation of anions in acid–base reactions.

3.9 Carbaboranes

Carboranes (or more correctly and less commonly called carbaboranes) are polyhedral boranes that contain framework C atoms as well as B atoms (Figure 3.27).

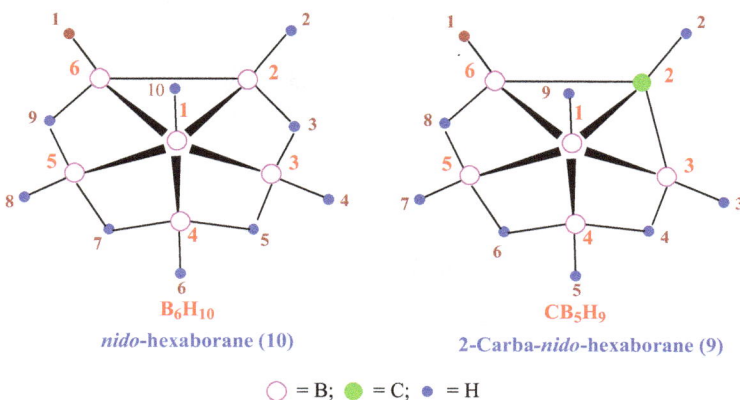

B_6H_{10}

nido-hexaborane (10)

CB_5H_9

2-Carba-*nido*-hexaborane (9)

$\bigcirc$ = B; ● = C; • = H

Figure 3.27: 2-Carbahexaborane (9) derived from *nido*-hexaborane (10).

Carbaboranes came onto the chemical scene in 1962–1963 when classified work that had been done in the late 1950s was cleared for publication. The following 25 years have been a tremendous growing of activity, and few other areas of chemistry have undergone such enormous development during this period. Much of the work has been carried out in USA and USSR, but significant contributions have also come from Czechoslovakia and the UK.

Carbaboranes and their related derivatives, the metallocarbaboranes, are now seen to occupy a strategic position in chemistry of the elements since they overlap and give coherence/consistency to several other large areas including the chemistry of polyhedral boranes, transition metal complexes, metal-cluster compounds and organometallicchemistry.

The field of carboranes has become so vast that it is only possible to give a few illustrative examples of the many thousands of known compounds, and to indicate the general structural unit a number of carbon and boron atoms arranged on the vertices of triangulated ordinary substances.

3.9.1 Structures of carbaboranes

The structures of carbaboranes are closely related to those of the isoelectronic boranes ($BH=B^- \equiv C$; $BH_2 \equiv BH^- \equiv CH$). This can be clearly seen from the carbaboranes derived from boranes.

For example, *nido*-hexaborane(10) [**nido-B$_6$H$_{10}$**] provides the basic structures for the 4-carbaboranes obtained from each successive replacement of a basal B atom by C being compensated by the removal of one H_μ (bridging hydrogen). These four carbaboranes are: CB_5H_9, $C_2B_4H_8$, $C_3B_3H_7$ and $C_4B_2H_6$, the structures of which are shown in Figure 3.28.

3.9.2 General formula of carbaboranes

Carbaboranes have the general formula $[(CH)_a(BH)_mH_b]^{c-}$ with a CH units and m BH units at the polyhedral vertices, plus b 'extra' H atoms which are either bridging (H_μ) or *endo* (i.e. tangential to surface of the polyhedron as distinct from the axial H_t atoms specified in the CH and BH groups; H_{endo} occur in BH_2 groups which are thus more precisely specified as BH_tH_{endo}).

It follows from the general formula that the number of electrons available for skeletal bonding is **3e** for each CH unit, **2e** from each BH unit, **1e** from each H_μ or H_{endo} and **ce** from the anionic charge.

Hence,

Total number of skeleton bonding electron pairs = $\frac{1}{2}(3a + 2m + b + c) = 3/2a + m + 1/2b + 1/2c = a + 1/2a + m + 1/2b + 1/2c = (a + m) + 1/2(a + b + c) = n + 1/2(a + b + c)$

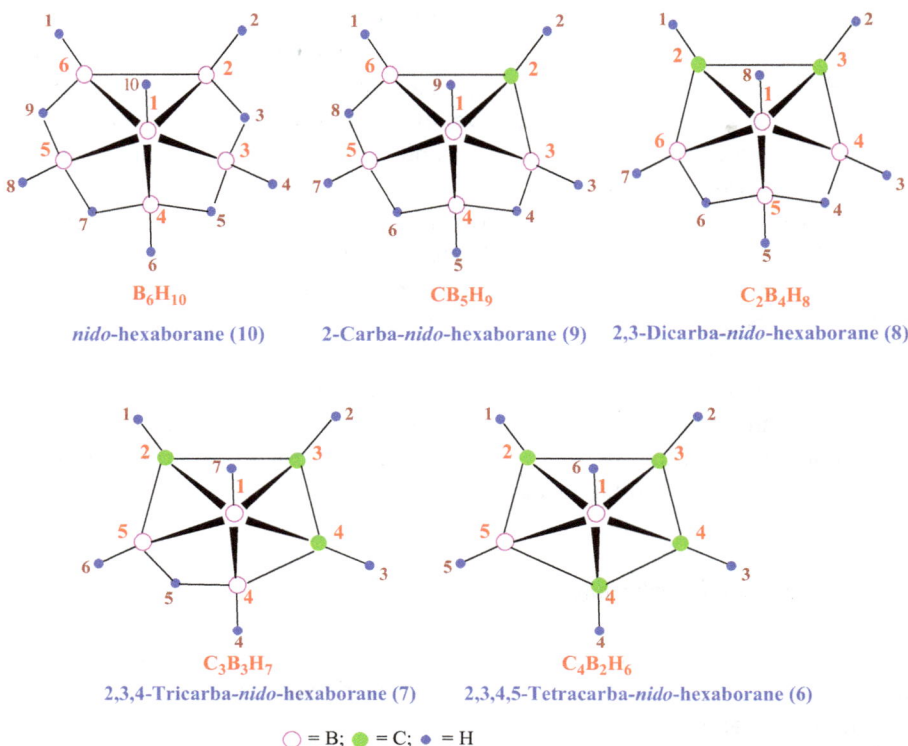

B_6H_{10}

nido-hexaborane (10)

CB_5H_9

2-Carba-*nido*-hexaborane (9)

$C_2B_4H_8$

2,3-Dicarba-*nido*-hexaborane (8)

$C_3B_3H_7$

2,3,4-Tricarba-*nido*-hexaborane (7)

$C_4B_2H_6$

2,3,4,5-Tetracarba-*nido*-hexaborane (6)

○ = B; ● = C; • = H

Figure 3.28: Four carbaboranes derived from *nido*-hexaborane(10) by successive replacement of a basal B atom by C.

Here, n = (a + m) is the number of occupied vertices of the polyhedron.

Based on the above skeleton bonding electron pair equation, the number of electron pairs of skeleton bonding electrons in different classes of carbaboranes can be worked out as follows:

(a) *Closo*-carbaborane structures have (n + 1) pairs of skeleton bonding electrons, that is, (a + b + c) = 2

(b) *Nido*-carbaborane structures have (n + 2) pairs of skeleton bonding electrons, that is, (a + b + c) = 4

(c) *Arachno*-carbaborane structures have (n + 3) pairs of skeleton bonding electrons, that is, (a + b + c) = 6

Let us again consider the general formula $[(CH)_a(BH)_m H_b]^{c-}$ of carbaboranes.

If a = 0 in the above formula (in case of no carbon atoms), the compound is a *borane or borane anion* rather than a carbaborane.

If b = 0, there are no H_μ or H_{endo}, this is the case of all *closo*-carbaboranes.

If c = 0, the compound is a *neutral carbaborane molecule* rather than an anion.

3.9.3 Nomenclature of carbaboranes

Nomenclature of carbaboranes follows the well-established oxa-aza convention of organic chemistry. Numbering starts with the apex atom of lowest coordination and successive rings or belts of polyhedral vertex atoms are numbered in a clockwise direction with C atoms being given the lowest possible numbers.

3.9.4 Synthesis of carbaboranes

Closo-carbaboranes are most numerous and most stable of carbaboranes. They are readily prepared from an alkyne and a borane by pyrolysis or by reaction in a silent electric discharge. This rout, which normally gives mixture of *closo*-carbaboranes (Figure 3.29), is particularly useful for small *closo*-carbaboranes (n = 5–7) and for intermediate *closo*-carbaboranes (n = 8–11). For example,

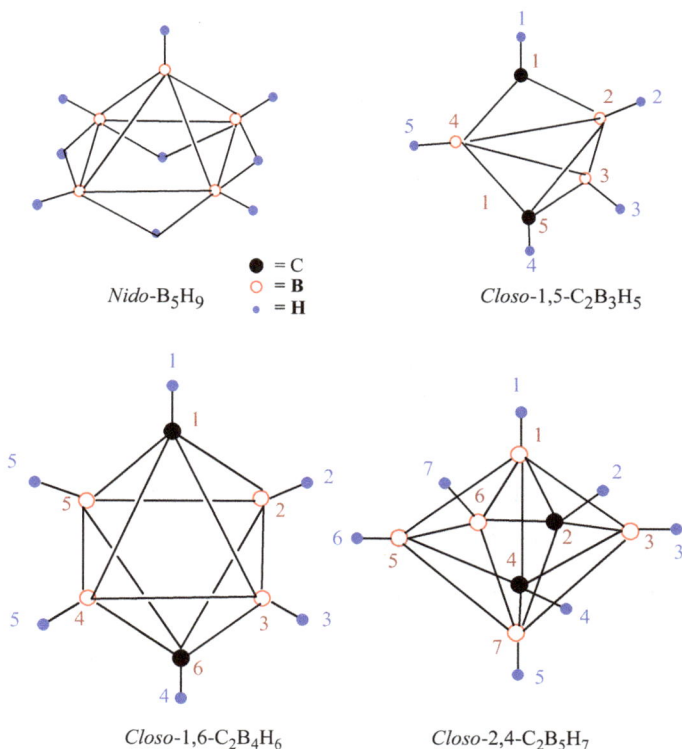

Nido-B_5H_9

● = C
○ = B
• = H

Closo-1,5-$C_2B_3H_5$

Closo-1,6-$C_2B_4H_6$

Closo-2,4-$C_2B_5H_7$

Figure 3.29: Different *closo*-dicarbaboranes derived from pyrolysis of *nido*-B_5H_9.

(i) $nido$-B_5H_9 $\xrightarrow[500-600°]{C_2H_2}$ $closo$-1,5-$C_2B_3H_5$ + $closo$-1,6-$C_2B_4H_6$ + $closo$-2,4-$C_2B_5H_7$

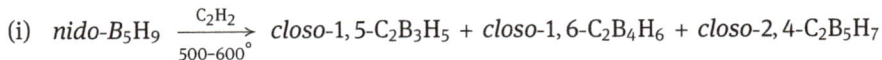

Under milder conditions, provide a route to $nido$-carbaboranes (Figure 3.30). For example,

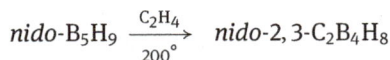

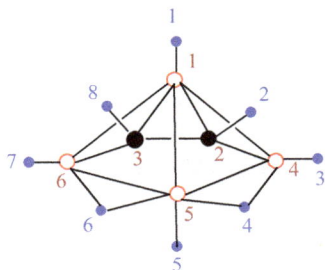

$$nido\text{-}B_5H_9 \xrightarrow[200°]{C_2H_4} nido\text{-}2,3\text{-}C_2B_4H_8$$

Figure 3.30: Nido-2,3-$C_2B_4H_8$ derived from nido-B_5H_9.

(ii) Pyrolysis of $nido$- or $arachno$-carbaboranes or their reactions in a silent electric discharge leads to $closo$-species either by loss of H_2 or disproportionation:

$$C_2B_nH_{n+4} \rightarrow C_2B_nH_{n+2} + H_2$$

$$2C_2B_nH_{n+4} \rightarrow C_2B_{n-1}H_{n+1} + C_2B_{n+1}H_{n+3} + 2H_2$$

For example,
(a) Pyrolysis of $nido$-2,3-$C_2B_4H_8$ gives the 3 $closo$-species 1,5-$C_2B_3H_5$, 1,6-$C_2B_4H_6$ and 2,4-$C_2B_5H_7$, whereas under the milder conditions of photolytic closure, the less-stable isomer $closo$-1,2-$C_2B_4H_6$ is obtained.
(b) Pyrolysis of alkyl boranes at 500–600 °C is a related route which is particularly useful to monocarboranes (Figure 3.31) though the yields are often poor; for example,

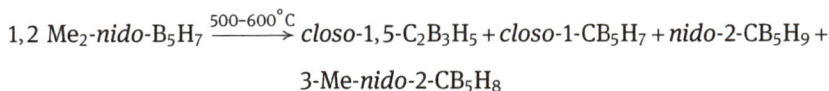

$$1,2\ Me_2\text{-}nido\text{-}B_5H_7 \xrightarrow{500-600°C} closo\text{-}1,5\text{-}C_2B_3H_5 + closo\text{-}1\text{-}CB_5H_7 + nido\text{-}2\text{-}CB_5H_9 +$$

$$3\text{-}Me\text{-}nido\text{-}2\text{-}CB_5H_8$$

(iii) Cluster expansion reactions with diborane provide an alternative route to intermediate $closo$-borane; for example,

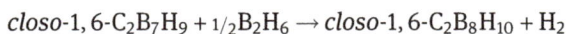

$$closo\text{-}1,7\text{-}C_2B_6H_8 + 1/2B_2H_6 \rightarrow closo\text{-}1,6\text{-}C_2B_7H_9 + H_2$$

$$closo\text{-}1,6\text{-}C_2B_7H_9 + 1/2B_2H_6 \rightarrow closo\text{-}1,6\text{-}C_2B_8H_{10} + H_2$$

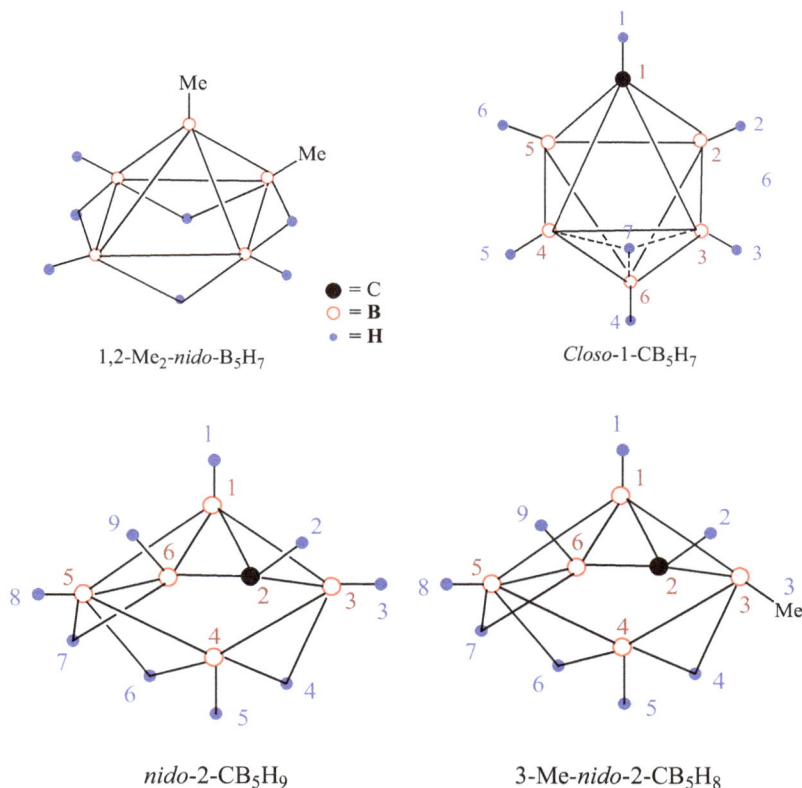

Figure 3.31: Structures of somemonocarboranes derived from pyrolysis of 1,2 Me$_2$-*nido*-B$_5$H$_7$.

(iv) Cluster degradation reactions lead to more open structures; for example,

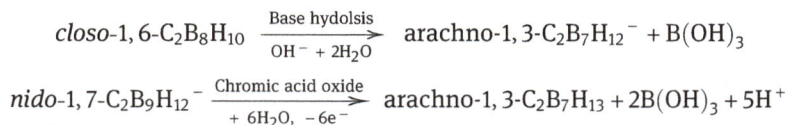

$$closo\text{-}1, 6\text{-}C_2B_8H_{10} \xrightarrow[\text{OH}^- + 2H_2O]{\text{Base hydolsis}} \text{arachno-1}, 3\text{-}C_2B_7H_{12}{}^- + B(OH)_3$$

$$nido\text{-}1, 7\text{-}C_2B_9H_{12}{}^- \xrightarrow[+ 6H_2O, \; -6e^-]{\text{Chromic acid oxide}} \text{arachno-1}, 3\text{-}C_2B_7H_{13} + 2B(OH)_3 + 5H^+$$

Closo-carboranes are most numerous and most stable of carboranes. They are colourless volatile liquids or solids (depending on molecular weights). Most *closo*-boranes are stable to at least 400 °C though they may undergo to rearrangement to more stable isomers in which the distance between carbon atoms is increased.

In general, *nido-* and *arachno*-carborane are less stable thermally than that of the corresponding *closo*-compounds. They are less stable to aerial oxidation and other reactions due to their more open structures and the presence of labile hydrogen atoms in the open face.

(v) The three isomeric icosahedral carbaboranes are unique in their case of preparation and their great stability in air, and consequently their chemistry has been the most fully studied. The 1,2-isomer in particular has been available for over decades on multikilogram scale. It is best prepared in bulk by the direct reaction of acetylene with decaborane in presence of a Lewis base, preferably Et_2S.

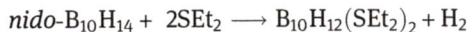

$$nido\text{-}B_{10}H_{14} + 2SEt_2 \longrightarrow B_{10}H_{12}(SEt_2)_2 + H_2$$

$$B_{10}H_{12}(SEt_2)_2 + C_2H_2 \longrightarrow closo\text{-}1,2\text{-}C_2B_{10}H_{12} + 2SEt_2 + H_2$$

The 1,7-isomer is obtained in 90% yield by heating the 1,2-isomer in the gas phase at 470 °C for several hours. In quantitative yield, it can also be obtained by flash pyrolysis of 1,2-isomer at 600 °C for 30 s. The 1,12-isomer is most efficiently prepared in 20% yield by heating (620 °C) the 1,7-isomer for a few seconds (Figure 3.32).

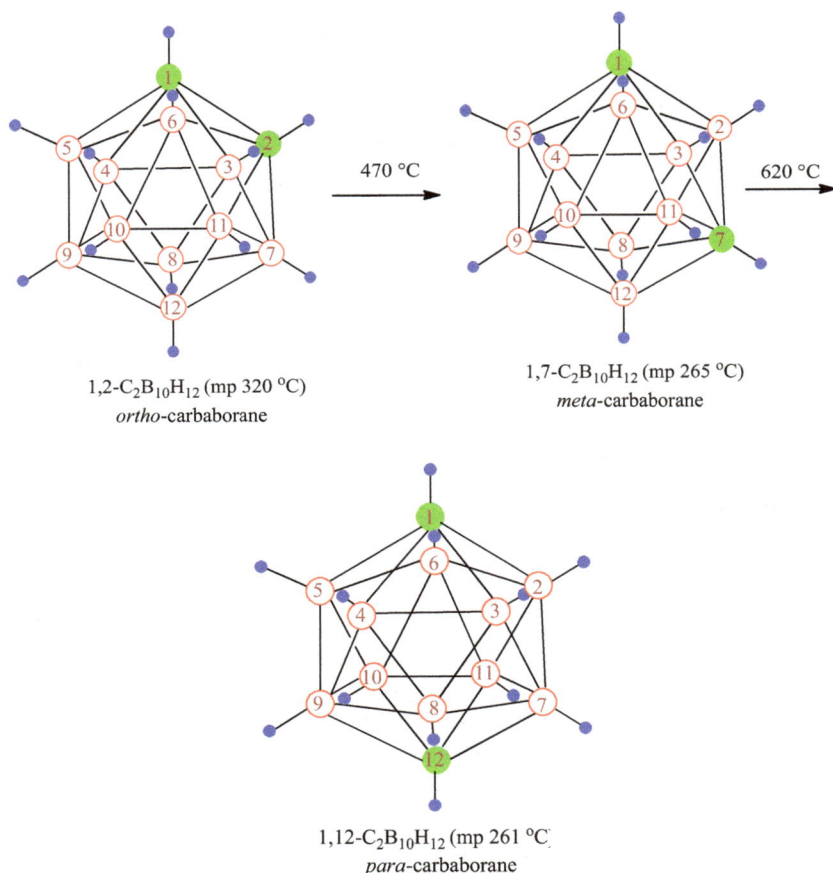

1,2-$C_2B_{10}H_{12}$ (mp 320 °C)
ortho-carbaborane

470 °C

1,7-$C_2B_{10}H_{12}$ (mp 265 °C)
meta-carbaborane

620 °C

1,12-$C_2B_{10}H_{12}$ (mp 261 °C)
para-carbaborane

Figure 3.32: Structure and isomerizations of the three isomers of dicarba-*closo*-dodecacarborane.

The mechanism of these isomerizations has been the subject of considerable speculation but definite experiments are hard to devise. A suitable mechanism which can lead to both 1,7- and 1,12-isomers is the successive concerted rotation of a three atoms on a triangular face as shown in Figure 3.33.

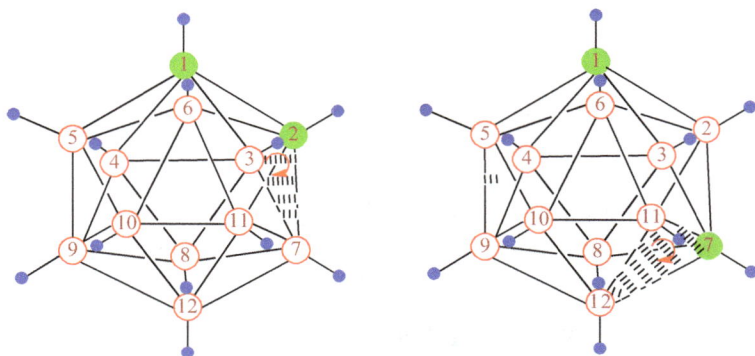

Figure 3.33: Successive concerted rotation of a three atoms on a triangular face shown leading to both 1,7- and 1,12-isomers.

3.10 Metallocarboranes or carborane complexes

When metals are incorporated in carborane cages, the resulting compounds are called metallocarboranes. The following are some examples of metallocarboranes:

$$(C_2B_9H_{11})Mo(CO)_3^{2-}$$

$$(C_2B_9H_{11})Re(CO)_3^{-}$$

$$(C_2B_9H_{11})Co(C_5H_5)$$

$$[(MeC)_2B_9H_9]Pd(C_4Ph_4)$$

$$[(MeC)(PhC)B_9H_9]RhH(PPh_3)_2$$

The large majority of known metallocarboranes have two carbon atoms per cage but there are now many examples of 1-, 3- and 4-carbon metallocarbaboranes. Complexes are known involving most transition metals, many lanthanides and numerous main group elements in groups 1, 2, 13, 14, 15 and 16. Moreover, combinations of these metals may occur in the same cage framework. Given this scope and the availability of many different carborane ligands, it is safe to say that metallocarboranes form by far the largest and most diverse class of cluster compounds known.

3.10.1 Synthetic routes to metallocarboranes

There are several synthetic routes to metallocarboranes and these are:
(i)   Coordination using *nido*-carborane anions as ligands
(ii) Polyhedral expansion
(iii) Polyhedral contraction
(iv) Polyhedral subrogation
(v)   Thermal metal transfer
(vi) Direct oxidative insertion of a metal centre

(i) Coordination using *nido*-carborane anions as ligands:

When $C_2B_9H_{11}^{2-}$ reacts with $FeCl_2$ in THF with rigorous exclusion of moisture and air, gives pink, diamagnetic bis-sandwich type complex of Fe(II) (Figure 3.34). This can be reversibly oxidized to the corresponding Fe(III) complex.

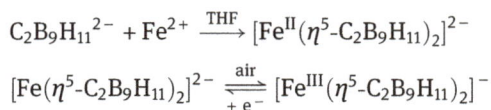

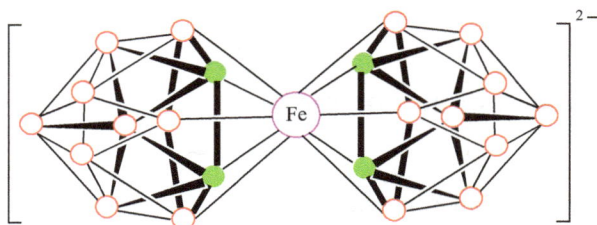

$$C_2B_9H_{11}^{2-} + Fe^{2+} \xrightarrow{\text{THF}} [Fe^{II}(\eta^5\text{-}C_2B_9H_{11})_2]^{2-}$$

$$[Fe(\eta^5\text{-}C_2B_9H_{11})_2]^{2-} \underset{+ e^-}{\overset{\text{air}}{\rightleftharpoons}} [Fe^{III}(\eta^5\text{-}C_2B_9H_{11})_2]^-$$

Figure 3.34: Structure of $[Fe(\eta^5\text{-}C_2B_9H_{11})_2]^{2-}$.

When the reaction is carried out in the presence of NaC_5H_5, the purple mixed sandwich complex (Figure 3.35) is obtained:

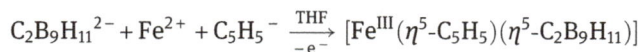

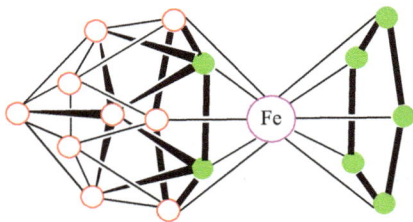

$$C_2B_9H_{11}^{2-} + Fe^{2+} + C_5H_5^- \xrightarrow[-e^-]{\text{THF}} [Fe^{III}(\eta^5\text{-}C_5H_5)(\eta^5\text{-}C_2B_9H_{11})]$$

Figure 3.35: Structure of $[Fe^{III}(\eta^5\text{-}C_5H_5)(\eta^5\text{-}C_2B_9H_{11})]$.

This reaction is general and can be applied to many transition metals. Metal carbonyls and other complexes may be used to supply the capping unit (Figure 3.36); for example,

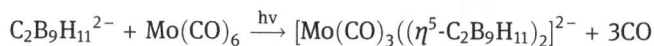

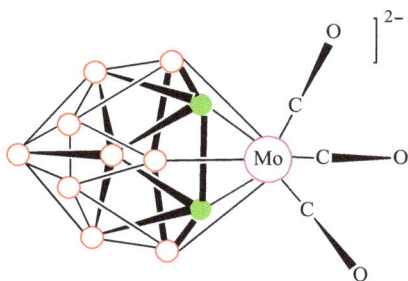

$$C_2B_9H_{11}{}^{2-} + Mo(CO)_6 \xrightarrow{hv} [Mo(CO)_3((\eta^5\text{-}C_2B_9H_{11})_2]^{2-} + 3CO$$

Figure 3.36: Structure of $[Mo(CO)_3((\eta^5\text{-}C_2B_9H_{11})_2]^{2-}$.

(ii) Polyhedral expansion:

This requires the 2-electron reduction of a *closo*-carbaborane with a strong reducing agent, such as sodium naphthalide in THF followed by reduction with a transition metal reagent.

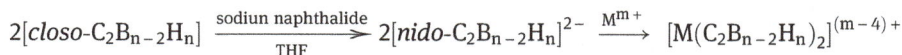

$$2[closo\text{-}C_2B_{n-2}H_n] \xrightarrow[\text{THF}]{\text{sodiun naphthalide}} 2[nido\text{-}C_2B_{n-2}H_n]^{2-} \xrightarrow{M^{m+}} [M(C_2B_{n-2}H_n)_2]^{(m-4)+}$$

The reaction, which is quite general for *closo*-carbaboranes, involves reductive opening of an *n*-vertex *closo*-cluster followed by metal insertion to give an (*n* + 1)-vertex *closo*-cluster. Numerous variants are possible including the insertion of a second metal centre into an existing metallocarboranes; for example,

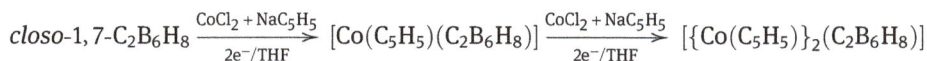

$$closo\text{-}1,7\text{-}C_2B_6H_8 \xrightarrow[\text{2e}^-/\text{THF}]{CoCl_2 + NaC_5H_5} [Co(C_5H_5)(C_2B_6H_8)] \xrightarrow[\text{2e}^-/\text{THF}]{CoCl_2 + NaC_5H_5} [\{Co(C_5H_5)\}_2(C_2B_6H_8)]$$

The structure of the bimetallic 10 vertex cluster was shown by X-ray diffraction studies to be as given in Figure 3.37.

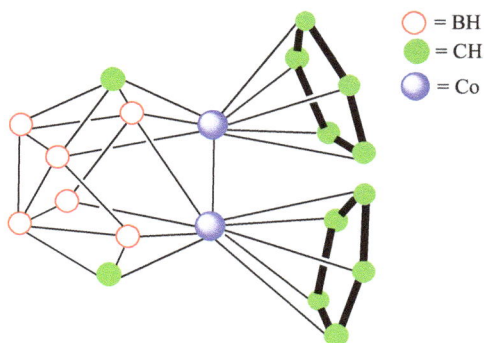

= BH
= CH
= Co

Figure 3.37: Structure of bimetallic 10 vertex cluster $[\{Co(C_5H_5)\}_2(C_2B_6H_8)]$.

When the icosahedral carborane $1,2\text{-}C_2B_{10}H_{12}$ was used in place of *closo*-1,7-$C_2B_6H_8$, the reaction led to the first supraicosahedral metallocarboranes with 13- and 14-vertex polyhedral structures (Figure 3.38). Facile isomerism of the 13-vertx monometallicdicarbaboranes was observed as indicated in the subjoined/attached scheme.

(iii) Polyhedron contraction:
This involves removal of one BH group from a *closo*-metallocarbaborane by nucleophilic base degradation, followed by oxidative closure of the resulting *nido*-metallocarbaborane complex to a *closo*-species with one vertex less than the original (Figure 3.39); for example,

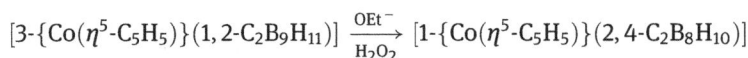

$$[3\text{-}\{Co(\eta^5\text{-}C_5H_5)\}(1,2\text{-}C_2B_9H_{11})] \xrightarrow[H_2O_2]{OEt^-} [1\text{-}\{Co(\eta^5\text{-}C_5H_5)\}(2,4\text{-}C_2B_8H_{10})]$$

Polyhedron contraction is not so a general method of preparing metallocarboranes as is polyhedron expansion since some metallocarboranes degrade completely under these conditions.

(iv) Polyhedron subrogation:
Replacement of a BH vertex by a metal vertex without changing the number of vertices in a cluster is termed as *polyhedron subrogation*. It is an offshoot of the polyhedron contraction route in which degradative removal of the BH unit is followed by reaction with a transition metal ion rather than with an oxidizing agent.

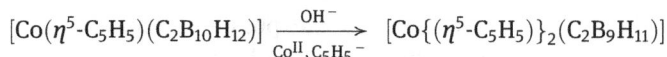

$$[Co(\eta^5\text{-}C_5H_5)(C_2B_{10}H_{12})] \xrightarrow[Co^{II}, C_5H_5^-]{OH^-} [Co\{(\eta^5\text{-}C_5H_5)\}_2(C_2B_9H_{11})]$$

The method is clearly of potential use in preparing mixed-metal clusters; for example, (Co + Ni) and (Co + Fe) and can be extended to prepare more complicated cluster arrays as shown in Figure 3.40. The subrogated B atom is being indicated as a broken circle.

(v) Thermal metal transfer:
This method is less general and less specific than the coordination of *nido*-anions or polyhedral expansion. It involves the pyrolysis of pre-existing metallocarbaboranes and consequent cluster expansion or disproportionation. Mixtures of products are usually obtained. For example,
(i)   Reaction shown in Figure 3.41 gives $(C_5H_5)_2Co_2C_2B_8H_{10}$ and five other isomers.

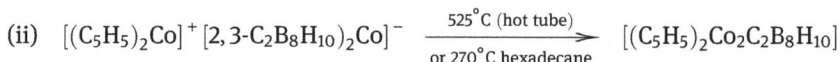

(ii) $[(C_5H_5)_2Co]^+[2,3\text{-}C_2B_8H_{10})_2Co]^- \xrightarrow[\text{or } 270°C \text{ hexadecane}]{525°C \text{ (hot tube)}} [(C_5H_5)_2Co_2C_2B_8H_{10}]$

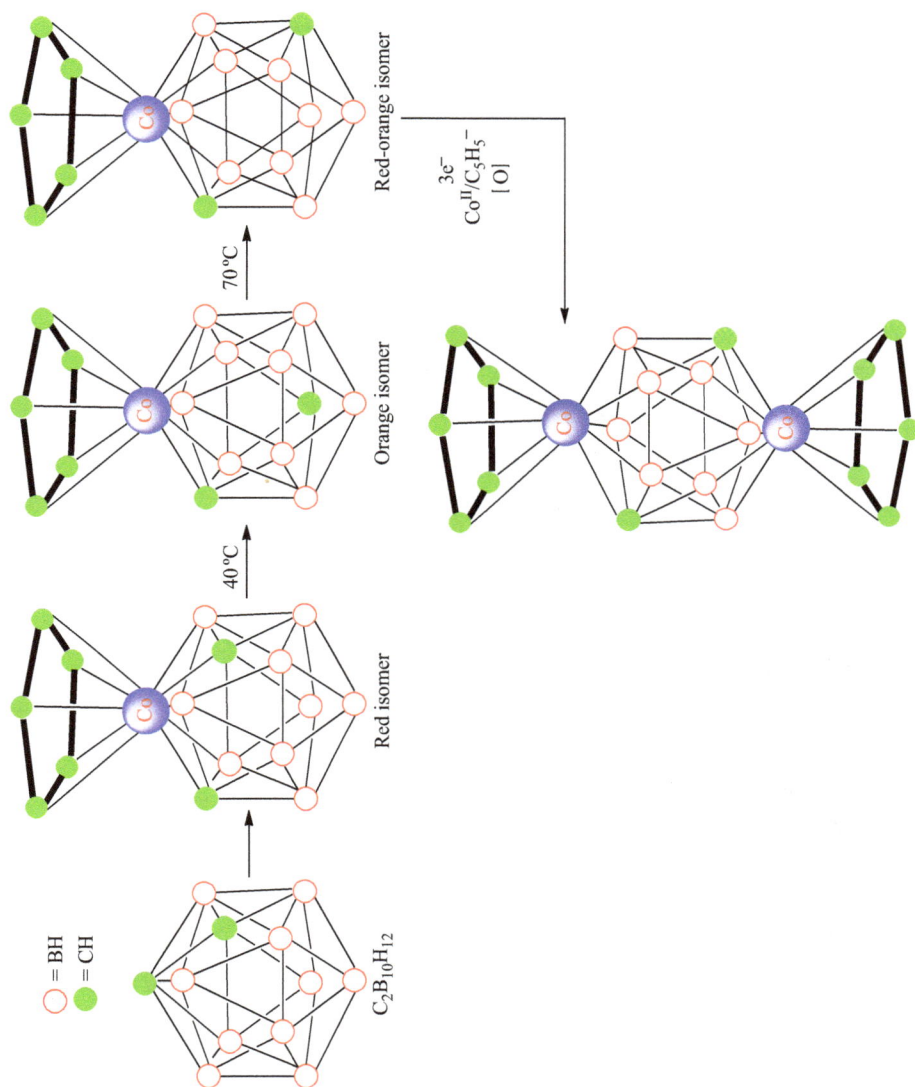

Figure 3.38: Supraicosahedral metallocarboranes with 13- and 14-vertex polyhedral structures.

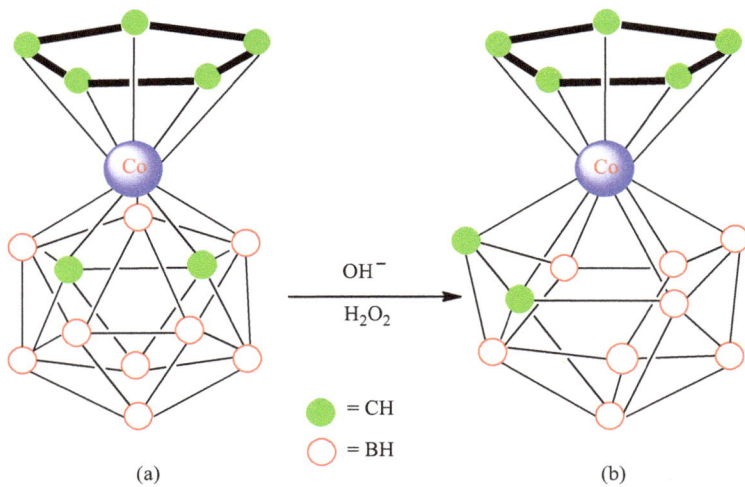

Figure 3.39: Structure of [3-{η^5-C$_5$H$_5$}(1,2-C$_2$B$_9$H$_{11}$)] (a) and [1-{η^5-C$_5$H$_5$}(2,4-C$_2$B$_8$H$_{10}$)] (b).

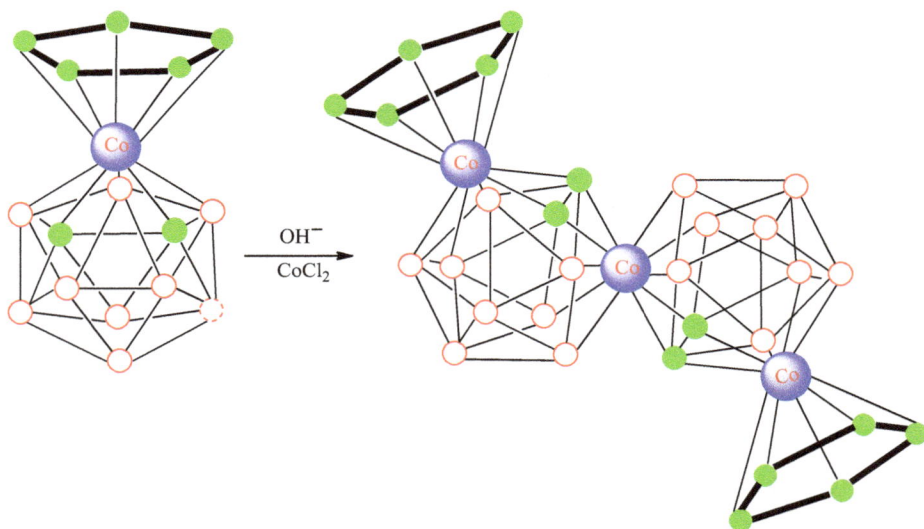

Figure 3.40: Formation of trimetallic carborane.

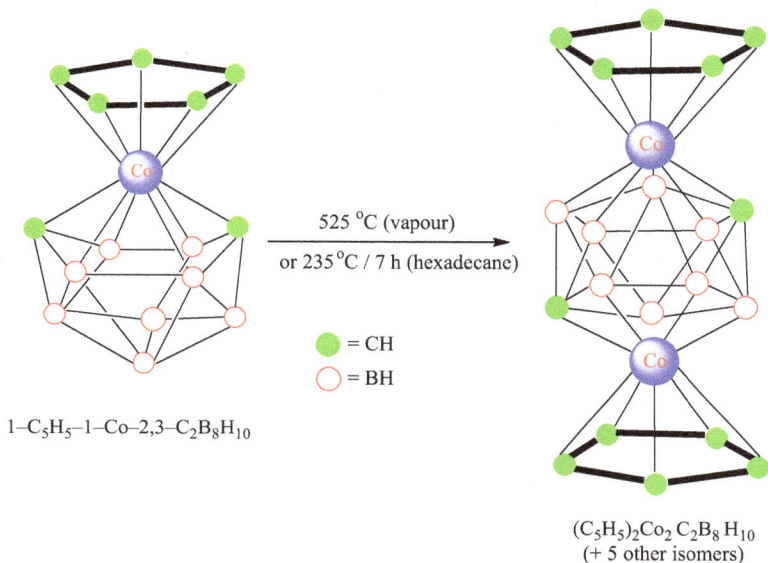

$1-C_5H_5-1-Co-2,3-C_2B_8H_{10}$

525 °C (vapour)

or 235 °C / 7 h (hexadecane)

● = CH

○ = BH

$(C_5H_5)_2Co_2\,C_2B_8\,H_{10}$
(+ 5 other isomers)

Figure 3.41: Formation of $(C_5H_5)_2Co_2C_2B_8H_{10}$ from $1-C_5H_5-1-Co-2,3-C_2B_8H_{10}$.

(vi) Direct oxidative insertion of a metal centre:

Nucleophilic zero-valent derivatives of Ni, Pd and Pt insert directly into *closo*-carborane clusters in a concerted process which involves a net transfer of electrons from the metal to the cage:

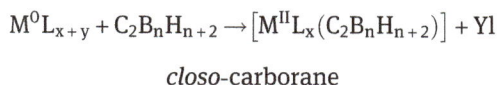

$$M^0L_{x+y} + C_2B_nH_{n+2} \rightarrow \left[M^{II}L_x(C_2B_nH_{n+2})\right] + Yl$$

closo-carborane

where L = PR_3, C_8H_{12}, RNC and so on.

A typical reaction is,

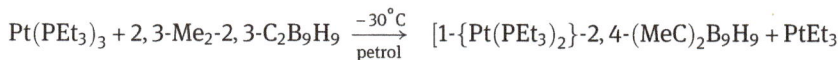

$$Pt(PEt_3)_3 + 2, 3\text{-Me}_2\text{-}2, 3\text{-}C_2B_9H_9 \xrightarrow[\text{petrol}]{-30°C} [1\text{-}\{Pt(PEt_3)_2\}\text{-}2, 4\text{-}(MeC)_2B_9H_9 + PtEt_3$$

Several novel cluster compounds have now been prepared in this way, including mixed metal clusters. But the structures sometimes have unexpectedly more open configurations than simple electron-counting rules would predict.

3.10.2 Reactivity of metallocarboranes

The following points are worth considering with regard to reactivity of metallo-carboranes:

(i) Normally, metallocarboranes are much less reactive (more stable) than the corre-sponding metallocenes and they tend to stabilize higher oxidation states of the later transition metals. For example, $[Cu^{II}(1,2\text{-}C_2B_9H_{11})_2]^{2-}$ and $[Cu^{III}(1,2\text{-}C_2B_9H_{11})_2]^{-}$ are known while cuprocene $[Cu^{II}(\eta^5\text{-}C_5H_5)_2]$ is not known. Similarly, Fe^{III} and Ni^{IV} carborane derivatives are extremely stable. Contrary to this, metallocarboranes tend to stabilize lower oxidation states of early transition elements, and complexes are well established for Ti^{II}, Zr^{II}, Hf^{II}, V^{II}, Cr^{II} and Mn^{II}. These do not react with H_2, N_2, CO or PPh_3 as do cyclopentadienyl derivatives of these elements.

(ii) Ferrocene can be protonated with strong acids to yield the cationic species $[Fe^{II}(\eta^5\text{-}C_5H_5)_2H]^+$, in which proton is associated with Fe atom. The carborane analogue $[Fe^{II}(\eta^5\text{-}1,2\text{-}C_2B_9H_{11})_2H]^-$ can be prepared similarly using HCl or $HClO_4$, and the protonated species is unique in promoting substitution at polyhedral boron (B) atom in good yields. For example, with weak Lewis base such as dialkyl sulphides:

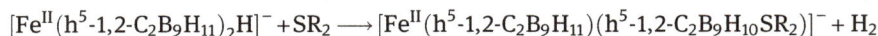

$$[Fe^{II}(h^5\text{-}1,2\text{-}C_2B_9H_{11})_2H]^- + SR_2 \longrightarrow [Fe^{II}(h^5\text{-}1,2\text{-}C_2B_9H_{11})(h^5\text{-}1,2\text{-}C_2B_9H_{10}SR_2)]^- + H_2$$

(iii) Bromination of the corresponding Co^{III} complex, in glacial acetic acid gives the hexabromo derivative $[Co^{III}(\eta^5\text{-}8,9,12\text{-}Br_3\text{-}1,2\text{-}C_2B_9H_8)_2]^-$ (Figure 3.42). Again, acid-

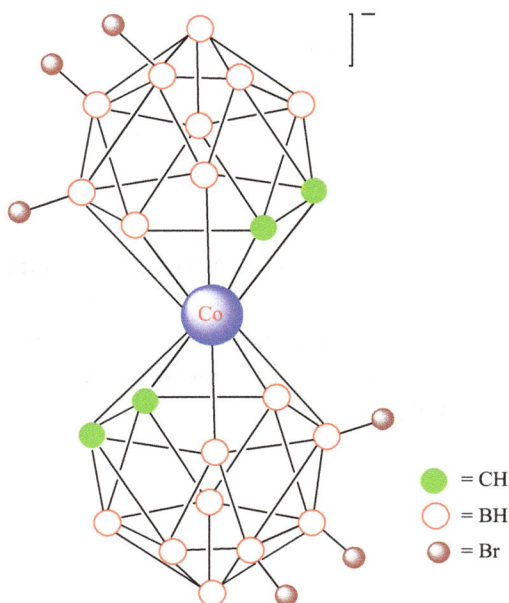

○ = CH
○ = BH
● = Br

Figure 3.42: Structure of $[Co^{III}(\eta^5\text{-}8,9,12\text{-}Br_3\text{-}1,2\text{-}C_2B_9H_8)_2]$.

catalyzed reaction of $K[Co((\eta^5\text{-}1,2\text{-}C_2B_9H_{11})_2]$ with CS_2 in the presence of $HCl/AlCl_3$ gives the novel neutral compound $[Co((\eta^5\text{-}1,2\text{-}C_2B_9H_{10})_2 (8,8'\text{-}S_2CH)]$ (Figure 3.43) in which the 8,8'-B atoms on the two icosahedra have been substituted by S and the polyhedral are linked by an S_2CH^+ bridge in a zwitterionic structure.

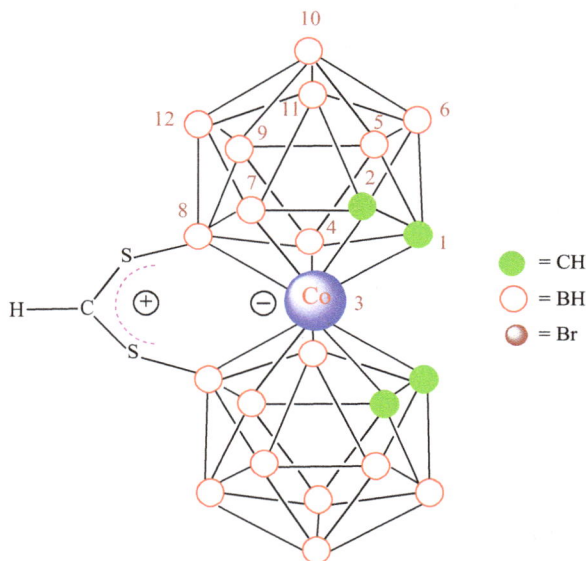

Figure 3.43: Structure of $[Co((\eta^5\text{-}1,2\text{-}C_2B_9H_{10})_2 (8,8'\text{-}S_2CH)]$.

Exercises

Multiple choice questions/fill in the blanks

1. Borane chemistry initiated in 1912 with the classic investigations of:
   (a) N. N. Greenwood   (b) W. N. Lipscomb   (c) A. Stock   (d) F. A. Cotton

2. Award of the 1976 Nobel Prize in Chemistry for studies on boranes was given to:
   (a) A. Stock   (b) W. N. Lipscomb   (c) G. Wilkinson   (d) None of these

3. *Nido*-boranes have structures in which the B_n cluster occupies n-corners of an:
   (a) $(n + 1)$-Cornered polyhedron   (b) $(n + 2)$-Cornered polyhedron
   (c) $(n + 3)$-Cornered polyhedron   (d) None of these

4. In *hypho*-boranes, the Greek word '*hyphe*' means:
   (a) a spiders web   (b) a nest   (c) a net   (d) None of these

5. In *closo*-boranes, name '*closo*' is derived from the word '*clovo*', which is a:
   (a) Latin word   (b) Greek word   (c) Japanese word   (d) All of these

6. In *hypho*-boranes, B atoms occupy n corners of a polyhedron having:
   (a) n Corners
   (b) (n + 2) Corners
   (c) (n + 3) Corners
   (d) (n + 4) Corners

7. Metalloboranes are borane cages containing metal atoms in number:
   (a) 1   (b) 2   (c) 3   (d) All of these

8. The basic idea of 3c-2e B–H–B and B–B–B types of bond was first time proposed by:
   (a) H. C. Longuet-Higgins
   (b) R. P. Bell
   (c) H. C. Longuet-Higgins and R. P. Bell
   (d) W. N. Lipscomb

9. In the formation of B_2H_6, B atom undergoes hybridization, namely:
   (a) sp   (b) sp^2 (c) sp^3 (d) None of these

10. The bond order of B–H bond in B–H–B bridging bond of diborane is:
    (a) 1.5   (b) 1   (c) 0.5   (d) 0

11. Which one of the following combinations is correct with regard to the orbitals involved from three B atoms in the formation of open B–B–B bond in higher boranes?
    (a) One sp^3 orbital of B_1, One sp^3 orbital of B_2, one p-orbital of B_3
    (b) One sp^3 orbital of B_2, One sp^3 orbital of B_3, one p-orbital of B_1
    (c) One sp^3 orbital of B_1, One sp^3 orbital of B_3, one p-orbital of B_2
    (d) All of these

12. The bond order of B–B bond in B–B–B bridging bond in higher boranes is:
    (a) 0.5   (b) 1   (c) 1.5   (d) 0

13. According to Lipscomb's rule, the number of atoms in a neutral boranes molecule is:
    (a) $(s + t + y + x)$
    (b) $2(s + t + y + x)$
    (c) $(s + t + y + x) + 2$
    (d) None of these

14. The number of electron pairs involved in the bonding of a borane cluster of n boron atoms is:
    (a) $n + (s + t + y + x)$
    (b) $n - (s + t + y + x)$
    (c) $2n + (s + t + y + x)$
    (d) None of these

15. The four-digit coding system of bonding in B_2H_6 is:
    (a) 2002   (b) 2020   (c) 2202   (d) None of these

16. The number of valance electrons for a borane of formula B_nH_m will be:
    (a) $(2n + m)$ (b) $(3n + m)$ (c) $(4n + m)$ (d) None of these

17. In *closo*-$B_6H_6^{2-}$, the total number of framework electrons will be:
    (a) 12   (b) 14   (c) 16   (d) None of these

18. The number of electron pairs of skeleton bonding electrons in *arachno*-carbaborane having n number of occupied vertices of the polyhedron will be:
    (a) $(n+1)$ pairs   (b) $(n+2)$ pairs   (c) $(n+3)$ pairs   (d) None of these

19. Wade's rules relating the structures of boranes with their composition can be extended to:
    (a) Carbaboranes   (b) metallocarbaboranes
    (c) Metalloboranes   (d) All on these

20. Replacement of a BH vertex by a metal vertex without changing the number of vertices in a cluster is termed as . . . .

21. The IUPAC name of $B_{20}H_{16}$ is . . . .

22. The four digit coding system of bonding in B_6H_{10} is . . . .

23. The IUPAC name of $B_5H_8^-$ is . . . .

24. When metals are incorporated in carborane cages, the resulting compounds are called . . . .

25. Carboranes are polyhedral boranes that contain framework C atoms as well as . . . . atoms.

Short answer type questions

1. What are boranes? Give names and formulae of five boranes.
2. Highlight the rules of nomenclature of boranes with suitable examples.
3. Briefly describe the importance of borane chemistry.
4. How are boranes classified?
5. What are *conjuncto*-boranes? Explain with suitable examples.
6. What are metalloboranes? Explain.
7. Highlight some important properties of boranes.
8. Draw the molecular orbital diagram related to the formation of B–H–B bridging bond in boranes. How bond order of B–H bond in B–H–B bridging bond is calculated?
9. Draw the molecular orbital diagram related to the formation of B–B–B bridging bond in boranes. How can its stability be ascertained?
10. Present the overlapping scheme of orbitals of three boron orbitals pictorially resulting bonding, non-bonding and anti-bonding MOs for the formation of open B–B–B bond.
11. What is Lispscomb rule? Give its utilities.
12. How are total number of valance electrons and number of bonds in boranes calculated without making use of Lipscomb's rule? Explain with a suitable example.

13. What are Wade's rules for relating the structures of boranes with their compositions?
14. What are carboranes or carbaboranes? Give the names and structures of some carbaboranes.
15. Present two synthetic routes to carboranes.
16. What are metallocarboranes? Give some examples of metallocarboranes.
17. Briefly highlight the reactivity of metallocarboranes.

Long answer type questions

1. Present a detailed view of classification of boranes with suitable examples of each type.
2. Discuss the formation of B–H–B bonds in diborane and higher boranes.
3. Present a detailed account of the formation of closed and open B–B–B bonds in higher boranes.
4. What is four digit coding system of bonding in boranes? Give its utility with suitable examples.
5. Describe in detail the Wade's rules pertaining to boranes of formula $B_nH_m{}^{x-}$, where $m \geq n$ and $x \geq 0$.
6. Present the structural and synthetic aspects of carboranes in detail.
7. What are metallocarboranes? Describe the various synthetic routes to metallo-carboranes.

Chapter IV
Synthesis and reactivity of metal clusters, and their bonding based on molecular orbital approach

4.1 Introduction

Inorganic chemistry encompasses numerous examples of molecular complexes containing metal–metal bonds. Such complexes are called metal clusters. **The name 'Metal Clusters' for metal–metal bonded compounds for the first time was coined by F. A. Cotton in 1964.** He defined metal clusters as

'The molecular complexes containing a finite group of metal atoms which are held together entirely or at least to a significant extent, by bonds between the metal atoms; some non-metal atoms may be associated with the cluster.' Certain metal clusters having metal–metal bonds are shown here (Figure 4.1).

Quite a few transition metals form metal–metal bonded compounds, and the structure and bonding of such compounds have been the subject of much research over the past several decades.

4.2 Metal cluster and catalysis

(i)   A principal reason for the extensive research on cluster compounds is their possible relevance to heterogeneous catalysis.
(ii)   Industrially important syntheses of organic compounds rely on catalysis.
(iii)   Most industrial reactions are catalyzed heterogeneously, usually by transition metals or their solid oxides.
(iv)   Practical considerations, which is ease of product separation often dictate a preference for heterogeneous catalysts.
(v)   Solid catalysts can be filtered off from a solution or can be suspended in a gaseous mixture. These solid catalysts are also stable at high temperatures requires for many reactions.

https://doi.org/10.1515/9783110727289-004

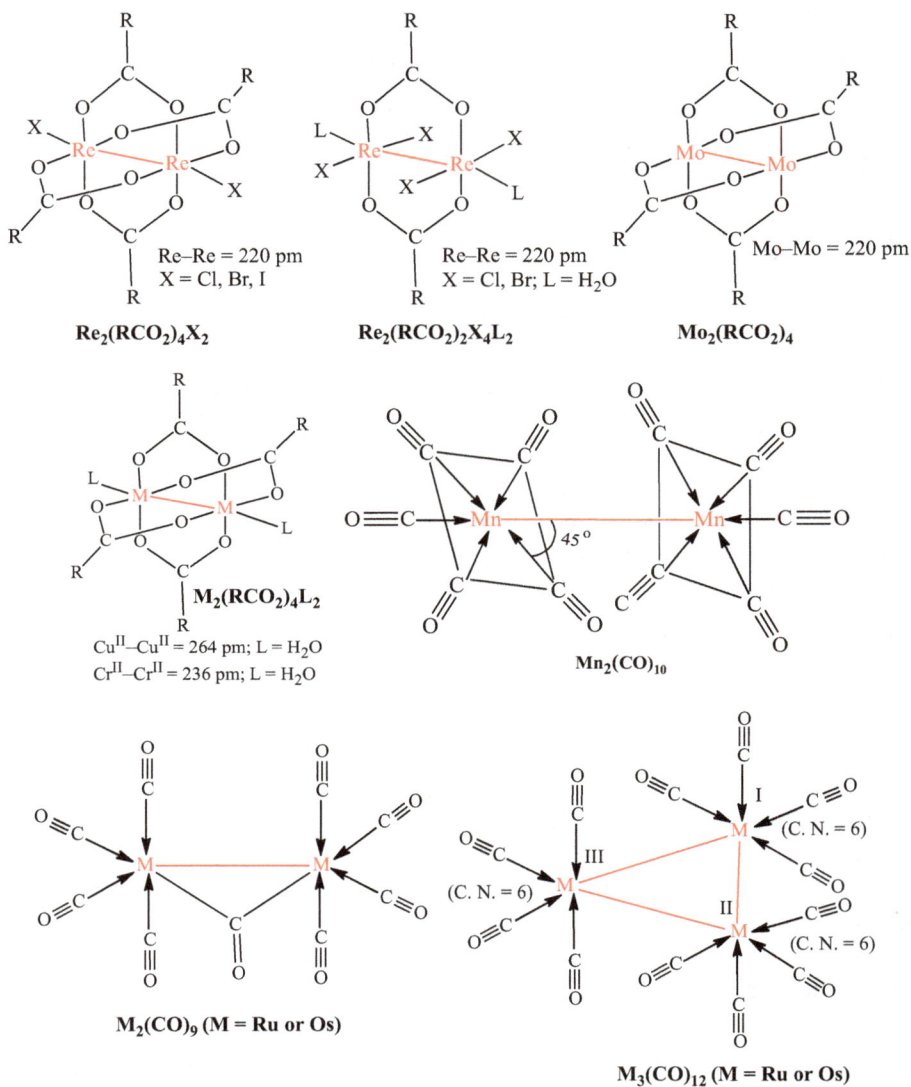

Re–Re = 220 pm
X = Cl, Br, I

Re₂(RCO₂)₄X₂

Re–Re = 220 pm
X = Cl, Br; L = H₂O

Re₂(RCO₂)₂X₄L₂

Mo–Mo = 220 pm

Mo₂(RCO₂)₄

M₂(RCO₂)₄L₂

Cu^{II}–Cu^{II} = 264 pm; L = H₂O
Cr^{II}–Cr^{II} = 236 pm; L = H₂O

45°

Mn₂(CO)₁₀

M₂(CO)₉ (M = Ru or Os)

(C. N. = 6)

(C. N. = 6)

I

III

II

(C. N. = 6)

(C. N. = 6)

M₃(CO)₁₂ (M = Ru or Os)

Figure 4.1: Structures of some metal clusters.

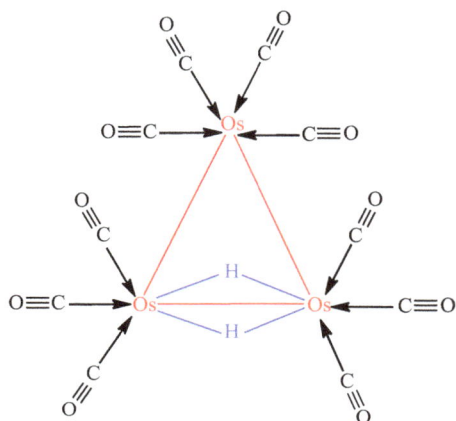

Cluster containing non-metal, H

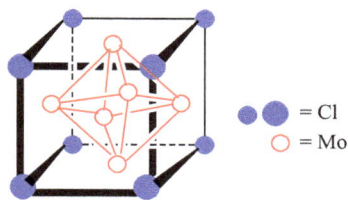

$Mo_6Cl_8^{4+}$ [6(2c-2e) bonds]

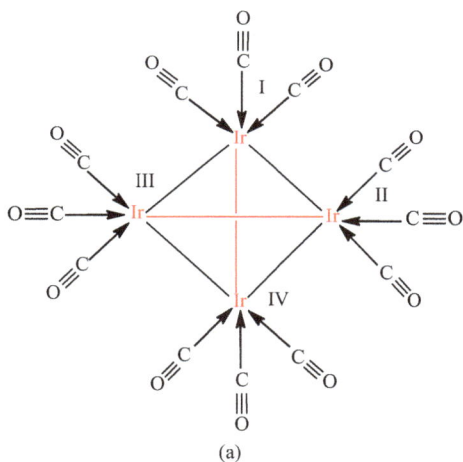

(a)

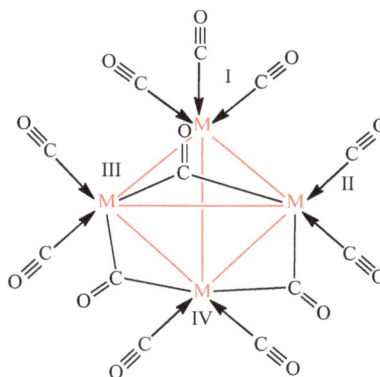

M = Co, Rh

(b)

$M_4(CO)_{12}$ (M = Co, Rh or Ir)

Figure 4.1 (continued)

4.3 Factors favouring for metal–metal bonding

There are four optimal conditions for the formation of M–M bonding in metal clusters:

4.3.1 Large energies of atomization

The tendency to retain clusters of metal atoms will predominate in those metals which have very large energies of atomization (and hence very high melting and boiling points). Thus, the most refractory metals (Zr, Nb, Mo, Tc, Ru, Rh, Hf, Ta, W, Re, Os, Ir and Pt) having large energies of atomization have the greatest tendency to form metal clusters.

4.3.2 Low oxidation states

The most important factor appears to be the low formal oxidation state. The extremely large majority of species containing M–M bonds have metal atoms in a formal oxidation state of +2 or less. The importance of this factor is probably connected with the dependence of M–M orbital overlap over the size of the d-orbitals.

The size of the d-orbitals is inversely related to the effective nuclear charge. Since effective overlap of d-orbitals appears necessary to stabilize metal clusters, excessive contraction of them (d-orbitals) will destabilize the clusters. Hence, large effective nuclear charges (resulting small size of d-orbitals) from high oxidation states are unfavourable. For the first transition series, the d-orbitals are relatively small and even in moderately low oxidation states (+2 and +3), they apparently do not extend sufficiently for good overlap.

4.3.3 Presence of only limited number of electrons in the valence shell

This is necessary to avoid the presence of too many electrons in the anti-bonding orbitals as that would affect the bond order adversely. It is for this reason that most of the transition metals which keenly form such complexes, fall in the first half of any transition series. Those with higher atomic numbers opt for such complexes only with ligands like CO which have the capacity to form back π-bonds.

4.3.4 Suitable valence shell configurations

A very favourable valence shell configuration (d^4) may result in strong M–M bonding even if the above three conditions are not suitable. Thus, $Re_2Cl_8^{2-}$ ions have very strong Re–Re bond (quadruple/fourfold bonds) despite of high oxidation states of +3 of metal atoms.

4.4 Evidence/identifying parameters for metal–metal bonding

4.4.1 Molecular structure

(a) In many instances, molecular structure itself provides surest indication of M–M bonding. For example, in Mn_2CO_{10}, each Mn atom is coordinated to five CO molecules and there is absolutely nothing to join the two $Mn(CO)_5$ halves except Mn–Mn bond as clear from its structure shown in Figure 4.2. The $Tc_2(CO)_{10}$ and $Re_2(CO)_{10}$ have similar structures.

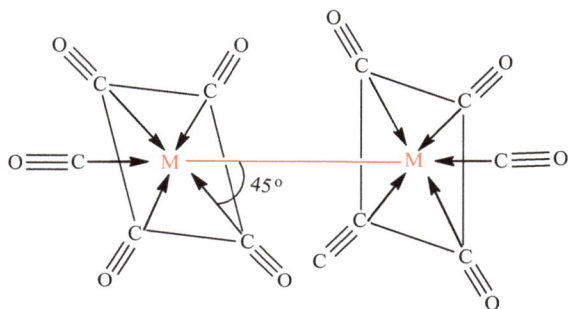

Figure 4.2: Structure of $M_2(CO)_{10}$ (M=Mn, Tc or Re).

(b) Another example is $W_2Cl_9^{3-}$ molecular ion (Figure 4.3). Here, each tungsten (W^{3+}, d^3 system) has approximately octahedral surrounding of six chloride ions and the three chlorides placed between them act as bridging unit. The molecular structure is appreciably distorted (compressed) so that the two tungstens are not exactly at the centre of their octahedra. Large compression brings the two tungstens appreciably closer and thus paves way for strong M–M interaction.

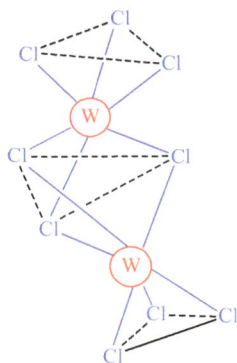

Figure 4.3: Structure of $W_2Cl_9^{2-}$.

If one considers the presence of three electrons in d_{xy}, d_{yz} and d_{xz} orbitals of tungsten, formation of two π-bonds through d_{xz}–d_{xz} and d_{yz}–d_{yz} overlaps, and one δ-bond through d_{xy}–d_{xy} overlap along the z-axis becomes apparent. This also makes the cluster anion diamagnetic.

(c) $Mo_2Cl_9^{3-}$ (Figure 4.4) too is very much similar to $W_2Cl_9^{3-}$, but since the molecular structure is less compressed, an overlap between d-orbitals of Mo is also less. This results in the M–M bonding of decreased strength.

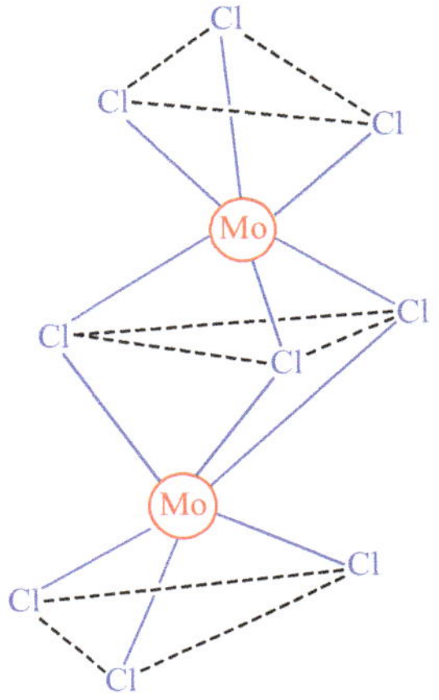

Figure 4.4: Structure of $Mo_2Cl_9^2$.

4.4.2 Magnetic susceptibility

Lowering of magnetic moments, even to zero, as in case of $W_2Cl_9^{3-}$, is another identifying parameter. This arises due to pairing of spins in the course of metal–metal interactions. As the magnetic susceptibilities are easy to measure, this parameter has often been used. However, this parameter may be misleading if the spin-orbit coupling constant is of high value, as in the case of heavier transition metals.

4.5 Classification of metal clusters

Metal cluster compounds can be conveniently grouped into two groups:
(i) Polynuclear carbonyl nitrosyl and related compounds containing organo-π-donors and other π-acceptor acids.
(ii) The lower halides and oxides.

4.6 Synthesis of metal clusters

The most active area in metal cluster chemistry currently is that of carbonyl-containing clusters. The synthesis of these species is presently in a rather unsatisfactory state. However, some preparative methods of metal clusters are given below:

4.6.1 Thermal condensation

A very important route to the synthesis of metal clusters is the thermal condensation reactions. In such reactions, carbonyl ligands are removed along with production of cluster compounds; for example,

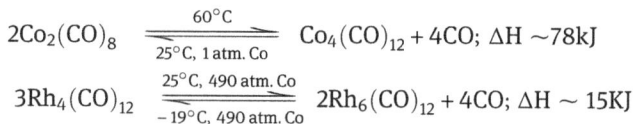

$$2Co_2(CO)_8 \underset{25°C, \, 1\,atm. \, Co}{\overset{60°C}{\rightleftharpoons}} Co_4(CO)_{12} + 4CO; \; \Delta H \sim 78kJ$$

$$3Rh_4(CO)_{12} \underset{-19°C, \, 490\,atm. \, Co}{\overset{25°C, \, 490\,atm. \, Co}{\rightleftharpoons}} 2Rh_6(CO)_{12} + 4CO; \; \Delta H \sim 15KJ$$

Both the above reactions are endothermic, as one would expect for processes that break strong M–CO bonds and substitute weak M–M bonds.

The greater strength of M–CO bonds as compared to M–M bonds for the first row transition elements requires that cluster synthesis may be carried out in absence of CO to avoid degradation of clusters. This suggests a preformed carbonyl compounds should be used in synthesis of clusters.

For second and third row transition elements, the reduction to carbonyl and condensation to metal clusters can be carried out in a single step; for example,

$$7K_3RhCl_6 + 48KOH + 28CO \xrightarrow[MeOH]{25°C, \, 1 \, atm. \, CO} K_3[Rh_7(CO)_{16}] + 42KCl + 12K_2CO_3 + 24H_2O$$

4.6.2 Redox condensation

The second method of general utility in preparing carbonyl clusters is redox condensations. These methods offer, particularly, attractive possibilities for making mixed-metal clusters. Redox condensation reactions often take place under very mild conditions; for example,

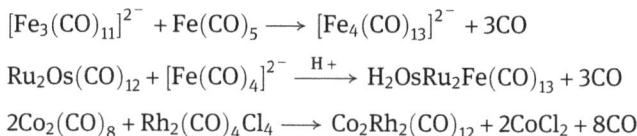

$$[Fe_3(CO)_{11}]^{2^-} + Fe(CO)_5 \longrightarrow [Fe_4(CO)_{13}]^{2^-} + 3CO$$

$$Ru_2Os(CO)_{12} + [Fe(CO)_4]^{2^-} \xrightarrow{H^+} H_2OsRu_2Fe(CO)_{13} + 3CO$$

$$2Co_2(CO)_8 + Rh_2(CO)_4Cl_4 \longrightarrow Co_2Rh_2(CO)_{12} + 2CoCl_2 + 8CO$$

4.7 Reactivity of metal clusters

Comparatively little is known of the reactivity of cluster species, and particularly nothing is known of reaction mechanism. Only some general reaction types are given below:

4.7.1 Reduction

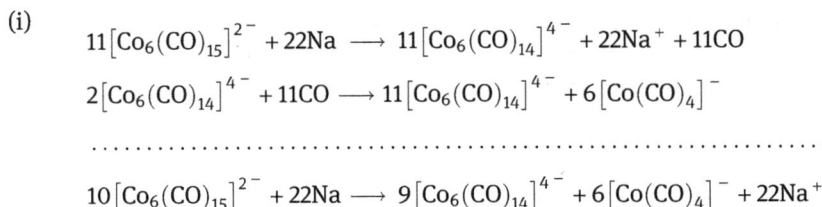

(i)

$$11\left[Co_6(CO)_{15}\right]^{2-} + 22Na \longrightarrow 11\left[Co_6(CO)_{14}\right]^{4-} + 22Na^+ + 11CO$$

$$2\left[Co_6(CO)_{14}\right]^{4-} + 11CO \longrightarrow 11\left[Co_6(CO)_{14}\right]^{4-} + 6\left[Co(CO)_4\right]^-$$

. . . .

$$10\left[Co_6(CO)_{15}\right]^{2-} + 22Na \longrightarrow 9\left[Co_6(CO)_{14}\right]^{4-} + 6\left[Co(CO)_4\right]^- + 22Na^+$$

It is well clear from the above reaction that the released CO in first step leads to cluster degradation, and thus the yield of the reduced product is decreased.
(ii) Use of strong base reduces coordinated CO, thus providing a better yield of the reduction product:

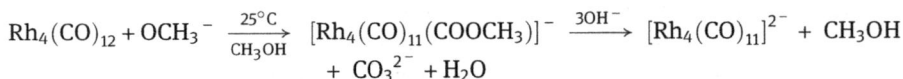

$$Rh_4(CO)_{12} + OCH_3^- \xrightarrow[CH_3OH]{25^\circ C} \left[Rh_4(CO)_{11}(COOCH_3)\right]^- \xrightarrow{3OH^-} \left[Rh_4(CO)_{11}\right]^{2-} + CH_3OH$$
$$+ CO_3^{2-} + H_2O$$

In this reaction, the intermediate carboalkoxy complex, $[Rh_4(CO)_{11}(COOCH_3)]$, has been isolated.

4.7.2 Oxidation

i. Electrochemical oxidation of $cp_4Fe_4(CO)_4$ to the +1 and +2 species is known.

ii. A second method of oxidation is by protonation of anionic cluster to give intermediate hydrido compound, which then react with additional proton eliminating H_2.

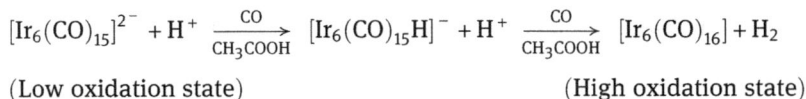

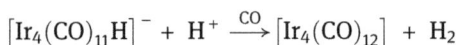

$$\left[Ir_6(CO)_{15}\right]^{2-} + H^+ \xrightarrow{CO}{CH_3COOH} \left[Ir_6(CO)_{15}H\right]^- + H^+ \xrightarrow{CO}{CH_3COOH} \left[Ir_6(CO)_{16}\right] + H_2$$

(Low oxidation state)   (High oxidation state)

$$\left[Ir_4(CO)_{11}H\right]^- + H^+ \xrightarrow{CO} \left[Ir_4(CO)_{12}\right] + H_2$$

iii. Some chemical oxidation can be affected with addition of oxidant, which is, I_2.

$$\left[Rh_6(CO)_{15}\right]^{2-} + I_2 \xrightarrow{THF} \left[Rh_6(CO)_{15}I\right]^- + I^-$$

4.7.3 Ligand substitution

(i) Carbonyl ligands in carbonyl clusters often can be replaced by Lewis bases. However, cluster fragmentation often occurs with first row transition metal species because of their weak metal–metal bonds.

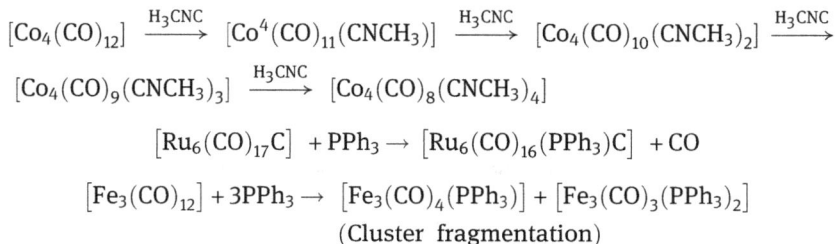

$$[Co_4(CO)_{12}] \xrightarrow{H_3CNC} [Co^4(CO)_{11}(CNCH_3)] \xrightarrow{H_3CNC} [Co_4(CO)_{10}(CNCH_3)_2] \xrightarrow{H_3CNC}$$

$$[Co_4(CO)_9(CNCH_3)_3] \xrightarrow{H_3CNC} [Co_4(CO)_8(CNCH_3)_4]$$

$$[Ru_6(CO)_{17}C] + PPh_3 \rightarrow [Ru_6(CO)_{16}(PPh_3)C] + CO$$

$$[Fe_3(CO)_{12}] + 3PPh_3 \rightarrow [Fe_3(CO)_4(PPh_3)] + [Fe_3(CO)_3(PPh_3)_2]$$
(Cluster fragmentation)

(ii) A cluster containing non-metals may display reactivity. The following scheme summarizes several reactions of $[(Cl)(C)Co_3(CO)_9]$:

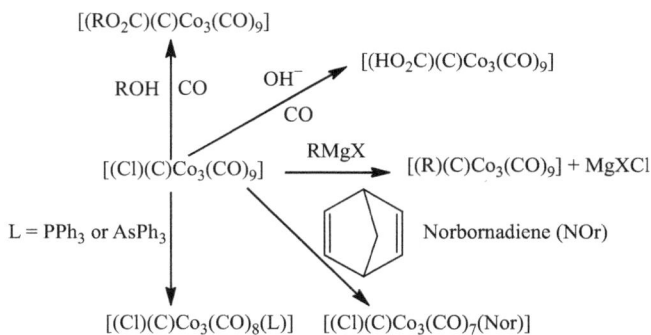

(iii) Substitution reactions occur readily in some Re–R and Mo–M quadruple bonded compounds

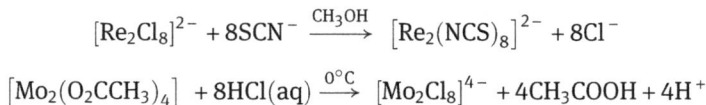

$$[Re_2Cl_8]^{2-} + 8SCN^- \xrightarrow{CH_3OH} [Re_2(NCS)_8]^{2-} + 8Cl^-$$

$$[Mo_2(O_2CCH_3)_4] + 8HCl(aq) \xrightarrow{0°C} [Mo_2Cl_8]^{4-} + 4CH_3COOH + 4H^+$$

At 60 °C, addition of HCl to $[Mo_2(O_2CCH_3)_4]$ results in oxidative addition to Mo–Mo quadruple bond,

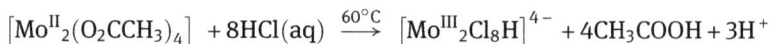

$$[Mo^{II}_2(O_2CCH_3)_4] + 8HCl(aq) \xrightarrow{60°C} [Mo^{III}_2Cl_8H]^{4-} + 4CH_3COOH + 3H^+$$

The structure of compound $[Mo^{III}_2Cl_8H]^{4-}$ is shown in Figure 4.5:

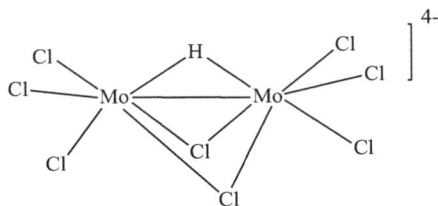

Figure 4.5: Structure of $[Mo^{III}_2Cl_8H]^4$.

4.8 Bonding in metal clusters: molecular orbital approach

The metal–metal bonding in metal clusters involves:
(i) Two metal atoms with localized single or multiple two centre homo- or hetero-nuclear covalent bonds, or
(ii) More than two metal atoms of same or different atomic numbers with considerable delocalization of covalent bonded electron density.

4.8.1 Two metal atoms with localized single or multiple bonds

(a) Metal clusters having M–M single bonds
In such cases M–M bonding in clusters can be adequately described in terms of 2c–2e bonds; for example, in $(CO)_5Mn–Mn(CO)_5$. Few clusters, such as, the $Os_3(CO)_{12}$ having a triangular arrangement of the three Os atoms in the molecule with four CO bonded to each Os atom, and $Mo_6Cl_8^{4+}$, having an octahedron of six Mo atoms can be treated satisfactorily as having a collection of 2c–2e M–M bonds (Figure 4.6).

(b) Metal clusters having M–M multiple bonds
$Re_2Cl_8^{2-}$ may be taken as an example of cluster compounds containing localized multiple bonding. This ion has been found present in the crystals of $K_2[Re_2Cl_8] \cdot 2H_2O$, and is X-ray characterized with the following unusual features:

(i) Re–Re distance:
It is only 224 pm against an average Re–Re distance of 275 pm in rhenium metal and 248 pm in Re_3Cl_9. This extreme closeness can be expected to counteract strong M–M bonding due to increased inter nuclear and inter-electronic repulsion.

(ii) Oxidation state of Re:
The oxidation state of Re in $Re_2Cl_8^{2-}$ is +3 and, hence, quite high to warrant strong M–M bonding.

Mn$_2$(CO)$_{10}$ [1(2C-2e) bond]

Os$_3$(CO)$_{12}$ [3(2c-2e) bonds]

= Cl
= Mo

Mo$_6$Cl$_8^{4+}$ [6(2c-2e) bonds]

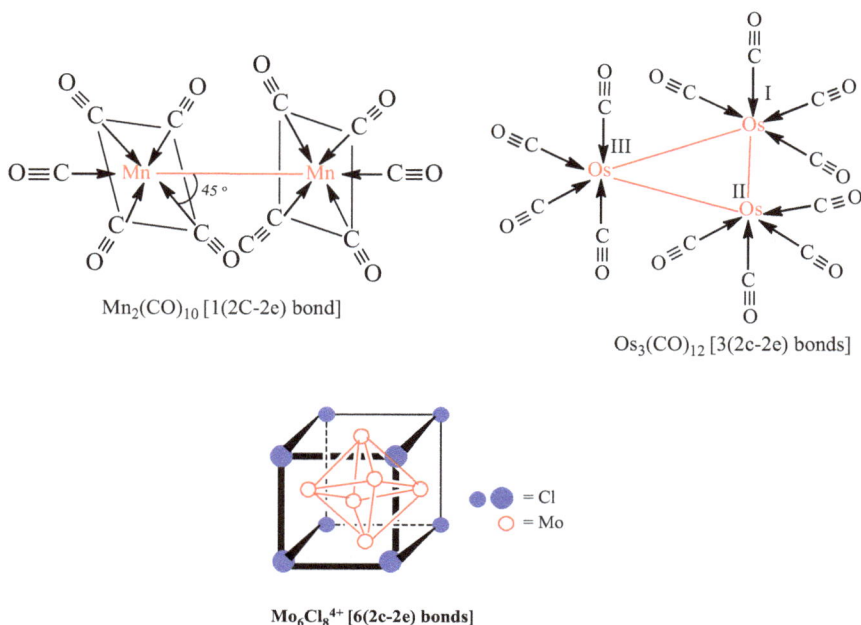

Figure 4.6: Metal clusters having collection of 2c–2e M–M bonds.

(iii) Eclipsed configuration of Cl$_4$ square:

Arrangement of four chlorines about each Re is square planar, and the two Cl$_4$ squares have a relative eclipsed configuration (Figure 4.7) instead of staggered. This speaks for undesirable Cl–Cl repulsion which should in turn tell upon the stability of this anion. This effect is aggravated by the fact that the Cl–Cl distance between two Cls of two different squares (3.32 Å) is much less than the distance between two Cls of the same square (3.35 Å). In spite of above, M–M bonding in this molecular ion is adjudged as extremely strong. This is possible only if the effect of these unusual features is somehow duly compensated. In fact, these unusual features facilitate a

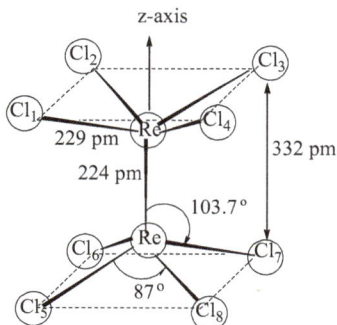

Figure 4.7: Octachlorodirhenate(III) ion, Re$_2$Cl$_8^{2-}$: eclipsed configuration.

strong M–M bonding in the ion as explained below by F. A. Cotton on the basis of MO theory (Cotton and Harris 1965).

Cotton's explanation

Cotton explained the above phenomena by invoking a quadruple (fourfold) bond which is unexpectedly short and unexpectedly strong. His explanation is as follows:

The z-axis may be taken as the line joining the Re atoms. Each Re atom is bonded to four Cl atoms almost in a square planar array (the Re atom is 50 pm out of the plane of the 4Cl$^-$). We may take the Re–Cl bonds to be approximate dsp^2 hybrids utilizing the $d_{x^2-y^2}$ orbital. The d_{z^2} and p_z orbitals on each Re lie along the Re–Re bond axis and may be hybridized to form one hybrid orbital on each Re atom directed to other Re atom, and other hybrid orbital on each Re atom directed in opposite direction. The former hybrid orbital can overlap with the similar hybrid orbital on second Re atom to from a σ-bond (or σ-bonding MO) while the second hybrid orbital on each Re forms an approximately **non-bonding orbital.** The formation of σ-**MO** with pure dz^2 is shown in Figure 4.8a.

The d_{xz} and d_{yz} orbitals of each Re are directed obliquely towards the other Re atom, overlapping to form dπ–dπ bonds (Figure 4.8b). This results in two π-bonds, one in xz plane and one in yz plane. A fourth bond can now form by overlap of the remaining d-orbital on each atom, that is, d_{xy}. The 'sideways' overlap of two d_{xy} orbitals results in a δ-bond (Figure 4.8c).

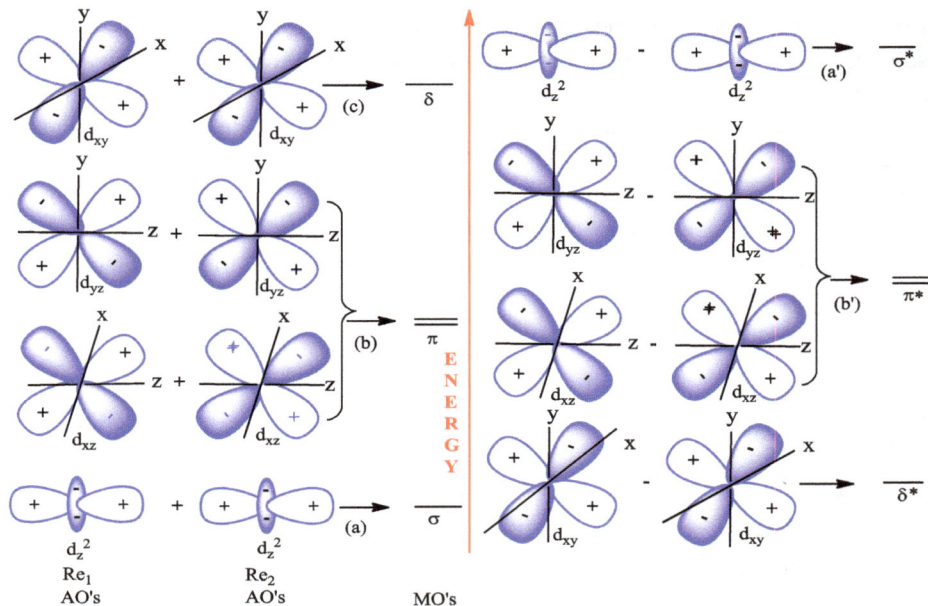

Figure 4.8: Overlap of d-orbitals of Re$_1$ and Re$_2$ and resulting MOs.

It is notable here that the maximum overlap can only occur if the Cl atoms are eclipsed (Figure 4.9(i)). If the Cl atoms are staggered, the two d_{xy} orbitals will likewise be staggered keeping resulting zero overlap (Figure 4.9(ii)).

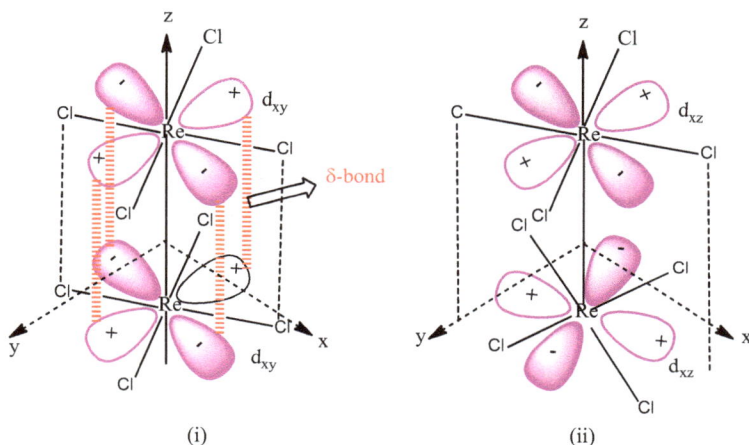

Figure 4.9: (i) Eclipsed configuration having maximum overlap of d_{xy}-orbital of each Re atom to form a δ-bond. (ii) Staggered configuration having zero overlap of d_{xy}-orbitals.

In $Re_2Cl_8^{2-}$, each Re (III) is a d^4 system ($Re_{75} = 5d^5 6s^2$). The Re–Cl bonds may be considered as dative bonds from Cl^- to Re^{3+} ions. After allowance for Re–Cl bonds and net charge, each Re atom will be left with four d-orbitals (one d-orbital, $d_{x^2-y^2}$ on each Re has already been utilized in dsp^2 hybridization for each $ReCl_4$ unit) and four electrons. The formation of one σ-, two π- and one δ-bonding MOs as shown and discussed above causes pairing of four electrons, and thus forming the quadruple (fourfold) bonds (Figure 4.10). The complex ion is, therefore, diamagnetic.

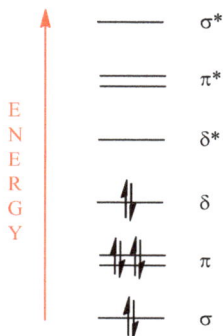

Figure 4.10: An energy level diagram of MOs involved in Re–Re bonding in $Re_2Cl_8^{2-}$.

The high strength (300–400 kcal mol^{-1}), short Re–Re distance and eclipsed configuration are accounted for by molecular orbital (MO) approach.

$Mo_2Cl_8^{4-}$ (Mo^{2+}, d^4) is isostructural with $Re_2Cl_8^{2-}$, and it has an eclipsed configuration and a very short (2.14 Å) Mo–Mo bond.

Re(III) and Mo(II) form a large series of carboxylate complexes of formulas, $Re_2(RCO_2)_4X_2$, $Re_2(RCO_2)_2X_4L_2$ and $Mo_2(RCO_2)_4$ (Figure 4.11).

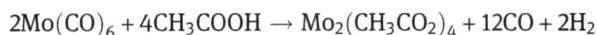

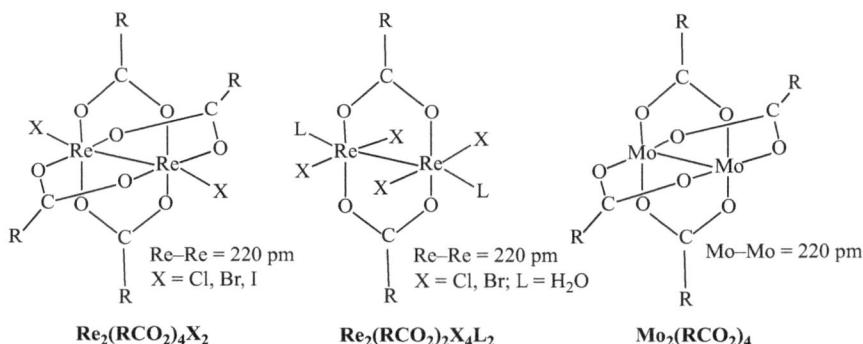

$$Re_2Cl_8^{2-} + 2CH_3COOH \rightarrow Re_2(CH_3CO_2)_2Cl_4 + 4Cl^- + 2H^+$$

$$Re_2Cl_8^{2-} + 4CH_3COOH \rightarrow Re_2(CH_3CO_2)_4Cl_2 + 6Cl^- + 4H^+$$

$$2Mo(CO)_6 + 4CH_3COOH \rightarrow Mo_2(CH_3CO_2)_4 + 12CO + 2H_2$$

Re–Re = 220 pm
X = Cl, Br, I

$Re_2(RCO_2)_4X_2$

Re–Re = 220 pm
X = Cl, Br; L = H_2O

$Re_2(RCO_2)_2X_4L_2$

Mo–Mo = 220 pm

$Mo_2(RCO_2)_4$

Figure 4.11: Molecular structures of some carboxylate complexes containing metal–metal bonding.

Structurally, these complexes are clearly related to $Re_2Cl_8^{2-}$, the only difference being the addition of a ligand to the non-bonding orbitals present in $Re_2(RCO_2)_2X_4L_2$. The absence of a ligand in the $Re_2Cl_8^{2-}$ is most likely due to steric hindrance of the large halide ions compared with the smaller oxygen atoms in the carboxylate group.

The structures of Cu(II) and Cr(II) carboxylates, $M(RCO_2)_4(H_2O)_2$ [M = Cu(II) or Cr(II)] (Figure 4.12) are also related to $Re_2Cl_8^{2-}$.

4.8.2 More than two metal atoms of same or different atomic numbers with considerable delocalization of covalent bonding

The formation of metal clusters through delocalized bonds/bonding means we can no longer distinguish electron pairs bonding pairs of atoms. Instead, we consider several electrons as belonging to a group of atoms and holding the clusters together.

The trinuclear $Re_3X_9L_3$ may be taken as a representative metal cluster to illustrate how the concept of delocalized MOs are used to explain the bonding in

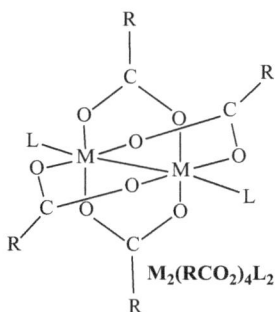

$M_2(RCO_2)_4L_2$

$Cu^{II}-Cu^{II} = 264$ pm; L = H_2O
$Cr^{II}-Cr^{II} = 236$ pm; L = H_2O

Figure 4.12: Molecular structures of $M(RCO_2)_4(H_2O)_2$ [M = Cu(II) or Cr(II)].

clusters having more than two metal atoms. The structure of the cluster is shown below (Figure 4.13).

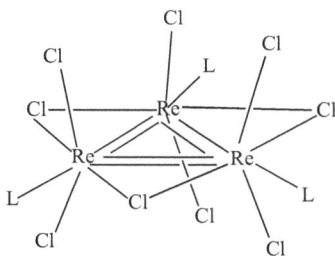

Figure 4.13: Structure of $Re_3Cl_9L_3$.

The simplest approach is to note that the metal atom in this compound is in +3 oxidation state (d^4 system), L is some neutral ligand and X usually some halide like Cl^-. Here, three Re form an equilateral triangle and each is coordinated to five ligands, one L and four Cls. Three Cls out of nine in the ion are present as bridging units for three Res. Thus, each Re is linked with two bridging Cls, two non-bridging Cls and one L. In addition, each Re is also linked with two other Re atoms.

The five ligands around any Re occupy five corners of a square pyramid. One can safely assume that the dsp^3 hybridization required for it involves a vacant d-orbital of Re, and its four electrons are accommodated in the remaining pure d-orbitals. The total of 12 such d-orbitals on three Re are somewhat twisted to involve themselves in mutual overlapping to give rise six delocalized bonding and six corresponding anti-bonding MOs (Figure 4.14).

It is seen that there are six bonding MOs, two of which (e′ and e″) are doubly degenerate. Thus altogether the bonding MOs can hold 12 electrons. As each Re atom has four electrons after allowance for metal–ligand bonds, hence the bonding MOs are fully occupied. The cluster is, therefore, diamagnetic with maximum M–M bonding.

$Re_{57} = 5d^5 6s^2$

$Re^{3+} = d^4 \rightarrow$

dsp^3 hybridization

= 2e donated by each ligand

$a_2'^*$

$a_2''^*$ ⎱ Six anti-bonding MOs

e'^*

e''^*

ENERGY

e''

e' ⎱ Six bonding MOs

a_2''

a_2'

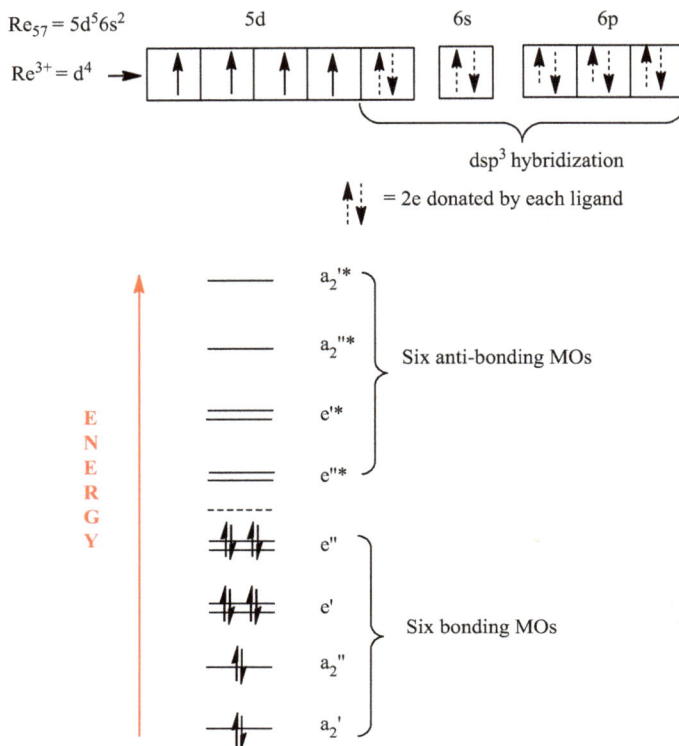

Figure 4.14: An energy level diagram of MOs involved in Re–Re–Re bonding in $Re_3Cl_9L_3$.

Since, there are six electron pairs distributed over three Re–Re edges of Re$_3$ tri-angle or two pair per Re–Re interaction, the MO picture is formally equivalent to describing each Re–Re bond as a double bond, consisting of one σ- and one π-bond ⊥r to Re$_3$ plane.

Exercises

Multiple choice questions/fill in the blanks

1. The name 'metal clusters' for metal–metal bonded compounds for the first time in 1964 was coined by:
   (a) N. N. Greenwood
   (b) W. N. Lipscomb
   (c) G. Wilkinson
   (d) F. A. Cotton

2. Sort out the metal cluster from the following:
   (a) B_2H_6 (b) $Fe(CO)_5$ (c) $Mn_2(CO)_{10}$ (d) $Cr(CO)_6$

3. Formation of metal cluster is favourable in metal of:
   (a) High formal oxidation state
   (b) Low formal oxidation state
   (c) Oxidation state is no bar
   (d) All of these

4. For the formation of stable metal cluster, effective nuclear charge of the metal should be:
   (a) Low
   (b) High
   (c) Effective nuclear charge low or high does not matter
   (d) Cannot be said

5. For the formation of stable metal cluster, the number of electrons in the valance shell of the transition metal show be:
   (a) Limited   (b) Moderate   (c) High   (d) All of these

6. Most of the transition metals which strongly form metal clusters fall in the first half of any transition series because they have electron in their anti-bonding orbitals in number:
   (a) Minimum including zero
   (b) Moderate
   (c) Much
   (d) Cannot be said

7. One of the identifying parameters in metal clusters is:
   (a) Low magnetic moment even to zero
   (b) High magnetic moment
   (c) Both low and high magnetic moment
   (d) All of these

8. In $W_2Cl_9{}^{3-}$ molecular ion cluster, strong W−W interaction is because of the compression of two tungsten octahedral units:
   (a) To some extent
   (b) Moderately
   (c) Appreciably
   (d) None of these

9. In $Mo_2Cl_9^{3-}$ cluster, Mo–Mo bonding of decreased strength is due to compression of two molybdenum octahedral units:
   (a) To less extent
   (b) To moderate extent
   (c) To appreciable extent
   (d) All of these

10. The metal–metal bonding in $Os_3(CO)_{12}$ having a triangular arrangement of the three Os atoms in the molecule with four CO bonded to each Os atom is can be described in terms of:
    (a) 1(2c–2e) bond
    (b) 2(2c–2e) bonds
    (c) 3(2c–2e) bonds
    (d) None of these

11. Which one of the following metal cluster forms quadruple (fourfold) metal–metal bonds?
    (a) $Mn_2(CO)_{10}$ (b) $Os_3(CO)_{12}$ (c) $Mo_6Cl_8^{4+}$ (d) $Mo_2Cl_8^{4-}$

12. In $Re_2Cl_8^{2-}$ cluster, maximum overlap can only occur to form quadruple bonds if the Cl atoms of two Re units are:
    (a) Eclipsed to each other
    (b) Staggered to each other
    (c) Gauche to each other
    (d) None of these

13. Sort out a metal cluster involving delocalized bonding:
    (a) $Os_3(CO)_{12}$ (b) $Fe_3(CO)_{12}$ (c) $Mo_2Cl_9^{2-}$ (d) $Re_3Cl_9L_3$

14. Synthesis of carbonyl clusters by redox condensation method offers possibility of mixed . . . . clusters.

15. Refractory metals having large energies of atomization have the greatest tendency to form . . . . clusters.

Short answer type questions

1. What are metal clusters? Give formulae and structures some metal clusters.
2. Highlight the factors favouring metal–metal bonding.
3. Briefly describe the identifying parameters for metal–metal bonding.
4. How are metal clusters synthesized?

5. Describe a method of mixed metal cluster synthesis.
6. Present a brief note on reactivity of metal cluster.
7. Highlight the bonding in metal clusters having metal–metal single bond with suitable examples.
8. Give the pictorial representations of overlapping of d-orbitals of Re_1 and Re_2 in $Re_2Cl_8^{2-}$ resulting molecular orbitals.
9. Draw the energy level diagram of molecular orbitals along with electron filling involved in Re–Re bonding in $Re_2Cl_8^{2-}$ and comments on its magnetic behaviour.
10. Draw the structure of trinuclear $Re_3Cl_9L_3$ and also draw the molecular orbital energy level diagram involved in Re–Re–Re bonding in this cluster.

Long answer type questions

1. Define the metal clusters with suitable examples. Present a detailed view of the synthesis and reactivity of metal clusters.
2. Give an explanatory note on factors favouring metal–metal bonding.
3. Present a detained view of evidences for metal–metal bonding in metal clusters.
4. Describe the metal–metal bonding in metal clusters with special reference to localized multiple two centre homo- or heteronuclear covalent bonds.
5. What do you mean by the formation of metal clusters through delocalized bonding? Taking a suitable example, explain the formation of delocalized bonding in metal cluster having more than two metal atoms.

Chapter V
Stability constants of metal complexes: some aspects

5.1 Introduction

The subject of stability of metal complexes is important in understanding the properties of complexes. Many variables associated with the central metal atom/ion and the ligand greatly complicate the study of this subject. The reasonable approach to the study of stability of complexes is to maintain as many variables as possible constant, and then examine a small area of the whole subject.

In a general sense, the stability of compounds means that the compounds existing under suitable conditions may be stored for a long period of time. However, when the formation of complexes in solution is studied, two types of stabilities, thermodynamic stability and kinetic stability, are considered.

In the language of thermodynamics, the equilibrium constants of a reaction are the measure of the heat released in the reaction and entropy change during reaction. The greater amount of heat evolved in the reaction, the most stable are the reaction products. Secondly, greater the increase in entropy during the reaction, greater is the stability of products. The greater heat evolved and increase in entropy during reaction for a stable complex can be illustrated by the following data:

$$[Cu(H_2O)_4]^{2+} + 2NH_3 \rightleftharpoons [Cu(H_2O)_2(NH_3)_2]^{2+} + 2H_2O$$

$$\Delta H = -46 \text{ kJ mol}^{-1}; \ \Delta S = -8.4 \text{ JK}^{-1}\text{mol}^{-1}$$

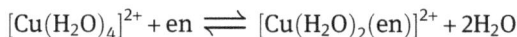

$$[Cu(H_2O)_4]^{2+} + en \rightleftharpoons [Cu(H_2O)_2(en)]^{2+} + 2H_2O$$

$$\Delta H = -54 \text{ kJ mol}^{-1}; \ \Delta S = 23 \text{ JK}^{-1}\text{mol}^{-1}$$

[$\Delta G = \Delta H - T\Delta S = -RT\ln K$, where ΔH is the enthalpy, ΔS the entropy and K is the equilibrium constant for the reaction and ΔG is the change in energy on going from reactants to products in the reaction].

The greater stability of ethylenediamine (en) chelate complex is primarily due to difference in entropy between chelate and non-chelate (NH_3) complex reactions. The formation of chelate complex results in greater disorder because of the formation of a larger number of free particles in the products whereas there is no change in the number of particles in the formation of comparable non-chelate complex. Moreover, heat evolved in chelate complex (-54 kJ mol^{-1}) is greater than the non-chelate complex (-46 kJ mol^{-1}). These data suggest that the complex $[Cu(H_2O)_2(en)]^{2+}$ is more stable than $[Cu(H_2O)_2(NH_3)_2]^{2+}$.

https://doi.org/10.1515/9783110727289-005

The kinetic stability of complexes refers to the speed with which transformation leading to the attainment of equilibrium will occur. Here, we are mainly concerned with the thermodynamic stability of the complex compound.

5.2 Types of stability of metal complexes

In studying the formation of coordination compounds in solutions, two types of stability have been distinguished, and these are:

5.2.1 Thermodynamic stability: stable and unstable complexes

This type of stability deals with the properties, like bond energies, stability constants and redox potentials that affect the equilibrium conditions. On the basis of thermodynamic stability of complexes in solution, Biltz has classified the complex compounds into stable and unstablecomplexes.

Stable complexes are those which possess sufficient stability to retain their identity in solution while unstable complexes are those which are reversibly dissociated in solution into their components.

5.2.2 Kinetic stability: labile and inert complexes

This type of stability deals with the rates of reactions (i.e. reactivity) of complexes in solution, the mechanism of chemical reactions, formation of intermediate complexes, activation energies for the process, etc. On the basis of the rates of reactions (i.e. kinetic stability) of the complex in solution, (Figure 5.1) has classified complexes into labile and inert complexes.

Figure 5.1: Prof. Henry Taube (1915–2005): Awardee of the Nobel Prize in Chemistry 1983, for his work '**On the mechanisms of electron transfer reactions, especially in metal complexes**'.

Labile complexes are those whose one or more ligands in the coordination sphere can be rapidly replaced by other ligands and the ability of a complex to replace its one or more ligands by other ligands is called its lability.

Contrary to the above, inert complexes are those whose one or more ligands can either not be replaced or can be replaced with difficulty by other ligands.

5.3 Thermodynamic stability versus kinetic stability

Very often the thermodynamic and kinetic stability terms are used incorrectly. This is due to the fact that stable complexes may be inert or labile and unstable complexes may be labile or inert. For example, CN^- forms a very stable complex with Ni^{2+}.

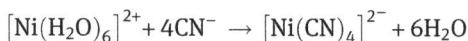

$$\left[Ni(H_2O)_6\right]^{2+} + 4CN^- \rightarrow \left[Ni(CN)_4\right]^{2-} + 6H_2O$$

Ni^{2+} prefers CN^- rather than H_2O as ligand. Thus, $[Ni(CN)_4]^{2-}$ is thermodynamically more stable than $[Ni(H_2O)_6]^{2+}$. However, when ^{14}C labelled CN^- ($^{14}CN^-$) is added to the solution of $[Ni(CN)_4]^{2-}$, it is almost instantaneously incorporated in the complex.

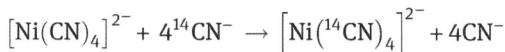

$$\left[Ni(CN)_4\right]^{2-} + 4\,^{14}CN^- \rightarrow \left[Ni(^{14}CN)_4\right]^{2-} + 4CN^-$$

This means that the complex $[Ni(CN)_4]^{2-}$ is kinetically labile. Thus, the stability of this complex does not ensure its inertness.

On the other hand, the complex $[Co(NH_3)_6]^{3+}$ is thermodynamically unstable but kinetically inert. The complex in aqueous solution remains undecomposed even over a period of several days, there being no exchange of the ligand NH_3 by the ligand H_2O. Thus, instability of a complex does not ensure lability.

$[Fe(H_2O)_6]^{3+}$ and $[Cr(H_2O)_6]^{3+}$ have roughly the same bond energy. This means that these two complexes are of equal stability from the thermodynamic point of view. But $[Fe(H_2O)_6]^{3+}$ is labile and exchanges its ligands with other ligands rapidly, whereas $[Cr(H_2O)_6]^{3+}$ is inert and hence exchanges its ligands very slowly.

Thus, these two types of stability are related to two different phenomena. The thermodynamic stability of a complex depends upon the difference in energy between the reactants and the products, called reaction energy. Greater is the reaction energy, greater will be the thermodynamic stability.

Contrary to this, the kinetic stability of a complex depends upon the difference in energy between the reactants and the activated complex, namely, the activation energy (Figure 5.2). Greater the activation energy, lesser will be reaction rate, implying that the complex is inert.

Activated complex

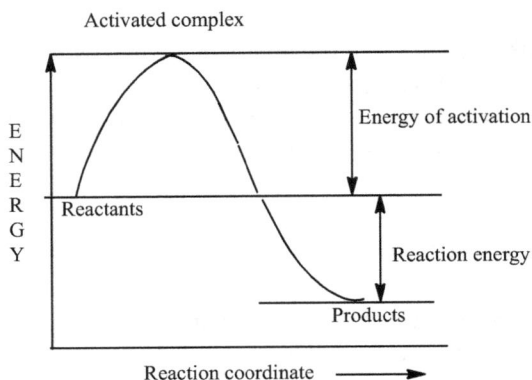

Figure 5.2: Graph showing the reaction energy and activation energy.

5.4 Dissociation of a complex in solution: dissociation constant (K_{diss}) or instability constant (K_i)

In aqueous solution, most of the complex ions are quite stable. However, they do dissociate in aqueous solution, to a slight extent, to establish equilibrium between the undissociated complex ion and the species obtained by dissociation of the complex ion.

Hence, the stability of the complex ion in solution is expressed in terms of equilibrium constant of the dissociation equilibrium. For example, in aqueous solution, the dissociation of $[Cu(NH_3)_4]^{2+}$ ion is represented by the equilibrium,

$$[Cu(NH_3)_4]^{2+} \rightleftharpoons Cu^{2+} + 4NH_3$$

The dissociation constant (K_{diss}) or instability constant (K_i) of the above equilibrium is given by

$$\text{Instability constant } (K_{diss}) = \frac{[Cu^{2+}][NH_3]^4}{[Cu(NH_3)_4]^{2+}} \tag{5.1}$$

Since the above equilibrium involves the formation of complex ion, the equilibrium constant of the above formation reaction is called formation (or stability) constant which is represented by K_{for}. Thus, K_{for} is given by

$$\text{Stability constant } (K_{for}) = \frac{[Cu(NH_3)_4]^{2+}}{[Cu^{2+}][NH_3]^4} \tag{5.2}$$

On comparing equations (5.1) and (5.2), we can write,

$$K_{for} \text{ or } (\beta) = \frac{1}{K_{diss}(\text{or } K_i)}$$

Thus, the formation constant (or stability constant), K_{for}, is reciprocal of dissociation constant (or instability constant), K_{diss}. The K_{for} or K_{diss} are also represented by β and K_i, respectively.

The values of K_{diss} and K_{for} of some of the complex ions in solution are given here:

Complex ion	K_{diss}	K_{for}
$[Cu(NH_3)_4]^{2+}$	1.0×10^{-12}	1.0×10^{12}
$[Co(NH_3)_6]^{3+}$	6.2×10^{-36}	1.6×10^{35}
$[Ag(CN)_2]^-$	1.8×10^{-19}	5.4×10^{18}
$[Hg(CN)_4]^{2-}$	4.0×10^{-42}	2.5×10^{41}
$[Fe(SCN)]^{2+}$	1.0×10^{-3}	1.0×10^3

Higher the value of stability constant (formation constant), for a complex ion, greater is the stability of the complex ion. Since $\beta\ \alpha\ 1/K_i$, we can say that smaller the value of instability constant, K_i of a complex ion, greater is the stability of the complex ion.

5.5 Stepwise formation/stability constants and overall formation/stability constants

5.5.1 Stepwise formation/stability constants

According to Niels J. Bjerrum (Figure 5.3), coordination compounds are assumed to be formed in aqueous solution by a stepwise displacement of coordinated water molecules by the ligand molecules or ion from the aqua complex of metal ion, $[M(H_2O)_n]^{n+}$.

Figure 5.3: Prof. Niels Janniksen Bjerrum (1879–1959).

If for simplicity we take a unidentate ligand L and ignore charges on the complex ion, then the different steps involved in the formation of a complex may be represented by the following reversible equations and respective equilibrium constants:

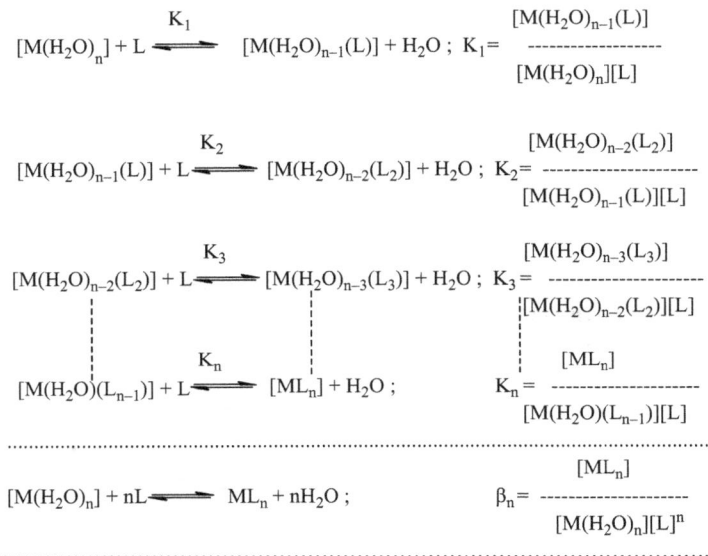

$$[M(H_2O)_n] + L \xrightleftharpoons{K_1} [M(H_2O)_{n-1}(L)] + H_2O \; ; \; K_1 = \frac{[M(H_2O)_{n-1}(L)]}{[M(H_2O)_n][L]}$$

$$[M(H_2O)_{n-1}(L)] + L \xrightleftharpoons{K_2} [M(H_2O)_{n-2}(L_2)] + H_2O \; ; \; K_2 = \frac{[M(H_2O)_{n-2}(L_2)]}{[M(H_2O)_{n-1}(L)][L]}$$

$$[M(H_2O)_{n-2}(L_2)] + L \xrightleftharpoons{K_3} [M(H_2O)_{n-3}(L_3)] + H_2O \; ; \; K_3 = \frac{[M(H_2O)_{n-3}(L_3)]}{[M(H_2O)_{n-2}(L_2)][L]}$$

$$[M(H_2O)(L_{n-1})] + L \xrightleftharpoons{K_n} [ML_n] + H_2O \; ; \qquad K_n = \frac{[ML_n]}{[M(H_2O)(L_{n-1})][L]}$$

$$[M(H_2O)_n] + nL \xrightleftharpoons{} ML_n + nH_2O \; ; \qquad \beta_n = \frac{[ML_n]}{[M(H_2O)_n][L]^n}$$

For each step in the above process, there is an equilibrium constant $K_1, K_2, K_3 \ldots \ldots .K_n$. These equilibrium constants are called stepwise stability constants or successive stability constants of the system.

By convention, displaced water is ignored in the each equilibrium constant equation since its concentration can be assumed to be constant or unchanged.

5.5.2 Overall formation/stability constants

The formation of the complex ML_n may also be represented by the following steps and equilibrium constants:

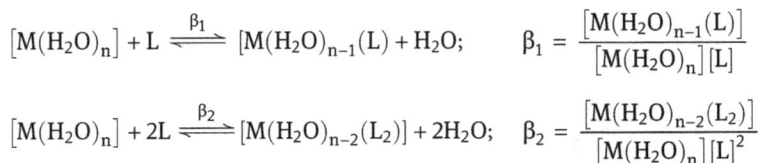

$$[M(H_2O)_n] + L \xrightleftharpoons{\beta_1} [M(H_2O)_{n-1}(L) + H_2O; \qquad \beta_1 = \frac{[M(H_2O)_{n-1}(L)]}{[M(H_2O)_n][L]}$$

$$[M(H_2O)_n] + 2L \xrightleftharpoons{\beta_2} [M(H_2O)_{n-2}(L_2)] + 2H_2O; \quad \beta_2 = \frac{[M(H_2O)_{n-2}(L_2)]}{[M(H_2O)_n][L]^2}$$

$$[M(H_2O)_n] + 3L \xrightarrow{\beta_3} [M(H_2O)_{n-3}(L_3)] + 3H_2O; \quad \beta_3 = \frac{[M(H_2O)_{n-3}(L_3)]}{[M(H_2O)_n][L]^3}$$

$$[M(H_2O)_n] + nL \xrightarrow{\beta_n} [ML_n] + nH_2O; \quad \beta_n = \frac{[ML_n)]}{[M(H_2O)_n][L]^n}$$

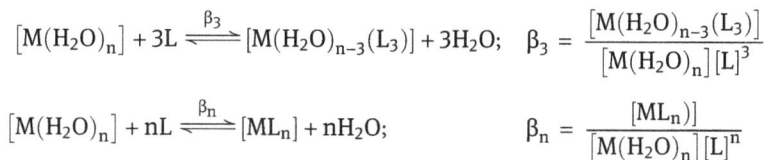

Here, the equilibrium constants β_1, β_2, β_3 . . . . β_n are called overall stability constants. β_n is termed as n^{th} overall (or cumulative) formation constant or overall stability constant.

5.5.3 Relationship between overall (β_n) and stepwise stability constants (K_1, K_2, K_3 . . . . . . K_n)

K's and β's are related to each other. Let us consider, for example, the equilibrium constant β_3 (the overall stability constant for the third-stage product), which is given by the expression:

$$\beta_3 = \frac{[M(H_2O)_{n-3}(L_3)]}{[M(H_2O)_n][L]^3}$$

On multiplying both numerator and denominator by $[M(H_2O)_{n-1}(L)]$ and $[M(H_2O)_{n-2}(L_2)]$, we get

$$\beta_3 = \frac{[M(H_2O)_{n-3}(L_3)] \cdot [M(H_2O)_{n-1}(L)] \cdot [M(H_2O)_{n-2}(L_2)]}{[M(H_2O)_n][L]^3 \cdot [M(H_2O)_{n-1}(L)] \cdot [M(H_2O)_{n-2}(L_2)]}$$

On rearranging the above expression, we get

$$\beta_3 = \frac{[M(H_2O)_{n-1}(L)] \cdot [M(H_2O)_{n-2}(L_2)] \cdot [M(H_2O)_{n-3}(L_3)]}{[M(H_2O)_n][L] \cdot [M(H_2O)_{n-1}(L)][L] \cdot [M(H_2O)_{n-2}(L_2)][L]}$$

The first term on the right-hand side of the above expression is equal to K_1, the second term is equal to K_2 and the third term is equal to K_3. Hence,

$$\beta_3 = K_1.K_2.K_3$$

For a system having n steps,

$$\beta_n = K_1.K_2.K_3. \ldots \ldots K_n \tag{5.3}$$

$$\text{or} \quad \beta_n = \sum_{n=1}^{n=n} K_n$$

From above relation, it is clear that the overall stability constant β_n is equal to the product of the successive/stepwise stability constants, K_1, K_2, K_3, ...K_n. This, in other words, means that the value of stability constants for a given complex is actually made up of a number of stepwise stability constants.

The overall stability constant is generally used as a guide to the stability of complexes. The value of stability constant may cover over a wide range. For extremely stable complexes, such as $[Fe(CN)_6]^{4-}$, the β_n is of the order of 10^{30} and for extremely unstable complexes, β_n may even be less than unity.

On account of the wide range, the overall stability constants are generally quoted on logarithm scale,

$$p\beta = \log_{10}\beta$$

Using equation (5.3) above [$\beta_n = K_1.K_2.K_3 \ldots \ldots K_n$ (i)], we get,

$$\log_{10}\beta_n = \log_{10} K_1 + \log_{10} K_2 + \log_{10} K_3 \ldots \ldots \log_{10} K_n$$

As a rough guide, $p\beta$ value greater than 8 represents a stable complex.

5.5.4 Trends in K values

The K values generally decrease with increasing substitution of H_2O from $[M(H_2O)_n]$ by L. This general decrease in K values is attributed to the following three factors:

(i) Statistical factor
As the coordinated water molecules are replaced by ligand molecules or ions (L), the number of water molecules in the complex formed decreases. Hence, the probability of replacing water molecules also decreases.

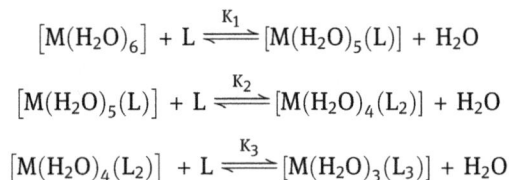

$$[M(H_2O)_6] + L \xrightarrow{K_1} [M(H_2O)_5(L)] + H_2O$$

$$[M(H_2O)_5(L)] + L \xrightarrow{K_2} [M(H_2O)_4(L_2)] + H_2O$$

$$[M(H_2O)_4(L_2)] + L \xrightarrow{K_3} [M(H_2O)_3(L_3)] + H_2O$$

As a result, K values generally decrease.

(ii) Steric factor
This arises only when the incoming ligands are bulkier in size than the coordinated water molecules. As the small-sized water molecules are replaced by bulkier ligands L, the steric crowing around the central metal increases. Consequently, the subsequent steps are retarded. Hence, the K values generally decrease.

(iii) Electrostatic factor

In the first step of complex formation, one ligand L replaces one coordinated water to give $[M(H_2O)_{n-1}(L)]$. In the second step, another ligand L of the same charge approaches the first step product. Now, there is an electrostatic repulsion between incoming ligand and a similar ligand already present in the complex.

As a result of this electrostatic repulsion between the ligands, the subsequent steps are retarded. Hence, K values decrease rapidly. Therefore, the general trend in K values are given below:

$$K_1 > K_2 > K_3 > K_4 . \ldots . . > K_n$$

Some examples illustrating such changes in stepwise stability constants are given below in Tables 5.1 and 5.2.

Table 5.1: Effect of ligand number on the stabilities of complexes.

System	log K_1	log K_2	log K_3	log K_4	log β_4
Cu^{2+}/NH_3	4.15	3.50	3.29	2.13	13.07
Ni^{2+}/NH_3	2.80	2.24	1.73	1.19	7.96

Table 5.2: Irregular successive stability constant.

System	log K_1	log K_2	log K_3	log K_4	log β_4
Cd^{2+}/Br	1.56	0.54	0.06	0.37	2.53

Any anomaly in the trend in K values suggests a major structural change. Generally, the aqua complexes are six coordinate whereas halo complexes are tetrahedral (4-coodinate). For instance, in the Cd^{2+}/Br system, the reaction of the fourth Br^- group and the complex with three Br groups is

$$[CdBr_3(H_2O)_3] + Br^- \longrightarrow [CdBr_4]^{2-} + 3H_2O$$

This step is forwarded by the release of $3H_2O$ molecules from the relatively restricted coordination sphere environment. The result is an increase of K values [Log K_4 > Log K_3].

5.6 Determination of stability/formation constant of binary complexes

The knowledge of stability constant is needed for computing quantitatively the concentration of free metal ion, ligand and any of its complexes formed in the system, under different conditions of pH. These data are extensively employed in analytical chemistry, stereochemistry and biochemistry and in the technology of non-ferrous and rare metals, solvent extraction, ion exchange and so on.

Let a metal ion (M^{n+}) combine with ligand (L) to form a complex, ML_n, then

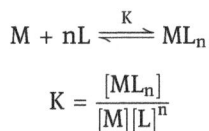

$$M + nL \xrightleftharpoons{K} ML_n$$

$$K = \frac{[ML_n]}{[M][L]^n}$$

Thus, by knowing the value of [M], [L] and [ML_n] the value of K, stability constant of the complex ML_n, can be determined.

Different methods are known for the determination of stability constants of complexes formed in aqueous medium. The experimental determination of stability constants is an important task but often it is difficult one. Some of the difficulties encountered in determination of stability constants are:

(i) Equilibrium constants depend on activities rather than on concentrations. Since activities and concentrations are equal in very dilute solutions, concentrations of all species (metal salts and ligands) have to be kept low.

(ii) During the measurement of equilibrium concentrations, the equilibrium may get disturbed.

(iii) A major problem in determination of stability constants is the difficulty in identification of complexes.

(iv) The selection of the best method to determine the stability constant of a complex is generally made on the basis of experimenter's experience. Results of more than one method are generally compared to get maximum reliability of the data.

There are so many techniques for the computation of stability constants. Here, only few methods will be taken up in the coming section, and these are known pH-metric method, spectrophotometric, polarographic method, ion exchange method and solubility method.

5.6.1 Determination of stepwise stability constants by pH-metric/ potentiometric/Bjerrum's method

As the majority of complexing agents used in analytical chemistry are moderately strong bases and get protonated in the pH range mostly applied in practice, the methods based on pH measurements are very often applicable for the determination of stability constants. The great advantage of these methods is that the pH measurements are not time consuming and the instruments necessary are not expensive.

The basis of this method is that during complex formation between metal ion and the protonated ligand, protons are liberated.

$$M + LH \rightleftharpoons ML + H^+$$

By determination of hydrogen ion (H^+) concentration potentiometrically, the degree of complex formation or the position of the equilibrium can be established. The magnitude of the observed pH range is then used for the determination of stepwise stability constants of the metal complexes.

This method was largely developed by J. Bjerrum. To facilitate the determination of the stepwise stability constants, he suggested certain formation functions such as $\bar{n}$, $\bar{n}_A$ and p_L.

These functions are employed to calculate the stepwise stability constants.

The formation function ($\bar{n}$) of a metal ligand (M, L) system can be mathematically represented as

$$\bar{n} = \frac{\text{Total conentration of L bound to the metal ion (M)}}{\text{Total conentration of the metal ion (M)}} \tag{5.4}$$

$$\bar{n} = \frac{[ML] + 2[ML_2] + 3[ML_3] + \text{--------}n[ML_n]}{[M] + [ML] + [ML_2] + [ML_3] + \text{---}[ML_n]} \tag{5.5}$$

As complexing processes are considered occurring by a series of stages, it is possible to express the formation constants referring specially to the addition of ligands in a stepwise manner as follows:

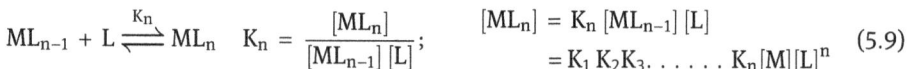

$$M + L \underset{}{\overset{K_1}{\rightleftharpoons}} ML \qquad K_1 = \frac{[ML]}{[M][L]}; \qquad [ML] = K_1[M][L] \tag{5.6}$$

$$ML + L \underset{}{\overset{K_2}{\rightleftharpoons}} ML_2 \qquad K_2 = \frac{[ML_2]}{[ML][L]}; \qquad \begin{aligned}[ML_2] &= K_2[ML][L] \\ &= K_1 K_2[M][L]^2\end{aligned} \tag{5.7}$$

$$ML_2 + L \underset{}{\overset{K_3}{\rightleftharpoons}} ML_3 \qquad K_3 = \frac{[ML_3]}{[ML_2][L]}; \qquad \begin{aligned}[ML_3] &= K_3[ML_2][L] \\ &= K_1 K_2 K_3[M][L]^3\end{aligned} \tag{5.8}$$

$$ML_{n-1} + L \underset{}{\overset{K_n}{\rightleftharpoons}} ML_n \qquad K_n = \frac{[ML_n]}{[ML_{n-1}][L]}; \qquad \begin{aligned}[ML_n] &= K_n[ML_{n-1}][L] \\ &= K_1 K_2 K_3 \ldots \ldots K_n[M][L]^n\end{aligned} \tag{5.9}$$

Putting the values of [ML], [ML$_2$], [ML$_3$] . . . . . . [ML$_n$] from equations (5.6), (5.7), (5.8) and (5.9) in equation (5.5), we have

$$\overline{n} = \frac{K_1[M][L] + 2K_1K_2[M][L]^2 + 3K_1K_2K_3[M][L]^3 + nK_1K_2K_3. \ldots \ldots K_n[M][L]^n}{[M] + K_1[M][L] + K_1K_2[M][L]^2 + \ldots \ldots K_1K_2K_3. \ldots \ldots K_n[M][L]^n}$$

Cancelling [M] throughout in numerator and denominator, we get

$$\overline{n} = \frac{K_1[L] + 2K_1K_2[L]^2 + 3K_1K_2K_3[L]^3 + nK_1K_2K_3. \ldots \ldots K_n[L]^n}{1 + K_1[L] + K_1K_2[L]^2 + K_1K_2K_3.[L]^3 + \ldots \ldots K_1K_2K_3. \ldots \ldots K_n[L]^n} \tag{5.10}$$

In terms of overall stability constant, the equation (5.10) may be represented as

$$\overline{n} = \frac{\beta_1[L] + 2\beta_2[L]^2 + 2\beta_3[L]^3 + \ldots \ldots + n\beta_n[L]^n}{1 + \beta_1[L] + \beta_2[L]^2 + \beta_3[L]^3 + \ldots \ldots \beta_n[L]^n}$$

$$\text{or} \quad \overline{n} = \frac{\displaystyle\sum_{i=0}^{i=n} i\beta_i[L]^i}{1 + \displaystyle\sum_{i=0}^{i=n} \beta_i[L]^i} \tag{5.11}$$

$$\text{or} \quad \overline{n}\left(1 + \sum_{i=0}^{i=n} \beta_i[L]^i\right) = \sum_{i=0}^{i=n} i\beta_i[L]^i$$

$$\text{or} \quad \overline{n} + \overline{n}\sum_{i=0}^{i=n} \beta_i[L]^i = \sum_{i=0}^{i=n} i\beta_i[L]^i \tag{5.12}$$

$$\text{or} \quad \overline{n} = \sum_{i=0}^{i=n} i\beta_i[L]^i - \overline{n}\sum_{i=0}^{i=n} i\beta_i[L]^i$$

$$\overline{n} = \sum_{i=0}^{i=n} i(1 - \overline{n})\beta_i[L]^i$$

In the similar way for ligand–proton (L-H) system, formation function ($\overline{n}_A$) is defined as

$$\overline{n}_A = \frac{\text{Total concentration of H bound to L}}{\text{Total concentration of L not bound to M}} \tag{5.13}$$

$$\overline{n}_A = \frac{[HL] + 2[H_2L] + 3[H_3L] + \ldots \ldots}{[L] + [HL] + [H_2L] + [H_3L] + \ldots \ldots} \tag{5.14}$$

$$\overline{n}_A = \frac{K_1^H[H][L] + 2K_1^HK_2^H[H]^2[L] + 3K_1^HK_2^HK_3^H[H]^3[L] + \ldots \ldots \ldots}{[L] + K_1^H[H][L] + 2K_1^HK_2^H[H]^2[L] + 3K_1^HK_2^HK_3^H[H]^3[L] + \ldots \ldots \ldots}$$

Cancelling [L] throughout in numerator and denominator, we get

$$\overline{n}_A = \frac{K_1^H[H] + 2K_1^HK_2^H[H]^2 + 3K_1^HK_2^HK_3^H[H]^3 + \ldots \ldots \ldots}{1 + K_1^H[H] + 2K_1^HK_2^H[H]^2 + 3K_1^HK_2^HK_3^H[H]^3 + \ldots \ldots \ldots}$$

$$\bar{n}_A = \frac{\beta_1{}^H[H] + 2\beta^H[H]^2 + 3\beta^H K_3{}^H[H]^3 + \dots\dots\dots}{1 + \beta_1{}^H[H] + 2\beta^H[H]^2 + 3\beta^H K_3{}^H[H]^3 + \dots\dots\dots} \tag{5.15}$$

$$\text{or} \quad \bar{n}_A = \frac{\sum\limits_{i=0}^{i=n} i\beta_i{}^H[H]^i}{1 + \sum\limits_{i=0}^{i=n} \beta_i{}^H[H]^i} \tag{5.16}$$

Now the formation function $(\bar{n})$ is

$$\bar{n} = \frac{T_{CL^\circ} - \text{concentration of L not bound to M}}{T_{CM^\circ}}$$

where $T_{CL}{}^\circ$ is the total concentration of L and $T_{CM}{}^\circ$ is the total concentration of M

or $\qquad \bar{n}\, T_{CM^\circ} = T_{CL^\circ} - \text{concentration of L not bound to M}$

or $\qquad \text{Concentration of L not bound to M} = T_{CL^\circ} - \bar{n}\, T_{CM^\circ}$
$$\tag{5.17}$$

From the value of $\bar{n}_A$ and using equations (5.15) and (5.16), we get

Total concentration of L not bound to $M = [L](1 + \beta_1{}^H[H] + + \beta_2{}^H[H]^2 +$
$$+ \beta_3{}^H[H]^3 + \dots\dots)$$

or Total concentration of L not bound to $M = [L]\sum\limits_{i=0}^{i=n} \beta_i{}^H[H]^i \tag{5.18}$

Putting the value of total concentration of L not bound to M from equation (5.18) to equation (5.17), we get

$$[L]\sum\limits_{i=0}^{i=n} \beta_i{}^H[H]^i = T_{CL^\circ} - \bar{n}\, T_{CM^\circ}$$

$$\text{or} \quad [L] = \frac{T_{CL^\circ} - \bar{n}\, T_{CM^\circ}}{\sum\limits_{i=0}^{i=n} \beta_i{}^H[H]^i} \tag{5.19}$$

$$\text{or} \quad [L]^{-1} = \frac{\sum\limits_{i=0}^{i=n} \beta_i{}^H[H]^i}{T_{CL^\circ} - \bar{n}\, T_{CM^\circ}} \tag{5.20}$$

Taking log of equation (5.20), we get

$$\log[L]^{-1} = \log_{10} \frac{\sum\limits_{i=0}^{i=n} \beta_i{}^H[H]^i}{T_{CL^\circ} - \bar{n}\, T_{CM^\circ}} \tag{5.21}$$

As log $[L]^{-1} = pL$

$$pL = \log_{10} \frac{\sum\limits_{i=0}^{i=n} \beta_i^H [H]^i}{T_{CL^\circ} - \bar{n} T_{CM^\circ}} \tag{5.22}$$

Calvin and Wilson (1945) have demonstrated that pH measurements made during titrations with alkali solution of ligand in the presence and absence of metal ion could be employed to calculate the formation functions $\bar{n}$, $\bar{n}_A$ and p_L and thereby stability constants can be calculated.

Irving and Rossotti (1954) titrated (pH titration) following solutions against standard sodium hydroxide solution (N°) keeping total volume (V°) constant:
1. Mineral acid (HClO₄) (concentration, E°)
2. Mineral acid + ligand solution
3. Mineral acid + ligand solution + metal ion solution

The ionic strength in each set is kept constant by adding appropriate quantities of a neutral electrolyte solution. The temperature of the solution in each case is kept constant. On plotting the observed pH against the volume of alkali, three curves (a) mineral acid titration curve, (b) mineral acid + ligand titration curve and (c) mineral acid + ligand + metal titration curve are obtained corresponding to the above titrations (Figure 5.4).

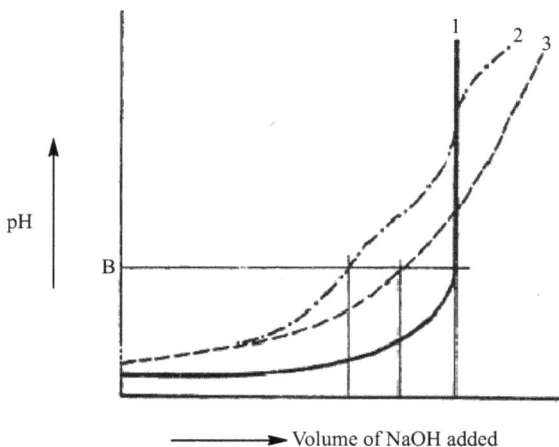

Figure 5.4: Curve 1: mineral acid alone; curve 2: mineral acid and ligand (oxine); curve 3: mineral acid, ligand (oxine) and metal.

The formation functions $\bar{n}$, $\bar{n}_A$ and p_L can be computed from the following equations:

$$\bar{n}_A = y - \frac{(V_1 - V_2)(N^\circ - E^\circ)}{(V^\circ + V_1) T_{Cl}^\circ} \tag{5.23}$$

$$\bar{n} = \frac{(V_3 - V_2)(N^\circ + E^\circ)}{(V^\circ + V_1)(\bar{n}_A)(T_M{}^\circ)} \tag{5.24}$$

$$pL = \log_{10} \frac{\sum_{n=0}^{n=n} \beta_n^H \cdot \frac{1}{(\text{anti} \log B)^n}}{T_{Cl^\circ} - \bar{n}T_{CM^\circ}} \times \frac{V^\circ + V_3}{V^\circ} \tag{5.25}$$

where Y = number of dissociable protons; V_1, V_2, V_3 = volume of NaOH used to bring the solution to same pH value B; $T_{CL}{}^\circ$ = total concentration of the ligands; and $T_{CM}{}^\circ$ = total concentration of the metal ion.

By the knowledge of $\bar{n}$, $\bar{n}_A$ and p_L, the stepwise stability constants of the complex can be computed by different methods as follows.

Computation of stability constants from experimental data
A large number of methods for computing stability constants from experimental data have been used by number of authors. Some of the more generally applicable computational methods are as follows:

(i) Least square method
From equations (5.10) and (5.11) developed above,

$$\bar{n} = \frac{K_1[L] + 2K_1K_2[L]^2 + 3K_1K_2K_3[L]^3 + nK_1K_2K_3 \ldots \ldots K_n[L]^n}{1 + K_1[L] + K_1K_2[L]^2 + K_1K_2K_3[L]^3 + \ldots \ldots \ldots K_1K_2K_3 \ldots \ldots K_n[L]^n} \tag{5.10}$$

$$\bar{n} = \frac{\beta_1[L] + 2\beta_2[L]^2 + 3\beta_3[L]^3 + \cdots \cdots \cdots + n\beta_n[L]^n}{1 + \beta_1[L] + \beta_2[L]^2 + \beta_3[L]^3 + \cdots \cdots \cdots \beta_n[L]^n}$$

$$\text{or} \quad \bar{n} = \frac{\sum_{i=0}^{i=n} i\beta_i[L]^i}{1 + \sum_{i=0}^{i=n} \beta_i[L]^i} \tag{5.11}$$

For i = 1
$$\bar{n} = \frac{K_1[L]}{1 + K_1[L]}$$

$$\text{or} \quad \bar{n}(1 + K_1[L]) = K_1[L]$$

$$\text{or} \quad \bar{n} + \bar{n}K_1[L] = K_1[L]$$

$$\text{or} \quad \bar{n} = K_1[L] - \bar{n} K1[L] = K_1[L](1 - \bar{n})$$

$$K_1 = \frac{\bar{n}}{(1 - \bar{n})(L)}$$

$$\text{or} \quad \log K_1 = \log \frac{\bar{n}}{(1 - \bar{n})} + \log [L]^{-1}$$

$$\text{or} \quad \log K_1 = \log \frac{\bar{n}}{(1 - \bar{n})} + pL \tag{5.26}$$

For i = 2

$$K_2 = \frac{1}{[L]} \cdot \frac{\bar{n}\,(\bar{n}-1)K_1[L]}{(2-\bar{n})K_1[L]}$$

$$\text{or} \quad \log K_2 = pL + \log \frac{\bar{n}\,(\bar{n}-1)K_1[L]}{(2-\bar{n})K_1[L]} \tag{5.27}$$

The term $(\bar{n}-1)K_1[L]$ is negligible when $\bar{n} > 0.5$

Hence,

$$\log K_2 = 2pL + \log \frac{\bar{n}}{(2-\bar{n})K_1} \tag{5.28}$$

The equations (5.26) and (5.27) are straight line equations. Thus, by plotting differ-ent values of $\bar{n}$ and [L] straight line will be obtained. Thus, the values of $\log K_1$ and $\log K_2$ can be obtained.

(ii) Half integral method

By putting the value $\bar{n}_A = 0.5$ in equation (5.26), we obtain $\log K_1 = pL$ as follows:

$$\log K_1 = \log \frac{\bar{n}}{(1-\bar{n})} + pL \tag{5.26}$$

when $\bar{n} = 0.5$, we get

$$\log K_1 = \log \frac{0.5}{(1-0.5)} + pL$$

$$= \log 1 + pL \qquad (\log 1 = 0)$$

$$= pL$$

Similarly by putting the value $\bar{n} = 1.5$ in the equation (5.27), we obtain $\log K_2 = pL$ as follows:

$$\log K_2 = pL + \log \frac{\bar{n}\,(\bar{n}-1)K_1[L]}{(2-\bar{n})K_1[L]} \tag{5.27}$$

when $\bar{n} = 1.5$

$$\log K_2 = pL + \log \frac{1.5 \times 0.5}{0.5}$$

$$\log K_2 = pL + \log 1.5 = pL + 0.1760$$

$$= \sim pL$$

It means if we plot a graph between $\bar{n}$ and pL then the corresponding values of pL at equal to 0.5 and 1.5 gives $\log K_1$ and $\log K_2$, respectively.

5.6.2 Spectrophotometric method

Principle

Spectrophotometric method is based on the principle that, in most of the cases, the absorption properties of solutions containing complexes differ from those of the constituent ions or molecules.

This change in absorption behaviour is closely related to the formation of coordinate bonds. Hence, the spectrophotometric method based on the measurement of light absorption can be used to study complex equilibria.

Procedure

This method for determining the stability constant of a complex can be exemplified by taking an example of complex formation between ferric ion and thiocyanate ion in aqueous medium as shown below:

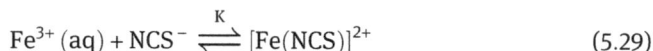

$$Fe^{3+}(aq) + NCS^- \overset{K}{\rightleftharpoons} [Fe(NCS)]^{2+} \tag{5.29}$$

Hence, the stability constant of this complex is given by

$$K = \frac{[[Fe(NCS)]^{2+}]}{[Fe^{3+}] \, [NCS^-]} \tag{5.30}$$

Here, the concentration terms represent the actual equilibrium concentration.

How, the absorption spectrum of a metal ion, M^{n+} changes on coordination with ligand L, has been shown in Figure 5.5 with references to Fe^{3+}–NCS system.

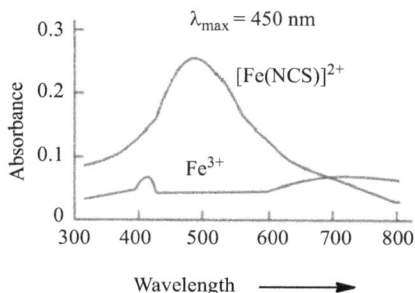

Figure 5.5: The absorption spectrum of F^{3+} and $[Fe(NCS)]^{2+}$.

It is seen from Figure 5.5 that the intensity of complex ion, $[Fe(NCS)]^{2+}$ absorption is increased and shifts to higher wavelength range compared to the weak absorption of Fe^{3+} ion. In fact, Fe^{3+} ion is almost colourless, whereas the complex $[Fe(NCS)]^{2+}$ is bright red. Its λ_{max} is 450 nm.

The concentration of the coloured complex ion, $[Fe(NCS)]^{2+}$ can be determined by measuring its absorbance in the visible region at a particular wavelength (~450 nm) using a spectrophotometer as per the Beer's law:

$$A = \varepsilon.c.l \qquad (5.31)$$

where ε (epsilon) is the molar extinction coefficient, c is the concentration of the complex in moles/litre and l is the path length of absorption cell/absorbing solution.

The value of ε for $[Fe(NCS)]^{2+}$ can be determined first by measuring the absorbance of a solution of the known concentration. Such a solution can be prepared by mixing a known amount of Fe^{3+} ion with a low molarity of NCS^- so that the complex formed is only monothiocyanataoiron(II), $[Fe(NCS)]^2$ in appreciable amount.

For calculating K, a solution of Fe^{3+} of known concentration, $[Fe^{3+}]_o$ is mixed with a solution of NCS^- of known concentration. The mixture in which the complex $[Fe(NCS)]^{2+}$ has been formed is equilibrated. Its absorbance A at 450 nm is measured. Then, K is evaluated as mentioned below:

$$A = \varepsilon. [[Fe(NCS)]^{2+}].l \qquad (5.32)$$

So, the equilibrium concentration of the complex, $[Fe(NCS)]^{2+}$ is,

$$[[Fe(NCS)]^{2+}] = \frac{A}{\varepsilon.l} \text{ (A, } \varepsilon \text{ and } l \text{ are known; } l = 1\,cm \text{ normally)} \qquad (5.33)$$

Now, the initial concentration, $[Fe^{3+}]_o$ of Fe^{3+} may be written as:

$$[Fe^{3+}]_o = [Fe^{3+}] + [Fe(NCS)] \text{ --------}$$

uncomplexed Fe^{3+} \qquad complexed Fe^{3+}

$$(5.34)$$

Therefore, equilibrium concentration of Fe^{3+} is:

$$[Fe^{3+}] = [Fe^{3+}]_o - [Fe(NCS)]^{2+} \qquad (5.35)$$

Similarly, the initial concentration, $[NCS^-]$, of NCS^- may be written as,

$$[NCS^-]_o = [NCS^-] + [Fe(NCS)]^{2+}$$

uncomplexed NCS^- \qquad complexed NCS^-

Then, equilibrium concentration of NCS^- is:

$$[NCS^-] = [NCS^-]_o - [Fe(NCS)]^{2+} \qquad (5.36)$$

Thus, if we put the value of $[[Fe(NCS)]^{2+}]$, $[Fe^{3+}]$ and $[NCS^-]$ from equations, (5.33), (5.35) and (5.36), respectively, in equation (5.30), we get the value of stability constant K of the Fe^{3+}–NCS system as

$$\text{Stability constant} = \frac{[[Fe(NCS)]^{2+}]}{[Fe^{3+}][NCS^-]}$$

The experiment is then repeated with three or more different initial Fe^{3+} and NCS^- concentrations to check the constancy of the equilibrium constant K. As this method does not disturb the equilibrium during absorbance measurement, it is a reliable method of determining K value.

5.6.2.1 Method of continuous variation (Job's method)

This method is a variation of the spectrophotometric method and is used to determine the composition (metal–ligand ration) of a complex as well its stability constant. It is also known as Job's method (Job 1928) although it is not originally due to him. This method is mainly used for solutions where only one complex is formed.

Principle

The principle of this method is that mole ratio of the metal ion and the ligand is varied between 0 and 1 at constant total concentration of the ligand and that of the metal ion, that is,

$$C_L + C_M = C$$

Here, C_L = total concentration of the ligand L and C_M = total concentration of the metal ion M.

The absorbances of the solutions of different compositions are measured, and then plotted against the mole fraction X_L (C_L/C) of the ligand. When only one complex species has been formed with the composition ML_n, and the absorbance is measured at a wavelength where neither the metal nor the ligand but only the complex absorbs, then n can be calculated from the abscissa of the maximum of the curve (X_{max}) as:

$$n = \frac{X_{max}}{1 - X_{max}}$$

Procedure

This involves the following steps:

(i) Make up 10 solutions of the complex containing equimolar concentration of metal ion and ligand in such proportions that the total volume of each solution is 10 mL as shown below:

Solution no.:	1	2	3	4	5	6	7	8	9	10
Vol. of metal ion (mL):	0	1	2	3	4	5	6	7	8	9
Vol. of ligand (mL):	10	9	8	7	6	5	4	3	2	1

Thus, we see that sum of the concentration of the ligand C_L and that of the metal ion, C_M is constant. Thus,

$$C_L + C_M = C \tag{5.37}$$

(ii) Now, determine the absorbances of the solutions as prepared in step (i) with the help of a spectrophotometer at such a wavelength of light where the complex absorbs strongly and the metal ion and the ligand do not.

(iii) Calculate the mole fractions of the ligand (**Vol. of the ligand/total vol.**) corresponding to 10 sets of solutions prepared in step (i) that will come out to be:

Solution no.:	1	2	3	4	5	6	7	8	9	10
Vol. of ligand (mL):	10	9	8	7	6	5	4	3	2	1
Mole fraction (X): of the ligand	1	0.9	0.8	0.7	0.6	0.5	0.4	0.3	0.2	0.1

$$10/10, 9/10 \ldots\ldots\ldots\ldots\ldots\ldots\ldots\ldots\ldots\ldots\ldots\ldots 1/10$$

(iv) Plot a graph between the mole fraction of the ligand ($X = C_L/C$) and absorbance. A curve of the type shown in Figure 5.6 is obtained. When the legs of the curve extrapolated, they cross each other at point at which absorbance is maximum. The mole fraction corresponding to this point is taken as X_{max}.

If the formula of the complex is ML_n, then

$$n = C_L/C_M \tag{5.38}$$

The equation (5.37) can be written as

$$C_L + C_M = C$$
$$\text{or} \quad C_L/C + C_M/C = 1 \tag{5.39}$$

$$\text{But } C_L/C = X \text{ (Mole fraction of the ligand)} \tag{5.40}$$

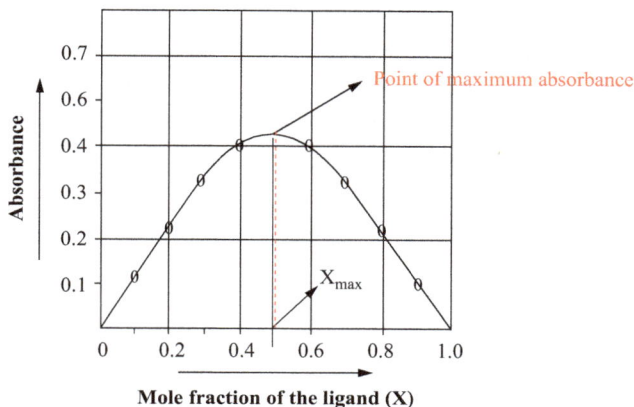

Figure 5.6: Plot of mole fraction of the ligand versus absorbance.

Equation (5.39) now becomes

$$X + C_M/C = 1$$
$$\text{or} \quad C_M/C = 1 - X$$

(5.41)

Dividing equations (5.40) and (5.41), we get

$$\frac{\frac{C_L}{C}}{\frac{C_M}{C}} = \frac{X}{1-X}$$

$$\text{or} \quad \frac{C_L}{C} \cdot \frac{C}{C_M} = \frac{X}{1-X} \quad \text{or} \quad \frac{C_L}{C_M} = \frac{X}{1-X}$$

$$\text{But} \quad \frac{C_L}{C_M} = n \quad [\text{see equation (2)}]$$

$$\text{Hence,} \quad n = \frac{X}{1-X}$$

(v) In fact, X in the above equation is X_{max} obtained from the graph corresponding to the point of intersection of the two legs of the curve [see point (iv)].

$$\text{Hence,} \quad n = \frac{X_{max}}{1 - X_{max}}$$

(5.42)

In the present case, X_{max} as obtained from the graph is 0.5

$$\text{Hence,} \quad n = \frac{0.5}{1 - 0.5} = \frac{0.5}{0.5} = 1$$

So, the hypothetical metal–ligand ratio is 1 in the present case. Thus, from the value of n as given by equation (5.42), we can determine the composition of the complex.

Determination of stability constant by Job's method

The stability constant of a complex in question can also be calculated from the curve by drawing tangents to the initial and final parts of the curve, and by using the coordinates of certain points on the tangents and the curve (Figure 5.7) without knowing the molar absorbance of the complex formed.

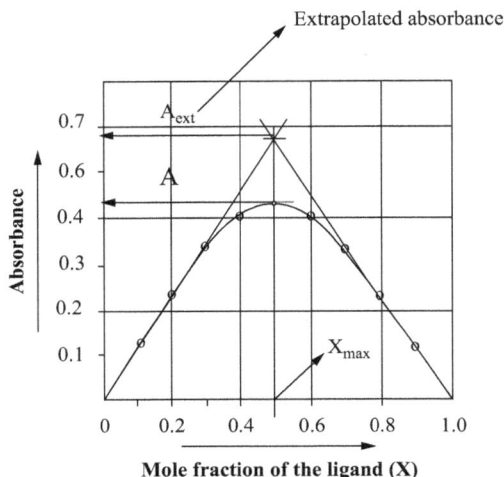

Figure 5.7: Plot of mole fraction of the ligand versus absorbance along with tangents drawn to the initial and final parts of the curve.

The points on the tangent drawn to the last part of the curve give the absorbances which would be measured if the metal were completely present in the form of the complex.

Before the intersection of the two straight lines, the complex formation is limited by the ligand concentration, and the absorbances of the points on the initial part on the ascending line are proportional to the concentration of complex equivalent to the total ligand concentration.

The ratio of the observed absorbance (A) to that indicated by the tangent (A_{ext}) (extrapolated absorbance) for the same value of mole fraction of the ligand (X) is equal to the mole fraction of the metal ion in the complex, when $X > X_{max}$ (when tangent intersection point is right-hand side to X_{max}), and the mole fraction of the ligand in the complex when $X < X_{max}$ (when tangent intersection point is left-hand side to X_{max})

$$\frac{A}{A_{ext}} = \frac{[ML_n]}{C_M} \text{ (mole fraction of the complex)} \tag{5.43}$$

$$\frac{A}{A_{ext}} = \frac{n[ML_n]}{C_L} \text{ (mole fraction of the complex)} \tag{5.44}$$

The concentration of the complex species can be expressed from these equations (5.43) and (5.44) as follows:

$$[ML_n] = \frac{A}{A_{ext}} \cdot C_M \tag{5.45}$$

$$[ML_n] = \frac{A}{A_{ext}} \cdot \frac{CL}{n} \tag{5.46}$$

In the above equations, if the multiplier of the ratio of absorbances (A/A_{ext}) is denoted by C_x, then

$$
\begin{aligned}
C_x &= C_M & \text{when } X > X_{max}, \\
C_x &= C_L/n & \text{when } X < X_{max}, \text{ and} \\
C_x &= C_M = C_L/n & \text{when } X = X_{max}
\end{aligned}
\tag{5.47}
$$

From these, the concentration of free metal ion and free ligandscan be expressed as follows:

$[M] =$	C_M	$-$	$[ML_n]$
(Free metal ion concentration)	(Total metal ion concentration)		(Complexed Metal ion concentration)
$[L] =$	C_L	$-$	$[ML_n]$
(Free ligand concentration)	(Total ligand concentration)		(Complexed ligand concentration)

or free metal ion concentration, [M] equals to

$$[M] = C_M - \frac{A}{A_{ext}} \cdot C_x \tag{5.48}$$

Because complexed ligand concentration will be n times of complexed metal ion concentration in the ML_n, free ligand concentration, [L] equals to

$$[L] = C_L - n. \frac{A}{A_{ext}} \cdot C_x \tag{5.49}$$

Looking into the formation of the complex shown below:

$$M + nL \overset{\beta_n}{\rightleftharpoons} ML_n$$

The overall stability constant is given by

$$\beta_n = \frac{[ML_n]}{[M][L]^n} \tag{5.50}$$

Putting the values of concentration $[ML_n]$ from equation (5.45)/(5.46) and the concentration of $[M]$ and $[L]$ from equations (5.48) and (5.49), respectively in equation (5.50), we get the value of β_n as:

$$\beta_n = \frac{\left[\frac{A}{A_{ext}} \cdot C_x \right]}{\left[C_M - \frac{A}{A_{ext}} \cdot C_x \right] \left[C_L - n. \frac{A}{A_{ext}} \cdot C_x \right]}$$

Here, C_x is the total concentration of the metal, C_M or total concentration of ligand/n (C_L/n) when $X = X_{max}$ as per equation (5.47).

Limitations of Job's method
(i)   This method can be used when only one complex is formed under the experimental conditions.
(ii) This method is used when there is no change in the total volume of the solution containing metal ion and ligand.
(iii) An important condition for the applicability of the method is that the metal and ligand do not react with other constituents present in solution.

5.6.2.2 Mole ratio method
This method is also a variation of the spectrophotometric method and is used to determine the composition (metal–ligand ratio) of the complex as well as its stability constant. This method is mainly used for solutions where only one complex is formed.

This method involves the following steps:
(i)   A series of solutions (usually 10) is prepared in which one component (usually concentration of the metal, C_M) is kept constant and that of the other (concentration of the ligand, C_L) is varied.
(ii) The absorbance of the solutions is measured at a suitable wavelength where the complex absorbs strongly and the metal ion and the ligand do not.
(iii) Plot a graph between the absorbance and ratio of the variable ligand concentration (C_L) and constant metal concentration (C_M), that is, C_L/C_M.
(iv) In case if only one stable complex is formed, which has selective light absorption, then the absorbance increases approximately linearly with mole ratio and then becomes constant.
(v)   Abscissa of the point of intersection of the two tangents gives the number of ligands in the complex (i.e. composition of the complex), if it was the ligand concentration that was varied.
(vi) The stability constant of the complex can be calculated from the coordinates of the points of the straight line (extrapolated line) and the curve (Figure 5.8) in a similar way to the case of the method of continuous variation.

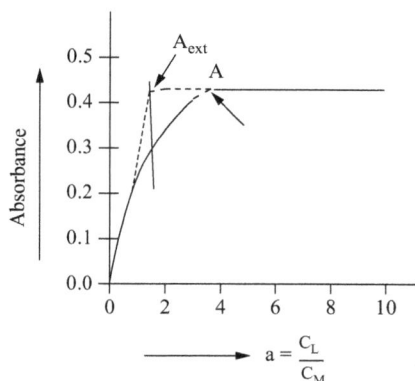

Figure 5.8: Plot of absorbance versus variable ligand concentration keeping the metal concentration constant.

(vii) In the simplest case, if only one complex of composition ML_n is formed, the mole fraction of the complex $(\phi_{MLn}) = [ML_n]/C_x = A/A_{ext}$ and hence,

$$\beta_n = \frac{\left[\frac{A}{A_{ext}} \cdot C_x\right]}{\left[C_M - \frac{A}{A_{ext}} \cdot C_x\right]\left[C_L - n.\frac{A}{A_{ext}} \cdot C_x\right]}$$

Here, 'A_{ext}' stands for extrapolated absorbance and 'A' for the actual absorbance at the same abscissa value. Before the intersection, $C_x = C_L$, and after that, $C_x = C_M$.

5.6.3 Polarographic method

Basic principle
If a metal ion (Lingane 1941) can be reversibly reduced to the metallic state at a dropping mercury electrode, the metal ion may behave in the presence of complexing agent in either of the following two ways:
(i) If the complex formation reaction is fast compared with the reduction of the metal ion, then one polarographic wave occurs, with a half wave potential more negative than that of the uncomplexed metal ion. By measuring this shift in half wave potential as a function of the concentration of the complex forming substance both the composition and the stability constant of the complex can be determined.
(ii) If the complex formation reaction is slow compared with the reduction of the metal ion, which occurs less frequently, then two polarographic waves are obtained, the first corresponding to the reduction of the free metal ion and second to that of the complexed metal ion. The amount of the free metal ion can be determined directly from the height of the first wave. From the dependence of the height of the waves on the ligand concentration, the composition and the stability constant of the complex formed can be calculated.

Taking into consideration of the first case, which occurs normally, let us consider the reaction

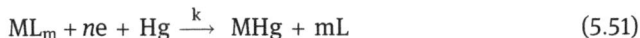

$$ML_m + ne + Hg \xrightarrow{k} MHg + mL \tag{5.51}$$

This equation may be imagined to occur in the following two steps:

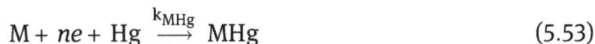

$$ML_m \xrightarrow{k_{ML}} M + mL \tag{5.52}$$

$$M + ne + Hg \xrightarrow{k_{MHg}} MHg \tag{5.53}$$

The term k corresponds to the respective formation constants of the above three reactions. In these equations, M is the metal, L the ligand, ML_m the complex compound and n is the number of electrons involved in the reaction.

Based on the above reactions, J. L. Lingane derived an equation connecting the potential of the dropping mercury electrode E, **the half wave potential $E_{0.5}$**, the current i and the limiting current i_d (the maximum of the polarographic wave), such as,

$$E_{25°C} = E_{0.5} - \frac{0.0591}{n} \log \frac{i}{i_d - i} \tag{5.54}$$

The plot of E against $\log i/(i_d - i)$ will, of course, be a straight line for a reversible potential, the slope of the line being $0.0591/n$, where n is the number of electrons transferred in the reduction.

By considering equations (5.51), (5.52) and (5.53), Lingane found that at 25 °C

$$E_{0.5} = E^0 - \frac{0.0591}{n} \log k - \frac{0.0591}{n} \log \frac{k_{ML}}{k_{MHg}} - m \frac{0.0591}{n} \log [L] \tag{5.55}$$

where E^0 is the standard potential of the metal amalgam electrode, the ratio k_{ML}/k_{MHg} depends on experimental conditions and electrode characteristics, m is the number of coordinated ligands and [L] is the ligand concentration.

On plotting the half wave potential $E_{0.5}$ against $\log [L]$, the slope of the straight line will be $0.0591 \, m/n$, from which m can be found. On the other hand, the formation or stability constant k of reaction (5.51) is found from the equation,

$$E_{0.5} - (E_{0.5})_s = - \frac{0.0591}{n} \log k - m \frac{0.0591}{n} \log [L] \tag{5.56}$$

where $(E_{0.5})_s$ is the half wave potential in the absence of ligands. The plot of the left-hand side of equation (5.56) against $\log [L]$ yields a straight line, the intercept of which on the ordinate axis gives $(0.0591/n)\log k$ and hence k can be calculated. It is also clear that the slope of the straight line is $(-m \times 0.0591)/n$, so that m can be calculated.

The validity of the foregoing relations is based on the following conditions:
(i) The potential must be reversible, that is, plot of E against $\log i/(i_d - i)$ must be a straight line.

(ii) The slowest, that is, the rate-determining step must be that involving diffusion of ions or molecules to and from the cathode.

(iii) The potential must be determined by the metal ions investigated.

(iv) The mercury must not participate in the electrolytic reduction process.

Procedure

(i) Prepare a number of solutions (eight in numbers), each containing 5×10^{-4} mol of cupric nitrate per litre of water and the following concentrations of potassium glycinate: 0.00, 0.01, 0.04, 0.08, 0.10, 0.20, 0.50 and 0.80 mol L^{-1} of solution.

(ii) Add 0.003% of methyl red as a maximum suppressor, and bring the solutions to an ionic strength of 1 with KNO_3.

(iii) Place the solution in the thermostatic bath (25 °C) for 30 min, and then find their pH.

(iv) Pass the nitrogen gas through the solutions for 10 min and then record their polarograms (the potential of the dropping mercury electrode should be checked at various voltages against a saturated calomel electrode).

(v) Find the half wave potential from its polarogram.

(vi) Calculate the concentrations of the free amino acid for each solution by subtracting the concentration of the combined glycine (complex CuL_2) from the total present. Hence, find the concentration of the free glycinate ion from the concentration of free acid, the pH (using pH meter) and the ionization constant ($pK_2 = 9.84$). Results are tabulated below:

Results

Total glycine (mol L^{-1})	0.00	0.01	0.04	0.08	0.10	0.20	0.50	0.80
pH	–	9.82	9.92	9.95	10.03	10.15	10.36	12.03
$-E_{0.5}$ (volt)		0.012	0.357	0.396	0.412	0.427	0.442	0.467	0.512

Calculate the values of $[E_{0.5} - (E_{0.5})_s]$ ($E_{0.5}$ = half wave potential in the presence of ligand and $(E_{0.5})_s$ = half wave potential in the absence of ligand). From the known values of pH and $pK_2 = 9.84$, find the values of [L] and then log [L] (can be calculated from the value of [L] using scientific calculator).

Calculations

$[E_{0.5} - (E_{0.5})_s]$ (volt)	−0.345	−0.384	−0.400	−0.415	−0.430	−0.455	−0.500
[L] (mol L^{-1})	0.0043	0.0210	0.0439	0.0711	0.1329	0.2560	0.7983
−log [L]	2.3665	1.6778	1.3575	1.1481	0.8766	0.5918	0.0978

(vii) Plot the values of $[E_{0.5} - (E_{0.5})_s]$ against $-\log [L]$ (plot shown below) (Figure 5.9) and find the intercept on the E (volt axis) at $\log [L] = 0$ and slop of the resulting straight line. Calculate m and k.

Two straight lines are obtained: first straight line at lower concentration of the ligand less than 0.2 moles [line (a)] and second straight line at higher concentration of the ligand higher than 0.2 moles [line (b)] (Figure 5.9)

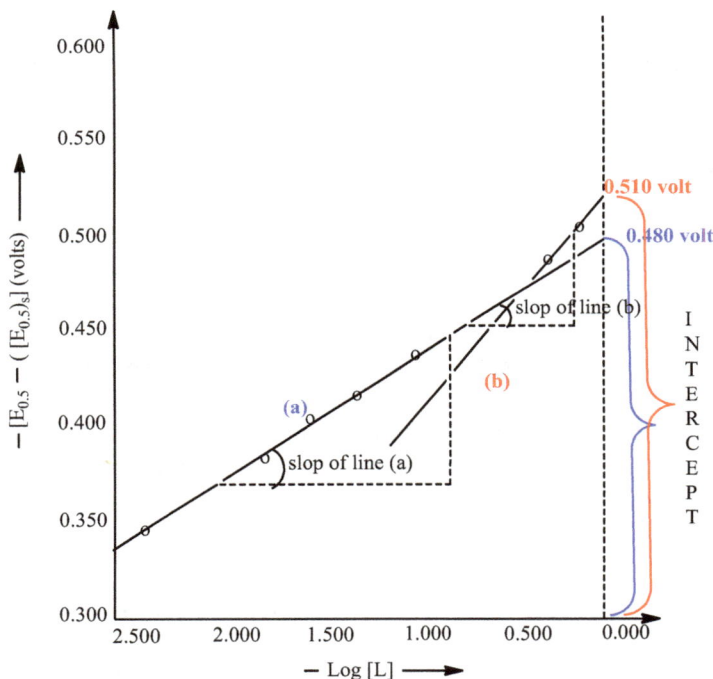

Figure 5.9: Plot the $[E_{0.5} - (E_{0.5})_s]$ against $-\log [L]$ giving two straight lines, (a) and (b).

(viii) For line (a), the intercept at $\log [L] = 0$ is -0.480 volt, so that

$$\text{Intercept} = -\frac{0.0591}{n} \quad \log k_a = -\frac{0.0591}{2} \quad \log k_a = -0.480$$

$$\text{Therefore,} \quad \log k_a = \frac{2 \times 0.480}{0.0591} = 16.2437$$

or $k_a \approx 1.8 \times 10^{16}$

For line (b), the intercept at log [L] = 0 is −0.510 volt, hence

$$-\frac{0.0591}{2} \log k_a = -0.480$$

$$\text{Therefore,} \quad \log k_b = \frac{2 \times 0.510}{0.0591} = 17.2589$$

$$\text{or } k_b \approx 1.8 \times 10^{17}$$

The slope of line (a) is −0.055, so that

$$\text{Slope} = -\frac{m \times 0.0591}{n} = -\frac{m \times 0.0591}{2} = -0.055$$

$$\text{Therefore,} \quad m = \frac{2 \times 0.055}{0.0591} = 1.86$$

Thus, the complex has composition [Cu(glycinate)$_2$] (Figure 5.10)

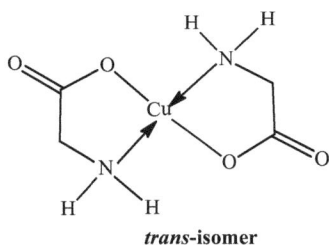

trans-isomer

Figure 5.10: Composition of the [Cu(glycinate)$_2$] complex.

For line (b), the slope is −0.096, hence

$$\text{Slope} = -\frac{m \times 0.0591}{n} = -\frac{m \times 0.0591}{2} = -0.096$$

$$\text{Therefore,} \quad m = \frac{2 \times 0.096}{0.0591} = 3.25$$

Thus, at higher concentrations of glycine ligand, the complex has composition [Cu(glycinate)$_3$]$^-$ (Figure 5.11)

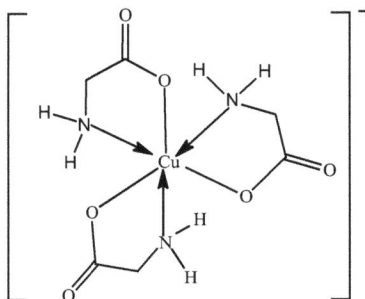

Figure 5.11: Composition of the [Cu(glycinate)$_3$]$^-$ complex.

5.6.4 Ion-exchange method

An ion-exchange resin is an insoluble organic polymeric solid having positive ions (called cation-exchange resin) or negative ions (anion-exchange resin). These ions (cation or anion) can be exchanged by some other ions of similar charge.

If a cation exchange resin is in contact with a solution of Na^+ and Ca^{2+} ions, an equilibrium constant for such a system may be represented as:

$$K_r = \frac{[Na^+]^2_{soln} [Ca^{2+}]_{resin}}{[Na^+]^2_{resin} [Ca^{2+}]_{resin}} \tag{5.57}$$

If the concentration of Na^+ is higher compared to Ca^{2+} concentration, then the Na^+ concentrations in two phases remain nearly constant. Consequently, equation (5.57) results:

$$K'_r = \frac{[Ca^{2+}]_{resin}}{[Ca^{2+}]_{sol}} \tag{5.58}$$

When a ligand such as citrate ion (Figure 5.12) is added, calcium citrate is formed

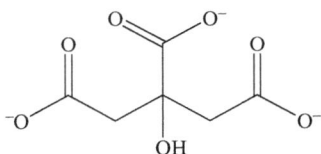

Figure 5.12: Structure of citrate ion.

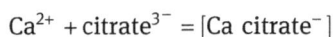

$$Ca^{2+} + citrate^{3-} = [Ca\ citrate^-]$$

Then, the formation constant for calcium citrate complex is given by the expression,

$$K = \frac{[Ca\ citrate^-]}{[Ca^{2+}]\ [citrate^{3-}]} \tag{5.59}$$

When citrate ion is added, the concentration of free Ca^{2+} decreases because of the formation of calcium citrate complex. Thus, the amount of Ca^{2+} in the resin decreases. The distribution quotient, D, is given by

$$D = \frac{[Ca^{2+}]_{resin}}{[Ca^{2+}]_{sol} + [Ca\ citrate^-]_{sol}} \tag{5.60}$$

or $\quad \dfrac{1}{D} = \dfrac{[Ca^{2+}]_{sol} + [Ca\ citrate^-]_{sol}}{[Ca^{2+}]_{resin}} \tag{5.61}$

Putting the value of $[Ca^{2+}]_{resin}$ from equation (5.58), we get

$$\frac{1}{D} = \frac{[Ca^{2+}]_{sol} + [Ca\,citrate^-]_{sol}}{K'_r \cdot [Ca^{2+}]_{sol}} \tag{5.62}$$

$$\text{or } \frac{1}{D} = \frac{[Ca\,citrate^-]_{sol}}{K'_r \cdot [Ca^{2+}]_{sol}} + \frac{[Ca^{2+}]_{sol}}{K'_r \cdot [Ca^{2+}]_{sol}}$$

$$= \frac{[Ca\,citrate^-]_{sol}}{K'_r \cdot [Ca^{2+}]_{sol}} + \frac{1}{K'_r} \tag{5.63}$$

Putting the value of $[Ca\,citrate^-]_{sol}$ from equation (5.59) to equation (5.63), we get

$$\frac{1}{D} = \frac{K[Ca^{2+}]\,[citrate^{3-}]}{K'_r \cdot [Ca^{2+}]_{sol}} + \frac{1}{K'_r}$$

$$\text{or } D = \frac{1}{\dfrac{K\,[citrate^{3-}]}{K'_r} + \dfrac{1}{K'_r}} \tag{5.64}$$

It is notable here that the distribution quotient, D, can be measured at different concentrations of citrate ion. Since the equation (5.64) is of the form $y = mx + c$, a plot of $1/D$ versus concentration of $[citrate^{3-}]$ will produce a straight line whose slope and intercept are:

$$\text{slope} = \frac{K}{K'_r}$$

$$\text{Intercept} = \frac{K}{K'_r}$$

From the slope and intercept, the formation constant for calcium citrate complex, K can be calculated as follows:

$$\frac{\text{slope}}{\text{Intercept}} = \frac{K}{K'_r} \times \frac{K'_r}{1} = K$$

This is fast and easy method for determining stability constant. If the metal ion is radioactive, then the concentration of the metal ion in the resin phase and solution phase can be more easily determined by measuring its radioactivity. If a suitable radioactive metal ion is available, then the method is preferred over the other methods.

5.6.5 Solubility method

If the metal ion forms a relatively insoluble salt whose solubility product (sp) is known, the stability constant of the more soluble complex may be determined by measuring the increased solubility of the salt due to the presence of the ligand.

Let us consider the reaction between CH_3COOAg and CH_3COO^-. When a solution of CH_3COONa of known high concentration is equilibrated at constant temperature with excess of CH_3COOAg, the equilibria given below are obtained:

$$CH_3COOAg = Ag^+ + CH_3COO^-$$

$$K_{sp} = [Ag^+][CH_3COO^-] \tag{5.65}$$

$$CH_3COONa = Na^+ + CH_3COO^-$$

$$Ag^+ + 2CH_3COO^- = [Ag(CH_3COO)_2^-]$$

$$\beta_2 = \frac{[Ag(CH_3COO)_2^-]}{[Ag^+][(CH_3COO)^-]^2} \tag{5.66}$$

After equilibration, the excess solid CH_3COOAg is removed by filtration and the solution is then analysed for total silver content. Now, the following relationship holds:

$$C_{Ag+} = [Ag^+] + [Ag(CH_3COO)_2^-] \tag{5.67}$$

or

$$[Ag(CH_3COO)_2^-] = C_{Ag+} - [Ag^+] \tag{5.68}$$

where C_{Ag+} is the Ag^+ concentration is solution.

From the solubility product expression (5.67), we have

$$[Ag^+] = \frac{K_{sp}}{[CH_3COO^-]} \tag{5.69}$$

Considering the excess CH_3COONa added, the above equation (5.69) becomes

$$[Ag^+] = \frac{K_{sp}}{C_{CH_3COONa}} \tag{5.70}$$

Putting the value of $[Ag(CH_3COO)_2^-]$ from equation (5.68) in equation (5.66), we get

$$\beta_2 = \frac{C_{Ag+} - [Ag^+]}{[Ag^+][(CH_3COO^-)]^2} \tag{5.71}$$

Now, putting the value of $[Ag^+]$ from equation (5.70) and considering the excess CH_3COONa added, the above equation becomes:

$$\beta_2 = \frac{C_{Ag^+} - \left(\frac{K_{sp}}{C_{CH_3COONa}}\right)}{\left(\frac{K_{sp}}{C_{CH_3COONa}}\right)[C_{CH_3COONa}]^2} \tag{5.72}$$

In the right-hand side of this expression, all the quantities are known, and so the overall stability constant β_2 can be calculated. It is necessary to vary the concentration of CH_3COONa to check whether β_2 is really a constant. Further refinement of this method would take into account of hydrolysis of acetate which is almost negligible in basic medium.

Exercises

Multiple choice questions/fill in the blanks

1. The thermodynamic stability deals with:
   (a) Bond energies
   (b) Stability constants
   (c) Redox potentials
   (d) All of these

2. Which one of the following comes under thermodynamic stability?
   (a) Rates of reactions
   (b) Formation of intermediate complexes
   (c) Activation energies for the process
   (d) Bond energies

3. Which one of the following comes in kinetic stability?
   (a) Redox potentials
   (b) Bond energies
   (c) Mechanism of chemical reactions
   (d) All of these

4. Sort out the term which comes in purview of kinetic stability:
   (a) Bond energies
   (b) Stability constants
   (c) Activation energies for the processes
   (d) Redox potentials

5. On the basis of kinetic stability of the complexes in solution, these are classified into labile and inert complexes first time by:
   (a) Ralph Gottfrid Pearson
   (b) Fred Pearson
   (c) Henry Taube
   (d) All of them

6. Out of $[Fe(H_2O)_6]^{3+}$ and $[Cr(H_2O)_6]^{3+}$, which is labile?
   (a) $[Fe(H_2O)_6]^{3+}$
   (b) $[Cr(H_2O)_6]^{3+}$
   (c) $[Cr(H_2O)_6]^{3+}$ and $[Fe(H_2O)_6]^{3+}$ both
   (d) Cannot be said

7. If the value of dissociation constant of a complex, $[Cu(NH_3)_4]^{2+}$ is 1.0×10^{-12}, the value of its stability constant will be:
   (a) 0.01×10^{12} (b) 0.1×10^{12} (c) 1.0×10^{12} (d) None of these

8. The concept of stepwise stability constant was first time given by:
   (a) S. M. Jorgensen   (b) A. Werner
   (c) N. J. Bjerrum   (d) None of them

9. For a system having 4 steps, the overall stability constant, β_4 is related to stepwise stability constants, K_1, K_2, K_3 and K_4 as:
   (a) $\beta_4 = K_1 + K_2 + K_3 + K_4$
   (b) $\beta_4 = K_1.K_2.K_3.K_4$
   (c) $\beta_4 = 2(K_1 + K_2 + K_3 + K_4)$
   (d) $\beta_4 = (K_1.K_2.K_3.K_4)/2$

10. For a system having n steps, which one of the following is correct?
    (a) $\log_{10}\beta_n = \log_{10} K_1 + \log_{10} K_2 + \log_{10} K_3 + \ldots \ldots + \log_{10} K_n$
    (b) $\log_{10}\beta_n = \log_{10} K_1 \times \log_{10} K_2 \times \log_{10} K_3 \times \ldots \ldots \times \log_{10} K_n$
    (c) $\log_{10}\beta_n = \log_{10} K_1 - \log_{10} K_2 + \log_{10} K_3 - \ldots \ldots - \log_{10} K_n$
    (d) None of the above

11. The basis of the Bjerrum's method for the determination of stepwise stability constants is that during complex formation between metal ion and the protonated ligand, protons:
    (a) Liberate   (b) Not liberate
    (c) Cannot predict   (d) None of these

12. Spectrophotometric method for the determination of stability constant is based on the principle that generally solutions containing complexes have:
    (a) The same absorption properties to those of the constituent ions or molecules
    (b) Different absorption properties to those of the constituent ions or molecules
    (c) Sometimes same and sometimes different absorption properties to those of the constituent ions or molecules
    (d) Nothing can be said

13. The half wave potential ($E_{1/2}$) for the polarographic reduction of a metal ion is . . . . when the metal is complexed.

14. The stepwise stability constants generally . . . . with increasing substitution of H_2O from $[M(H_2O)_n]$ by ligand L.

15. Job's method for the determination of metal-ligand ratio in complexes is mainly used for solutions where only . . . . is formed.

Short answer type questions

1. Differentiate between thermodynamic and kinetic stability of complexes.
2. Justify the two statements with suitable example that (i) thermodynamic stability of a complex does not ensure its inertness and (ii) thermodynamic instability of a complex does not ensure its lability.
3. Explain with suitable examples along with relevant data that formation constant (or stability constant) is reciprocal of dissociation constant.
4. Explain the formation of coordination compounds in aqueous solution through successive stability constants of the system taking a suitable example of an aqua complex.
5. Describe the formation of an ML_n complex in aqueous solution through overall stability constants of the system taking a suitable example of an aqua complex.
6. How are overall stability constant and stepwise stability constants of a complex formation system related?
7. The K values generally decrease with increasing substitution of H_2O from $[M(H_2O)_n]$ by ligand L. Justify this statement.
8. What are the limitations of Job's method for the determination of stability constant of complexes in aqueous solutions?
9. Briefly highlight the various steps involved in the determination of metal–ligand ratio as well stability constant of complexes by mole ratio method.
10. How is overall stability constant of a system is determined by solubility method?

Long answer type questions

1. Differentiate between stepwise and overall stability constants. How are they related to each other?
2. Mathematically develop the formation functions $\bar{n}$, $\bar{n}_A$ and p_L used to calculate the stepwise stability constants by Bjerrum's method.
3. Present a detained view of determination of stepwise stability constants by pH-metric/potentiometric method.
4. How is metal–ligand ration determined of a complex formation in aqueous solution by Job's method of continuous variation?
5. Describe the determination of stability constant by spectrophotometric method taking example of a complex formation between ferric ion and thiocyanate ion in aqueous medium.
6. Present a detailed view of determination of stability constant by polarographic method.
7. Describe the ion exchange method for the determination of stability constant taking a suitable example.
8. Describe the method for the determination of stability constant of a complex by Job's method. Highlight the limitations of this method also.

Chapter VI
Principles of magnetochemistry and its multiple applications in coordination compounds

6.1 Introduction

Measurements of magnetic properties have been used to characterize a wide range of systems from oxygen, metallic alloys, solid state materials and coordination complexes containing metals. Most organic and main group element compounds have all the electrons paired and these are diamagnetic molecules with very small magnetic moments. All of the transition metals have at least one oxidation state with an incomplete d subshell. Magnetic measurements, particularly for the first row transition elements, give information about the number of unpaired electrons. The number of unpaired electrons provides information about the oxidation state and electron configuration. The determination of the magnetic properties of the second and third row transition elements is more complex.

The magnetic properties of metal complexes in terms of unpaired electrons and their magnetic or spin properties are useful in determining structural features in transition metal compounds. Complexes that contain unpaired electrons are paramagnetic and are attracted into magnetic fields. Diamagnetic compounds are those with no unpaired electrons that are repelled by a magnetic field. All compounds, including transition metal complexes, possess some diamagnetic component which results from paired electrons moving in such a way that they generate a magnetic field that opposes an applied field. A compound can still have a net paramagnetic character because of the large paramagnetic susceptibility of the unpaired electrons. The number of unpaired electrons can be determined by the magnitude of the interaction of the metal compound with a magnetic field. This is directly the case for 3d transition metals but not always for 4d or 5d transition metals, whose observed magnetic properties may arise not only from the spin properties of the electrons, but also from the orbital motion of the electrons (Figure 6.1).

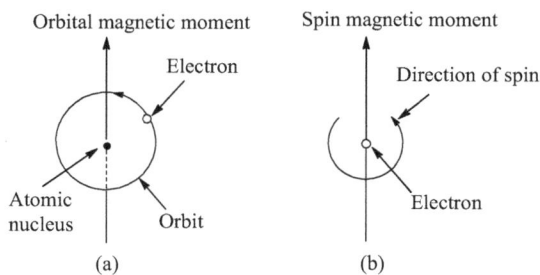

Figure 6.1: (a) Magnetic moment of the electron due to orbital motion, (b) magnetic moment of the electron due to spin motion.

https://doi.org/10.1515/9783110727289-006

Chemical useful information of transition metal complexes is obtained by proper interpretation of measured value of magnetic moments. However, magnetic moments are not measured directly. Instead, one measures magnetic susceptibility of a material, from which, it is possible to calculate the magnetic moment of the paramagnetic ion or atom therein. The meaning of magnetic susceptibility may be understood in the following section.

6.2 Principle of magnetism

Since the time of Michael Faraday (Figure 6.2) it had been realized that all substances possess magnetic properties. That is to say, all substances are affected in some way by the application of a magnetic field. The general expression of this fact is as follows:

Michael Faraday
(1791–1867)

Hans Christian Oersted
(1777–1851)

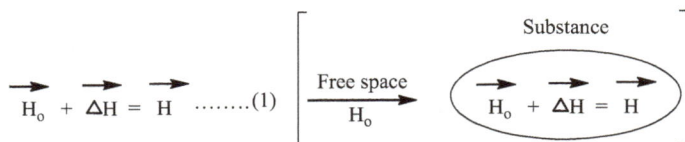

$$H_o + \Delta H = H \quad \cdots \cdots (1)$$

Free space
H_o

Substance
$H_o + \Delta H = H$

Figure 6.2: Difference in the magnitude of magnetic field in free space and inside a substance.

When a substance is placed in a magnetic field of magnitude H_o oersteds (in the name of Hans Christian Oersted (Figure 6.2)), the magnitude inside a substance differs from the free space value of the applied field, that is, H_o. The difference of the field strength as shown in Figure 6.2 can be represented in the form of following equation:

$$H_o + \Delta H = H \tag{6.1}$$

Where H_o is the free space applied field and ΔH is the field produced by the magnetic polarization.

The relationship given in equation (6.1) is usually expressed as follows:

$$B = H_o + 4\pi I \tag{6.2}$$

Here B is identical with H and is known as magnetic induction or density of lines of force within the substance, I is the intensity of magnetization or magnetic moment per unit volume.

Dividing the equation (6.2) both side by H_o, we get

$$B/H_o = 1 + 4\pi I/H_o$$

$$\text{or} \quad P = 1 + 4\pi k \tag{6.3}$$

$$\text{or} \quad k = P - 1/4\pi \tag{6.4}$$

The ratio B/H_o is called magnetic permeability, P and k are called magnetic suscep-tibility per unit volume or volume susceptibility, and it is a measure of how suscep-tible the substance is to magnetic polarization.

The quantity that is usually measured is not the volume susceptibility (k), but another quantity known as specific, weight or gram magnetic susceptibility (χ_g). χ_g is magnetic susceptibility per gram of the material. The two quantities (k and χ_g) are related to density (d) of the material, such as,

$$\chi_g = \frac{k}{d} (cm^3/gram) \tag{6.5}$$

Substituting the value of k from equation (6.4), we get

$$\chi_g = \frac{P-1}{4\pi d} \tag{6.6}$$

The molar susceptibility, χ_M is given as:

$$\chi_M = \text{gram susceptibility} \times \text{molecular weight}$$

$$= \chi_g \times M = \frac{k}{d} \times M \, (cm^3/mole)$$

For normal paramagnetic and diamagnetic substances, k, χ_g and χ_M are constants and independent of field strength.

6.3 Types of magnetic substances

From equation (6.3)–(6.6), we get

$$\chi_g = \frac{P-1}{4\pi d} = \frac{k}{d} = \frac{1}{Ho.d}$$

This equation leads to the most fundamental classification of substances.

6.3.1 Diamagnetic substances

If P < 1, that is, χ_g, k and I are negative, then in that case substance is said to be diamagnetic. Such a substance allows a smaller number of lines of force to pass through it as compared to that *in vacuum* (Figure 6.3) and, therefore, prefers to move to the region of the lowest magnetic field. In other words such substances are repelled by the magnetic field.

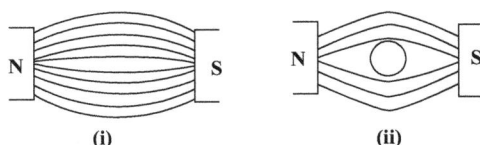

(i)   (ii)

Figure 6.3: (i) Magnetic field lines of flux in vacuum and (ii) magnetic field lines of flux for a diamagnetic substance in a field.

Diamagnetic substances have the following characteristics:

(a) When a diamagnetic substance is put into the magnetic field, small magnetic moments (or magnetic field) are induced into the substance. These induced magnetic moments are in opposition to the inducing field. Consequently, the substance is repelled by the magnetic field.

(b) Substances having no unpaired electrons or having all the electrons in the paired state show diamagnetic character. When an orbital contains two electrons with opposite spins (↑↓), the magnetic moment (or magnetic field) generated by one electron is cancelled by that generated by the other electron as the magnetic moments of the two electrons are equal and opposite to each other (Figure 6.4). Thus, an atom, ion or a molecule having paired electrons will be diamagnetic.

(c) The diamagnetic nature of the substance can be found by weighing the substance first in air and then after suspending it between the poles of the magnetic field. If the substance weighs less in the magnetic field, it shows that the substance is repelled by the magnetic field and hence is diamagnetic.

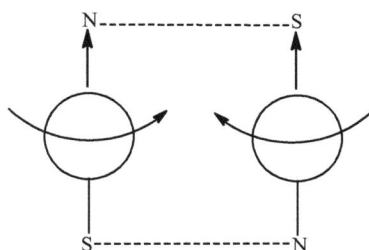

Figure 6.4: Magnetic field generated by two electrons with opposite spins cancelling each other.

6.3.2 Paramagnetic substances

If P > 1, that is, χ_g, k and I are positive, then in that case substance is said to be para-magnetic. Such a substance allows more of lines of force to pass through it, and therefore prefers to move to the region of the highest magnetic field (Figure 6.5).

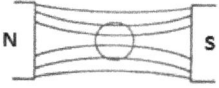

Figure 6.5: Magnetic field lines of flux for a paramagnetic substance in a field allowing more of lines of force to pass through it.

Paramagnetic substances have the following characteristics:
(a) Substances having one or more unpaired electrons show paramagnetic charac-ter. A single electron (unpaired) spinning on its axis generates a magnetic field and behaves like a magnet as shown in Figure 6.6. Hence, when a substance whose orbital contains an unpaired electron is placed in an electromagnet, it is attracted towards the poles of the electromagnet. It follows that an atom, ion or molecule containing one or more unpaired electrons will be paramagnetic.

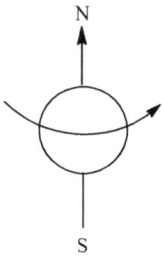

Figure 6.6: An electron spinning on its axis behaves like a small magnet.

(b) When a paramagnetic substance is placed in an external magnetic field, the indi-vidual atoms or molecules of the substance behave as permanent magnets and align themselves in the same direction as that of external magnetic field. Hence, the paramagnetic substances are attracted towards magnetic field. Consequently, when a paramagnetic substance is placed in a magnetic field, the number of lines of force passing through paramagnetic substance would be more than that would pass through air.
(c) The paramagnetic nature of the substance can be found by weighing the sub-stance first in air and then after suspending it between the poles of the magnetic field. If the substance weighs more in the magnetic field, it shows that it is at-tracted by the magnetic field and hence is paramagnetic.

6.4 Types of paramagnetic materials

Paramagnetic substances are subdivided into two classes: (i) Ferromagnetism and (II) anti-ferromagnetism materials.

6.4.1 Ferromagnetism

Substances which show permanent magnetism even in the absence of magnetic field are called ferromagnetic substances. Such substances remain permanently magnetized, once they have been magnetized. This type of magnetism arises due to spontaneous alignment of magnetic moments due to unpaired electrons in the same direction, as shown in Figure 6.7. This is the case of large amount of paramagnetism.

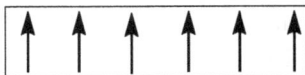

Figure 6.7: Alignment of neighbouring spins due to unpaired electrons in the same direction in ferromagnetic materials.

Atoms or ions with incomplete d or f subshells exhibit ferromagnetism. Examples of substances showing ferromagnetism are: Fe, Co, Ni, Gd, Dy, number of alloys of Cu, Al and Mn, and oxides of metals of first transition series such as CrO_2, Fe_3O_4, etc. Ferromagnetic substances have their magnetic permeability greater, P ≫ 1.

6.4.2 Effect of temperature on magnetic susceptibly of ferromagnetic substances

Ferromagnetic behaviour of a substance exists up to a certain temperature which is called Curie temperature or Curie point (T_C). A plot of temperature (T) and magnetic susceptibility (χ_g) of a ferromagnetic substance along with a paramagnetic substance, for the sake of comparison, is given in Figure 6.8.

It is seen from the plot that below T_C, say at T_1, χ_g for a ferromagnetic substance is much higher than for a normal paramagnetic substance. Thus, ferromagnetic substance does not obey Curie–Weiss law in this range, but above Curie temperature (T_C) ferromagnetic substance follows Curie–Weiss. As the temperature is lowered below, T_C, magnetic susceptibility sharply increases and attains the highest value at T = 0.

Such behaviour of ferromagnetic substance can be explained as follows: We know that in ferromagnetic substances all electron spins are parallel and hence generate a collective magnetic moment. As the temperature is raised, parallel arrangement of electron spins tends to become partially anti-parallel arrangement.

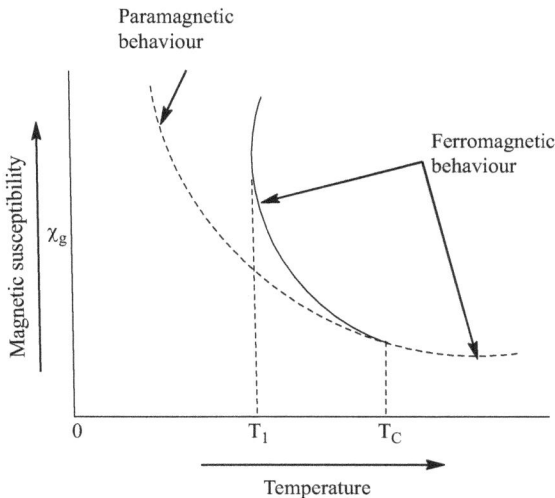

Figure 6.8: Variation of magnetic susceptibility (χ_g) with temperature for a Ferromagnetic substance.

Above T_C, all arrangement of electron spins breaks down and ferromagnetic substance behaves as normal paramagnetic substance. As χ_g decreases above Tc, that is, with increasing temperature, Curie law is being followed.

6.4.3 Magnetically dilute and magnetically concentrated substances

Substances in which the individual magnetic centres present in the crystal lattice are sufficiently far apart to minimize any further magnetic interaction between nearest neighbours, are called magnetically dilute substances. On the other hand substances, in which the individual magnetic centres present in the crystal lattice are close enough to initiate further magnetic interaction between nearest neighbours, are called magnetically concentrated substances. For example, $[Ni(H_2O)_6]Cl_2$ is an example of magnetically dilute substance while $Cu_2(CH_3COOH)_4 \cdot 2H_2O$ is magnetically concentrated substance as in this the unpaired electrons of the two Cu^{2+} ions interact with each other.

6.4.4 Anti-ferromagnetism

Substances which are expected to exhibit paramagnetism or ferromagnetism on the basis of unpaired electrons but actually they possess zero net magnetic moment are called Anti-ferromagnetic substances. Anti-ferromagnetism is due to the presence of equal number of magnetic moments in the opposite directions as shown in Figure 6.9.

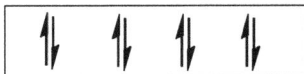

Figure 6.9: Neighbouring spins (magnetic moments) are in opposite direction in anti-ferromagnetic substances.

In first transition series, many oxides of the metals, such as, V_2O_3, CrO_3, MnO, Mn_2O_3, MnO_2, FeO, Fe_2O_3, CoO, Co_3O_4 and NiO exhibit anti-ferromagnetic behaviour.

Magnetically concentrated systems have magnetic exchange interaction between neighbouring paramagnetic metal ion centres. This affects the magnetic properties of the complexes and sometimes swamps the ligand field effect. The exchange interaction occurs between the spins of the neighbouring paramagnetic ions when the paramagnetic centres are close enough for direct or indirect orbital overlaps. In fact, all most all the paramagnetic compounds are involved in exchange interaction to a certain extent and the interaction is dominant only at very low temperatures. If the exchange interaction is greater than kT, we get cooperative phenomenon of ferromagnetism and anti- ferromagnetism.

In case of anti- ferromagnetic substances these exchange interactions lead to lowering of magnetic moment compared to that expected on monomer and no interaction basis. Thus, in anti-ferromagnetic substances the neighbouring spins are opposed to each other.

6.5 Explanation of anti-ferromagnetic behaviour of substances

Anti-ferromagnetic behaviour of substances arises either due to metal–metal interaction which takes place through the overlap of suitable orbitals on metal atom or due to spin exchange interactions taking place through diamagnetic O^{2-} or F^- ions bridging between paramagnetic metal ion.

6.5.1 Metal–metal interaction

Metal–metal interaction that takes place through overlapping of suitable orbitals on metal atoms can be explained by considering the intramolecular anti-ferromagnetic character in Cu(II) acetate monohydrate dimer, $Cu_2(CH_3COO)_4 \cdot 2H_2O$. In this compound the two Cu atoms are 2.64 Å apart. These two Cu atoms are bridged by four CH_3COO^- groups and two water molecules are attached with two Cu atoms along Cu–Cu-axis (Figure 6.10(a)).

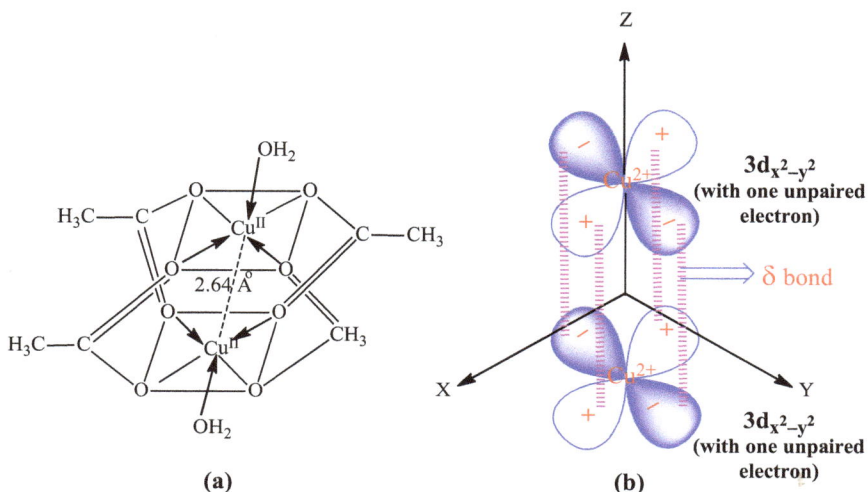

Figure 6.10: (a) Structure of $Cu_2(CH_3COO)_4 \cdot 2H_2O$ (dimer). (b) Formation of δ bond by the overlap of two $3d_{x^2-y^2}$ ions in the dimer.

Cu^{2+} ion is a d^9 system, and the distribution of nine electrons in different d-orbital is as: $3d^2_{xy}$ $3d^2_{yz}$ $3d^2_{xz}$ $3d^2_{z^2}$ $3d^1_{x^2-y^2}$. The $3d_{x^2-y^2}$ of two Cu^{2+} ions each containing one unpaired electron make a lateral overlap and produce a δ bond (Figure 6.10(b)). Because of the formation of δ bond the two unpaired electrons of two Cu^{2+} ions get paired, and consequently magnetic susceptibility lowers ($\mu = 1.43$ B.M.). The lowering of magnetic susceptibility continues to drop below T_N, Neel temperature). The anti-ferromagnetism occurring in $Cu_2(CH_3COO)_4 \cdot 2H_2O$ is called intramolecular anti-ferromagnetism as the interacting paramagnetic centres are present in the same molecule.

6.5.2 Spin exchange interaction through bridging of O^{2-} or F^- ions

This type of interaction can be explained by considering anti-ferromagnetic character of MnO. This oxide contains Mn^{2+} and O^{2-} ions. The valence shell electronic configurations of these two ions are: Mn^{2+} (d^5): $3d^1_{xy}$ $3d^1_{yz}$ $3d^1_{xz}$ $3d^1_{z^2}$ $3d^1_{x^2-y^2}$ and O^{2-}: $2s^2 2p_x^2\ 2p_y^2\ 2p_z^2$ ($2s^2 2p^{4+2}$). The spins of the five electrons in one Mn^{2+} ion are parallel (↑↑↑↑↑) while those in other neighbouring Mn^{2+} ion are anti-parallel (↓↓↓↓↓). In O^{2-} ion all thee 2p orbitals have paired electrons. Thus, O^{2-} ion is diamagnetic.

The $2p_z$ orbital of the bridging O^{2-} ion having paired spins (↑↓) overlaps with two similar 3d-orbitals (*viz.*, $3d_{xz}$ orbitals) of two Mn^{2+} ions. Each of the two $3d_{xz}$-orbitals in Mn^{2+} ion has one unpaired electron with opposite spins (↑ and ↓) (Figure 6.11). This type of overlap between orbitals yields stable MnO.

Movement of one of the two electrons from
$2p_z$ orbital of $O_2{}^{2-}$ into $3d_{xz}$ orbital of
each Mn^{2+}

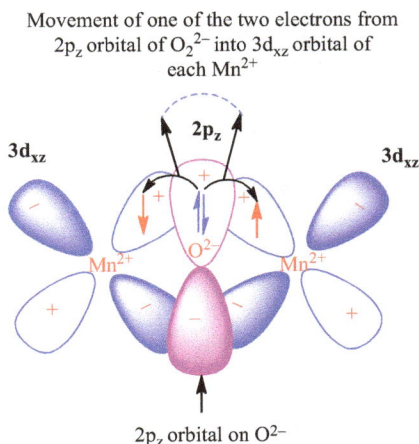

Figure 6.11: Overlap between 2p-orbital on O^{2-} ion with two $3d_{xz}$ orbitals on two Mn^{2+} ions in MnO.

In the present overlap, one of the two electrons present in the $2p_z$-orbital of the bridging O^{2-} ion moves into the $3d_{xz}$ orbital of one Mn^{2+} ion which carries one electron with opposite spin. Similarly, the other electron of the $2p_z$-orbital migrates to $3d_{xz}$ orbital of other Mn^{2+} ion. It is due to the spin exchange interaction so that both Mn^{2+} ions get paired electrons in 3d-orbitals.

It is clear from the above discussion on anti-ferromagnetic substances that pairing of total electron spins of atoms or ions of a substance are responsible for its anti-ferromagnetic behaviour.

6.5.3 Effect of temperature on magnetic susceptibility of anti-ferromagnetic substances

Magnetic susceptibility of anti-ferromagnetic substances is very much dependent on temperature. The variation of χ_g with temperature (T) for an anti-ferromagnetic substance is shown in Figure 6.12. The χ_g versus T graph for a paramagnetic substance is also shown in this figure for the sake of comparison.

For anti-ferromagnetic substances also, there is a characteristic temperature, known as **Neel temperature** or **Neel point**, above which the anti-ferromagnetic substance behaves as a normal paramagnetic substance. But below the **Neel point**, χ_g decreases with decrease in temperature as shown in Figure 6.12. The reason for this type of behaviour is discussed below.

We have seen in the earlier discussion on anti-ferromagnetism that the coupling of total electron spins (also called electron spin pair interactions) of atoms or ions of a substance are responsible for its anti-ferromagnetic behaviours. Bur the thermal energy tends to decouple and randomize these coupled total electron spins.

$\uparrow\downarrow \ \uparrow\downarrow \ \uparrow\downarrow \ \uparrow\downarrow \ \uparrow\downarrow$

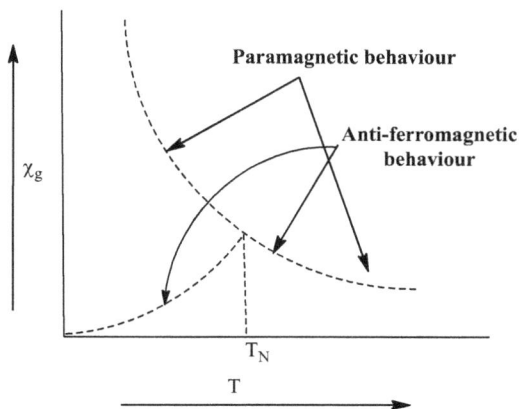

Figure 6.12: Variation of χ_g with temperature for an anti-ferromagnetic substance.

At the Neel point, the thermal energy available to the anti-ferromagnetic system is comparable to the energy of the electron spin pair interaction (the two energies have opposite sign). Electron spin pair interactions decrease the energy of the system. Hence, at the Neel point, both the coupled total electron spins and the uncoupled randomized total electron spins of atoms or ions of the substance are equally plausible.

Above the Neel point, no coupling of total electron spins (or no electron spin pair interactions) is possible, and only the uncoupled and completely randomized total electron spins of atoms or ions exist as happens in the case of a paramagnetic substance. Therefore, at temperature above the Neel point, the anti-ferromagnetic substances behave as paramagnetic substances because the randomized total electron spins tend to align along the applied magnetic field. Similar to other paramagnetic substances, χ_g in these substances is also inversely proportional to temperature measured above Neel temperature.

Below the Neel point, thermal energy available to the anti-ferromagnetic system is less than the energy of the electron spin pair interactions. Therefore, below the Neel point as we go on decreasing the temperature, the uncoupling and randomized total electron spins goes on decreasing and the extent of coupling of total electron spins goes on increasing. Consequently, below the Neel point anti-ferromagnetic character of a substance, which is directly related to the extent of electron spin pair interactions (or to the extent of coupling of total electron spins), goes on increasing with decreasing temperature. In the other words, below the Neel point, χ_g decreases with decrease in temperature (that is, $\chi_g \propto T$) in anti-ferromagnetic substances.

Intramolecular anti-ferromagnetism occurs when the electron spins of the metal ions belong to the same molecule of a polynuclear compound get coupled. An example of this type of anti-ferromagnetic material is provided by Cu(II) acetate monohydrate dimer, $Cu_2(CH_3COO)_4 \cdot 2H_2O$ (Figure 6.13).

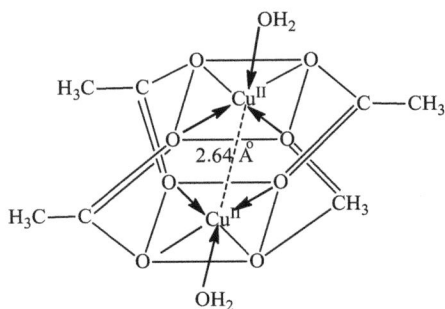

Figure 6.13: Anti-ferromagnetic Cu_2 $(CH_3COO)_4 \cdot 2H_2O$.

6.6 Magnetic moment from magnetic susceptibility

Paramagnetic molecules are free to orient themselves ideally in the gas phase, but in practice also in solution and even in the solid state if magnetically dilute. They are subjected to two opposing effect: (i) the magnetic field H, which tends to align the molecular magnets in the same direction as that of the field and (ii) the thermal agitation or thermal energy, kT (called kT effect) that tends to orient the molecular magnets randomly. The effect of the magnetic field becomes more pronounced with the decrease in the value of absolute temperature T.

Based on quantum mechanical calculations, the paramagnetic susceptibility (χ_p) is given by

$$\chi_p = \frac{N\mu^2\beta^2}{3kT} \tag{6.7}$$

where N = Avogadro's number: 6.023×10^{23}, k = Boltzmann's constant: 1.381×10^{-16} ergs/deg., T = temperature in absolute, β = Bohr magneton: 0.9273×10^{-20} ergs/gauss and μ is the magnetic moment.

The value of χ_M calculated from the experimental value of χ_g is the algebraic sum of χ_p, χ_{TIP} and χ_{dia}, that is,

$$\chi_M = \chi_p + \chi_{TIP} + \chi_{dia.} \tag{6.8}$$

As we do not know the value of χ_{TIP}, corrected molar susceptibility, χ_M corr., which is equal $\chi_M - \chi_{dia.}$ is also used in place of χ_p in equation (6.7),

$$\chi_p = \frac{N\mu^2\beta^2}{3kT} \tag{6.7}$$

The value of magnetic moment obtained by putting $\chi_M^{corr.}$ in place of χ_p in equation (6.7) is called effective magnetic moment and hence is represented as μ_{eff}. Thus, equation (6.7) becomes:

$$\chi_M^{corr.} = \frac{N\beta^2\mu^2eff}{3kT} \tag{6.9}$$

$$\text{or } \mu_{eff}^2 = \frac{3k\,\chi_M corr.}{N\beta^2} \cdot T$$

$$\text{or } \mu_{eff} = \sqrt{\frac{3k}{N\beta^2}} \ \sqrt{\chi_M^{corr.} \cdot T} \ \text{ B.M.}$$

$$= \sqrt{\frac{3 \times 1.381 \times 10^{-16}}{6.023 \times 10^{23} \times (0.9273 \times 10^{-20})^2}} \cdot \sqrt{\chi_M^{corr.} \cdot T} \ \ \text{B.M.}$$

$$= \sqrt{\frac{4.143 \times 10^{-16}}{6.023 \times 10^{23} \times (0.9273)^2 \times 10^{-40}}} \cdot \sqrt{\chi_M^{corr.} \cdot T} \ \ \text{B.M.}$$

$$= \sqrt{\frac{4.143 \times 10^{-16}}{5.1790 \times 10^{17}}} \cdot \sqrt{\chi_M^{corr.} \cdot T} \ \ \text{B.M.}$$

$$= \sqrt{\frac{4.143 \times 10}{5.1790}} \cdot \sqrt{\chi_M^{corr.} \cdot T} \ \ \text{B.M.}$$

$$= \sqrt{7.999613825} \cdot \sqrt{\chi_M^{corr.} \cdot T} \ \ \text{B.M.}$$

$$\mu_{eff} = 2.828\sqrt{\chi_M^{corr.} \cdot T} \ \ \text{B.M.} \tag{6.10}$$

Magnetic moments are expressed in Bohr magneton (B.M.). 1 B.M. = 0.9273×10^{-20} ergs/gauss

6.7 Curie and Curie–Weiss law: effect of temperature on χ_M^{corr} of paramagnetic substances

χ_M^{corr} depends on temperature since the thermal agitation randomize the atomic or molecular magnetic dipoles of the paramagnetic substances when get aligned in the direction of the magnetic field.

From equation (6.11) given above, and the same presented below,

$$\chi_M^{corr.} = \frac{N\beta^2\mu^2_{eff}}{3kT} \tag{6.11}$$

it is well evident that $\chi_M{}^{corr}$ is inversely proportional to absolute temperature (T) since thermal agitation opposes the paramagnetic alignment. As it was observed by the Pierre Curie while investigating the effect of temperature on magnetic properties of paramagnetic substances at the end of the nineteenth century, this is called the Curie law. Thus,

$$\chi_M{}^{corr} \; \alpha \; \frac{1}{T}$$

$$\text{or} \quad \chi_M{}^{corr} = \frac{C}{T} \tag{6.12}$$

Here C is a constant equal to:

$$\frac{N\beta^2\mu^2_{eff}}{3k}$$

This constant is characteristic of the paramagnetic substance and is called Curie's constant and the equation (6.12) is called Curie equation. From this equation it is evident that the plot of $\chi_M{}^{corr}$ values against 1/T (or $1/\chi_M{}^{corr}$ against T) will be a straight line with slope equal to C which passes through the origin (Figure 6.14 (a)).

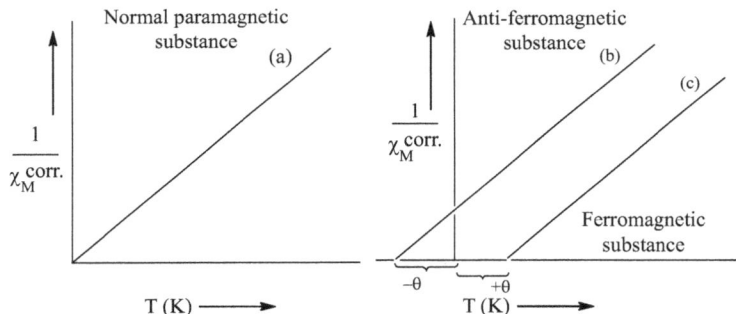

Figure 6.14: (a) Perfect plot of Curie law, (b, c) Curie–Weiss plots.

6.7.1 Limitation of Curie law and Curie–Weiss law

Curie law represented by equation (6.12) is obeyed by those paramagnetic substances in which magnetic dipoles of the substance are completely independent of each other. If there are some cooperative interactions between the neighbouring dipoles or if the applied magnetic field induces some temperature-independent paramagnetism, Curie law is not obeyed, that is, the straight line does not pass through the origin but cut the T-axis at temperature below 0° (Figure 6.14(b)) or at temperature above 0° (Figure 6.14(c)). For such paramagnetic substances, a modified law known as Curie–Weiss law has been suggested. This law is represented by the following relationship:

$$\chi_M{}^{corr} = \frac{C}{T - \theta} \tag{6.13}$$

Here θ is the intercept on the temperature axis and is called **Curie–Weiss constant.** In fact, θ takes into account of cooperative interactions mentioned above. For ferromagnetic substances θ is positive while for anti-ferromagnetic materials θ is negative. The magnitude and sign of θ can be obtained from the plot of $1/\chi_M{}^{corr}$ versus T.

6.7.2 Calculation of μ_{eff}

The equation (6.11) given above gives the value of $\chi_M{}^{corr.}$ in terms of Curie constant, C as follows:

$$\chi_M{}^{corr.} = \frac{N\beta^2\mu^2 eff}{3kT}$$

$$\text{or } \mu_{eff}{}^2 = \frac{3k\,\chi_M{}^{corr.}}{N\beta^2} \cdot T$$

$$\text{or } \mu_{eff} = \sqrt{\frac{3k}{N\beta^2}} \; \sqrt{\chi_M{}^{corr.} \cdot T} \quad \text{B.M.}$$

Putting the value of Avogadro's number (N): 6.023×10^{23}, = Boltzmann's constant (k): 1.381×10^{-16} ergs.deg^{-1}, β = Bohr magneton: 0.9273×10^{-20} ergs/gauss in the above equation, we get

$$\mu_{eff} = 2.83 \; \sqrt{\chi_M{}^{corr.} \cdot T} \quad \text{B.M.} \tag{6.14}$$

If in place of Curie equation, we make use of Curie–Weiss equation (6.13), then the μ_{eff} is given by

$$\mu_{eff} = 2.83 \; \sqrt{\chi_M{}^{corr.} (T - \theta)} \quad \text{B.M.} \tag{6.15}$$

The value of θ can be known from the variable temperature susceptibility data. The value of θ for a lot of complexes is close to zero and hence equation (6.15) reduces to equation (6.14).

In the above equations (6.14) and (6.15), we use the corrected value of molar magnetic susceptibility ($\chi_M{}^{corr.}$) of the compound which is given by

$$\chi_M{}^{corr.} = \chi_c \times \text{molar mass of the compound}$$

Here χ_c is the magnetic susceptibility of the compound.

The corrected value, $\chi_M^{corr.}$ is given by:

$$\chi_M^{corr.} = \chi_{exp} - \chi_{dia}$$

Here, χ_{exp} is the experimental value of magnetic susceptibility of the paramagnetic compound and χ_{dia} is diamagnetic susceptibility contribution which is equal to sum of the contributions made by all individual atoms or ions and constitutive corrections for certain bonds.

6.8 Temperature-independent paramagnetism

Some of the substances like $Cr^{VI}O4$ and $Mn^{VII}O_4^-$, which do not contain unpaired electrons, show week magnetism, when the substance is placed in a magnetic field. This paramagnetism is independent of temperature and is referred to as temperature-independent paramagnetism (TIP). This paramagnetism is produced by mixing of the ground state (which is diamagnetic) with excited states of high energy (which are paramagnetic) under the influence of the applied field. So, ground state assumes some of the magnetic properties of the excited states. Thus, TIP is not due to presence of paramagnetic centres in the sample but is induced in the sample when it is placed in a magnetic field.

6.9 Diamagnetic correction: Pascal's constants and constitutive corrections

The measured magnetic susceptibility for a given substance will consist of contribution from paramagnetic and diamagnetic susceptibilities, the former being much greater. Thus, $\chi_M^{corr.}$ for a paramagnetic compound is obtained by subtracting the susceptibility of a diamagnetic group or ligands present from the molar susceptibility (χ_M).

$$\chi_M^{corr.} = \chi_M - \chi_{dia}$$

P. Pascal carried out much precise susceptibility determinations on a host of diamagnetic compounds. Out of these determinations he was able to express diamagnetic susceptibility of a compound as follows:

$$\chi_{dia} = \Sigma n_A \chi_A + \Sigma \lambda$$

Here χ_A represents the gram atomic susceptibility of the atom A and n_A is the number of atoms of A present in the molecule and λ is a constitutive corrections for certain bond type.

The values for χ_A and λ are referred to as Pascal's constants for the atoms/ion and the bonds concerned.

The diamagnetic contributions (Pascal's constants) for some common atoms/ions molecule and bonds are given in Tables 6.1–6.5.

Table 6.1: Pascal's constant for atoms (all values in 10^{-6} cgs units/gram atom).

Atom	χ_A	Atom	χ_A
H	−2.93	Cl	−20.1
C	−6.00	Br	−30.6
N (open chain)	−5.57	I	−44.6
N (ring)	−4.61	S	−15.0
N (monoamide)	−1.54	P	−26.3
N (diamide, imide)	−2.11	As(III)	−20.9
O (alcohol, ether)/O	−4.61	As(V)	−43.0
O (aldehyde, ketone)	+1.73	Sb(III)	−74.0
O_2 (carboxylate)	−7.95	Se	−23.0
F	−6.3		

Table 6.2: Pascal's constitutive corrections (λ) (all values in 10^{-6} cgs units).

Atom/Bond	χ_{dia}	Atom/Bond	χ_{dia}
C (in ring)	−0.24	C=N	+8.15
C (shared by two rings)	−3.07	C≡N	+0.8
C=C	+5.5	N=N	+1.8
C≡C	+0.8	N=O	+1.7

Since, in diamagnetic substances, the magnetic field induced is in the direction opposite to the direction of the magnetic field applied, χ_{dia} is always negative. However, χ_{dia} for bonds is positive.

Table 6.3: Diamagnetic susceptibilities of some cations (all values in 10^{-6} cgs units/mole).

Cations	χ_{dia}	Cations	χ_{dia}
Li^+	−1.0	Ag^+	−24.0
Na^+	−6.8	Ba^{2+}	−32.0
K^+	−14.9	Be^{2+}	−0.4
Rb^+	−22.5	Co^{2+}	−12.8
Cs^+	−35.5	Cd^{2+}	−22.0
NH_4^+	−13.3	Cu^{2+}	−12.8
Mg^{2+}	−5.0	Cr^{3+}	−11.0
Ca^{2+}	−10.4	Mn^{2+}	−14.0
Zn^{2+}	−15.0	Mn^{3+}	−10.0
Hg_{2+}	−40.0	Mo^{3+}	−23.0
V^{4+}	−12.0	Mo^{5+}	−12.0
Eu^{3+}	−20.0	Fe^{2+}	−12.8
Nd^{3+}	−20.0	Ni^{2+}	−12.8
Pr^{3+}	−20.0	Sn^{2+}	−20.0
Sm^{3+}	−20.0	Th^{4+}	−23.0
Yb^{3+}	−18.0		

Table 6.4: Diamagnetic susceptibilities of some anions (all values in 10^{-6} cgs units/mole).

Anions	χ_{dia}	Anions	χ_{dia}
F^-	−9.1	ClO_4^-	−32.0
Cl^-	−23.4	NO_2^-	−10.0
Br^-	−34.6	NO_3^-	−18.9
I^-	−50.6	OH^-	−12.0
CN^-	−13.0	O^{2-}	−7.0
CNS^-	−31.0	$PtCl_6^{2-}$	−148
CO_3^{2-}	−28.0	SO_4^{2-}	−40.1

Table 6.5: Diamagnetic susceptibilities of some common ligands (all values in 10^{-6} cgs units/mole).

Ligands	χ_{dia}	Ligands	χ_{dia}
Water	−13	Acetate	−30
Ammonia	−18	Glycinate	−37
Hydrazine	−20	Ethylenediamine	−46
Urea	−34	Oxalate	−24
Thiourea	−42	Acetylacetone	−52
Ethylene	−15	Pyridine	−49
Oxinate	−86	Dipyridyl	−105
Salene	−182	o-phenanthroline	−128
Diarsine	−194	Phthalocyanine	−422

6.9.1 Calculations of χ_{dia} of some compounds

The use of Pascal's constants can be illustrated by the following examples:

(i) Pyridine (C_5H_5N)

Atom correction ($\Sigma\chi_A$)

5 C atoms	$= 5 \times (-6.0 \times 10^{-6})$	$= -30.0 \times 10^{-6}$
5H atoms	$= 5 \times (-2.93 \times 10^{-6})$	$= -14.65 \times 10^{-6}$
1N (ring)	$= 1 \times (-4.61 \times 10^{-6})$	$= -4.61 \times 10^{-6}$

$$\Sigma\chi_A = -49.26 \times 10^{-6}$$

Constitutive correction ($\Sigma\lambda$)
$5 \times \text{ring } C = 5 \times (-0.24 \times 10^{-6}) = -1.20 \times 10^{-6}$

$$\chi_{dia} = \Sigma n_A \chi_A + \Sigma\lambda$$
$$= -49.26 \times 10^{-6} - 1.20 \times 10^{-6} = -50.46 \times 10^{-6} \text{ cgs units}$$

The experimental value of $\chi_{dia} = -49 \times 10^{-6}$ cgs units

(ii) Acetone, (CH₃)₂CO

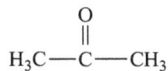

$$
\begin{array}{c}
\overset{\displaystyle O}{\underset{\displaystyle \|}{}} \\
H_3C-C-CH_3
\end{array}
$$

Atom correction ($\Sigma\chi_A$)

3C atoms	$= 3 \times (-6.0 \times 10^{-6})$	$= -18.0 \times 10^{-6}$	
6H atoms	$= 6 \times (-2.93 \times 10^{-6})$	$= -17.4 \times 10^{-6}$	
1O (Ketone)	$= 1 \times (+1.73 \times 10^{-6})$	$= +1.73 \times 10^{-6}$	

$$\Sigma\chi_A = -33.67 \times 10^{-6}$$

Constitutive correction ($\Sigma\lambda$)

Contributions from C–C and C–H and C=O bonds are taken to be zero (Table 6.2).

Therefore, χ_{dia} for acetone $= -33.67 \times 10^{-6}$ cgs units

(iii) CuSO₄ · 5H₂O

Atom correction ($\Sigma\chi_A$)

One Cu^{2+}	$= -12.8 \times 10^{-6}$
One SO_4^{2-}	$= -40.1 \times 10^{-6}$
Five H_2O molecules	$= 5 \times (-13 \times 10^{-6}) = -65 \times 10^{-6}$

$$\Sigma\chi_A = -117.9 \times 10^{-6} \text{ cgs units}$$

(iv) 1,10 or o-Phenanthroline (C₁₂H₈N₂)

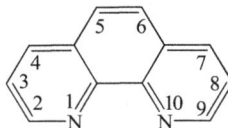

Atom correction ($\Sigma\chi_A$)

12 C atoms	$= 5 \times (-6.0 \times 10^{-6})$	$= -72.0 \times 10^{-6}$	
8H atoms	$= 8 \times (-2.93 \times 10^{-6})$	$= -23.44 \times 10^{-6}$	
2N (ring)	$= 2 \times (-4.61 \times 10^{-6})$	$= -9.22 \times 10^{-6}$	

$$\Sigma\chi_A = -104.66 \times 10^{-6}$$

Constitutive correction (ΣΛ)

8 ring C	$= 8 \times (-0.24 \times 10^{-6}) = -\ 1.92 \times 10^{-6}$
4 shared C	$= 4 \times (-3.07 \times 10^{-6}) = -12.28 \times 10^{-6}$

$$\Sigma\,\lambda = -14.20 \times 10^{-6}$$

Thus, χ_{dia} for o-Phenanthroline $= -104.66 \times 10^{-6} + (-14.20 \times 10^{-6})$
$$= -118.86 \times 10^{-6} \text{ cgs units}$$
The experimental value of $\chi_{dia} = -128 \times 10^{-6}$ cgs units

(v) Salicylideneglycine

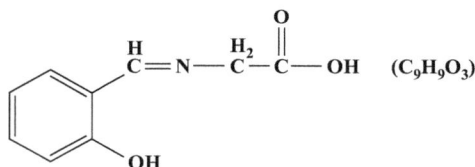

Atom correction (ΣχA)

9C atoms	$= 9 \times (-6.0\ \times 10^{-6}) = -54.0\ \times 10^{-6}$
9H atoms	$= 9 \times (-2.93 \times 10^{-6}) = -26.\,37 \times 10^{-6}$
O_2 (carboxylate)	$= 1 \times (-7.95 \times 10^{-6}) = -\ 7.95 \times 10^{-6}$
O	$= 1 \times (-4.61 \times 10^{-6}) = -\ 4.61 \times 10^{-6}$

$$\Sigma\chi_A = -92.93 \times 10^{-6} \text{ cgs units}$$

Constitutive correction (ΣΛ)

6 ring C	$= 6 \times (-0.24 \times 10^{-6}) = -1.44 \times 10^{-6}$
1C=N	$= 1 \times (+8.15 \times 10^{-6}) = +8.15 \times 10^{-6}$

$$\Sigma\lambda = +6.71 \times 10^{-6} \text{ cgs units}$$

$\chi_{dia} = -92.93 \times 10^{-6} + 6.71 \times 10^{-6}$
$= -86.22 \times 10^{-6}$ cgs units

(vi) Ethyl benzoylacetate/benzoylacetic acid ethyl ester ($C_6H_5COCH_2COOC_2H_5$)

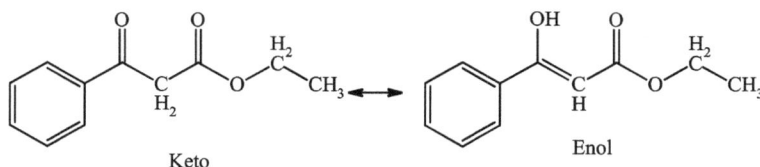

Keto

Enol

Ethyl benzoylacetate may exist in either ketone or enol form shown above. The dia-magnetic susceptibility (χ_{dia}) of **keto form** may be calculated as follows:

Atom correction ($\Sigma\chi_A$)

11C atoms	$= 11 \times (-6.00 \times 10^{-6})$	$= -66.00 \times 10^{-6}$
12H atoms	$= 12 \times (-2.93 \times 10^{-6})$	$= -35.16 \times 10^{-6}$
O (carboxylate)	$= 1 \times (-7.95 \times 10^{-6})$	$= -7.95 \times 10^{-6}$
O (alcohol)	$= 1 \times (-4.61 \times 10^{-6})$	$= -4.61 \times 10^{-6}$
1O atom (ketone)	$= 1 \times (+1.73 \times 10^{-6})$	$= +1.73 \times 10^{-6}$

$$\Sigma\chi_A = -111.99 \times 10^{-6} \text{ cgs units}$$

Constitutive correction ($\Sigma\lambda$)

6 ring C $\qquad = 6 \times (-0.24 \times 10^{-6}) = -1.44 \times 10^{-6}$ cgs units

χ_{dia} **(Total) (Keto form):** $= -111.99 \times 10^{-6} -1.44 \times 10^{-6}$
$$= -113.43 \times 10^{-6} \text{ cgs units}$$

Enol form

Atom correction ($\Sigma\chi_A$)

The diamagnetic susceptibility (χ_{dia}) of **enol form** may be calculated as follows:

11C atoms	$= 11 \times (-6.0 \times 10^{-6})$	$= -66.0 \times 10^{-6}$
12H atoms	$= 12 \times (-2.93 \times 10^{-6})$	$= -35.16 \times 10^{-6}$
O (carboxylate)	$= 1 \times (-7.95 \times 10^{-6})$	$= -7.95 \times 10^{-6}$
2O (alcohol)	$= 2 \times (-4.61 \times 10^{-6})$	$= -9.22 \times 10^{-6}$

$$\Sigma\chi_A = -118.33 \times 10^{-6} \text{ cgs units}$$

Constitutive correction ($\Sigma\lambda$)

6 ring C	$= 6 \times (-0.24 \times 10^{-6}) =$	-1.44×10^{-6}
1C=C	$= 1 \times (+5.5 \times 10^{-6}) =$	$+5.5 \times 10^{-6}$

$$\Sigma\lambda = +4.06 \times 10^{-6} \text{ cgs units}$$

χ_{dia} **(Total) Enol** $\qquad = -118.33 \times 10^{-6} + 4.06 \times 10^{-6}$
$$= -114.27 \times 10^{-6} \text{ cgs units}$$

The measured molar susceptibility for ethyl benzoylacetate is −115.2 cgs units.

6.9.2 Utility of Pascal's constants

We often come across a ligand which can be directly synthesized and isolated. The diamagnetic susceptibility of such a ligand can be measured directly. However, there are ligands which cannot be synthesized and isolated. Such a ligand can be synthesized in the form of a metal complex by the metal ion-promoted template reaction. The Schiff base salicylidenglycine has not been isolated. However, its complex can be synthesized by reacting metal acetate, salicylaldehyde and glycine in aqueous alcohol. To calculate the diamagnetic susceptibility of such ligands, the only alternative way is to take help from Pascal's constants of atoms and Pascal's constitutive corrections.

6.10 Origin of magnetism: orbital and spin motion

The study of the magnetic properties of coordination compounds has contributed a lot to our understanding of the chemistry of transition metal complexes, the stereochemistry of the central metal ion and the nature of metal–ligand bonding. It is of interest, therefore, to describe the origin of the type of magnetic behaviour by transition metal complexes.

An electron has two motions associated with it. One of these is its motion around the nucleus in an orbit, called the orbital motion. Since the electron is itself charged, this orbital motion is equivalent to the movement of a charge along a loop. This generates a magnetic field perpendicular to the plane of the orbit. The other motion of the electron is the spinning of the electron about its own axis, which too is equivalent to the movement of a charge and generates a magnetic field.

The overall magnetic moment of an atom is the resultant of the magnetic moment generated by the orbital motion of the electrons, called orbital magnetic moment (μ_l), and the magnetic moment generated by the spin motion of the electrons, called spin magnetic moment (μ_s). The generation of these two types of magnetic moments are pictorially shown in Figure 6.15.

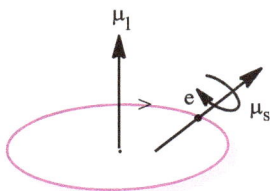

Figure 6.15: Orbital and spin motion of electrons generating orbital and spin magnetic moments.

6.10.1 Orbital magnetic moments (μ_l)

A negatively charged electron moving in a circular orbit is similar to the flow of the electric current in a circular wire. The electron motion in an orbit, therefore, generates a magnetic field in a direction perpendicular to the plane of the orbit containing the electron. The moment of the magnetic field so generated is given by the product of the strength of the current and surface area of the orbit. Thus,

Orbital magnetic moment (μ_l) = strength of the current × surface area of the orbit

$$\mu_l = \frac{ew}{2\pi c} \times \pi r^2 = \frac{ewr^2}{2c} \tag{6.16}$$

where r is the radius of the orbit, w is the angular velocity of the electron, c is the velocity of light and e is the electronic charge.

An electron moving in an orbit generates orbital angular momentum, (l), which is given by

$$(l) = \sqrt{l(l+1)} \cdot \frac{h}{2\pi} \tag{6.17}$$

where l is the **orbital angular momentum quantum number**, which is distinctly different from the orbital angular momentum, (l).

Since the orbital angular momentum of a particle of mass m moving in a circle of radius r is given by mwr^2. Hence,

$$(l) = mwr^2 = \sqrt{l(l+1)} \cdot \frac{h}{2\pi} \tag{6.18}$$

$$\text{or} \qquad wr^2 = \sqrt{l(l+1)} \cdot \frac{h}{2\pi m}$$

Substituting the value of wr^2 in equation (6.16), we get

$$\text{Orbital angular momentum, } \mu_l = \frac{e}{2c}\sqrt{l(l+1)} \cdot \frac{h}{2\pi m}$$
$$= \sqrt{l(l+1)} \cdot \frac{eh}{4\pi mc} \tag{6.19}$$
$$= \sqrt{l(l+1)} \text{ B.M.}$$

The magnetic moment (μ_l) is measured in the unit of Bohr Magneton (B.M.), which is equal to $eh/4\pi mc$. When the appropriate values are substituted, we get 1 B.M. = 0.9723×10^{-20} erg.gauss^{-1}. Thus, for an electron moving in an orbit around the nucleus, μ_l in the units of Bohr magneton and orbital angular momentum (l) in the units of $h/2\pi$ have the same magnitude, $viz.,$. $\sqrt{l(l+1)}$.

6.10.2 Spin magnetic moment (μ_s)

The negatively charged electron in its spin motion produces magnetic field whose moment is directed along the spin axis. It is postulated that the ratio of spin magnetic moment μ_s to the spin angular momentum (s) is twice the ration of μ_l and (l) in their respective unit of B.M. and $h/2\pi$. Since the ration of μ_l to (l) in their respective unit is 1, the ratio of μ_s to (s) should be 2. The ratio of spin magnetic moment μ_s to the spin angular momentum (s) is called **Lande splitting factor**, often called simply the 'g factor'. Thus,

$$\frac{\mu_s}{(s)} = \frac{\mu_s}{\sqrt{s(s+1)}} = g$$

$$\text{or} \quad \mu_s = g\sqrt{s(s+1)} = 2\sqrt{s(s+1)} \text{ B.M.} \quad \left(\text{Because } \frac{\mu_s}{(s)} = 2\right)$$

$$= \sqrt{4s(s+1)}$$

The spin angular momentum, (s) is equal to:

$$\sqrt{s(s+1)} \cdot h/2\pi$$

Again s and (s) represent different quantity. While (s) represents **spin angular momentum, s** represents spin angular momentum quantum number.

When we apply an external magnetic field, both μ_l and μ_s will interact with the applied magnetic field. Like small bar magnet, they tend to align themselves with the applied magnetic field and thereby reinforcing the applied magnetic field.

In a multi-electron atom, which is of great interest to us, the spin magnetic moments of the individual electrons will interact with one another to give a resultant or an overall magnetic moment. In such a case, we shall have to use the resultant spin angular momentum quantum number quantum number **S (Capital S)** for calculating resultant spin magnetic moment μ_S (**capital S**). Similarly, we shall have to use the resultant orbital angular momentum quantum number L for calculating resultant orbital magnetic moment μ_L.

The overall magnetic moment μ_{S+L} for a multi-electron system will be the resultant of μ_S and μ_L provided (L) and (S) vectors do not couple or mix with each other. Thus,

$$\mu_{S+L} = \sqrt{L(L+1) + 4S(S+1)} \text{ B.M.}$$

In case (L) and (S) vectors couple, the magnetic moment of the system will be calculated from the resultant angular momentum quantum number J. Thus,

$$\mu_J = g\sqrt{J(J+1)} \text{ B.M.}$$

$$\text{where} \quad g = 1 + \frac{J(J+1) + S(S+1) - L(L+1)}{2J(J+1)}$$

In this case, **g** is known as Lande splitting factor since Lande derived this equation.

6.11 Quenching of orbital angular momentum: orbital contribution to magnetic moment of O_h, T_d and square planar complexes

6.11.1 What is quenching of orbital angular momentum?

In a number of coordination compounds the average magnetic moment associated with its orbital angular momentum (because of its orbital motion of electrons) is reduced to zero in presence of certain electrostatic ligand fields is called quenching of orbital angular momentum. In fact, the presence of certain electrostatic ligand fields in coordination compounds causes the orbit of electrons to precess rapidly so that the average magnetic moment associated with its orbital angular momentum is reduced to zero.

6.11.2 Quenching of orbital angular momentum: explanation

There are two types of angular momenta of an unpaired electron: (i) spin angular momentum and (ii) orbital angular momentum. The spin angular momentum of an unpaired electron is not affected by change in stereochemistry of the complexes. However, orbital angular momentum of an unpaired electron is quenched under certain conditioned that are explained as follows:

6.11.3 Octahedral and tetrahedral complexes

Five d-orbitals in a free ion are degenerate and hence the electrons present in d-orbitals have magnetic moment. According to crystal field theory, when the metal ion is surrounded octahedrally by six or tetrahedrally by four ligands, the degeneracy of the d-orbitals is lifted, wherein the metal ion are split into two sets of orbitals which have different energy as shown in Figure 6.16. All the orbitals in each set are degenerate.

According to the wave mechanical model of atom, an electron will possess an orbital angular momentum along a given axis if it is possible, by rotation about this axis, to transfer the orbital that it occupies into another orbital which is equivalent to it in shape, size and energy. Let us now observe the fate of these two sets of orbitals with regard to orbital angular momentum.

Although $d_{x^2-y^2}$ and d_{z^2} orbitals of e_g set are degenerate in octahedral (O_h) complexes, but due to different shape and size (Figure 6.17), these orbitals cannot be

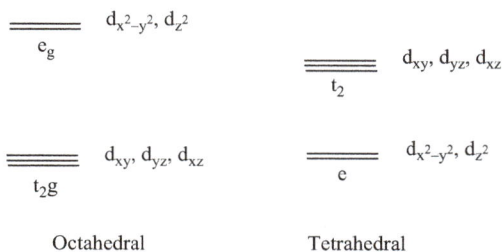

Figure 6.16: Splitting of d-orbitals in octahedral and tetrahedral ligand fields.

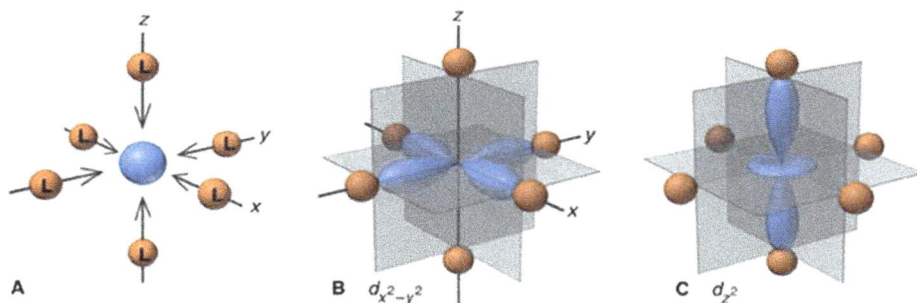

Figure 6.17: Different shape and size of $d_{x^2-y^2}$ and d_{z^2} orbitals (e_g set).

transferred in to one another by rotation. Thus, the electron present in the orbital of e_g set cannot generate any orbital magnetic moment to the overall magnetic moment. Hence $d_{x^2-y^2}$ and d_{z^2} orbitals are said to be non-magnetic doublet.

Contrary to the above, each of the three orbitals of lower energy t_{2g} set (d_{xy}, d_{yz}, d_{xz} orbitals) (of equivalent shape, size and energy) (Figure 6.18) in O_h complexes is transferred into the other two orbitals through rotation by an angle of 90° about the relevant axis. Hence, the electrons present in these orbitals will generate orbital

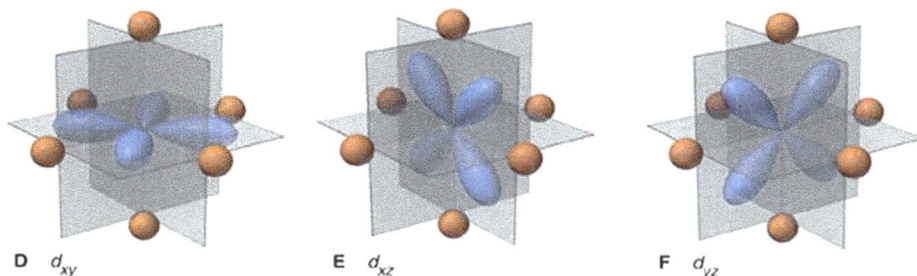

Figure 6.18: t_{2g} Set, d_{xy}, d_{yz} and d_{xz} orbitals of equivalent shape, size and energy.

contribution to the magnetic moment provided that these orbitals do not contain electrons in the same spin. Thus, the electrons present in $t_{2g}3$ and $t_{2g}6$ configurations do not contribute any orbital magnetic moment to the overall magnetic moment.

The Table 6.6(a) given below provides which of the of the d^n configuration of metal ion in high-spin octahedral and tetrahedral complexes has orbital magnetic moment contribution and which does not have orbital contribution to the overall magnetic moment (μ_{eff}). Orbital contribution of different d^n configuration of metal ion in low-spin octahedral complexes is also given in the another Table 6.6(b). It is notable here that the tetrahedral crystal field being weak cannot produce any low-spin complex. Both the tables also include the ground state term symbol for each d^n ion. Ground state terms for high-spin complexes (d^1 to d^4 and d^6 to d^9) and d^5 high-spin compleces can be seen from Orgel diagrams given in Figures 6.19 and 6.20, respectively. Also Ground state terms for low-spin complexes can be seen from Tanabe–Sugano digrams given in Figures 6.21 and 6.22.

Table 6.6(a): Orbital contribution (expected/non-expected) in high-spin octahedral and tetrahedral complexes.

d^n ion (n = 1 to 9)	High-spin octahedral complexes		Tetrahedral complexes	
	$t_{2g}{}^p e_g{}^q$ configuration (Ground state)	Orbital contribution	$e^p t_2{}^q$ configuration (Ground state)	Orbital contribution
d^1	$t_{2g}{}^1 e_g{}^0$ ($^2T_{2g}$)	Expected	$e^1 t_2{}^0$ (^2E)	Not Expected
d^2	$t_{2g}{}^2 e_g{}^0$ ($^3T_{1g}$)	Expected	$e^2 t_2{}^0$ ($^3A_2$)	Not Expected
d^3	$t_{2g}{}^3 e_g{}^0$ ($^4A_{2g}$)	Not Expected	$e^2 t_2{}^1$ ($^4T_1$)	Expected
d^4	$t_{2g}{}^3 e_g{}^1$ (^5E_g)	Not Expected	$e^2 t_2{}^2$ ($^5T_2$)	Expected
d^5	$t_{2g}{}^3 e_g{}^2$ ($^6A_{1g}$)	Not Expected	$e^2 t_2{}^3$ ($^6A_1$)	Not Expected
d^6	$t_{2g}{}^4 e_g{}^2$ ($^5T_{2g}$)	Expected	$e^3 t_2{}^3$ (^5E)	Not Expected
d^7	$t_{2g}{}^5 e_g{}^2$ ($^4T_{1g}$)	Expected	$e^4 t_2{}^3$ ($^4A_2$)	Not Expected
d^8	$t_{2g}{}^6 e_g{}^2$ ($^3A_{2g}$)	Not Expected	$e^4 t_2{}^4$ ($^3T_1$)	Expected
d^9	$t_{2g}{}^6 e_g{}^3$ (^2E_g)	Not Expected	$e^4 t_2{}^5$ ($^2T_2$)	Expected

Table 6.6(b): Orbital contribution in low-spin octahedral and tetrahedral complexes.

d^n ion	Low-spin octahedral complexes		Tetrahedral complexes
	$t_{2g}^p e_g^q$ configuration (Ground state)	Orbital contribution	
d^4	$t_{2g}^4 e_g^0$ ($^3T_{1g}$)	Expected	Low-spin tetrahedral complexes are not known
d^5	$t_{2g}^5 e_g^0$ ($^2T_{2g}$)	Expected	
d^6	$t_{2g}^6 e_g^0$ ($^1A_{1g}$)	Not Expected	
d^7	$t_{2g}^6 e_g^1$ (^2E_g)	Not Expected	
d^8	$t_{2g}^6 e_g^2$ ($^3A_{2g}$)	Not Expected	
d^9	$t_{2g}^6 e_g^3$ (^2E_g)	Not Expected	

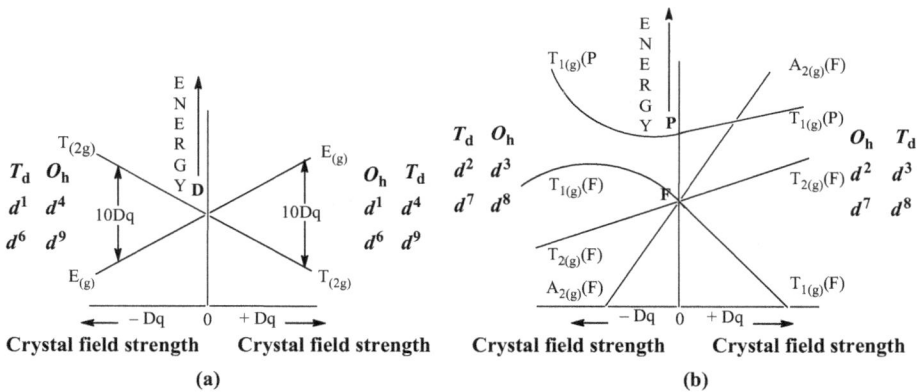

Figure 6.19: Orgel diagrams showing splitting of the terms of d^1 to d^9 for O_h and T_d complexes: (a) d^1, d^9; high-spin d^4 and d^6 and (b) d^2, d^3, d^8 and high-spin d^7.

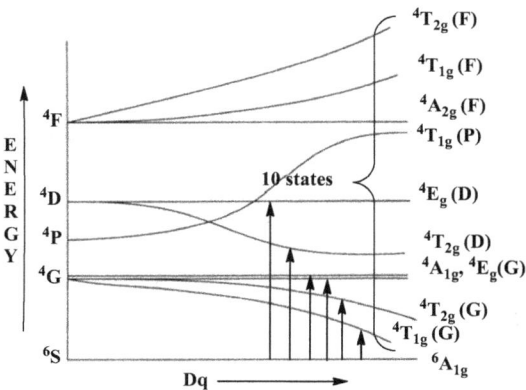

Figure 6.20: Orgel diagram for d^5 configuration showing only quartet terms.

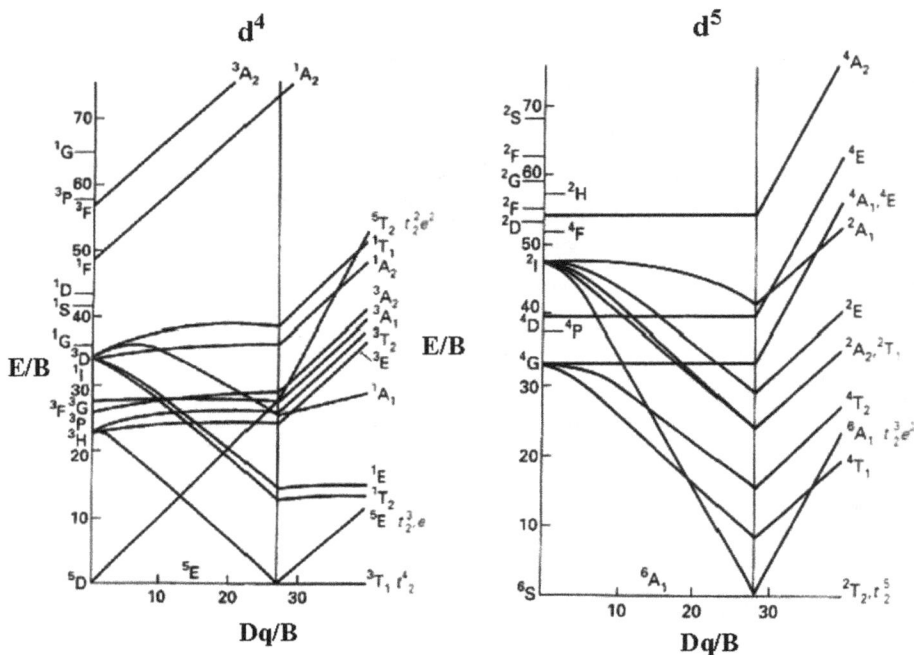

Figure 6.21: Tanabe–Sugano diagram for low-sin complexes of d^4 and d^5 configurations.

In both the tables:

(i) The $t_{2g}^p e_g^q$ and $e^p t_2^q$ are the electronic configurations and the respective ground states in high-spin and low-spin complexes are given in bracket. The ground state of d^1 to d^9 systems can be obtained from the respective Orgel diagram of high-spin complexes (Figures 6.19 and 6.20) and Tanabe–Sugano diagram of low-sin complexes of d^4–d^7 (Figures 6.21 and 6.22). For d^8 and d^9 configurations, the ground state remains the same in high-spin and low-spin complexes.

(ii) The ground state term A is singlet, E is doublet and T is triplet. The left hand superscript is spin multiplicity of the ground state. The right hand subscript g indicates *gerade*, that is, even orbitals which retain the same sign of orbital wave function on inversion about the centre of octahedron. A tetrahedron has no inversion centre and hence g subscript for the ground states of the tetrahedron complexes is dropped.

(iii) Tables 6.6(a) reveals that in both types of complexes (octahedral and tetrahedral), the configurations having T as the ground state have orbital contribution but the configurations having A and E as their ground state is quenched. The same situation is with Table 6.6(b) also in octahedral complexes.

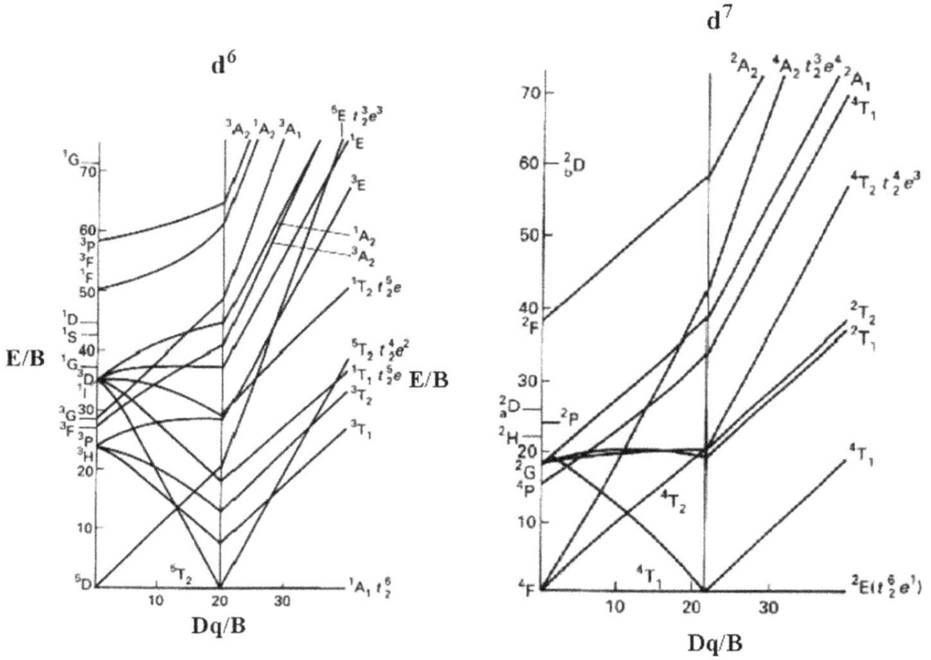

Figure 6.22: Tanabe–Sugano diagram of low-sin complexes of d^6 and d^7 configurations.

(iv) We can thus generalize that the orbital angular momentum is completely quenched when the metal ion in its complexes has A or E as the ground state, and it is partially quenched when the metal ion in its complexes has T as the ground state. This explains why in octahedral Co(II) (d^7), octahedral Fe(II) (d^6) and tetrahedral Ni(II) (d^8), the magnetic moment is higher than the spin only magnetic moment, (μ_s) [$\mu_s = \sqrt{n(n+2)}$]. The experimental and spin only magnetic moments of some Co(II), Fe(II) octahedral and Ni(II) tetrahedral complexes are given in Table 6.7.

Table 6.7: Observed magnetic moments and spin only magnetic moments of some complexes.

Compound	No. of unpaired electrons (n)	Spin only magnetic moment (μ_s) (B.M.)	μ_{eff} (Experimental) (B.M.)
[Fe^{II}(NH$_3$)$_4$(Cl)$_2$] (d^6) ($^5T_{2g}$)	4	4.90	5.45
[Co^{II}(en)$_3$]SO$_4$ (d^7) ($^4T_{1g}$)	3	3.87	4.56
[Co^{II}(aniline)$_2$(Cl)$_2$(EtOH)$_2$](d^7) ($^4T_{1g}$)	3	3.87	5.0
[Co^{II}(aniline)$_2$(Br)$_2$(EtOH)$_2$] (d^7) ($^4T_{1g}$)	3	3.87	5.0
(Et$_4$N)$_2$[Ni^{II}Cl$_4$] (d^8) e^{4}t$_2$^{4} ($^3T_1$)	2	2.83	3.89

(v) Although in most of the ions with A and E as ground state, the experimentally observed μ_{eff} is very close to μ_s and is temperature independent, yet there are some metal ions for which experimental value of μ_{eff} are different from their μ_s values. For example, octahedral Ni(II) and tetrahedral Co(II) complexes exhibit magnetic moment substantially higher than μ_s as shown in Table 6.8. This can be explained as follows:

Table 6.8: Observed magnetic moments and spin-only magnetic moments of some complexes.

Compound	No. of unpaired electrons (n)	Spin only magnetic moment (μ_s) (B.M.)	μ_{eff} (Experimental) (B.M.)
$[Ni^{II}(NH_3)_6]Cl_2$ (d^8) ($^3A_{2g}$)	2	2.83	3.32
$[Ni^{II}(H_2O)_6]SO_4$(d^8) ($^3A_{2g}$)	2	2.83	3.23
$[Co^{II}Cl_4]^{2-}$ (d^7) ($^4A_2$)	3	3.87	4.59
$[Co^{II}Br_4]^{2-}$ (d^7) ($^4A_2$)	3	3.87	4.69
$[Co^{II}I_4]^{2-}$ (d^7) ($^4A_2$)	3	3.87	4.77

In fact, in such cases, a low-lying excited T-state (of the same multiplicity as the ground state) mixes up due to spin-orbit coupling with A ground state and so A state will have some character of the T-state. In other words, the ground state A state in such cases will retain some orbital angular momentum. In case of octahedral Ni(II) and tetrahedral Co(II) complexes such low-lying excited T state of multiplicity similar to ground state are given in Table 6.9: The Orgel diagrams showing ground state and low-lying excited state of d^8 octahedral and d^7 tetrahedral systems are shown in Figures 6.23 and 6.24.

The contribution of orbital magnetic moment to overall μ_{eff} in such cases depends upon the extent of mixing of ground state with the excited state and is, therefore, inversely proportional to $\Delta = 10\ Dq$, the energy difference between the ground and excited state. Taking this into account, the following equation has been suggested for μ_{eff}.

$$\mu_{eff} = \mu_s\left(1 - \frac{\alpha\lambda}{10Dq}\right) \quad (6.19)$$

Here α is a constant depending the spectroscopic ground term and the number of d-electron. The $\alpha = 2$ for an E ground term, $\alpha = 4$ for A_2 ground term and $\alpha = 0$ for A_1 ground term. Values of α for some metal ions are given as: d^1 (Ti^{3+}) = 2; d^2 (V^{3+}) = 4; d^3 (Cr^{3+}) = 4; d^4 (Cr^{2+}, Mn^{3+}) = 4; d^5 (Fe^{3+}, Mn^{2+}) = 0; d^6 (Fe^{2+}) = 2; d^7 (Co^{2+}) = 4; d^8 (Ni^{2+}) = 4; d^9(Cu^{2+}) = 2. λ is the spin–orbit coupling parameter which is +ve for d^n configurations with n < 5 and −ve for d^n configurations with n > 5.

Table 6.9: Ground state and low-lying excited state and of octahedral Ni(II) and tetrahedral Co(II) complexes.

Compound	Ground state	Low-lying excited state
Octahedral Ni(II) (d^8)	$^3A_{2g}$	$^3T_{2g}$
Tetrahedral Co(II) (d^7)	$^4A_2$	$^4T_2$

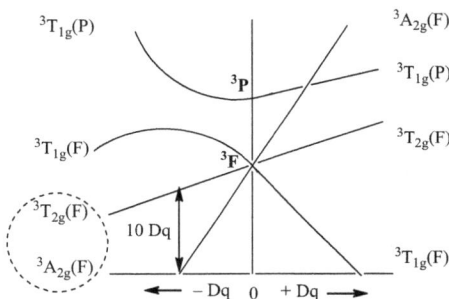

Figure 6.23: Orgel diagram for d^8 metal ion in O_h crystal field.

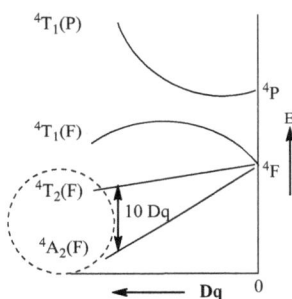

Figure 6.24: Orgel diagram for d^7 metal ion in T_d ligand field.

Let us consider the following cases:

Case of d^3 and d^4 ions in high-spin octahedral complexes

For d^3 ion, since ground state is $^4A_{2g}$, $\alpha = 4$ and λ is positive, equation (6.19) can be written as

$$\mu_{eff} = \mu_s \left(1 - \frac{4\lambda}{10Dq}\right) \tag{6.20}$$

For d^4 ion, since ground state is ^5E_g, $\alpha = 2$ and λ is positive, equation (6.19) becomes,

$$\mu_{eff} = \mu_s \left(1 - \frac{2\lambda}{10Dq}\right) \tag{6.21}$$

Above discussion suggests that for high-spin octahedral complexes of d^3 and d^4 ions, the experimental value of μ_{eff} is lower than the μ_s. Therefore, the value of for d^3 and d^4 ions should be calculated with the help of equation (6.20) and (6.21), respectively.

Case of d^8 and d^9 ions in high-spin octahedral complexes

For d^8 ion, since ground state is $^3A_{2g}$, $\alpha = 4$ and λ is negative, equation (6.19) becomes,

$$\mu_{eff} = \mu_s \left(1 + \frac{4\lambda}{10Dq}\right) \tag{6.22}$$

For d^9 ion, since ground state is ^2E_g, $\alpha = 2$ and λ is negative, equation (6.19) becomes,

$$\mu_{eff} = \mu_s\left(1 + \frac{2\lambda}{10Dq}\right) \qquad (6.23)$$

Above discussion suggests that for high-spin octahedral complexes of d^8 and d^9 ions, the experimental value of μ_{eff} is higher than the μ_s. Therefore, the value of for d^8 and d^9 ions should be calculated with the help of equation (6.22) and (6.23), respectively.

The values of μ_{eff}, for d^3, d^4, d^8 and d^9 ions are independent of temperature as the energy gap between (A/E) and the first excited state T is far greater than kT.

Case of d^5 ion in high-spin octahedral complexes

For d^5 ion, since ground state is $^6A_{1g}$, $\alpha = 0$ and λ is positive, equation (6.19) becomes,

$$\mu_{eff} = \mu_s\left(1 + \frac{0 \cdot \lambda}{10Dq}\right) \qquad (6.24)$$

or $\mu_{eff} = \mu_s$

This shows that in case of d^5 ion, μ_{eff} and μ_s are the same. In fact, the ground state of high-spin (HS) octahedral complexes of d^5 is $^6A_{1g}$, and they do not have any excited state of the same multiplicity as shown in Figure 6.25. Consequently, the mixing of ground state with excited state due to spin–orbit coupling is not possible in this case. Thus, there is no orbital contribution to μ_{eff} in this ion. Hence, HS octahedral complexes of d^5 ion have μ_{eff} (experimental) equal to μ_s.

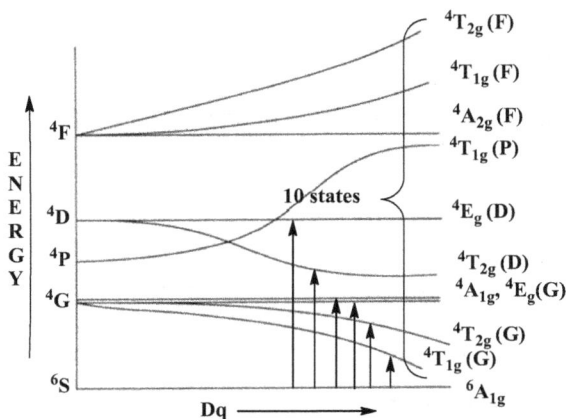

Figure 6.25: Orgel diagram for d^5 configuration showing only quartet terms.

6.11.4 Orbital contribution in square planar complexes

We know that since $d_{x^2-y^2}$ and d_{z^2} orbitals (e_g set) in octahedral complexes cannot be transformed into one another and so these orbitals cannot contribute to the overall magnetic moment value. It is only d_{xy}, d_{yz} and d_{xz} orbitals (t_{2g} set) which contribute to the overall μ_{eff} value. In square planar complexes, there is further splitting of the t_{2g} set (Figure 6.26), and hence the scope of orbital contribution from t_{2g} set is minimized.

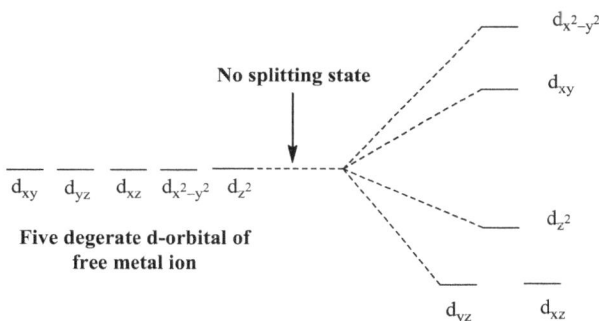

Figure 6.26: Splitting of d-orbitals of the central metal ion in square planar geometry.

Now, since d_{xy}, d_{yz} and d_{xz} orbitals are no longer of the same energy in square planar complexes, there is genuine energy barrier from the movement of electrons among these orbitals. Thus, in square planar complexes, μ_{eff} value is likely to be closed to μ_s value. In other words, we can say that orbital contribution in square planar complexes is not more likely than in octahedral complex.

6.12 Magnetic properties of paramagnetic substances

We are aware of the origin of the Russell–Saunders terms for the ground state and the excited states. The magnetic property of a paramagnetic substance is dictated by the population of the molecules in both these types of states. Here, we shall consider this population of molecules in the ground state and the excited states and derive the magnetic moment equations for paramagnetic compounds.

6.12.1 Thermal energy and magnetic property

According to Russell–Saunders coupling scheme, a particular term with given spin multiplicity and orbital multiplicity, can assume several J values which range from

(L+S) to (L–S) or (S–L) whichever is positive. Thus, a multi-electron system possesses several J levels, the ground state J being decided by Hund's rules. The J levels of a given Russell–Saunders term constitute a multiplet, and a given J level of the multiplet is called a component. The energy gap or separation (ΔE) between two successive J levels is known as multiplet width.

Normally, the magnetic property of a substance originates from its ground state. But this is not so when its excited states lie close to its ground state, that is, when the multiplet width is in the neighbourhood of kT, the thermal energy. In such a situation, the excited state or states may be significantly populated. Therefore, when deriving the magnetic moments equation, we need to consider also the population of the excited states. Such a consideration leads to three distinct situations, namely: (i) multiplet widths large as compared to kT, the energy separation (ΔE) between J states are greater than kT (ii) multiplet widths small as compared to kT, that is, the energy separation (ΔE) between J states are very small compared to kT and (iii) multiplet widths comparable to kT, that is, the energy separation (ΔE) between J states are comparable to kT (Figure 6.27).

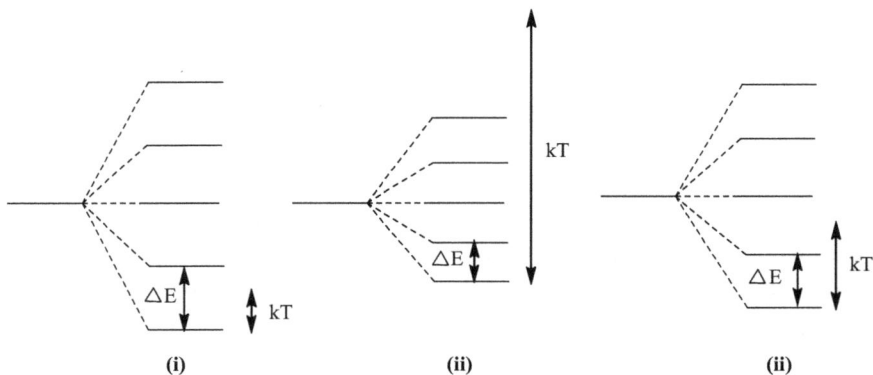

Figure 6.27: Multiplet widths may be wide, narrow or comparable to kT.

Now, we will derive the equation of magnetic moment for different multiplet widths.

6.12.2 Multiplet widths large as compared to kT

In this case the spin-orbit coupling constant λ is quite high so that the L and S vectors interact strongly. This means that these vectors precess rapidly about the direction of the resultant J vector. In this situation J becomes a good quantum number such that the quantum numbers L and S no longer dictate the ultimate magnetic properties. It then follows that the lowest lying component, that is, lowest lying J states alone is populated by metal ions. The coupling of the L and S vectors and the resulting magnetic moments are shown in Figure 6.28(a). It should be noted that the direction of momentum and the resulting magnetic moment are opposite to each other because the latter is generated by the flow of the positive current, and the momentum by the motion of the negatively charged electrons. It is also to be noted that the behaviour of the spin being anomalous, the spin moment vector is taken as double of the spin momentum vector.

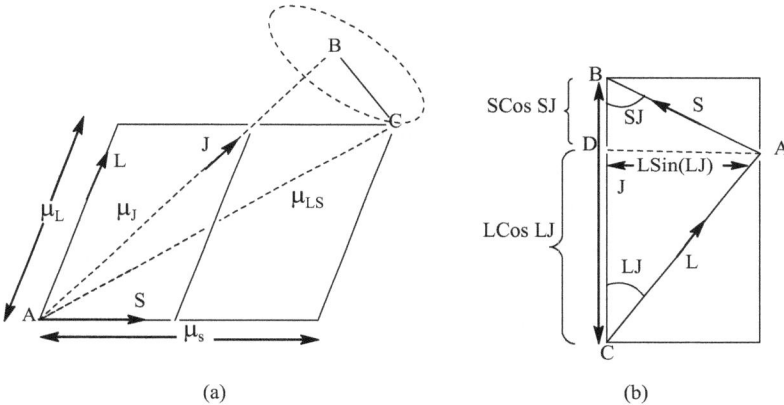

Figure 6.28: (a) Coupling of L and S vectors. (b) Components of L and S vectors along J vector.

The resulting magnetic moment μ_{LS} is given by the vector AC. Since the vectors L and S are precessing rapidly about the direction of J, we can assume that μ_L and μ_S are precessing rapidly about J. Effectively, μ_{LS} is precessing about J, so that, in any finite time, the vector BC averages out to zero. Thus, finally, μ_{LS} is approximately given by μ_J, that is, the AB vector.

With the help of Figure 6.28(b), we may derive the following two equations:

In $\triangle$ ABD: $S^2 = AD^2 + BD^2$ or $S^2 = AD^2 + (J - DC)^2 = [L\,\mathrm{Sin}\,(LJ)]^2 + [J - L\,\mathrm{Cos}(LJ)]^2$

$$= L^2\,\mathrm{Sin}^2(LJ) + J^2 + L^2\,\mathrm{Cos}^2(LJ) - 2LJ\,\mathrm{Cos}(LJ)$$

$$= L^2\left[\mathrm{Sin}^2(LJ) + \mathrm{Cos}^2(LJ)\right] + J^2 - 2LJ\,\mathrm{Cos}(LJ)$$

$$S^2 = L^2 + J^2 - 2LJ\,\mathrm{Cos}(LJ) \quad \left[\mathrm{As}\left\{\mathrm{Sin}^2(LJ) + \mathrm{Cos}^2(LJ)\right\} = 1\right]$$

(6.25)

Similarly,

In $\triangle$ ADC: $L^2 = AD^2 + DC^2$ or $L^2 = AD^2 + (J - BD)^2 = [S\mathrm{Sin}(SJ)]^2 + [J - S\,\mathrm{Cos}(SJ)]^2$

$$= S^2\,\mathrm{Sin}^2(SJ) + J^2 + S^2\,\mathrm{Cos}^2(SJ) - 2JS\,\mathrm{Cos}(SJ)$$

$$= S^2\left[\mathrm{Sin}^2(SJ) + \mathrm{Cos}^2(SJ)\right] + J^2 - 2JS\,\mathrm{Cos}(SJ)$$

$$\text{or } L^2 = S^2 + J^2 - 2JS\,\mathrm{Cos}(SJ) \quad \left[\mathrm{As}\left\{\mathrm{Sin}^2(SJ) + \mathrm{Cos}^2(SJ)\right\} = 1\right]$$

(6.26)

From equations (6.25) and (6.26), we have

$$L\,Cos(LJ) = L^2 + J^2 - S^2/2J \text{ and}$$

$$S\,Cos(SJ) = S^2 + J^2 - L^2/2J$$

Here, L Cos(LJ) is the component of L in the direction of J and S Cos(SJ) is the component of S in the direction of J.

Referring to Figure 6.28(a), we see that

Effective magnetic moment (μ_J) = Vector AB

Vector AB = Moment corresponding to the component of L in the direction of J

+ Twice the moment corresponding to the component of S in the direction of J (due to anomalous spin behaviour)

$= \beta L Cos(LJ) + 2\beta S Cos(SJ)$ (β = Bohr magneton, which is a unit of magnetic)

$= [\{L^2 + J^2 - S^2/2J\} + 2\{S^2 + J^2 - L^2/2J\}]\beta$

$= [L^2 + J^2 - S^2 + 2J^2 + 2S^2 - 2L^2]\beta/2J$

$= [3J^2 + S^2 - L^2]\beta/2J$

$= [3J^2/2J.J + (S^2 - L^2)/2J.J]\,\beta J$ (J is multiplied in numerator and denominator

$= [3/2 + (S^2 - L^2)/2J^2]\beta J$ and then taken commom outside with β)

$= [1 + 1/2 + (S^2 - L^2)/2J^2]\beta J$

$= [1 + (J^2 + S^2 - L^2)/2J^2]\beta J$

or $\mu_J = g\beta J$ (6.27)

$$[\text{Here } g = 1 + (J^2 + S^2 - L^2)/2J^2] \qquad (6.28)$$

Substituting $S = [S(S+1)]^{1/2} \cdot h/2\pi$, $L = [L(L+1)]^{1/2} \cdot h/2\pi$, and $J = [J(J+1)]^{1/2} \cdot h/2\pi$ (S, L and J indicate the respective quantum number in the unit of h/2π) in equation (6.28), we get

$$g = 1 + [J(J+1) + S(S+1) - L(L+1)/2J(J+1)] \qquad (6.29)$$

Putting the value of J $\{J = [J(J+1)]^{1/2} \cdot h/2\pi\}$ in equation (6.27), we get

$$\mu_J = g\beta J = g\beta[J(J+1)]^{1/2} \text{ e.m.u.}$$

or $\mu_J = g[J(J+1)]^{1/2}$ B.M.   (6.30)

$$\mu_{eff} = g\sqrt{J(J+1)} \text{ B.M.}$$

Here, h/2π is included in β, the unit of magnetic moment.

From equation (6.30), we can arrive at the equations for μ_L and μ_S easily.

For orbital magnetic moment only, $S = 0$, so that $J = L$, and hence $g = 1$ as clarified below:

$$g = 1 + [J(J+1) + S(S+1) - L(L+1)]/2J(J+1)]$$
$$= 1 + [L(L+1) + 0 - L(L+1)]/2L(L+1)$$
$$= 1$$

$$\therefore \mu_L = 1 \times [L(L+1)]^{1/2} \text{ B.M.} = [L(L+1)]^{1/2} \text{ B.M.}$$

or $\mu_L = \sqrt{L(L+1)}$ B.M.

For spin only magnetic moment, $L = 0$, so that $J = S$, and hence $g = 2$ as clarified below:

$$g = 1 + [S(S+1) + S(S+1) - 0]/2S(S+1)]$$
$$= 1 + 1$$
$$= 2$$

$$\therefore \mu_s = 2 \times [S(S+1)]^{1/2} \text{ B.M.} = 2[S(S+1)]^{1/2} \text{ B.M.}$$

We know that $S = n.1/2$ where n stands for number of unpaired electron.

$$\therefore \mu_s = 2[S(S+1)]^{1/2} = [4S(S+1)]^{1/2} = [4 \times n/2(n/2+1)]^{1/2}$$
$$= [2n(n/2+1)]^{1/2} = [2n(n+2)/2)]^{1/2}$$
$$= n(n+2)^{1/2} \text{ B.M.}$$

or $\mu_s = \sqrt{n(n+2)}$

This category of paramagnetism is found in tripositive lanthanide ions in which electrons in incomplete d or f shells are well shielded by other electrons present in outer shells. The separation (ΔE) between the adjacent J states of tripositive lanthanide ions are comparable to separation between adjacent energy terms arising due to interelectronic repulsion energies (~ 1000 cm^{-1}). The J states in such cases are well separated from one another so that at ordinary temperatures only the lowest J state is occupied by such metal ions, that is, the energy separation (ΔE) between J states in such cases are greater than kT (≈ 200 cm^{-1} at room temperature).

The magnetic moments, μ_{eff}, obtained for lanthanide ions are in close agreement with μ_{eff} calculated by the relation,

$$\mu_{eff} = g\sqrt{J(J+1)} \quad \text{B.M.}$$

except for Sm^{3+} and Eu^{3+} in which the ground state is closed to the excited state so that $\Delta E \approx kT$. Hence, a fraction of these ions will be present in higher J states also at room temperature as per Boltzman's distribution law. In such cases, μ_{eff} calculated only by taking J of the ground state (that is, by assuming that all the ions occupy only in the ground J state) will not give the correct magnetic moment of the system.

The μ_{eff} values calculated and actually observed in case of lanthanide tripositive ions are given in Table 6.10.

Table 6.10: Calculate and experimentally observed values for tripositive lanthanide ion.

No. of f electrons	Lanthanide ion (M^{3+})	Ground state	Lande splitting factor (g)	μ_{eff} (B.M.)	
				Calculated	exptl.
0	La	$^1S_0$	–	0	Diamagnetic
1	Ce	$^2F_{5/2}$	6/7	2.54	2.3–2.5
2	Pr	$^3H_4$	4/5	3.58	3.4–3.6
3	Nd	$^4I_{9/2}$	8/11	3.62	3.5–3.6
4	Pm	$^5I_4$	3/5	2.68	–
5	Sm	$^6H_{5/2}$	2/7	0.84	1.5–1.6
6	Eu	$^7F_0$	–	0	3.4–3.6
7	Gd	$^8S_{7/2}$	2	7.94	7.8–8.0
8	Tb	$^7F_6$	3/2	9.72	9.4–9.6
9	Dy	$^6H_{15/2}$	4/3	10.63	10.4–10.5
10	Ho	$^5I_8$	5/4	10.60	10.3–10.5
11	Er	$^4I_{15/2}$	6/5	9.57	9.4–9.6
12	Tm	$^3H_6$	7/6	7.63	7.1–7.4
13	Yb	$^2F_{7/2}$	8/7	4.50	4.4–4.9
14	Lu	$^1S_0$	–	0	Diamagnetic

6.12.3 Multiplet widths small as compared to kT

In the present case, the energy separation between various states of the ground term is smaller than kT so that the J states are almost equally occupied by the metal ions at room temperature. This means that the coupling of L and S vectors in this case is so poor that these vectors maintain their individuality. Hence, L and S vectors are said to interact independently with the magnetic field. Thus, J is no longer a good quantum number.

Generally the magnetic moment of a substance containing unpaired electrons is the resultant of spin magnetic moment, μ_S and orbital magnetic moment, μ_L. In the present case, the effective magnetic moment (μ_{eff}) is the resultant of both type of magnetic moments mentioned above. Since magnetic susceptibility has additive property, total magnetic susceptibility is given by

$$\chi_M = \frac{N\mu_L^2}{3kT} + \frac{N\mu_S^2}{3kT}$$

$$\text{or} \quad \chi_M = \frac{N}{3kT}(\mu_L^2 + \mu_S^2) \tag{6.31}$$

$$\text{and} \quad \mu_{eff} = \mu_{L+S} = \sqrt{L(L+1) + 4S(S+1)} \text{ B.M.} \tag{6.32}$$

The μ_{eff} expressed by equation (6.32) is called spin plus orbit magnetic moment and hence is represented as μ_{L+S}. This equation is applicable to those paramagnetic metal ions of the first transition series wherein the spin angular momentum (generating spin magnetic moment) and orbital angular momentum (producing orbital magnetic moment) both are effective. This means that this equation is applicable to those paramagnetic metal ions in which orbital magnetic moment is not quenched.

It has been seen that in case of metal ions present in the first half of the first transition series, orbital angular momentum is quenched, $L = 0$ for these metal ions. Thus, the value of μ_{eff} for these metal ions is given by equation shown below which is obtained by putting $L = 0$ in equation (6.32)

$$\mu_{eff} = \mu_S = \sqrt{0\,(0+1) + 4S(S+1)} \quad \text{B. M.}$$

$$\text{or} \quad \mu_S = \sqrt{4S(S+1)} \quad \text{B. M.} \tag{6.33}$$

As $S = n.1/2$ where n stands for number of unpaired electron, equation (6.33) can be written as

$$\mu_S = \sqrt{n(n+2)} \quad \text{B. M.} \tag{6.34}$$

The equations (6.33) and (6.34) are called spin only formulae and the μ_{eff} calculated from these formulae are called spin only magnetic moment, μ_S or $\mu_{S.O.}$.

The values of μ_{eff} calculated from equation (6.33) or (6.34) of some ions of first transition series are given in Table 6.11. The experimental values of μ_{eff} are also given for comparison. It may be seen from the table that the experimental values differ slightly from the calculated values. The difference between the calculated and experimental values depends on the extent to which the orbital angular momentum of electrons in metal ions is quenched.

Although, the values of μ_{eff} calculated from spin only formula of most of the ions of first row transition metals are in good agreement with the experimental values (μ_{exp})

Table 6.11: Calculated (μ_S) and experimental values of magnetic moments of the metal ions of first transition series.

Meta ion (d^n)	No. of unpaired electron (n)	S = n/2	$\mu_s = 4S(S+1)^{1/2}$ $= n(n+2)^{1/2}$ (B.M.)	Experimental value (B.M.)
Ti^{3+}, V^{4+} ($3d^1$)	1	1/2	1.73	1.6–1.8
Ti^{2+}, V^{3+} ($3d^2$)	2	1	2.83	2.7–2.9
V^{2+}, Cr^{3+} ($3d^3$)	3	3/2	3.87	3.7–3.8
Cr^{2+}, Mn^{3+} ($3d^4$)	4	2	4.90	4.8–4.9
Mn^{2+}, Fe^{3+} ($3d^5$)	5	5/2	5.92	5.7–6.0
Fe^{2+}, Co^{3+} ($3d^6$)	4	2	4.90	5.0–5.6
Co^{2+} ($3d^7$)	3	3/2	3.87	4.1–5.2
Ni^{2+} ($3d^8$)	2	1	2.83	2.8–3.5
Cu^{2+} ($3d^9$)	1	1/2	1.73	1.8–2.1

of these ions, yet in some cases μ_{exp} values are higher than μ_s values. For example, μ_{exp} value of Co^{2+} ($3d^7$, $t_{2g}^5 e_g^2$) (4.1–5.2) is higher than μ_s value (3.87). The higher experimental value in this case is because of some orbital contribution from Co^{2+} ion to the value of μ_s. The orbital contribution of Co^{2+} ion is due to its $t_{2g}^5 e_g^2$ configuration with $^4T_{1g}$ ground state.

As orbital contribution is not quenched in case of 2nd and 3rd row transition metals, the spin-only formula cannot be used for calculating μ_{eff} of the ions of these metals. In such cases, equation (6.32) (given below) is used for calculating the μ_{eff}. These metal ions also show

$$\mu_{eff} = \mu_{L+S} = \sqrt{L(L+1) + 4S(S+1)} \ \text{B. M.} \tag{6.32}$$

extensive temperature-dependent paramagnetism. This is due to the spin–orbit coupling which removes the degeneracy from the lowest energy level in the ground state.

6.12.4 Multiplet widths comparable to kT

When the multiple widths are comparable to kT, the situation is a mixture of the two extremes already dealt with. All or at least several components of the lowest lying multiplet will be significantly populated but to different extents. The average susceptibility of a sample of the material is the result of contribution from each component of the multiplet, weighted according to its population.

The expression for the molar susceptibility (χ_M) due to complete population of a single component will be,

$$\chi_M = \frac{N\beta^2 g^2 J(J+1)}{3kT} \tag{6.35}$$

In the present case, the contribution to the susceptibility of each level is given by substituting N_J for N, where N_J is the number of molecules per gram mole with that particular value of J. Thus the value of χ_M in the present case will be given by:

$$\chi_M = \frac{\sum N_J \beta^2 g^2 J(J+1)}{3kT} \tag{6.36}$$

Remembering that J level has as many as (2J + 1) orientations, we have for a J level having an energy E above the ground state J:

$$N_J \; \alpha \; (2J+1)e^{-E/kT}$$

or

$$N_J = C\,(2J+1)e^{-E/kT}$$

Putting the value of N_J in equation (6.36), we have

$$\chi_M = C\,\frac{\sum \beta^2 g^2 J(J+1)\,(2J+1)e^{-E/kT}}{3kT} \tag{6.37}$$

$$\chi_M = \frac{N\beta^2\mu^2}{3kT} = \frac{N}{3kT}\,\frac{\sum g^2\beta^2 J(J+1)\,(2J+1)e^{-E/kT}}{\sum (2J+1)e^{-E/kT}}$$

$$\text{or} \quad \mu^2 = \frac{\sum g^2 J(J+1)\,(2J+1)e^{-E/kT}}{\sum (2J+1)e^{-E/kT}}$$

$$\text{or} \quad \mu = \sqrt{\frac{\sum g^2 J(J+1)\,(2J+1)e^{-E/kT}}{\sum (2J+1)e^{-E/kT}}} \tag{6.38}$$

Since $N = \Sigma\, N_J = C\,\Sigma\,(2J+1)e^{-E/kT}$ [because $N_J = C(2J+1)e^{-E/kT}$]

Hence, $$C = \frac{N}{\sum (2J+1)e^{-E/kT}} \tag{6.39}$$

Putting the value of C from equation (6.39) in equation (6.37), the final expression for χ_M is given by:

$$\chi_M = \frac{N}{3kT}\,\frac{\sum g^2\beta^2 J(J+1)\,(2J+1)e^{-E/kT}}{\sum (2J+1)e^{-E/kT}} \tag{6.40}$$

We know that

$$\chi_M = \frac{N\beta^2\mu^2}{3kT}$$

Therefore,

$$\chi_M = \frac{N\beta^2\mu^2}{3kT} = \frac{N}{3kT}\frac{\sum g^2\beta^2 J(J+1)(2J+1)e^{-E/kT}}{\sum(2J+1)e^{-E/kT}}$$

$$\text{or}\quad \mu^2 = \frac{\sum g^2 J(J+1)(2J+1)e^{-E/kT}}{\sum(2J+1)e^{-E/kT}}$$

Therefore, the magnetic moment (μ) is given by

$$\mu = \sqrt{\frac{\sum g^2 J(J+1)(2J+1)e^{-E/kT}}{\sum(2J+1)e^{-E/kT}}} \tag{6.41}$$

As is evident from the equation (6.40), χ_M is not directly proportional to 1/T. Such systems, therefore, do not obey the Curie law.

The results for the different multiple widths, on the assumption of *LS* coupling, are summarized in Table 6.12.

Table 6.12: Different multiple widths and temperature dependence of χ_M.

Multiple width	μ in Bohr magnetons	Temperature dependence of χ_M
Large as compared to *kT*	$g\sqrt{J(J+1)}$	$\chi_M \propto 1/T$ (Curie law obeyed)
Small as compared to *kT*	$\sqrt{L(L+1)+4S(S+1)}$	$\chi_M \propto 1/T$ (Curie law obeyed)
Comparable to *kT*	Complicated function of *J* and *T*	Curie law obeyed

6.13 Anomalous magnetic moments

Anomalous magnetic moment for a metal ion is the magnetic moment in a discrete molecular species having a value which falls outside the range of values predicted on the basis of the spin angular and orbital angular momenta of the electrons in the metal ion in the ligand fields of given strength and symmetry. The discrete molecular species means that there is no secondary magnetic interaction between the neighbouring molecules. In other words, ferromagnetic or anti-ferromagnetic interaction in these systems is absent, that is, the systems are magnetically dilute. The meaning of 'anomalous' magnetic moment may be understood by taking some examples of complexes of different stereochemistry.

6.13.1 Octahedral complexes

(i) Ni(II) complexes
In regular octahedral geometry, the $^3A_{2g}$ ground state term of Ni(II) ion [see Orgel diagram of d^8 metal ion (Figure 6.29)] acquires some orbital contribution from the next

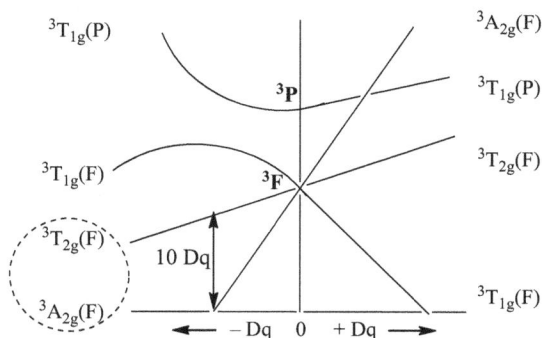

Figure 6.29: Orgel diagram of d^8 metal ion.

higher $^3T_{2g}$ state term due to spin–orbit coupling. The overall magnetic moment is dependent of $\Delta/10Dq$ value and λ as expected from the following expression:

$$\mu_{\text{eff}} = \mu_s \left[1 - \frac{\alpha\lambda}{10Dq}\right]$$

Assuming $10\,Dq = 8000$ cm^{-1} and λ (spin–orbit coupling parameter) = 75 % of the free ion value (~315 cm^{-1}), we find that

$$\mu_{\text{eff}} = \mu_s \left[1 - \frac{\alpha\lambda}{10Dq}\right]$$

$$\mu_{\text{eff}} = 2.83 \left[1 - \frac{4 \times (-315)}{8000}\right] \quad (\mu_s = 2.83; \ \alpha = 4 \text{ for } A_2 \text{ ground term})$$

$$= 2.83\,[1 + 0.1575] \ = \ 2.83 \times 1.1575$$

$$= 3.275 = \text{~}3.3$$

The calculated μ_{eff} is about 3.3 B.M., which is fairly above the spin-only moment of 2.83 B.M. This also includes a small temperature-independent paramagnetism contribution. Generally, μ_{eff} is found in the range of 3.0–3.3 B.M. However, observed values are slightly higher than this value. It is notable that Fe(II), Fe(III) and Co(II) octahedral complexes also show similar trends. However, in some instances orbital angular momentum for the ground state for these ions in octahedral symmetry are not quenched and the expression for μ_{eff} is more complicated. For these systems, there also exists the possibility of 'high-spin' and 'low-spin' complexes with variation in the ligand field strength.

(ii) Fe(II) and Fe(III) complexes

The free ion term for Fe(II) (d^6) is a ^5D state and the absence of adjacent higher terms of the same multiplicity simplifies the calculations. The ground state in octahedral symmetry is $^5T_{2g}$ and μ_{eff} is expected to have a value of 5.6 B.M. at room temperature. Experimental values of the moments fall in the range 5.1–5.7 B.M. The Fe(III) (d^5)

under weak ligand fields (high-spin complexes) has a $^6A_{1g}$ ground state arising from a free ion ^6S state and the magnetic moments are in the order of 5.9 B.M.

(iii) Cobalt (II) complexes

For high-spin octahedral Co(II) complexes, the ground-state configuration is $^4T_{1g}$ resulting from a free ion ^4F ground state. The following two cases are considered for the calculation of μ_{eff}: (a) The ligand field is small compared with the inter-electronic interactions (b) The field is large compared with inter-electronic interactions. The calculated μ_{eff} is in the range 4.7–5.1 B.M. Depending on the ligand field strength and the amount of distortion from octahedral symmetry, experimental values lies in the range of 4.7–5.2 B.M.

6.13.2 Tetrahedral complexes

Relatively large contributions to the observed magnetic moments from orbital angular momentum are expected and observed for Ni(II) in tetrahedral fields with ground state in an orbitally degenerate $^3T_1$ state. For regular tetrahedral complexes, μ_{eff} should be within the range of 3.45–4.0 BM. Tetrahedral Co(II) complexes usually exhibit moments closer to the spin-only values and these are of the order 4.2–4.4 B.M. Tetrahedral Fe(II) and Fe(III) complexes are to some extent rare.

6.13.3 Square planar complexes

Square planar Ni(II) complexes have no unpaired electron and essentially exhibit zero moments. Square planar Co(II) species having slightly large orbital contributions than low-spin octahedral complexes are paramagnetic and reported to exhibit magnetic moments in the range 2.1–2.9 B.M.

The normal and anomalous magnetic moments for the Ni(II), Fe(II), Fe(III) and Co(II) complexes are summarized in Table 6.13.

From the above table, it is notable that that range of anomalous magnetic moments for the Ni(II) complexes is ~0.0–2.8 B.M., Co(II) complexes is ~2.9–4.2 B.M., Fe(II) complexes is 0.0–5.1 B.M. and Fe(III) complexes is 2.3–5.9 B.M.

6.13.4 Factors responsible for anomalous magnetic moments

A variety of models has been suggested to account for the anomalous magnetic moments of various metal complexes. These are:
(i)   Equilibrium between two spin states
(ii) Magnetically non-equivalent sites in the unit cell

Table 6.13: Normal and anomalous magnetic moments for some metal complexes.

Ion	Octahedral	5-Coordinate (trigonal bipyramidal/ square pyramidal)	Tetrahedral	Square planer	Anomalous magnetic moment Range
Ni^{2+}					
High spin	3.0–3.3	3.0–3.45	3.45–4.0	–	} 0.0–2.8
Low spin	–	0	–	0	
Co^{2+}					
High spin	4.7–5.2	4.2–4.6	4.2–4.8	–	} 2.9–4.2
Low spin	1.8–2.0	1.7–2.1	–	2.1–2.9	
Fe^{2+}					
High spin	5.1–5.7	5.1–5.5	5.0–5.2	5.4	} 0.0–5.1
Low spin	0	2.9–3.1	–	–	
Fe^{3+}					
High spin	5.9	–	–	–	} 2.3–5.9
Low spin	2.3	–	–	–	

(iii) Solute–solvent interaction
(iv) Solute–solute interaction
(v) Configurational equilibrium

Let us briefly understand these models by taking appropriate examples.

6.13.4.1 Equilibrium between two spin states: spin cross-over

In cases of d^4, d^5, d^6 and d^7 electronic configurations, a spin state equilibrium may arise in the octahedral geometry provided the crystal field strength is in the region of the critical 10Dq. Thus, if the energy required for promoting an electron from the lower to higher set of orbitals (that is, 10Dq) greater than the spin-pairing energy (SPE) then the electron will prefer to pair and be in the lower set of orbitals, resulting in a low-spin complex. Contrary to this If the spin pairing energy is greater than the energy gap between the orbital sets (10Dq) then a high-spin complex will result. In other words

$$10Dq \gg SPE \qquad Low\ Spin$$

$$SPE \gg 10Dq \qquad High\ Spin$$

However, if the ligand field splitting 10Dq is approximately equal to the spin pairing energy, known as the cross-over point, then small changes in conditions such as temperature or pressure can result in the complex switching from high spin to low spin, or vice versa. This situation is well known for Iron(II) complexes ($3d^6$ electron configuration), which are ideal, as the low-spin state is diamagnetic and the high-spin state has four unpaired electrons as shown in Figure 6.30. Consequently, the change in magnetic moment is large.

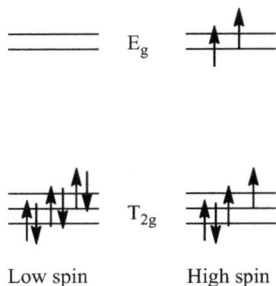

Low spin   High spin   **Figure 6.30:** Low-spin and high-spin state in d^6 Fe(II) complexes.

Let us consider the case of two octahedral iron(II) phenanthroline complexes shown in, and their magnetic behaviour with change in temperature (Figure 6.31).

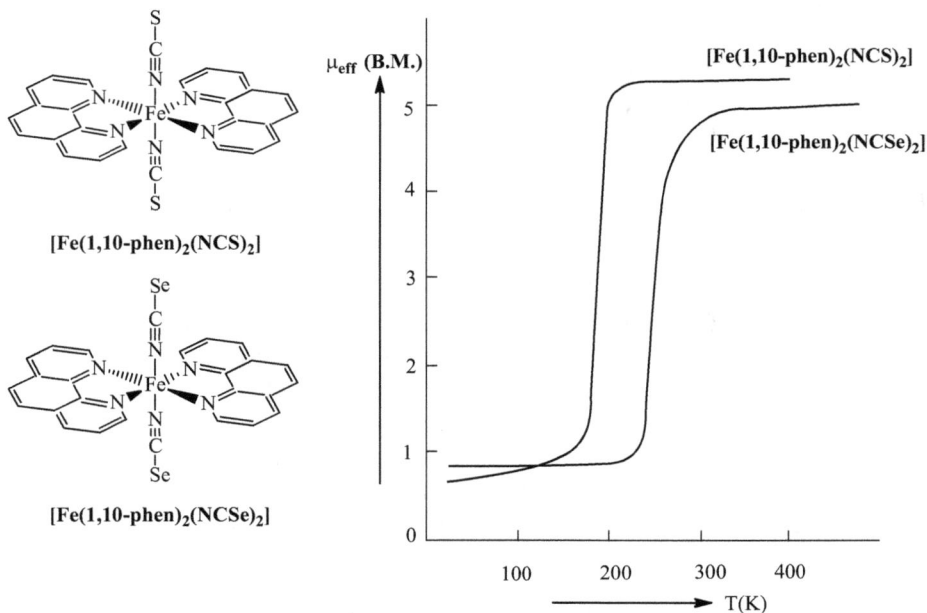

Figure 6.31: Fe(II) phenanthroline complexes and their magnetic behaviour.

It is well clear from the magnetic curves that the magnetic moments of both the complexes at room temperature are around 5 B.M. Both the NCS and NCSe complexes exist in a $^5T_2$ ground state and therefore have magnetic moments which are greater than the spin only value (4.90 B.M.) for four unpaired electrons. At 174 K the magnetic moment of the SCN complex falls dramatically to below 1 B.M. indicative of a low-spin, diamagnetic compound, consistent with the expected $^1A_{1g}$ ground state. It is notable here that the selenium analogue undergoes transition at a much higher temperature of 232 K.

A range of other species also exhibit such behaviour, including iron(II) complexes of tris(pyrazolyl)borate ligands. Iron(III) complexes in a sulphur coordination sphere, such as the tris(dithiocarbamate) complexes $[Fe(S_2C = NR_2)_3]$ also show this type of spin equilibrium. Here, the two states have five and one unpaired electrons, respectively.

Let us now consider the case of d^8 Ni(II) complexes and see whether spin state equilibrium arises in this case or not. For Ni^{2+} complexes under O_h symmetry, there is only one way in which the eight d electrons can be arranged for states of lowest energy. All regular octahedral complexes of Ni^{2+} must, therefore, show paramagnetism corresponding to the presence of two unpaired spins. The situation changes, however, whenever an axial tetragonal distortion is applied to the regular octahedral field. Such distortion lowers the symmetry from O_h to D_{4h} and is accompanied by a further loss of degeneracy of the d orbitals. When this tetragonal distortion is strong, the energy separation between the $d_{x^2-y^2}$ and d_{z^2} orbitals (Δ_1) or between $d_{x^2-y^2}$ and d_{xy} orbitals (Δ_2) (Figure 6.32) may exceed the electron-pairing energy. In this case there will be a change in the magnetic moment from 3.0 B.M. to zero.

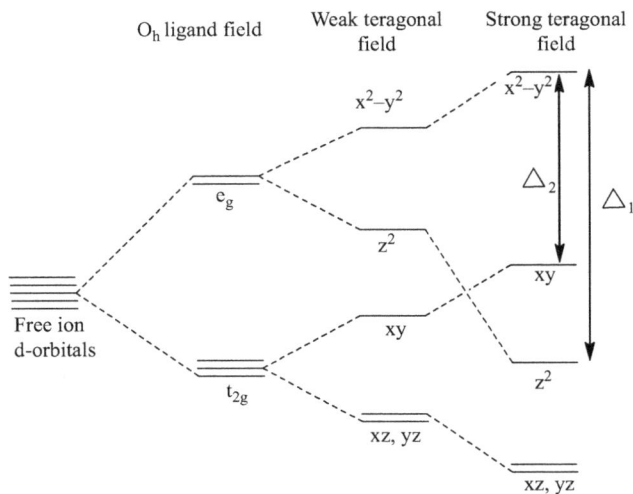

Figure 6.32: Effect of tetragonal distortion in octahedral complex on energy of orbitals.

Conceptually, tetragonal distortion can be envisaged as arising by moving the two ligands on the z-axis to greater distances from the metal ion. In fact this occurs only for certain electronic configurations and is an example of Jahn–Teller distortion. Alternatively, this effect will be generated if we have a mixed-ligand complex, Nia_4b_2, where bs are the axial ligands. If a and b have a similar crystal field strength, it is likely that Δ_1 or Δ_2 will be small and $\mu = \sim 3$. On the other hand, if b has a very low position as compared to a in the spectrochemical series, Δ_1 or Δ_2 may exceed the spin pairing energy (SPE), resulting in diamagnetism. Thus, when Δ_1 or Δ_2 is closed to SPE, we will have spin state equilibrium in a tetragonal nickel(II) complex.

The complex dichlorotetrakis(diethythiourea)nickel(II) (Figure 6.33) is spin paired ($\mu = 0$) below 194 K but attain partial paramagnetism as the temperature is raised. It is believed that Ni(II) has a weak tetragonal field and that the magnetic property is dictated only by the thermal population of the two spin states. For the equilibrium, singlet (low spin) $\leftrightarrow$ Triplet (high spin), the equilibrium constant, K is give by

$$K = \frac{\text{Triplet}}{\text{Singlet}} \tag{6.42}$$

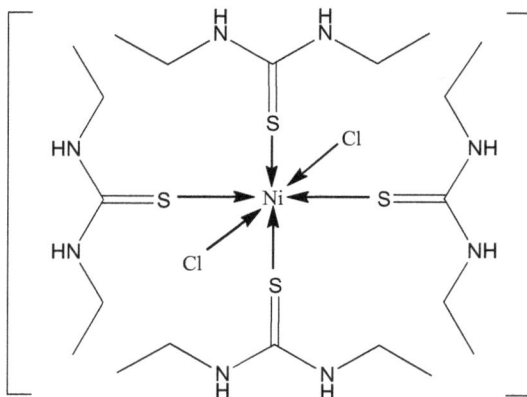

Figure 6.33: Structure of dichlorotetrakis(diethythiourea)nickel(II).

K can be calculated from the knowledge of the mole fractions γ_{HS} (mole fraction of high-spin species) and γ_{LS} (mole fraction of low-spin species). The relation between χ_m and mole fractions is given by

$$\chi_{m(exp)} = \gamma_{HS}\, \chi_{m(HS)} + \gamma_{LS}\, \chi_{m(LS)} \tag{6.43}$$

$$\left(\gamma_{HS} + \gamma_{LS} = 1\right)$$

This relation can be converted in terms of magnetic moment and mole fractions, and it is given by

$$\mu^2_{(exp)} = \gamma_{HS}\,\mu^2_{(HS)} + \gamma_{LS}\,\mu^2_{(LS)} \tag{6.44}$$

$$\gamma_{HS} = \frac{\mu^2_{(exp)} - \mu^2_{(LS)}}{\mu^2_{(HS)} - \mu^2_{(LS)}} \tag{6.45}$$

$$\gamma_{LS} = \frac{\mu^2_{(HS)} - \mu^2_{(exp)}}{\mu^2_{(HS)} - \mu^2_{(LS)}} \tag{6.46}$$

and

$$K = \frac{\gamma_{HS}}{\gamma_{LS}} = \frac{\mu^2_{(exp)} - \mu^2_{(LS)}}{\mu^2_{(HS)} - \mu^2_{(exp)}} \tag{6.47}$$

The γ_{HS} and γ_{LS} can be calculated at different temperatures by using the above equations. The γ_{HS}, γ_{LS}, μ_{eff} and K for the complex, dichlorotetrakis(diethythiourea) nickel(II) at different temperatures are given in Table 6.14.

Table 6.14: Various parameters for the complex dichlorotetrakis(diethythiourea)nickel(II) at different temperature.

T (K)	μ_{exp} (B.M.)	γ_{LS}	γ_{HS}	K
<194	0.0	1.00	0.00	–
280	1.10	0.88	0.12	0.14
300	1.33	0.82	0.18	0.22
320	1.57	0.76	0.24	0.32
360	2.05	0.58	0.42	0.72

These data suggest that effective magnetic moment and equilibrium constant K is increasing with temperature. That means low-spin complex is converting into high spin at higher temperatures. This suggests that the magnetic cross-over is the main cause for anomalous magnetic moment of the complex, dichlorotetrakis(diethythiourea)nickel(II).

Another interesting series of examples showing spin-state isomerism are the nickel(II) complexes (Figure 6.34) of the planar quadridentate macrocyclic ligand TAAB (= tetrakisanhydroaminobenzaldehyde). Such a macrocyclic nickel(II) complex is obtained by way of self-condensation of o-aminobenzaldehyde in the presence of suitable nickel(II) salt in the alcoholic medium. Table 6.15 summarizes the data obtained for these systems.

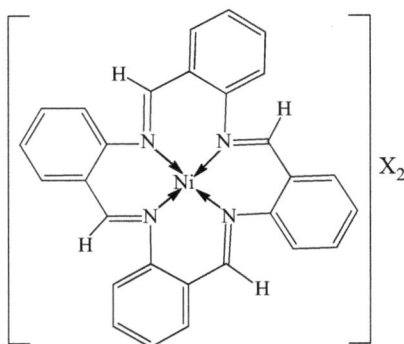

Figure 6.34: Structure of [Ni(TAAB)]X$_2$.

Table 6.15: Magnetic moment data of [Ni(TAAB)]X$_2$ complexes.

X$^-$	μ_{eff} (B.M.)	X$^-$	μ_{eff} (B.M.)
ClO$_4^-$	0.0	NO$_3^-$	3.2
BF$_4^-$	0.0	I$^-$	3.2
BPh$_4^-$	0.0	Cl$^-$	1.7
NCS$^-$	3.2	Br$^-$	1.5

The anhydrous perchlorate, tetraflouroborate and tetraphenylborate salts of [Ni(TAAB)]$^{2+}$, are diamagnetic, whereas its anhydrous thiocyanate, nitrate and iodide are paramagnetic ($\mu = \sim 3.2$). The chloride and bromide salts of this complex form very stable hydrates and exhibit anomalous room temperature magnetic moments of 1.7 and 1.5 B.M., respectively. The K values of these halide salts have been evaluated at different temperatures. The values thus obtained show that spin-state equilibrium exists in these salts.

It has been demonstrated that the order of crystal field strength in the halide salts is anomalous, that is, I$^-$ > Cl$^-$ > Br$^-$. The water molecule may form strong hydrogen bond with the singlet chloride/bromide salt but not with triplet chloride/bromide salt. This is because the residual charge on halide is maximum for the singlet state since nickel–halide bond is weak in this state. Thus, the contribution of hydrate water to the energy balance system is very crucial.

6.13.4.2 Magnetically non-equivalent sites in the unit cell

In a unit cell, (i) the metal ions may have the same coordination number, the same set of ligands but different geometries, and (ii) the metal ions may have different coordination numbers and thereby different geometries. The former class is best demonstrated by the green coloured complex dibromobis(benzyldiphenylphosphine)nickel(II),

[Ni{P(Ph–CH$_2$)Ph$_2$}$_2$Br$_2$], which exhibits an anomalous magnetic moment 2.7 B.M. The X-ray crystallographic study of this compound has revealed that the unit cell has three Ni(II) complexes: one square planar and two tetrahedral. Remembering the additivity of the square of the magnetic moment and taking into account of the mole fractions of the two forms, we have

$$\mu^2 = 0.33 \times \mu^2_{Ni^{2+} \text{ (square planar)}} + 2 \times 0.33 \times \mu^2_{Ni^{2+} \text{ (tetrahedral)}}$$

or $\quad (2.7)^2 = 0.33 \times 0^2 + 0.66\mu^2_{Ni^{2+} \text{ (tetrahedral)}}$

or $\quad 7.29 = 0.66\mu^2_{Ni^{2+} \text{ (tetrahedral)}}$

or $\quad \mu^2_{Ni^{2+} \text{ (tetrahedral)}} = 7.29/0.66 = 11.045$

or $\quad \mu_{Ni^{2+} \text{ (tetrahedral)}} = \sqrt{11.045} = 3.32$

Thus, the magnetic moment of each tetrahedral Ni(II) complex comes out to be 3.32 B.M., a value which is not that bad. Such spin isomers as just discussed differ only in bond angles and are termed as intrallogons (*allos* means different; *gonia* means angles).

An example of class (ii) is found in the yellow form of the Lifschitz compound, bis(*meso*-stilbeneiamine)nickel(II) dichloroacetae · (2/3)C$_2$H$_5$OH · (4/3)H$_2$O. This compound has a magnetic moment 2.58 B.M. and the unit cell consists of one four coordinate square planar nickel(II) complex and two six-coordinate pseudooctahedral nickel(II) complexes. The square planar nickel(II) complex has two bidentate stilbeneiamine coordinated to nickel(II), and the pseudooctahedral nickel(II) complex has two stilbeneiamine molecules along with two anions or one anion and a solvent. Using the room temperature magnetic susceptibility data, we get $\mu = 3.16$ B.M. for the pseudooctahedral nickel(II), a reasonably good value indeed. The colour of Lifschitz compound varies from shades of yellow to blue, depending on the nature of the anion, the coordinating solvent and the diamine used.

6.13.4.3 Solute–solvent interaction

An anomalous magnetic moment may also occur when a particular species interacts with a coordinating solvent. This is because many square planar diamagnetic nickel(II) complexes become partially paramagnetic due to the equilibrium of the type

Square planar (diamagnetic) + solvent $\rightleftharpoons$ pseudo – octahedral (paramagnetic)

This is best illustrated by the equilibrium,

$$[Ni(CRH)]^{2+} + xH_2O \rightleftharpoons [Ni(CRH)(H_2O)_x]^{2+}$$
$$\text{Diamagnetic} \qquad\qquad \text{Paramagnetic}$$

which has been studies in detail using the magnetic and spectroscopic methods. The complex [Ni(CRH)]$^{2+}$ is macrocyclic and is obtained by catalytic hydrogenation of $[Ni(CR)]^{2+}$, which in turn is obtained by the condensation of 2,6-diacetylpyridine and 3,3′-diaminodipropylamine in the presence of the nickel(II) ion.

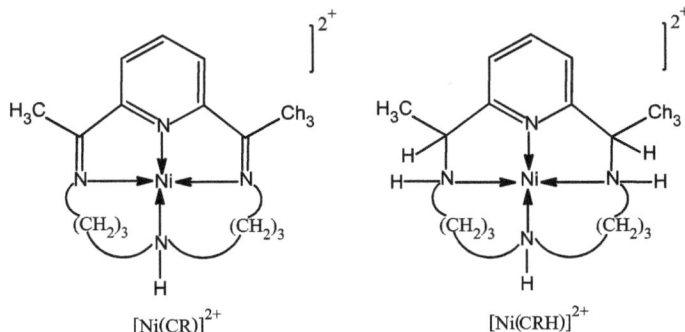

$[Ni(CR)]^{2+}$ $[Ni(CRH)]^{2+}$

The thermodynamic parameters, ΔH and ΔS, are concentration dependent because of the changes in solvent activity. The solvent activity apparent determines the position of equilibrium and hence influences magnetic moment in solution. The addition of salt from outside to an aqueous solution of the complex lowers the activity of the solvent water and thus influences the relative amounts of spin free and spin paired complexes. This has actually been observed in diaquabis(meso-2,3-diaminobutane)nickel(II) and bis((meso-2,3-diaminobutane)nickel(II), the latter being diamagnetic.

6.13.4.4 Solute–solute interaction

Indeed, the solute–solute interaction means an association of two or more molecules of a complex, leading to an increase in the coordination number of the metal ion. This interaction is found to change the spin state of the metal ion. The complex, bis(N-methyl-salicylaldiminato)nickel(II) (Figure 6.35) is diamagnetic in the solid state but shows an anomalous magnetic moment in the range 1.9–2.3 B.M. in solution, depending upon the nature of the non-coordinating solvent.

The dipole moment of this complex in either of the solvents benzene and 1,4-dioxane is practically zero, thus ruling out a planar tetrahedral equilibrium. If the concentration of the complex in any such solvent is increased, the molecular weight and the magnetic moment also increase. This indicates the presence of solute–solute interaction. If instead, a steric barrier is introduced via substitution at the position 3 of the aromatic ring (Figure 6.33), the complex is no longer paramagnetic. On the other hand, if the substitution is carried out at the position 5, the reduction in paramagnetism is much less. Evidently, the substitution at the position 3 does not allow the phenolic oxygen to get attached axially to a second nickel(II), thus preventing the change of the spin state.

The complex bis(acetylacetonato)nickel(II) (Figure 6.36) is linear trimeric, each nickel(II) being six-coordinate with $\mu = 3.23$ B.M. It attains the octahedral structure

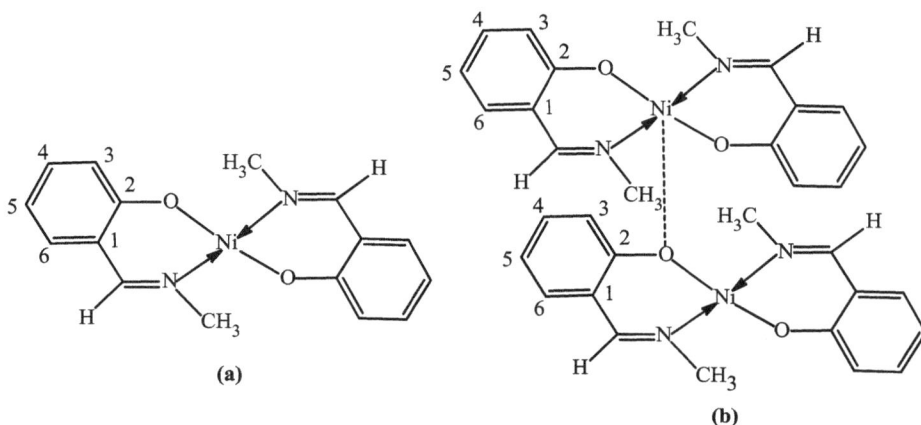

Figure 6.35: (a) Monomeric bis(N-methyl-salicylaldiminato)nickel(II) [**diamagnetic**]. (b) Dimeric bis(N-methyl-salicylaldiminato)nickel(II) [**paramagnetic**].

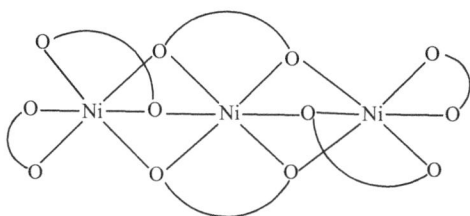

Figure 6.36: Linear trimeric structure of bis(acetylacetonato)nickel(II).

through the oxygen atom functioning as bridges between the nickel(II) ions. The trimeric unit persists in solution (as evidenced from the molecular weight and spectral measurements) but becomes reversibly dissociated into the red monomer at high temperature (~200 °C). The anomalous magnetic moment arises due to a square planar monomer-octahedral trimer equilibrium. Solvent such as pyridine destroy the trimeric structure and give the solvent coordinated six-coordinate monomer.

It is notable here that the substitution of the CH_3 groups of acetylacetone by the very bulky $C(CH_3)_3$ groups prevents trimerization and produces only the diamagnetic square planar monomer (Figure 6.37).

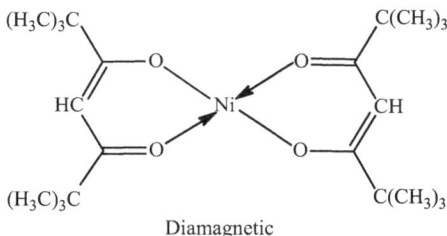

Diamagnetic

Figure 6.37: Monomeric square planar structure of bis(acetylacetonato)nickel(II) wherein CH_3 groups of acetylacetone is replaced by $C(CH_3)_3$.

It is significant to note that the substituent groups which are intermediate between CH_3 and $C(CH_3)_3$ provide a monomer-trimer equilibrium in a non-coordinating solvent, giving an anomalous magnetic moment.

6.13.4.5 Congfigurational equilibrium

Bis(N-*sec*-alkylsalicylaldiminato)nickel(II) complex in an inert solvent at room temperature has been demonstrated to provide an anomalous magnetic moment. This has been shown due to the presence of both square planar diamagnetic and tetrahedral paramagnetic forms in it in comparable proportions. The assignment of the geometry to each of such complex in the solid state has been made on the basis of magnetic moment and reflectance spectrum and also in solution on the basis of dipole moment and molecular weight determinations. Although, there is evidence for some association in solution, it has been shown that above 37 °C it is essentially monomeric and appreciably paramagnetic.

A square planar-tetrahedral equilibrium has also been recognized for the bis (2-hydroxy-1-naphthalaldiminato)nickel(II) complexes. For instance, when the aldehyde is 2-hydroxy-1-naphthaldehyde and the amine is $(C_2H_5)_2CH-NH_2$ (pentan-3-amine), the nickel(II) Schiff base complex is diamagnetic in the solid state but develop a magnetic moment 1.80 B.M. at 50 °C in the chloroform solution. A square planar-tetrahedral equilibrium has also been demonstrated for bis(β-ketoiminato) cobalt(II) complexes.

The nickel(II) complexes NiL_2 (L = Figure 6.38) has been demonstrated to show a monomeric octahedral-square planar equilibrium in a non-coordinating solvent. The equilibrium appears

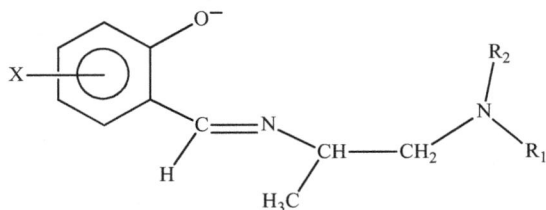

Figure 6.38: A tridentate ONN donor ligand L.

only when R_1=H and R_2=C_6H_5 or substituted phenyl or α-naphthyl. The nature of the substituents determines the solid state behaviour. In solution, an increase in temperature shifts the equilibrium towards the four-coordinate form, giving a low magnetic moment. Then, it is obvious that the tridentate ONN-donor behaviour of the ligand L in the octahedral complex changes to bidentate ON-donor behaviour in the square planar complex.

6.14 Experimental methods of determination of magnetic susceptibility

The techniques normally used for the determination of magnetic susceptibility of coordination compounds are:
(i) Gouy's method (ii) Faraday method and (iii) NMR method

6.14.1 Gouy method

This is the simplest of all and most common. It has the advantage that the apparatus is simple and robust and can be used to measure a wide range of susceptibilities.

Basic principle
This is the simplest of all and most common. It has the advantage that the apparatus is simple and robust and can be used to measure a wide range of susceptibilities.

In this method, a tube filled up to a certain height with the magnetic sample is suspended from an arm of a sensitive balance between the poles of a magnet (Figure 6.39) so that the bottom part of the sample is in a strong magnetic field and the top part in a zero filed. The whole set up is housed inside an enclosure such that there is no vibrational and air disturbance. Generally, an electromagnet giving a constant magnetic field in the range of 5,000 to 20,000 gauss is used.

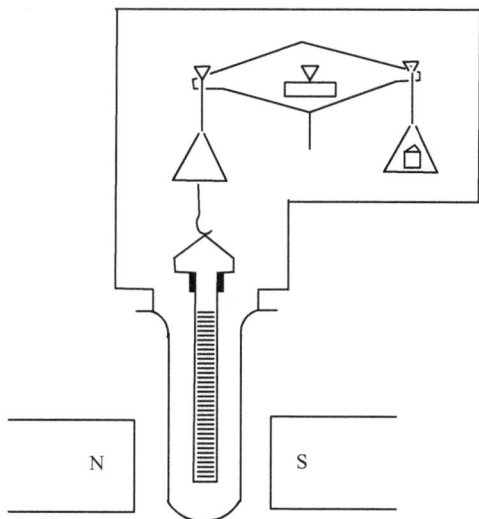

Figure 6.39: A set-up of Gouy magnetic balance.

A small volume of sample, dv, of volume magnetic susceptibility k will experience a force dF given by

$$dF = H \times k \times dv(dH/dx) \tag{6.48}$$

where dH/dx is the magnetic field gradient.

Taking A as the cross-sectional area of the sample and dx as the small height of the sample, we have

$$dF = H \times k \times A \, dx(dH/dx)$$
$$\text{or} \quad dF = H \times k \times A \, dH \tag{6.49}$$

Considering the magnetic field between H and H_o and integrating equation (6.49), we get

$$F = \int dF = \int_{H_0}^{H} H. k. A \, dH = A(k) \frac{1}{2} (H^2 - H_0)$$

$$\text{or} \quad F = A \chi_g \cdot d \, \frac{1}{2} (H^2 - H_0) \quad (d = density)$$

$$\text{or} \quad F = A \chi_g \cdot \left(\frac{m}{v}\right) \frac{1}{2} (H^2 - H_0) \quad \left(d = \frac{m}{v}\right)$$

$$\text{or} \quad F = \chi_g \cdot \left(\frac{m}{l}\right) \frac{1}{2} (H^2 - H_0) \quad (v = A.l)$$

where m and l represent the mass and the length of the sample, respectively. If H_0 is made negligible, then

$$F = \chi_g \cdot \left(\frac{m}{2l}\right) H^2 \tag{6.50}$$

The force experienced by the sample is measured by the change in its weight in the magnetic field. Thus, gram magnetic susceptibility (χ_g) in equation (6.50) may be expressed as

$$\chi_g = \frac{2l F}{m H^2} = \frac{2l \Delta W.g}{(\text{mass of the sample}) \, H^2} = \frac{2l \Delta W.981 \, cm/second^2}{(\text{mass of the sample}) \, H^2} \tag{6.51}$$

where ΔW is the change in mass magnetic material due to imposition of the magnetic field H.

The mass of the sample is determined with and without the magnetic field on. Similarly, the mass of a standard of known gram magnetic susceptibility is determined with and without the same magnetic field on. Under these conditions a change in weight of the sample and standard will be noticed. With the help of this change in weight of the sample and standard, one can determine the magnetic susceptibility using the equation (6.51), details of which are given below under procedure.

Standard/calibrant

$Hg[Co(NCS)_4]$ and $[Ni(en)_3]S_2O_3$ are two very good solid standards or calibrants. They can be easily prepared in pure state. Moreover, they do not decompose or absorb moisture, and pack well. Their magnetic susceptibilities are 16.44×10^{-6} and 11.03×10^{-6} cgs units, decreasing by 0.05×10^{-6} and 0.04×10^{-6} per degree temperature rise, respectively, near room temperature. The $Hg[Co(NCS)_4]$, besides having the higher susceptibility, also packs rather densely and is suitable for calibrating low magnetic fields, while the nickel compound $[Ni(en)_3]S_2O_3$ with lower susceptibility and density is suitable for higher magnetic fields.

For solution measurements, a liquid calibrant eliminating any possible packing errors is to be preferred. The susceptibility of pure H_2O at 20 °C is -0.720×10^{-6} cgs units (χ increases by $\sim 0.0009 \times 10^{-6}$ per degree rise of temperature near 20 °C), but oxygen should be excluded, otherwise this value will be appreciably reduced. Solutions of $NiCl_2$ may be used if higher susceptibility is required. For solution of about 30% by weight of $NiCl_2$, χ is given by

$$\chi = \left[\frac{10030\,p}{T} - 0.720(1-p)\right] \times 10^{-6}$$

Here p is the weight fraction of $NiCl_2$ in solution, and T is the temperature in °K. The disadvantage of this is that p must be determined by chemical analysis which may introduce a significant error. This can be avoided by using solutions of Cs_2CoCl_4 which can be made up directly from known weights of the solid without analysis of solution. In this case χ is given by

$$\chi = \left[\frac{6867\,p}{T+18} - 0.720(1-p)\right] \times 10^{-6}$$

Where p is now the weight fraction of Cs_2CoCl_4 in the solution.

Procedure

It involves the following steps:

(a) Experiment with standard

(i) A given tube is first weighed without and with magnetic field, H on. Let weight of the empty tube without magnetic field = a, and weight of the empty tube with magnetic field on = b. Then, the difference in weights = (b − a).

(ii) The tube is then filled up to a height of 10–11 cm with a standard and weighed it without and with magnetic field on. Suppose, weight of the tube plus standard without magnetic field = c, and weight of the tube plus standard with magnetic field on = d. Then, the difference in weights = (d − c).

Based on the above data:

The weight of the standard $(W_s) = c - a$

and

Change in the weight of the standard $(\Delta W_s) = (d - c) - (b - a)$

(b) Experiment with compound

(iii) The tube is next cleaned and dried. Now, the cleaned tube is first weighed without magnetic field and then with magnetic field on. Suppose, weight of the empty tube without magnetic field = e, and weight of the empty tube with magnetic field on = f. Then, the difference in weights = (f–e).

(iv) The tube is then filled up to with the same height with the compound (whose magnetic susceptibility is to be determined) and weighed it without and with magnetic field on. Let, weight of the tube plus compound without magnetic field = g, and weight of the tube plus compound with magnetic field on = h. Then, the difference in weights = (h–g).

Based on the above data:

The weight of the compound $(W_c) = g - e$

and

Change in the weight of the compound $(\Delta W_c) = (h - g) - (f - e)$

When Hg[Co(NCS)$_4$] with gram magnetic susceptibility value of 16.44×10^{-6} cgs units at 293 K is used as the standard, then $\chi_{g(s)}$ of the standard using equation (6.51) will be

$$\chi_{g(s)} = \frac{2l\,\Delta W_s \cdot g}{W_s\,H^2} \tag{6.52}$$

Similarly, $\chi_{g(c)}$ of the compound in question using equation (6.51) will be

$$\chi_{g(c)} = \frac{2l\,\Delta W_c \cdot g}{W_c\,H^2} \tag{6.53}$$

Dividing equation (6.53) by equation (6.52), we get

$$\frac{\chi_{g(c)}}{\chi_{g(s)}} = \frac{2l\Delta W_c \cdot g \; W_s H^2}{W_c H^2 \cdot 2l\Delta W_s \cdot g} \tag{6.54}$$

$$\text{or} \quad \frac{\chi_{g(c)}}{\chi_{g(s)}} = \frac{\Delta W_c \cdot W_s}{W_c \cdot \Delta W_s} \tag{6.55}$$

$$\text{or } \chi_{g(c)} = \chi_{g(s)} \cdot \frac{\Delta W_c \cdot W_s}{W_c \cdot \Delta W_s} \tag{6.56}$$

Putting the value of $\chi_{g(s)}$ (gram magnetic susceptibility of the standard, 16.44×10^{-6} cgs units) and other experiment values, *viz.*, W_s, ΔW_s, W_c and ΔW_s in the above equation, we can calculate the magnetic susceptibility, $\chi_{g(c)}$ of the compound.

The molar susceptibility $[\chi_{M(c)}]$ of the compound is obtained from the expression:

$$\chi_{M(c)} = \chi_{g(s)} \times \text{Molecular weight of the compound}$$

The value of $\chi_{M(c)}$ so obtained is corrected for diamagnetism of ligands, anions, solvents of crystallization and metal ion, and temperature-independent paramagnetism (TIP). Thus,

$$\chi_{M}\text{corr} = \chi_M - \text{diamagnetic correction} - \text{TIP}$$

The μ_{eff} of the compound is obtained from the Curie's equation,

$$\mu_{eff} = 2.828 \sqrt{\chi_{M}\text{corr.} \cdot T} \quad \text{B. M.}$$

On putting the value of $\chi_M{}^{corr}$ as given above and T and T = 293 K, we get the value of μ_{eff}.

Source of error
The main error in the Gouy method for solid samples arises from the inhomogeneous packing of the sample. This can be reduced by repeating measurements on repacked samples until relatively constant values of χ are obtained. Agreement within 1% is to be considered good and, in some cases, much higher discrepancies are unavoidable.

6.14.2 Faraday's method

In this method, a very small volume of the magnetic sample is packed in a quartz ampoule and suspended from a sensitive balance is placed in a region of fairly strong magnetic field. In this situation, the product H(dH/dx), where dH/dx is the field gradient, is constant over the volume of the sample. The whole set-up is housed in an enclosure which can be flushed with nitrogen or helium. The ampoule has an internal diameter of ~1 mm and the balance used is a quartz fibre torsion balance.

The region of uniform H(dH/dx) is determined by placing a small volume of a calibrant of mass m and of known magnetic susceptibility at different point along the field. The value of H(dH/dx) is obtained from the relation:

$$dF = m\,\chi_g\,H\left(\frac{dH}{dx}\right) \tag{6.57}$$

$$\text{or } H\left(\frac{dH}{dx}\right) = \frac{dF}{m\,\chi_g} \qquad (6.58)$$

Where dF is the force experienced by the sample due to the magnetic field and is measured using a cathetometer.

With the help of the measurements first with calibrant and then with the sample, we can write the relation,

$$\chi_s = \chi_c \frac{d_s}{d_c} \cdot \frac{m_c}{m_s} \qquad (6.59)$$

Here χ_s and χ_c are the gram susceptibilities of the sample and calibrant, m_s and m_c are the respective masses and d_s and d_c are the respective deflections at constant H (dH/dx).

In the above equation (6.59), all the parameters on the right hand side are known, and hence gram susceptibility χ_s of sample in question can be worked out.

6.14.3 NMR method

This method is based on the principle that the position of proton resonance lines of a compound is dependent on the bulk susceptibility of the medium in which the compound is placed. In this method, a concentric cell (Figure 6.40) of length > diameter is used. The inner tube of the cell contains the aqueous solution of the paramagnetic substance whose magnetic susceptibility is to be measured and ~3% *tert*-butanol as internal standard. An identical solvent (H_2O) without the paramagnetic substance is placed in the annular section of the cell.

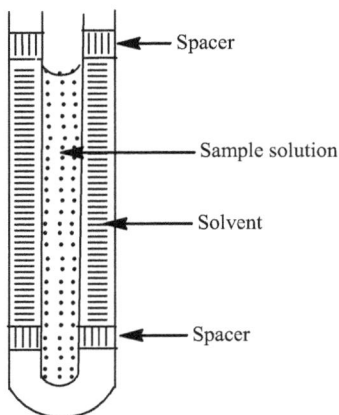

Figure 6.40: Concentric NMR tube for measuring magnetic susceptibility in solution.

The paramagnetic substance shifts the proton resonance lines of the standard (that is *tert*-butanol) and consequently two resonance lines are observed for the methyl protons of the *tert*-butanol.

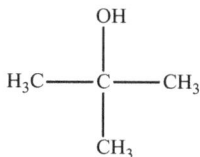

tert-Butanol

The shift of the proton resonance lines is given by the equation:

$$\Delta H = \frac{2\pi}{3} \Delta V \cdot H \tag{6.60}$$

where H is the applied magnetic field and $\Delta V = V_{soln.} - V_{solv}$, the V_{soln} and V_{solv} being the volume susceptibility of the solution and the solvent, respectively.

The gram magnetic susceptibility of the paramagnetic substance can be calculated from the equation,

$$X_g = X_{solv} + \frac{X_{solv}\,(d_{solv} - d_{soln})}{m} \cdot \frac{3}{2\pi m} \cdot \frac{\Delta H}{H}$$

Here X_{solv} is the gram susceptibility of the solvent (H_2O) (-0.72×10^{-6} cgs units for 3% aqueous solution of *tert*-butanol), m is the mass of the paramagnetic substance contained in 1 mL of the solution, and d_{solv} and d_{soln} are the density of the solvent and the solution, respectively. It is notable here that 1% teramethylsilane or 1% benzene can also be used as standard in this method.

6.14.4 Advantages and disadvantages of different techniques

The advantages and disadvantages of the three techniques discussed above for the determination of magnetic susceptibility of paramagnetic substances are given below:

(a) Gouy method

Advantages:
(i) It requires very simple equipment of reasonable cost, easy to assemble, and it is very simple to operate.
(ii) Since the amount of sample taken in this technique is quite large, even an ordinary chemical balance can measure the change in mass.

Disadvantages:
(i)   Large amount of sample is required.
(ii) It requires uniform and compact packing of the sample in Gouy tube for correct results. Packing error is ~3–5%.
(iii) Magnetic susceptibly of sample in the form solution can be determined conveniently.

(b) Faraday method

Advantages:
(i)   It requires small amount of sample compared to Goy method.
(ii) Good sensitivity is additional advantage of this technique.

Disadvantages:
Delicate equipment, fragile suspension devices, constructional difficulty, inconvenient solution measurements and small weight changes are some of the disadvantages of this technique.

(c) NMR method

Advantages:
(i)   It requires small amount of solution (~0.2 mL).
(ii) The measurement is simple and speedy.
(iii) Temperature can be controlled and varied.

Disadvantages:
(i)   Only sample in the form of solution can be handled.
(ii) Costly NMR spectrometer is required to record proton shifts.

6.15 Uses of magnetic moments data of complexes

The calculated spin only and experimental μ_{eff} of some low-spin (inner orbital orbital) octahedral, high-spin (outer orbital) octahedral, square planar and tetrahedral complexes of metal ions of 3d series are given in Table 6.16. We will make use of these magnetic data to know the following:

Table 6.16: Calculated spin only and experimental μ_{eff} values of some metal complexes of 3d series.

(i) Low-spin (inner orbital) octahedral complexes (d^2sp^3 hybridization):

Complex	Metal ion with d^n configuration	No. of unpaired electrons	μ_{eff} (B.M.)	
			μ_s	μ_{exp}
1. $K_3[TiF_6]$	Ti^{3+} ($3d^1$)	1	1.73	1.70
2. $[V(acac)_3]$	V^{3+} ($3d^2$)	2	2.83	2.80
3. $NH_4[V(SO_4)_2] \cdot 12H_2O$	V^{3+} ($3d^2$)	2	2.83	2.80
4. $[Cr(NH_3)_6]Br_3$	Cr^{3+} ($3d^3$)	3	3.87	3.77
5. $[Cr(BigH)_3]Cl_3$	Cr^{3+} ($3d^3$)	3	3.87	3.86
6. $[Cr(dipy)_3]Br_2 \cdot 4H_2O$	Cr^{2+} ($3d^4$)	2	2.83	3.27
7. $K_4[Mn(CN)_6] \cdot 3H_2O$	Mn^{2+} ($3d^5$)	1	1.73	2.18
8. $[Mn(CN)_6]^{3-}$	Mn^{3+} ($3d^4$)	2	2.83	3.2
9. $K_3[Fe(CN)_6]$	F^{3+} ($3d^5$)	1	1.73	2.25
10. $[Fe(o\text{-}phen)_3]^{3+}$	F^{3+} ($3d^5$)	1	1.73	2.50
11. $K_4[Fe(CN)_6]$	F^{2+} ($3d^6$)	0	0	0
12. $[Fe(dipy)_3]_{2+}$	F^{2+} ($3d^6$)	0	0	0
13. $Na_2[Fe(CN)_5(NO^+)]$	F^{2+} ($3d^6$)	0	0	0
14. $[Fe(NO)(H_2O)_5]^{2+}$	F^+ ($3d^7$)	3	3.87	3.90
15. $K_2Ba[Co(NO_2)_6]$	Co^{2+} ($3d^7$)	1	1.73	1.88
16. $[Co(NH_3)_6]Cl_3$	Co^{3+} ($3d^6$)	0	0	0

acacH = acetylacetone; BigH = protonated biguanide; dipy = dipyridyl; o-phen = orthophenanthroline.

(ii) High-spin (outer orbital) octahedral complexes (sp^3d^2 hybridization):

Complex	Metal ion with d^n configuration	No. of unpaired electrons	μ_{eff} (B.M.)	
			μ_s	μ_{exp}
$[Cr(H_2O)_6](ClO_4)_2$	Cr^{2+} ($3d^4$)	4	4.90	4.97
$[Cr(H_2O)_6]SO_4$	Cr^{2+} ($3d^4$)	4	4.90	4.80
$[Mn(py)_6]Br_2$	Mn^{2+} ($3d^5$)	5	5.92	6.00
$[Mn(SCN)_6]^{4-}$	Mn^{2+} ($3d^5$)	5	5.92	6.10

Table 6.16 (continued)

Complex	Metal ion with d^n configuration	No. of unpaired electrons	μ_{eff} (B.M.)	
			μ_s	μ_{exp}
[Mn(acac)$_3$]	$Mn^{3+}(3d^4)$	4	4.90	5.0
Na[Mn(salgly)$_2$]	$Mn^{3+}(3d^4)$	4	4.90	4.80
Na$_3$[FeF$_6$]	$Fe^{3+}(3d^5)$	5	5.92	5.85
[Fe(NH$_3$)$_6$]Cl$_2$	$Fe^{2+}(3d^6)$	4	4.90	5.45
[Fe(en)$_3$]$^{2+}$	$Fe^{2+}(3d^6)$	4	4.90	5.5
[Fe(NO$^+$)(H$_2$O)$_5$]$^{2+}$	$Fe^{+}(3d^7)$	3	3.87	3.90
Na$_3$[CoF$_6$]	$Co^{3+}(3d^6)$	4	4.90	5.39
[Co(H$_2$O)$_6$]SO$_4$	$Co^{2+}(3d^7)$	3	3.87	5.10
[Co(NH$_3$)$_6$]$^{2+}$	$Co^{2+}(3d^7)$	3	3.87	5.0
[Co(aniline)$_2$(Cl)$_2$(EtOH)$_2$]	$Co^{2+}(3d^7)$	3	3.87	5.0
[Co(en)$_3$]SO$_4$	$Co^{2+}(3d^7)$	3	3.87	4.56
[Co(py)$_4$Cl$_2$]	$Co^{2+}(3d^7)$	3	3.87	5.15
[Ni(NH$_3$)$_6$]Cl$_2$	$Ni^{2+}(3d^8)$	2	2.83	3.32
[Ni(H$_2$O)$_6$]SO$_4$	$Ni^{2+}(3d^8)$	2	2.83	3.23
[Cu(phen)$_3$]ClO$_4$	$Cu^{2+}(3d^9)$	1	1.73	1.96

salglyH$_2$ = salicylideneglycine.

(iii) Tetrahedral complexes (sp^3 hybridization):

Complex	Metal ion with d^n configuration	No. of unpaired electrons	μ_{eff} (B.M.)	
			μ_s	μ_{exp}
(Et$_4$N)$_2$[MnCl$_4$]	$Mn^{2+}(3d^5)$	5	5.92	5.94
[Fe(Ph$_3$P)$_2$I$_2$]	$Fe^{2+}(3d^6)$	4	4.90	5.10
(Et$_4$N)$_2$[FeCl$_4$]	$Fe^{2+}(3d^6)$	4	4.90	5.40
[Co(py)$_2$Cl$_2$]	$Co^{2+}(3d^7)$	3	3.87	4.42
Cs$_2$[CoCl$_4$]	$Co^{2+}(3d^7)$	3	3.87	4.59
(Et$_4$N)$_2$[NiCl$_4$]	$Ni^{2+}(3d^8)$	2	2.83	3.89
[Ni(CO)$_4$]	$Ni^{0}(3d^8 4s^2 = 3d^{10})$	0	0	0
Cs$_2$[CuCl$_4$]	$Cu^{2+}(3d^9)$	1	1.73	2.0

(iv) Square planar complexes (dsp^2 hybridization):

Complex	Metal ion with d^n configuration	No. of unpaired electrons	μ_{eff} (B.M.)	
			μ_s	μ_{exp}
[Fe(phthalocyanine)]	Fe^{2+}(3d^6)	2	2.83	3.96
[Co(phthalocyanine)]	Co^{2+}(3d^7)	1	1.73	2.73
[Co(BigH)$_2$]SO$_4$	Co^{2+}(3d^7)	1	1.73	2.49
[Ni(Et$_3$P)$_2$Cl$_2$]	Ni^{2+}(3d^8)	0	0	0
[Ni(CN)$_4$]$^{2-}$	Ni^{2+}(3d^8)	0	0	0
[Ni(BigH)$_2$]Cl$_2$	Ni^{2+}(3d^8)	0	0	0
[Ni(Et$_3$P)Br$_3$]	Ni^{3+}(3d^7)	1	1.73	1.80
[Cu(BigH)$_2$]Cl$_2$	Cu^{2+}(3d^9)	1	1.73	1.79
[Cu(DMG)Cl$_2$]	Cu^{2+}(3d^9)	1	1.73	1.85

(a) To know the type of hybridization in complexes

(i) Na$_2$[Fe(CN)$_5$(NO)] · 2H$_2$O: sodium nitroprusside

Magnetic measurement of this octahedral compound has shown that it is diamagnetic ($\mu_{eff} = 0$) [see Table 6.16 (i)] in nature. This compound contains iron as Fe^{2+} (d^6) and nitric oxide NO$^+$. This suggests that all the six electrons in the metal ion are paired. According to valance bond theory (VBT), this can happen only if Fe^{2+} metal ion rearrange its six electrons in such a manner that it vacates two of the five d-orbitals for the formation of 3d^{2}4sp^3 hybrid orbitals which can accommodate the six ligand electron pairs. In other words Na$_2$[Fe(CN)$_5$(NO)] · 2H$_2$O is an inner orbital (low-spin) complex of Fe(II) involving d^{2}sp^3 hybridization. This is pictorially shown in the following hybridization scheme (Figure 6.41).

(ii) Pentaaquanitrosyliron(I) ion, [Fe(NO)(H$_2$O)$_5$]$^{2+}$

The experimental magnetic moment of octahedral [Fe(NO)(H$_2$O)$_5$]$^{2+}$ ion is 3.90 B.M. [see Table 6.16 (i)], and this suggests the presence of three unpaired electrons on the Fe atom in this ion. Thus, Fe is present as Fe$^+$ (Fe = 3d^{6}4s^2; Fe$^+$ = 3d^{6}4s^1 = 3d^7 with n = 3) and nitric oxide as NO$^+$ in this complex ion. In order to explain the presence of three unpaired electrons, the valance bond theory (VBT) suggests that one of the 4s electron shifts to 3d orbitals, and that 3d-orbitals with three unpaired electrons do not participate in hybridization. Consequently, the formation of complex

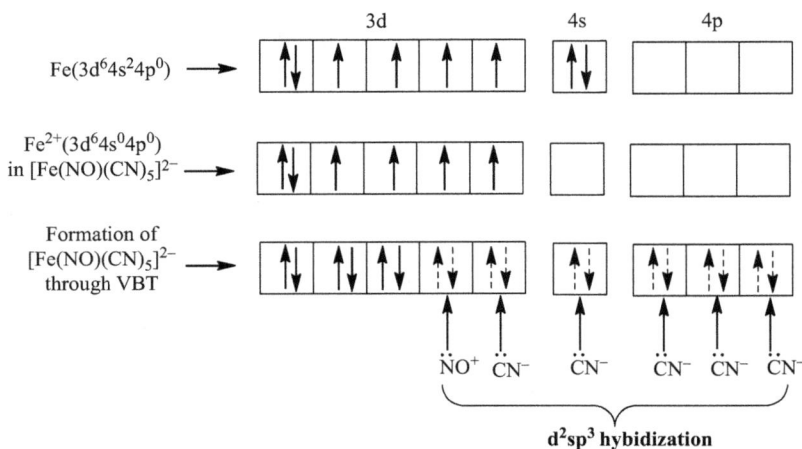

Figure 6.41: d^2sp^3 Hybridization scheme for the formation of sodium nitroprusside.

ion results from $4s4p^34d^2$ (sp^3d^2) hybridization of Fe^+. Hence, the complex ion in question, is an outer orbital octahedral ion of Fe(I) involving sp^3d^2 hybridization (Figure 6.42).

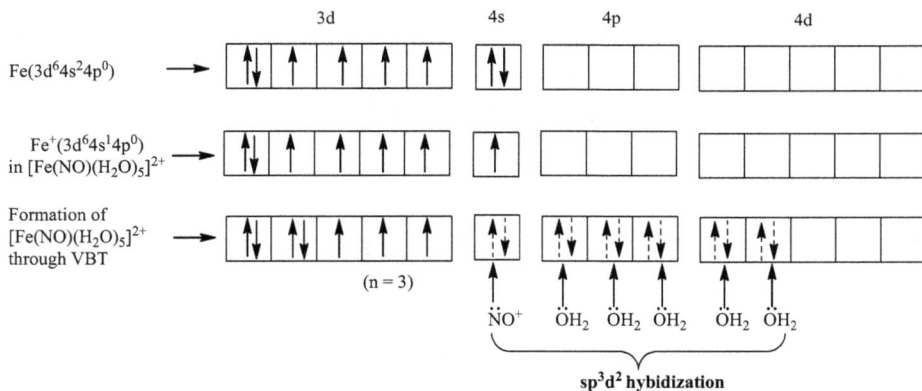

Figure 6.42: sp^3d^2 Hybridization scheme for the formation of $[Fe(NO)(H_2O)_5]^{2+}$.

(iii) $[Co(NH_3)_6]^{3+}$ and $[CoF_6]^{3-}$

The observed magnetic moment of $[Co(NH_3)_6]^{3+}$ ion (d^6) is zero (0) which corresponds to having no unpaired electron, that is, n = 0, while $[CoF_6]^{3-}$ ion has magnetic moment value of 5.39 B.M. corresponding to four unpaired electrons (n = 4). These data suggest that $[Co(NH_3)_6]^{3+}$ is inner orbital complex involving d^2sp^3 hybridization while $[CoF_6]^{3-}$ is outer orbital complex having sp^3d^2 hybridization.

(iv) $[Fe(o\text{-phen})_3]^{3+}$ and $[FeF_6]^{3-}$

The μ_{exp} (2.50 B.M.) for $[Fe(o\text{-phen})_3]^{3+}$ (d^5) correspond to $n = 1$ while μ_{exp} (5.85 B.M.) for $[FeF_6]^{3-}$ corresponds to $n = 5$. These magnetic data indicate that $[Fe(o\text{-phen})_3]^{3+}$ is inner orbital complex involving d^2sp^3 hybridization whereas $[CoF_6]^{3-}$ is outer orbital complex undergoing sp^3d^2 hybridization.

(b) Determination of number of unpaired (n) in complexes

This use of magnetic moment data can be understood by taking some examples as follows:

(i) $Na_3[CoF_6]$

This compound contains Co as Co^{3+} (d^6) and the observed magnetic moment (μ_{exp}) is found to be 5.39 B.M. We know that spin only magnetic moment (μ_s) for $n = 4$ and $n = 5$ are 4.90 and 5.92 B.M., respectively. As $\mu_{exp} = 5.39$ B.M. is closed to 4.90 B.M. rather than 5.92 B.M, hence the number of unpaired electrons in $Na_3[CoF_6]$ is 4.

(ii) $[Fe(Ph_3P)_2I_2]$

This tetrahedral compound (Figure 6.43) contains Fe as Fe^{2+} (d^6) and the experimental magnetic moment (μ_{exp}) is 5.10 B.M. The spin only magnetic moment (μ_s) for compounds having four and five unpaired electrons is 4.90 and 5.92 B.M., respectively. Since, $\mu_{exp} = 5.10$ B.M. of this compound is closed to 4.90 B.M. compared to 5.92 B.M, it has four unpaired electrons.

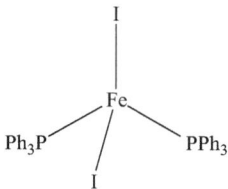

Figure 6.43: Tetrahedral structure of $[Fe(Ph_3P)_2I_2]$.

(iii) $[Cu(DMG)Cl_2]$

This square planar complex (Figure 6.44) has Cu as Cu^{2+}(d^9). Its magnetic moment experimentally found is 1.85 B.M, which is closed to spin only value (μ_s) 1.73 B.M. for one unpaired electron. Hence, $[Cu(DMG)Cl_2]$ has one unpaired electron.

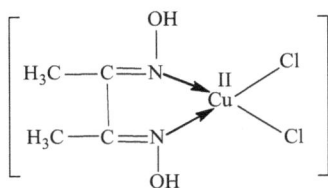

Figure 6.44: Square planar structure of [Cu(DMG)Cl$_2$].

(c) Prediction of the geometry of complexes

If the value of μ_{exp} of a given coordination compound or the number of unpaired electrons present in it is known, the geometry of that compound can be known. This can be understood through the following examples:

(i) Geometry of [NiCl$_4$]$^{2-}$

This complex ion contains Ni as Ni^{2+} with valance shell electronic configuration $3d^84s^04p^0$ and the number of unpaired electrons (n) equals to 2. As the magnetic moment value 3.89 B.M. (including orbital contribution also because of $^3T_1$ ground state) of [NiCl$_4$]$^{2-}$ is found corresponding to two unpaired electrons (n = 2), the two unpaired electrons present in $3d^84s^04p^0$ configuration do not pair up. Thus, the formation of this complex ion results from $4s4p^3$ or sp^3 hybridization of Ni^{2+} ion (Figure 6.45), and hence its geometry is tetrahedral.

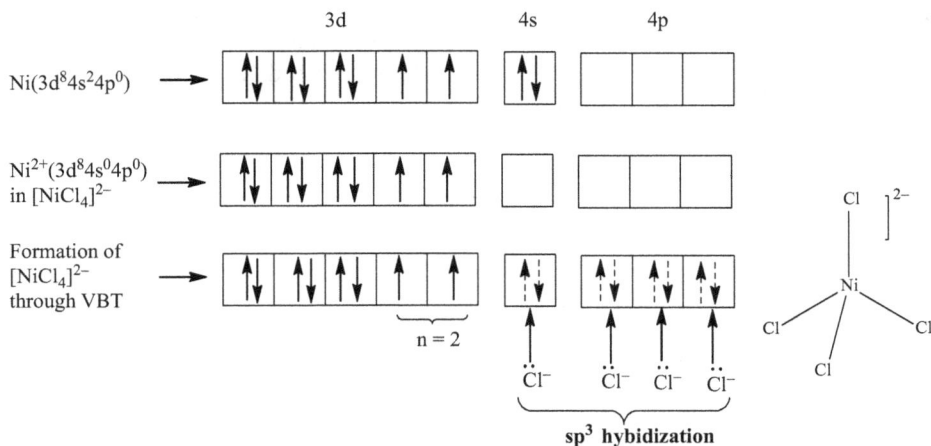

Figure 6.45: sp^3 Hybridization resulting tetrahedral geometry of [NiCl$_4$]$^{2-}$.

(ii) Geometry of [Ni(CN)$_4$]$^{2-}$

This complex ion has Ni as Ni^{2+} with valance shell electronic configuration $3d^84s^04p^0$ and the number of unpaired electrons (n) equals to 2. As the magnetic moment value [Ni(CN)$_4$]$^{2-}$ is found to be zero, the two unpaired electrons present in $3d^84s^04p^0$ configuration paired up in four 3d-orbitals. Therefore, the formation of this complex ion results from 3d4s4p^2 or dsp^2 hybridization of Ni^{2+} ion (Figure 6.46), and thus its geometry is square planar.

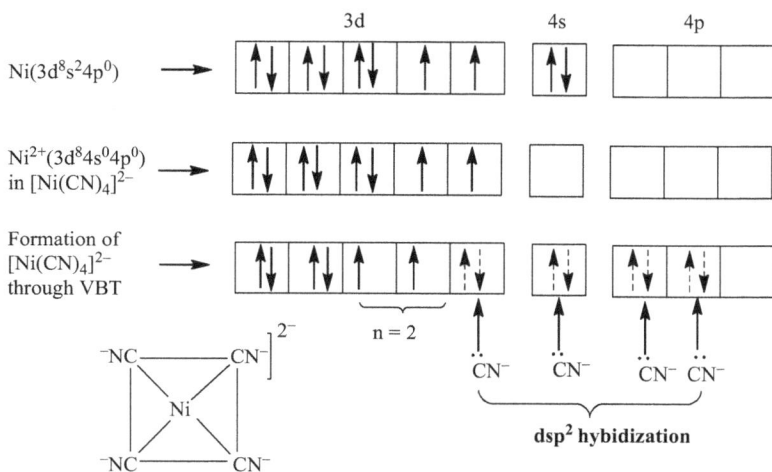

Figure 6.46: dsp^2 Hybridization resulting square planar geometry of [Ni(CN)$_4$]$^{2-}$.

(iii) Geometry of [Fe(phthalocyanine)]

This complex compound contains Fe as Fe^{2+} with valance shell electronic configuration $3d^64s^04p^0$ and the number of unpaired electrons (n) equals to 2. The magnetic moment value 3.96 B.M. of [Fe(phthalocyanine)] (although high because of some unknown reasons) corresponds to two unpaired electrons (n = 2). Hence, out of four unpaired electrons present in $3d^64s^04p^0$ only two paired up and two do not, and thus vacating one d-orbital. So, the formation of this compound result from 3d4s4p^2 or dsp^2 hybridization of Fe^{2+} ion (Figure 6.47), and thus its geometry is square planar (Figure 6.48).

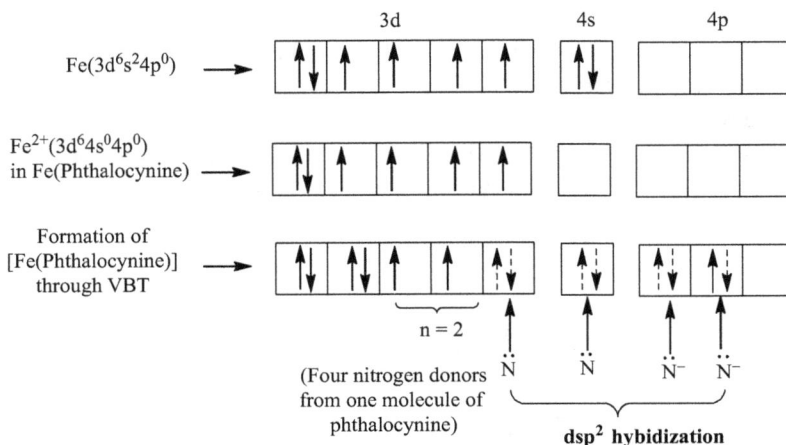

Figure 6.47: dsp^2 Hybridization resulting square planar geometry of [Fe(phthalocyanine)].

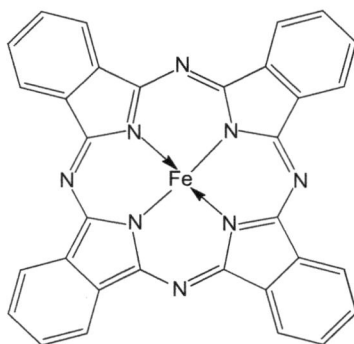

Figure 6.48: Square planar geometry of [Fe(phthalocyanine)].

6.16 Limitations of valance bond theory: magnetic behaviour of complexes

(i) The explanation for the magnetic behaviour of complexes presented by va-lance bond theory is purely qualitative. This theory fails to explain why in some cases the observed μ_{eff} values are higher and in some cases are lower than the spin only (μ_s) value.

(ii) The valance bond theory fails to explain why the μ_{eff} values of tetrahedral com-plexes of a metal ion are different from the μ_{eff} values of octahedral complexes of the same metal ion containing the same number of unpaired electrons. For example, the μ_{eff} values for tetrahedral Ni(II) complexes containing two un-paired electrons ranges between 3.2 and 4.0 B.M., whereas the μ_{eff} values for octahedral Ni(II) complexes, also containing two unpaired electrons ranges between 2.9 and 3.3 B.M. Likewise, the μ_{eff} values for tetrahedral Co(II) complexes containing three unpaired electrons ranges between 4.4 and 4.8 B.M., whereas the μ_{eff} values for outer orbital octahedral Co(II) complexes, also containing three unpaired electrons ranges between 4.8 and 5.2 B.M.

(iii) The valance bond theory (VBT) does not explain why $[Co(NH_3)_6]^{3+}$ is diamagnetic and $[CoF_6]^{3-}$ is paramagnetic, although both the complex ions have the same geometry (octahedral) and the oxidation state of Co atom is the same (= +3). In order to explain this discrepancy in magnetic behaviour of the two complexes, VBT says that Co^{3+} ion in $[Co(NH_3)_6]^{3+}$ uses 3d orbital for $3d^2 4s 4p^3$ (d^2sp^3) hybridization, whereas Co^{3+} ion in $[CoF_6]^{3-}$ utilizes 4d orbital for $4s 4p^3 4d^2$ (sp^3d^2) hybridization. VBT does not explain why Co^{3+} ion utilizes d-orbital of different principal quantum shells in $[Co(NH_3)_6]^{3+}$ and $[CoF_6]^{3-}$ ions.

(iv) The correlation between geometry and magnetic behaviour of a complex is sometimes misleading. For instance, square planar complex of Ni(II), such as, $[Ni(CN)_4]^{2-}$ is diamagnetic because this complex is formed by $3d 4s 4p^2 (dsp^2)$ hybridization of Ni^{2+} ion and has n = 0. We may thus conclude that all square complexes of Ni(II) should be diamagnetic. But in reality, it is not so. In fact, there are few square planar complexes of Ni(II) which are paramagnetic due to the presence of two unpaired electrons. This observation is explained by VBT as follows:

Square planar paramagnetic complexes of Ni(II) resulting from $3d 4s 4p^2 (dsp^2)$ hybridization of Ni^{2+} ion, has two unpaired electrons: one electron present in 3d-orbital and one electron is shifted to 4p-orbital (see the hybridization scheme shown in Figure 6.49). VBT cannot explain why the promotion of an electron from 3d to 4p orbital occurs in the present case but in the case of $[Ni(CN)_4]^{2-}$ ion this type of promotion of an electron does not take place.

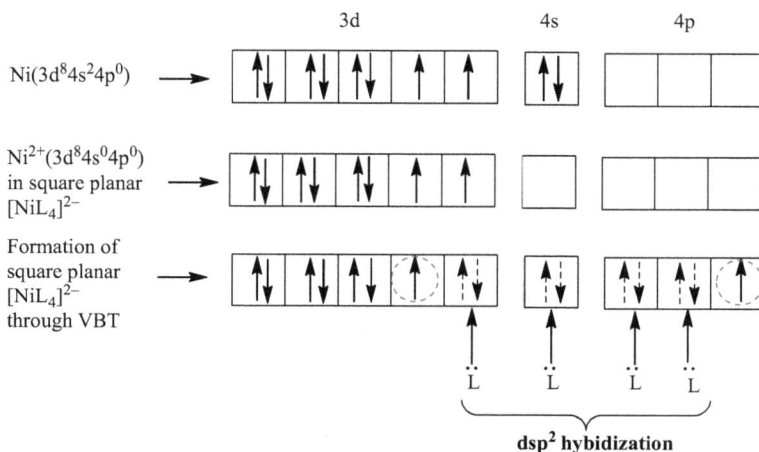

Figure 6.49: Promotion of one electron from 3d orbital to 4p orbital and dsp² hybridization resulting paramagnetic square planar complexes.

(v) In most of the complexes of first transitional series, the μ_{eff} values are in good agreement with μ_s values, but μ_{eff} values for tetrahedral complexes of Ni(II) and for octahedral Co(II) are higher than μ_s values. VBT is failed to explain the enhanced values of μ_{eff} values in tetrahedral Ni(II) and octahedral Co(II) complexes.

(vi) It has been established that the colour and magnetic moment of the complexes of transition metal ions are due to the presence of unpaired electrons present in d-orbitals of metal ions. So, there must be a quantitative correlation between spectra and magnetic moment of the complexes. Unfortunately, there is no such connection according to VBT.

(vii) The μ_{eff} value for outer orbital complex of Cu(II) ($Cu^{2+} = 3d^9$ with $n = 1$) is 1.96 B.M. This vale is greater than its μ_s value of 1.73 B.M. VBT has no explanation for such a difference in magnetic moment.

6.17 Crystal field theory and magnetic behaviour of complexes

The application of crystal field theory to explain the magnetic behaviour of transition metal complexes meets with much greater success than the valance bond theory. Unlike the valance bond theory, the crystal field theory does not have to assume the energetically unfavourable excitation of electrons from lower energy orbitals to higher energy d, s, or p orbitals in octahedral complexes of metal ions with d^7, d^8 and d^9 configurations. Nor does the crystal field theory have to assume the same ground state for free and complexes metal ion as is done in the case of valance bond theory. Apart from this, the crystal field theory satisfactorily explains all those features of magnetochemistry of transition metal complexes, which the crystal field theory fails to, explain. Some of the features of magnetochemistry of metal complexes satisfactorily explained by crystal filed theory are illustrated below:

(i) $[Co(NH_3)_6]^{3+}$ and $[CoF_6]^{3-}$ are octahedral complexes of the same metal ion, Co^{3+}, but $[Co(NH_3)_6]^{3+}$ is diamagnetic, and $[CoF_6]^{3-}$ is paramagnetic with respect to three unpaired electrons. Valance bond theory is failed to explain such a difference in magnetic behaviour of these complexes. However, the crystal field theory (CFT) explains the different magnetic behaviour of these two complexes. According to CFT, the five d-orbitals in the isolated Co^{3+} ion are degenerate. But in the presence of six ligands, d-orbitals are split into two sets of orbitals-t_{2g} and e_g. The t_{2g} set contains d_{xy}, d_{yz} and d_{xz} (triply degenerate orbitals) while e_g set has $d_{x^2-y^2}$ and d_{z^2} orbitals (doubly degenerate orbitals). The t_{2g} set has lower energy than e_g set (Figure 6.50).

In $[CoF_6]^{3-}$, because of the weak F^- ligand, the energy separation (Δ_o) between t_{2g} and e_g is small. Hence, distribution of six electrons of Co^{3+}ion taking place according to Hund's rule gives $t_{2g}^4 e_g^2$ electronic configuration and thus four unpaired electrons. This is why $[CoF_6]^{3-}$ is paramagnetic. Contrary to this, due to the presence of strong NH_3 ligands in $[Co(NH_3)_6]^{3+}$ ion, the energy difference (Δ_o) between

t_{2g} and e_g is large. Therefore, distribution of six electrons of Co^{3+} ion takes place in such a way that all the six electrons occupy in t_{2g} set $(t_{2g}{}^6 e_g{}^0)$ only which gives $n = 0$. This makes $[Co(NH_3)_6]^{3+}$ ion diamagnetic. The above two situations of energy difference between t_{2g} and e_g sets are pictorially shown in Figure 6.50.

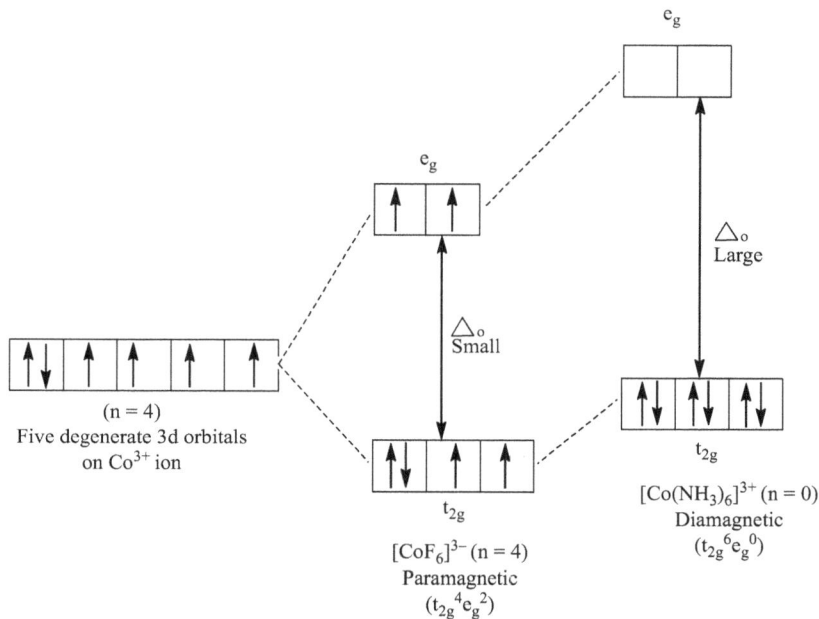

Figure 6.50: CFT based splitting of d-orbitals in weak (F^-) and strong (NH_3) ligand field in Co(III) octahedral complexes resulting paramagnetic and diamagnetic complexes, respectively.

(ii) The explanation for quenching of orbital magnetic momentum based on the crystal field theory is highly logical as presented below.

6.17.1 Octahedral complexes

It is convenient to visualize an octahedral complex as being formed by six ligands (Ls), or donor atoms (Figure 6.51), approaching the central metal ion along its x, y and z axes. The relationship between these axes and five orbitals is shown in Figure 6.52.

In the free metal ion, the d-orbitals are energetically identical or degenerate. However, as the ligands approach, any electrons in the d-orbitals is repelled, that is, energy of the d-orbitals is increased. In addition, their *degeneracy is removed*. Since, the $d_{x^2-y^2}$ and d_{z^2} orbitals lie along x, y and z axes and are thus destabilized

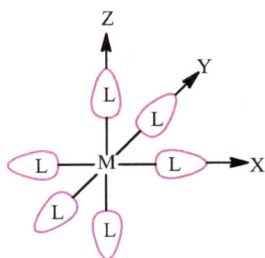

Figure 6.51: Central metal ion surrounded by six ligands (Ls) in octahedral complexes.

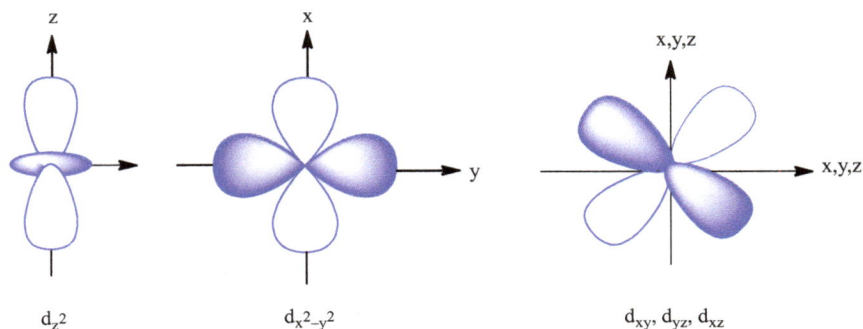

Figure 6.52: Pictorial representation of d-orbitals.

more than d_{xy}, d_{yz} and d_{xz} orbitals which lie between the axes. The two sets of orbitals are referred to as e_g set and t_{2g} set, respectively. The difference in energy between e_g and t_{2g} is Δ_o or 10Dq, which is the measure of the intensity of the electrostatic field acting on the metal ion. The more usual representation of the splitting of d-orbitals showing removal of degeneracy is given in Figure 6.53.

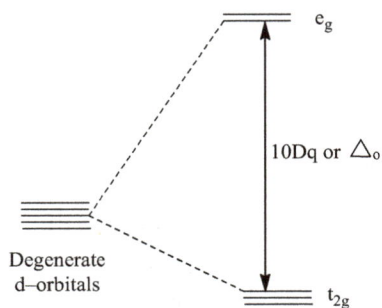

Figure 6.53: Splitting of d-orbitals into doublet and triplet sets in an octahedral complex.

It is this removal of the degeneracy of the d-orbitals which is responsible for the quenching of the orbital contribution and consequently applicability of the spin-only formula for the magnetic moment. This may be understood as follows:

It has been established that orbital angular momentum arises from rotation of the electron about the nucleus as in the Bohr model of the atom. This simple picture is complicated in the wave mechanical model of the atom where the orbital angular momentum may be pictorially associated with the interchange or transformation of one orbital with another by rotation about the appropriate axis. For this transformation or interchange idea of an orbital to be possible, *the d-orbitals must be degenerate* and *of the same shape* and *must not contain* both, *electrons of the same spin*. If these conditions are satisfied, then an electron in one of the orbitals will be able to rotate about the axis, being effectively an electric current producing a magnetic field.

The d_{xy}, d_{yz} and d_{xz} orbitals can transform into one another by 90° rotations about the relevant axes while, even more effectively, 45° rotations of the d_{xy} orbital about the z-axis transform it into the $d_{x^2-y^2}$ orbital. For a single electron *in a free ion* all these rotations are possible and full orbital angular momentum is developed. For d^5 and d^{10} configurations (Figure 6.54), $L = 0$ and no orbital contribution to the magnetic moment is expected. These are the cases in which every orbital contains electrons of the same spin, thus preventing the development of any orbital angular momentum.

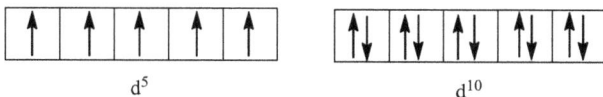

d^5 d^{10}

Figure 6.54: D^5 and d^{10} electron configurations in free ion having same spin in every orbital.

The effect of the imposing of an octahedral filed on to the free ion can now be seen.
(i)   In the first step the interchange of d_{xy} and $d_{x^2-y^2}$ orbitals, which was an importance source of orbital contribution on magnetic moment in the free ion, is no longer possible.
(ii) Secondly, although interchange with the t_{2g} set is still energetically feasible, the condition that the orbitals in O_h complexes must not contain electrons of the same spin is now more preventive/limiting. In the free ion this precludes orbital contribution only in two configurations d^5 and d^{10}, whereas now the orbital contribution is quenched in all cases involving t_{2g}^3 or t_{2g}^6 configurations irrespective of the number of electrons in the e_g set.
(iii) It can be seen from Figure 6.52 that the d_{z^2} and $d_{x^2-y^2}$ orbitals, being of different shapes cannot be interchanged and hence have no orbital angular momentum associated with them. It is for this reason that e_g set is known as a 'non-magnetic'

doublet. This expression implies nothing about the spin angular momentum of electrons in the e_g set, which remain unaffected.

(iv) Electronic configurations involving t_{2g}^1, t_{2g}^2, t_{2g}^4 and t_{2g}^5 will give rise to some orbital contribution leading to moments in excess of the spin only values.

For d^1, d^2 and d^3 configurations of the electrons, the electron enter the lower energy/more stable t_{2g} set, while, for d^8 and d^9, the unpaired electrons must be in the higher e_g set. However, for each of the intervening configurations- d^4, d^5, d^6 and d^7, there are two distinct possibilities, shown in Figure 6.55 along with orbital contribution expected/not expected in respective configuration. Which of these actually occurs will depend upon the relative magnitudes of the two opposing effects.

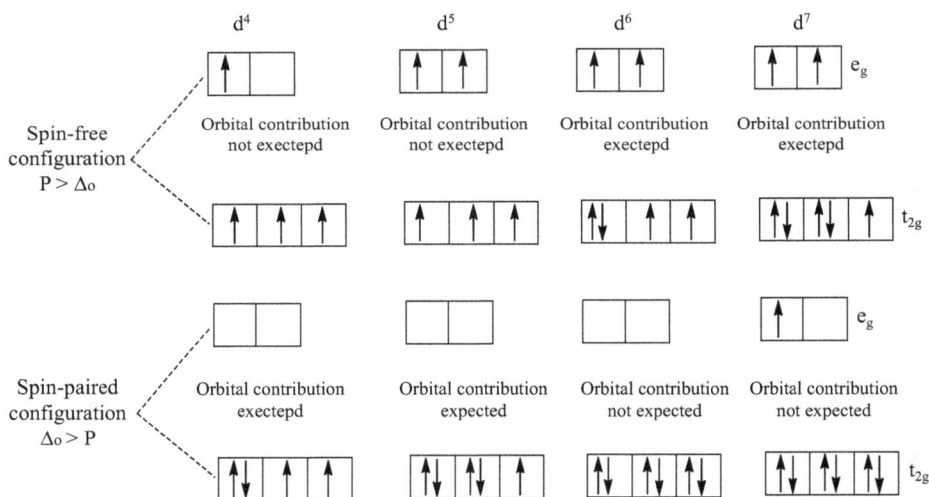

Figure 6.55: Spin-free and spin-paired configurations in case of d^4, d^5, d^6 and d^7 electron configurations.

6.17.2 Tetrahedral complexes

In case of tetrahedral complexes, the situation in many ways is reverse of the octahedral case. The ligands (Ls) are now non-axial and their distribution around the central metal atom (M) is shown in Figure 6.56. Comparison with the shapes of the d-orbitals (Figure 6.18) shows that, because of their closer proximity to the ligands, it is the non-axial orbitals-d_{xy}, d_{yz} and d_{xz} which are destabilized more (Figure 6.56). The d-orbitals are now split into an upper t_2 set of three and a lower e set of two orbitals.

Unlike to the e_g set in the octahedral case, the t_2 orbitals, although now closest to the ligands, do not point directly at the ligands. For this reason, and also because there are less number of ligands compared to the octahedral case, the tetrahedral

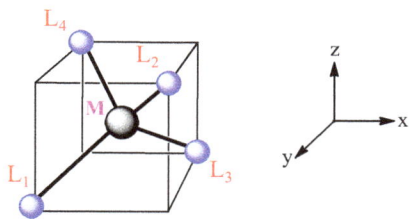

Figure 6.56: Non-axial ligands environment in a tetrahedral complex.

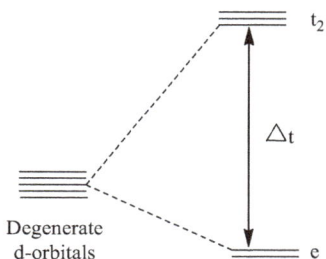

Figure 6.57: The splitting of d-orbitals in a tetrahedral complex.

orbital splitting, Δ_t is less than the octahedral. In fact, for the same metal and ligands and the same internuclear distances

$$\Delta_t = 4/9\Delta_o$$

Because of this relatively small orbital splitting, spin pairing in tetrahedral complexes is rather unlikely. Although, the diamagnetism of $RbReCl_4$ was ascribed due to spin-pairing, it has finally been shown to arise from metal–metal bonding in a trimeric anion (Figure 6.58).

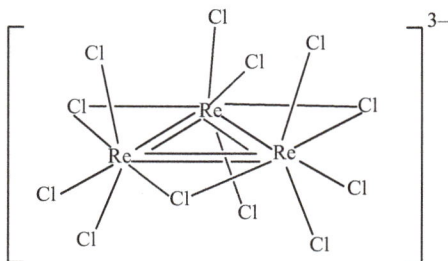

Figure 6.58: Trimeric structure of $[Re_3Cl_{12}]^{3-}$ having Re–Re bonding responsible for diamagnetic nature of anion.

If the possibility of spin-pairing is ignored, the configuration d^3, d^4, d^8 and d^9 should retain orbital contribution while in the remaining d-configurations (d^1, d^2, d^5, d^6, d^7 and d^{10}) the orbital contribution should be completely quenched. The spin-free configurations of d^3, d^4, d^8 and d^9 along with their orbital contribution are shown in Figure 6.59. Similarly, the spin-free configurations- d^1, d^2, d^5, d^6, d^7 and d^{10} and their orbital contribution in magnetic moments are also shown in Figure 6.60.

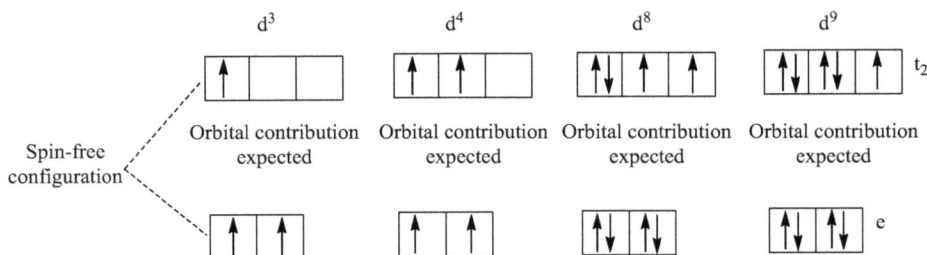

Figure 6.59: Spin-free configuration of d^3, d^4, d^8 and d^9 and their expected orbital contribution on the basis of electron(s) in t_2 orbitals of not the same spin.

The occurrence of orbital contribution/no orbital contribution in octahedral and tetrahedral transition metal complexes are summarized in Table 6.17.

(a) Other stereochemistries
The orbital splitting in square planar complexes can be derived from the octahedral case by assuming a tetragonal distortion caused by withdrawing the two ligands from the z-axis. In an octahedral complex the d_{z^2} is the orbital most strongly repelled by the ligands on the z-axis, followed by the d_{xz} and d_{yz} orbitals. Removal of the z-axis will, therefore, stabilize these orbitals, the most marked effect on the d_{z^2} as shown in Figure 6.61.

Whether or not the d_{z^2} orbital actually lies lowest in a square planar complex will depend on the particular metal and ligands involved.

The d-orbital splittings for some other Stereochemistries, such as, square pyramidal, trigonal bipyramidal and dodecahedral are given in Figure 6.62.

Reported work on these stereochemistries is very meagre compared with the more highly symmetrical octahedral and tetrahedral arrangement and so exact predictions would be rather hasty. However, it is evident that the lower the symmetry, the more the d-orbital lose their degeneracy, and greater the probability of the orbital contribution being quenched.

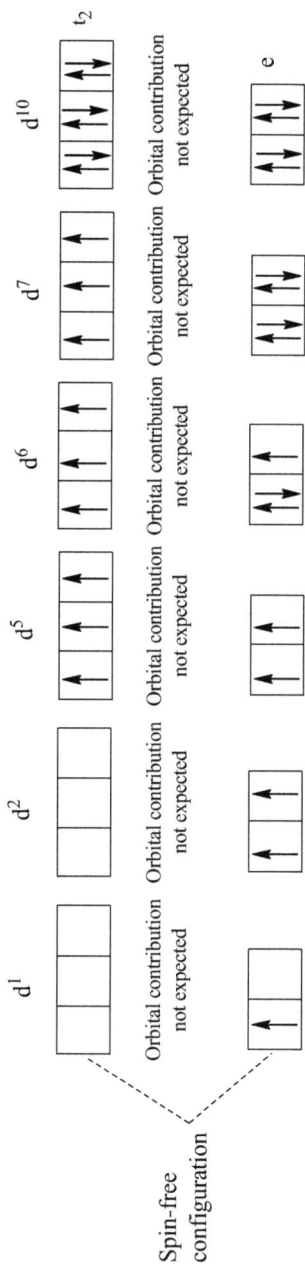

Figure 6.60: Spin-free configuration of d^1, d^2, d^5, d^6, d^7 and d^{10} and their no orbital contribution on the basis of electron(s) in t_2 orbitals of the same spin.

Table 6.17: Orbital contribution/No orbital contribution in octahedral and tetrahedral complexes of d^n configurations.

d^n Configurations	Spin-free octahedral		Spin-paired octahedral		Spin-free tetrahedral	
	Orbital contribution	No orbital contribution	Orbital contribution	No orbital contribution	Orbital contribution	No orbital contribution
d^1	t_{2g}^1					e^1
d^2	t_{2g}^2					e^2
d^3		t_{2g}^3			$e^2 t_2^1$	
d^4		$t_{2g}^3 e_g^1$	t_{2g}^4		$e^2 t_2^2$	
d^5		$t_{2g}^3 e_g^2$	t_{2g}^5			$e^2 t_2^3$
d^6	$t_{2g}^4 e_g^2$			t_{2g}^6		$e^3 t_2^3$
d^7	$t_{2g}^5 e_g^2$			$t_{2g}^6 e_g^1$		$e^4 t_2^3$
d^8		$t_{2g}^6 e_g^2$		$t_{2g}^6 e_g^2$	$e^4 t_2^4$	
d^9		$t_{2g}^6 e_g^3$		$t_{2g}^6 e_g^3$	$e^4 t_2^5$	

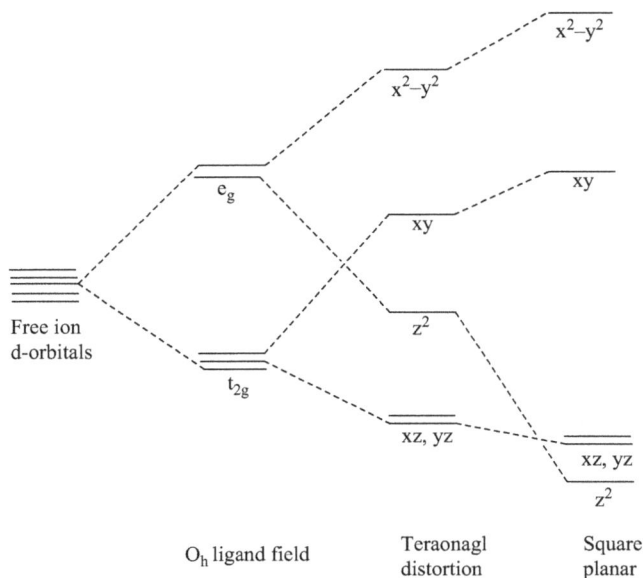

Figure 6.61: The effect of an increasing tetragonal distortion in an octahedral complex leading finally to a square planar stereochemistry.

── x^2–y^2	── z^2	── x^2–y^2
	══ x^2–y^2, xy	══ xz, yz
── z^2		
══ xz, yz		── z^2
── xy	══ xz, dyz	── xy
Square pyramidal	Trigonal pyramidal	Dodecahedral

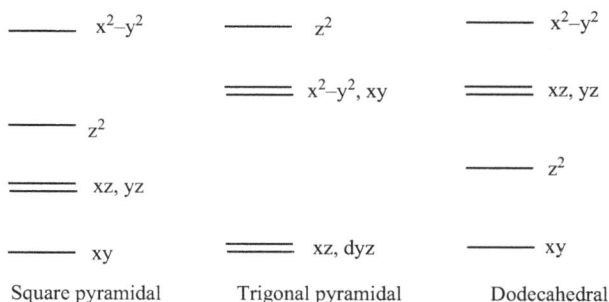

Figure 6.62: Splitting of d-orbitals in some non-cubic fields.

Exercises

Multiple choice questions/fill in the blanks

1. Magnetic properties of 4d or 5d transition metal compounds arise from one of the following:
   (a) Orbital motion of the electrons   (b) Spin properties of the electrons
   (c) Both (a) and (b)   (d) All of these

2. When symbols have their usual meanings, sort out the correct statement for diamagnetic substances:
   (a) $P < 1$; χ_g, k and I are negative   (b) $P > 1$; χ_g, k and I are positive
   (c) $P = 1$; χ_g, k and I are positive   (d) None of these

3. Considering symbols have their usual meanings, which one is correct statement for paramagnetic substances:
   (a) $P < 1$; χ_g, k and I are negative   (b) $P > 1$; χ_g, k and I are positive
   (c) $P = 1$; χ_g, k and I are positive   (d) All of these

4. Substances which show permanent magnetism even in the absence of magnetic field are known as:
   (a) Paramagnetic substances   (b) Anti-ferromagnetic substances
   (c) Ferromagnetic substances   (d) None of these

5. Sort out a metal oxide from the following which shows anti-ferromagnetic behaviour:
   (a) CrO_2 (b) Fe_3O_4 (c) CrO_3 (d) All of these

6. In ferromagnetic materials, below a particular temperature, magnetic suscepti-bility sharply increases and attains the highest value at $T = 0$. This particular temperature is called:
   (a) Neel temperature   (b) Curie temperature
   (c) Room temperature   (d) None of these

7. For anti-ferromagnetic substances, there is a characteristic temperature above which the anti-ferromagnetic substance behaves as a normal paramagnetic substance. This characteristic temperature is called:
   (a) Room temperature   (b) Curie temperature
   (c) Neel temperature   (d) None of these

8. For a normal paramagnetic substance, plot of $1/\chi_M^{corr}$ against T gives a straight line which:
   (a) Passes through the origin
   (b) Cuts the T-axis at temperature below $0°$
   (c) Cuts the T-axis at temperature above $0°$
   (d) None of them

9. Curie–Weiss law is represented by equation:
   (a) $\chi_{Mcorr} \dfrac{C}{T + \theta}$ (b) $\chi_{Mcorr} \dfrac{C}{T - \theta}$
   (c) $\chi_{Mcorr} \dfrac{2C}{T - \theta}$ (d) $\chi_{Mcorr} \dfrac{C}{2(T + \theta)}$

10. Corrected molar magnetic susceptibility ($\chi_M^{corr.}$) of a paramagnetic compound is obtained by the relationship:
    (a) Molar susceptibility (χ_M) + susceptibility of diamagnetic group/ligands present
    (b) Molar susceptibility (χ_M) – susceptibility of diamagnetic group/ligands present
    (c) Molar susceptibility (χ_M) × susceptibility of diamagnetic group/ligands present
    (d) None of these

11. In the equation for diamagnetic susceptibility, $\chi_{dia} = \Sigma\, n_A\, \chi_A + \Sigma\lambda$ of a compound, Pascal constants are:
    (a) Value of λ for certain bond type   (b) Value of χ_A for atom A
    (c) Both (a) and (b)   (d) None of these

12. Sort out a high- spin octahedral complex with expected orbital contribution to the magnetic moment from the following electronic configurations:
    (a) $t_{2g}^3 e_g^0$ (b) $t_{2g}^3 e_g^2$ (c) $t_{2g}^6 e_g^2$ (d) $t_{2g}^1 e_g^0$

13. Which one of the following low-spin octahedral complexes with the electronic configurations (along with ground state terms) given is expected to show orbital contribution to the magnetic moment?
    (a) $t_{2g}^6 e_g^0 (^1A_{1g})$ (b) $t_{2g}^6 e_g^1 (^2A_g)$ (c) $t_{2g}^6 e_g^2 (^3A_{2g})$ (d) $t_{2g}^5 e_g^0 (^2T_{2g})$

14. In high-spin Mn(II) (d^5) complexes, orbital contribution to the magnetic moment is not expected because its ground state term is:
    (a) E   (b) A   (c) T   (d) None of these

15. Octahedral Ni(II) and tetrahedral Co(II) complexes with ground state term A in each exhibit magnetic moments substantially higher than the spin only value because of the mixing of A with:
    (a) First excited state T
    (b) Second excited state T
    (c) Third exited state T
    (d) None of these

16. In case of d^3 high-spin octahedral complexes, experimental value of μ_{eff} is lower than the μ_s, and hence μ_{eff} is calculated using the equation:

    (a) $\mu_{eff} = \mu_s \left(1 - \dfrac{\alpha\lambda}{10Dq} \right)$
    (b) $\mu_{eff} = \mu_s \left(1 - \dfrac{2\alpha\lambda}{10Dq} \right)$

    (c) $\mu_{eff} = \mu_s \left(1 - \dfrac{\alpha\lambda}{10Dq} \right)$
    (d) $\mu_{eff} = \mu_s \left(1 - \dfrac{2\lambda}{10Dq} \right)$

    (λ = spin–orbit coupling parameter; α is constant depending upon the spectroscopic term and 10 Dq = energy difference between the ground and excited state)

17. Orbital contribution towards magnetic moment in square planar complexes is no more likely than in octahedral complex because of the reason:
    (a) $d_{x^2-y^2}$ and d_{z^2} orbitals are non-magnetic doublet.
    (b) T_{2g} set (d_{xy}, d_{yz}, d_{xz}) orbitals of equivalent shape, size and energy generate orbital contribution
    (c) Splitting of T_{2g} set orbitals in two groups of different energy make the orbital contribution hindered
    (d) None of these

18. When multiplet widths is large as compared to kT, the energy separation (ΔE) between J states are greater than kT, then at ordinary temperatures metal ions occupy J state of
    (a) Highest energy
    (b) Medium energy
    (c) Lowest energy
    (d) All of these

19. The magnetic moments, μ_{eff}, obtained for most of lanthanide ions (multiplet widths large compared to kT) are in close agreement with μ_{eff} calculated by the equation:
    (a) $\mu_L = \sqrt{L(L+1)}$
    (b) $\mu_s = \sqrt{n(n+2)}$
    (c) $\mu_{eff} = g\sqrt{J(J+1)}$
    (d) All of these

20. In case of Sm^{3+} and Eu^{3+} ions, the experimental magnetic moment values are higher than the calculated magnetic moment using the formula $\mu_{eff} = g\sqrt{J(J+1)}$. This is due to:
    (a) Ground state is closed to the first excited state
    (b) Ground state is not so closed to the first excited state
    (c) Ground state is far away to the first excited state
    (d) None of these

21. When in a compound or metal ion (of 2nd and 3rd transition series) multiplet widths are small as compared to kT, the magnetic moment is calculated using the formula:
    (a) $\mu_{eff} = \sqrt{L(L+1) + 4S(S+1)}$ (b) $\mu_{eff} = \sqrt{4S(S+1)}$
    (c) $\mu_{eff} = \sqrt{L(L+1)}$ (d) None of these

22. The diamagnetic square planar complex bis(acetylacetonato)nickel(II) in solution exhibits anomalous magnetic moment (μ = 3.23 B.M.). This is due to:
    (a) A square planar monomer-octahedral trimer equilibrium
    (b) A square planar monomer-octahedral dimer equilibrium
    (c) A square planar monomer-octahedral tetramer equilibrium
    (d) All of these

23. Sort out suitable model(s) that account for anomalous magnetic moments in different metal complexes:
    (a) Solute–solvent interaction   (b) Solute–solute interaction
    (c) Configurational equilibrium   (d) All of these

24. Which one of the following is not a calibrant to be used in magnetic susceptibility measurement?
    (a) H_2O (b) $[Ni(en)_3]S_2O_3$ (c) $Hg[Co(NCS)_4]$ (d) C_6H_6

25. E_g set of orbitals are degenerate in O_h complexes, but due to different shape and size, these orbitals cannot be transferred in to one another by . . . . Hence, electron present in the e_g set orbital cannot generate any . . . . magnetic moment.

26. T_{2g} set orbitals of equivalent shape, size and energy in O_h complexes is transferred into the other two orbitals through rotation by an angle . . . . about the relevant axis. Hence, the electrons present in these orbitals will generate . . . . contribution to the magnetic moment when these orbitals do not contain electrons in the . . . . spin.

Short answer type questions

1. Derive an equation for molar magnetic susceptibility χ_M.
2. Differentiate between diamagnetic and paramagnetic substances.
3. What are ferromagnetic substances? Explain with suitable examples.
4. What are magnetically dilute and magnetically concentrated substances? Explain.
5. Briefly describe the effect of temperature on magnetic susceptibility of anti-ferromagnetic substances.
6. What is temperature-independent paramagnetism (TIP)? How and in what type of substances it appear?

7. What is Curie equation? Highlight its limitation.
8. Briefly describe the utility of Pascal's constants.
9. How corrected molar susceptibility ($\chi_M^{corr.}$) is related to Pascal's constants and constitutive corrections? Explain.
10. What is quenching of orbital angular momentum? How does it occur?
11. Orbital contribution in square planar complexes is not more likely than in octahedral complex. Justify this statement.
12. Briefly explain how thermal energy affects the magnetic property of materials.
13. What will be the fate of magnetic moment when multiplet widths small as compared to kT ?
14. What do you understand by anomalous magnetic moments? Explain it taking suitable magnetic moment data of metal complexes in different stereochemistries.
15. Highlight the advantages and disadvantages of different techniques for the determination of magnetic susceptibility of different samples.
16. Briefly describe the prediction of the geometry of complexes on the basis of magnetic moment data.

Long answer type questions

1. What is ferromagnetism in certain substances? High light the effect of temperature on magnetic susceptibly of ferromagnetic substances.
2. What is anti-ferromagnetism in certain substances? Explain the anti-ferromagnetic behaviour of substances due to metal–metal and spin exchange interactions with suitable examples.
3. Derive the standard equation of magnetic moment from magnetic susceptibility.
4. How the equation $\mu_{eff} = 2.828\sqrt{\chi_M^{corr.} \cdot T}$ is derived?
5. Present a detailed view of Curie and Curie–Weiss law. Also derive the equation of magnetic moment in terms of Curie and Curie–Weiss law.
6. Describe the role of Pascal's constants and constitutive corrections to compute corrected molar magnetic susceptibility ($\chi_M^{corr.}$) from molar susceptibility (χ_M). Taking two suitable examples of your choice, calculate the diamagnetic susceptibility (χ_{dia}) using Pascal's constants and constitutive corrections.
7. Present a detailed account of origin of magnetism via orbital and spin motion of electrons generating orbital and spin magnetic moments.
8. What is quenching of orbital angular momentum? Present a detailed mechanism of quenching of orbital angular momentum in low-spin and high-spin octahedral and tetrahedral complexes.
9. In high-spin and low-spin octahedral and tetrahedral complexes, the ground state electronic configurations having T as the ground state have orbital contribution but the configurations having A and E as their ground state is quenched. Justify this statement with suitable examples.

10. Explain the effect of mixing of ground state with excited state due to spin–orbit coupling on magnetic moment (μ_{eff}) in high-spin octahedral complexes of d^n systems with suitable examples. Also explain why in d^5 high-spin octahedral complexes, μ_{eff} and μ_s are the same.

11. Derive the equation for effective magnetic moment when multiplet widths are large as compared to kT, the thermal energy.

12. Develop the equation for magnetic moment (μ) when multiplet widths are comparable to kT (thermal energy).

13. What are anomalous magnetic moments? Describe the occurrence of anomalous magnetic moments in some complexes via solute–solvent and solute–solute interactions taking suitable examples.

14. Present a detailed view of the factors responsible for anomalous magnetic moments in complexes.

15. Describe the basic principle and experimental procedure for determination of magnetic susceptibility of a sample by Gouy method.

16. Discuss the utility of Faraday and NMR methods for the determination of magnetic susceptibility of different samples. Also highlight their advantages and disadvantages.

17. Describe the various uses of magnetic moment data in complexes.

18. Taking suitable examples highlight the limitations of valance bond theory to explain the magnetic behaviour of complexes.

19. Explain the role of crystal field theory in describing the magnetic behaviour of complexes in different stereochemistries.

Chapter VII
Mechanism of inorganic reactions: a study of metal complexes in solution

7.1 Introduction

Reactions in inorganic chemistry are mostly ionic in character. As such, they take place almost instantaneously and so no reaction mechanism is involved. However, reactions of metal complexes are slow and proceed via definite pathways. The mechanism involved in some of the reactions of coordination compounds, particularly substitution reactions and electron transfer reactions, are being described in detail in this chapter.

7.2 Substitution reactions in octahedral metal complexes

Substitution reactions in octahedral metal complexes are the reactions in which a ligand present in the coordination sphere of the complex compound is substituted by another ligand (nucleophile) or the metal in a complex compound is replaced by another metal (electrophile).

Depending on whether a ligand is substituted by another ligand or a metal is replaced by another metal, substitution reactions are of two types:

(i) Nucleophile or ligand substitution reactions (S_N reactions)
When a ligand present in the coordination sphere of a complex compound is replaced by another ligand, such substitution reactions are called nucleophilic substitution (S_N) reactions as ligands are all nucleophiles. For example:

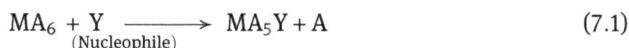

$$MA_6 + \underset{\text{(Nucleophile)}}{Y} \longrightarrow MA_5Y + A \tag{7.1}$$

(ii) Electrophile (or metal) substitution reactions (S_E reactions)
In some reactions, the central metal ion may get replaced by other metal ions. Such substitution reactions are called electrophilic substitution (S_E) reactions since metals are all electrophiles. For example:

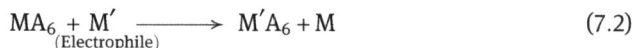

$$MA_6 + \underset{\text{(Electrophile)}}{M'} \longrightarrow M'A_6 + M \tag{7.2}$$

https://doi.org/10.1515/9783110727289-007

Here, we shall discuss only nucleophilic substitution reactions in octahedral complexes in detail, since these reactions proceed so slowly that their kinetics can be followed by conventional techniques.

7.3 Mechanism: nucleophilic substitution reaction in O_h complexes

Let us consider the substitution reaction wherein one of the ligands A is replaced by another ligand Y:

$$[MA_6] + Y \longrightarrow [MA_5Y] + A$$

There are two mechanisms to fully explain the occurrence of ligand substitution reactions in O_h complexes, and these are:

(a) Unimolecular nucleophilic substitution (S_N^1) or dissociative S_N^1 mechanism

(b) Bimolecular nucleophilic substitution (S_N^2) or associative S_N^2 mechanism

7.3.1 Unimolecular nucleophilic substitution (S_N^1) or dissociative S_N^1 mechanism

According to this mechanism, the complex in question first undergoes dissociation losing a ligand to be replaced and changes into a five-coordinate such as square pyramidal (SP) intermediate complex, which then readily adds the new incoming ligand Y. This is illustrated as follows:

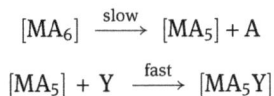

$$[MA_6] \xrightarrow{\text{slow}} [MA_5] + A$$

$$[MA_5] + Y \xrightarrow{\text{fast}} [MA_5Y]$$

Since the dissociation step is the slow step, it is the rate-determining step. Hence, the rate law for such a reaction may be represented as:

$$\text{Rate}(r) = k[MA_6]$$

Since the rate of the reaction depends only on the concentration of one species, $[MA_6]$, the mechanism is referred to as unimolecular nucleophilic substitution, that is S_N^1 mechanism. Since in this mechanism the complex first gets dissociated, it is also termed as dissociative S_N^1 mechanism.

7.3.2 Bimolecular nucleophilic substitution (S_N^2) or associative S_N^2 mechanism

According to this mechanism, the new incoming ligand first add on to the complex in question to form a seven-coordinate activated complex, which then readily undergoes dissociation to yield the final product. This may be shown as follows:

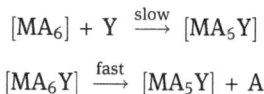

$$[MA_6] + Y \xrightarrow{slow} [MA_5Y]$$

$$[MA_6Y] \xrightarrow{fast} [MA_5Y] + A$$

As the step first in this case is slow, it is the rate-determining step. Hence, the rate law may be expressed as:

$$Rate(r) = k\,[MA_6][Y]$$

Since the rate of this substitution reaction depends on the concentration of two species, the mechanism is called bimolecular nucleophilic substitution, that is S_N^2. As in this mechanism, the incoming ligand first gets associated with the complex, it is called associative S_N^2 mechanism.

7.4 Types of intermediate/activated complex formed during S_N^1 and S_N^2 mechanism

7.4.1 S_N^1 mechanism

Let us consider an octahedral complex MA_5X being attacked by a nucleophile Y. If the reaction proceeds through S_N^1 mechanism, two types of intermediates can be expected:

(i) In the first type, the bond M–X dissociates causing least disturbance to the remaining MA_5 intermediate which has a SP geometry. This intermediate is then attacked by Y to give MA_5Y as shown in Figure 7.1.

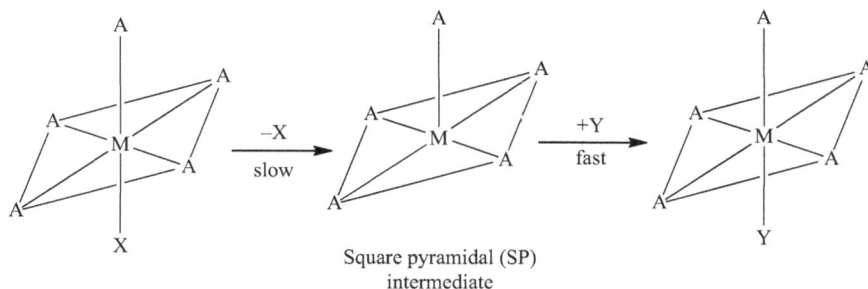

Square pyramidal (SP) intermediate

Figure 7.1: Formation of square pyramidal intermediate in S_N^1 mechanism.

(ii) In the second type, the M–X dissociates and the remaining MA_5 species immediately adjust the bond angles to produce trigonal bipyramidal (TBP) intermediate, which is then attacked by Y to produce MA_5Y as given in Figure 7.2.

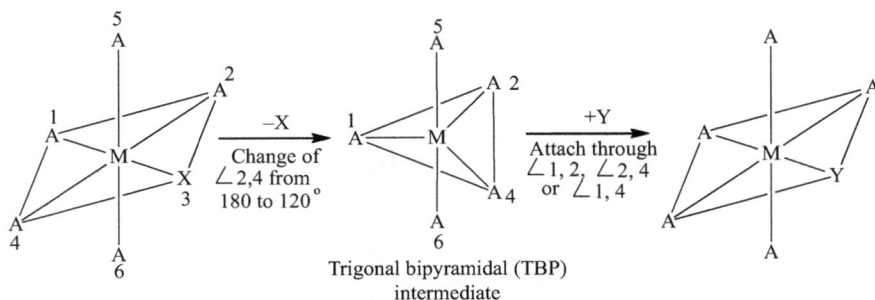

Figure 7.2: Formation of trigonal bipyramidal intermediate in S_N^1 mechanism.

It is very clear that the formation of TBP intermediate requires the movement of at least two metal–ligand bonds through change in angles whilst no such movement is required in the formation of SP intermediate. Thus, the S_N^1 mechanism generally proceeds through more stable SP intermediate. If TBP geometry is much more stabilized by π-bonding, then S_N^1 mechanism proceeds through TBP intermediate.

7.4.2 S_N^2 mechanism

If the reaction proceeds through S_N^2 mechanism, again two types of intermediate can be formed:

(i) The first type of intermediate is formed if the nucleophile Y attacks through one of the edges of the octahedron which leads to the formation of pentagonal bipyramidal (PBP), as shown in Figure 7.3.

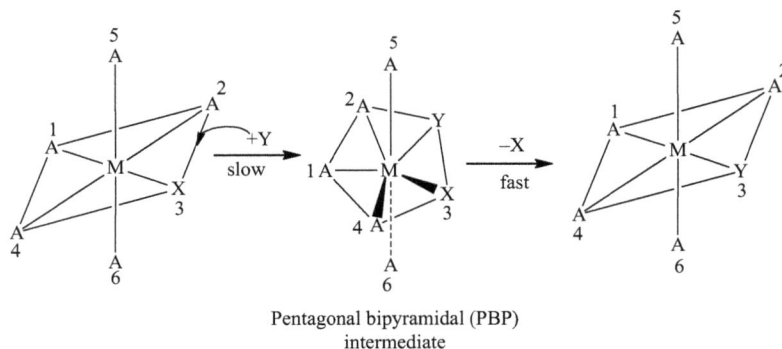

Figure 7.3: Formation of pentagonal bipyramidal intermediate in S_N^2 mechanism.

It is evident that the formation of PBP intermediate requires the movement of at least four ligands to adjust the nucleophile Y. Moreover, the ligand–ligand repulsion also increases the energy of PBP intermediate. This is because the decrease in A–M–A bond angles brings the electron pairs of metal–ligand bonds nearer to one another in this intermediate.

(ii) The second type of intermediate is formed if the molecule Y attacks through the middle of one of the triangular face of the octahedron. As soon as Y starts approaching the central metal, M, the outgoing ligand X starts moving towards the middle of another triangular face so that the 'octahedral wedge' (OW) intermediate formed has both X and Y ligands in equivalent position. The formation of an OW intermediate is shown in Figure 7.4.

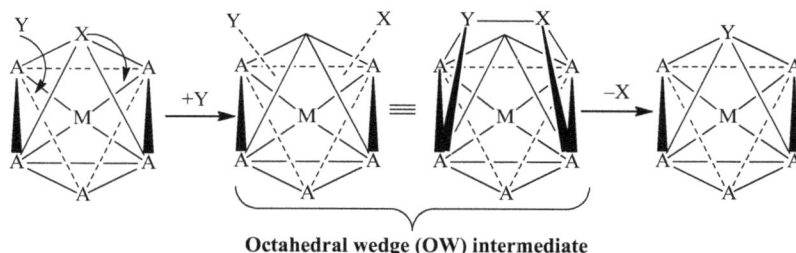

Octahedral wedge (OW) intermediate

Figure 7.4: Formation of octahedral wedge intermediate in S_N^2 mechanism.

The formation of OW intermediate requires minimum movement of ligands and the ligand–ligand repulsions are less compared to PBP intermediate. The S_N^2 reaction, therefore, would preferably proceed through an OW intermediate rather than PBP intermediate.

7.5 Lability and inertness in octahedral complexes: kinetically labile and kinetically inert complexes

Depending on whether the rate of substitution reaction in complexes is slow or fast, the complexes have been classified as labile and inert complexes by Taube (Figure 7.5). Since the term labile and inert refers to the rate of substitution reaction, labile and inert complexes are called kinetically labile and kinetically inert complexes, respectively.

Figure 7.5: Henry Taube (1915–2005): Nobel Prize (1983) for his work on the 'mechanisms of electron transfer reactions, especially in metal complexes'.

7.5.1 Kinetically labile complexes

Such complexes are those in which one or more ligands present in coordination sphere can be replaced by other ligands quickly or rapidly. Thus, the complexes in which the ligand substitution is fast are labile complexes. The rate of substitution of labile complexes is difficult to measure and hence special techniques are used to study the kinetics of such complexes.

7.5.2 Kinetically inert complexes

Such complexes are those in which the ligands present in the coordination sphere can be replaced slowly but not unable to be replaced. Thus, the complexes in which the ligand substitution is slow are called inert complexes. The rate of substitution of inert complexes can be measured easily by conventional techniques.

7.6 Lability/internees versus thermodynamically stable/ unstable complexes

A thermodynamically stable complex has high value of its formation constant whilst a thermodynamically unstable complex has low value of its formation constant. It should be understood clearly that a thermodynamic stable complex may be kinetically labile (fast reacting complex). Similarly, a thermodynamically unstable complex may be kinetically inert (slow reacting complex). Thus, thermodynamic stability and kinetic stability are different from each other. Thermodynamically stable complex may be labile or inert. For example:

(i) $[Fe(H_2O)_6]^{3+}$ (bond energy, 116 Kcal/mol) and $[Cr(H_2O)_6]^{3+}$ (bond energy, 122 Kcal/mol) have the almost the same thermodynamic stability, but $[Fe(H_2O)_6]^{3+}$ is kinetically labile (exchanges its ligands with other ligands rapidly) whilst $[Cr(H_2O)_6]^{3+}$ is kinetically inert (exchanges ligands slowly).

(ii) $[Co(NH_3)_6]^{3+}$ is thermodynamically unstable because high value of dissociation constant, $[K_{diss} = 10^{25}]$ (or low β value, 1×10^{-25}) but is kinetically inert because it remains undecomposed in acidic solution (or its reaction with H_2O in acidic medium is very slow).

$$\left[Co(NH_3)_6\right]^{3+} + 6H_3O^+ \rightleftharpoons \left[Co(H_2O)_6\right]^{3+} + 6NH_4^+ ; K_{diss} = 10^{25}$$

(iii) $[Ni(CN)_4]^{2-}$ is thermodynamically stable because of low value of its dissociation constant $[K_{diss} = 10^{-22}]$ but is kinetically labile, since it exchanges CN^- ion very rapidly when added with isotopically labelled cyanide ion ($^{14}CN^-$).

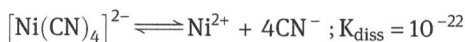

$$\left[Ni(CN)_4\right]^{2-} \rightleftharpoons Ni^{2+} + 4CN^- ; K_{diss} = 10^{-22}$$

$$\left[Ni(CN)_4\right]^{2-} + {}^{14}CN^- \longrightarrow \left[Ni({}^{14}CN)_4\right]^{2-} + CN^-$$

Thus, the kinetic nature of the complexes, that is, lability and inertness, is not decided by the thermodynamic stability alone. In fact, it is influenced by other factors such as the nature of the other reactant, temperature, solvent and so on.

7.7 Interpretation of lability and inertness of transition metal complexes

Attempts have been made to explain lability and inertness of transition metal complexes on the basis of **valence bond theory** (VBT) as well as **crystal field theory** (CFT). The VBT is simple in approach but offers only partial explanation for the lability and inertness of complexes. However, the CFT explains the phenomenon more satisfactorily.

7.7.1 The valence bond theory approach

(i) According to the VBT, transition metal complexes undergoing substitution reactions **through the dissociative $S_N{}^1$ mechanism** would be **labile** if the metal–ligand bonds **are comparatively weak** and would **be inert** if such bonds are comparatively **strong**.

(ii) Since the outer nd orbitals are associated with higher energy than the inner $(n-1)d$ orbitals, the metal–ligand bonds utilizing metal hybrid orbitals containing contribution from outer nd orbitals would be less stable than the metal–ligand bonds utilizing metal hybrid orbitals containing contribution from inner $(n-1)d$ orbitals.

(iii) Consequently, the outer orbital complexes in which metal ion utilizes $nsnp^3nd^2$ hybrid orbitals for metal–ligand bonding would permit easier delinking of outgoing ligands compared to the inner orbital octahedral complexes in which metal ion utilizes $(n-1)d^2nsnp^3$ hybrid orbitals for metal–ligand bonding. It is notable here that two $(n-1)d$ orbitals are the orbitals of e_g set and the three orbitals of t_{2g} set remains empty or contains the electrons of the metal ion.

(iv) Thus, according to VBT, **all outer orbital (high-spin) complexes would be labile** and all **inner orbital complexes would be inert** if substitution reactions **proceed through S_N^1 mechanism.**

(v) It is notable here that the metal–ligand bonds in inner orbital complexes are quite strong and hence require considerable energy for their dissociation. Consequently, **the inner orbital (low-spin) complexes would prefer to follow an associative S_N^2 mechanism** rather than dissociative S_N^1 mechanism in their substitution reaction.

Outcome of the VBT for lability and inertness

(a) Inner orbital complexes containing metal ions with at least one empty $(n-1)d$ t_{2g} orbital [left after hybridization] which can accommodate electrons form the incoming ligand (nucleophile) which would facilitate the formation of seven-coordinated intermediate. **Such complexes would, therefore, be labile.** Examples of such **labile** inner orbital complexes are:

(i) O_h complexes of Sc^{3+}, Y^{3+}, Zr^{4+}, Ce^{4+} (all d^0 system) (Ce_{58}: $4f^25d^06s^2$)

(ii) O_h complexes of Ti^{3+}, V^{4+}, Mo^{5+} (all d^1 system)

(iii) O_h complexes of Ti^{2+}, V^{3+}, Nb^{3+}, Mo^{4+} (all d^2 system)

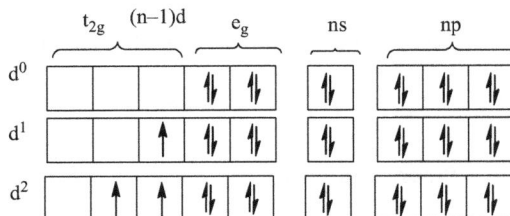

(b) Inner orbital complexes in which the metal ions have **at least one electron in each of the $(n-1)d$ t_{2g} orbital** [left after hybridization] will have to accommodate electrons from incoming nucleophile into an outer higher energy nd orbital. Such complexes would, therefore, form 7-coodinate intermediate with difficulty and would thus be inert. Examples of such inner orbital complexes are the O_h complexes with d^3, d^4, d^5 and d^6 metal ion configurations.

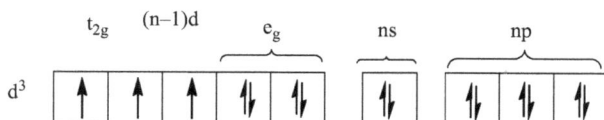

(c) Some exceptions to the above generalizations are also known. For example, O_h complexes of Ni(II) with d^8 configuration are predicted by VB theory to be *labile* because according to this theory Ni(II) will utilize $nsnp^3nd^2$ hybrid metal orbitals for metal ligand bonding (outer orbital complex), but experimentally they are found to be inert. This and some other exceptions are, however, well explained by the CFT.

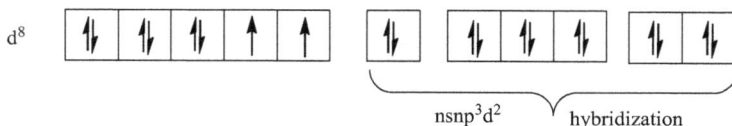

7.7.2 Crystal field theory

In order to explain lability and inertness of octahedral complexes, it is worthwhile to differentiate between activation energy and crystal field activation energy (CFAE).

Activation energy

Activation energy is defined as the energy required for transforming the reacting complex into the transition state, that is, the intermediate. The activation energy is made up of energy changes in metal–ligand bond lengths, bond angles, ligand–ligand repulsions and crystal field stabilization energies (CFSEs), etc.

Crystal field activation energy

CFAE is defined as the change inCFSE when the reacting complex is transferred into the transition state, that is, the intermediate. Thus,

$$\text{CFAE} = \text{CFSE of intermediate} - \text{CFSE of reacting complex}$$

The CFAE is only a part of the overall activation energy. Since the geometry of the reacting complex and the intermediate are different, the order and extent of splitting of d-orbitals of metal ion in the two species would be different. Hence, CFSE of the reacting complex would be different from the CFSE of the intermediate.

Evaluation of CFAE

The following assumptions are made for calculating CFSEs from which CFAEs are easily calculated:

(i) The geometry of the reacting complex is assumed to be O_h even if all the six ligands are not identical.

(ii) Interelectronic repulsions amongst the d-electrons are assumed to be negligible.

(iii) The Dq for the reacting complex is assumed to be the same as the Dq for its intermediate. Actually the two Dqs are quite different due to different geometries of the complex and intermediate.

(iv) The Jahn–Teller effect which causes distortions in octahedral geometries is assumed to be negligible.

Because of the above drastic assumptions, some of the calculated CFAEs come out to be negative. However, such CFAEs, when calculated by employing more exact but very tedious calculations come out to be either zero or small but are never negative. Therefore, negative CFAEs can be roughly taken as zero.

CFAE: lability and inertness

According to CFT, if the calculated CFAE is negative or zero, the reacting complex would require less energy for its transformation into the intermediate. On the other hand, if the calculated CFAE is high, the reacting complex would require more energy for its transformation into the intermediate.

It is notable here that since CFAE is only a part of the activation energy, a valid comparison of lability and inertness on the basis of CFAE can be made only if all the other factors which contribute to the activation energy are more or less the same for the reacting complex and the intermediate.

We know that the ligand substitution reaction in O_h complexes proceed either through S_N^1 or S_N^2 mechanism. In order to explain lability and inertness of O_h complexes by CFT, we calculate the value of CFAE for various d^n ions of the complexes in both the mechanisms. If the value of CFAEs is negative or zero, the complex is labile. If the value of CFAE is positive, then the complex is inert.

CFAE and S_N^1 substitution reaction

We are aware that under normal conditions, the S_N^1 reactions proceed through lower energy SP intermediate instead of higher energy TBP intermediate. CSAEs of O_h complexes undergoing S_N^1 substitution reactions through SP intermediates for different d^n configurations are given in Table 7.1.

Table 7.1: CSAEs of O_h complexes undergoing S_N^1 substitution reactions through SP for different d^n configurations.

d^n ion	Weak field complexes (HS)/spin free			Strong field complexes (LS)/spin paired		
	CFSE for:		CFAE (b–a)	CFSE for:		CFAE (b-a)
	O_h complexes (a)	SP intermediate (b)		O_h complexes (a)	SP intermediate (b)	
d^0	0	0	0	0	0	0
d^1	−4	−4.57	−0.57	−4	−4.57	−0.57
d^2	−8	−9.14	−1.14	−8	−9.14	−1.14
d^3	−12	−10.00	2.00	−12	−10.00	2.00
d^4	−6	−9.14	−3.14	−16	−14.57	1.43
d^5	0	0	0	−20	−19.14	0.86
d^6	−4	−4.57	−0.57	−24	−20.00	4.00
d^7	−8	−9.14	−1.14	−18	−19.14	−1.14
d^8	−12	−10.00	2.00	−12	−10.00	2.00
d^9	−6	−9.14	−3.14	−6	−9.14	−3.14
d^{10}	0	0	0	0	0	0

From Table 7.1, the following conclusions may be drawn:

(i) Amongst the weak field/high spin O_h complexes, metal ion with $d^0, d^1, d^2, d^4, d^5, d^6, d^7, d^9$ and d^{10} configurations have negative or zero CFAEs, and would therefore be labile whereas those containing d^3 and d^8 configurations of metal ion have positive CFAEs, and would, therefore, be slow to react or inert.

(ii) Amongst the strong field/low spin O_h complexes, metal ion with d^0, d^1, d^2, d^7, d^9 and d^{10} configurations have negative or zero CFAEs, and would, therefore, be labile whereas those containing d^3, d^4, d^5, d^6 and d^8 configurations of metal ion have positive CFAEs, and would, therefore, be slow to react. Due to decreasing order of positive CFAE values, the order of reactivity of the complexes is $d^6 < d^3 \sim d^8 < d^4 < d^5$.

CFAE and S_N^2 substitution reaction

As we know that an OW intermediate requires less energy for its formation than the PBP intermediate, the S_N^2 reactions proceed through lower energy OW intermediates rather than higher energy PBP intermediate. CSAEs of O_h complexes undergoing S_N^2 substitution reactions through OW intermediates for d^n configurations are given in Table 7.2.

Table 7.2: CSAEs of O_h complexes undergoing S_N^2 substitution reactions through OW for different d^n configurations.

d^n ion	Weak field complexes (HS)			Strong field complexes (LS)		
	CFSE for:		CFAE (b-a)	CFSE for:		CFAE (b-a)
	O_h complexes (a)	OW intermediate (b)		O_h complexes (a)	OW intermediate (b)	
d^0	0	0	0	0	0	0
d^1	−4	−6.08	−2.08	−4	−6.08	−2.08
d^2	−8	−8.68	−0.68	−8	−8.68	−0.68
d^3	−12	−10.20	1.80	−12	−10.20	1.80
d^4	−6	−8.79	−2.79	−16	−16.26	−0.26
d^5	0	0	0	−20	−18.86	1.14
d^6	−4	−6.08	−2.08	−24	−20.37	3.63
d^7	−8	−8.68	−0.68	−18	−18.98	−0.98
d^8	−12	−10.20	1.80	−12	−10.20	1.80
d^9	−6	−8.79	−2.79	−6	−8.79	−2.79
d^{10}	0	0	0	0	0	0

We draw the following conclusions from Table 7.2:

(i) Amongst the weak field/high spin O_h complexes, metal ion with $d^0, d^1, d^2, d^4, d^5,$ d^6, d^7, d^9 and d^{10} configurations have negative or zero CFAES, and would therefore be labile. The complexes with d^3 and d^8 configurations of metal ion have positive CFAEs. They would, therefore, be slow to react.

(ii) Amongst the strong field/low spin O_h complexes, metal ion with $d^0, d^1, d^2, d^4,$ d^7, d^9 and d^{10} configurations have negative or zero CFAES, and would, therefore, be labile whereas those containing d^3, d^5, d^6 and d^8 configurations of metal ion have positive CFAEs, and would, therefore, be slow to react. Due to decreasing order of positive CFAE values, the order of reactivity of the complexes is $d^6 < d^3 \sim d^8 < d^5$.

From the combined CFAE data given in Tables 7.1 and 7.2, the following conclusions can be drawn:

(a) Octahedral complexes of metal ion involving d^3 and d^8 configurations and spin paired octahedral complexes of d^5 and d^6 metal ions are inert either by S_N^1 or by S_N^2 mechanism.

(b) Octahedral complexes with d^0, d^1, d^2, d^9 and d^{10} configurations and high spin complexes of d^4, d^5, d^6 and d^7 configurations would be labile whatever the mechanism of their substitution reactions is.

7.8 Substitution reactions in octahedral complexes: acid and base hydrolysis

Most of the kinetics studies on substitution reactions of octahedral complexes have been carried out in aqueous medium. Consider a substitution reaction of an O_h complex $[MA_5X]^{n+}$ in presence of ligand Y^- in aqueous medium. Since in aqueous medium much more water is present compared to Y^-, it is always the aquation of the complex that occurs first.

$$[MA_5X]^{n+} + H_2O \longrightarrow [MA_5(H_2O)]^{(n+1)} + X^- \qquad (7.3)$$

The Y^- would then replace the coordinated water to give $[MA_5Y]^{n+}$ provided Y^- is present in appreciable amount.

$$[MA_5(H_2O)]^{(n+1)} + Y^- \longrightarrow [MA_5Y]^{n+} + H_2O \qquad (7.4)$$

Since all substitution reactions in aqueous medium proceed through the reaction of the complex with water, it is imperative to study the mechanism of substitution reactions of complexes with water.

The substitution reaction in aqueous medium in which water molecule replaces a coordinated ligand from the complex species is termed as **aquation reaction or acid hydrolysis** (reaction 7.3).

The substitution reaction in aqueous medium in which anion of the water, that is, OH^- replaces a coordinated ligand from a complex species is known as **base hydrolysis** of an octahedral complex.

$$[MA_5X]^{n+} + OH^- \longrightarrow [MA_5(OH)]^{n+} + X^-$$

Since some OH- is always present due to auto ionization of H_2O, some $[MA_5(OH)]^{n+}$ is always formed along with $[MA_5(H_2O)]^{(n+1)}$ during the hydrolysis of $[MA_5X]^{n+}$ even in neutral aqueous medium.

7.9 Mechanism of acid hydrolysis (aquation) of different types of O_h complexes

Mechanistic studies on the hydrolysis of a number of octahedral complexes of Co (III) and other metal ions are reported in the literature. We shall, however, discuss only few representative cases.

7.9.1 Mechanism of acid hydrolysis when no inert ligand in the complex is a π-donor or π-acceptor

Let us consider the acid hydrolysis of $[MA_5X]^{n+}$ complex in which As are inert ligands (ligands which remain attached with the metal ion in the products also), and none of these ligands is a π-donor or π-acceptor. The acid hydrolysis of this complex may be represented as:

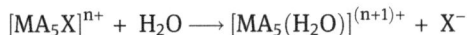

$$[MA_5X]^{n+} + H_2O \longrightarrow [MA_5(H_2O)]^{(n+1)+} + X^-$$

S_N^1 mechanism

In S_N^1 mechanism of acid hydrolysis, the slow, that is, the rate-determining step is the one in which the M–X bond dissociate to form a five-coordinated complex as intermediate. This then quickly reacts with water to give the hydrolysis product, as shown below:

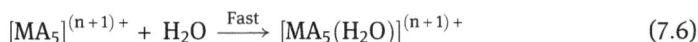

$$[MA_5X]^{n+} \xrightarrow{\text{Slow}} [MA_5]^{(n+1)+} + X^- \qquad (7.5)$$

$$[MA_5]^{(n+1)+} + H_2O \xrightarrow{\text{Fast}} [MA_5(H_2O)]^{(n+1)+} \qquad (7.6)$$

Hence, Rate of the reaction $= k_1\{[MA_5X]^{n+}\}$

S_N^2 mechanism

In S_N^2 mechanism of acid hydrolysis, the slow, that is, the rate-determining step is the one in which a seven-coordinated intermediate is formed which then readily gives up the leaving group X to yield the hydrolysis product, as shown below:

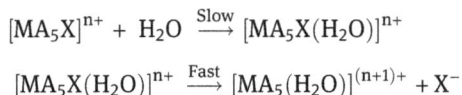

$$[MA_5X]^{n+} + H_2O \xrightarrow{\text{Slow}} [MA_5X(H_2O)]^{n+}$$

$$[MA_5X(H_2O)]^{n+} \xrightarrow{\text{Fast}} [MA_5(H_2O)]^{(n+1)+} + X^-$$

Hence, Rate of the reaction $= k_2\{[MA_5X]^{n+}\}[H_2O]]$

Since water is present in large abundant quantity, its concentration, which is $[H_2O]$, remains constant. Hence,

$$\text{Rate of the reaction} = k_2\{[MA_5X]^{n+}\} \times \text{constant}$$

or $\qquad\qquad$ Rate of the reaction $= k_{2'}.\{[MA_5X]^{n+}\}$

It is clear from the above discussion that both the mechanisms predict that the rate of hydrolysis is dependent only on the concentration of the complex $[MA_5X]^{n+}$. This has been verified experimentally. Thus, the kinetic measurement of hydrolysis process is not able to decide whether the hydrolysis reaction proceeds through

S_N^1-mechanaism or through S_N^2 mechanism. Consequently we have to look over other factors to decide the type of mechanism. These factors are:

(i) Charge on the substrate

It has been observed during the hydrolysis of several octahedral complexes of Co (III) and other metal ions that the rate of hydrolysis of a complex decreases with increase in the charge of the complex/substrate. For example:

(a) The rate of hydrolysis of cis-[Co(en)$_2$Cl$_2$]$^+$ is some 100 times faster than that of cis-[Co(en)$_2$Cl(H$_2$O)]$^{2+}$.

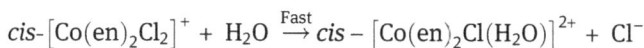

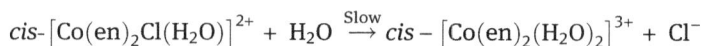

$$cis\text{-}\left[Co(en)_2Cl_2\right]^+ + H_2O \xrightarrow{Fast} cis - \left[Co(en)_2Cl(H_2O)\right]^{2+} + Cl^-$$

$$cis\text{-}\left[Co(en)_2Cl(H_2O)\right]^{2+} + H_2O \xrightarrow{Slow} cis - \left[Co(en)_2(H_2O)_2\right]^{3+} + Cl^-$$

(b) Similarly, the rate constants for the aquation of complexes [RuCl$_6$]$^{3-}$, [RuCl$_3$ (H$_2$O)$_3$]^0 and [RuCl (H$_2$O)$_5$]$^{2+}$ decrease regularly with increase in the positive charge, the rate constant being $1.0\ s^{-1}$, $2.1 \times 10^{-6}\ s^{-1}$ and $\sim 10^{-8}\ s^{-1}$, respectively.

The above observations favour a dissociative S_N^1 path for the aquation process since the increase in +ve charge on the substrate would obviously render the dissociation of leaving group from M more difficult, resulting in slower reaction rate by this mechanism.

    If the hydrolysis is supposed to proceed through S_N^2 mechanism, the rate of hydrolysis would remain unchanged with increase in the charge on the complex. This is because the increase in the charge would not only make the breaking of the M–X bond more difficult but would also make the formation of the M–H$_2$O bond easier. Thus, the two processes kinetically affect in opposite directions. Consequently, the overall aquation process would remain practically unchanged with any change in the charge on the substrate.

(ii) Strength of metal-leaving group bond: basicity of the leaving group

The observed rate constants (k) for hydrolysis of [Co(NH$_3$)$_5$X] are as given in Table 7.3.

Table 7.3: Observed rate constants (k) for hydrolysis of [Co(NH$_3$)$_5$X] with different leaving group X.

Leaving group (X$^-$)	Dissociation constant of M–X bond	Rate constant ($k_{hydrolysis}$)
CF$_3$COO$^-$	2.0×10^{-14}	5.5×10^{-3}
CCl$_3$COO$^-$	5.0×10^{-14}	5.4×10^{-3}

Table 7.3 (continued)

Leaving group (X⁻)	Dissociation constant of M–X bond	Rate constant ($k_{hydrolysis}$)
$CHCl_2COO^-$	2.0×10^{-13}	1.6×10^{-3}
CH_2ClCOO^-	7.1×10^{-12}	0.6×10^{-3}
$CH_3CH_2COO^-$	6.6×10^{-10}	0.3×10^{-3}

Basicity of the of the leaving group (X⁻) increases as:

$$CF_3COO^- < CCl_3COO^- < CHCl_2COO^- < CH_2ClCOO^- < CH_3CH_2COO^-$$

Rate of hydrolysis decreases as:

$$CF_3COO^- > CCl_3COO^- > CHCl_2COO^- > CH_2ClCOO^- > CH_3CH_2COO^-$$

It is evident from the data shown above that the rate of hydrolysis goes on decreasing with increase in basicity of the leaving group X⁻.

Since the strength of the M–X bond is directly proportional to the basicity of the group, it can be said that the rate of hydrolysis goes on deceasing with increase of the strength of M–X bond. This indicates that the rate-determining step in the aquation process does involve the dissociation of M–X bond. The data thus clearly supports a dissociative S_N^1 mechanism for the hydrolysis of the octahedral complex.

(iii) Inductive effect of the inert ligands/group

It has been observed that the rate constants of acid hydrolysis reaction,

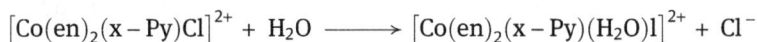

$$[Co(en)_2(x-Py)Cl]^{2+} + H_2O \longrightarrow [Co(en)_2(x-Py)(H_2O)l]^{2+} + Cl^-$$

increase with the CH_3 substitution in pyridine. Here, x-Py stands for various derivatives of pyridine obtained by removing one of the H atoms of pyridine by CH_3 group. The increase in the rate constants is due to inductive effect caused by increasing CH_3 substitution which results in distorting/repelling electron density towards Co-centre and thus helps in dissociation of Cl^- ion (leaving group). This again confirms that the acid hydrolysis reaction given above occurs through dissociative S_N^1 mechanism.

The values of the rate constants (in s^{-1}) for the acid hydrolysis reaction:

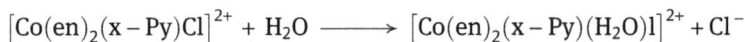

$$[Co(en)_2(x-Py)Cl]^{2+} + H_2O \longrightarrow [Co(en)_2(x-Py)(H_2O)l]^{2+} + Cl^-$$

in which x-Py = pyridine, 3-methyl pyridine and 4-methyl pyridine are 1.1×10^{-5}, $1.3 \times 10^{-5} s^{-1}$ and $1.4 \times 10^{-5} s^{-1}$, respectively.

(iv) Steric effects

In the complexes of *trans*-$[Co(LL)_2Cl_2]^+$ type, if the bidentate ligand LL is $NH_2-CH_2-CH_2-NH_2, NH_2-CH_2-CH(CH_3)-NH_2$ or $dl-NH_2-CH(CH_3)-CH(CH_3)-NH_2$ then the ligand LL becomes more bulky. Due to the increase in the bulk of the ligand, the steric overcrowding of the ligand around the central metal ion (CO^{3+}) also increases. Because of the increase of the overcrowding around the CO^{3+}, another ligand cannot be taken by the complex, that is, there is no possibility of S_N^2 mechanism. Contrary to this, the removal of the leaving group Cl^- will reduce the overcrowding of the ligand around the central metal ion CO^{3+}. This gives the evidence of the S_N^1 mechanism.

The increase in the bulk of the ligand (LL) also increases the value of the rate constant for the acid hydrolysis of the complexes as established from the results of following acid hydrolysis reactions:

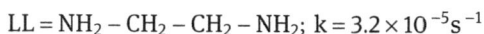

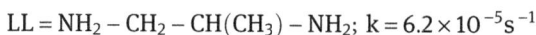

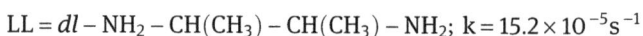

$$trans - \left[Co(LL)_2Cl_2\right]^+ + H_2O \longrightarrow \left[Co(LL)_2Cl\ (H_2O)\right]^{2+} + Cl^-$$

$$LL = NH_2 - CH_2 - CH_2 - NH_2;\ k = 3.2 \times 10^{-5}s^{-1}$$

$$LL = NH_2 - CH_2 - CH(CH_3) - NH_2;\ k = 6.2 \times 10^{-5}s^{-1}$$

$$LL = dl - NH_2 - CH(CH_3) - CH(CH_3) - NH_2;\ k = 15.2 \times 10^{-5}s^{-1}$$

Let us compare the rate of hydrolysis of $[Co(NH_2-CH_2-CH_2-CH_2-NH_2)_2Cl_2]^+$ (I, propylenediamine) and $[Co(NH_2-CH_2-CH_2-NH_2)_2Cl_2]^+$ (II, ethylenediamine). Complex (I) contains 6-membered chelate ring whilst complex (II) has 5-membered chelate ring. Now, since the 6-membered chelate ring in complex (I) produces greater steric strain around the central metal ion than 5-membered chelate ring present in complex (II), the aquation of (I) will be faster than that of (II) through dissociative S_N^1 mechanism. Experimentally, this prediction has been found true $[k = 1,000 \times 10^{-5}\ s^{-1}$ (complex I) and $k = 3.2 \times 10^{-5}\ s^{-1}$ (complex II)].

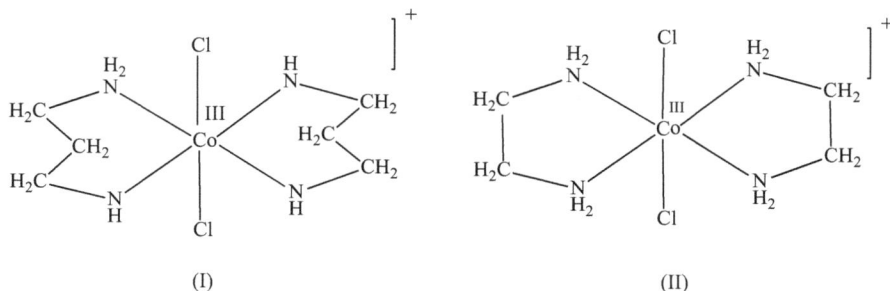

(I)   (II)

Similarly, the rate of constants of other two Co(III) complexes containing the bulkier ligand mentioned above can be explained on the basis of steric stain through dissociative S_N^1 mechanism.

(v) Solvation effects

Solvation theory is based on the following facts:

(i)   In aqueous medium, the reacting species, the intermediate and the products are present in the hydrated state.

(ii) The hydration of any species decreases its energy and thus stabilizes it. The extent of stabilization is related to the extent of hydration of the species.

(iii) The greater the charge and smaller the size of the species, the greater would be the extent of its hydration, and hence extent of its stabilization.

Let us consider the rate of aquation of the complexes: (a) $[Co(NH_3)_5Cl]^{2+}$, (b) $[Co(en)(NH_3)_3Cl]^{2+}$, (c) $[Co(en)(diene)Cl]$ and (d) $[Co(tetraene)Cl]$. It has been observed that the rate of aquation of these complexes decreases in the order of their mention.

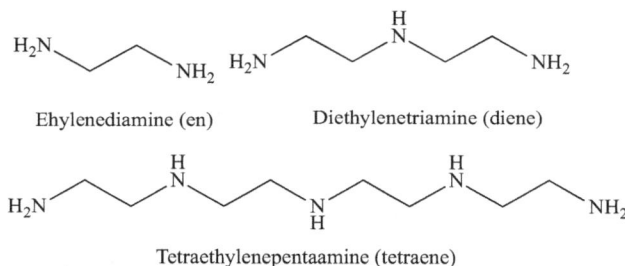

Ehylenediamine (en)   Diethylenetriamine (diene)

Tetraethylenepentaamine (tetraene)

It is noticeable from the above-mentioned order that the rate of aquation goes on decreasing as the extent of chelation in the reacting complex goes on increasing. Even though, the above-mentioned order of aquation can be explained by S_N^2 (trans) attack of the water to some extent, yet it cannot explain satisfactorily the relative rate of aquation of highly chelated complexes (c) and (d).

The salvation theory, however, explains the relative rates of aquation of all the chelated complexes through S_N^1 mechanism in a satisfactory manner as follows:

(i)   The five-coordinated intermediate formed during aquation through S_N^1 mechanism is smaller in size than the seven-coordinated intermediate formed through S_N^2 mechanism.

(ii) The five-coordinated intermediate being smaller in size, would get hydrated and thus get stabilized to a greater extent than the seven-coordinated intermediate.

(iii) The greater stability of the five-coordinated intermediate makes the S_N^1 mechanism more feasible than the S_N^2 mechanism.

(iv) The presence of chelating molecules, such as en, diene, tetraene and so on, in place of NH_3 causes an increase in the size of the complex ion without causing any appreciable increase in the steric overcrowding around the central Co(III) ion.

Let us consider the aquation/hydrolysis of (a) $[Co(NH_3)_5Cl]^{2+}$ (smaller size) and (b) $[Co(en)(NH_3)_3Cl]^{2+}$ (bigger size). Assuming that the process involves S_N^1 mechanism, the aquation reactions can be represented as follows:

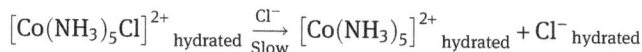

$$\left[Co(NH_3)_5Cl\right]^{2+}_{hydrated} \xrightarrow[\text{Slow}]{Cl^-} \left[Co(NH_3)_5\right]^{2+}_{hydrated} + Cl^-_{hydrated}$$

(a)   (I)

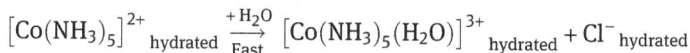

$$\left[Co(NH_3)_5\right]^{2+}_{hydrated} \xrightarrow[\text{Fast}]{+H_2O} \left[Co(NH_3)_5(H_2O)\right]^{3+}_{hydrated} + Cl^-_{hydrated}$$

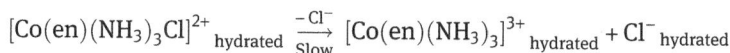

$$[Co(en)(NH_3)_3Cl]^{2+}_{hydrated} \xrightarrow[\text{Slow}]{-Cl^-} [Co(en)(NH_3)_3]^{3+}_{hydrated} + Cl^-_{hydrated}$$

(b)

$$[Co(en)(NH_3)_3]^{3+}_{hydrated} \xrightarrow[\text{Fast}]{+H_2O} [Co(en)(NH_3)_3(H_2O)]^{3+}_{hydrated} + Cl^-_{hydrated}$$

(II)

Now, since the intermediate (II) of the chelated complex is bigger in size than the intermediate (I) of the complex containing unidentate NH_3 ligands, the former (II) is stabilized by hydration to a lesser extent and thus requires more energy than the latter (I). Consequently, complex (b) would be aquated at a slower rate than complex (a).

Similarly, it can be inferred that a chelate complex of a bigger size would be aquated at a slower rate than a chelate complex of a smaller size provided the chelate rings which increase the sizes of the chelate complexes do not bring about any appreciable change in the steric environments around the central metal ions.

7.9.2 Mechanism of acid hydrolysis in octahedral complexes in which the inert ligand is a π-donor

Let us consider the aquation reaction of *cis*-$[Co(en)_2(OH)Cl]^+$ and *cis*-$[Co(en)_2(NH_3)Cl]^{2+}$ represented in Figure 7.6.

It may be seen from the values of rate constant (k) that the rate of aquation of *cis*-$[Co(en)_2(OH)Cl]^+$ is much higher than that of *cis*-$[Co(en)_2(NH_3)Cl]^{2+}$. The difference in the values of k of the two complexes can be explained on the basis of the capacity of OH^- and NH_3 ligands to form π-bonding. The OH^- has filled p-orbitals ($O^- = 2s^2 2p^5$) whilst NH_3 has no such orbital. The lone pair of electrons on N atom in NH_3 molecule is used up in coordination. The central metal ion Co^{3+} of SP intermediate formed during aquation has one empty d^2sp^3 hybrid orbital. This empty hybrid orbital overlaps with a filled p-orbital on OH^- to form a π-bond as shown in Figure 7.7.

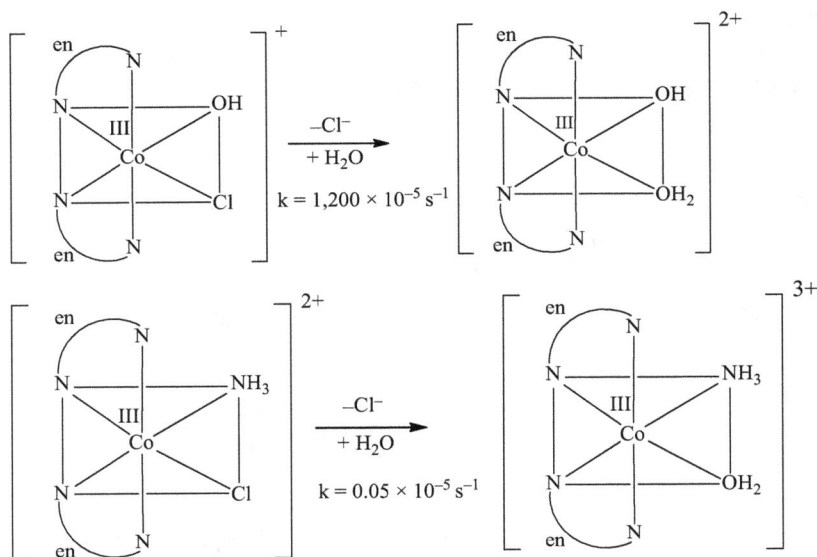

Figure 7.6: Aquation reaction of *cis*-[Co(en)$_2$(OH)Cl]$^+$ and *cis*-[Co(en)$_2$(NH$_3$)Cl]$^{2+}$ complexes.

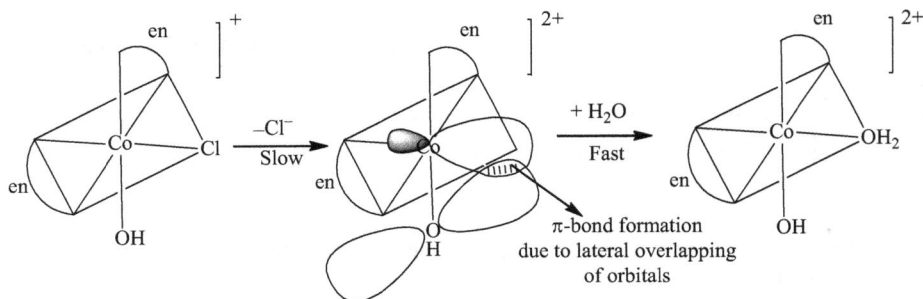

Figure 7.7: Formation of a π-bond in square pyramidal intermediate by lateral overlapping of one empty d^{2}sp^3 hybrid orbital of Co^{3+} ion with a filled p-orbital on OH.

Because of stabilization of SP intermediate through π-bonding, the aquation of *cis*-[Co(en)$_2$(OH)Cl]$^+$ complex becomes much easier than the aquation of *cis*-[Co(en)$_2$ (NH$_3$)-Cl]$^{2+}$ whose SP intermediate cannot be stabilized by π-bonding.

Let us now consider the aquation reaction of *trans*-[Co(en)$_2$(OH)Cl]$^+$ and *trans*-[Co(en)$_2$(NH$_3$)Cl]$^{2+}$ complexes (Figure 7.8) through S$_N^1$ mechanism represented as follows:

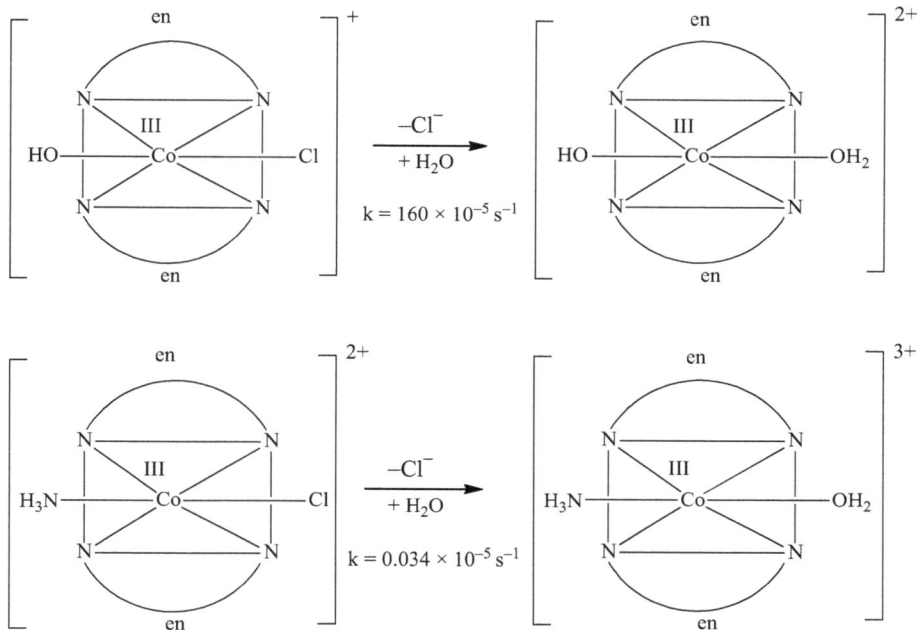

Figure 7.8: Aquation reaction of *trans*-[Co(en)$_2$(OH)Cl]$^+$ and *trans*-[Co(en)$_2$(NH$_3$)Cl]$^{2+}$ complexes through S$_N^1$ mechanism.

The values of rate constants of aquation reactions indicate that the aquation of *trans*-[Co(en)$_2$(OH)Cl]$^+$ complex is also much higher than the aquation of *trans*-[Co(en)$_2$(NH$_3$)Cl]$^{2+}$ complex. However, the empty d^{2}sp^3 hybrid orbital of the central metal ion in SP intermediate of the *trans*-[Co(en)$_2$(OH)Cl]$^+$ complex is incapable of forming π-bond with any filled p-orbitals of the coordinated OH$^-$ ligand because of lack of symmetry as shown in Figure 7.9:

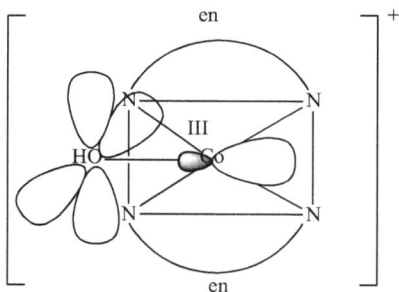

Figure 7.9: Square pyramidal intermediate incapable of forming π-bond due to lack of symmetry.

The above observation rules out the formation of SP intermediate during the aquation of *trans*-[Co(en)$_2$(OH)Cl]$^+$ complex. The fact that still remains is how to explain the faster aquation rate of *trans*-[Co(en)$_2$(OH)Cl]$^+$ complex compared to the aquation rate of *trans*-[Co(en)$_2$(NH$_3$)Cl]$^{2+}$ complex. The formation of TBP intermediate is another possibility to be considered for the aquation of hydroxo *trans* complex which can be stabilized by π-bonding. Looking over the TBP intermediate, it is well clear that it is being stabilized by π-bonding involving the overlap of the empty d-orbital of the metal with the filled p-orbital of the coordinated OH$^-$ ligand as shown in Figure 7.10:

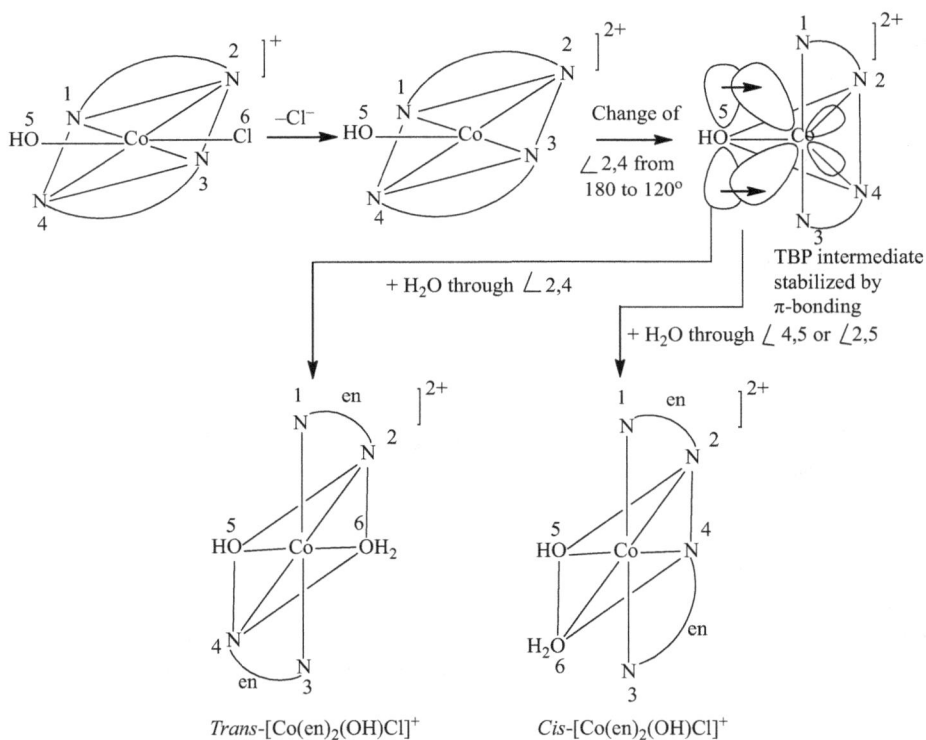

Figure 7.10: Representation of aquation of *trans*-[Co(en)$_2$(OH)Cl]$^+$ complex through S$_N$^{1} mechanism involving the formation of a TBP intermediate.

It is notable here that the formation of TBP intermediate from a SP intermediate requires some energy because some bond angles have to change from 180° to 120°. However, this extra energy is more than compensated by the energy released as a result of π-bonding. Thus, a π-bonded TBP intermediate formed during the aquation of *trans*-[Co(en)$_2$(OH)Cl]$^+$ is more stable than an SP intermediate. This explains the faster rate of aquation of *trans*-[Co(en)$_2$(OH)Cl]$^+$ compared to that of *trans*-[Co(en)$_2$(NH$_3$)Cl]$^{2+}$

complex. It also explains the formation of a mixture of *cis* and *trans* products during the aquation of *trans*-$[Co(en)_2(OH)Cl]^+$. The attack of H_2O through $\angle$s 2,5 and 4,5 of TBP intermediate leads to *cis* product and through $\angle$ 2,4 to *trans* product. An SP intermediate would have given the product having the same geometry as that of the reacting complexes. For example, *trans*-$[Co(en)_2(NH_3)Cl]^{2+}$ complex on aquation through the formation of SP intermediate gives only the *trans* product *trans*-$[Co(en)_2(NH_3)(H_2O)]^{3+}$.

From the above discussion, the following points are notable:

(i)   When a π-bonding inert ligand is absent in a complex, both *cis* and *trans* isomers undergo aquation through the formation of SP intermediate giving products which have the same geometries as those of the parent reacting complexes.

From the above discussion, the following points are notable:

(ii)   When a π-bonding inert ligand is absent in a complex, both *cis* and *trans* isomers undergo aquation through the formation of SP intermediate giving products which have the same geometries as those of the parent reacting complexes.

(iii) When a π-bonding inert ligand is present, *cis*-isomers aquate through the formation of a π-bonded SP intermediate whereas the *trans*-isomers aquate through the formation of a π-bonded TBP.

(iv)   When SP intermediate is formed, the aquation reaction always leads to the retention of geometry. On the other hand when aquation reactions occur through the formation of a TBP intermediate, a mixture of *cis* and *trans* products is obtained.

(v)   Like OH^-, NH_2^-, Cl^-, Br^- and so on are π-bonding inert ligands containing filled p-orbitals. The mechanism of aquation of complexes containing these ligands is the same as that of the complexes containing OH^- ligand.

(vi)   The aquation reaction in all the case discussed above involves a dissociative S_N^1 mechanism and not an associative S_N^2 mechanism.

(vii) In all case, aquation proceeds through a dissociative S_N^1 mechanism.

Other π-bonding inert ligands containing filled p-orbitals are NH_2^-, Cl^-, Br^- and so on. The mechanism of aquation of complexes containing these ligands is the same as that of complexes containing OH^-. The rates of aquation of such complexes are directly related to the π-bonding capability of the inert ligand.

7.9.3 Mechanism of acid hydrolysis in octahedral complexes in which the inert ligand is a π-acceptor

The mechanism of acid hydrolysis of octahedral complexes containing an inert π-acceptor ligand lying *cis* or *trans* to the leaving group is different from the mechanism of aquation when inert ligands are either π-donors or are capable of forming π-bonding.

7.9.3.1 When inert π-acceptor ligand is *trans* to the leaving group

Let us consider the aquation of *trans*-$[O_2NCo(en)_2Cl]^+$ in which the inert π-acceptor ligand NO_2 is *trans* to the leaving group Cl. Here, in this complex, one of the filled t_{2g} orbitals of the central Co^{3+} ion can overlap with an empty p-orbital of NO_2 group to form π-bond. Thus, the electronic charge of t_{2g} orbital would shift towards NO_2 group to maximize π-overlap. Consequently, there is decrease of the electronic charge from and around the leaving group Cl^-. The lone pair of electrons of the attacking H_2O will, therefore, experience lesser repulsion from t_{2g} electrons from the direction *cis* to the leaving group (or *trans* to the NO_2 group). As a result, the attack of H_2O on the central Co atom from a position *cis* to the leaving group Cl^- or *trans* to the NO_2 group becomes easier in the presence of such π-bonding. This is illustrated in Figure 7.11.

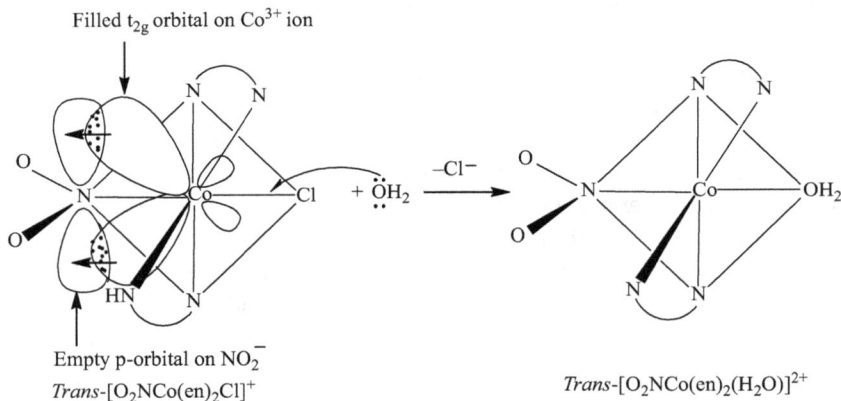

Figure 7.11: Formation of π-bond by overlapping of a filled t_{2g} orbital of Co^{3+} ion with an empty p-orbital of NO_2 group and attack of H_2O from a position *cis* to the leaving group Cl.

The presence of a π-acceptor ligand like NO_2 in the complex would make the formation of the Co–H_2O bond easier, and this would favour an associative S_N^2 mechanism for aquation. Because of the withdrawing inductive effect of NO_2 group, the bonding electrons of Co–Cl are pulled towards Co making the dissociation of the Co–Cl bond difficult to release Cl^-. Hence, a dissociative S_N^1 path for the aquation of this *trans* complex is unlikely.

7.9.3.2 When inert π-acceptor ligand is *cis* to the leaving group

When NO_2 group is *cis* to the leaving group Cl^-, the extent of overlap between one of the filled t_{2g} orbitals of Co^{3+} ion with an empty p-orbital of NO_2 group is less than when NO_2 group is placed *trans* to the leaving group Cl^-. As a result, the withdrawal of t_{2g} electronic charge from and around the leaving group Cl^- would be less. Therefore, the formation of Co–H_2O bond would not be as easy in the *cis* complex as it is in *trans* complex. As a result, the aquation of *cis*-$[O_2NCo(en)_2Cl]^+$ would be slower than the aquation of *trans* isomer, *trans*-$[O_2NCo(en)_2Cl]^+$. However, it would still be faster than the aquation of, which is, *cis* or *trans*-$[H_3NCo(en)_2Cl]^+$, in which the inert ligands are incapable of forming π-bonds with the filled t_{2g} orbitals of the central metal ion. These facts are well clear by the data given in Table 7.4.

Table 7.4: Rate of aquation of complexes containing π-acceptor inert ligands.

Complex	Rate constant k(s^{-1})
trans-$[O_2NCo(en)_2Cl]^+$	98×10^{-5}
trans-$[H_3NCo(en)_2Cl]^+$	0.034×10^{-5}
cis-$[O_2NCo(en)_2Cl]^+$	11×10^{-5}
cis-$[H_3NCo(en)_2Cl]^+$	0.05×10^{-5}

Aquation of other metal complexes involving inert π-acceptor ligands, which are CO, CN$^-$ and so on, also proceed through an associative S_N^2 mechanism rather than a dissociative S_N^1 path.

7.9.3.3 Intermediate formed during aquation of complexes containing π-acceptor ligands

Experimentally, it is observed that aquation of *cis*-$[LCo(en)_2X]^{n+}$ complex always yields 100% *cis*-$[LCo(en)_2(H_2O)]^{(n+1)+}$ complex, where L is an inert π-acceptor ligand, which are NO_2^-, CN$^-$, CO and so on. Similarly, aquation of *trans*-$[LCo(en)_2X]^{n+}$ complex always yields 100% *trans*-$[LCo(en)_2(H_2O)]^{(n+1)+}$ complex.

The above-mentioned observations can be explained by assuming that H_2O attacks the reacting species from a position *cis* to the leaving group resulting in the formation of a PBP intermediate. But, the formation of this intermediate is energetically unfavourable because it requires the adjustment of four metal–ligand bonds to accommodate the incoming H_2O ligand.

Another intermediate requiring lesser energy for its formation and having incoming H_2O and leaving group X$^-$ in equivalent position is the OW intermediate. The formation of such a symmetric intermediate in which leaving group X$^-$ and incoming group H_2O occupy equivalent sites require minimum movement of ligands and is of lower energy compared to PBP intermediate.

The experimental observation that aquation of complexes containing π-acceptor ligands proceeds with the retention of geometry can also be explained satisfactorily through OW intermediate as illustrated in Figure 7.12.

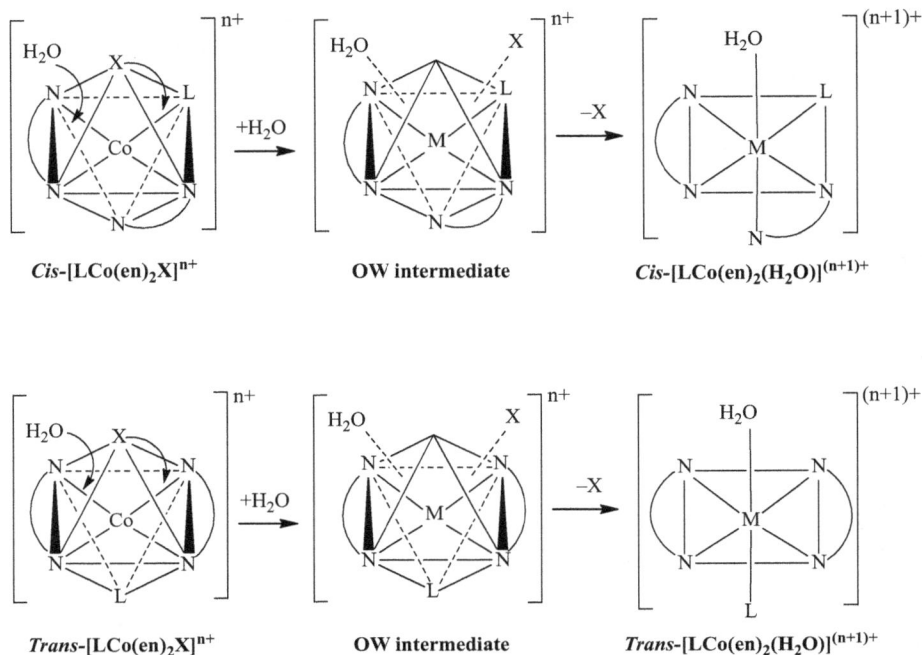

Figure 7.12: Aquation of *cis*-[LCo(en)$_2$X]$^{n+}$ and *trans*-[LCo(en)$_2$X]$^{n+}$ through OW intermediate retaining geometry of the respective aquated product.

7.10 Base hydrolysis

The reaction in aqueous medium in which OH$^-$ replaces/substitutes a coordinated ligand from the complex species is known as base hydrolysis of the complex species. For instance, the base hydrolysis of an octahedral complex [MA$_5$ L] may be represented as

$$[MA_5X] + OH^- \longrightarrow [MA_5(OH)] + X^- \tag{7.7}$$

In this reaction, the coordinated ligand X$^-$ is replaced by OH$^-$ ion. Thus, X$^-$ is the leaving group and OH$^-$ is the entering group.

The base hydrolysis is known to proceed with a faster rate compared to the acid hydrolysis of octahedral complexes. The rate constant for hydrolysis in basic solutions is often a million times of that found for acidic solutions.

The rate of base hydrolysis is a second-order reaction, being first order with respect to the complex ion and first order with respect to OH$^-$ ion. Thus, the rate of base hydrolysis may be represented as

$$r = k[\text{complex}][\text{OH}^-]$$

Since, the base hydrolyses of pentaamminecobalt(III) complexes have been extensively studied, we shall discuss here the mechanism of base hydrolysis by taking into consideration the octahedral ammine complexes of cobalt(III).

7.11 Mechanism of base hydrolysis

In order to explain the base hydrolysis of octahedral complexes, two mechanisms have been reported in the literature, and these are S_N^2 and S_N^1(CB) mechanisms.

7.11.1 Associative S_N^2 mechanism for base hydrolysis

According to this mechanism, the ligand OH^- ion, being a strong nucleophile, attacks the ammine complex $[Co(NH_3)_5X]^{2+}$ as given below:

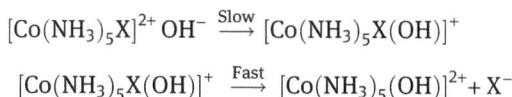

$$[Co(NH_3)_5X]^{2+} OH^- \xrightarrow{Slow} [Co(NH_3)_5X(OH)]^+$$

$$[Co(NH_3)_5X(OH)]^+ \xrightarrow{Fast} [Co(NH_3)_5(OH)]^{2+} + X^-$$

Accordingly, the rate of hydrolysis (r) is given as

$$r = k[Co(NH_3)_5X]^{2+}[OH^-]$$

In general, the rate of hydrolysis (r) is given as

$$r = k[complex\ ion][OH^-]$$

Hence, the rate of base hydrolysis is a second-order reaction, being first order with respect to the complex ion and first order with respect to OH^- ion.

Shortcomings of the S_N^2 mechanism
The above-mentioned mechanism is unable to explain the following two experimental observations:
(i) At high concentration of OH^-, the reaction rate becomes almost independent of OH^- concentration, and appears to be first order with respect to the complex ion only. This observation cannot be explained by a simple associative S_N^2 mechanism.
(ii) As per the associative S_N^2 mechanism, the rate of base hydrolysis should be directly related to the strength of the nucleophilic character of the attacking ligands. However, there are several equally strong nucleophiles other than OH^-, which are NCS^-, NO_2^- and N_3^-, whose concentration do not affect the rates of hydrolysis of the ammine complexes. Hence, the rates of hydrolysis of cobalt(III) ammine complexes are independent of the concentration of these nucleophiles

and are dependent on the concentration of the complex ion only. The fact that why OH^- alone and no other strong nucleophile affect the rate of hydrolysis cannot be explained by a simple S_N^2 mechanism.

7.11.2 S_N^1 (CB) mechanism for base hydrolysis

The S_N^1 (CB) mechanism was proposed by Garrick (1937). The above-mentioned shortcomings and several other experimental observations can be explained successfully by this mechanism. The symbol S_N^1(CB) stands for substitution (S), nucleophilic (N), unimolecular (1) and conjugate base (CB). The main features of this mechanism are given below:

(i)   In organic chemistry, there are several reactions in which OH^- serves as a catalyst but the rates of these reactions depend on the concentration of the OH^- added. Such reactions normally involve the acid–base equilibrium as shown below:

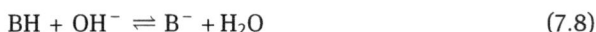

$$BH + OH^- \rightleftharpoons B^- + H_2O \qquad (7.8)$$

where BH is an organic compound with removable hydrogen, and B^- is its CB.

Since the ammine complexes of cobalt(III) also contain removable hydrogens, occurrence of an acid–base equilibrium similar to equation (7.10) was proposed by Garrick as the first step in the base hydrolysis. For example,

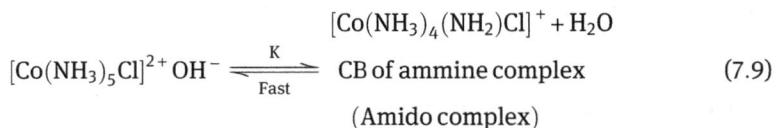

$$[Co(NH_3)_4(NH_2)Cl]^+ + H_2O$$

$$[Co(NH_3)_5Cl]^{2+} OH^- \underset{Fast}{\overset{K}{\rightleftharpoons}} \text{CB of ammine complex} \qquad (7.9)$$

$$\text{(Amido complex)}$$

(ii)   The CB produced in equation (7.11) contains a coordinated NH_2^- ligand, which is capable of forming π-bonding with Co(III) in the five-coordinated intermediate more efficiently than in seven-coordinated intermediate. Hence, the better stabilized five-coordinated intermediate was proposed to be formed from the CB in the second step.

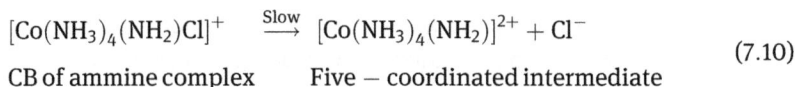

$$[Co(NH_3)_4(NH_2)Cl]^+ \overset{Slow}{\longrightarrow} [Co(NH_3)_4(NH_2)]^{2+} + Cl^-$$

$$\text{CB of ammine complex} \qquad \text{Five – coordinated intermediate} \qquad (7.10)$$

The π-bonding NH_2^- ligand also helps in the dissociation of Cl^- from Co(III). The effect of π-bonding on the stability of five-coordinated intermediate has already been explained in the section dealing with the acid hydrolysis of octahedral complexes.

(iii) The five-coordinated intermediate $[Co(NH_3)_4(NH_2)]^{2+}$ then quickly reacts with H_2O to give the final product of base hydrolysis.

$$[Co(NH_3)_4(NH_2)]^{2+} + H_2O \xrightarrow{\text{fast}} [Co(NH_3)_5(OH)]^{2+} \tag{7.11}$$

7.12 Observed reaction rates in different cases: explanation through $S_N^1(CB)$ mechanism

7.12.1 Second-order kinetics of base hydrolysis

Experimentally observed second-order kinetics of base hydrolysis of Co(III) ammine complexes can be explained satisfactorily by $S_N^1(CB)$ mechanism provided the equilibrium (ii) is established quickly and the amount of the CB present at equilibrium is small, that is, K is small.

Since the reaction (7.12) involves the dissociation of Cl^- from the CB, it is supposed to be slower than the reactions (7.11) and (7.13). The reaction (7.12) thus constitutes the rate-determining step of the reaction. Reaction (7.13) involves the acceptance of a proton from H_2O by the basic NH_2^- group. This reaction is assumed to be fastest. Therefore, the rate of hydrolysis (r) is given by the rate-determining step (7.10). Thus,

$$r = k\,[Co(NH_3)_4(NH_2)Cl]^+ \tag{7.12}$$

From reaction (7.9),

$$K = \frac{[Co(NH_3)_4(NH_2)Cl]^+}{[Co(NH_3)_5Cl]^{2+}\,[OH^-]} \tag{7.13}$$

or $\quad [Co(NH_3)_4(NH_2)Cl]^{2+} = K\,[Co(NH_3)_5Cl]^{2+}\,[OH^-] \tag{7.14}$

Putting the value of $[Co(NH_3)_4(NH_2)Cl]^{2+}$ from equation (7.14) in equation (7.12), we get

$$r = k\,K\,[Co(NH_3)_5Cl]^{2+}\,[OH^-]$$

or $\quad r = k'\,[Co(NH_3)_5Cl]^{2+}\,[OH^-] \tag{7.15}$

Thus, the rate of base hydrolysis is a second-order reaction, being first order with respect to the complex ion and first order with respect to OH^- ion.

7.12.2 Base hydrolysis of [Co(NH₃)₅Cl]²⁺ at high concentration of OH⁻

As mentioned in shortcomings at point (i), the rate of base hydrolysis tend to be almost independent of [OH⁻] at very high concentration of OH⁻. This observation can be explained by S_N^1(CB) mechanism as follows:

When the amount of OH⁻ added is very large, there would be very little change in the concentration of OH⁻ as a result of the acid–base equilibrium reaction (7.9). In this situation, the concentrate ion of OH⁻ can be taken as constant so that the rate of hydrolysis as given by equation (7.15), becomes

$$r = k' \, [Co(NH_3)_5Cl]^{2+} \times \text{constant}$$

$$\text{or } r = k''[Co(NH_3)_5Cl]^{2+}$$

(7.16)

This rate equation shows that at very high concentration of OH⁻, the rate of base hydrolysis would depend only on the concentration of the complex ion, and is independent of OH⁻ concentration.

7.12.3 Base hydrolysis of ammine complexes by nucleophiles/ bases NO₂⁻, NCS⁻ and N₃⁻

The anions NO_2^-, NCS^- and N_3^- are strong nucleophiles as OH⁻ but do not affect the rate of base hydrolysis of ammine complexes. This observation can be explained by S_N^1(CB) mechanism as follows:

Although the aforesaid anions are strong nucleophile as OH⁻, yet they are not as strong bases as the OH⁻. In fact, these are weak bases than OH⁻, and hence none of them are capable of abstracting a proton from ammine complex, $[Co(NH_3)_5Cl]^{2+}$, to yield a CB of the complexes by reaction (7.9).

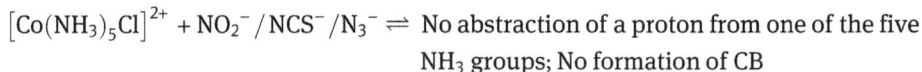

$$[Co(NH_3)_5Cl]^{2+} + NO_2^- / NCS^- / N_3^- \rightleftharpoons \text{No abstraction of a proton from one of the five}$$
NH₃ groups; No formation of CB

Because there is no formation of a CB, the hydrolysis of the complexes cannot proceed through S_N^1(CB) mechanism.

7.12.4 Base hydrolysis of [M(en)₂Cl₂]⁺ [M=Co(III) or Rh(III)]

The chelating complex, $[M(en)_2Cl_2]^+$ [M=Co(III) or Rh(III)], exists in *cis* and *trans* isomers as shown in Figure 7.13.

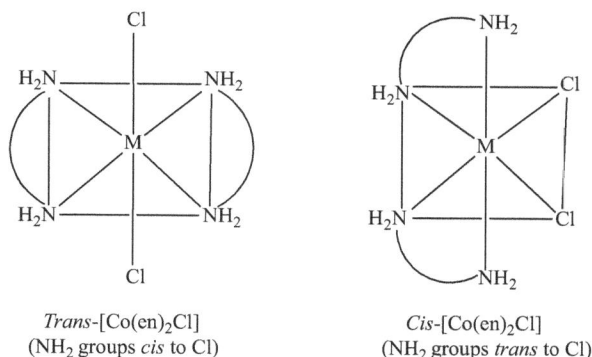

Trans-[Co(en)$_2$Cl]
(NH$_2$ groups *cis* to Cl)

Cis-[Co(en)$_2$Cl]
(NH$_2$ groups *trans* to Cl)

Figure 7.13: *cis* and *trans* isomers of [M(en)$_2$Cl$_2$]$^+$ [M=Co(III) or Rh(III)].

It is well clear from the structure of *trans* isomer that it contains NH$_2$ groups of en *cis* to Cl$^-$ ions whilst *cis*-isomer contains NH$_2$ groups trans to Cl$^-$ ions.

It has been established by ^{1}H−NMR that the amine hydrogens *trans* to Cl$^-$ ions in *cis*-isomer are 100 times more acidic than those of amine hydrogens *cis* to Cl$^-$ ions in *trans* isomer of [M(en)$_2$Cl$_2$]$^+$ [M = Co(III) or Rh(III)]. Both metal ions have the same configuration, t_{2g}^6. Hence, removal of protons from NH$_2$ groups *trans* to Cl$^-$ ions in *cis* isomer is easier than its removal from NH$_2$ groups *cis* to Cl$^-$ ions in *trans* isomer. Consequently, the formation of CB in *cis*-isomer should be easier than the formation of CB in *trans*-isomer. Therefore, the base hydrolysis of *cis*-[M(en)$_2$Cl$_2$]$^+$ should be faster than the *trans*-[M(en)$_2$Cl$_2$]$^+$. Experimentally, it is found to be so.

7.12.5 Hydrolysis of octahedral complex ions having no acidic hydrogen

For the hydrolysis of any complex ion to proceed through an S_N^1(CB) mechanism, the formation of its CB is essential. But this is possible only if the complex contains at least one acidic hydrogen. There are several complex ions which do not contain any acidic hydrogen but undergo hydrolysis. The rate of their hydrolysis is independent of the concentration of OH$^-$ ions. Thus, the hydrolysis of such complexes does not occur through S_N^1(CB) mechanism. Their hydrolysis proceed through a mechanism other than S_N^1(CB) mechanism. This is found to be so experimentally. For example:

(a) Hydrolysis of [Co(CN)$_5$Br]$^{3-}$ and [Co(CN)$_5$I]$^{3-}$

It has been found experimentally that the complex ion [Co(CN)$_5$Br]$^{3-}$ and [Co(CN)$_5$I]$^{3-}$ hydrolyze slowly to form [Co(CN)$_5$(OH)]$^{3-}$ and the rate of their hydrolysis is independent of the concentration of OH$^-$ ions. This indicates that their hydrolysis does not proceed through S_N^1(CB) mechanism. Moreover, neither of these complexes contains any acidic hydrogen, the hydrolysis of these complex ions is not possible through S_N^1(CB) mechanism.

(b) Hydrolysis of [Co(py)₄Cl₂]⁺ and [Co(dipy)₂(CH₃COO)₂]⁺

The complex ions $[Co(py)_4Cl_2]^+$ and $[Co(dipy)_2(CH_3COO)_2]^+$ (py = pyridine and dipy = 2,2′-dipyridine) hydrolyze in the pH range 6–12 and the rates of their hydrolysis are independent of the concentration of OH^- ions. As neither of py or dipy contains any acidic H atom, the hydrolysis of these complex ions does not occur through S_N^1 (CB) mechanism.

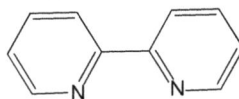

Pyridine   2,2'-Dipyridine

(c) Hydrolysis of [Fe(CN)₅NH₃]³⁻

This complex ion contains removable acidic hydrogen which is present in the co-ligand NH_3. But this acidic H atom is removed with great difficulty due to high negative charge (−3) on the complex ion, $[Fe(CN)_5NH_3]^{3-}$. Consequently, it is very difficult for this complex ion to form its CB. So, the hydrolysis of this complex ion does not occur through S_N^1(CB) mechanism and the rate of hydrolysis of the complex is independent of the concentration of OH^- ions. Experimentally, this is found to be so.

7.12.6 Catalytic role of OH⁻ ions in hydrolysis of octahedral [Co(en)₂(NO₂)Cl]⁺ ion in non-aqueous medium

In non-aqueous dimethyl sulphoxide medium, the reaction of $[Co(en)_2(NO_2)Cl]^+$ with NO_2^- yields $[Co(en)_2(NO_2)_2]^+$. The rate of reaction is slow ($t_{1/2} = 5$–6 h) and is independent of the concentration of NO_2^-. The following mechanism has been proposed for the reaction:

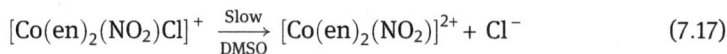

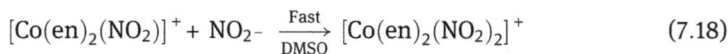

$$[Co(en)_2(NO_2)Cl]^+ \xrightarrow[DMSO]{Slow} [Co(en)_2(NO_2)]^{2+} + Cl^- \qquad (7.17)$$

$$[Co(en)_2(NO_2)]^+ + NO_2{}^- \xrightarrow[DMSO]{Fast} [Co(en)_2(NO_2)_2]^+ \qquad (7.18)$$

The addition of small amount of OH^- reduces the $t_{1/2}$ to a few seconds, that is, it increases the reaction rate substantially. The action of OH^- in non-aqueous medium can be explained satisfactorily by S_N^1(CB) mechanism as follows:

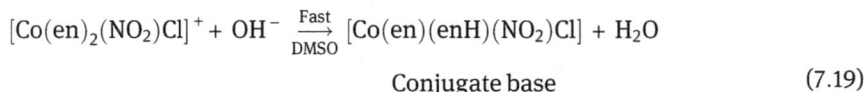

$$[Co(en)_2(NO_2)Cl]^+ + OH^- \xrightarrow[DMSO]{Fast} [Co(en)(enH)(NO_2)Cl] + H_2O$$

$$\text{Conjugate base} \qquad (7.19)$$

[en − H = ethylenediamine from which one proton (H^+) is removed.]

$$[Co(en)(en-H)(NO_2)Cl] \xrightarrow[DMSO]{Slow} [Co(en)(enH)(NO_2)]^+ + Cl^- \tag{7.20}$$

Conjugate base

$$[Co(en)(en-H)(NO_2)]^+ + NO_2- \xrightarrow[DMSO]{Fast} [Co(en)(enH)(NO_2)_2] \tag{7.21}$$

$$[Co(en)(en-H)(NO_2)_2] + H_2O \xrightarrow[DMSO]{Fast} [Co(en)_2(NO_2)_2]^+ + OH^- \tag{7.22}$$

The dissociation of Co–Cl bond in the CB $[Co(en)(en\text{-}H)(NO_2)Cl]$ is easier because of the presence of π-bonding NH^- group compared to the dissociation of Co–Cl bond in the complex $[Co(en)_2(NO_2)Cl]^+$ in the absence of OH^-. This explains why the rate-determining step (7.17) is comparatively slower with $t_{1/2}$ = 5–6 h than the step (7.20) with $t_{1/2}$ = a few seconds although both of them are kinetically slow step.

It is notable here that the OH^- consumed in step (7.19) is regenerated in step (7.22). This shows that the catalytic role of OH^- is in the present hydrolysis.

Other bases such as piperdine (hexahydropyridine, $C_5H_{11}N$) (B) also act as catalyst for the above-mentioned reaction in non-aqueous medium yielding the same product. Their catalytic role can be similarly explained on the basis of $S_N^1(CB)$ mechanism. The reactions involving consumption and regeneration of the base (B) are as follows:

$$[Co(en)_2(NO_2)Cl]^+ + B \xrightarrow[DMSO]{Fast} [Co(en)(en-H)(NO_2)Cl] + BH^+ \tag{7.23}$$

Conjugate base

$$[Co(en)(en-H)(NO_2)Cl] + BH^+ + NO_2^- \xrightarrow[DMSO]{Fast} [Co(en)_2(NO_2)_2Cl] + B \tag{7.24}$$

A simple S_N^1 or S_N^2 mechanism is incapable of explaining the catalytic role of OH^- or other bases (B) in the above reaction in non-aqueous medium.

7.12.7 Base hydrolysis of Co(III) ammine complexes in presence of H₂O₂

If base hydrolysis of Co(III) ammine complexes occurs through an associative S_N^2 mechanism, then the conversion of OH^- to a better nucleophile (but a weaker base) would obviously increase the reaction rate. On the other hand, if base hydrolysis occurs through $S_N^1(CB)$ mechanism for which the presence of OH^- is must, then the conversion of OH^- to a better nucleophile (but a weaker base) would retard the reaction.

It has been seen that all the reaction proceeding through S_N^2 mechanism, HO_2^- is a better nucleophile than OH^-. The anion HO_2^- can be generated by the action of H_2O_2 on OH^-.

$$H_2O_2^+ \; OH^- \overset{K}{\rightleftharpoons} HO_2- + H_2O; \qquad (K > 150)$$

Hence, if the rate of base hydrolysis increases with the addition of H_2O_2, an S_N^2 mechanism would be operative because H_2O_2 converts OH^- to a better nucleophile, HO_2^-. However, if the rate of base hydrolysis decreases with the addition of H_2O_2, an $S_N^1(CB)$ mechanism would be operative because the concentration of OH^- necessary for the formation of the CB would decrease since OH^- gets converted to HO_2^-.

It has been observed experimentally that the addition of H_2O_2 actually decreases the rate of base hydrolysis of ammine complexes of Co(III). This observation strongly favours an $S_N^1(CB)$ mechanism for hydrolysis of these complexes.

The isotope exchange studies on base hydrolysis using $^{18}OH^-$ unambiguously support the $S_N^1(CB)$ mechanism for base hydrolysis of amine complexes of Co(III) and not a pure S_N^1 and S_N^2 mechanism.

7.13 Stereochemistry of intermediates formed during base hydrolysis of Co(III) ammine complexes

Let us consider the base hydrolysis of the complex ion $[Co(en)_2AX]^+$. Assuming that the hydrolysis occurs through $S_N^1(CB)$ mechanism, the rate-determining step yields a five-coordinated intermediate $[Co(en)(en-H)A]^+$ as given below:

$$[Co(en)_2AX]^+ + OH^- \rightleftharpoons \begin{matrix} [Co(en)(en-H)AX] + H_2O \\ \text{Conjugate base} \end{matrix}$$

$$[Co(en)(en-H)AX] \xrightarrow{\text{Slow}} [Co(en)(en-H)A]^+ + X^-$$

$$[Co(en)(en-H)A]^+ + H_2O \xrightarrow{\text{Fast}} [Co(en)_2A(OH)]^+$$

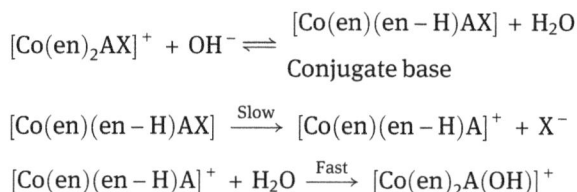

The five-coordinated intermediate can have the following two possible geometries:

(i) Square pyramidal

The formation of five-coordinated SP intermediate requires no movement of ligands which remain attached with Co(III) after the dissociation of leaving group X from the CB. Hence, under normal conditions, that is, in the absence of π-bonding, SP intermediate is more stable than the other possible intermediate (TBP). The formation of an SP intermediate from the CB is shown in Figure 7.14.

Obviously, if the base hydrolysis occurs through the formation of a SP intermediate, the geometry of the product and of the reacting ammine complex would remain the same. This is due to the fact that the attacking H_2O molecule would preferentially approach the metal centre through the site vacated by the leaving group.

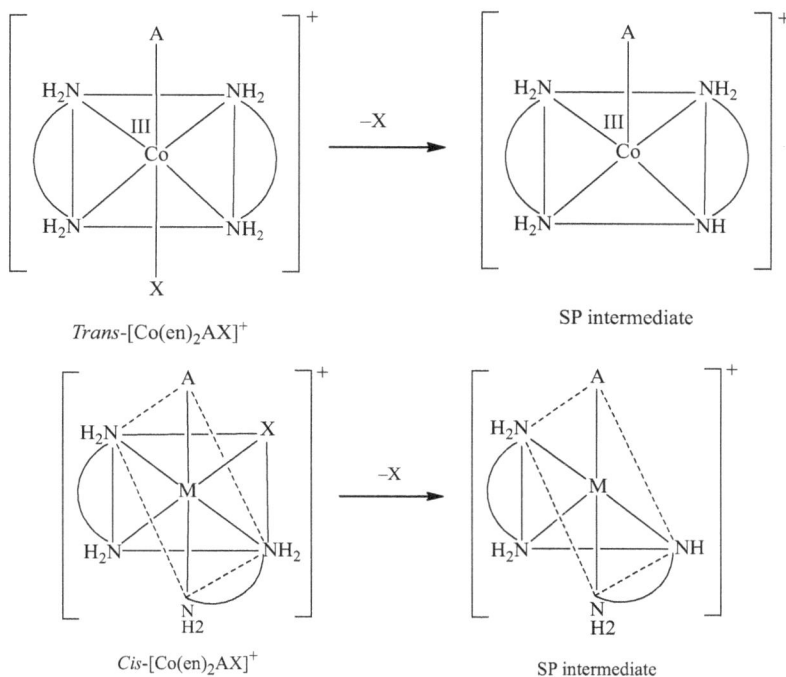

Trans-[Co(en)$_2$AX]$^+$

SP intermediate

Cis-[Co(en)$_2$AX]$^+$

SP intermediate

Figure 7.14: Formation of square pyramidal (SP) intermediate from the conjugate base.

(ii) Trigonal bipyramidal

The formation of a five-coordinated TBP intermediate requires movement of certain Co-ligand bonds. This obviously requires some energy. Therefore, under normal conditions, a TBP intermediate is unlikely to form. On the other hand, when coordinated ligands, $-NH_2^-$ or $>NH^-$, are present in the intermediate, which is capable of forming π-bond with the Co(III) ion, the TBP intermediate becomes more stable than the SP intermediate. This in due to the fact that amount of energy released on account of π-bonding far exceeds the energy required for the formation of TBP intermediate.

As the intermediate formed during base hydrolysis of ammine complexes of Co (III) contain π-bonding ligands such as $-NH_2^-$ or $>NH^-$, a TBP intermediate is expected to be more stable than an SP intermediate.

A TBP intermediate produces a mixture of *cis* and *trans* products irrespective of the fact that whether the reacting complex has a *cis*- or *trans*-geometry. Therefore, one can predict the geometry of the intermediate form during base hydrolysis. If the product has the same geometry as that of the reacting complex, the intermediate form would be SP. Contrary to this, if the product is a mixture of *cis* and *trans* forms, irrespective of the geometry of the reacting complex, the intermediate formed would be TBP.

Experimental studies on the geometries of the products formed during base hydrolysis of ammine complexes of Co(III) reveal that the product formed in each case is a mixture *cis* and *trans* forms irrespective of the geometry (*cis* or *trans*) of the reacting complexes. This observation clearly proves that the five-coordinated intermediate formed during the base hydrolysis of ammine complexes of Co(III) through an $S_N^1(CB)$ mechanism is TBP.

The stereochemistry of base hydrolysis of *trans*- and *cis*-[Co(en)$_2$AX]$^+$ complexes are shown in Figure 7.15(a) and (b), respectively.

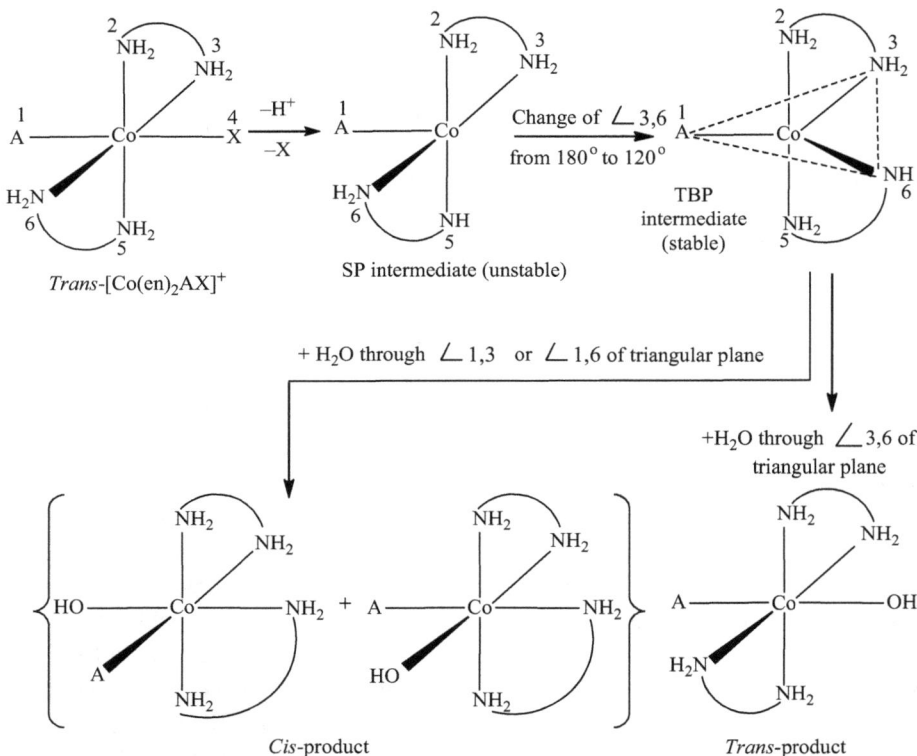

Figure 7.15: (a) Stereochemistry of base hydrolysis of *trans*-[Co(en)$_2$AX]$^+$ complex and (b) stereochemistry of base hydrolysis of *cis*-[Co(en)$_2$AX]$^+$ complex.

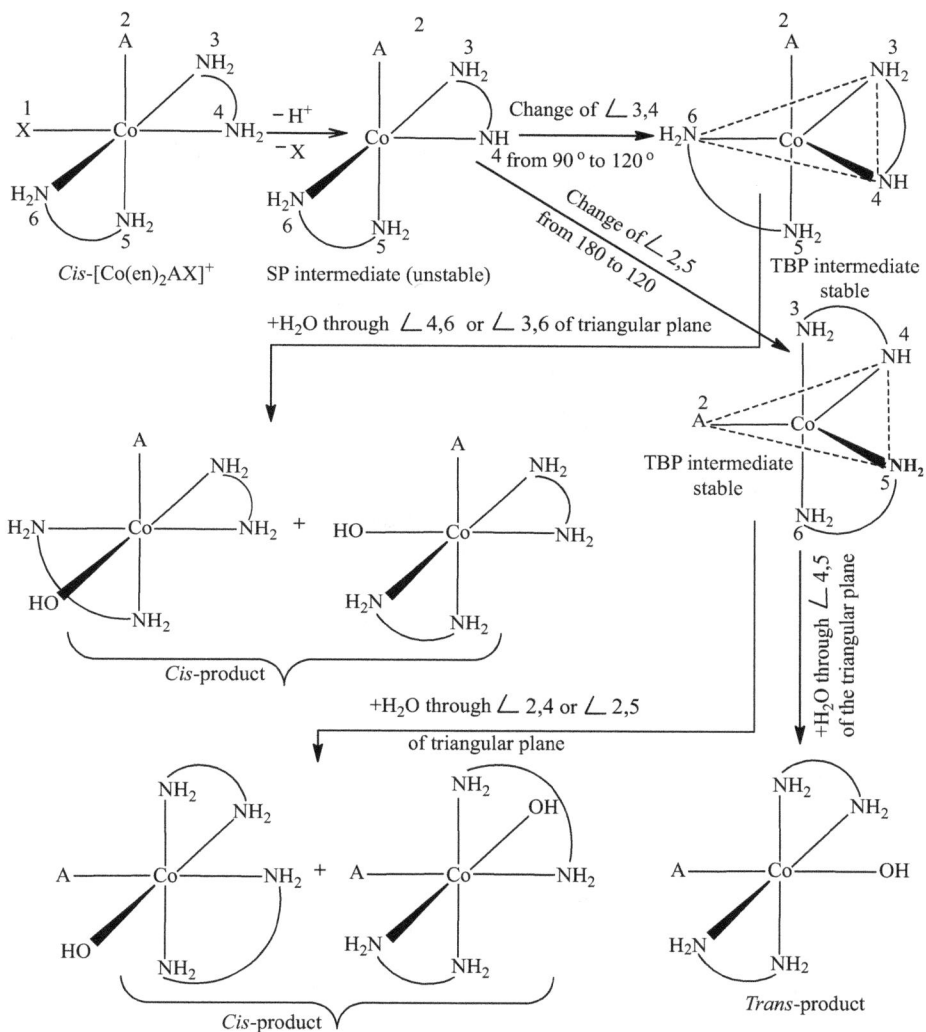

Figure 7.15 (continued)

7.14 Reaction involving replacement of coordination water: anation reactions

This is a kind of ligand substitution reaction. In such reactions, one or more H_2O ligands in a complex are replaced by an anion, and is sometimes called the **anation reaction**. For example, the reaction,

$$[Co(NH_3)_5(H_2O)]^{3+} + X^- \longrightarrow [Co(NH_3)_5X]^{2+} + H_2O$$

is an anation reaction in which H_2O is the leaving group and X^- is the entering ligand.

We have seen that in acid hydrolysis (aquation) reaction, a ligand present in the coordination sphere is replaced by H_2O molecule. For instance, the reaction,

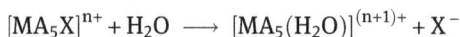

$$[MA_5X]^{n+} + H_2O \longrightarrow [MA_5(H_2O)]^{(n+1)+} + X^-$$

is the acid hydrolysis reaction. Here, X^- is the leaving group and H_2O is the entering ligand. Thus, it is obvious from the above discussion that anation reaction is the reverse of aquation (acid hydrolysis) reaction.

An specific example of anation reaction is

$$[Co(NH_3)_5(H_2O)]^{3+} + Cl^- \longrightarrow [Co(NH_3)_5Cl]^{2+} + H_2O$$

The rate of this reaction depends on the concentration of the complex as well as a nucleophile.

Mechanism of anation reactions

Let us consider the anation reaction,

$$[Co(NH_3)_5(H_2O)]^{3+} + X^- \longrightarrow [Co(NH_3)_5X]^{2+} + H_2O$$

Kinetic studies of these reactions in **aqueous solutions** after suitable corrections, often find them to be second order with a rate dependant on the concentration of anion, that is, $[X^-]$ and the concentration of the parent complex (Adell 1941). However, from this information alone, it cannot be concluded that these reactions are bimolecular. The same second-order kinetics would be observed for a unimolecular process, such as,

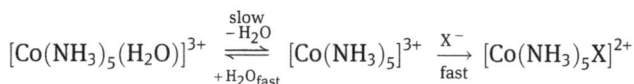

$$[Co(NH_3)_5(H_2O)]^{3+} \underset{+H_2O \text{ fast}}{\overset{\substack{\text{slow} \\ -H_2O}}{\rightleftharpoons}} [Co(NH_3)_5]^{3+} \xrightarrow[\text{fast}]{X^-} [Co(NH_3)_5X]^{2+}$$

In such a mechanism/scheme, a pseudo equilibrium exists between the aqua complex, $[Co(NH_3)_5(H_2O)]^{3+}$, and the five-coordinated intermediate, $[Co(NH_3)_5]^{3+}$. As X^- is expected to compete with the solvent H_2O for the formation of active intermediate, the rate of formation of $[Co(NH_3)_5X]^{2+}$ would depend on $[X^-]$. If the concentration of X^-, that is $[X^-]$, is very high, then the rate of replacement of H_2O would no longer be dependent upon the concentration of X^-. At such a high concentration of X^-, the rate of formation of $[Co(NH_3)_5X]^{2+}$ should be equal to the rate of formation of $[Co(NH_3)_5]^{3+}$.

A number of anionic ligands X^- result anation reactions with aqua complexes of Co(III) and Fe(III). The rate constants of some such reactions are listed in Table 7.5.

Table 7.5: Rate constants for some anation reactions shown below at 25 °C and $\mu = 0.5$.

$[Co(NH_3)_5(H_2O)]^{3+} + X^- \xrightarrow{k} [Co(NH_3)_5X]^{2+} + H_2O$	
X^-	$k, M^{-1}s^{-1}$
Cl^-	2.1×10^{-6}
Br^-	2.5×10^{-6}
NO_3^-	2.3×10^{-6}
NCS^-	1.3×10^{-6}
SO_4^{2-}	1.5×10^{-6}
$H_2PO_4^-$	2.0×10^{-6}

7.15 Substitution reactions without breaking metal–ligand bond

We have so far studied the ligand substitution reactions wherein the metal–ligand bond is broken. However, there are a quite few ligand substitution reactions in which the metal–ligand bond is not broken. The following three examples substantiate this:

(i) In acidic solution, carbonato complex, such as $[Co(NH_3)_4(CO_3)]^+$ (here CO_3^{2-} is serving as a bidentate ligand), is converted to an aqua complex with release of CO_2. When the reaction is carried out in the presence of $H_2{}^{18}O$, no ^{18}O is found in the aqua complex. Hence, Co–O bond is retained during the reaction. It is suggested that Co-bonded O of CO_3^{2-} is protonated and CO_2 is then removed as shown in Figure 7.16(a).

In acidic solution, $[Co(NH_3)_4(CO_3)]^+$, having bidentate CO_3^{2-}, is converted to cis-$[Co(NH_3)_4(H_2O)_2]^{3+}$. In $H_2{}^{18}O$, half the oxygen in the product is found to have come from solvent H_2O. The first step involves breaking the chelate ring, with an H_2O or H_3O^+ substituting the one carbonato O. The second steps involves removal of CO_2 without rupture of Co–O bond as in the case of $[Co(NH_3)_4(CO_3)]^+$ (Dasgupta and Harris 1969).

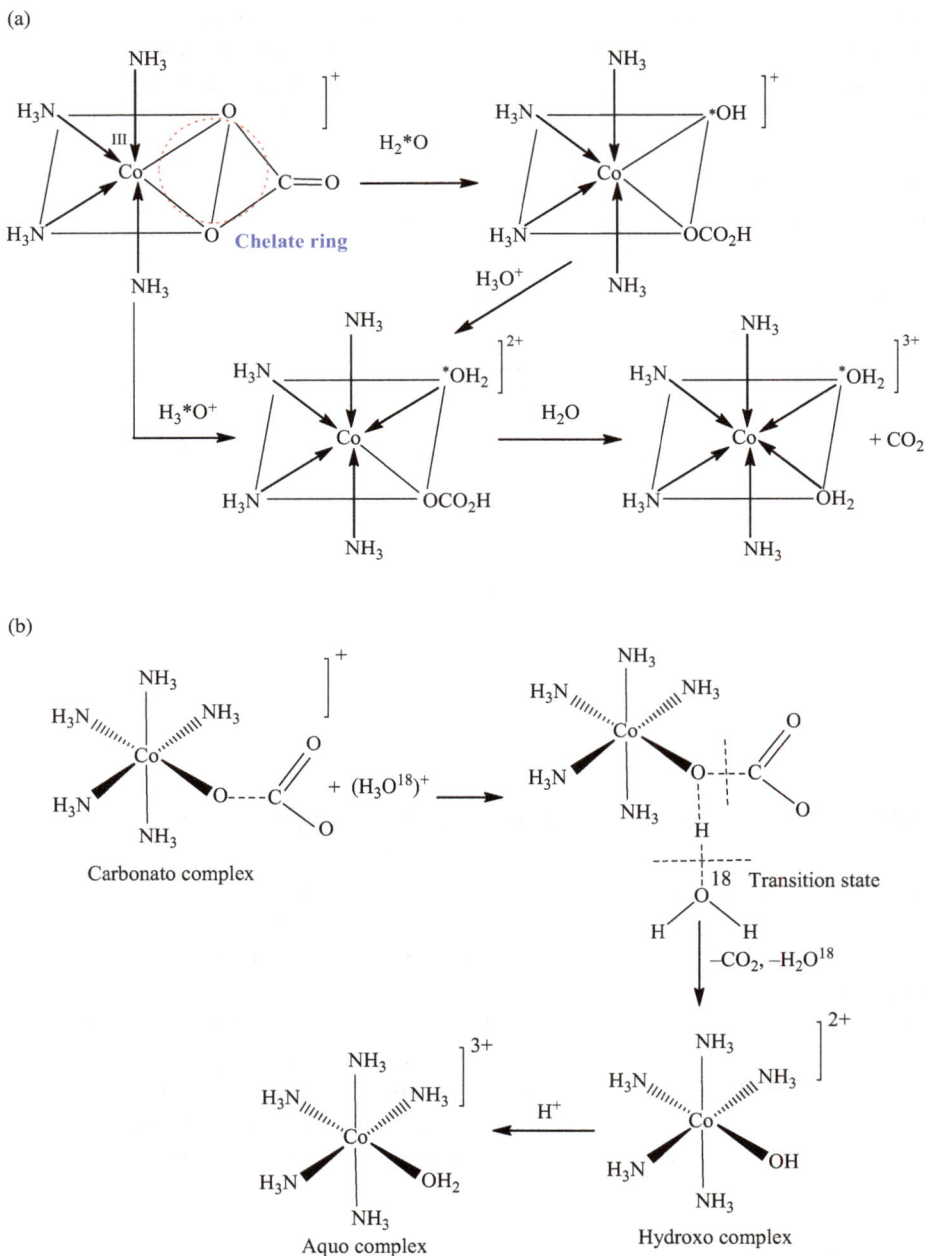

(a)

(b)

Figure 7.16: (a) Decarboxylation of a carbonato complex without the cleavage of a metal–ligand bond and (b) decarboxylation of a carbonato complex without the cleavage of a metal–ligand bond.

(ii) Another example of decarboxylation of a carbonato complex to give an aquo complex is

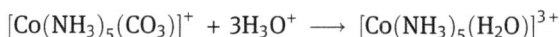

$$[Co(NH_3)_5(CO_3)]^+ + 3H_3O^+ \longrightarrow [Co(NH_3)_5(H_2O)]^{3+}$$

Primarily, this is an acid hydrolysis. However, the isotopic labelling of oxygen of a nucleophile by (O^{18}) revealed that the oxygen atom from the nucleophile does not enter into the product. The most probable mechanism for this reaction is shown in Fig. 7.16(b).

The mechanism shows that upon the attack of the proton on the oxygen atom bonded to the metal ion, carbon dioxide is eliminated, accompanied by the release of a water molecule. This forms a hydroxo complex, which quickly takes up a proton from the medium to give the resultant aquo complex. The absence of isotopic oxygen (O18) can be easily justified by the above mechanism.

(iii) The anation reaction of a cobalt(III) ammine aquo complex using (NO_2^-) as an anionic ligand also occurs without the cleavage of metal–ligand bond.

$$[(NH_3)_5Co - O^{18}H_2]^{3+} + NO_2^- \longrightarrow [(NH_3)_5Co - O^{18}NO]^{2+} + H_2O$$

The retainment of the O^{18} isotope on the aquo ligand in the product complex confirms that the M–(O^{18}) bond is not broken.

7.16 Substitution reactions in square planar complexes

7.16.1 Introduction

Metal ions with the d^8 configuration such as Au(III), Ni(II), Pt(II), Pd(II), Rh(I) and Ir (I) usually form four-coordinated square planar complexes, especially with strong field ligands. Ligand substitution reactions of square planar geometry are mainly confined to Pt(II) complexes because these complexes are slow towards substitution and thus are easier to study. More recently, the availability of methods for faster reactions has opened the area to study the highly labile Pd(II) and other d^8 systems.

7.16.2 Mechanism of ligand substitution reactions in square planar complexes

Ligand substitution reactions of Pt(II) complexes appear to proceed generally by associative S_N^2 rather than dissociative S_N^1 mechanism. It is due to lack of steric crowding and availability of an empty 'p' orbital perpendicular to the molecular plane. This may be inferred from the following observations:

(i) In square planar complexes of Ni(II), Pd(II) and Pt(II), the five empty metal orbitals (one d, one s and three p) of comparable energies can be made available for σ-bonding. As only four ligands are to be σ-bonded with the metal, only four of the five orbitals of the metal ion are used up for this purpose for dsp^2 hybridization. The fifth, the empty p-orbital can accommodate electrons from the attacking ligand. In other words, a five-coordinated intermediate is formed from a four-coordinated square planar complex through an associative S_N^2 mechanism.

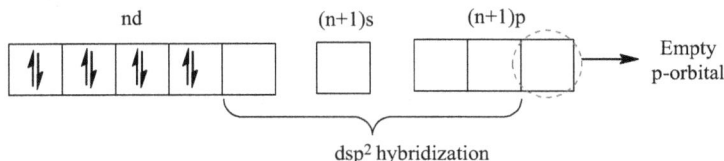

(ii) The rates of substitution reactions of square planar complexes are sensitive to the nature of reacting complex. For example, among the following reactions:

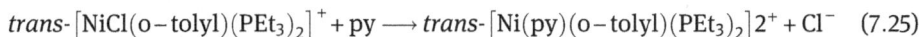

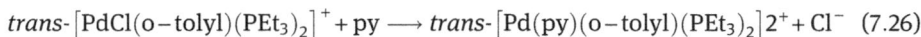

$$trans\text{-}[NiCl(o-tolyl)(PEt_3)_2]^+ + py \longrightarrow trans\text{-}[Ni(py)(o-tolyl)(PEt_3)_2]2^+ + Cl^- \quad (7.25)$$

$$trans\text{-}[PdCl(o-tolyl)(PEt_3)_2]^+ + py \longrightarrow trans\text{-}[Pd(py)(o-tolyl)(PEt_3)_2]2^+ + Cl^- \quad (7.26)$$

$$trans\text{-}[PtCl(o-tolyl)(PEt_3)_2]^+ + py \longrightarrow trans\text{-}[Pt(py)(o-tolyl)(PEt_3)_2]2^+ + Cl^- \quad (7.27)$$

The reaction (i) is 50 times faster than the reaction (ii), which, in turn, is about 100,000 times faster than the reaction (iii).

It is recognized that the square planar Ni(II) complex expands its coordination number with greater ease than the square planar Pd(II) complex, and the square planar Pd(II) complex expands its coordination number with much greater ease than the square planar Pt(II) complex. Thus, there exists a parallelism between the reactivity of square planar complexes of these metal ions with the ease with which they expand their coordination number. Such a relationship clearly indicates the formation of an intermediate with a higher coordination number during the substitution reactions of square planar complexes of Ni(II), Pd(II) and Pt(III). This type of intermediate is formed in such reactions only through associative S_N^2 mechanism.

(iii) Consider the substitution reaction of square planar complex, $[PtL_nCl_{4-n}]$, in aqueous medium as given below:

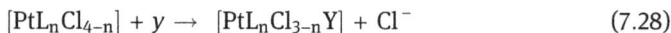

$$[PtL_nCl_{4-n}] + y \rightarrow [PtL_nCl_{3-n}Y] + Cl^- \quad (7.28)$$

Here L is a unidentate ligand and n can vary from 0 to 3. The complexes $[PtL_nCl_{4-n}]$ and $[PtL_nCl_{3-n}Y]$ can be neutral or charged species depending upon the nature of L.

The rate of equation (r) in such cases is found to be represented by the expression:

$$r = k_1\{[PtL_nCl_{4-n}]\}[Y] + k_2\{[PtL_nCl_{4-n}]\} \quad (7.29)$$

Here $\{[PtL_nCl_{4-n}]\}$ is the concentration of the parent complex, $[PtL_nCl_{4-n}]$ and $[Y]$ is the concentration of the entering group Y.

The above rate equation suggests that the substitution of Cl^- by Y in reaction (7.28) follows two independent pathways. The reaction rate for the first pathways is given by

$$r = k_1\{[PtL_nCl_{4-n}]\}[Y] \tag{7.30}$$

This rate equation suggests that the rate-determining step is bimolecular and the mechanism involved is associative S_N^2.

As the concentration of the ligand Y is much less than that the concentration of H_2O which acts as medium as well as a ligand, the fraction of molecules of the sub-strate $[PtL_nCl_{4-n}]$ reacting through this path (first path) is far less than the fraction of molecules reacting through second path in which H_2O also acts as a ligand. The second reaction path, therefore, consists of a slow aquation reaction, such as,

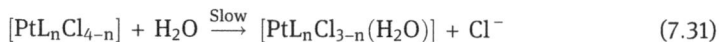

$$[PtL_nCl_{4-n}] + H_2O \xrightarrow{\text{Slow}} [PtL_nCl_{3-n}(H_2O)] + Cl^- \tag{7.31}$$

followed by a fast replacement of coordinated water by the ligand Y as follows:

$$[PtL_nCl_{3-n}(H_2O)] + Y \xrightarrow{\text{Fast}} [PtL_nCl_{3-n}Y] + Cl^- \tag{7.32}$$

Reaction (7.32), being slow, is the rate-determining step of the second reaction path. The two mechanisms (S_N^1 and S_N^2) are possible for this step. Let us consider these mechanisms one by one.

S_N^1 mechanism

This mechanism for step (7.32) involves the following reactions:

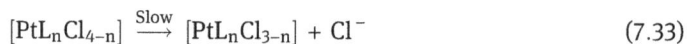

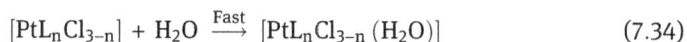

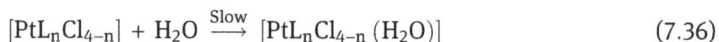

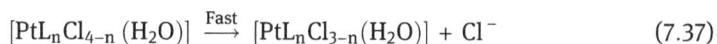

$$[PtL_nCl_{4-n}] \xrightarrow{\text{Slow}} [PtL_nCl_{3-n}] + Cl^- \tag{7.33}$$

$$[PtL_nCl_{3-n}] + H_2O \xrightarrow{\text{Fast}} [PtL_nCl_{3-n}(H_2O)] \tag{7.34}$$

In the present case, the rate of reaction (r) will be given by

$$r = k_2\{[PtL_nCl_{4-n}]\} \tag{7.35}$$

S_N^2 mechanism

This mechanism for step (7.32) involves the following reactions:

$$[PtL_nCl_{4-n}] + H_2O \xrightarrow{\text{Slow}} [PtL_nCl_{4-n}(H_2O)] \tag{7.36}$$

$$[PtL_nCl_{4-n}(H_2O)] \xrightarrow{\text{Fast}} [PtL_nCl_{3-n}(H_2O)] + Cl^- \tag{7.37}$$

In this case, the rate of reaction is given by

$$r = k_2\{[PtL_nCl_{4-n}]\}[H_2O] \qquad (7.38)$$

Now, since H_2O is present in large excess, its concentration remains practically constant. Hence, the reaction rate (r) is given by

$$r = k_2\{[[PtL_nCl_{4-n}]\} \qquad (7.39)$$

Thus, both S_N^1 and S_N^2 mechanisms for reaction (7.31) give the same expression for the rate equation, that is, $r = k_2\{[[PtL_nCl_{4-n}]\}$ as given in the rate law equations (7.35) and (7.39).

Experimentally, it is found that there is very little variation in the rates of aquation (substitution of Cl^- by H_2O) of complexes such as $[PtCl_4]^{2-}$, $[PtCl_3(NH_3)]^-$, $[PtCl_2(NH_3)_2]$ and $[PtCl(NH_3)_3]^+$. In case aquation of these complexes occurs through S_N^1 mechanism, it would involve breaking of Pt–Cl bond which becomes increasingly difficult in going from $[PtCl_4]^{2-}$ to $[PtCl(NH_3)_3)]^+$. This would happen due to increase in the positive charge on the complexes from –2 to +1. Hence, the rate of aquation should decrease considerably in going from a complex of lower positive charge to a complex of higher positive charge. Since this is contrary to the experimental observations, an S_N^1 mechanism is ruled out in the aquation reactions of square planar complexes.

The increase in positive charge on the complexes would make the breaking of Pt–Cl bond difficult but it would make the formation of Pt–OH$_2$ bond easier. In the S_N^2 mechanism, the two effects are equally important. As a result, the reaction rates of the above-mentioned complexes via S_N^2 mechanism should be virtually the same irrespective of the charge on the complex, as is actually the case. The above observations clearly indicate that the aquation reaction (7.31) and general substitution reaction (7.28) follow the S_N^2 mechanism.

The two path mechanism of the substitution reaction (7.28) taking $[Pt(NH_3)Cl_3]$ as a substrate can be summarized in a single diagram as shown in Figure 7.17.

7.16.3 Stereochemistry of S_N^2 mechanism in substitution reaction of square planar complexes

The stereochemistry of S_N^2 mechanism in substitution reaction of square planar complexes may be illustrated as shown in Figure 7.18.

The incoming ligand Y attacks M from positive z-direction (possibly utilizing the empty p_z orbital of the central metal) to form a five-coordinated SP intermediate (ii) which then transform into a more stable TBP intermediate (iii). This again gets changed to another SP intermediate (iv) containing the leaving group X along the negative z-direction. The removal of X from this intermediate yields the substituted square planar complex (v).

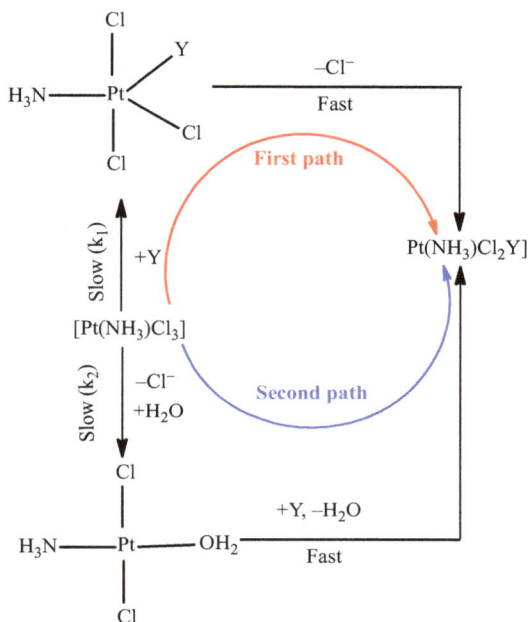

Figure 7.17: The two pathways proposed for the substitution reaction, $[Pt(NH_3)Cl_3] + Y \rightarrow [Pt(NH_3)Cl_2Y]$.

It is notable here that the TBP intermediate (iii) should be more stable than the two SP intermediates (ii) and (iv) in the above mechanism. However, the energy of conversion of the SP intermediate to TBP and vice-versa should be very small for this mechanism to be operative.

7.17 Role of some ligands preferentially direct the substitution of a ligand *trans* to themselves in square planar complexes: *trans* effect

7.17.1 Introduction

The *trans* effect was first time recognized by Ilya Ilich Chernyaev, a Russian Chemist in square planar complexes of Pt(II), and named in 1926 (Chernyaev 1926). It has been observed by him that during the substitution reactions of square planar complexes, some coordinated ligands preferentially direct the substitution *trans* to themselves, that is, the choice of leaving group is determined by the nature of ligand *trans* to it. Such a ligand has a marked influence on the rate of a reaction. This type of observation is more pronounced in the square planar complexes formed by d^8 metal ions, such as Ni(II), Pd(II), Pt(II) and Au(III). The following example of a substitution reaction makes it clear.

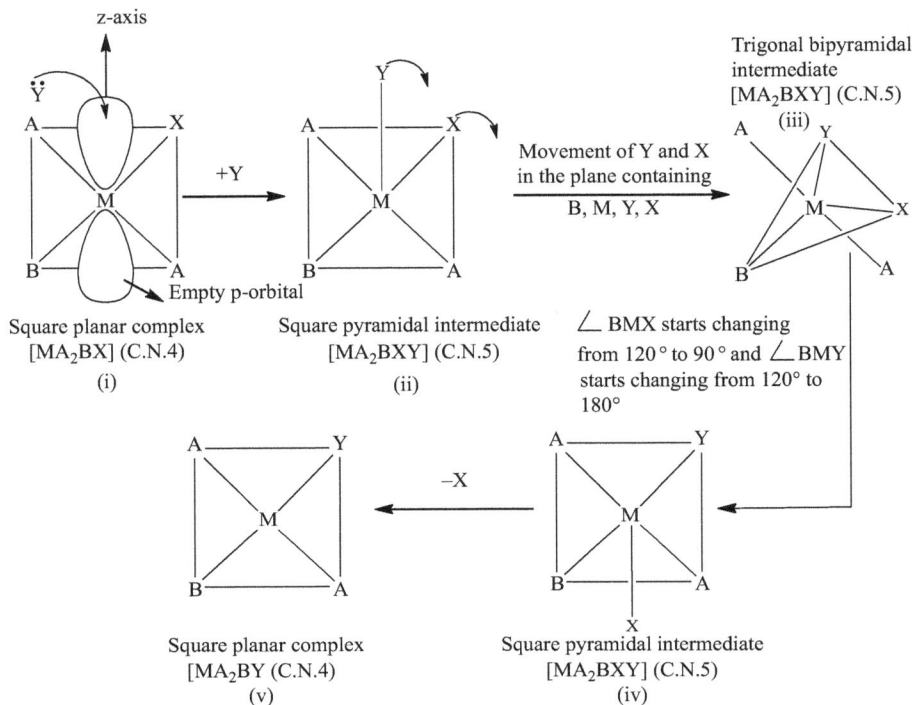

Figure 7.18: Stereochemistry of S_N^2 mechanism in substitution reaction of square planar complexes.

Let us consider the action of NH_3 on the square planar complex, $[Pt(NO_2)Cl_3]^{2-}$. Theoretically, there are two possible reaction products (I) and (II) as shown in Figure 7.19.

However, experimental studies have shown that only the *trans*-product (I) is formed by the replacement of Cl^- lying *trans* to NO_2^- in $[Pt(NO_2)Cl_3]^{2-}$. The formation of *trans*-product (I) is explained by saying that Cl^- lying *trans* to NO_2^- in $[Pt(NO_2)Cl_3]^{2-}$ is far easily replaced by NH_3 than either of the two Cl^- lying *cis* to NO_2^- ion. The phenomenon of such type of replacement is called *trans* effect. Groups like NO_2^- which direct the entering ligand to occupy the position *trans* to them are called *trans*-directing groups. The property of these groups due to which ligands lying *trans* to them are replaced far more readily by the entering ligand is called *trans*-directing character or *trans*-directing influence or simply *trans* effect.

Thus, the *trans* effect may be defined in different ways as follows:

(i) It is the labilization of ligands *trans* to other *trans*-directing ligands.

(ii) It is the effect of a ligand over rate of substitution of another ligand positioned *trans* to it in the square planar complexes.

(iii) It is the effect of a coordinated ligand upon the rate of substitution of ligands opposite to it.

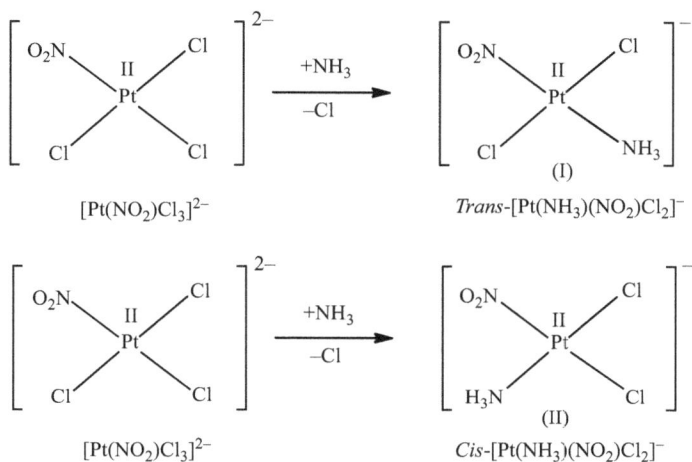

Figure 7.19: Action of NH_3 on the square planar complex, $[Pt(NO_2)Cl_3]^{2-}$ and formation of two possible reaction products (I) and (II).

7.17.2 *Trans*-directing series

On the basis of kinetic and equilibrium data of a large number of reactions, various ligands can be arranged in an order of their *trans*-directing power and are called *trans*-directing series.

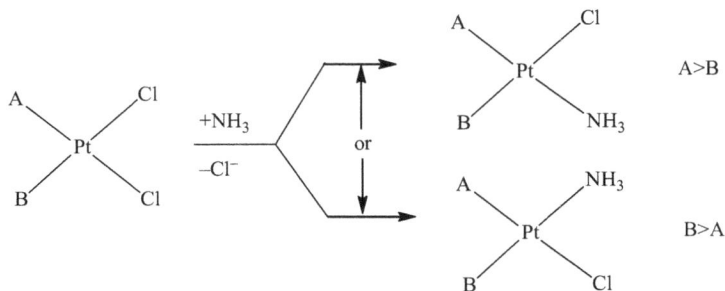

The approximate order of decreasing *trans* effect of some common ligands in this series is as follows:

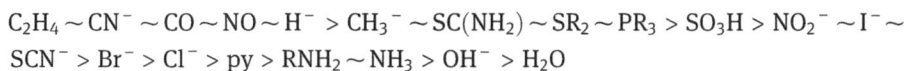

$$C_2H_4 \sim CN^- \sim CO \sim NO \sim H^- > CH_3^- \sim SC(NH_2) \sim SR_2 \sim PR_3 > SO_3H > NO_2^- \sim I^- \sim$$
$$SCN^- > Br^- > Cl^- > py > RNH_2 \sim NH_3 > OH^- > H_2O$$

The ligands lying at the high end (left hand side) of the series have vacant π of π^* orbitals which can accept electrons from metal orbital to form metal–ligand π-bond ($d\pi$–$d\pi$ or $d\pi$–$p\pi$ bond). Hence, these ligands are called π-bonding ligands. The *trans*-directing ability of these π-bonding ligands increases with the increase of their ability to form metal–ligand π-bond. The *trans* effect of the ligands which are unable to form metal–ligand π-bonds increases with increase of their polarizability. For example,

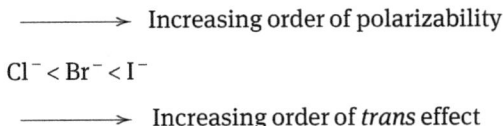

————————→ Increasing order of polarizability

$$Cl^- < Br^- < I^-$$

————————→ Increasing order of *trans* effect

Trans effect is more pronounced when the central metal ion is more polarizable. The observed ordering of susceptibility to *trans* effect is:

$$Pt(II) > Pd(II) > Ni(II)$$

7.17.3 Representative examples of synthesis of *cis*- and *trans*-isomers

Let us consider the synthesis of *cis*- and *tran*-isomers of [Pt(NH₃)₂Cl₂], a square planar complex of Pt(II).

(i) *cis*-[Pt(NH₃)₂(Cl₂)]
This isomer can be prepared taking $[PtCl_4]^{2-}$ as follows:

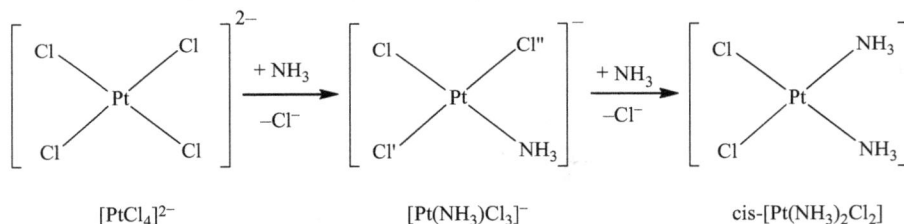

[PtCl₄]²⁻   [Pt(NH₃)Cl₃]⁻   cis-[Pt(NH₃)₂Cl₂]

In step (i), any of the four Cl⁻ of $[PtCl_4]^{2-}$ can be replaced by NH_3 molecule to form $[Pt(NH_3)Cl_3]^-$. Now in the step (ii), the *trans*-directing effect of Cl⁻ > NH₃, Cl″ which is *trans* to Cl′ in [Pt(NH₃)Cl₃], is replaced by NH₃ to give *cis*-product.

(ii) *trans*-[Pt(NH₃)₂(Cl₂)]
This isomer can be prepared taking parent compound, $[Pt(NH_3)_4]^{2+}$, as follows:

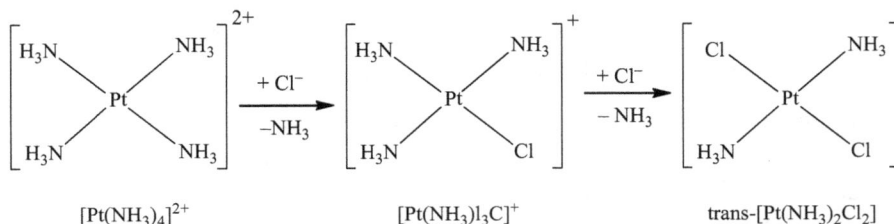

[Pt(NH₃)₄]²⁺   [Pt(NH₃)l₃C]⁺   trans-[Pt(NH₃)₂Cl₂]

In the second step, the *trans*-directing effect of $Cl^- > NH_3$, therefore *trans*-isomer is formed.

7.17.4 Utilities of *trans* effect

(i) In the synthesis of square planar Pt(II) complexes
By ordering the sequence of addition of substituents, the *trans* effect may be utilized to get the desired isomer in an otherwise complicated system. For example:

(a) Synthesis of three geometrical isomers of [Pt((NH₃)(Py)(Cl)(Br)]
By utilizing the knowledge of *trans* effect, the following sequence of reactions were suggested and carried out by Hel'man et al. for the synthesis of first isomer (Hel'man and Essen 1948)

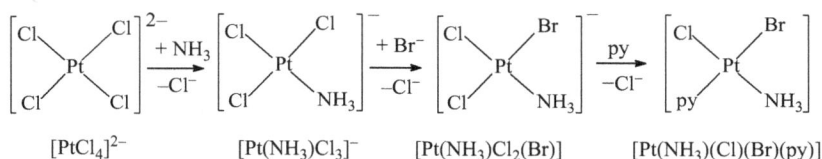

$$
[PtCl_4]^{2-} \xrightarrow[-Cl^-]{+NH_3} [Pt(NH_3)Cl_3]^- \xrightarrow[-Cl^-]{+Br^-} [Pt(NH_3)Cl_2(Br)] \xrightarrow[-Cl^-]{py} [Pt(NH_3)(Cl)(Br)(py)]
$$

In step (i), any of the four Cl^- of $[PtCl_4]^{2-}$ can be replaced by NH_3 molecule to form $[Pt(NH_3)Cl_3]^-$. In the second step, the chloride ion *trans* to another chloride ion is replaced more readily than the NH_3 group trans to a chloride ion. This is because a metal–halogen bond is more labile than a metal–nitrogen bond. In the third step, Cl^- is replaced by py because Br^- is a better *trans* director than Cl^-.

The synthesis of the second isomer follows the following sequence of reactions:

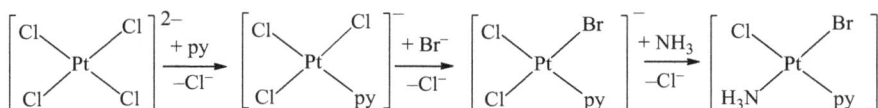

$$
[PtCl_4]^{2-} \xrightarrow[-Cl^-]{+py} [Pt(Cl_3)(py)]^- \xrightarrow[-Cl^-]{+Br^-} [Pt(Cl_2)(Br)(py)]^- \xrightarrow[-Cl^-]{+NH_3} [Pt(Cl)(Br)(NH_3)(py)]
$$

Here, in this case too, the lability of metal–chlorine bond results in preferential replacement of Cl^- instead of py in the second step, and the *trans* influence of Br^- determines the final geometry.

The third isomer may be formed with the following sequence of reactions:

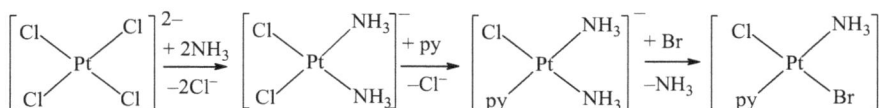

$$
[PtCl_4]^{2-} \xrightarrow[-2Cl^-]{+2NH_3} [Pt(Cl_2)(NH_3)_2] \xrightarrow[-Cl^-]{+py} [Pt(Cl)(NH_3)_2(py)]^+ \xrightarrow[-NH_3]{+Br} [Pt(Cl)(NH_3)(py)(Br)]
$$

In this synthesis, the *trans* effect predicts the formation of the *cis*-derivative in the first step and the inherent lability of the Pt–Cl bond directs the second step. In the third step, the replacement of NH_3 molecule *trans* to Cl^- takes place rather than the NH_3 molecule *trans* to pyridine. Thus, in the third step, this inherent lability of Cl^- runs counter (opposite) to the labilizing *trans* effect. Hence, the entering Br^- ion displaces an ammonia molecule rather than the Cl^- *trans* to NH_3 group. This is a noble example of the fact that *trans* effect provides us with qualitative information concerning which ligands will be more labile (than they would be in the absence of a *trans* director) but no information on the absolute lability of various ligands. Nevertheless, considerable ordering is given by application of the *trans* effect and empirical observations, which is inherent lability of the M–Cl bonds.

(b) Synthesis of *cis* and *trans*-[Pt (NH₃)(NO₂)Cl₂]⁻

Reaction (i)

In step one, any of the four Cl^- of $[PtCl_4]^{2-}$ can be replaced by a NH_3 molecule to form $[Pt(NH_3)Cl_3]^-$. In step two, eventually, *cis*-isomer is formed because Cl^- is a better *trans* director than NH_3.

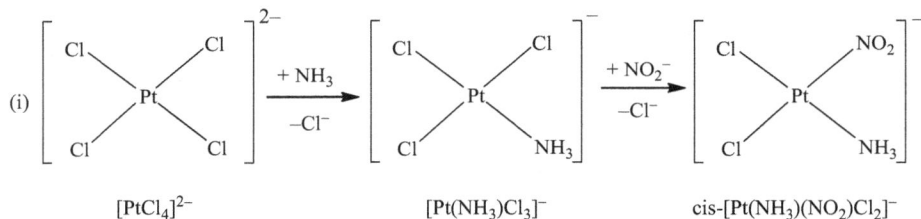

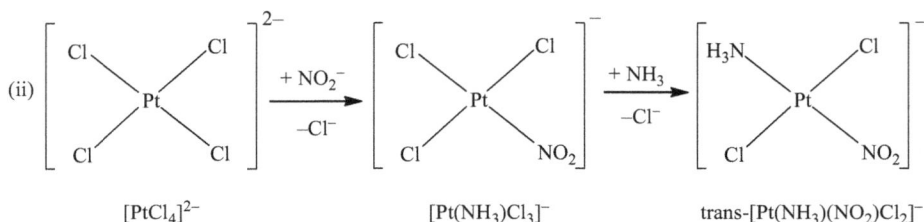

Reaction (ii)

In step one, any of the four Cl^- of $[PtCl_4]^{2-}$ can be replaced by a NO_2^- to form $[Pt(NO_2)Cl_3]^-$.

Now, in step two, since NO_2^- ion present in $[Pt(NO_2)Cl_3]^-$ has greater trans effect than Cl^-, Cl^- which is *trans* to NO_2^- will be replaced by NH_3 molecule to form the *trans*-isomer [Pt (NH₃)(NO₂)Cl₂].

(c) Synthesis of *trans*-[Pt^{II} (R$_3$P)$_2$I$_2$]

The *trans*-[Pt (R$_3$P)$_2$I$_2$] can be formed with the following sequence of reactions:

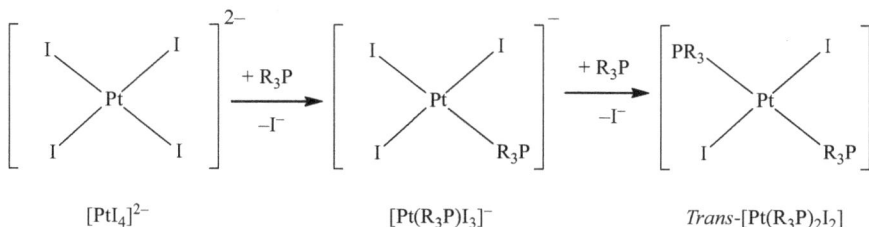

$[PtI_4]^{2-}$ $[Pt(R_3P)I_3]^-$ *Trans*-$[Pt(R_3P)_2I_2]$

In step one, any of the four I$^-$ of [PtI]$^{2-}$ can be replaced by a R$_3$P to form [Pt(R$_3$P)I$_3$]$^-$. In the second step, since, the *trans*-directing effect of R$_3$P > I, therefore [Pt (R$_3$P)$_2$I$_2$] is formed.

(d) Synthesis of *cis*-and *trans*-isomer of [Pt^{II}(PPh$_3$)(NH$_3$)Cl$_2$]

These isomers can be synthesized by the following two sequences of reactions:

Reaction (i)

In step one, any of the four Cl$^-$ of [PtCl$_4$]$^{2-}$ can be replaced by a PPh$_3$ molecule to form [Pt(PPh$_3$)Cl$_3$]$^-$. In step two, eventually, *trans*-isomer is formed because PPh$_3$ is a better *trans* director than Cl$^-$.

Reaction (ii)

In step one, any of the four Cl$^-$ of [PtCl$_4$]$^{2-}$ can be replaced by a NH$_3$ to form [Pt(NH$_3$)Cl$_3$]$^-$. Now, in step (ii), finally, *cis*-[Pt(PPh$_3$)(NH$_3$)Cl$_2$] is formed because Cl$^-$ is a better *trans* director than NH$_3$.

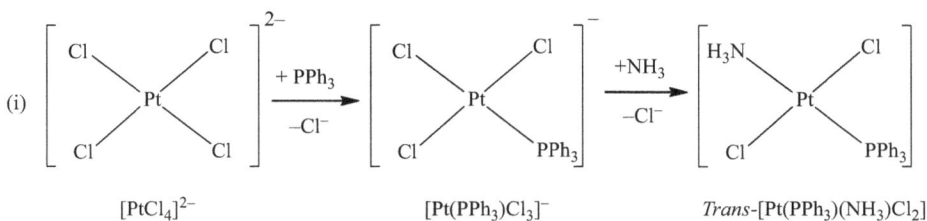

$[PtCl_4]^{2-}$ $[Pt(PPh_3)Cl_3]^-$ *Trans*-$[Pt(PPh_3)(NH_3)Cl_2]$

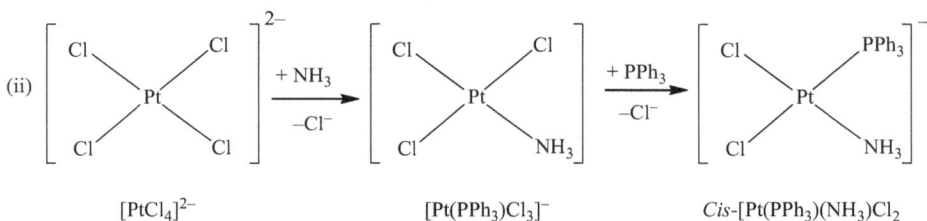

$[PtCl_4]^{2-}$ $[Pt(PPh_3)Cl_3]^-$ *Cis*-$[Pt(PPh_3)(NH_3)Cl_2$

7.18 Distinction between *cis*-and *trans*-isomers of complexes of the type [Pt(A)₂X₂]

Kurnakov's test

An interesting application of the *trans* effect is to distinguish between *cis*-and *trans*-isomers of the complexes of the type $[Pt(A)_2X_2]$ (where A = ammine and X = halide). This test was devised by Russia's one of most distinguished and versatile Chemists, Nikolai Semenovich Kurnakov (Figure 7.20), in 1893. The test procedure is as follows:

Figure 7.20: Nikolai Semenovich Kurnakov (1860–1941).

cis-$[Pt(NH_3)_2(Cl_2]$ when added with thiourea $[(NH_2)_2CS]$ (tu) gives the final product $[Pt(tu)_4]^{2+}$ by replacement of all the former ligands of the parent complex.

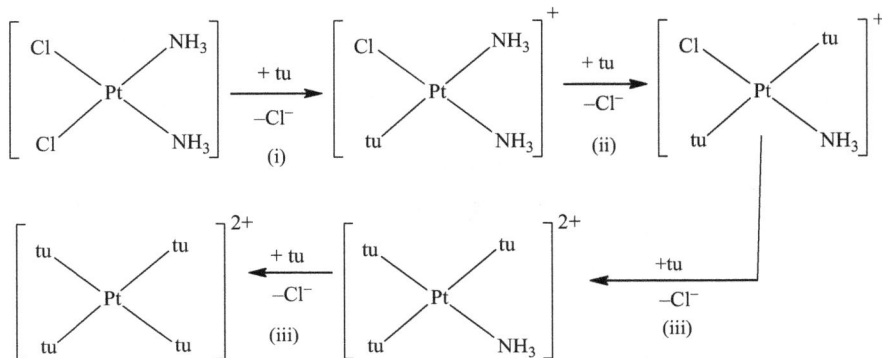

In step (i), one of the Cl *trans* to NH_3 is replaced by tu because of the inherent lability of the M–Cl bond, whilst in step (ii), NH_3 *trans* to tu is replaced by another tu

molecule because the *trans* effect of tu > NH₃. In step (iii), the product shown is
formed because of the inherent lability of the M–Cl bond. In step (iv), NH$_3$ *trans* to tu
is replaced by another tu molecule because the *trans* effect of tu is greater than NH$_3$.

Contrary to the above, in the *trans* isomer of [Pt(NH$_3$)$_2$(Cl$_2$)], replacement stops after
the two chloride ions have been replaced. The explanation behind this is as follows:

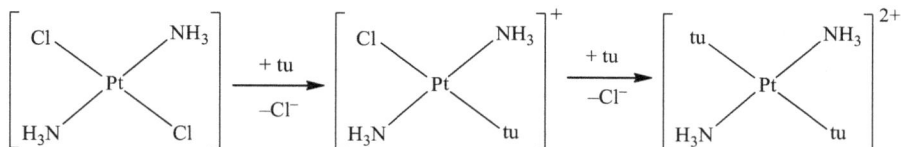

In step (i), any of the two Cls, *trans* to each other is replaced by tu molecule. In step
(ii), Cl *trans* to tu is replaced by another tu molecule because *trans* effect of tu > Cl⁻.
Since, the *trans* ammonia molecules do not labilize each other, no replacement is
further possible by any number of tu molecules.

7.19 Mechanism of the *trans* effect

Kinetic control is must in the *trans* effect because the thermodynamically most stable
isomer is not always produced. This is obvious since it is possible to form two different isomers or three different isomers (vide the synthetic applications of the *trans* effect) depending upon the reaction sequence, and only one of the isomers can be most
stable in the thermodynamic sense. Furthermore, we observe that best *trans* directors
(say T) are often those that are the best π-acceptor. Hence, in the formation of bis-
isomers of the type/sort [PtCl$_2$T$_2$] form [PtCl$_4$]$^{2-}$, the *trans*-isomer is always observed
even though *cis*-isomers are favoured thermodynamically in general.

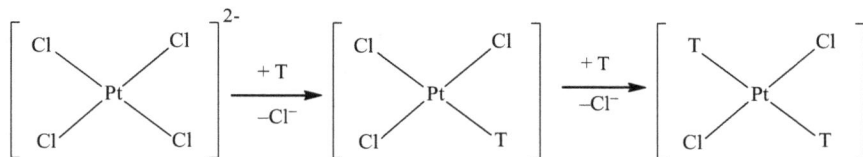

Such kinetically controlled reactions are common both in organic and inorganic
chemistry and represent examples of reactions in which the energy of activation of
activated complex is more important in determining the course of reaction than the
energy of the product.

Two opinions have been advanced with regard to the mechanism of *trans* effect.
The first is essentially a static one emphasizing weakening of *trans* bond, the second
is the lowering of the activation energy of *trans* replacement. The earlier literature

made no attempt to distinguish between the two effects but in view of the above discussion, it is imperative to separate them as follows:

The *trans* effect is a kinetic factor and it is the effect of a coordinated group (T) upon the rate of substitution reactions of ligands opposite to it by the stabilization of the transition state. The strong π-acceptors like NO^+, C_2H_4, CO and CN^- stabilize *the transition state* by accepting electron density that the incoming nucleophilic ligand donates to the metal through π-interaction. In contract, the *trans* influence is a thermodynamic factor, and is defined as the extent to which that ligand weakens the metal–ligand bond *trans* to itself in the equilibrium state of a substrate. For example, strong σ-donor ligands, such as H^-, I^-, Me^- and PR_3, destabilize/weaken the metal–ligand bond *trans* to themselves and thereby bringing the easy substitution of that ligand.

7.20 Theory of *trans* effect

The *trans* effect is a kinetic phenomenon affecting the magnitude of activation energy of a reaction. The stability of both the ground state (i.e. the square planar complex before substitution) and of the activated complex can, in principle, affect the activation energy required for the substitution reaction. Hence, any factor that changes the stability of the ground state and/or which changes the stability of the activated complex would be a contributor to the *trans* effect shown by the attached ligand. There are two theories of *trans* effect, one relates to the ground state and the other to the activated complex. These are discussed further in the chapter.

7.21 The polarization theory

The earliest theory that still has current application is the polarization theory of *trans* effect proposed by Grinberg (1935). This theory is primarily concerned with the effects on the ground state. According to this theory, if the square planar complex is symmetrical, which is $[MCl_4]^{2-}$, the bond dipoles to the various ligands will be identical and cancel *trans* to each other (Figure 7.21). Consequently, no *trans* effect is operative in such complexes.

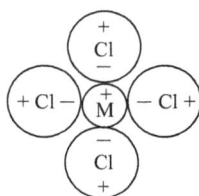

Figure 7.21: *Trans* effect is not operative in $PtCl_4^{2-}$ due to cancellation of bond dipoles.

Contrary to this, if a more polarizable and more polarizing ligand L (such as I⁻) is introduced in the square planar complex, such as M LA$_2$X having two bonds L–M and M–X *trans* to each other (Figure 7.22(a)), the primary charge on the metal ion polarizes the electron charge cloud on L and thus induces a dipole in L. The dipole in L, in turn, induces a dipole in M as shown below (Figure 7.22(b)).

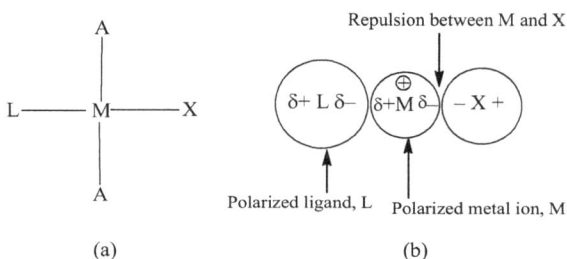

Figure 7.22: Pictorial representation of polarization theory of *trans* effect.

The orientation of the induced dipole on the metal ion is such that it opposes/repels the natural dipole of the ligand X *trans* to the polarizing ligand L. So, the attraction of ligand X for the metal ion is decreased because of the polarizable and more polarizing ligand L. Consequently, M–X bond becomes weaker. The weakening of the M–X bond facilitates the replacement of X by the entering ligand. Accordingly, the polarization of ligand L is directly proportional to the *trans* effect of this ligand.

The polarization theory is supported by the following two facts:
(i) It should be most important when the central atom is large and polarizable, and indeed the ordering of the *trans* effect is Pt(II) > Pd(II) > Ni(II) with the decreasing trend of size in the order of their mention.
(ii) The *trans*-directing series (C_2H_4 ~ CN⁻ ~ CO ~ NO ~ H⁻ > CH$_3$⁻ ~ SC(NH$_2$) ~ SR$_2$ ~ PR$_3$ > SO$_3$H > NO$_2$⁻ ~ I⁻ ~ SCN⁻ > Br⁻ > Cl⁻ > py > RNH$_2$ ~ NH$_3$ > OH⁻ > H$_2$O) should also be polarizability series. In general, *trans*-directing groups are more polarizable either because they are large in size (iodide) or multiple bonded (C_2H_4, CN⁻, CO). Nevertheless, the emphasis on weakening of the *trans* bond limits its (polarization theory) application to the *trans* influence.

7.22 The π-bonding theory

A second approach that involves the weakening of *trans* bond is the static π-bonding theory This theory satisfactorily explains the *trans* effect of those attached ligands in square planar complexes which are π-acceptors or π-acids, which are PPh$_3$, CN⁻, CO, C_2H_4, PR$_3$ and so on.

Let us consider the substitution reaction given below occurring in a square planar complex, MA_2LX, in which L is a π-acceptor ligand whose *trans* effect is to be investigated, and X is *trans* to L. This substitution reaction takes in the steps as shown in Figure 7.23.

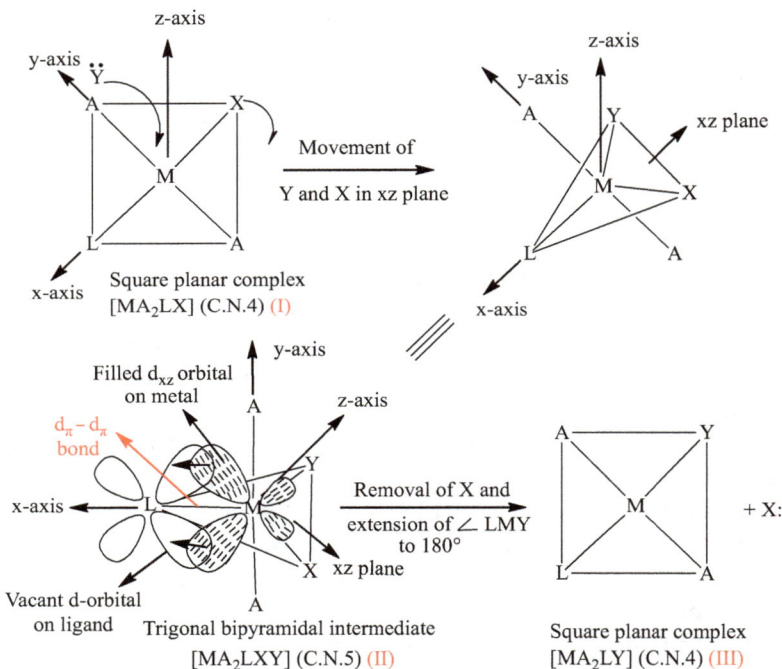

$$MA_2LX + :Y \longrightarrow MA_2LY + :X$$

Figure 7.23: Pictorial representation of π-bonding theory of *trans* effect.

(i) In step one, the entering ligand Y attacks the square planar complex $[MA_2LX]$ (I) and a five coordinate TBP intermediate (transition state), $[MA_2LXY]$ (II) is obtained. This TBP intermediate is formed due to the movement of X and Y ligands in the xz trigonalplane of TBP geometry. In this geometry, the two A ligands occupy the *trans* position along the y axis. It may be seen from the structure of TBP intermediate that the leaving group X, entering group Y and π-bonding ligand L form the trigonal xz plane of the TBP geometry.

(ii) In step two, the transition state (II) loses X and square planar (SP) complex, $[MA_2LY]$ (III) is formed. In the formation of this SP complex, the $\angle LMY$ equals to $120°$ in TBP geometry is extended to $180°$ in SP complex, $[MA_2LY]$ (III). In this SP complex, the entering group Y is placed *trans* to the π-acid ligand L. It is notable here that the position of the entering group Y in [MA$_2$LY] (III) is the same with respect to L as that of leaving group X in [MA$_2$LX] (I) with respect to L.

(iii) Chatt et al. (1955) and Orgel (1956) independently pointed out that consideration of activated complex improves the π-bonding theory. They suggested that the TBP intermediate (II) is stabilized by the formation of M–L π-bond between the π-bonding ligand L and metal M. This π-bond is formed in xz plane of TBP transition state, [MA$_2$LXY] (II) by the overlap between the empty π-orbital (p, d or π*) of ligand L and filled d$_{xz}$ orbital of the metal M. Orgel has pointed out that the filled d$_{xz}$ orbital of the metal involved in the overlap lie in xz trigonal plane formed by LXY. Due to formation of M–L π-bond, the electrons are back donated from the metal d-orbital to the empty p/d/π*-orbital of ligand L. Consequently, the electron density between the metal M and the ligand X which is *trans* to L in the [MA$_2$LX] is decreased, that is, the electron density in M–X bond is decreased. The decrease in electron density in M–X bond facilitates the incoming ligand Y to occupy a position *trans* to L to form substituted product [MA$_2$LY]. The formation of a dπ–dπ M–L bond in TBP intermediate (II) is shown in Figure 8.23.

Concluding remarks

1. Though the π-bonding theory proposes stabilization of the TBP intermediate/ transition state, there is evidence that the M–X bonds are longer when they are placed *trans* to a ligand with a strong *trans* effect than they are placed *cis* to that such a group even in the ground state. Therefore, a ligand with strong *trans* effect affects both the ground as well as the transition state.

2. The present view is, therefore, that both the theory, namely, polarization theory which weakens the M–X bond in the ground state, and π-bonding theory which stabilizes TBP activated complex/intermediate contributes towards the *trans* effect shown by an attached *trans*-directing ligand. The extent of contribution by each depends upon the nature of the ligand.

7.23 *Trans* effect in octahedral complexes

7.23.1 Introduction

It has long been accepted that a coordinated ligand can exert a profound influence upon the metal-to-ligand bonding and lability of other ligands within a complex, particularly those in a *trans* position. A *trans* effect is the effect of a coordinated group on the rate of substitution reactions of ligands *trans* to itself. Hence, it is a kinetic phenomenon which depends upon both ground state and transition state factors, and is distinct from purely ground-state properties such as bond length.

The term **'trans-influence'** was originally used to describe the tendency of a ligand to selectively weaken the bond *trans* to itself, and is commonly used to describe ground-state phenomena. However, although bond weakening is often assumed to be synonymous with bond lengthening, it should always be remembered that this is not necessarily the case. There is no logical reason why an 'effect' should describe a kinetic property, whilst an 'influence' refers to a thermodynamic property and ambiguous use of the term *trans* effect is a common source of confusion.

In view of the above, *trans* effect is now used in a general sense to cover both kinetic and equilibrium phenomena, and the specific terms **'structural trans-effect'** (**STE**) to refer to the effect of a ligand on the *bond distance* to a *trans* ligand and **'kinetic trans-effect'** (**KTE**) to describe the effect on the *lability* of a *trans* ligand, are now adopted for *trans* effect.

7.23.2 STEs in octahedral metal complexes

One might expect the *trans* effect to be operable in octahedral complexes. However, evidence for its presence in octahedral complexes is not abundant. There is definite evidence for a *trans* influence (now, **STE**) on the ground state. For example:

(i) Direct structural evidence for STEs of phosphite ligands is limited. However, X-ray studies on complexes, *trans*-$[Cr^0(CO)_4(PPh_3)X]$ [X = $P(^nBu)_3$, $P(OMe)_3$, $P(OPh)_3$, or CO] (Figure 7.24) have afforded the bond distance data given in Table 7.5. These data show that the Cr–PPh$_3$ bond lengthens gradually as the π-acceptor strength of X increases in the order $P(^nBu)_3 < P(OMe)_3 < P(OPh)_3 < CO$, the difference between the strong σ-donor $P(^nBu)_3$ and the strong π-acceptor CO being ca. 0.07 Å. In keeping with the competitive π-bonding theory, the Cr–P(X) bond is strengthened and shortened at the expense of the *trans* Cr–PPh$_3$ bond.

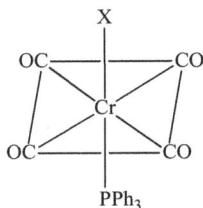

X = $P(^nBu)_3$, $P(OMe)_3$, $P(OPh)_3$ or CO **Figure 7.24:** Structure of *trans*-$[Cr^0(CO)_4(PPh_3)X]$.

Table 7.5: Lengthening of Cr–PPh$_3$ and shortening of Cr–P(X) bond distances in *trans*-[Cr^0(CO)$_4$(PPh$_3$) X] complexes [X = P(^{n}Bu)$_3$, P(OMe)$_3$, P(OPh)$_3$ or CO].

X	Cr–PPh$_3$ (Å)	Cr–P(X) (Å)
P(^{n}Bu)$_3$	2.349	2.344
P(OMe)$_3$	2.364	2.261
P(OPh)$_3$	2.395	2.228
CO	2.422	–

(ii) Nitric oxide coordinates most often as a linearly bonded terminal ligand, but can also adopt a bent bonding mode. When coordinated in the latter fashion, NO behaves as 'NO$^-$' and is a very strong σ-donor which may also engage in some π-back-bonding. Although octahedral bent nitrosyl complexes are rare and comparatively little structural information is available, crystal structure data available (given in Table 7.6) on some five bent nitrosyl complexes (Figure 7.25) clearly show that NO$^-$ has a particularly strong STE on a range of ligands.

Table 7.6: Crystal structure data of some bent nitrosyl complexes showing NO$^-$ having strong STE on a range of ligands.

Compound	L	M–L$_{trans}$ (Å)	M–L$_{cis}$ (Å)
[Co^{III}(NO)(NH$_3$)$_5$]Cl$_2$	NH$_3$	2.220(4)	1.981(8)
trans-[Co^{III}(NCS)(NO)(diars)$_2$]NCS	NCS$^-$	2.12(1)	
trans-[Co^{III}(ClO$_4$)(NO)(en)$_2$]ClO$_4$	ClO$_4^-$	2.360(4)	
mer, *trans*-[Rh^{III}(NO)(MeCN)$_3$(PPh$_3$)$_2$](PF$_6$)$_2$	MeCN	2.308(8)	2.02(1)
mer, *trans*-[Ir^{III}(NO)(MeCN)$_3$(PPh$_3$)$_2$](PF$_6$)$_2$	MeCN	2.36(3)	1.94(2)

(iii) The linearly bonded nitrosyl ligand is a poor σ-donor, but a very strong π-acceptor. The STEs of NO$^+$ are particularly interesting since they depend very much upon the bonding properties of the ligand being affected. Some of the linearly bonded nitrosyl complexes showing STE of NO$^+$ are illustrated below.

In the unusual complex, [W^{II}O(Me)(^{i}Pr$_3$tacn)(NO)] · EtOH · 2H$_2$O (^{i}Pr$_3$tacn=N, N', N''-tri(isopropyl)triazacyclononane) (Figure 7.26), the W–N(^{i}Pr$_3$tacn) bond distances are 2.422(9) Å, (*trans* to O$_2^-$), 2.42(1) Å, (*trans* to NO$^+$) and 2.297(8) Å, (*trans* to Me$^-$). In this case, the STE of NO$^+$ on an amine donor is as large as that of an oxo ligand and much larger than that of a methyl group.

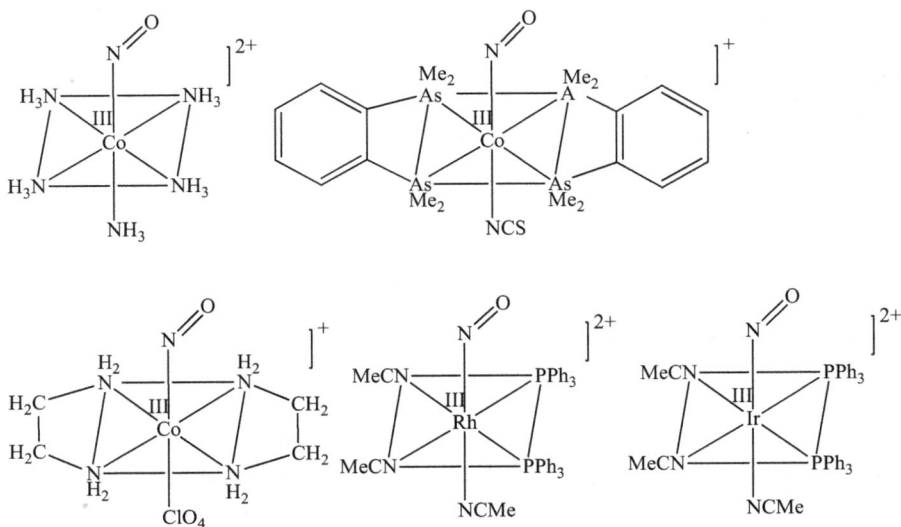

Figure 7.25: Structures of some bent nitrosyl complexes.

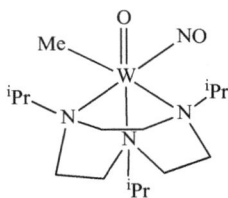

Figure 7.26: [W^{II}O(Me)(^{i}Pr$_3$tacn)(NO)].

Selected crystallographic data for some linear nitrosyl complexes showing STEs are given in Table 7.7.

Table 7.7: Crystallographic data for some linear nitrosyl complexes showing structural trans-effects (STEs).

Compound	L	M–L$_{trans}$ (Å)	M–L$_{cis}$ (Å)
1. *trans*-[V^{-I}(CO)$_4$(PMe$_3$)(NO)]	PMe$_3$	2.543(1)	
2. *mer*-[Cr^0(CN^{t}Bu)$_3$(dppm)(NO)]PF$_6$	dppm	2.478(2)	2.347(2)
3. *trans, mer*-[Cr^I(ONO)$_2$(py)$_3$(NO)] · py	py	2.17(1)	2.096(8)
4. [Co^{III}(en)$_3$][Cr^I(CN)$_5$(NO)] · 2H$_2$O	CN$^-$	2.08(1)	2.033(7)
5. *trans*-[Mo^{0}Cl$_2$(bpy)(NO)$_2$]	bpy	2.20(1)	
6. *cis*-(AsPh$_4$)$_2$[Re^I(CN)$_4$(tu)(NO)]	CN$^-$	2.23(1)	2.12(1)
7. *trans*-Et$_4$N[Re^{II}Cl$_4$(py)(NO)]	py	2.218(6)	

From the data given in Table 7.7, it is notable that when NO^+ is *trans* to another π-acceptor ligand, it does exhibit STEs which vary from small to moderate. Although compound **4** demonstrates STE on the weakly π-accepting CN^-, certain related complexes, such as, that in $Na_2[Ru^{II}(CN)_5(NO)] \cdot 2H_2O$ do not show such an effect. This may be due to subtle differences in electronic structure and bonding caused by changing the nature of the metal ion.

(iv) By thermogravimetric studies made on some mixed-ligand cyanonitrosyl complexes of chromium(I) of composition, $[Cr^I(NO)(CN)_2(L)_2(H_2O)]$ (L = aniline and its derivatives) (Figure 7.27), Maurya et al. (1986) have demonstrated that the weight loss observed at 95–110 °C corresponds to elimination of one coordinated water molecule *trans* to lineally bonded NO^+. The low decomposition temperature of coordinated water molecule in these complexes is most probably due to strong STE of NO^+ on water molecule *trans* to it.

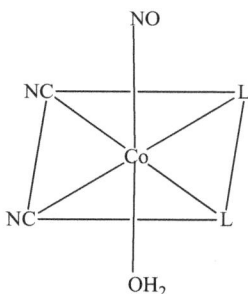

Figure 7.27: $[Cr^I(NO)(CN)_2(L)_2(H_2O)]$ (L = aniline and its derivatives) showing strong STE of NO^+ on water molecule *trans* to it.

(v) STEs in metal–metal bonded octahedral complexes

Since the presence of a quadruple Re–Re bond in $[Re^{III}_2Cl_8]^{2-}$ recognized in 1964, numerous complexes containing metal–metal bonds have been studied, many of which feature two (or three) octahedrally coordinated metal ions. A number of the structural studies on such complexes have featured a consideration of STEs, both those of the axial ligands and those of the metal–metal bonded units themselves. For example, the Pt–Pt single bond distances in the compounds $[Pt^{III}_2X_2(pyd)_2(NH_3)_4](NO_3)_2 \cdot nH_2O$ (pyd = α-pyridonato) are as follows: 2.547(1) Å, (X = ONO_2, n = 0.5), 2.568(1) Å, (X = Cl^-, n = 0), 2.576(1) (X = NO_2^-, n = 0.5) and 2.582(1) (X = Br^-, n = 0.5.)

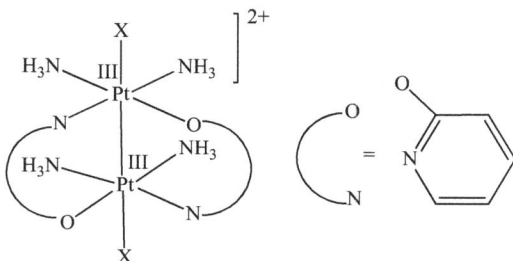

Furthermore, the axial Pt–X bonds in these complexes are ca. 0.10–0.15 Å, longer than the expected values, showing that the $\{Pt_2\}^{6+}$ unit exerts a large STE on the axial ligands (Hollis and Lippard 1983; Hollis et al. 1983).

(vi) The hydride ion behaves as a strong σ-donating ligand which can exert pronounced STEs due to polarization effects. This has been observed experimentally. X-ray crystallographic data for representative octahedral hydrido complexes (Figure 7.28) featuring a range of metal ions and *trans* ligands are given in Table 7.8.

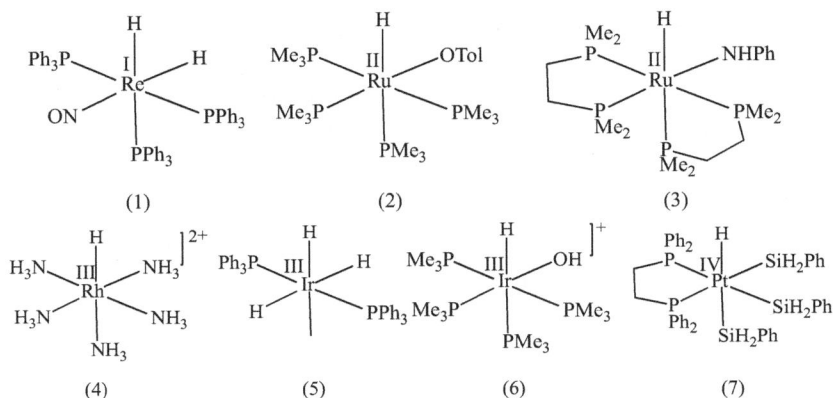

Figure 7.28: Structure of some octahedral hydrido complexes.

Table 7.8: STEs due to polarization effects of hydride ion in some octahedral hydrido complexes.

Compound	L	M–L$_{trans}$ (Å)	M–L$_{cis}$ (Å)
1. *cis, mer*-[Re^{I}(H)$_2$(PPh$_3$)$_3$(NO)]·0.5C$_6$H$_6$	PPh$_3$	2.47(1)	2.39(1)
2. *cis*-[Ru^{II}H(OTol)(PMe$_3$)$_4$]	PMe$_3$	2.364(2)	2.328(3) (*trans* P)
			2.229(2) (*trans* O)
3. *cis*-Ru^{II}H(NHPh)(dmpe)$_2$]	dmpe	2.335(3)	2.286(4) (*trans* P)
			2.244(3) (*trans* N
4. [Rh^{III}H(NH$_3$)$_5$](ClO$_4$)$_2$	NH$_3$	2.24(1)	2.079(7)
5. *mer*-[Ir^{III}(H)$_3$(PPh$_3$)$_3$]·0.5C$_6$H$_6$	PPh$_3$	2.347(3)	2.286(4)
6. *cis*-[Ir^{III}H(OH)(PMe$_3$)$_4$]PF$_6$	PMe$_3$	2.369(2)	2.337(1) (*trans* P)
			2.259(2) (*trans* O)
7. *fac*-[Pt^{IV}H(SiH$_2$Ph)$_3$(dppe)]	SiH$_2$Ph$^-$	2.406(5)	2.374(7)

dmpe = 1,2-bis(dimethylphospino)ethane; dppe = 1,2-bis(diphenylphospino) ethane

7.23.3 KTEs in octahedral metal complexes

Though a STE destabilizes a complex in the ground state, this will enhance reactivity towards ligand substitutions only if there is no effect on the energy of the transition state. This is sometimes the case for dissociatively activated substitutions, and the STE and KTE ligand series of a particular complex will then be similar. However, if the transition state energy is affected by the ligand in question, then there may be no correlation between STEs and KTEs. This is commonly observed in substitution reactions of square planar complexes which react via associative mechanisms. A general KTE sequence for such complexes has been established as follows:

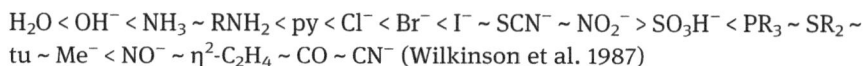

$$H_2O < OH^- < NH_3 \sim RNH_2 < py < Cl^- < Br^- < I^- \sim SCN^- \sim NO_2^- > SO_3H^- < PR_3 \sim SR_2 \sim$$
$$tu \sim Me^- < NO^- \sim \eta^2\text{-}C_2H_4 \sim CO \sim CN^- \text{ (Wilkinson et al. 1987)}$$

Kinetic data for ligand substitution reactions are most often obtained via spectrophotometric methods, but NMR (nuclear magnetic resonance) and other techniques may also be used. Substantial studies of KTEs in octahedral complexes are relatively scarce compared to the reports concerning STEs. This is largely because ground-state phenomena are more readily understood, and the complete explanation of kinetic properties requires knowledge of ligand substitution reaction mechanisms. Some illustrative examples showing KTEs in octahedral metal complexes are given below:

(i) In metal carbonyl compounds (Figure 7.29), KTE seems to be operative to the extent that substitution proceeds readily as long as there are carbonyl groups *trans* to each other (presumably activating each other).

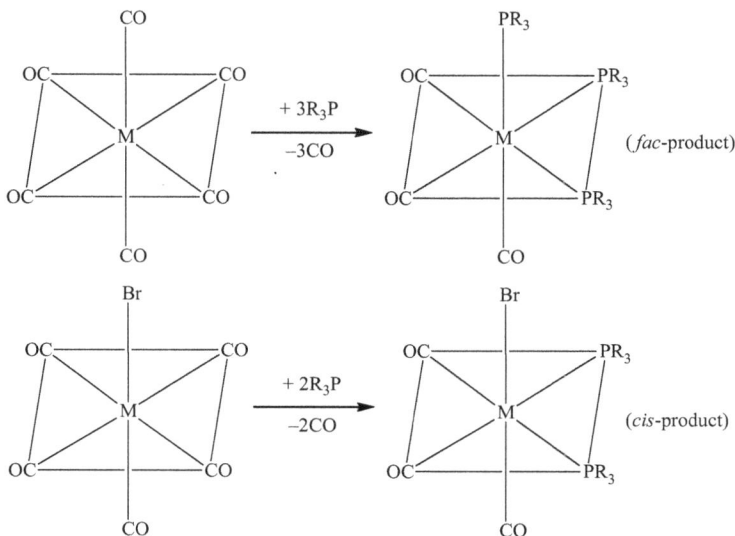

Figure 7.29: Metal carbonyls showing KTE.

These products are invariably *facial* (or *fac*) if trisubstituted (when three identical ligands occupy one face of an octahedron, the isomer is said to be *facial* or *fac*) or *cis* if reaction is stopped at the disubstituted products. If steric hindrance exists between the *cis*-substituents, isomerization to a *trans* product may occur. Otherwise, greater stability is found with the carbonyl groups *trans* to the weaker π-bonding phosphines.

(ii) The first-order kinetics and activation parameters for the reactions of chromum(0) complexes shown in Figure (7.30) indicate that dissociation of L is the rate-determining step. A comparison of the k_1 (first-order rate constant for ligand dissociation) values yields the following KTE series for loss of either PPh$_3$ or P(OPh)$_3$ (L):

$$CO < P(OMe)_3 \sim P(OPh)_3 < P(nBu)_3 < PPh_3$$

In the case of L = P(OPh)$_3$, the relative rate enhancement moving from X = CO to X = PPh$_3$ is ca. 2×10^4. It was suggested that these widely differing KTEs arise from electronic rather than steric factors (Wovkulich and Atwood 1982).

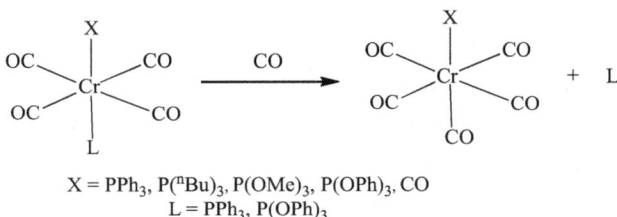

X = PPh$_3$, P(^{n}Bu)$_3$, P(OMe)$_3$, P(OPh)$_3$, CO
L = PPh$_3$, P(OPh)$_3$

Figure 7.30: Chromum(0) complexes showing KTE.

(iii) The reactivity studies between the complex *cis*, *mer*-[Ru^{II}Cl$_2$(CO)(PMe$_2$Ph)$_3$] (Figure 7.31) and I$^-$ showed that the chloride ligand *trans* to PMe$_2$Ph is replaced much more rapidly than that *trans* to CO. The greater KTE of the phosphine ligand was attributed to its stronger σ-donor strength when compared to the very weakly basic CO (Lupin and Shaw 1968).

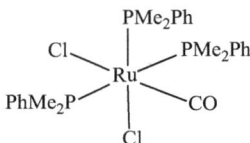

Figure 7.31: *cis*, *mer*-[Ru^{II}Cl$_2$(CO)(PMe$_2$Ph)$_3$] exhibiting KTE due to σ-donating phosphine ligand.

The ease of selective substitution by phosphine or arsine ligands of the PR$_2$Ph ligand *trans* to H$^-$ in the complexes *cis*, *mer*-[Ru^{II}Cl(H)(CO)(PR$_2$Ph)$_3$] (R = Et, ^{n}Pr, ^{n}Bu) (Figure 7.32) provides clear evidence for the pronounced KTE of the strongly σ-donating hydrido ligand (Douglas and Shaw 1970).

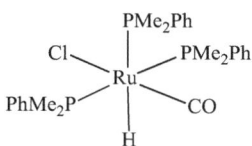

Figure 7.32: *cis*, *mer*-[Ru^{II}Cl(H)(CO)(PR$_2$Ph)$_3$] showing KTE due to strongly σ-donating hydrido ligand.

7.24 Redox reactions: electron transfer reactions

7.24.1 Introduction

Electron transfer reactions, also known as oxidation–reduction reactions or redox reactions, involve a transfer of electrons from one atom to the other atom. Redox reactions or oxidation–reduction are of wide importance in chemistry. Many classical analytical methods are based on rapid redox reactions, including Fe^{II} determination by titration with HCrO$_4^-$, as well as oxidation of substances such as C$_2$O$_4^{2-}$ by MnO$_4^-$. The role of transition metal ions in life processes depends on their ability to participate selectively in electron transfer reactions. Redox reactions involving two transition metal complexes generally occur very rapidly. Thus, values of Eo are a rather good guide to the chemistry that actually occurs on a convenient timescale.

7.24.2 Categories of redox reactions

Redox reactions can occur in two ways and these are: (i) reaction through atom or group transfer and (ii) reaction through electron transfer. Reactions of both categories are discussed.

7.24.2.1 Redox reactions through atom or group transfer
Particularly in aqueous solution, it is usually possible to imagine atom or group transfer, rather than electron transfer, as occurring in redox reaction. For example, Fe(II) ion in aqueous solution may act as a reducing agent by transferring H atom from its hydration shell to an organic free radical. Thus,

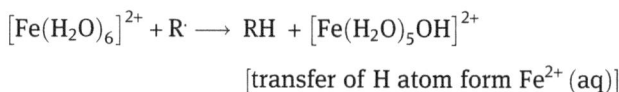

$$[Fe(H_2O)_6]^{2+} + R^{\cdot} \longrightarrow RH + [Fe(H_2O)_5OH]^{2+}$$

[transfer of H atom form Fe^{2+} (aq)]

On the other hand, Fe(III) ion in aqueous solution may act as a oxidizing agent by transferring OH radical from its inner hydration shell to an organic substrate. Thus,

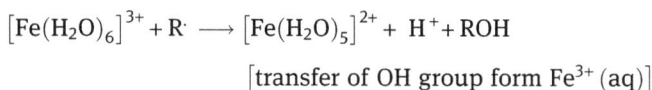

$$[Fe(H_2O)_6]^{3+} + R^{\cdot} \longrightarrow [Fe(H_2O)_5]^{2+} + H^+ + ROH$$

[transfer of OH group form Fe^{3+} (aq)]

In general, transfer of a positive group or atom is equivalent to transfer of electron(s), and transfer of a negative group or atom is equivalent to the taking up of electrons(s).

7.24.2.2 Redox reactions through electron transfer

The quantum mechanical picture of an electron in an orbital is of a charge with charge density falling off with distance from the nucleus and the probability of the existence of electron even far away from the nucleus being non zero. This means that the electron may be occasionally found even far from the nucleus. In other words, the electron may momentarily extend its orbital and come very close to another nucleus which may accept that electron subject to certain conditions.

Consider atom A and A^+ of a monoatomic gas. Since the range of extension of electron orbital is more than the accepted diameters of A and A^+, the electron from A may get transferred to A^+ even before the actual collision for electron transfer occurs between A and A^+ (all the collision, however, do not lead to electron transfer).

The situation for electron transfer from a metal ion in solution is more difficult. First, the extension of orbital is likely to be hindered by the solvent molecules. Second, the ligand of the complex would be likely to insulate the electron from the metal ion and thus inhibit the extension of orbital of metal ion.

(a) Electron transfer: law of conservation of energy

Let us consider the reaction,

$$\left[Fe(H_2O)_6\right]^{3+} + \left[Fe(H_2O)_6\right]^{2+} \longrightarrow \left[Fe^*(H_2O)_6\right]^{2+} + \left[Fe^*(H_2O)_6\right]^{3+}$$

The reacting complex ion, $[Fe(H_2O)_6]^{3+}$, in its ground state (or stable state) has an equilibrium Fe–O distance of 2.05 Å. Likewise, the equilibrium Fe–O distance in [Fe(H_2O)_6]^{2+} in its ground state is 2.21 Å. Any change in this Fe–O distance would increase the energy of the respective reacting complex ion.

According to Frank–Condon principle, the nuclear positions and hence the inter-nuclear distances in a molecule remain unchanged during electron transfer reaction. This is because electrons move too much faster than the nuclei in the specie. Hence, the Fe–O distance in $[Fe^*(H_2O)_6]^{2+}$ produced from $[Fe(H_2O)_6]^{3+}$ by electron transfer would at least be 2.05 Å. This distance is shorter than Fe–O distance of 2.21 Å in $[Fe(H_2O)_6]^{2+}$ in the stable or ground state. Thus, $[Fe^*(H_2O)_6]^{2+}$ produced would be in higher energy or excited state. It would come to the lower energy or stable state by re-adjusting the Fe–O distance to 2.21 Å. In doing so, it has to lose the excess energy to the system.

Similarly, $[Fe^*(H_2O)_6]^{3+}$ produced from $[Fe(H_2O)_6]^{2+}$ by electron transfer would at first, have the Fe–O distance of 2.05 Å in its stable state. Hence, the $[Fe^*(H_2O)_6]^{3+}$ produced would also be in excited state and would come to the ground state by re-adjusting the Fe–O distance to 2.21 Å. Whilst doing so, it also loses the excess energy to the system.

It is notable here that after the products $[Fe^*(H_2O)_6]^{2+}$ and $[Fe^*(H_2O)_6]^{3+}$ come to the stable state, these are in no way different from reacting species $[Fe(H_2O)_6]^{3+}$ and $[Fe(H_2O)_6]^{2+}$. Hence, they can again participate in electron transfer and can again liberate energy. Thus, electron transfer between two ions of the same metal in aqueous solution would lead to constant generation of energy. But, this is contrary to the law of conservation of energy. However, the law is not violated if the metal–ligand distances in both the complex ions are, somehow, made the same prior to electron transfer between them.

In the above example, the ions $[Fe(H_2O)_6]^{2+}$ and $[Fe(H_2O)_6]^{3+}$ can have the same Fe–O distance falling in between 2.05 Å and 2.21 Å. It has been calculated that minimum energy is required to bring Fe–O distance in both the species to the same value of 2.09 Å. Let E_a be the energy required to produce the $[Fe\#(H_2O)_6]^{2+}$ and $[Fe\#(H_2O)_6]^{3+}$ species having the same distance of 2.09 Å in each. Then

$$E_a = \text{Energy of} \left\{ \left[Fe\#(H_2O)_6\right]^{2+} + \left[Fe\#(H_2O)_6\right]^{3+} \right\}$$
$$- \text{Energy of} \left\{ \left[Fe(H_2O)_6\right]^{2+} + \left[Fe(H_2O)_6\right]^{3+} \right\}$$

The electron transfer process when all the four species have the same Fe–O distance of 2.09 Å may be represented as

$$\left[Fe\#(H_2O)_6\right]^{2+} \text{ and } \left[Fe\#(H_2O)_6\right]^{3+} \longrightarrow \left[Fe\#(H_2O)_6\right]^{3+} + \left[Fe\#(H_2O)_6\right]^{2+}$$

After the electron transfer, $[Fe\#(H_2O)_6]^{3+}$ and $[Fe\#(H_2O)_6]^{2+}$ species produced would come to the ground state by releasing energy exactly equal to the energy E_a which was required for the formation of reacting species $[Fe\#(H_2O)_6]^{2+}$ and $[Fe\#(H_2O)_6]^{3+}$. In this way, no net energy would occur during electron transfer between ions to yield different oxidation states of the same metal.

In the case of electron transfer between ions of different metals M and M′, the probability of electron transfer form M^{2+} to M'^{3+} may be different, which is, more than the probability of electron transfer form M'^{3+} to M^{2+}. In that situation, the backward reaction may not occur. This case is different from the one involving electron transfer between the ions of the same metal in which there is equal probability of electron transfer between forward and backward reactions. Thus, electron transfer between ions of different metals does not violate the law of conservation of energy.

(b) Electron transfer: spin state of metal ions
Another constraint placed on electron transfer between ions of the same metal in solution phase is that, leaving aside the electrons that are to be transferred, the two ions should have essentially the same electron spin. Let us consider the case of

[Co(NH$_3$)$_6$]$^{2+}$–[Co(NH$_3$)$_6$]$^{3+}$ system. The NH$_3$ ligand is of moderate crystal field strength. It causes electron spin pairing in [Co(NH$_3$)$_6$]$^{3+}$ (d^6) but cannot do so in [Co(NH$_3$)$_6$]$^{2+}$ (d^7). Consequently, Co in the former complex, [Co(NH$_3$)$_6$]$^{3+}$ (d^6) is in the low-spin state (t$_{2g}^6$) and in the latter complex, [Co(NH$_3$)$_6$]$^{2+}$ (d^7), it is in the high-spin state (t$_{2g}^5$eg^2). As such, the electron transfer from [Co(NH$_3$)$_6$]$^{2+}$ to [Co(NH$_3$)$_6$]$^{3+}$ is not allowed if both the ions are in their ground spin states. For electron transfer to occur, excitation of one or both the ions is required to such an extent that the electron spins of both the ions becomes the same (barring the spin of the electron being transferred). This would obviously require

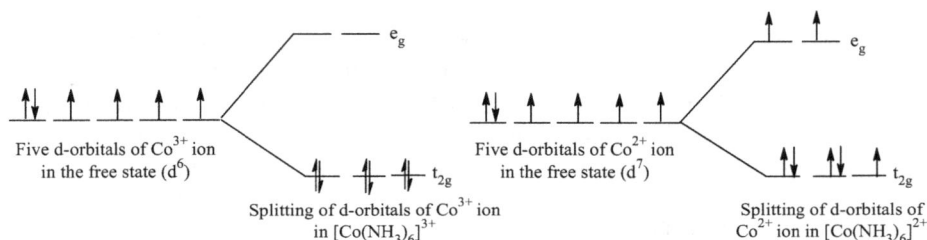

Five d-orbitals of Co^{3+} ion in the free state (d^6)

Splitting of d-orbitals of Co^{3+} ion in [Co(NH$_3$)$_6$]$^{3+}$

Five d-orbitals of Co^{2+} ion in the free state (d^7)

Splitting of d-orbitals of Co^{2+} ion in [Co(NH$_3$)$_6$]$^{2+}$

considerable energy. Thus, the electron transfer from [Co(NH$_3$)$_6$]$^{2+}$ to [Co(NH$_3$)$_6$]$^{3+}$ is more difficult than, which is, the electron transfer from [Fe(H$_2$O)$_6$]$^{2+}$ to [Fe(H$_2$O)$_6$]$^{3+}$. Both Fe(II) and Fe(III) are in high-spin state in these complexes.

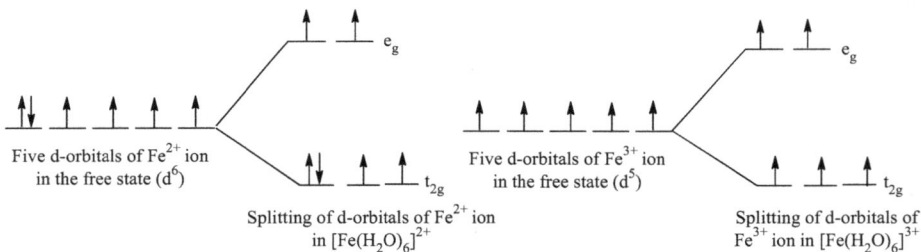

Five d-orbitals of Fe^{2+} ion in the free state (d^6)

Splitting of d-orbitals of Fe^{2+} ion in [Fe(H$_2$O)$_6$]$^{2+}$

Five d-orbitals of Fe^{3+} ion in the free state (d^5)

Splitting of d-orbitals of Fe^{3+} ion in [Fe(H$_2$O)$_6$]$^{3+}$

7.25 Mechanism of electron transfer reaction in solution

The possibility of an electron leaking through a potential energy barrier (that would be impenetrable according to classical mechanics) is a well-known quantum mechanical phenomenon. As already mentioned, electron can be transferred from distances which are considerably larger than the distance essential for actual collision of the reactants. Thus, 'electron tunnelling effect' is directly related to the phenomenon of extension of electronic orbitals in space.

Electron transfer reactions are generally faster than those involving the exchange of ligands on the metal ions. These, however, can be conveniently studied by radioactive exchange as well as NMR techniques.

Two types of transition states or intermediates are generally formed during electron transfer reactions. These are: (i) outer sphere transition states and (ii) inner sphere transition states

(i) Outer sphere transition states: outer sphere mechanism

When the inner coordination spheres of two reacting complex ions remain intact in the transition states of an electron transfer reaction, it is called *outer sphere transition state*. The mechanism of electron transfer reaction through such a transition state is called *outer sphere mechanism.*

(ii) Inner sphere transition states: inner sphere mechanism

When the coordinated ligand of one complex ion forms a bridge with the other complex ion in the transition state of an electron transfer reaction, it is called the *inner sphere transition state*. The mechanism of electron transfer reaction through such a transition state is called *inner sphere mechanism.*

Both the mechanisms of electron transfer reactions are presented below in detail with suitable examples.

7.25.1 Outer sphere mechanism

If the reductant and the oxidant complex ions are both inert and if their coordination shells are saturated, the rate of ligand exchange would be very small and also the probability of the formation of ligand bridge would be negligible. However, the observed rate of electron transfer between such pair of complex ions is quite fast. This cannot be explained by any mechanism involving the dissociation of an atom or group, or exchange of coordinated ligand, since both the complex ions are inert. Such electron transfer reaction would obviously occur through outer sphere mechanism. We shall now consider some examples of electron transfer reaction occurring through this mechanism.

(a) $[Co(NH_3)_6]^{2+}-[Co(NH_3)_6]^{3+}$ system

Let us consider the reaction,

$$\left[Co(NH_3)_6\right]^{2+} + \left[Co(NH_3)_6\right]^{3+} \longrightarrow \left[Co(NH_3)_6\right]^{3+} + \left[Co(NH_3)_6\right]^{2+}$$

The following points are notable with regard to the above reaction:

(i) Since the oxidant, $[Co(NH_3)_6]^{3+}$ is an inert complex and Co^{3+} ion has in it a saturated inner coordination shell, the exchange of NH_3 ligands between the two reacting complexes $[Co(NH_3)_6]^{2+}$ and $[Co(NH_3)_6]^{3+}$ is found to be extremely slow.

(ii) The coordinated NH_3 ligands do not have any other lone pair through which they can form a bridge between the two cobalt ions. Therefore, an inner transition state of the type $[(NH_3)_5Co-(NH_3)-Co(NH_3)_5]^{5+}$ is not possible.

(iii) In view of the above, electron transfer between $[Co(NH_3)_6]^{2+}$ and $[Co(NH_3)_6]^{3+}$ **must occur** through **outer sphere mechanism** in which inner coordination cells of the reductant, $[Co(NH_3)_6]^{2+}$ and oxidant $[Co(NH_3)_6]^{3+}$ remain intact in the transition state.

The outer sphere electron transfer process is **slow** because of following reasons:

(i) Since the Co–N distances in the oxidant and reductant complex ions are significantly different, considerable energy of activation is required to excite the two ions to have identical Co–N distance so that electron transfer reaction between the reacting complex ions may occur.

(ii) The electron to be transferred from the reductant $[Co(NH_3)_6]^{2+}$ to the oxidant $[Co(NH_3)_6]^{3+}$ resides in e_g orbital of $t_{2g}^5 e_g^2$ configuration of Co^{2+} ion in reductant. Now, since d_z^2 and $d_{x^2-y^2}$ orbitals of e_g set along with s- and p-orbitals of the Co^{2+} ion are already engaged in forming Co–N sigma bonding, the delinking of this electron form Co(II) would, therefore, require more energy.

(iii) The spin state of Co(II) present in $[Co(NH_3)_6]^{2+}$ ($t_{2g}^5 e_g^2$; high-spin) and Co(III) present in $[Co(NH_3)_6]^{3+}$ ($t_{2g}^6 e_g^0$; low-spin) are different.

(b) $[Fe(CN)_6]^{4-}-[Fe(CN)_6]^{3-}$ system

Let us consider the reaction,

$$\left[Fe(CN)_6\right]^{4-} + \left[Fe(CN)_6\right]^{3-} \longrightarrow \left[Fe(CN)_6\right]^{3-} + \left[Fe(CN)_6\right]^{4-}$$

The following points are noteworthy for this reaction.

(i) In this reaction, an electron is transferred from hexacyanoferrate(II), $[Fe(CN)_6]^{4-}$ to hexacyanoferrate(III),$[Fe(CN)_6]^{3-}$. Thus, the former is reductant (electron loser) and the latter is oxidant (electron gainer).

(ii) Both the reacting complexes are inert towards substitution and their coordination shells are fully saturated.

(iii) The isotopic exchange studies confirm that the coordinated CN^- ligands of both the reacting complex ions are not exchangeable. Hence, the formation of inner sphere bridged transition state is ruled out. Thus, the electron transfer in the above reaction proceeds through the outer sphere mechanism.

The observed rate of electron transfer in the above reaction has been found to be fast. This is because of the following reasons:

(i) Since Fe^{2+}–C bond distance in reductant is slightly larger than Fe^{3+}–C bond distance in oxidant, very little energy is required to make these bond distances in the two complex ions identical so that electron transfer may occur.

(ii) Leaving aside the electrons that are to be transferred, both the metal ions have the same electron spin (both are in low-spin state). This makes the electron transfer easy.

(iii) The electron to be transferred from the reductant to the oxidant resides in t_{2g} orbital of $t_{2g}^6 e_g^0$ configuration of Fe^{2+} ion in reductant. Now since none of t_{2g}^6 electron is engaged in Fe–CN σ-bonding, the transfer of an electron t_{2g}^6 configuration is easy.

(iv) The ligand CN^- is unsaturated and the presence of unsaturated ligands on the reductant facilitates the electron tunnelling.

(v) Since CN^- is a π-acceptor ligand, it can form a π-bond by accepting electron from the from the t_{2g} orbitals of the metal ion. Such ligands always stabilize the lower oxidation state ion. This implies that such ligands stabilize the product, $[Fe(CN)_6]^{4-}$ formed due to acceptance of electron by $[Fe(CN)_6]^{3-}$ (oxidant).

(c) $[Os(dipy)_3]^{2+}$–$[Os(dipy)_3]^{3+}$ system

Let us consider the reaction,

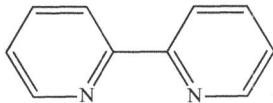

$$[Os(dipy)_3]^{2+} + [Os(dipy)_3]^{3+} \longrightarrow [Os(dipy)_3]^{3+} + [Os(dipy)_3]^{2+}$$

2,2'-dipyridine (dipy)

This reaction cannot occur through inner sphere mechanism because the complex ions are inert and their coordination shells are fully saturated.

The observed rate of the above reaction is found to be fast and this may be due to the following reasons:

(i) The Os–N bod distances in the reacting complex ions are almost the same. Hence, very little energy is required to make the Os–N bod distances equal in both the complex ions so that electron transfer between the complex ions may occur.

(ii) The ligand dipyridine is a π-acceptor and is highly conjugated. This facilitates electron tunnelling.

(iii) The electron to be transferred from the reductant to the oxidant is present in the t_{2g} orbital $[Os^{2+}(d^6)$: electronic configuration $t_{2g}^6 e_g^0]$ which is not involved in Os–N σ-bonding. Hence, this electron can be easily delinked from the Os^{2+} ion.

(iv) Except for the electron which is being transferred, both the metal ions have the same electron spin (both are in low-spin state). This facilitates the electron transfer.

(d) Some other systems

Some other systems and their electron transfer reactions occurring through outer sphere mechanism are presented below:

(i) $[Fe(dipy)_3]^{2+} - [Fe(dipy)_3]^{3+}$ **System:**

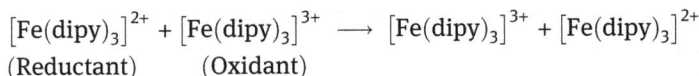

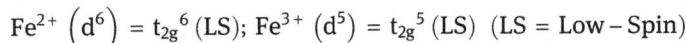

$$[Fe(dipy)_3]^{2+} + [Fe(dipy)_3]^{3+} \longrightarrow [Fe(dipy)_3]^{3+} + [Fe(dipy)_3]^{2+}$$

(Reductant)   (Oxidant)

$Fe^{2+} \left(d^6\right) = t_{2g}^{6}$ (LS); $Fe^{3+} \left(d^5\right) = t_{2g}^{5}$ (LS) (LS = Low – Spin)

(ii) $[Fe(phen)_3]^{2+} - [Fe(phen)_3]^{3+}$ **System:**

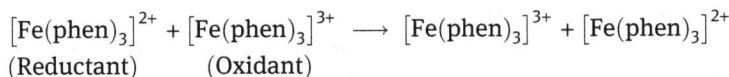

$$[Fe(phen)_3]^{2+} + [Fe(phen)_3]^{3+} \longrightarrow [Fe(phen)_3]^{3+} + [Fe(phen)_3]^{2+}$$

(Reductant)   (Oxidant)

$Fe^{2+} \left(d^6\right) = t_{2g}^{6}$ (LS); $Fe^{3+} \left(d^5\right) = t_{2g}^{5}$(LS)

(iii) $[Mn(CN)_6]^{4-} - [Mn(CN)_6]^{3-}$ **System:**

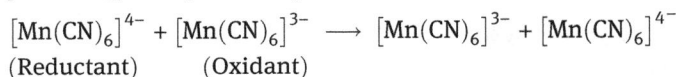

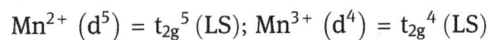

$$[Mn(CN)_6]^{4-} + [Mn(CN)_6]^{3-} \longrightarrow [Mn(CN)_6]^{3-} + [Mn(CN)_6]^{4-}$$

(Reductant)   (Oxidant)

$Mn^{2+} (d^5) = t_{2g}^{5}$ (LS); $Mn^{3+} (d^4) = t_{2g}^{4}$ (LS)

(iv) $[Mo(CN)_6]^{4-} - [Mo(CN)_6]^{3-}$ **System:**

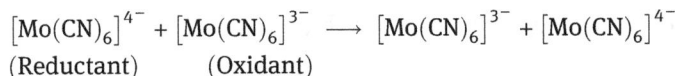

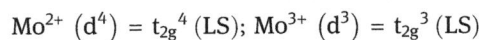

$$[Mo(CN)_6]^{4-} + [Mo(CN)_6]^{3-} \longrightarrow [Mo(CN)_6]^{3-} + [Mo(CN)_6]^{4-}$$

(Reductant)   (Oxidant)

$Mo^{2+} (d^4) = t_{2g}^{4}$ (LS); $Mo^{3+} (d^3) = t_{2g}^{3}$ (LS)

(v) $[IrCl_6]^{3-} - [IrCl_6]^{4-}$ **System:**

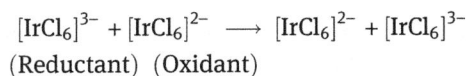

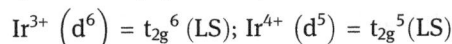

$$[IrCl_6]^{3-} + [IrCl_6]^{2-} \longrightarrow [IrCl_6]^{2-} + [IrCl_6]^{3-}$$

(Reductant) (Oxidant)

$Ir^{3+} \left(d^6\right) = t_{2g}^{6}$ (LS); $Ir^{4+} \left(d^5\right) = t_{2g}^{5}$(LS)

(vi) $[Fe(H_2O)_6]^{2+} - [Fe(H_2O)_6]^{3+}$ **System:**

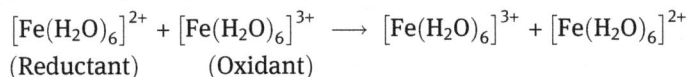

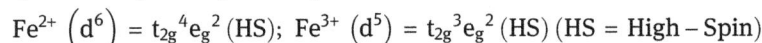

$$[Fe(H_2O)_6]^{2+} + [Fe(H_2O)_6]^{3+} \longrightarrow [Fe(H_2O)_6]^{3+} + [Fe(H_2O)_6]^{2+}$$

(Reductant)   (Oxidant)

$Fe^{2+} \left(d^6\right) = t_{2g}^{4}e_g^{2}$ (HS); $Fe^{3+} \left(d^5\right) = t_{2g}^{3}e_g^{2}$ (HS) (HS = High – Spin)

(vii) $[Co(en)_3]^{2+} - [Co(en)_3]^{3+}$ **System:**

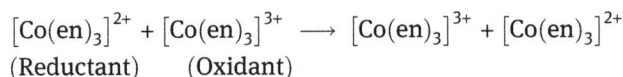

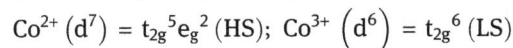

$$[Co(en)_3]^{2+} + [Co(en)_3]^{3+} \longrightarrow [Co(en)_3]^{3+} + [Co(en)_3]^{2+}$$

(Reductant)   (Oxidant)

$Co^{2+} (d^7) = t_{2g}^{5}e_g^{2}$ (HS); $Co^{3+} \left(d^6\right) = t_{2g}^{6}$ (LS)

(viii) $\left[Co(C_2O_4)_3\right]^{4-} - \left[Co(C_2O_4)_3\right]^{3-}$ **System:**

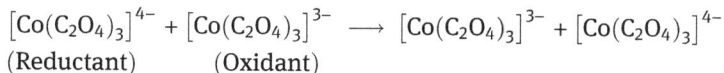

(Reductant)   (Oxidant)

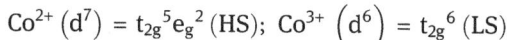

(ix) $\left[MnO_4\right]^{2-} - \left[MnO_4\right]^{3-}$ **System:**

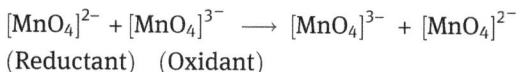

(Reductant) (Oxidant)

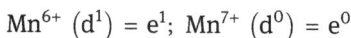

Following points are notable with regard to above reactions:

(i) Reactions (i) to (v) are very fast (rate constant ~10^4–10^6 L mol^{-1} s^{-1}), because the metal–ligand bond distances in the two reacting complexes differ very slightly. Thus, very little energy of bond stretching and bond compressing is needed to achieve symmetrical transition state.

(ii) The complex ions $\left[Fe(H_2O)_6\right]^{2+}$ and $\left[Fe(H_2O)_6\right]^{3+}$ in reaction (vi) react very slowly, since there exists a considerable difference between Fe–O bond lengths in the two species.

(iii) For reactions (vii) and (viii), there is a considerable difference in two reactants in metal–ligand bond distance and so reaction is rarely slow (rate constant ~10^{-4} L mol^{-1}s^{-1}).

(iv) In the reaction (ix), both the reacting ions are inert towards substitution. Although the MnO_4^{2-}/MnO_4^- pair, the bond length difference is somewhat greater, the transfer of electron from MnO_4^{2-} to MnO_4^- takes place rapidly (rate constant ~3×10^3 L mol^{-1} s^{-1}).

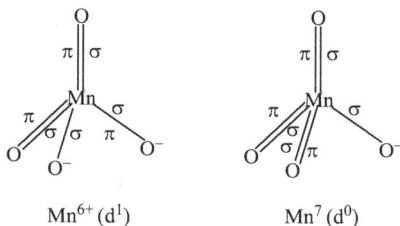

$Mn^{6+} (d^1)$ $Mn^7 (d^0)$

Main characteristics of electron transfer reactions through outer sphere mechanism

Based on the above observations, main characteristics of electron transfer reactions occurring through outer sphere mechanism are:

(i) Electron transfer reactions occur between those metal ions which are either both inert with their coordination shell saturated or one of the complex ions is inert with is coordination shell saturated. In the latter case, none of the coordinated ligands should be a bridging ligand.

(ii) Electron transfer between ions of the same metal is fast if metal–ligand distances in the reactant and oxidant are nearly the same.

(iii) Electron transfer is fast if electrons are able to reach the surface of the complex either through unsaturation or through conjugation.

(iv) Electron transfer is fast if the electron being transferred is present in t_{2g} orbital and slow if present in e_g-orbital.

(v) The rate of electron transfer depends upon the polarizing power of the central metal ion.

(vi) Electron transfer is fast if the coordinated ligands present on the system are π-acceptors.

(vii) Electron transfer is fast if the electron spins of the two complex ions (leaving aside the electrons being transferred) are the same.

Electron transfer: molecular orbital theory

According to molecular orbital theory, the highest occupied molecular orbital of the reductant is the donor orbital and the lowest unoccupied molecular orbital of the oxidant is the receptor orbital. If either of these orbitals does not meet the symmetry requirements for effective electron transfer, then higher chemical activation involving strong deformation and electron configuration change is necessary for reaction. This factor is a qualitative guide to speculate the speed of electron transfer reactions. The electron transfer reactions given below involve symmetric molecular orbitals:

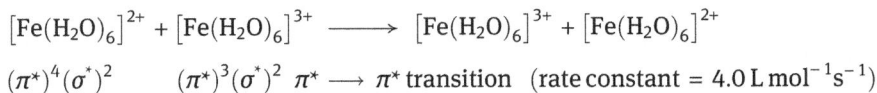

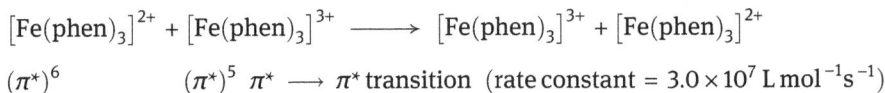

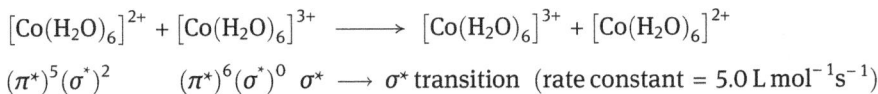

$$\left[Fe(H_2O)_6\right]^{2+} + \left[Fe(H_2O)_6\right]^{3+} \longrightarrow \left[Fe(H_2O)_6\right]^{3+} + \left[Fe(H_2O)_6\right]^{2+}$$

$(\pi^*)^4(\sigma^*)^2 \qquad (\pi^*)^3(\sigma^*)^2 \; \pi^* \longrightarrow \pi^*$ transition $\;$ (rate constant $= 4.0\,L\,mol^{-1}s^{-1}$)

$$\left[Fe(phen)_3\right]^{2+} + \left[Fe(phen)_3\right]^{3+} \longrightarrow \left[Fe(phen)_3\right]^{3+} + \left[Fe(phen)_3\right]^{2+}$$

$(\pi^*)^6 \qquad\qquad (\pi^*)^5 \; \pi^* \longrightarrow \pi^*$ transition $\;$ (rate constant $= 3.0 \times 10^7 \, L\,mol^{-1}s^{-1}$)

$$\left[Co(H_2O)_6\right]^{2+} + \left[Co(H_2O)_6\right]^{3+} \longrightarrow \left[Co(H_2O)_6\right]^{3+} + \left[Co(H_2O)_6\right]^{2+}$$

$(\pi^*)^5(\sigma^*)^2 \qquad (\pi^*)^6(\sigma^*)^0 \; \sigma^* \longrightarrow \sigma^*$ transition $\;$ (rate constant $= 5.0\,L\,mol^{-1}s^{-1}$)

Cross reaction: the Marcus–Hush principle

Electron transfer reactions between two completely different complex ions or molecules are known as *cross reactions*. For example,

$$\left[Fe(CN)_6\right]^{4-} + \left[Mo(CN)_8\right]^{3-} \longrightarrow \left[Fe(CN)_6\right]^{3-} + \left[Mo(CN)_8\right]^{4-}$$

A cross reaction involves a large chemical activation. However, it is fast because a large $E_a^{\#}$ requirement is partly compensated by an attendant negative free energy change, ΔG. For cross reactions, $\Delta G < 0$, the products are more stable than reactants.

The rate constant for a cross reaction involving the **outer sphere mechanism** can be predicted by an equation proposed by R. A. Marcus and N. S. Hush,

$$k_{1,2} = [k_1 k_2 K_{1,2} f]^{1/2} \qquad (7.40)$$

Here $k_{1,2}$ is the rate constant of the cross reaction, k_1 and k_2 each is the rate constant for electron exchange species of the same metal (self-exchange reaction) and $K_{1,2}$ is the equilibrium constant for the cross reaction.

The term f is close to 1 and is given by:

$$\log f = \frac{(\log K_{1,2})^2}{4} \log \frac{K_1 K_2}{z^2} \qquad (7.41)$$

where z is the measure of collision frequency.

The above equation (**Marcus–Hush equation**) can now be applied for prediction of rate constant for electron transfer in a cross reaction.

Consider the cross reaction,

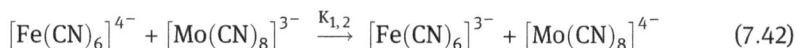

$$\left[Fe(CN)_6\right]^{4-} + \left[Mo(CN)_8\right]^{3-} \xrightarrow{K_{1,2}} \left[Fe(CN)_6\right]^{3-} + \left[Mo(CN)_8\right]^{4-} \qquad (7.42)$$

The equilibrium constant $K_{1,2}$ for this reaction is 1.0×10^2, and the electrochemical potential is $E = 0.12$ V. The self-exchange reactions that apply are given below:

$$\left[Fe(CN)_6\right]^{4-} + \left[Fe(CN)_6\right]^{3-} \xrightarrow{K_1} \left[Fe(CN)_6\right]^{3-} + \left[Fe(CN)_6\right]^{4-} \qquad (7.43)$$

$$\left[Mo(CN)_8\right]^{4-} + \left[Mo(CN)_8\right]^{3-} \xrightarrow{K_2} \left[Mo(CN)_6\right]^{3-} + \left[Mo(CN)_8\right]^{4-} \qquad (7.44)$$

The value of self-exchange rate constant for equation (7.43) is $k_1 = 7.4 \times 10^4 \, L \, mol^{-1} s^{-1}$ and for equation (7.44) is $k_2 = 3.0 \times 10^4 \, L \, mol^{-1} \, s^{-1}$. The calculated value of f using equation (7.41) for equation (7.42) is 0.85.

Substituting the values of k_1, k_2, $K_{1,2}$ and f in equation (7.40) yields the prediction value of rate constant $k_{1,2}$ as

$$k_{1,2} = [7.4 \times 10^4 . 3.0 \times 10^4 . 1.0 \times 10^2 . 0.85]^{1/2}$$

$$= [18.87 \times 10^{10}]^{1/2}$$

$$= [18.87]^{1/2} \times 10^5$$

$$= 4.34 \times 10^5 \, Lmol^{-1}s^{-1}$$

$$= 43.4 \times 10^4 \, Lmol^{-1}s^{-1}$$

7.25.2 Inner sphere (or ligand bridged) mechanism

Several experimental studies carried out by **H. Taube and his students** have established that electron transfer reactions occurring through inner sphere mechanism normally have one inert and one labile reactant. The inert reactant has a coordinated ligand which is capable of forming a bridge between the two reacting complex ions. We shall now discuss some examples of electron transfer reaction occurring through this mechanism.

Example 1

Let us consider the following reactions:

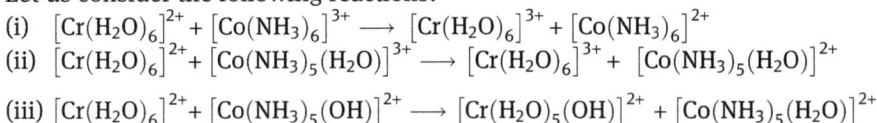

(i) $[Cr(H_2O)_6]^{2+} + [Co(NH_3)_6]^{3+} \longrightarrow [Cr(H_2O)_6]^{3+} + [Co(NH_3)_6]^{2+}$

(ii) $[Cr(H_2O)_6]^{2+} + [Co(NH_3)_5(H_2O)]^{3+} \longrightarrow [Cr(H_2O)_6]^{3+} + [Co(NH_3)_5(H_2O)]^{2+}$

(iii) $[Cr(H_2O)_6]^{2+} + [Co(NH_3)_5(OH)]^{2+} \longrightarrow [Cr(H_2O)_5(OH)]^{2+} + [Co(NH_3)_5(H_2O)]^{2+}$

In all three reactions, the reductant is $[Cr(H_2O)_6]^{2+}$. This complex ion is labile and can easily lose a H_2O molecule. The order of the rate of electron transfer in these reactions is: (i) $\ll$ (ii) $<$ (iii).

In reaction (i), the oxidant $[Co(NH_3)_6]^{3+}$ is an inert complex ion with no coordinated ligand capable of forming a bridge with the reductant. Therefore, in this reaction, electron transfer cannot occur through inner sphere mechanism and hence the rate of electron transfer reaction would be slow. In reaction (ii), the oxidant $[Co(NH_3)_5(H_2O)]^{3+}$ is also inert, but coordinated H_2O is capable of forming a bridge, though a very weak one, with the reductant $[Cr(H_2O)_6]^{2+}$. This leads to the formation of the intermediate bridged complex $[(H_2O)_5Cr-(H_2O)-[Co(NH_3)_5]^{5+}$. This happens because the oxygen of the coordinated H_2O has one unused pair of electrons through which it can link with the reductant. The formation of an aqua bridge has been established by H_2O^{18} studies. Thus, the reaction (ii) proceeds through inner sphere mechanism and is much faster than reaction (i). The mechanism of reaction (ii) may be represented as:

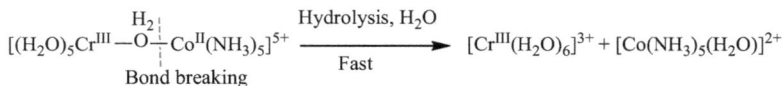

$$[Cr^{II}(H_2O)_6]^{2+} + [Co^{III}(NH_3)_5(H_2O)]^{3+} \longrightarrow [(H_2O)_5Cr^{II}-(H_2O)-[Co^{III}(NH_3)_5]^{5+} + H_2O$$

$$\Big\updownarrow \begin{array}{l} \text{Electron} \\ \text{transfer} \end{array}$$

$$[(H_2O)_5Cr^{III}-(H_2O)-Co^{II}(NH_3)_5]^{5+}$$

$$[(H_2O)_5Cr^{III}-O\underset{\substack{|\\ \text{Bond breaking}}}{\overset{\substack{H_2\,|\\|}}{-}}Co^{II}(NH_3)_5]^{5+} \xrightarrow[\text{Fast}]{\text{Hydrolysis, } H_2O} [Cr^{III}(H_2O)_6]^{3+} + [Co(NH_3)_5(H_2O)]^{2+}$$

In reaction (iii), the inert oxidant $[Co(NH_3)_5(OH)]^{2+}$ contains a coordinated OH⁻ ligand capable of forming a fairly strong bridge with the reductant, $[Cr(H_2O)_6]^{2+}$.

This leads to the formation of a bridged intermediate $[(H_2O)_5Cr-(HO)-[Co(NH_3)_5]^{4+}$. Since, OH-bridge is stronger/more stable than the aqua bridge, the energy of activation required to from OH bridge is less than required for the formation of aqua bridge intermediate. Reaction (iii), therefore, proceeds through inner sphere mechanism, and is more facile than reaction (ii). The mechanism of reaction (iii) may be represented as:

$$[Cr^{II}(H_2O)_6]^{2+} + [Co^{III}(NH_3)_5(OH)]^{2+} \longrightarrow [(H_2O)_5Cr^{II}-(HO)-Co^{III}(NH_3)_5]^{4+} + H_2O$$

$$\updownarrow \begin{matrix} \text{Electron} \\ \text{transfer} \end{matrix}$$

$$[(H_2O)_5Cr^{III}-(HO)-Co^{II}(NH_3)_5]^{4+}$$

$$[(H_2O)_5Cr^{III} \underset{\text{Bond breaking}}{-O\dot{+}Co^{II}(NH_3)_5}]^{4+} \quad \xrightarrow[\text{Fast}]{\text{Hydrolysis, } H_2O} \quad [Cr^{III}(H_2O)_5(OH)]^{2+} + [Co(NH_3)_5(H_2O)]^{2+}$$

Example 2

Let us consider the reaction,

$$\left[Cr^{II}(H_2O)_6\right]^{2+} + \left[Co^{III}(NH_3)_5X\right]^{2+} \xrightarrow{H_2O} \left[Cr^{III}(H_2O)_5X\right]^{2+} + \left[Co^{II}(NH_3)_5(H_2O)\right]^{2+} \quad (7.45)$$

This reaction has been studied in aqueous medium in presence of radioactive free X^- ($X^- = F^-$, Cl^-, Br^- or I^-), and has been found that the product $[Cr^{III}(H_2O)_5X]^{2+}$ shows an absence of radioactivity. This observation confirms that it is the X^- ligand initially attached to the inert oxidant $[Co^{III}(NH_3)_5X]^{2+}$ gets transferred to the reductant $[Cr^{II}(H_2O)_6]^{2+}$ forming $[Cr^{III}(H_2O)_5-X]^{2+}$. This cannot happen if X^- completely dissociates from the oxidant and becomes free. Therefore, the transition state/ intermediate must have X^- associated with both the oxidant and reductant.

The above observations provide an indirect proof for the formation of inner sphere or bridged complex in the transition state. Thus, the mechanism of reaction in question may be represented as given below:

$$\text{(a) } [Cr^{II}(H_2O)_6]^{2+} + [Co^{III}(NH_3)_5X]^{2+} \xrightarrow{\text{Slow}} [(H_2O)_5Cr^{II}-X-Co^{III}(NH_3)_5]^{4+} + H_2O$$

$$\updownarrow \begin{matrix} \text{Electron} \\ \text{transfer} \end{matrix}$$

$$[(H_2O)_5Cr^{III}-X-Co^{II}(NH_3)_5]^{4+}$$

$$\text{(b) } [(H_2O)_5Cr^{III}\underset{\text{Bond breaking}}{-X\dot{+}Co^{II}(NH_3)_5}]^{5+} \quad \xrightarrow[\text{Fast}]{\text{Hydrolysis, } H_2O} \quad [Cr^{III}(H_2O)_5X]^{2+} + [Co(NH_3)_5(H_2O)]^{2+}$$

The step (a), being the slow step, is the rate-determining step. The rate of electron transfer in reaction (iv) is thus given by:

$$R = k \left(\text{Concn. of } \left[Cr^{II}(H_2O)_6 \right]^{2+} \right) \times \left(\text{Concn. of } \left[Co^{III}(NH_3)_5X \right]^{2+} \right)$$

$$= k \times [\text{Reductant}][\text{Oxidant}]$$

It has been observed that the rate of electron transfer depends upon the nucleophilic character (not the basicity) of the halide ion X^-. The rates are thus in the order $I^- > Br^- > Cl^- > F^-$. It is obvious that the stronger the nucleophilic character of the halide ion, the greater would be the capability to form $Cr-X-Co$ bridge.

Example 3
Let us consider the reaction,

Obviously, the COOR group forms a bridge between reductant and oxidant so that the reaction proceeds through inner sphere mechanism. This explains why the rate of electron transfer in this reaction is fast compared to reaction (i) in **Example 1**. The rate of reaction is further increased if R is unsaturated or conjugate. This is due to the fact that this factor known to facilitate electron transfer.

Example 4
The reactions given below are worth considering as electron transfer reactions.

(i) $[Cr^{II}(H_2O)_6]^{2+} + [Cr^{III}(NH_3)_5(NCS)]^{2+} \longrightarrow [(H_2O)_5Cr^{II} - NCS - Cr^{III}(NH_3)_5]^{4+}$
    Reductant   Oxidant

$[(H_2O)_5 Cr^{II} - NCS - Cr^{III}(NH_3)_5]^{4+} \xrightleftharpoons{\text{Electron transfer}} [(H_2O)_5Cr^{III} - NCS - Cr^{II}(NH_3)_5]^{4+}$

$$[(H_2O)_5 Cr^{III} - NCS - Cr^{II}(NH_3)_5]^{4+} \xrightarrow{\text{Hydrolysis, } H_2O} [Cr^{III}(H_2O)_5(NCS)]^{2+} +$$
$$[Cr^{II}(NH_3)_5(H_2O)]^{2+}$$

(ii) $[Cr^{II}(H_2O)_6]^{2+} + [Cr^{III}(NH_3)_5(N_3)]^{2+} \longrightarrow [(H_2O)_5 Cr^{II} - N = N - N = Cr^{III}(NH_3)_5]^{4+}$

$$[(H_2O)_5 Cr^{II} - N = N - N = Cr^{III}(NH_3)_5]^{4+} \xrightleftharpoons{\text{Electron transfer}} [(H_2O)_5 Cr^{III} - N = N - N = Cr^{II}(NH_3)_5]^{4+}$$

$$[(H_2O)_5 Cr^{III} - N = N - N = Cr^{II}(NH_3)_5]^{4+} \xrightarrow{\text{Hydrolysis, } H_2O} [Cr^{III}(H_2O)_5(N_3)]^{2+} +$$
$$[Cr^{II}(NH_3)_5(H_2O)]^{2+}$$

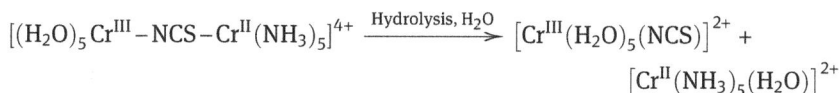

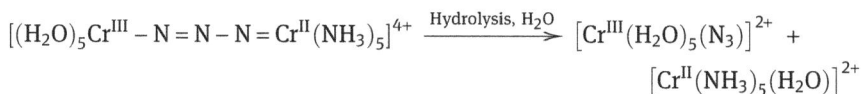

The electron transfer reaction (ii) is found to be faster than reaction (i). This can be explained if we assume inner sphere mechanism for both these reactions.

In inner sphere mechanism, the bridged transition state of reaction (i) is less stable and so would require more energy for its formation compared to the energy required for the formation of bridged transition state of reaction (ii). Consequently, electron transfer reaction (ii) is faster than reaction (i). In fact, the bridge Cr–NCS–Cr bridge is less stable because NCS–Cr linkage is weaker than Cr–NCS linkage. On the other hand Cr–N–N–N–Cr bridge is more stable as it acquires greater stability through resonance as follows:

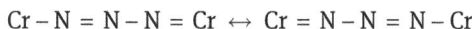

$$Cr - N = N - N = Cr \leftrightarrow Cr = N - N = N - Cr$$

The observed energy of activation (Ea) for the entire above mentioned electron transfer reactions is found to be very low (about 3–4 kcal mol^{-1}) and the entropy change (ΔS) is found to be negative.

In electron transfer reactions occurring through inner sphere mechanism, an atom or group is also transferred from one complex to another. However, the transfer of electron is not caused from atom or group which is transferred. In the electron transfer process, the oxidant and the reductant first form a bridge intermediate/transition state which then breaks up to give the products. *The transfer or non-transfer of the bridging atom or group would depend on the relative stability of the products* that are possible from the bridged intermediate. This is well illustrated from the following example.

Example 5
Let us consider the reaction,

$$[Cr^{II}(H_2O)_6]^{2+} + [Ir^{IV}Cl_6]^{2-} \longrightarrow [(H_2O)_5 Cr^{II} - Cl - Ir^{IV}Cl_5]^{0}$$
$$\text{Reductant} \qquad \text{Oxidant}$$

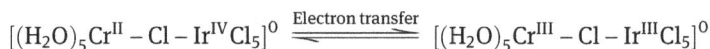

$$[(H_2O)_5 Cr^{II} - Cl - Ir^{IV}Cl_5]^{0} \xrightleftharpoons{\text{Electron transfer}} [(H_2O)_5 Cr^{III} - Cl - Ir^{III}Cl_5]^{0}$$

$$[(H_2O)_5Cr^{III} \!\!-\!\! Cl - \!\!|\!\!- Ir^{III}Cl_5]^0 \atop [(H_2O)_5Cr^{III} - \!\!|\!\!- Cl \!\!-\!\! Ir^{III}Cl_5]^0 \Bigg\} \quad \xrightarrow{\text{Hydrolysis, } H_2O}$$

$\longrightarrow [Cr^{III}(H_2O)_5Cl]^{2+} \;+\; [Ir^{III}Cl_5(H_2O)]^{2-}$

Less stable

$\longrightarrow [Cr^{III}(H_2O)_6]^{3+} \;+\; [Ir^{III}Cl_6]^{3-}$

Most stable

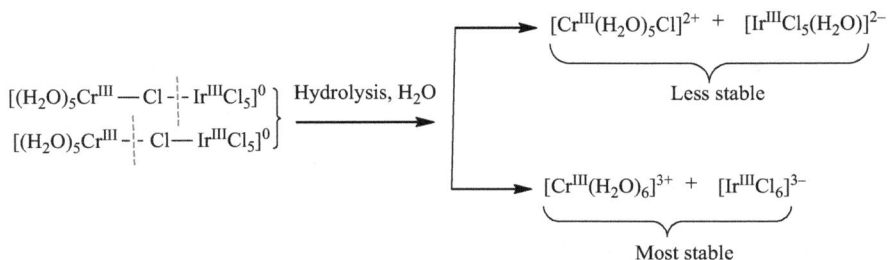

There is possibility of formation of two sets of products, one by transfer of bridging Cl atom and other by no transfer of bridging Cl atom from the bridged intermediate. As the products $[Cr^{III}(H_2O)_6]^{3+}$ and $[Ir^{III}Cl_6]^{3-}$ are found to be more stable than the products $[Cr^{III}(H_2O)_5Cl]^{2+}$ and $[Ir^{III}Cl_5(H_2O)]^{3-}$ in the above reaction, it appears that no transfer of Cl atom has taken place from the bridged intermediate $[(H_2O)_5Cr^{III}-Cl-Ir^{III}Cl_5]^0$, although an electron has been transferred from the reductant to oxidant.

Main characteristics of electron transfer reactions through inner sphere mechanism

Based on the observations given above through different examples, the main characteristics of electron transfer reactions occurring through inner sphere mechanism are given below:

(i)   For electron transfer to occur through inner sphere mechanism one of the reacting complexes should be inert and the other should be labile. Moreover, the inert complex ion should have a ligand capable of bridging both the complex ions during the formation of intermediate.

(ii) The rate of electron transfer increases with increase in the nucleophilic character of the bridging ligand.

(iii) The rate of electron transfer also increases if the bridging ligand has unsaturation or extended conjugation in its structure.

(iv) Although with transfer of electron from reductant to oxidant, the bridging ligand is also transferred from one complex ion to other. But it is not always necessary. The transfer or non-transfer of the bridging ligand depends upon the relative stabilities of the products that are possible from the bridged intermediate.

7.26 Two-electrons transfer reactions

It is well known that some post-transition elements exist in two stable oxidation states differing by two electrons. For example, Tl^+ and Tl^{3+}, the oxidation state Tl^{2+} in between these two being unstable. Thus, when a complex of Tl^+ reacts with a complex of Tl^{3+}, two electrons are transferred from reductant to oxidant. *Such a reaction is called a two electron transfer reaction.* An example for such a reaction is

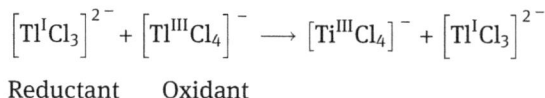

$$\left[Tl^ICl_3\right]^{2-} + \left[Tl^{III}Cl_4\right]^{-} \longrightarrow \left[Ti^{III}Cl_4\right]^{-} + \left[Tl^ICl_3\right]^{2-}$$

Reductant   Oxidant

A mechanism involving an activated complex containing seven Cl^- ions has been suggested for this two-electron transfer.

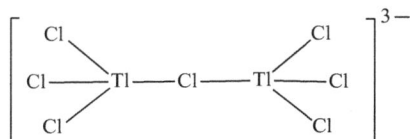

Likewise, two-electron transfer reactions occur in some Pt(II) and Pt(IV) complexes.
     The two electron transfer reactions are further divided into two groups:
(i)   Complementary reactions
(ii) Non-complementary reactions

(i) Complementary reactions

Such reactions are those in which reductant loses and the oxidant gains two electrons. For example, in the Tl^+–Tl^{3+} system in aqueous medium containing ClO_4^- ion, the two electrons are transferred from $Tl(aq)^+$ to $Tl(aq)^{3+}$. The observed rate of the two-electron transfer reaction is given by

$$r = k_1[Tl^+]\,[Tl^{3+}] + k_2[Tl^+]\,[TlOH^{2+}]$$

The above rate law can be explained when one assumes the following two reactions occurring independently:

(i) $\left[Tl(H_2O)_n\right]^+ + \left[Tl(H_2O)_n\right]^{3+} \longrightarrow \left[Tl(H_2O)_n\right]^{3+} + \left[Tl(H_2O)_n\right]^+$

(ii) $\left[Tl(H_2O)_n\right]^+ + \left[Tl(H_2O)_{n-1}(OH)\right]^{2+} \longrightarrow \left[Tl(H_2O)_n\right]^{3+} + \left[Tl(H_2O)_{n-1}(OH)\right]^0$
                     (Present in small quantity)

Several other two electron transfer reactions are known. Some of these are:

$$V^{2+} + Tl^{3+} \longrightarrow V^{4+} + Tl^+$$

$$Sn^{2+} + Hg^{2+} \longrightarrow Sn^{4+} + Hg^0$$

$$Sn^{2+} + Tl^{3+} \longrightarrow Sn^{4+} + Tl^+$$

(ii) Non-complementary reactions

Such reactions are those in which the number of electrons released by the reactant is not equal to the number of electrons accepted by oxidant. Such reactions generally proceed through a multi-step path. For example, the reduction of Tl^{3+} with Fe^{2+} is believed to involve the following steps:

$$Fe(aq)^{2+} + Tl(aq)^{3+} \longrightarrow Fe(aq)^{3+} + Tl(aq)^{2+}$$

$$Fe(aq)^{2+} + Tl(aq)^{2+} \longrightarrow Fe(aq)^{3+} + Tl(aq)^{+}$$

$$\cdots\cdots\cdots\cdots\cdots\cdots\cdots\cdots\cdots\cdots\cdots$$

$$2Fe(aq)^{2+} + Tl(aq)^{3+} \longrightarrow 2\,Fe(aq)^{3+} + Tl(aq)^{+}$$

The non-complementary reactions are generally slower than complementary reactions because in the former termolecular collision is required – that is, more difficult than the bimolecular collision in a complementary reaction.

Exercises

Multiple choice questions/fill in the blanks

1. In nucleophilic substitution reaction in a complex compound:
    (a) Central metal ion is replaced by other metal ion
    (b) Ligand is replaced by another ligand
    (c) Both metal ion and ligand are replaced by other metal ion and ligand
    (d) None of these

2. When a metal ion is replaced by other metal ion in a complex, the substitution reaction is called:
    (a) Nucleophilic substitution
    (b) Electrophilic substitution
    (c) Electromeric substitution
    (d) None of these

3. Bimolecular nucleophilic substitution is called:
    (a) Dissociative S_N^1 mechanism
    (b) Associative S_N^1 mechanism
    (c) Associative S_N^2 mechanism
    (d) Dissociative S_N^1 mechanism

4. Complexes have been classified as labile and inert in 1952 by:
   (a) G. M. Harris
   (b) F. J. Garrick
   (c) Henry Taube
   (d) None of these

5. Sort out an octahedral inert complex system from the following inner orbital systems with d^n electronic configuration:
   (a) All d^0 systems
   (b) All d^1 systems
   (c) All d^2 systems
   (d) All d^3 systems

6. Crystal field stabilization energy (CFSE) of four low-spin octahedral complexes undergoing S_N^1 substitution reactions is given below. Point out a labile complex from them:
   (a) 2.00   (b) −1.14   (c) 1.43   (d) 0.86

7. CSAEs of some four high-spin O_h complexes undergoing S_N^2 substitution reactions are given. Which one is the inert complex from them?
   (a) −2.08   (b) −0.68   (c) −2.79   (d) 1.80

8. The base hydrolysis is the substitution reaction in which:
   (a) H_2O substitutes a coordinated ligand from the complex
   (b) An anion substitutes one or more water molecules from the complex
   (c) OH^- substitutes a coordinated ligand from the complex
   (d) All of above

9. The $S_N^1(CB)$ mechanism is for:
   (a) Anation reaction
   (b) Acid hydrolysis
   (c) Base hydrolysis
   (d) None of these

10. The $S_N^1(CB)$ mechanisms was proposed in 1937 by:
   (a) F. Basolo   (b) R. G. Pearson   (c) F. J. Garrick   (d) G. M. Harris

11. The correct trend of rates of aquation of some octahedral Co(III) complexes on the basis of the solvation effect is:

    (a) $[Co(NH_3)_5Cl]^{2+} < [Co(en)(NH_3)_3Cl]^2$
    $$+ < [Co(en)_2(NH_3)Cl]^{2+} < [Co(en)(diene)Cl]^{2+}$$

    (b) $[Co(NH_3)_5Cl]^{2+} > [Co(en)(NH_3)_3Cl]^{2+}$
    $$> [Co(en)_2(NH_3)Cl]^{2+} > [Co(en)(diene)Cl]^{2+}$$

    (c) $[Co(NH_3)_5Cl]^{2+} > [Co(en)(NH_3)_3Cl]^{2+}$
    $$> [Co(en)_2(NH_3)Cl]^{2+} = [Co(en)(diene)Cl]^{2+}$$

    (d) $[Co(NH_3)_5Cl]^{2+} < [Co(en)(NH_3)_3Cl]^{2+} = [Co(en)_2(NH_3)Cl]^{2+}$
    $$< [Co(en)(diene)Cl]^{2+}$$

12. The actual mechanism of acid hydrolysis of octahedral complexes when no inert ligand in the complex is a π-donor or π-acceptor is:

    (a) $S_N{}^2$ (b) $S_N{}^1$ (c) $S_N{}^1(CB)$ (d) None of these

13. In the aquation of *cis*- and *trans*-$[Co(en)_2(OH)Cl]^+$ complexes, sort out the correct statement with regard to the formation of intermediate product:

    (a) Both form SP intermediate
    (b) Both form TBP intermediate
    (c) *cis*-forms SP and *trans*-forms TBP intermediate
    (d) *cis*-forms trigonal pyramidal and *trans*-forms square bipyramidal intermediate

14. The mechanism involved in the aquation reaction/acid hydrolysis of *cis*-[Co $(en)_2(NH_2)$- Cl]$^+$ with filled p-orbital on nitrogen of $NH_2{}^-$ ($N^- = 2s^2 2p^4$) is:

    (a) $S_N{}^1$ (b) $S_N{}^2$ (c) $S_N{}^1(CB)$ (d) None of these

15. The mechanism involved in the aquation reaction of *cis*- and *trans*-$[O_2NCo(en)_2Cl]^+$ is:

    (a) $S_N{}^1$ (b) $S_N{}^2$ (c) $S_N{}^1(CB)$ (d) None of these

16. Ligand substitution reaction in which one or more H_2O ligand(s) in a complex is replaced by an anion is called:

    (a) Acid hydrolysis
    (b) Base hydrolysis
    (c) Anation reaction
    (d) None of these

17. Ligand substitution reactions in Pt(II) complexes proceed by associative $S_N{}^2$ rather than dissociative $S_N{}^1$ mechanism. It is due to:

    (a) Lack of steric crowding and availability of an empty 'd' orbital
    (b) Lack of steric crowding and availability of an empty 'p' orbital

(c) Lack of steric crowding only

(d) All of these

18. The *trans* effect was first time reported in 1926 by:

 (a) I. I. Chernyaev

 (b) A. D. Hel'man

 (c) L. N. Essen

 (d) E. F. Karandashova

19. Electron transfer reaction through outer sphere mechanism is fast if the electron spins of the two complex ions (leaving aside the electrons being transferred) are the . . . .

20. Electron transfer reaction through outer sphere mechanism is . . . . if the electron being transferred is present in t_{2g} orbital and . . . . if present in e_g- orbital.

21. The rate constant for a cross reaction involving the outer sphere mechanism can be predicted by an equation proposed by . . . . and . . . .

22. The rate of electron transfer through inner sphere mechanism increases if the bridging ligand has . . . . or extended . . . . in its structure.

Short answer type questions

1. What are nucleophilic substitution reactions in octahedral complexes? Derive the rate equations for S_N^1 and S_N^2 mechanisms.

2. A thermodynamic stable complex may be kinetically labile and a thermodynamically unstable complex may be kinetically inert. Justify this statement taking suitable examples.

3. What is the basic principle of interpretation of lability and inertness of transition metal complexes on the basis of valance bond theory (VBT)?

4. Differentiate between activation energy and crystal field activation energy (CFAE).

5. What are the basic assumptions for evaluation of CFAE?

6. What do you understand by acid hydrolysis? Justify that both S_N^1- and S_N^2- mechanism of acid hydrolysis of an octahedral complex (when no inert ligand in the complex is a π-donor or π-acceptor) predict that the rate of hydrolysis is dependent only on the concentration of the complex only.

7. Explain why the rate of aquation of *cis*-$[Co(en)_2(OH)Cl]^+$ is much higher than that of *cis*-$[Co(en)_2(NH_3)Cl]^{2+}$.

8. Present the salient features of associative S_N^2 mechanism for base hydrolysis along with its shortcomings.

9. Taking two suitable examples, explain how the solvation effect justify the order of rates of aquation in Co(III) complexes through S_N^1 mechanism.
10. Highlight the intermediate formed during aquation of complexes containing π-acceptor ligands with justification.
11. What do you understand by base hydrolysis? Show that rate of base hydrolysis is a second order reaction.
12. Explain why the base hydrolysis of $[Co(NH_3)_5Cl]^{2+}$ tend to be almost independent of $[OH^-]$ at very high concentration of OH^-.
13. The anions, which are NO_2^-, NCS^- and N_3^- are strong nucleophiles as OH^- but do not affect the rate of base hydrolysis of ammine complexes. Explain.
14. The base hydrolysis of cis-$[M(en)_2Cl_2]^+$ is faster than the trans-$[M(en)_2Cl_2]^+$ (M = Co(III) or Rh(III). Explain, why?
15. Highlight the effect of H_2O_2 in base hydrolysis of Co(III) ammine complexes in aqueous medium.
16. What is anation reaction? Explain with suitable examples along with its mechanism.
17. Decarboxylation of a carbonato complex to give an aqua complex is given below:

$$[Co(NH_3)_5(CO_3)]^+ + 3H_3O^+ \longrightarrow [Co(NH_3)_5(H_2O)]^{3+}$$

Justify through a suitable mechanism that this reaction completes without cleavage of metal–ligand bond.
18. Draw the two mechanistic pathways proposed for the substitution reaction in square planar complex, which is $[Pt(NH_3)Cl_3] + Y \longrightarrow [Pt (NH_3)Cl_2Y]$.
19. Make a sketch of stereochemistry of S_N^2 mechanism in substitution reaction of square planar complexes with brief note.
20. What is Kurnakov's test for distinction between cis-and trans- isomers of the complexes of the type $[Pt(A)_2X_2]$ (where A = ammine and X = halide)? Explain.
21. Differentiate between 'structural trans-effect' (STE and 'kinetic trans-effect' (KTE).
22. Highlight the main characteristics of electron transfer reactions through inner Sphere mechanism.
23. Briefly describe the main characteristics of electron transfer reactions through outer sphere mechanism.
24. What are cross reactions? Highlight the Marcus–Hush Principle for cross reactions involving the outer sphere mechanism.

Long answer type questions

1. Present an explanatory note on types of intermediate/activated complex formed during S_N^1 and S_N^2 mechanism.
2. What are inert and labile complexes? How does valance bond theory explain lability and inertness in octahedral complexes?

3. Present a detailed view of interpretation of lability and inertness in low-spin and high-spin octahedral complexes. In what way it is superior to valance bond theory?

4. Describe the factors deciding S_N^1 mechanism rather than S_N^2 for acid hydrolysis when no inert ligand in the complex is a π-donor or π-acceptor.

5. Discuss the type of mechanism operative for acid hydrolysis of octahedral complexes when no inert ligand in the complex is a π-donor or π-acceptor by the following factors:
   (a) Inductive effect of ligands
   (b) Charge of the substrate
   (c) Strength of metal-leaving group bond
   (d) Steric effects

6. Explain the following order (decreasing) of the rates of aquation of complexes on the basis of the solvation effect:

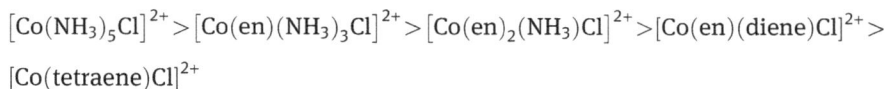

$$[Co(NH_3)_5Cl]^{2+}>[Co(en)(NH_3)_3Cl]^{2+}>[Co(en)_2(NH_3)Cl]^{2+}>[Co(en)(diene)Cl]^{2+}>$$
$$[Co(tetraene)Cl]^{2+}$$

7. Discuss the mechanism of aquation of *cis*-and *trans*-[Co(en)$_2$(OH)Cl]$^+$ complexes. Explain why their rates of aquation are much faster than cis-[Co(en)$_2$(NH$_3$)Cl]$^{2+}$.

8. Present the S_N^1-mechanism of acid hydrolysis in octahedral complexes in which the inert ligand is a π-donor.

9. The mechanism of acid hydrolysis of octahedral complexes containing an inert π-acceptor ligand lying *cis* or *trans* to the leaving group is different from the mechanism of aquation when inert ligands are either π-donors or are capable of forming π-bonding. Justify this statement.

10. Explain why the rate of aquation of *cis*-[O$_2$NCo(en)$_2$Cl]$^+$ would be slower than the aquation of trans isomer, *trans*-[O$_2$NCo(en)$_2$Cl]$^+$. Here inert NO$_2^-$ is a π-acceptor ligand.

11. In what way the acid hydrolysis of *cis*-[Co(en)$_2$(OH)Cl]$^+$ complex differ from that of the *trans*-[O$_2$NCo(en)$_2$Cl]$^+$ complex? Explain.

12. Justify that the metal complexes involving inert π-acceptor ligands, which are CO and CN$^-$, also proceed through an associative S_N^2 mechanism rather than a dissociative S_N^1 path.

13. Present two arguments against the S_N^2 mechanism for the base hydrolysis of octahedral ammine complex of Co(III).

14. Highlight all points of evidence to establish that base hydrolysis of octahedral ammine complexes of Co(III) proceeds through S_N^1(CB) mechanism.

15. Present a detailed view of the stereochemistry of the intermediates formed during base hydrolysis of octahedral ammine complexes of Co(III).

16. Reaction of $[Co(en)_2(NO_2)Cl]^+$ with NO_2^- yields $[Co(en)_2(NO_2)_2]^+$. Explain the catalytic role of OH^- ions in the reaction of $[Co(en)_2(NO_2)Cl]^+$ with NO_2^- in non-aqueous medium on the basis of $S_N^1(CB)$ mechanism.

17. Present an explanatory note on substitution reactions without breaking metal–ligand bond.

18. Discuss the mechanism of substitution reaction in square planar complexes taking a suitable example.

19. What is *trans* effect? Describe it various utilities.

20. What is *trans* effect? Which theory of trans effect satisfactorily explains the following order of trans effect of inert ligands: $F^- < Cl^- < Br^- < I^-$

21. Discuss the π-bonding theory of *trans* theory. Out of the two theories of *trans* effect which one is better to explain trans effect of CO compared to that of pyridine.

22. Present a detailed view of *trans* effect in octahedral complexes.

23. Explain the terms 'structural trans-effect' (STE) and 'kinetic trans-effect' (KTE) with suitable examples.

24. In electron transfer reaction through inner sphere mechanism, the transfer or non-transfer of the bridging atom or group would depend on the relative stability of the products that are possible from the bridged intermediate. Justify this statement with suitable example.

25. Present the outer sphere mechanism of electron transfer reactions. Explain why the electron transfer reaction in the system, $[Co(NH_3)_6]^{2+}-[Co(NH_3)_6]^{3+}$, is slower than that in the system, $[Fe(CN)_6]^{4-} - [Fe(CN)_6]^{3-}$.

26. Present a detailed view of electron transfer reactions in aqueous medium occurring through inner sphere mechanism.

27. Explain why the transfer of electron from $[Cr(H_2O)_6]^{2+}$ to $[Co(NH_3)_6]^{3+}$ in aqueous solution is slower than the transfer of electron from $[Cr(H_2O)_6]^{2+}$ to $[Co(NH_3)_5(OH)]^{2+}$.

28. Explain why the transfer of electron from $[Cr(H_2O)_6]^{2+}$ to $[Cr^{III}(NH_3)_5(N_3)]^{2+}$ in aqueous solution is faster than the transfer of electron from $[Cr^{II}(H_2O)_6]^{2+}$ to $[Cr^{III}(NH_3)_5(NCS)]^{2+}$.

29. Discuss the complimentary and non-complimentary two electron transfer reactions citing suitable examples.

Chapter VIII
Bonding in transition metal complexes: molecular orbital theory approach

8.1 Introduction

In order to explain the nature of bonding between metal atom and ligands in coordination compounds/transition metal complexes, the following theories have been suggested:

(i)   Valence bond theory (VBT) (due to L. Pauling and J. L. Slater)
(ii) Crystal field theory (CFT) (due to H. Bethe and J. H. Van Vleck)
(iii) Ligand field theory (LFT) or molecular orbital theory (MOT) (due to J. H. Van Vleck)

According to Werner's theory, it seemed reasonable that ligands donate electron pairs to metal ions or atoms to form coordinate linkage. This approach was first applied to coordination compounds by Linus Pauling and Slater in 1931. Hence, this theory is called *Pauling's theory of complexes* or *valance bond theory* (*VBT*). According to VBT, the metal–ligand bonding in complexes is covalent only because it assumes that ligand electrons are partially donated to metal orbitals. It deals with the ground state electronic structure of the central metal atom and is primarily concerned with the kind of bonding, stereochemistry and gross magnetic properties present in complexes. The orbital in the complexes are designated only in terms of central atom orbitals and hybridization of these to produce bonding orbitals. It is somewhat successful in describing and predicting some of the physical and chemical properties of complexes known at that time. But at the end of 1950s, some facts which were not easily explained by this theory became known. At this stage, another theory called the CFT revived from the early work of J. H. Van Vleck and H. Bethe, both physicists, was proposed.

CFT is very much different from the VBT. According to VBT, the metal–ligand bonding is purely covalent, while CFT involves an electrostatic approach to the bonding in complexes. Advanced by H. Bethe and J. H. Van Vleck in 1929, it was first applied to the ionic-type crystalline substances to explain their colours and magnetic properties. Hence, it is very often called CFT. This theory considers the metal ion as being placed in an electrostatic field created by surrounding ligand molecules or ions. This electric field changes the energies of the d-electrons in the case of transition metal ions. Many of the properties of the complexes are related to these energy changes. This theory considers the bonding in complexes to be entirely electrostatic. The CFT was proposed a little earlier than the VBT, but it took about 20 years for this theory to be used by inorganic chemists. In 1951, many theoretical

https://doi.org/10.1515/9783110727289-008

chemists working independently used the CFT to explain the spectra of complexes of the d-block elements.

Although CFT adequately accounts for a surprisingly large amount of data, it has many limitations as pointed out below:

(i)   CFT considers only the metal ion d-orbitals and gives no consideration to other metal orbitals, which are s, p_x, p_y and p_z orbitals and the ligand π-orbitals. Therefore, to explain all the properties of the complexes dependent on π-ligand orbitals will be outside the scope of the CFT. In fact, CFT does not consider the formation of π-bonding in complexes, although it is very common and particularly great in complexes of metals in unusually low and high oxidation states, and in complexes of alkenes, alkynes, cyclopentadienides, aromatics and so on.

(ii) CFT is unable to account satisfactorily for the relative strength of ligands. For example: (i) it offers no explanation of why H_2O appears in the spectrochemical series as a stronger ligand than OH^- and (ii) it gives no account as to why CO, with no ionic charge and almost no dipole moment ($\mu = 0.374 \times 10^{-30}$ Cm), is placed at the right-hand side of the spectrochemical series. Thus, the interpretation of the spectrochemical series in terms of a point charge model of CFT is extremely tenuous/weak and unconvincing.

(iii) According to CFT, the bonds between the metal and ligands are purely ionic. It gives no account of the partly covalent nature of the metal–ligand bonds. Thus, the effect is directly dependent on covalency cannot be explained by CFT.

(iv) It offers no explanation for charge transfer bands normally appearing in ultraviolet region.

Because of the above-mentioned limitations of CFT and to explain these anomalies, a modified form of CFT came into picture in which some parameters are empirically adjusted to allow for covalence in complexes, without clearly introducing orbital overlap. This modified form of CFT (or LFT) is called *adjusted CFT* by Cotton and Wilkinson. As the adjusted CFT makes use of the concept of the MOT to allow for covalence in complexes, the modified form of CFT is also termed as *MOT*.

8.2 Evidence of covalent bonding in complexes

Before coming to MOT, it is proper to include the evidence for covalent bonding in complexes first.

8.2.1 Electron spin resonance spectra

Most direct evidence of covalent bonding is obtained from electron spin resonance (ESR) spectra of complexes. For example, ESR spectrum of $[Ir^{IV}Cl_6]^{2-}$ shows that it has a complex pattern of sub-bands called the hyperfine structure. This hyperfine structure has been satisfactorily explained by assuming that certain of the iridium orbitals and certain orbitals of the surrounding Cl^- ions overlap to such an extent that the single unpaired electron is not localized entirely on the metal ion but instead is about 5% localized on each Cl^-ion. The hyperfine structure is caused by the nuclear magnetic moments of the chloride ions, and the hyperfine splittings are proportional to the fractional extent to which the unpaired electron occupies the orbitals of these chloride ions. The electron is thus only 70% [$100-5\% \times 6$ $(Cl^-) = 70\%$] an 'Iridium(IV) 4d electron', instead of 100% that is assumed in the purely electrostatic CFT.

Another related example is that of $[Mo(CN)_8]^{3-}$ ion. The ESR spectrum of this ion when it is enriched in ^{13}C, which has a nuclear spin (^{12}C does not has nuclear spin), displays marked hyperfine structure showing that unpaired electron is significantly delocalized on the carbon atom of CN^- ions.

8.2.2 Nuclear magnetic resonance spectra

Sometimes in nuclear magnetic resonance (NMR) experiments of metal complexes, the nuclear resonances of atoms in ligands are found to be affected by unpaired electrons of the metal in a manner which can only be explained by assuming that electron spin is transferred from metal orbitals into orbitals of the ligand atoms. For example, the resonance frequency of the ring protons H_α in tris(acetylacetonato)vanadium(III) [$V(acac)_3$] (acacH = acetylacetone) (Figure 8.1) is considerably shifted from its position in a comparable diamagnetic compound, say Al^{III} analogue, tris (acetylacetonato)aluminium(III). In order to account for the magnitude of the shift, it is necessary to assume that the spin density of unpaired electrons, formally restricted to t_{2g} metal orbitals in crystal field treatment, actually moves out into the π-electron system of the ligand to a significant extent and eventually into the 1s orbitals of the hydrogen atoms.

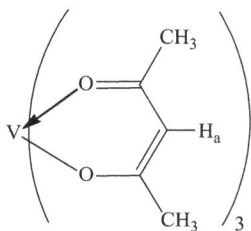

Figure 8.1: Tris(acetylacetonato)vanadium(III) indicating the ring hydrogen, H_α.

Similar but much more extensive studies of aminotropoponeiminate complexes of Ni(II) (Figure 8.2) have revealed the transfer of unpaired electron density from

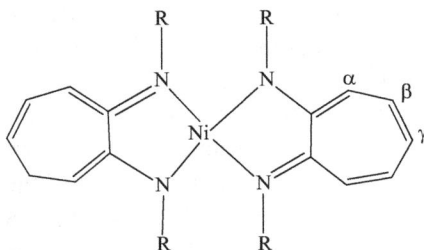

Figure 8.2: Bis(N,N'-dialkylaminotropoponeiminato)nickel(II), showing the three different proton positions, α, β and γ, for which large nuclear resonance shifts caused by Ni(II) d-electrons, have allowed the estimation of non-zero spin densities at these positions.

nickel atoms into the ligand π-system. Indeed, the various large shifts in the positions of the different (α,β,γ) proton NMRs has permitted determination of the spin density residing on each different carbon atom. For example, in the analogue in which R groups are each C_2H_5, a calculation which assumes the transfer of 0.10 electrons from Ni(II) to the ligands yields spin densities at the α, β and γ carbon atoms of +0.033, −0.023 and +0.057, respectively, whereas the experimental spin densities are +0.041, −0.021 and +0.057, respectively.

8.2.3 Nuclear quadrupole resonance studies

The nuclear quadrupole resonance spectra of square planar complexes of Pt(II) and Pd(II) such as $[Pt^{II}X_4]^{2-}$ and $[Pd^{II}X_4]^{2-}$ (X = halide ions) suggest that there is considerable amount of covalency in the metal–ligand bonds, that is, Pt–X or Pd–X bonds.

8.2.4 Intensities of d–d transitions

The unusually large absorption band intensities due to d–d transitions in tetrahedral complexes like $[CoCl_4]^{2-}$ where there is no centre of symmetry have been explained by saying that the metal–ligand bonds have appreciable covalent character.

8.2.5 Interelectronic repulsion: the nephelauxetic effect

Electrons in the partly filled d-orbitals of a metal ion repel one another and give rise to a number of energy levels depending upon the arrangement of these electrons in the d-orbitals. The energy of each of these levels can be represented in terms of some interelectronic repulsion parameters called Racah repulsion parameters B and C. The energy gap between two such energy levels which have the same multiplicity can be expressed in terms of Dq and B and the energy gap between two energy levels having different spin multiplicities is expressed in terms of Dq, B and C. It is experimentally observed that the magnitude of these interelectronic repulsion parameters always decreases on complexation of the metal ion. This is possible only if interelectronic repulsion between the d-electrons of the metal ion decreases on complexation.

The magnitude of interelectronic repulsion is inversely proportional to the distance r between the regions of maximum charge density of the d-orbitals which are occupied by the electrons. This repulsion would decrease only if the distance r increases or if lobes of d-orbitals containing electrons extend in space, as illustrated in Figure 8.3.

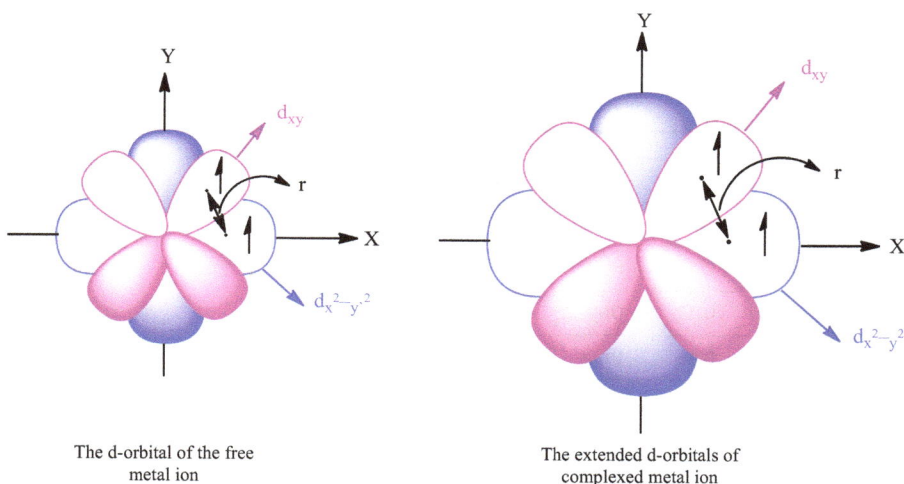

The d-orbital of the free
metal ion

The extended d-orbitals of
complexed metal ion

Figure 8.3: Extension of d-orbital in space in complexed metal ion.

The extension of the lobes of d-orbitals which means the expansion of d-electron charge cloud is known as *nephelauxetic effect* (from Greek, meaning 'cloud expanding').

The extension of d-orbitals of the complexed metal ion in space occurs obviously to maximize the overlap of these orbitals with the orbitals of the ligand. This is an essential condition for covalent bonding. In other words, it is the covalent bonding in metal–ligand bond which decreases the interelectronic repulsion parameters when a free metal ion gets complexed. The larger the decrease in the interelectronic repulsion parameters, the greater is the extent of covalent bonding in metal–ligand bond of the complex. The capability of d-orbitals of different metal ions to extend themselves in space forms the basis of *nephelauxetic series*. Since the interelectronic repulsion parameters almost always decrease on complexation of a metal ion, the metal–ligand bonds in all the complexes must contain some contribution from covalent bonding.

Nephelauxetic series

It is possible to place metal ion and ligands in a series (called nephelauxetic series) based on the order in which another factor, ($B_{\text{Free metal ion}} - B_{\text{Complexed metal ion}}$), goes on increasing. In this factor, B stands for Racah interelectronic repulsion parameter which measures interelectronic repulsion amongst d-electrons due to an increase in the size of the d-orbitals which invariably occurs when the metal ion gets complexes. The approximate value of B is calculated using the empirical relation,

$$B = B_0(1 - hk) \times 10^3 \text{cm}^{-1}$$

Here, B_0 = Racah repulsion parameter for free metal ion and h and k are arbitrary parameters which are assigned empirical values for various metal ions and ligands. These empirical values are given in Table 8.1.

Table 8.1: Empirical values of h and k assigned to various metal ions and ligands.

Metal ions	Empirical value of (h)	Ligand	Empirical value of (k)
Mn^{2+}	0.07	F^-	0.8
V^{2+}	0.1	H_2O	1.0
Ni^{2+}	0.12	DMF	1.2
Mo^{3+}	0.15	Urea	1.2
Cr^{3+}	0.20	NH_3	1.4
Fe^{3+}	0.24	en	1.5
Rh^{3+}	0.28	$C_2O_4^{2-}$	1.5
Ir^{3+}	0.28	Cl^-	2.0
Tc^{4+}	0.3	CN^-	2.1
Co^{3+}	0.33	Br^-	2.3
Mn^{4+}	0.5	N_3^-	2.4
Pt^{4+}	0.6	I^-	2.7
Pd^{4+}	0.7		
Ni^{4+}	0.8		

According to the above data, approximate value of B for $[Ni(H_2O)_6]^{2+}$ comes out to be:

$$B = 1,080(1 - 0.12 \times 1.0) \times 10^3 cm^{-1} = 1,080 \times 0.88 = 950.4 \times 10^3 cm^{-1}$$

The experimentally observed value of B is 905×10^3 cm^{-1}. B_0 for $Ni^{2+} = 1,080$ cm^{-1}.

Based on the values of the factor ($B_{Free\ metal\ ion} - B_{Complexed\ metal\ ion}$), the metal ions and ligands are arranged in the following series:

Metal ions:

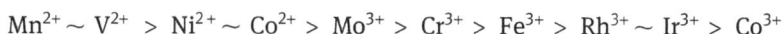

$$Mn^{2+} \sim V^{2+} > Ni^{2+} \sim Co^{2+} > Mo^{3+} > Cr^{3+} > Fe^{3+} > Rh^{3+} \sim Ir^{3+} > Co^{3+}$$

Ligands:

$$F^- > H_2O > Urea > NH_3 > en \sim C_2O_4{}^{2-} > SCN^- > Cl^- > CN^- > Br^- > I^-$$

8.3 Molecular orbital theory: transition metal complexes

8.3.1 Introduction

Although more complicated than VBT and CFT, MOT explains the nature of bonding in coordination compounds more satisfactorily. For appropriate understanding of MOT, some basic concept of group theory is desirable. Since group theory is an independent topic whose approach is outside the scope of the present discussion, we will discuss the MOT using minimum possible reference to the group theory. It would be desirable to explain the symmetry symbols which would be used in the subsequent discussion of MOT. These symmetry symbols are *a*, *e* and *t*. Here, *a* stands for non-degenerate orbital, *e* stands for doubly degenerate orbitals and *t* stands for triply degenerate orbitals. Degenerate means orbitals of equal energy.

The symmetry symbol for a particular orbital of the metal may be different in different environments. Thus, a set of p-orbitals of a metal has t_{1u} symmetry in octahedral environment whereas it has t_2 symmetry in tetrahedral environment. Details of some other symmetry symbols will be explained wherever required.

8.3.2 Qualitative aspect of MOT

We shall now consider the qualitative aspects of the application of MOT to complexes. This theory starts with the proposition that overlap of atomic orbitals (AOs) of the central metal ion and those of the ligands will occur, to some degree, whenever symmetry permits. It thus includes the electrostatic situation (no overlap) as one extreme, maximal overlapping of orbitals as the other extreme, and all intermediate degrees of overlap in its scope.

The first task in working out the molecular orbital treatment for a particular type of complex is to find out which AOs of the central metal and the ligands can combine. This can be done quite elegantly and systematically using some principles of group theory. As the group theory is very complex, we shall simply present the results obtained for complexes of different geometries. The molecular orbitals (MOs), we shall be using here will be of the linear combination of AOs type.

We shall apply the above-mentioned principles to explain the nature of bonding in coordination compounds of octahedral, tetrahedral and square planar complexes involving σ (sigma)- and π (pi)-bonding both.

8.4 Sigma (σ) bonding in octahedral complexes

Consider a metal ion M^{n+} surrounded octahedrally by six ligands, which are, L_1, L_2, L_3, L_4, L_5 and L_6 (Figure 8.4) and suppose that each ligand has a filled pσ orbital on it which is capable of forming σ-bond with the metal ion M^{n+}. Thus, we have six

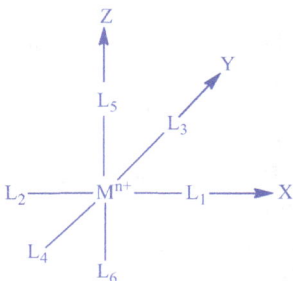

Figure 8.4: A metal ion M^{n+} surrounded octahedrally by six ligands, L_1, L_2, L_3, L_5 and L_6.

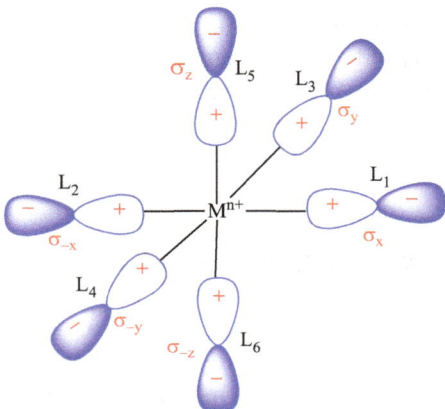

Figure 8.5: Filled pσ–orbitals lying along x-, y- and z-axes in an octahedral complex.

ligand pσ orbitals. These pσ orbitals are lying along x-, y- and z-axes. Hence, these six ligand pσ orbitals can be represented as σ_x, σ_{-x}, σ_y, σ_{-y}, σ_z and σ_{-z} (Figure 8.5).

The formation of six metal-ligand σ-bonds (or six MOs) in an octahedral complex, ML_6 takes place through the following steps:

Step I: The central metal cation of 3d series contains in all nine valence AOs: 4s, $4p_x$, $4p_y$, $4p_z$, $3d_{xy}$, $3d_{yz}$, $3d_{xz}$, $3d_{x^2-y^2}$ and $3d_{z^2}$. All these nine orbitals have been grouped into four symmetry classes as follows:

(i) $4s \rightarrow A_{1g}$ or a_{1g} (ii) $4p_x$, $4p_y$, $4p_z \rightarrow T_{1u}$ or t_{1u} (iii) $3d_{x^2-y^2}$, $3d_{z^2} \rightarrow E_g$ or e_g (iv) $3d_{xy}$, $3d_{yz}$, $3d_{xz} \rightarrow T_{2g}$ or t_{2g}

The symbols A_{1g}/a_{1g}, E_g/e_g, T_{1u}/t_{1u} and T_{2g}/t_{2g} are derived from group theory and their meanings are as follows: The letter A or a denotes an energy level which is singly degenerate, corresponding to single orbital having full symmetry of the system. E or e denotes an energy level which is doubly degenerate and corresponds to a pair of orbitals differing only in directional properties. T or t denotes a triply degenerate energy level corresponding to a set of three orbitals differing only in directional properties. Subscript 1 means the wave functions do not change sign on rotation about the Cartesian axis. Subscript 2 means they do not change sign on rotation about axis diagonal to the Cartesian axis. The Subscripts g and u indicate whether the orbitals are centrosymmetric (g from the German word gerade = even) or antisymmetric (u from ungerade = uneven), respectively. A set of three p-orbitals corresponds to a triply degenerate energy level T_{1u} or t_{1u}, a set of two $3d_{x^2-y^2}$ and $3d_{z^2}$ orbitals corresponds to a doubly degenerate energy level E_g or e_g and a set of three $3d_{xy}$, $3d_{yz}$ and $3d_{xz}$ orbitals corresponds to a set of triply degenerate energy level T_{2g} or t_{2g}.

We are aware that in an octahedral complex six σ-orbitals of the six ligands are approaching along the axes (Figure 8.5), in which ligand σ-orbitals along the +X, −X, +Y, −Y, +Z and −Z axes have been represented as σ_x, σ_{-x}, σ_y, σ_{-y}, σ_z and σ_{-z}, respectively. In order to form six metal−ligand σ-orbitals (or six MOs), these σ-orbitals, will overlap more effectively, with only those metal ion valence AOs that are having their lobes along the axes, that is, along the metal−ligand directions. Quite evidently, such AOs are 4s, $4p_x$, $4p_y$, $4p_z$, $3d_{z^2}$ and $3d_{x^2-y^2}$ since these AOs have their lobes lying along the axes (Figure 8.6).

The remaining three AOs namely, $3d_{xy}$, $3d_{yz}$ and $3d_{xz}$, do not participate in σ-bonding process because these have their lobes oriented in space between the axes (Figure 8.7). These orbitals remain non-bonding and hence are called non-bonding orbitals.

These orbitals can, however, overlap sidewise with filled or unfilled orbitals of the same six ligands to form metal−ligand π-bonds (or π-MOs). Thus, these orbitals are generally referred to as πd-orbitals.

Step II: In this step, the σ_x, σ_{-x}, σ_y, σ_{-y}, σ_z and σ_{-z} orbitals of the ligands combine together linearly to form such LGOs that should be capable of overlapping with the

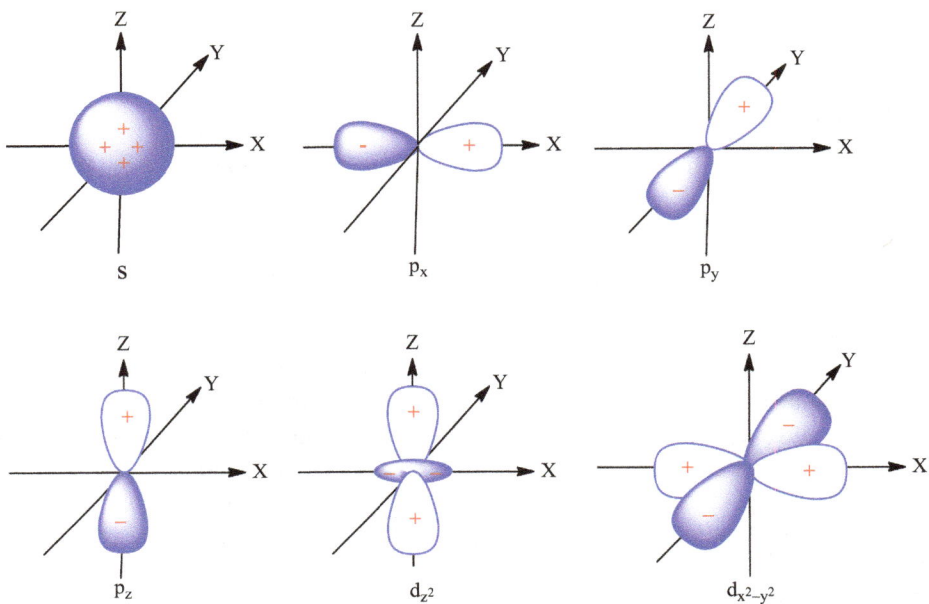

Figure 8.6: Metal atomic orbitals having their lobes along the axes.

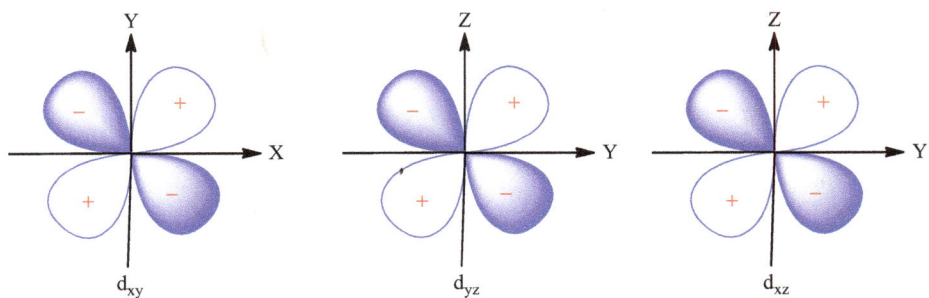

Figure 8.7: Metal atomic orbitals having their lobes between the axes.

central metal ion six AOs, namely $4s$, $4p_x$, $4p_y$, $4p_z$, $3d_{z^2}$ and $3d_{x^2-y^2}$. The determination of such linear combination of ligand σ-orbitals is made by inspection method which is straight forward and common sense method which is shown below:

(a) Since $4s$ orbital has the same sign in all the directions, the linear combination of ligand σ-orbitals which can overlap with $4s$ orbital is: $\sigma_x + \sigma_{-x} + \sigma_y + \sigma_{-y} + \sigma_z + \sigma_{-z}$.

This linear combination is represented by Σ_s, which in the normalized form is given by

$$\Sigma_s = 1/\sqrt{6}(\sigma_x + \sigma_{-x} + \sigma_y + \sigma_{-y} + \sigma_z + \sigma_{-z})$$

(Group symmetry symbol: A_{1g} or a_{1g})

(b) Since one lobe of $4p_x$ orbital has positive sign and the other has negative sign, the linear combination of ligand σ-orbitals that can overlap with $4p_x$ orbital is: $\sigma_x - \sigma_{-x}$. It is represented by Σ_x, which in the normalized form is given by

$$\Sigma_x = 1/\sqrt{2}(\sigma_x - \sigma_{-x}) \quad \text{(Group symmetry symbol: } T_{1u} \text{ or } t_{1u})$$

Similarly, $\sigma_y - \sigma_{-y}$ and $\sigma_z - \sigma_{-z}$ is the linear combination of ligand σ-orbitals that can overlap with $4p_y$ and $4p_z$ AOs, respectively. Thus,

$$\Sigma_y = 1/\sqrt{2}(\sigma_y - \sigma_{-y}) \quad \text{(Group symmetry symbol: } T_{1u} \text{ or } t_{1u})$$

$$\Sigma_z = 1/\sqrt{2}(\sigma_z - \sigma_{-z}) \quad \text{(Group symmetry symbol: } T_{1u} \text{ or } t_{1u})$$

(c) Since, one opposite pair of lobes of $3d_{x^2-y^2}$ orbital has +ve sign and the other has −ve sign, the linear combination of ligand σ-orbitals for this orbital is $\sigma_x + \sigma_{-x} - \sigma_y - \sigma_{-y}$. Thus,

$$\Sigma_{x^2-y^2} = 1/2(\sigma_x + \sigma_{-x} - \sigma_y - \sigma_{-y}) \quad \text{(Group symmetry symbol: } E_{1g} \text{ or } e_{1g})$$

(d) There is some difficulty in finding the ligand σ-orbital combination for $3d_{z^2}$ orbital. The analytical function for $3d_{z^2}$ orbital is proportional to $3z^2-r^2$. The proper σ-orbital combination is easily written down by substituting $x^2 + y^2 + z^2$ for r^2 in $3z^2-r^2$.
Thus,

$$3z^2 - r^2 = 3z^2 - (x^2 + y^2 + z^2) = 2z^2 - x^2 - y^2$$

Consequently, the proper combination for $3d_{z^2}$ orbital is:

$$2(\sigma_z + \sigma_{-z}) - (\sigma_x + \sigma_{-x}) - (\sigma_y - \sigma_{-y}) = (2\sigma_z + 2\sigma_{-z} - \sigma_x - \sigma_{-x} - \sigma_y - \sigma_{-y})$$

Thus, the linear combination of the ligand σ-orbitals for $3d_{z^2}$ AO in the normalized form is given by

$$\Sigma_z^2 = 1/2\sqrt{3}(2\sigma_z + 2\sigma_{-z} - \sigma_x - \sigma_{-x} - \sigma_y - \sigma_{-y})$$

(Group symmetry symbol: E_{1g} or e_{1g})

The group symmetry symbols given above (A_{1g}/a_{1g}, T_{1u}/t_{1u} and E_g/e_g) show that there is no combination of ligand σ-orbitals of T_{2g}/t_{2g} group symmetry symbol. Thus, $3d_{xy}$, $3d_{yz}$ and $3d_{xz}$ AOs of T_{2g} or t_{2g} symmetry will not combine with any of the above

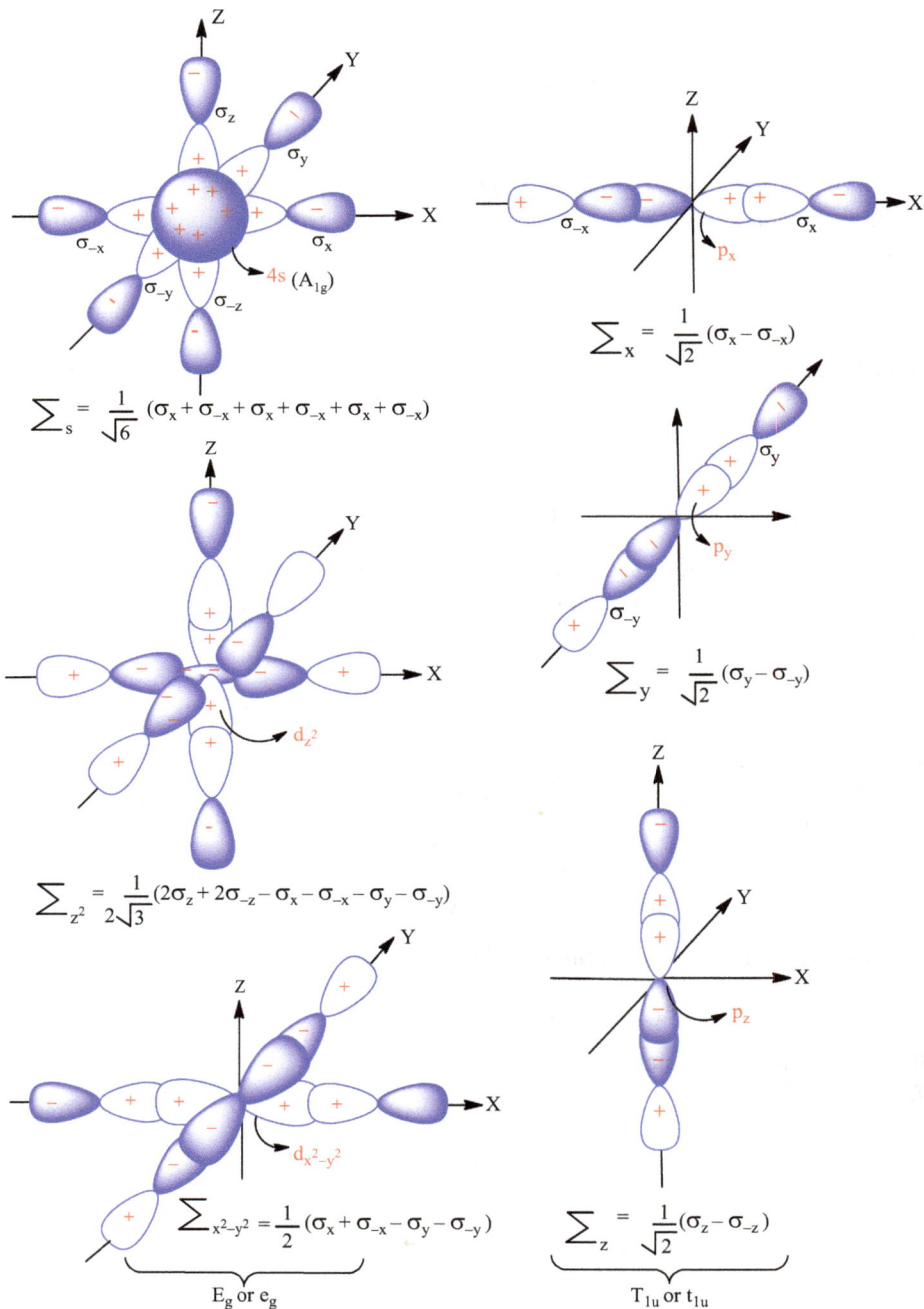

$$\Sigma_s = \frac{1}{\sqrt{6}} (\sigma_x + \sigma_{-x} + \sigma_x + \sigma_{-x} + \sigma_x + \sigma_{-x})$$

$$\Sigma_x = \frac{1}{\sqrt{2}} (\sigma_x - \sigma_{-x})$$

$$\Sigma_y = \frac{1}{\sqrt{2}} (\sigma_y - \sigma_{-y})$$

$$\Sigma_{z^2} = \frac{1}{2\sqrt{3}} (2\sigma_z + 2\sigma_{-z} - \sigma_x - \sigma_{-x} - \sigma_y - \sigma_{-y})$$

$$\Sigma_{x^2-y^2} = \frac{1}{2} (\sigma_x + \sigma_{-x} - \sigma_y - \sigma_{-y})$$

$$\underbrace{}_{E_g \text{ or } e_g}$$

$$\Sigma_z = \frac{1}{\sqrt{2}} (\sigma_z - \sigma_{-z})$$

$$\underbrace{}_{T_{1u} \text{ or } t_{1u}}$$

Figure 8.8: Pictorial representation of ligand group orbitals (LGOs) and matching atomic orbitals (AOs) of the same symmetry to form bonding σ-MOs in octahedral complex.

group ligand σ-orbitals. These AOs are, therefore, called non-bonding orbitals in σ-bonding complexes.

The LGOs and matching AOs of the same symmetry are given in Figure 8.8.

Step III: In this step, the six AOs of the central metal cation, which are $4s$, $4p_x$, $4p_y$, $4p_z$, $3d_{z^2}$ and $3d_{x^2-y^2}$, overlap with six LGOs, which are Σ_s, Σ_x, Σ_y, Σ_z, Σ_{z^2} and $\Sigma_{x^2-y^2}$, respectively, to form six σ-bonding and six σ-antibonding MOs. It is notable here that the metal AOs and LGOs which are overlapping together to form MOs are of the same symmetry. Thus,

   (i) $4s$ and Σ_s which have the same symmetry (a_{1g}) overlap to form one $\sigma_s{}^b$ MO and one $\sigma_s{}^*$ MO.

  (ii) $4p_x$ and Σ_x (both with t_{1u} symmetry) overlap to form one $\sigma_x{}^b$ MO and one $\sigma_x{}^*$ MO.

 (iii) $4p_y$ and Σ_y (both with t_{1u} symmetry) overlap to form one $\sigma_y{}^b$ MO and one $\sigma_y{}^*$ MO.

 (iv) $4p_z$ and Σ_z (both with t_{1u} symmetry) overlap to form one $\sigma_z{}^b$ MO and one $\sigma_z{}^*$ MO.

  (v) $3d_{z^2}$ and Σ_{z^2} (both with e_g symmetry) overlap to form one $\sigma_{z^2 b}$ MO and one σ_{z^2*} MO.

 (vi) $3d_{x^2-y^2}$ and $\Sigma_{x^2-y^2}$ (both with e_g symmetry) overlap to form one $\sigma_{x^2-y^2 b}$ MO and one $\sigma_{x^2-y^2*}$ MO.

Thus, we have a total of 12 MOs, six σ^b MOs and six σ^* MOs. It is thus obvious that on adding 12 MOs to three non-bonding AO's, which are $3d_{xy}$, $3d_{yz}$ and $3d_{xz}$, we get in all 15 orbitals potentially available for electron filling.

It is remarkable to note that σ^b MOs are generated when the metal AOs and the ligand group σ-orbitals combine together with maximum positive overlap and σ^* MOs are formed when the orbitals overlap together with maximum negative overlap.

The overlap of metal AOs: $4s$, $4p_x$, $4p_y$, $4p_z$, $3d_{z^2}$ and $3d_{x^2-y^2}$ with group ligand σ-orbitals (LGOs): Σ_s, Σ_x, Σ_y, Σ_z, Σ_{z^2} and $\Sigma_{x^2-y^2}$, respectively, which are of the same symmetry to produce σ^b MOs is shown pictorially in Figure 8.8. Taking negative overlap of AOs with respective LGOs into consideration, formation of σ^* MOs can also be shown pictorially.

8.4.1 Energy order of orbitals and their filling with electrons

The order of energy of different orbitals formed in an octahedral complex depend on the nature of ligands, that is, whether the ligands are strong or weak. Thus, two cases arise.

8.4.1.1 When the ligands are strong

Strong ligands such as NH_3 molecules split the σ-bonding MOs, namely, $\sigma_s{}^b$, $\sigma_x{}^b = \sigma_y{}^b = \sigma_z{}^b$, $\sigma_{x2-y2b} = \sigma_{z2b}$ more widely and the energy difference between Δ_o between the t_{2g} set of non-bonding AOs ($3d_{xy}$, $3d_{yz}$ and $3d_{xz}$) and $e_g{}^*$ set of MOs (σ_{x2-y2^*} and σ_{z2^*}) is greater than the electron pairing energy, P (i.e. $\Delta_o > P$). In this case, the order of energy of different orbitals is shown as follows:

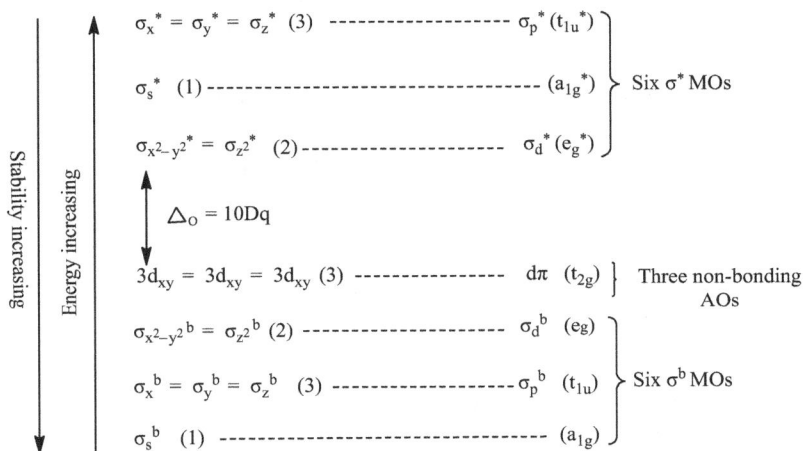

The Arabic numerals given in the bracket indicate the degeneracy of the MOs or AOs as the case may be. For example, the MOs, which are $\sigma_x{}^b$, $\sigma_y{}^b$ and $\sigma_z{}^b$, are triply degenerate (3). Quite evidently, the σ^b MOs have the lowest energy, σ^* MOs have the highest energy while the non-bonding orbitals, $3d_{xy}$, $3d_{yz}$ and $3d_{xz}$, have an energy level intermediate between the bonding and antibonding MOs.

Let us see now how the distribution of electrons in $[Co(NH_3)_6]^{3+}$ (containing strong NH_3 ligands) takes place in various orbitals according to the order of energy. This complex has 18 electrons in it (12e from six NH_3 ligands, plus 6e from 3d-orbitals in Co^{3+} ion). The distribution of 18e in various orbitals in shown in Figure 8.9, which is commonly called MOT energy level diagram.

In connection with the MOT energy level diagram shown in Figure 8.9, the following points are noteworthy:

(i) It is certain that the overlap of 4s and 4p orbitals with ligands is considerably better than that of the 3d orbitals (in general, d-orbitals tend to be large and diffuse and as a result overlap of d-orbitals may be quantitatively poor even when qualitatively favourable). As a result, a_{1g} and t_{1u} MOs are the lowest in energy and the corresponding $a_{1g}{}^*$ and $t_{1u}{}^*$ antibonding orbitals are the highest in energy. The e_g and $e_g{}^*$ orbitals arising from the 3d orbitals are displaced less from their barycentre because of the poorer overlap. The t_{2g} orbitals are non-bonding (in a σ-only system) and not displaced.

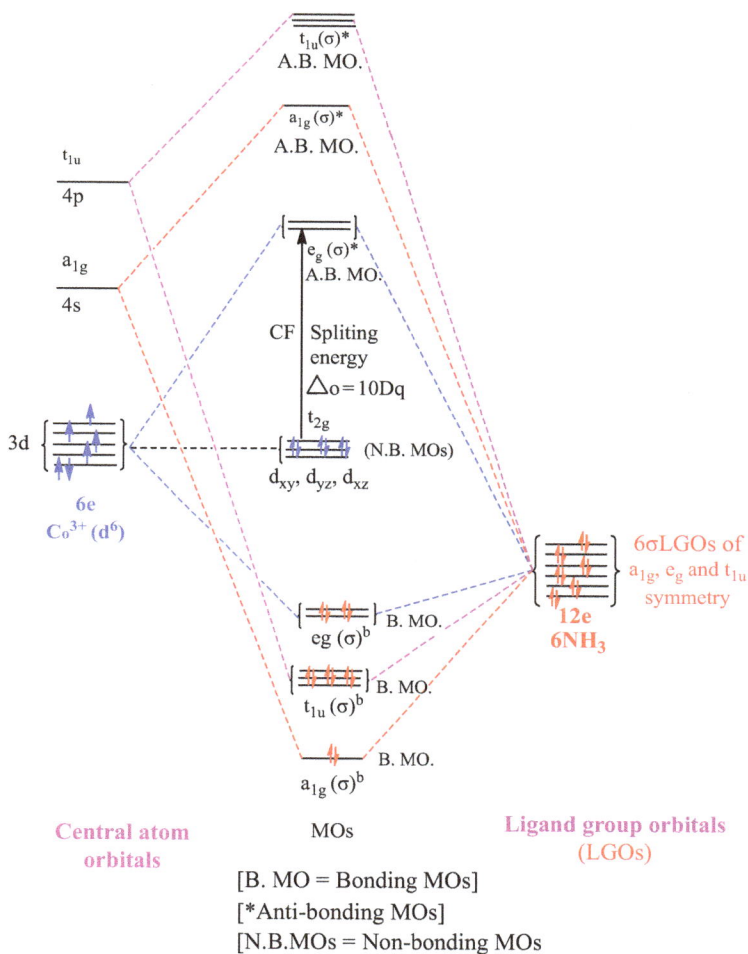

Figure 8.9: MOT energy level diagram of $[Co(NH_3)_6]^{3+}$ (drawn not to scale).

(ii) In general, it may be assumed that if an MO is much nearer to one of the AOs or LGOs used in its construction than to the other one, it will have much more characteristics of the nearer one. Thus, the six bonding MOs [$a_{1g} (\sigma)^b$ (1), $t_{1u} (\sigma)^b$ (3) and $e_g (\sigma)^b$ (2) and or σ_s^b, σ_x^b, σ_y^b, σ_z^b, $\sigma_{x^2-y^2b}$, σ_{z^2b}] are considered to have the character of LGOs than metal AOs. We, therefore, consider electrons in these MOs to be mainly ligand electrons. Hence, 12e from $6NH_3$ ligands ($6 \times \uparrow\downarrow = 12$) are filled in these orbitals as shown by **orange** arrows ($\uparrow\downarrow$) in the MOT diagram.

Likewise, any electrons occupying any of the antibonding MOs are considered to be predominantly 'metal electrons'. Similarly, any electrons in the non-σ-bonding t_{2g} orbitals ($3d_{xy}$, $3d_{yz}$ and $3d_{xz}$) will be purely metal electrons, provided there exist no ligand π-orbitals to overlap with these t_{2g} orbitals. This is why electrons occupying t_{2g} orbitals are shown by **blue** arrows ($\uparrow\downarrow$). In fact, **6e** of Co^{3+} ion occupying 3d-orbitals are already shown by **blue** arrows ($\uparrow\downarrow$) as parity.

(iii) It is evident from the MO diagram that the t_{2g} and $e_g{}^*$ levels, both containing only metal atom orbitals, are split apart (qualitatively) in the same manner as they were by the purely electrostatic argument from the CFT. All that has changed in this limited portion of the energy level diagram is that in MOT $e_g{}^*$ orbitals are not pure metal atom d-orbitals. Furthermore, in the CFT, the splitting arises from only *electrostatic* and symmetry considerations whereas in the MO the splitting arises from *covalent bonding* and symmetry considerations.

(iv) It is evident from the MO diagram that the $[Co(NH_3)_6]^{3+}$ is a diamagnetic complex.

8.4.1.2 When the ligands are weak

In case of weak ligands, such as, F^- ion, the energy different, Δ_o between the t_{2g} set and $e_g{}^*$ set is smaller than the electron pairing energy, P (that is $\Delta_o < P$) and hence the lowest energy anti-bonding Mos, which are $\sigma_{x^2-y^2*}$ and σ_{z^2*}, have approximately the same energy as the non-bonding AOs: $3d_{xy}$, $3d_{yz}$ and $3d_{xz}$. Consequently, the order of energy of different orbitals becomes as shown below:

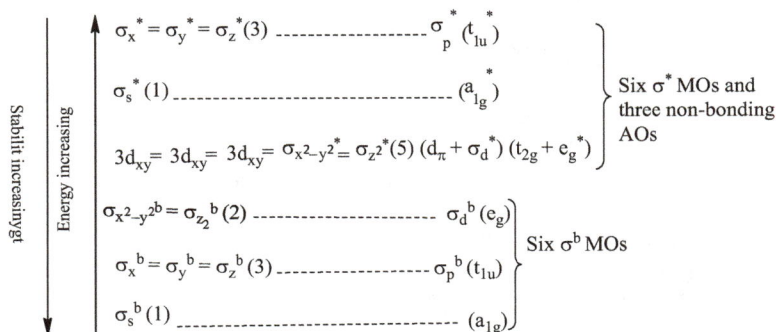

$$\sigma_x{}^* = \sigma_y{}^* = \sigma_z{}^*(3) \text{-----------------} \sigma_p{}^* \, (t_{1u}{}^*)$$

$$\sigma_s{}^*(1) \text{----------------------------} (a_{1g}{}^*)$$

$$3d_{xy} = 3d_{xy} = 3d_{xy} = \sigma_{x^2-y^2}{}^* = \sigma_{z^2}{}^*(5) \, (d_\pi + \sigma_d{}^*) \, (t_{2g} + e_g{}^*)$$

Six σ^* MOs and three non-bonding AOs

$$\sigma_{x^2-y^2}{}^b = \sigma_{z^2}{}^b (2) \text{-------------------} \sigma_d{}^b \, (e_g)$$

$$\sigma_x{}^b = \sigma_y{}^b = \sigma_z{}^b (3) \text{------------------} \sigma_p{}^b \, (t_{1u})$$

$$\sigma_s{}^b (1) \text{--------------------------} (a_{1g})$$

Six σ^b MOs

(left margin: Stabilit increasingt; Energy increasing)

The distribution of 18 electrons in $[CoF_6]^{3-}$ (6 + 12 = 18), which contains weak ligand field, F^-, takes place in various orbitals according to the above energy order scheme as shown in MOT energy level diagram given in Figure 8.10. This energy level diagram clearly shows that in case of weak field complexes (high spin complexes), Hund's rule is obeyed and the compound is paramagnetic with respect to four unpaired electrons.

Electronic configurations of octahedral complexes of d^1 to d^8 ions having strong (low-spin) and weak ligand (high-spin) ligands can be worked out using MOT energy level diagram shown in Figures 8.9 and 8.10, respectively, and thus the magnetic behaviour.

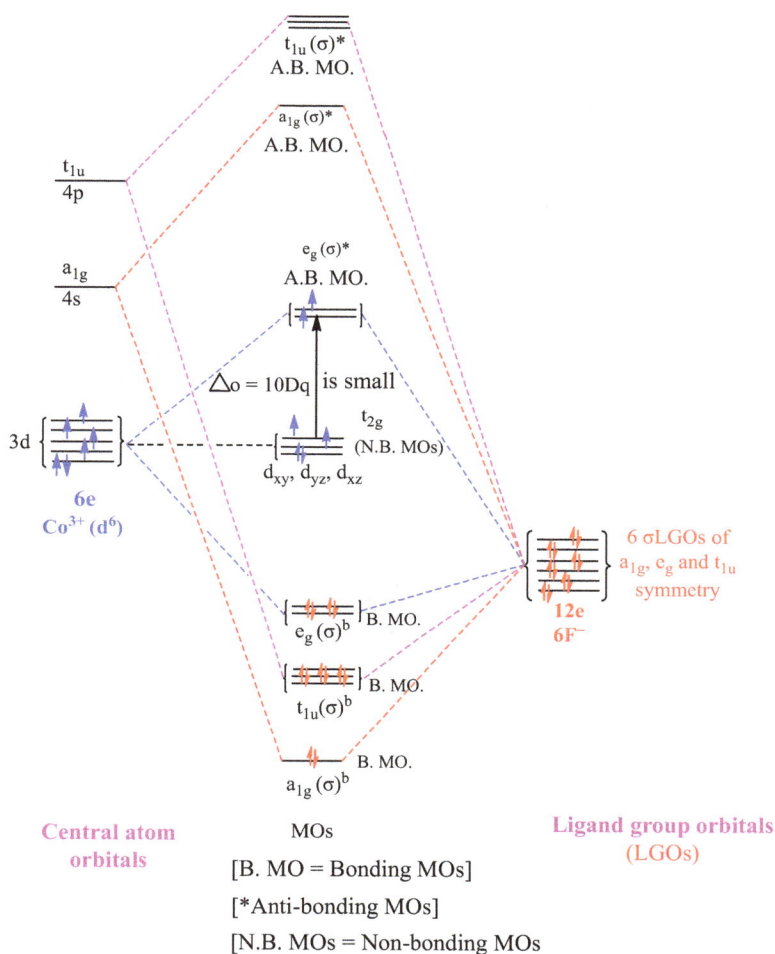

Figure 8.10: MOT energy level diagram of $[CoF_6]^{3-}$ (drawn not to scale).

8.5 Pi (π)-bonding in octahedral complexes

From the MO diagram of ML_6 (M from 3d transition series) octahedral complexes involving only σ-bonding, it is seen that d_{xy}, d_{yz} and d_{xz} (t_{2g}) orbitals remain non-bonding. If the ligand atoms have π-orbitals, they can combine to form MOs of π-symmetry.

Most of the ligand atoms possess π-orbitals which may be filled or empty (unfilled) and which, therefore, may interact with metal t_{2g} d-orbitals, that is, the set d_{xy}, d_{yz} and d_{xz}. Considering the octahedral case, we may consider each ligand to possess a pair of mutually perpendicular π-orbitals, giving rise to a 12 p-π orbitals. The directions of the AOs of the ligands L_1, L_2 . . ... L_6, which can form π-bonds with AOs of the same symmetry on the metal atom M, are represented in Figure 8.11.

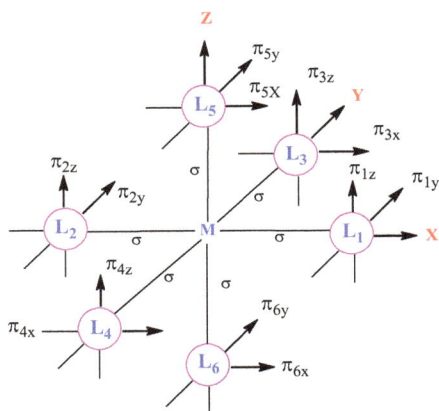

Figure 8.11: Ligand π-orbitals in octahedral complexes, ML_6.

It is found from the group theoretical calculations that these 12 p-π orbitals may be combined into four triply degenerate set belonging to four symmetry classes t_{1g}, t_{2g}, t_{1u} and t_{2u}. The t_{1g}/T_{1g} and t_{2u}/T_{2u} classes orbitals must remain non-bonding for the simple reason that metal atoms do not possess orbitals corresponding to these symmetries with which interaction might occur. This is well clear from the character table (Table 8.2.) of O_h point group when we see the columns 3 and 4 corresponding

Table 8.2: Character table of O_h point group.

O_h	E	$8C_3$	$3C_2$	$6C_4$	$6C_2$	i	$6S_4$	$8S_6$	$3\sigma_h$	$6\sigma_d$		
A_{1g}	1	1	1	1	1	1	1	1	1	1		$x^2+y^2+z^2$
A_{2g}	1	1	1	−1	−1	1	−1	1	1	−1		
E_g	2	−1	2	0	0	2	0	−1	2	0		$(2z^2-x^2-y^2, x^2-y^2)$
T_{1g}	3	0	−1	1	−1	3	1	0	−1	−1	(R_x, R_y, R_z)	
T_{2g}	3	0	−1	−1	1	3	−1	0	−1	1	(xz, yz, xy)	
A_{1u}	1	1	1	1	1	−1	−1	−1	−1	−1		
A_{2u}	1	1	1	−1	−1	−1	1	−1	−1	1		
E_u	2	−1	2	0	0	−2	0	1	1	0		
T_{1u}	3	0	−1	1	−1	−3	−1	0	0	1	(x, y, z)	
T_{2u}	3	0	−1	−1	1	−3	1	0	0	−1		

The overall results are tabulated below:

T_{1g}	T_{2g}	T_{1u}	T_{2u}
None	d_{xz}, d_{yz}, d_{xy}	p_x, p_y, p_z	None

to T_{1g} and T_{2u} symmetry, which are blank. However, the t_{1u}/T_{1u} class ligand orbitals have the metal orbitals of same symmetry, that is, the metal p-AOs: $4p_x$, $4p_y$ and $4p_z$ metal (see column 3 of the character table corresponding to t_{1u}/T_{1u} symmetry). But this interaction cannot be very important since these same metal atom orbitals are already engaged in σ-bonding with the ligand atoms. The remaining t_{2g} LGOs may, however, combine with t_{2g}/T_{2g} AOs ($3d_{xy}$, $3d_{yz}$, $3d_{xz}$) (see column 4 of the character table corresponding to T_{2g} symmetry), and if this happens π-bonding results.

8.5.1 Types of π-bonds

We may classify the types of π-bonds according to the nature of the ligand π-orbitals as follows:

(i) Simple pπ-orbitals are always filled, as in O_2^-, RO^-, F^-, Cl^-, Br^-, I^-, RS^- and so on. In the present case, the ligand L is both σ- and π-donor, and the metal acts as receptor with suitable empty orbital on itself. Metals early in each transition series and those in higher oxidation states have few d-electrons and can thus serve as receptors. Ligands stated above, while engaging in σ-bonding to the metal, have electrons in essentially non-bonding pπ-orbitals of the donor atoms, and act as good π-donors forming π-bonds (Figure 8.12).

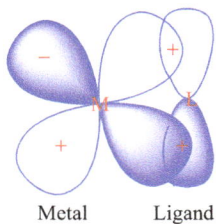

Metal   Ligand

Figure 8.12: Formation of dπ–pπ bonds between a metal d-orbital and ligand p-orbitals.

(ii) Simple dπ-orbitals, always empty, as in phosphines, arsines, sulphides and so on. While engaging in σ-bonding to the metal, the ligands with P, As and S donor atoms have empty low-lying d-orbitals of π-symmetry leading to dπ–dπ bonding (Figure 8.13)

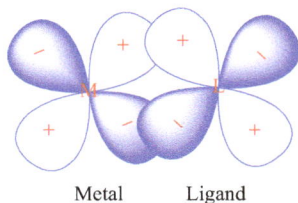

Metal   Ligand

Figure 8.13: Formation of dπ–dπ bonds between a metal d-orbital and ligand d-orbitals.

(iii) Empty MOs of certain polyatomic ligands such as NO_2^-, CN^-, CO, NO^+, py, *o-phen*, *dipy*, *acac*$^-$ and unsaturated organic molecules. Ligands such as CN^-, CO and NO^+ possess empty π^*-anti-bonding orbitals favouring back donation of electrons from the filled t_{2g}-orbitals of the metal ion. Metals late in each transition series and those in zero or low oxidation states acts as π-donors.

Some of the heterocyclic aromatics, dipyridine (*dipy*) and o-phenanthroline (*o-phen*), are also good π acids in that they contain empty low-lying π^*-orbitals for M–L π-bonding. Formation of π-bonds between a filled metal d-orbital and ligand empty π^*-antibonding orbital is pictorially shown in Figure 8.14.

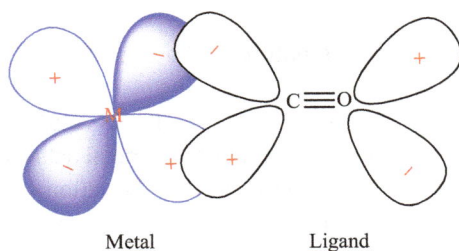

Metal   Ligand

Figure 8.14: Formation of π-bonds between a filled metal d-orbital and ligand π^*-antibonding orbital.

8.5.2 Formation of LGOs for π-bond formation

The formation of LGOs of t_{2g} symmetry taking part in π-bond formation with metal t_{2g} orbitals can be understood by taking example of ligands having simple $p\pi$-orbitals belonging to category (i) above.

Simple $p\pi$-orbitals taking part in π-bond formation are not directed along the line joining the ligand and the metal ion (that is, along the axis) but are perpendicular to one another and also to the metal–ligand bond. Thus, there are two such $p\pi$-orbitals on each of the six ligands in an octahedral complex as shown in Figure 8.11.

Table 8.3 lists the $p\pi$-orbitals on each of the ligands $L_1, L_2 \ldots \ldots L_6$.

Table 8.3: The $p\pi$-orbitals on each of the ligands $L_1, L_2 \ldots \ldots L_6$.

Ligands	pπ-orbitals on the ligands
L_1	π_{1y}, π_{1z}
L_2	π_{2y}, π_{2z}
L_3	π_{3x}, π_{3z}
L_4	π_{4x}, π_{4z}
L_5	π_{5x}, π_{5y}
L_6	π_{6x}, π_{6y}

The letter x in π_{3x} to π_{6x} indicates that these pπ-orbitals are along the x-axis and the letter y in π_{1y} to π_{6y} indicates that these pπ-orbitals are along the y-axis. Similarly, the letter z in π_{1z} to π_{4z} denotes that these pπ-orbitals are along the z-axis. The numbers 1, 2, 3, 4, 5 and 6 used as coefficient of x, y and z in different pπ-ligand orbitals are of ligand, namely, L_1, L_2, L_3, L_4, L_5 and L_6, respectively. The arrow heads of these pπ-orbitals points in the direction of the positive lobe of the orbitals.

Before combining/overlapping with AOs of the central metal cation, these ligand pπ-orbitals combine together to form LGOs of t_{1g}, t_{2g}, t_{1u} and t_{2u} symmetries, but for the reasoning mentioned above, we will give here the combination of ligand pπ-orbitals in the formation of LGOs of t_{2g} symmetry only.

Taking help of Figure 8.11, it is quite easy to see that ligand pπ-orbitals (not in the normalized from) which can overlap with d_{xz} metal AO is:

$$(\pi_{1z} - \pi_{2z} + \pi_{5x} - \pi_{6x})$$

Table 8.4 gives the appropriate combination of pπ-orbitals of the ligands (LGOs) in the normalized form that can overlap with d_{xy}, d_{yz} and d_{xz} AOs of t_{2g} symmetry.

Table 8.4: The appropriate combination of pπ-orbitals of the ligands (LGOs) suitable for overlapping with d_{xy}, d_{yz} and d_{xz} AOs.

Group theory symmetry	Central ion orbitals	Appropriate LGOs
t_{2g}	$3d_{xy}$	$\pi_{xy} = \frac{1}{2}(\pi_{1y} - \pi_{2y} + \pi_{3x} - \pi_{4x})$
	$3d_{yz}$	$\pi_{yz} = \frac{1}{2}(\pi_{3z} - \pi_{4z} + \pi_{5y} - \pi_{6y})$
	$3d_{xz}$	$\pi_{xz} = \frac{1}{2}(\pi_{1z} - \pi_{2z} + \pi_{5x} - \pi_{6x})$

The overlap of $3d_{xy}$, $3d_{yz}$ and $3d_{xz}$ AOs of the central metal cation with corresponding LGO is shown in Figure 8.15.

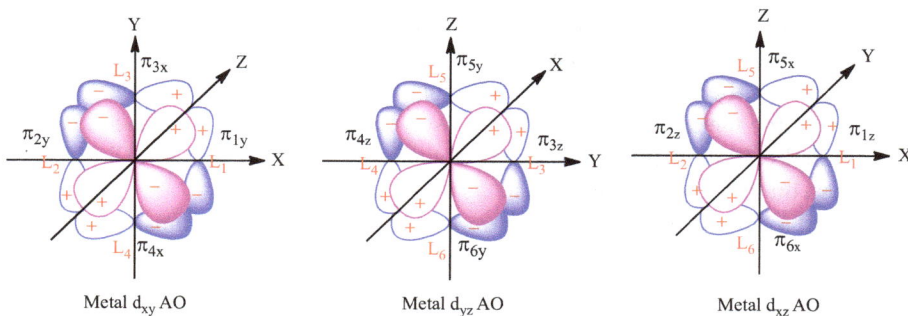

Metal d_{xy} AO   Metal d_{yz} AO   Metal d_{xz} AO

Figure 8.15: The overlap of metal d_{xy}, d_{yz} and d_{xz} AOs with LGOs π_{xy}, π_{yz} and π_{xz}, respectively.

The three LGOs π_{xy}, π_{yz} and π_{xz} combine with the central atom orbitals d_{xy}, d_{yz}, d_{xz} forming three bonding and three anti-bonding MOs as given in Table 8.5.

Table 8.5: Formation ofbonding and antibonding orbitals.

Symmetry	Bonding MOs	Anti-bonding MOs
t_{2g}	$\psi_{xy} + k\pi_{xy}$	$\psi_{xy} - k\pi_{xy}$
	$\psi_{yz} + k\pi_{yz}$	$\psi_{yz} - k\pi_{yz}$
	$\psi_{xz} + k\pi_{xz}$	$\psi_{xz} - k\pi_{xz}$

Here, the notation k used in the present table represents the ratio of the coefficients of the central atom and the LGO. Since, ψ_{xy}, ψ_{xz} and ψ_{yz} are degenerate and π_{yz}, π_{xz} and π_{yz} are also degenerate, the coefficients k is same resulting in three degenerate t_{2g} MOs.

8.5.3 Effect of π-bonding on the magnitude of Δ_o: construction of MO energy level diagram/types of π-bonding complexes

The involvement of π-bonding in complexes affect the energy level of metal t_{2g} orbitals just as the formation of σ-bonding affects the energy of the e_g orbitals of the central metal atom/ion.

The effect on the energy level of metal t_{2g} orbitals due to π-bonding depends on the following two factors:
(i) Whether the ligand π-orbitals of t_{2g} symmetry are of higher or lower energy than the metal t_{2g} orbitals
(ii) Whether the ligand π-orbitals are filled or empty

Based on the above two factors, following types of π-bonding complexes are known:

8.5.3.1 Complexes having filled (i.e. donor) ligand π-orbitals of lower in energy than metal t_{2g} orbitals

Let us consider the case of a ligand involving filled pπ-orbitals with energy lower than that of the metal t_{2g} orbital. This is observed in complexes of 3d-metal ions in their normal oxidation states with ligands such as O^{2-} (H_2O), RO^-, F^-, Cl^-, Br^-, I^-, RS^- and so on, in which the ligands act as π-bases.

Since the ligand ions are more electronegative that the metal atoms, the np-orbitals (2p in case of F^-) lie at a lower energy than the corresponding 3d-metal orbitals of 3d-series. A qualitative MO energy level diagram applicable for such complexes is shown in Figure 8.16. This energy level diagram supplements to the information already given in σ-bonding diagram.

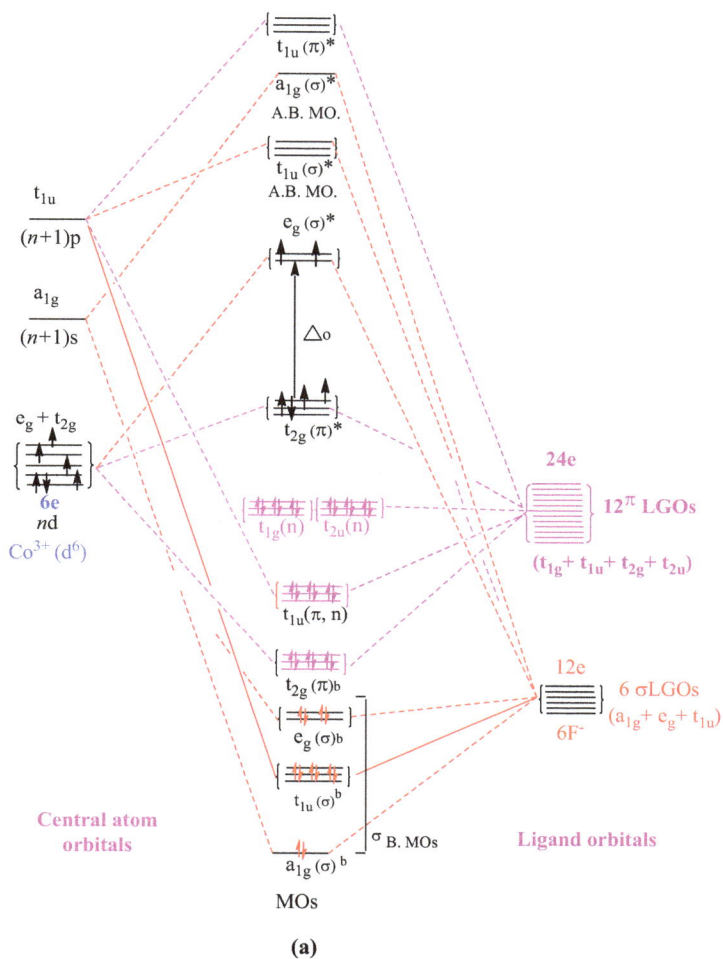

Figure 8.16: (a) MO diagram of $[CoF_6]^{3-}$ involving σ- and π-bonding. (b) Simplified MO energy level diagram showing the effect of π-bonding on Δ_o in low energy filled ligand orbitals.

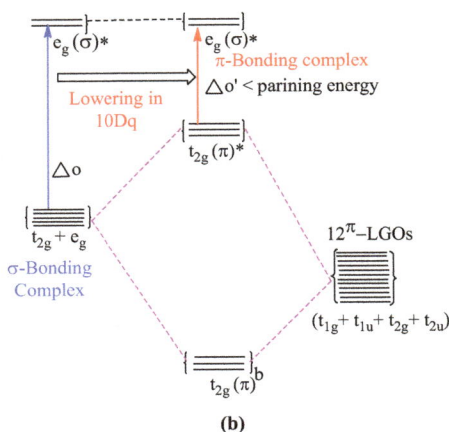

(b)

Figure 8.16 (continued)

Some important features of this diagram are highlighted below:

(i) The order of the energy of the various orbitals is: σ-LGO < π-LGO < nd < $(n+1)s$ < $(n+1)p$.

(ii) The metal t_{2g}-set of orbitals which remained non-bonding in pure σ-bonding case is now involved in π-bonding. Consequently, it interacts with the t_{2g} LGOs to give a triply degenerate pair of MOs: $t_{2g}(\pi)^b$ and $t_{2g}(\pi)^*$.

(iii) As the energy-wise π-LGO < nd, the bonding $t_{2g}(\pi)^b$ MO is basically of ligand in character and the anti-bonding $t_{2g}(\pi)^*$ is of mostly metal in character. In this process, $t_{2g}(\pi)^*$ is slightly raised in energy [destabilized relative $E_g(\sigma)^*$] compared to the position of T_{2g} non-bonding MO in pure σ-bonding case. Since, the level of $e_g(\sigma)^*$ orbitals in σ-bonding case is unaffected by the π-interaction, the ligand field parameter Δ_o (10Dq) is reduced to Δ_o' as a result of π-bonding.

(iv) It is worthwhile to recall here that 10Dq is the separation between t_{2g} orbitals occupied by the metal electrons [which is now $t_{2g}(\pi)^*$] and the metal orbitals, which is $e_g(\sigma)^*$ with respect to σ-bonding. Thus, the 10Dq in case of filled ligand π-orbital is reduced. This explains why the position of F^- (and other halide ions), OH^-, H_2O, etc., lie at a weak field stream in the spectrochemical series.

(v) The t_{1u}-πLGO will not remain totally non-bound as this set has the right symmetry to that of the metal t_{1u} set which had primarily participated in σ-bonding. However, we cannot rule out a slight π-interaction between t_{1u}-πLGO of the ligand and metal t_{1u}-orbitals and that is reflected in slight lowering of t_{1u}-LGO forming t_{1u}-(π,n) MO showing π-character in addition to its non-bonding character because t_{1u} set has already been involved in σ-bonding. The remaining π-LGO, t_{1g} and t_{2u} are shown as pure non-bonding as they have no suitable metal orbitals for π-interaction.

Taking $[CoF_6]^{3-}$ as an example of such a complex, the molecular electronic configuration as shown in the MO diagram can be written as:

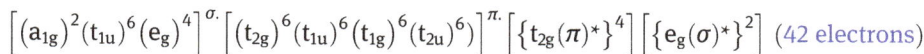

$$\left[(a_{1g})^2(t_{1u})^6(e_g)^4\right]^{\sigma.}\left[(t_{2g})^6(t_{1u})^6(t_{1g})^6(t_{2u})^6)\right]^{\pi.}\left[\{t_{2g}(\pi)^\star\}^4\right]\left[\{e_g(\sigma)^\star\}^2\right] \text{ (42 electrons)}$$

Evidently, such ligands favour the formation of high-spin complexes. The π-bonding is usually represented as L→M and the ligands are termed as π-donors.

8.5.3.2 Complexes having vacant/empty ligand dπ-orbitals of higher energy than metal t_{2g} orbitals

An important case of π-bonding under this category occurs with ligands such as PR_3, PX_3, AsR_3, R_2S and so on, where X is a halide and R is a phenyl group. In these molecules, the ligating atom can form σ-bond with the metal through an approximately sp^3 hybrid orbital in a manner similar to NH_3.

Both phosphorus and sulphur, however, have empty 3d-orbitals ([15]P; $1s^2\ 2s^22p^6\ 3s^23p^3\ 3d^0$; [16]S: $1s^2\ 2s^22p^6\ 3s^23p^4\ 3d^0$) which can receive electron density from the metal. A ligand of this type is referred to as π-acceptor (π-acid) ligand because of the presence of empty dπ-orbitals in it, and the π-bonding established in such a case is called M→L π-bonding or back bonding. These orbitals have fairly low electron density compared to a positively charged metal ion. So, the **dπ-LGOs** will lie at a higher energy than the corresponding metal orbitals.

The net result of π-interaction is that the metal t_{2g} orbital is stabilized relative to the e_g (σ)*. Consequently, the separation (10Dq) between t_{2g} orbitals occupied by the metal electrons [which is now t_{2g} $(\pi)^b$] and the metal orbitals, which is e_g (σ)* with respect to σ-bonding will be increased.

Based on the above consideration, the MO energy level diagram of $[(Cl_3P)_3Mo^0(CO)_3]$ involving empty ligand π-orbitals of higher energy is shown in Figure 8.17(a).

Some important features of MO diagram

(i) The order of the energy of the various orbitals is: dπ-LGO > np > ns > (n−1)d > σ-LGO.

(ii) The metal t_{2g}-set of orbitals which remained non-bonding in pure σ-bonding case is now involved in π-bonding. Consequently, it interacts with the T_{2g} dπ-LGOs to give a triply degenerate pair of MOs: $t_{2g}(\pi)^b$ and $t_{2g}(\pi)^\star$.

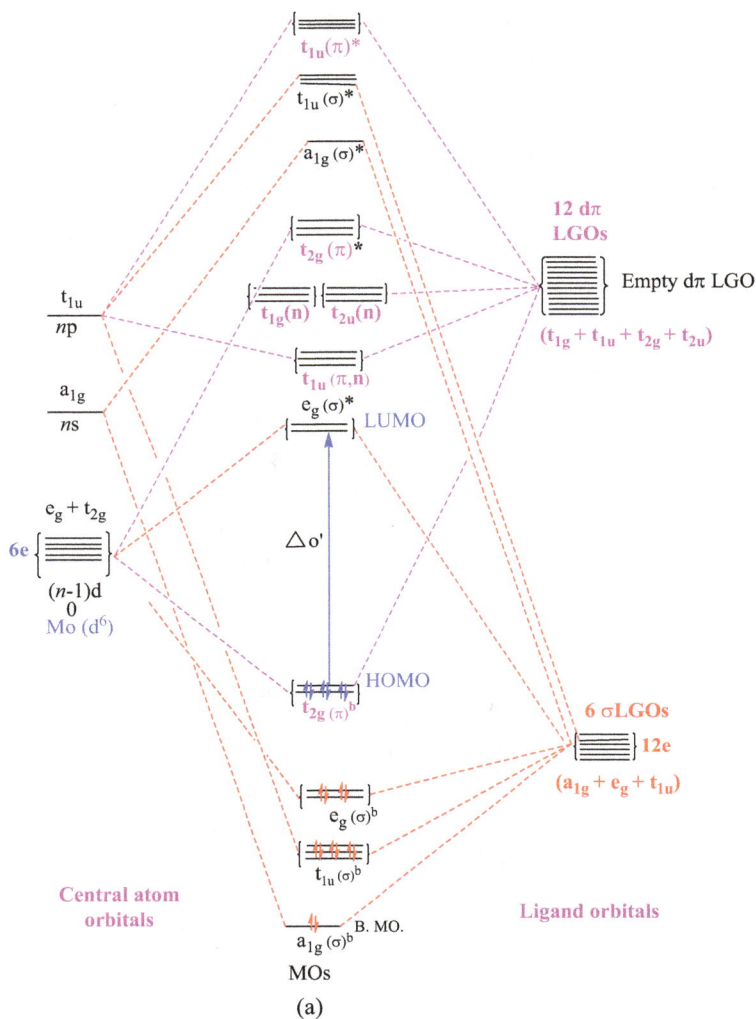

Figure 8.17: (a) MO diagram of O_h $[(Cl_3P)_3Mo^0(CO)_3]$ involving σ- and π-bonding. (b) Simplified MO energy level diagram of $[(Cl_3P)_3Mo^0(CO)_3]$ showing the effect of π-bonding on Δ_o in empty ligand π-orbitals of higher energy.

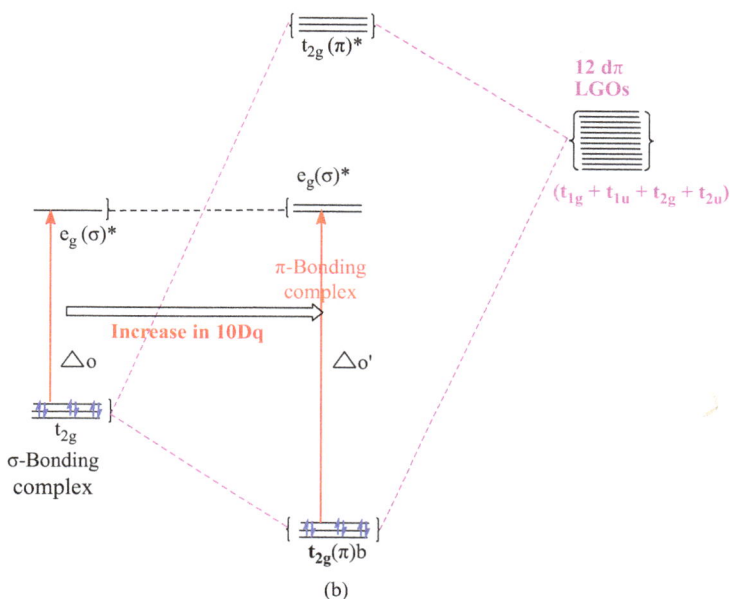

Figure 8.17 (continued)

(iii) As the energy-wise $d\pi$-LGO $>$ (n–1)d, the bonding $t_{2g}(\pi)^b$ MO is basically of metal in character. In this process, $t_{2g}(\pi)^b$ is significantly lowered in energy compared to the position of t_{2g} non-bonding MO in pure σ-bonding case. Since, the level of $e_g(\sigma)^*$ orbitals in σ-bonding case is unaffected by the π-interaction, the ligand field parameter, Δ_o (10Dq), is increased.

(iv) The t_{1u}-LGO will not remain totally non-bonding as this set has the right symmetry to that of the metal t_{1u} set which had primarily participated in σ-bonding. However, we cannot rule out a slight π-interaction between t_{1u}-LGO and metal t_{1u}-orbitals and that is reflected in slight lowering of t_{1u}-LGO forming t_{1u}-(π,n) MO showing π-character in addition to its non-bonding character because t_{1u} set has already been involved in σ-bonding.

(v) The remaining π-LGO, t_{1g} and t_{2u} are shown as pure non-bonding as they have no suitable metal orbitals for π-interaction.

Taking $[(Cl_3P)_3Mo(CO)_3]$ [molybdenum is zero O.S. in this compound] as an example of such a complex, the molecular electronic configuration as shown in the MO diagram can be written as:

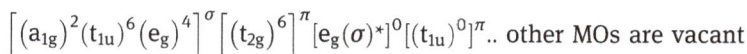

$$\left[(a_{1g})^2(t_{1u})^6(e_g)^4\right]^{\sigma}\left[(t_{2g})^6\right]^{\pi}[e_g(\sigma)^*]^0[(t_{1u})^0]^{\pi}.. \text{ other MOs are vacant}$$

Evidently, such ligands favour the formation of low-spin (strong field) complexes. The π-bonding is usually represented as M→L and the ligands are termed as π-acceptors. *It is to be mentioned here that the effect of π-bonding of CO is not considered here.*

A very simplified MO diagram showing the effect of π-interaction is shown in Figure 8.17(b). This diagram very clearly shows how the π-bonding interaction in the complex increases the ligand field parameter from Δ_o to Δ_o'.

8.5.3.3 Complexes having filled and empty ligand π-orbitals of higher energy than metal t_{2g} orbitals

Another important case of π-bonding arises in octahedral complexes with ligands such as halides (barring F⁻), which have filled pπ-orbitals and vacant dπ-orbitals along with σ-orbitals, and also with ligands, which are CN⁻, CO, pyridine (py), *o*-phenanthroline (*o-phen*), dipyridyl (*dipy*), acetylacetonate (*acac⁻*) and so on having filled π-MOs and vacant π*-MOs along with suitable σ-orbitals.

Obviously, there is competition between the two effects (i.e. effect of filled and empty π-orbitals on bonding), and hence the prediction of the effect of π-bonding on the magnitude of 10Dq is difficult.

There are ample reasons to believe that in the CO or CN⁻ ligand, the high energy anti-bonding (π*) orbitals are predominantly operative. As for the case of halides, effect of low energy filled pπ-orbitals dominates in some of their complexes, while in others the effect of high energy vacant dπ-orbitals.

In the halide complexes of the metals of 3d series, the low energy filled pπ orbitals are predominantly operative, while those of the 4d- and 5d-series metal ions, and particularly Pt(II), Pd(II), Hg(II) and Au(III), it is the vacant *d*-orbitals that plays the dominant role.

A detailed construction of MO energy level diagram of an O_h complex involving CO/CN⁻ as a ligand along with its salient features of π-bonding effect is being discussed here as one of the representative examples under the present category.

The ligand contains filled π-MOs and vacant π*-MOs along with suitable σ-orbitals having energy order:

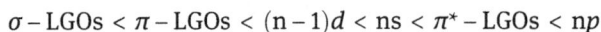

$$\sigma - \text{LGOs} < \pi - \text{LGOs} < (n-1)d < \text{ns} < \pi^* - \text{LGOs} < np$$

The vacant π*-MOs is more localized on carbon (say in CO) and has suitable symmetry to overlap with filled T_{2g}-orbitals of the central metal (Figure 8.18). The filled π-MOs are localized on electronegative oxygen and are not available for overlap with T_{2g}-orbitals of the metal. The acidic π*-MOs on the ligand are higher in energy than the metal T_{2g}-orbitals.

Metal d_{xz}-orbital   Ligand molecule π^*–MO
(e.g. CO or CN$^-$)

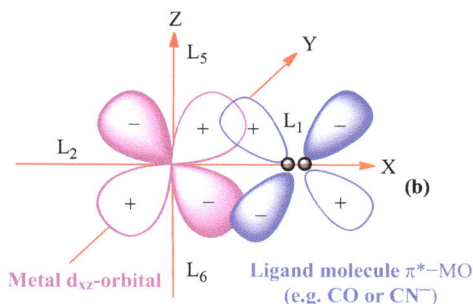

Figure 8.18: Overlap of π^*-MOs of CO having suitable symmetry with filled T_{2g}-orbitals of the central metal.

Based on the above consideration, the MO energy level diagram for an O_h [ML$_6$] complex, where L is ($\sigma + \pi + \pi^*$) type, is given in Figure 8.19.

The vacant π^*-LGOs of T_{2g} symmetry overlap with the metal T_{2g} orbitals to give a pair of triply degenerate π^*-MOS, the T_{2g} $(\pi^*)^b$ and T_{2g} $(\pi^*)^*$.

As the energy-wise π^*-LGO > (n–1)d, the bonding $T_{2g}(\pi^*)^b$ MO is basically of metal in character. In this process, $T_{2g}(\pi^*)^b$ is significantly lowered in energy compared to the position of T_{2g} non-bonding MO in pure σ-bonding case.

Since, the level of $e_g(\sigma)^*$ orbitals in σ-bonding case is unaffected by the π^*-interaction, the ligand field parameter, Δ_o (10Dq) is significantly increased. It is worthwhile to recall here that 10Dq is the separation between T_{2g} orbitals occupied by the metal electrons [which is now $T_{2g}(\pi^*)^b$] and the metal orbitals, which is $e_g(\sigma)^*$ with respect to σ-bonding.

The t_{1u}-π^*LGO will not remain totally non-bonding as this set has the right symmetry to that of the metal t_{1u} set, which had primarily participated in σ-bonding. However, we cannot rule out a slight π^*-interaction between t_{1u}-π^*LGO and metal t_{1u}-orbitals and that is reflected in slight lowering of t_{1u}-*LGO forming t_{1u} (π^*, n) MO showing π-character in addition to its non-bonding character because metal t_{1u} set has already been involved in σ-bonding. The remaining π^*-LGO, t_{1g} and t_{2u} are shown as pure non-bonding as they have no suitable metal orbitals for π^*-interaction.

Taking [Cr(CO)$_6$] [Cr^0] as an example of such a complex, the molecular electronic configuration as shown in the MO diagram can be written as:

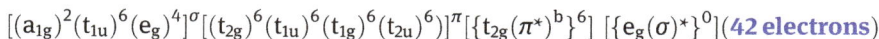

$$[(a_{1g})^2(t_{1u})^6(e_g)^4]^\sigma[(t_{2g})^6(t_{1u})^6(t_{1g})^6(t_{2u})^6)]^\pi[\{t_{2g}(\pi^*)^b\}^6]\ [\{e_g(\sigma)^*\}^0](\textbf{42 electrons})$$

Evidently, such ligands favour the formation of low-spin complexes. The π-bonding in such complexes is usually represented as M→L and the ligands are termed as π-acid ligands. As the transfer of electron density takes place from metal to CO in its $T_{2g}(\pi^*)^b$ MOs in metal carbonyls, the π-bonding in it is usually termed as π-back bonding.

The ligand CO may be predicted to bond increasingly strongly with electron releasing metal atoms. The bond order of CO decreases progressively in metal carbonyls as

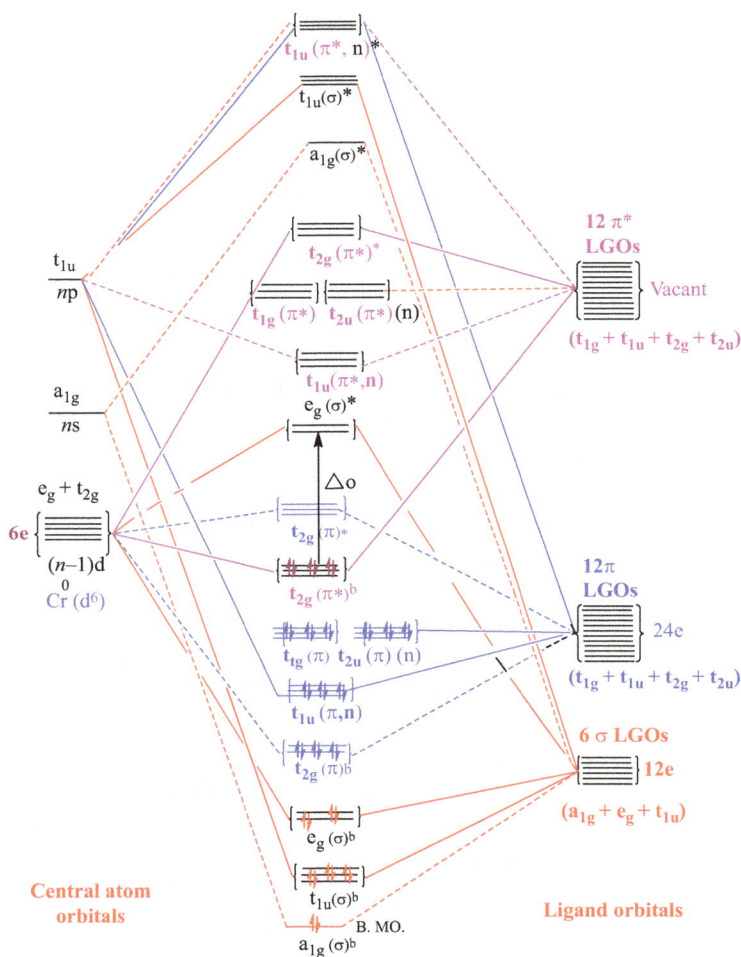

Figure 8.19: MO diagram of an octahedral [Cr(CO)$_6$] involving high energy vacant π*MO.

the $T_{2g}(\pi\star)^b$ MOs are increasingly populated by M→L donation. The decrease of CO bond order of coordinated CO in metal carbonyls is clearly reflected from the $\nu(CO)$ data in the infrared spectra of metal carbonyls.

Free CO in the gas phase has a stretching frequency of 2,143 cm^{-1}. The $\nu(CO)$ in tetrahedral metal carbonyls, say, [Ni(CO)$_4$], [Co(CO)$_4$]$^-$ and [Fe(CO)$_4$]$^{2-}$ is found, respectively, at 2,057, 1,886 and 1,786 cm^{-1}.

This indicates CO coordination with increasingly strong electron releasing atoms, in going from 0 to −1 to −2 oxidation states. The bond order of CO decreases progressively as the $T_{2g}(\pi\star)^b$ MOs become increasingly populated.

Similarly, when we move from [Mn(CO)$_6$]$^+$ to isoelectronic [Cr(CO)$_6$] and [V(CO)$_6$]$^-$, the $\nu(CO)$ decreases from 2,090 to 2,000 and 1,860 cm^{-1}, respectively. Thus, infrared spectroscopy provides a method for determining the tendency of a metal to participate in d-π^*donation.

The simplified MO energy level diagram showing the effect of π^*-interaction is given in Figure 8.20. This diagram very clearly shows how the empty high energy π^*-bonding interaction in the complex increases the ligand field parameter from Δ_o to Δ_o'. This explains the position of CO and CN$^-$ at the high field stream in the spectrochemical series. Here, Δ_o' is shown as Highest occupied molecular orbital (HOMO)-Lowest unoccupied molecular orbital (LUMO) separation which can be determined experimentally.

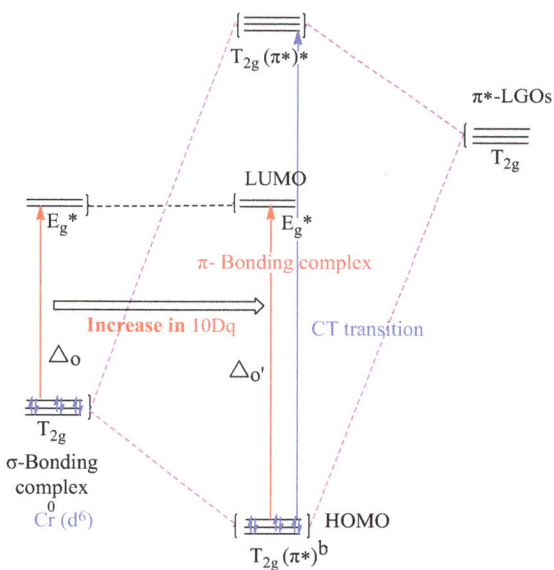

Figure 8.20: Simplified MOT energy level diagram showing the effect of π^*-bonding in high energy empty ligand orbitals.

Thus, besides six σ-bonds, there will be three effective π-bonds between Cr and six CO groups. The transfer of electrons from $T_{2g}(\pi^*)^b \rightarrow E_g^*$ corresponds to d–d transition that occurs in the visible region. But the $T_{2g}(\pi^*)^b \rightarrow T_{2g}(\pi^*)^*$ transition corresponds to transfer of electron from the MOs essentially belonging to metal to the MOs belonging to the ligands. Thus, this is an M $\rightarrow$ L charge transfer transition.

8.6 Tetrahedral complexes

A regular tetrahedron can be inscribed in a cube by taking the alternate vertices. If each of the four vertices of a tetrahedron is occupied by an atom/group surrounding

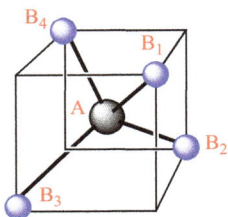

Figure 8.21: A tetrahedral compound/complex.

an atom at the centre, it gives a tetrahedral molecule (Figure 8.21). CH_4, CCl_4, $NiCl_4^{2-}$, $Ni(CO)_4$, $CoCl_4^{2-}$ and $Zn(CN)_4^{2-}$ belong to this group. Tetrahedral AB_4 complexes can involve both σ- and π-bonds between the metal ion and the ligands.

8.6.1 Sigma (σ) bonding in tetrahedral complexes

As a typical example, let us consider ML_4 tetrahedral complex as being inscribed in a cube (Figure 8.22) where L is only a σ-bonding type ligand. The σ-bonds are formed by the interaction of suitable AOs of the ligand L with suitable AOs of the metal M. The four σ-bonds can be represented as four vectors pointing towards the central metal atom M.

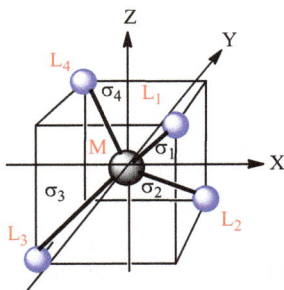

Figure 8.22: The molecular geometry of a tetrahedral ML_4 complex inscribed in a cube showing the coordinate axes and the σ-vectors representing the metal–ligand bonds.

For the formation of MOs in tetrahedral complexes, the σ-orbitals of the ligand atoms combine to form a certain numbers of LGOs. These composite orbitals or LGOs will combine with the central atom orbitals of the same symmetry to form bonding and anti-bonding orbitals.

In order to find out the number and nature of symmetries of the resulting composite orbitals, the concept of group theory is applied using the character table of T_d point group. Accordingly, the symmetries of the LGOs are:

$$\Gamma_\sigma = A_1 + T_2$$

Thus, there are four LGOs formed, one of non-degenerate A_1 symmetry and the other three of triply degenerate T_2 symmetry. These composite orbitals/LGOs will combine with the central atom orbitals of the same symmetry.

The central atom orbitals corresponding to A_1 and T_2 symmetries can be found out from the character table (Table 8.6) of the T_d point group given below:

Table 8.6: Character table of T_d point group.

T_d	E	$8C_3$	$3C_2$	$6S_2$	$6\sigma_d$		
A_1	1	1	1	1	1		$x^2 + y^2 + z^2$
A_2	1	1	1	-1	-1		
E	2	-1	2	0	0		$(2z^2 - x^2 - y^2, x^2 - y^2)$
T_1	3	0	-1	1	-1	(R_x, R_y, R_z)	
T_2	3	0	-1	-1	1	(x, y, z)	(xz, yz, xy)

Accordingly,

$$A_1 \longrightarrow s \text{ and}$$

$$T_2 \longrightarrow p_x, p_y, p_{z;} d_{xy}, d_{yz}, d_{xz}$$

We, therefore, see that to form a complete set of tetrahedrally directed σ bonds, central atom A will have to provide an s-orbital, as well as a set of p-orbitals or a set of d-orbitals. In the elements of the second period, the tetrahedral hybridization (as in the case of CH_4 or NH_4^+) will obviously be sp^3 as d-orbitals are available only in the upper energy level.

In CrO_4^{2-} and MnO_4^-, the type of hybridization is most probably d^3s, in which 3d- and 4s-orbitals are used. Of course, we do not discard completely the possibility of some mixing amongst 3d- and 4p-orbitals. Taking example of CrO_4^{2-}, MnO^{4-} and MnO_4^{2-}, pictorial representation of d^3s hybridization in this anion is shown in Figure 8.23.

Case I: Let us assume that the identity of the central atom M is such that a set of p-orbitals (p_x, p_y, p_z) of the ligands is appropriate. The complexes, $[AlCl_4]^-$, $[ZnCl_4]^{2-}$, $[NiCl_4]^{2-}$, $[MnCl_4]^{2-}$, $[CoCl_4]^{2-}$, $[Ni(CO)_4]$, $[Ni(NH_3)_4]^{2+}$, $[Zn(NH_3)_4]^{2+}$, $[Cu(py)_4]^{2+}$ may be considered under sp^3 category.

The compositions of the LGOs can be worked out by looking over the overlap of the ligand σ orbitals with the central atom orbitals. An alternate approach is symmetry adopted linear combinations of σ_1, σ_2, σ_3 and σ_4 making composite orbitals or LGOs of required symmetry. But it is a time-consuming process, so we will make use of the former method.

Let us consider first the composition of LGO of A_1 symmetry suitable to overlap with s-orbital. It can be seen that all the four σ orbitals have equivalent overlap with s-orbital (Figure 8.24). Hence, the composition of this LGO in normalized form will be:

$$\sum_s (A_1) = 1/2(\sigma_1 + \sigma_2 + \sigma_3 + \sigma_4)$$

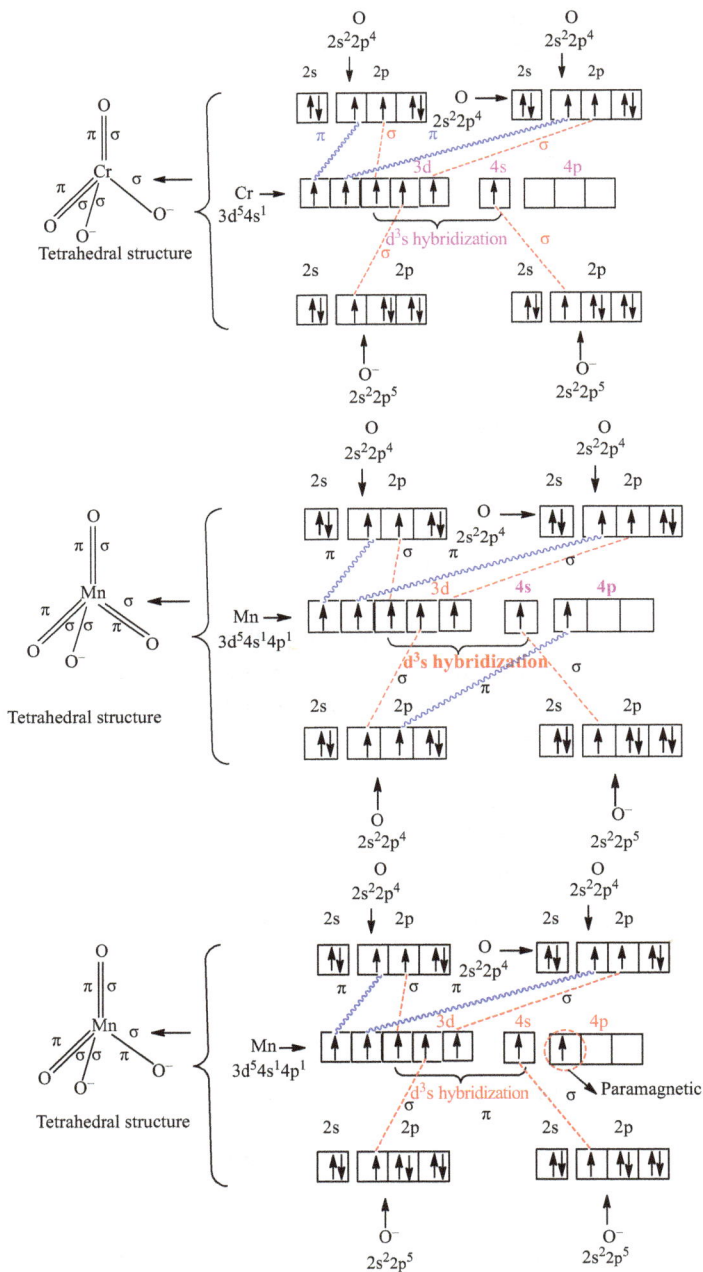

Figure 8.23: d^{3}s Hybridization scheme in CrO_4^{2-}, MnO_4^- and MnO_4^{2-}.

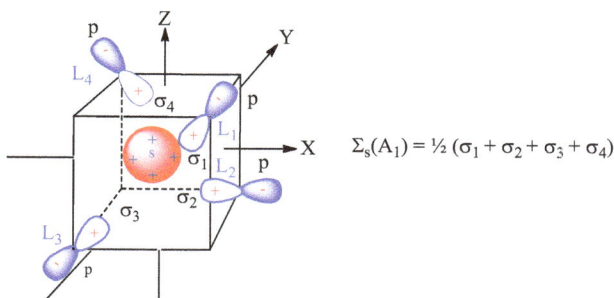

$\Sigma_s(A_1) = \frac{1}{2}(\sigma_1 + \sigma_2 + \sigma_3 + \sigma_4)$

Figure 8.24: Overlap of the metal s-orbital with ligand pσ-orbitals.

Similarly, compositions of LGOs of T_2 symmetry to overlap with metal p_x, p_y and p_z orbitals (Figure 8.25) can be worked out, which are shown below:

$$\sum_x (T_2) = 1/2\,(\sigma_1 + \sigma_2 - \sigma_3 - \sigma_4)$$

$$\sum_y (T_2) = 1/2\,(-\sigma_1 + \sigma_2 - \sigma_3 + \sigma_4)$$

$$\sum_x (T_2) = 1/2\,(\sigma_1 + \sigma_2 - \sigma_3 - \sigma_4)$$

In Figures 8.24 and 8.25, the metal ion is shown at the centre of a cube and the alternate apices of which form the tetrahedron. Cartesian coordinate axes (X, Y, Z) are shown as passing through the mid points of pairs of opposite phases of the cube.

The resulting composite orbitals/LGOs, $\Sigma_s(A_1)$, $\Sigma_x(T_2)$, $\Sigma_y(T_2)$ and $\Sigma_z(T_2)$ combine with the corresponding metal atom orbitals to form bonding and anti-bonding orbitals (Table 8.7) as follows:

Based on the above discussion, the MO energy level diagram of AB_4/$[Zn(NH_3)_4]^{2+}$ tetrahedral complexes can be shown (Figure 8.26) as follows:

In case of the tetrahedral $[Zn(NH_3)_4]^{2+}$ complex, eight electrons from the $4NH_3$ ligands are accommodated in the four σ-bonding MOs, and hence the bond between Zn^{2+} and each NH_3 molecule is a single σ-bond. All the d-orbitals remain as non-bonding MOs.

Case II: Let us assume that the identity of the central atom M is such that a set of d-orbitals $(3d_{xy}, 3d_{yz}, 3d_{xz})$ of the ligands is appropriate. The complexes, $[CrO_4]^{2-}$ $[MnO_4]^-$, $[RuO_4]$ and $[OsO_4]$ may be considered under d^3s category in which 3d-$(3d_{xy}, 3d_{yz}, 3d_{xz})$ and 4s-orbitals are used.

Again, the compositions of the LGOs can be worked out by looking over the overlap of the ligand σ orbitals with the central atom orbitals.

$\Sigma_x (T_2) = \frac{1}{2} (\sigma_1 + \sigma_2 - \sigma_3 - \sigma_4)$

$\Sigma_y (T_2) = \frac{1}{2} (-\sigma_1 + \sigma_2 - \sigma_3 + \sigma_4)$

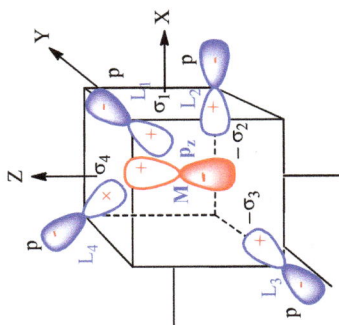

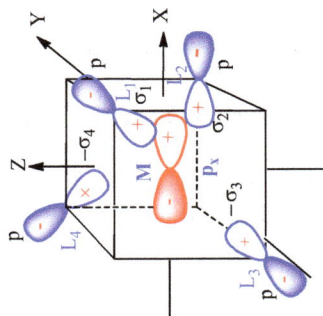

$\Sigma_z (T_2) = \frac{1}{2} (\sigma_1 - \sigma_2 - \sigma_3 + \sigma_4)$

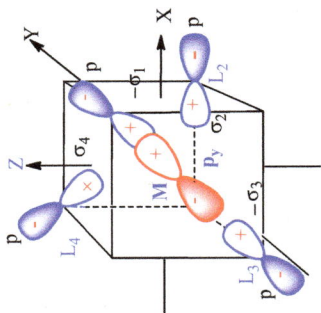

Figure 8.25: Overlap of the metal $p_x/p_y/p_z$-orbital with ligand pσ-orbitals.

Table 8.7: Formation of bonding and anti-bonding orbitals.

Symmetry	Bonding MOs	Anti-bonding MOs
A_1	$\psi_s + \lambda\Sigma_s$	$\psi_s - \lambda\Sigma_s$
T_2	$\psi_{px} + \lambda\Sigma_x$	$\psi_{px} - \lambda\Sigma_x$
	$\psi_{py} + \lambda\Sigma_y$	$\psi_{py} - \lambda\Sigma_y$
	$\psi_{pz} + \lambda\Sigma_z$	$\psi_{pz} - \lambda\Sigma_z$

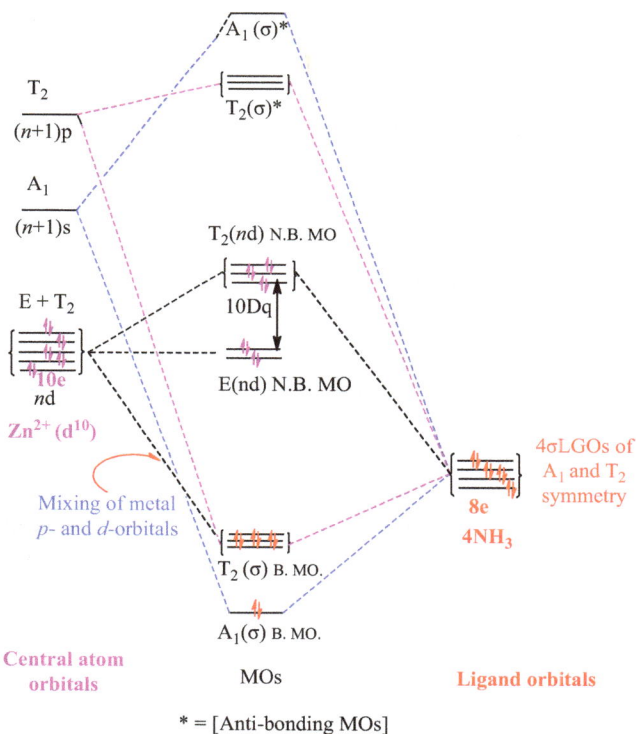

Figure 8.26: MO diagram of $AB_4/[Zn(NH_3)_4]^{2+}$.

Let us consider first the composition of LGO of A_1 symmetry suitable to overlap with s-orbital. It can be seen that all the four σ orbitals have equivalent overlap with s-orbital of the metal (Figure 8.27). Hence, the composition of this LGO in normalized form will be:

$$\sum_s (A_1)\,1/2\,(\sigma_1 + \sigma_2 + \sigma_3 + \sigma_4)$$

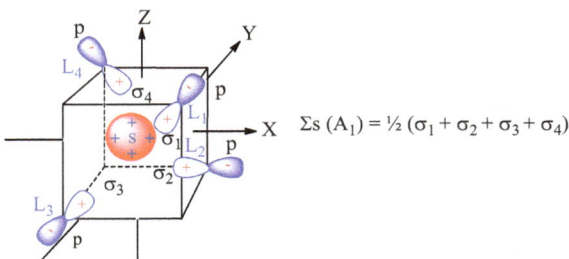

$\Sigma s\ (A_1) = \frac{1}{2}\ (\sigma_1 + \sigma_2 + \sigma_3 + \sigma_4)$

Figure 8.27: Overlap of the metal s-orbital with ligand pσ-orbitals.

Likewise, compositions of LGOs of T_2 symmetry to overlap with metal d_{xy}, d_{yz}, d_{xz} orbitals can be worked out, which are pictorially shown in Figure 8.28.

$$\sum_{xy}(T_2) = 1/2\,(-\sigma_1 - \sigma_2 + \sigma_3 + \sigma_4)$$

$$\sum_{xz}(T_2) = 1/2\,(\sigma_1 - \sigma_2 - \sigma_3 + \sigma_4)$$

$$\sum_{yz}(T_2) = 1/2\,(-\sigma_1 + \sigma_2 - \sigma_3 + \sigma_4)$$

The LGOs, $\Sigma_s(A_1)$, $\Sigma_{xy}(T_2)$, $\Sigma_{xz}(T_2)$ and $\Sigma_{yz}(T_2)$ so obtained combine with the corresponding metal atom orbitals to form bonding and anti-bonding orbitals as shown in Table 8.8

Based on the above discussion, the MO energy level diagram of $AB_4/[CrO_4]^{2-}$ tetrahedral complexes may be shown as given in Figure 8.29.

In the present case, eight electrons from the four oxygens are accommodated in the four σ-bonding MOs, and this accounts for four σ-bonds in CrO_4^{2-}. As all the bonding electrons are paired, the anion is diamagnetic. Similarly, the MO diagrams for MnO_4^-, MnO_4^{2-}, RuO_4 and OsO_4 can be generated. MO diagram of MnO_4^{2-} is presented in Figure 8.30 showing its **paramagnetic behaviour**.

The presence of 8e in four σ-bonding MOs [$A_1(\sigma)$ and $T_2(\sigma)$] accounts for four σ-bonds in MnO_4^{2-}. Beside this, the presence of one metal electron in non-bonding MO E suggest the paramagnetic behaviour of this oxyanion.

8.6.2 π-Bonding in tetrahedral complexes

When a tetrahedral complex, ML_4, is inscribed in a cube (Figure 8.31), the directions of valence shell orbitals of the central metal ion can be visualized perspectively with the X, Y, Z axes passing through centres of cube faces. The lobes of the d_{xy}, d_{yz}

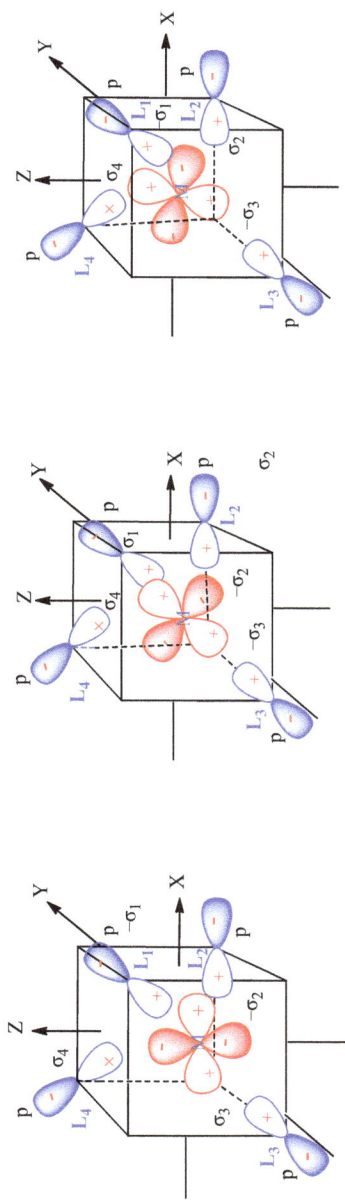

$$\Sigma_{xy}(T_2) = \tfrac{1}{2}(-\sigma_1 - \sigma_2 + \sigma_3 + \sigma_4)$$

$$\Sigma_{xz}(T_2) = \tfrac{1}{2}(\sigma_1 - \sigma_2 - \sigma_3 + \sigma_4)$$

$$\Sigma_{yz}(T_2) = \tfrac{1}{2}(-\sigma_1 + \sigma_2 - \sigma_3 + \sigma_4)$$

Figure 8.28: Overlap of the metal $d_{xy}/d_{yz}/d_{xz}$-orbital with ligand $p\sigma$-orbitals.

Table 8.8: Formation of bonding and anti-bonding orbitals.

Symmetry	Bonding MOs	Anti-bonding MOs
A_1	$\psi_s + \lambda\Sigma_s$	$\psi_s - \lambda\Sigma_s$
T_2	$\psi_{dxy} + \lambda\Sigma_{xy}$	$\psi_{dxy} - \lambda\Sigma_{xy}$
	$\psi_{dyz} + \lambda\Sigma_{xz}$	$\psi_{dyz} - \lambda\Sigma_{xz}$
	$\psi_{xz} + \lambda\Sigma_{yz}$	$\psi_{xz} - \lambda\Sigma_{yz}$

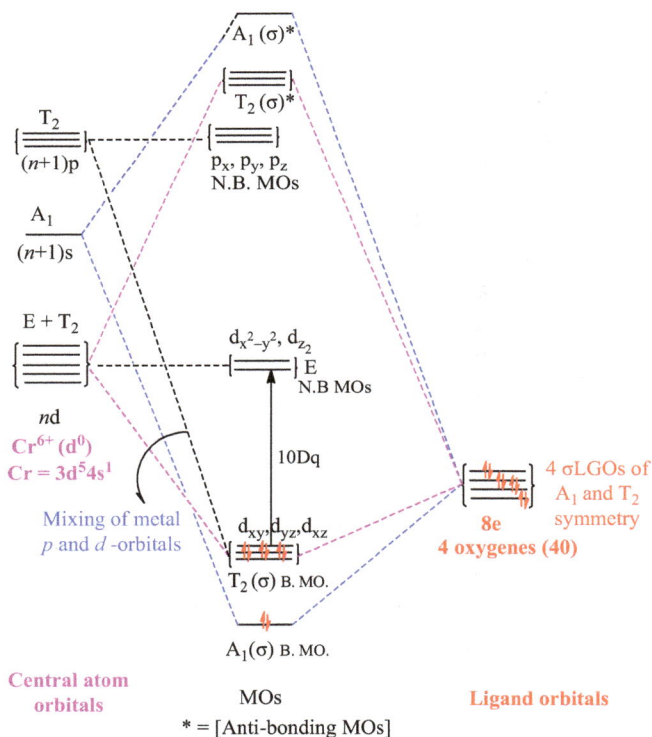

Figure 8.29: MO diagram of AB_4/$CrO_4{}^2$.

and d_{yz} orbitals (t_2 set) are directed towards the midpoints of the cubic edges at an angle of 35.3° with respect to the ligands, while the lobes of the remaining pairs of d-orbitals, d_{z^2} and $d_{x^2-y^2}$, are directed towards the cubic faces and bisects the angle between the pairs of ligands, thus making an angle of 54.7° with the ligands (half of the tetrahedral angle, i.e. 109.4°/2).

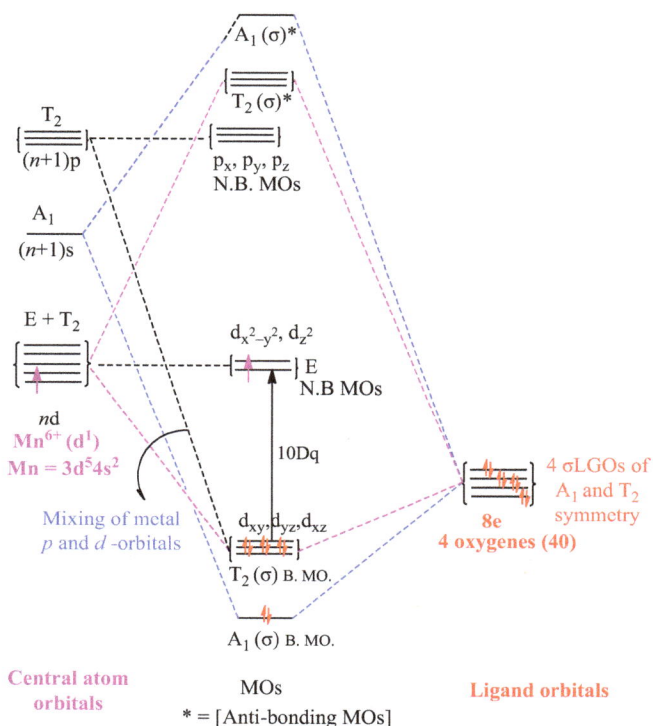

Figure 8.30: MO diagram of $AB_4/MnO_4{}^{2-}$.

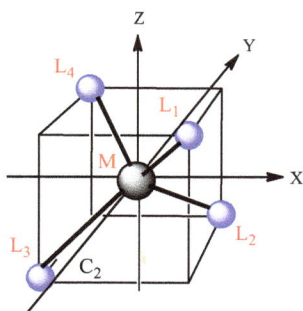

Figure 8.31: A tetrahedral complex, ML_4, inscribed in a cube.

In the σ-bonding case, one pσ-orbital on each ligand has been utilized for bonding to the metal. This leaves another two mutually perpendicular p-orbitals on each ligand which is capable of π-bonding to the metal. Thus, we can assign a pair of vectors to represent these two π-orbitals for each of the four ligands in ML_4 molecule. The directions of the vectors on the L atoms are given in Figure 8.32.

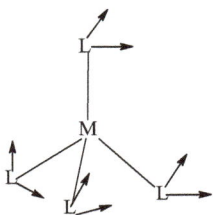

Figure 8.32: Directions of p-π orbitals in a tetrahedral ML₄ molecule.

Taking these vectors as the basis and applying the concept of group theory, one can get the number and nature of symmetries of the resulting LGOs. Accordingly, the symmetries of the LGOs are:

$$\Gamma_\pi = E + T_1 + T_2$$

Thus, there are eight LGOs formed, two of doubly degenerate E symmetry, three of triply-degenerate T_1 symmetry and three of triply degenerate T_2 symmetry. These LGOs will combine with the central atom orbitals of the same symmetry. The central atom orbitals corresponding to E, T_1 and T_2 symmetries can be found out from the character table (Table 8.9) of the T_d point group.

Table 8.9: Character table of the T_d point group.

T_d	E	$8C_3$	$3C_2$	$6S_2$	$6\sigma_d$		
A_1	1	1	1	1	1		$x^2 + y^2 + z^2$
A_2	1	1	1	-1	-1		
E	2	-1	2	0	0		$(2z^2 - x^2 - y^2, x^2 - y^2)$
T_1	3	0	-1	1	-1	(R_x, R_y, R_z)	
T_2	3	0	-1	-1	1	(x, y, z)	(xz, yz, xy)

Accordingly,

$$E \rightarrow d_z{}^2, x^2 - y^2$$

$$T_1 \rightarrow \text{none and}$$

$$T_2 \rightarrow p_x,\ p_y,\ p_{z;}\ d_{xy},\ d_{yz},\ d_{xz}$$

We, therefore, see that to form a complete set of tetrahedrally directed π- bonds, all the d-orbitals and all the p-orbitals of the metal can be utilized for π-bonding. The p- and d-orbitals of T_2 symmetry have well enough overlapping orientation for σ-bonding, but are poor for π type overlap because of their improper orientations with respect to pπ orbitals of ligands (Ls). However, the E set of d_{z^2} and $d_{x^2-y^2}$ orbitals can make overlap to some extent with pπ-orbitals of the ligands. Hence, because of their improper orientation and poor overlap with the pπ-orbitals of the ligands, the wave functions for their LGOs are difficult to evaluate.

A qualitative MO energy level diagram incorporating the above information can be generated as shown in Figure 8.33. In a π-bonding tetrahedral complex, AB₄ [such as $[Zn(CN)_4]^{2-}$ or $[Ni(CO)_4]$], each ligand contribute two σ- and four π-electrons (six electrons). Thus, the four ligands all together give $6 \times 4 = 24$ electrons for bonding. They occupy $A_1(\sigma)(2e)$, $T_2(\sigma)$ (6e),$T_2(\pi)$ (6e), $E(\pi)(4e)$ and $T_1(n)(6e)$. It is notable here that due to the right symmetry, the T_2 (σ^*) MO of the σ-bonding case is now raised little bit to be finally assigned as a mixed MO character, that is, T_2 (σ^*,π^*) because of the mixed character of π-bonding. Likewise, $T_2(\pi^*)$ is also assigned as $T_2(\pi^*,\sigma^*)$ because of the mixed character of σ-bonding. The metal electrons are distributed between $E(\pi^*)$ and $T_2(\pi^*,\sigma^*)$ The molecular energy order is σ-LGO < π-LGO < (n)d < $(n + 1)s < (n + 1)p$.

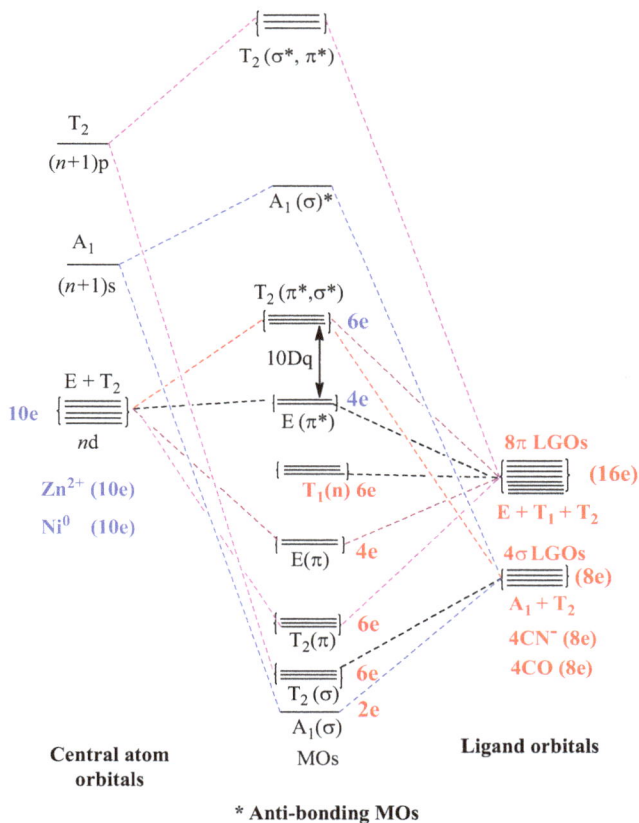

Figure 8.33: MO diagram of AB₄ (tetrahedral) involving σ- and π-bonding both; $[Zn(CN)_4]^{2-}$, $[Ni(CO)_4]$.

The presence of 8e in four σ-bonding MOs [$A_1(\sigma)$ and $T_2(\sigma)$] and 10 electrons in five π-bonding MOs [$T_2(\pi)$ and $E(\pi)$] are responsible for increasing stability of π-bonding tetrahedral complexes [$Zn(CN)_4]^{2-}$ and [$Ni(CO)_4$]. As all the electrons present in MOs are in paired state; both the complexes are diamagnetic. The π-bonding in tetrahedral complexes becomes very important as evidences by the occurrence of oxyanions like, chromate (CrO_4^{2-}), permanganate (MnO_4^-) and manganate (MnO_4^{2-}). The MO diagram of manganate (MnO_4^{2-}) involving σ- and π-bonding is shown in Figure 8.34, and this clearly indicates the paramagnetic behaviour of this anion with respect to one unpaired electron. Again, the presence of 8e in four σ-bonding MOs [$A_1(\sigma)$ and $T_2(\sigma)$] and 10 electrons in five π-bonding MOs [$T_2(\pi)$ and $E(\pi)$] are responsible for increasing stability of π-bonding MnO_4^{2-}.

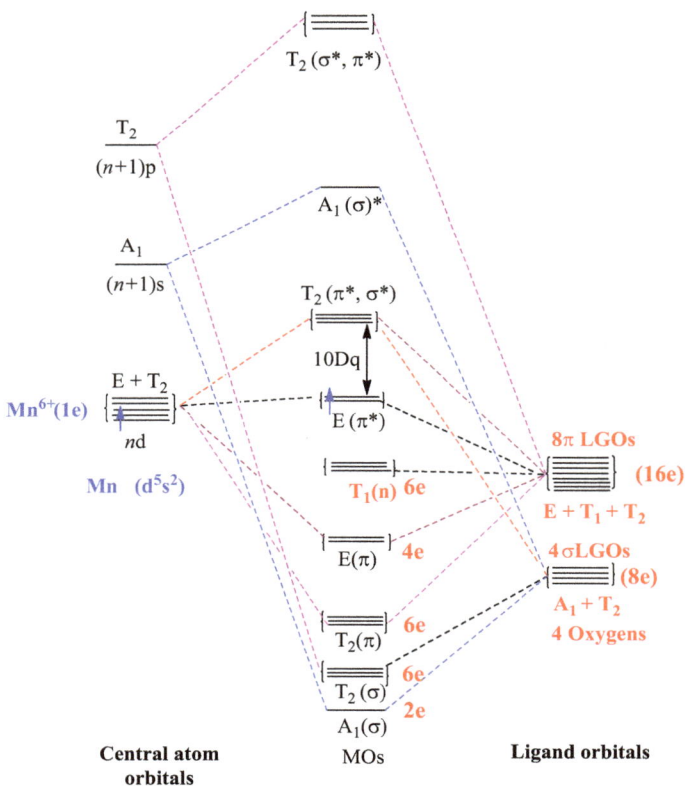

Figure 8.34: MO diagram of MnO_4^{2-} involving σ- and π-bonding.

8.7 Square planar complexes

Square planar AB_4 complexes can involve both σ- and π-bonds between the metal ion and the ligands.

8.7.1 σ-Bonding in square planar complexes

Let us consider the case of a pure σ-bonding in ML_4 square planar complexes. In case of a square planar complex, the ligand atom orbitals are at the corners of a square plane. The molecule belongs to D_{4h} point symmetry. Let us choose a Cartesian coordinate system with the central metal ion M at the origin and the molecule is disposed in the XY plane with its four σ-bonds (M–L) along X and Y axes as shown in Figure 8.35.

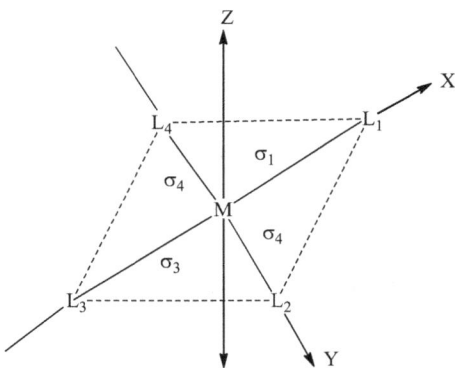

Figure 8.35: The orientation of ML_4 square planar complex in Cartesian coordinate system and transformation of σ-vectors in D_{4h} point group.

Similar to the octahedral case, for the formation of MOs in square planar complexes, the σ-orbitals of the ligand atoms combine to form certain numbers of LGOs. These LGOs combine with the central atom orbitals of the same symmetry to form bonding and anti-bonding orbitals.

Taking these vectors as the basis and applying the concept of group theory, one can get the number and nature of symmetries of the resulting LGOs. Accordingly, the symmetries of the LGOs are:

$$\Gamma\sigma = A_{1g} + B_{1g} + E_u$$

Thus, there are four LGOs formed, one is singly degenerate of A_{1g} symmetry, another is singly degenerate of B_{1g} symmetry and the rest two are doubly degenerate of E_u symmetry. These LGOs will combine with the central atom orbitals of the same symmetry. The central atom orbitals corresponding to A_{1g}, B_{1g} and E_u symmetries can be found out from the character table (Table 8.10) of the D_{4h} point group.

Table 8.10: Character table of the D_{4h} point group.

D_{4h}	E	$2C_4$	C_2	$2C_2$	$2C_2$	i	$2S_4$	σ_h	$2\sigma_v$	$2\sigma_d$		
A_{1g}	1	1	1	1	1	1	1	1	1	1		x^2+y^2, z^2
A_{2g}	1	1	1	−1	−1	1	1	1	−1	−1	R_z	
B_{1g}	1	−1	1	1	−1	1	−1	1	1	−1		(x^2-y^2)
B_{2g}	1	−1	1	−1	1	1	−1	1	−1	1		xy
E_g	2	0	−2	0	0	2	0	−2	0	0	(R_x, R_y)	(xz, yz)
A_{1u}	1	1	1	1	1	−1	−1	−1	−1	−1		
A_{2u}	1	1	1	−1	−1	−1	−1	−1	1	1	z	
B_{1u}	1	−1	1	1	−1	−1	1	−1	−1	1		
B_{2u}	1	−1	1	−1	1	−1	1	−1	1	−1		
E_u	2	0	−2	0	0	−2	0	2	0	0	(x, y)	

Accordingly,

$$A_{1g} \rightarrow s, d_{z^2}; \quad B_{1g} \rightarrow d_{x^2-y^2}; \quad E_u \rightarrow p_x \text{ and } p_y$$

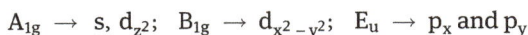

The composition of the composite orbitals or LGOs can be worked out by looking at the way the 4σ-orbitals of the ligands overlap with the central atom orbitals. Accordingly, the compositions of the composite orbitals based of the way of the overlaps are given in Figure 8.36.

The overlap of d_{z^2} orbital with ligand σ-hybrid orbital needs special attention since the large lobes of +ve sign of d_{z^2} are oriented along the Z-axis where there are no ligands whereas it small annular ring with −ve sign in the XY plane. Obviously, ligands present on X and Y axis will interact with this annular planar ring rather weakly. *Therefore, we will not include this LGO in the generation of MO energy level diagram.*

The remaining four LGOs combine with the central atom orbitals of the same symmetry to form bonding and anti-bonding orbitals as given in Table 8.11

The metal ion orbitals, d_{xy}, d_{xz}, d_{yz} and d_{z^2} and p_z remain as non-bonding MOs.

A qualitative MO energy level diagram incorporating the above information can be built up as shown in Figure 8.37.

The metal atom orbitals d_{xy} (B_{2g}), d_{xz}, d_{yz} (E_g) and d_{z^2} (A_{1g}) remain as non-bonding orbitals (nb). In the complex compound $[Pt(NH_3)_4]^{2+}$, there are 8e in the metal d-orbitals and 8e are available from the ligands, $4NH_3$. The 8e of the ligands are placed in the four σ-bonding orbitals and 8e of the metal d-orbitals are filled up in four non-bonding MOs. This supports/explains the presence of four σ bonds between Pt(II) and four NH_3 ligands in $[Pt(NH_3)_4]^{2+}$.

The B_{2g} (nb) (HOMO) and $B_{1g}(\sigma^\star)$ (LUMO) separation is the actual ligand field parameter (10Dq or Δ_1) which can be estimated from the experiments.

$$\Sigma_s = \frac{1}{2}(\sigma_x + \sigma_{-x} + \sigma_y + \sigma_{-y})$$

$$\Sigma_{x^2-y^2} = \frac{1}{2}(\sigma_x + \sigma_{-x} - \sigma_y - \sigma_{-y})$$

$$\Sigma_{z^2} = \frac{1}{2}(-\sigma_x - \sigma_{-x} - \sigma_y - \sigma_{-y})$$

$$\Sigma_x = \frac{1}{\sqrt{2}}(\sigma_x - \sigma_{-x})$$

$$\Sigma_y = \frac{1}{\sqrt{2}}(\sigma_y - \sigma_{-y})$$

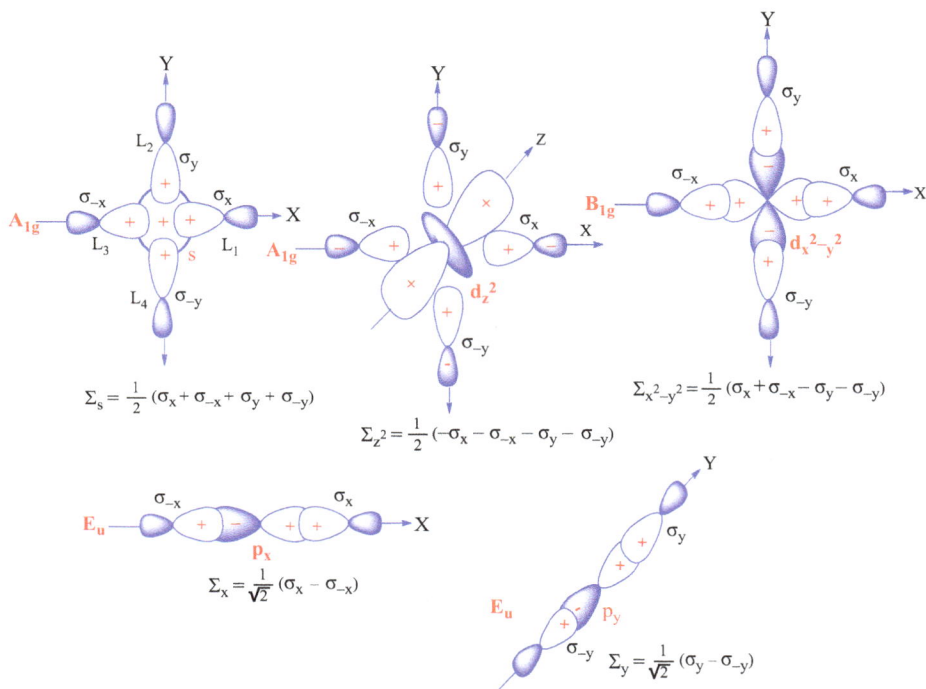

Figure 8.36: A pictorial representation of σ-LGO formation for s, d_{z^2}, $d_{x^2-y^2}$, p_x and p_y metal orbitals combining with σ-hybrid orbital (dsp^2) of same symmetry.

Table 8.11: Formation of bonding and anti-bonding MOs in σ-bonding square planar complexes.

Symmetry	Bonding MO	Anti-bonding MO
A_{1g}	$A_{1g}{}^b = \psi_s + \lambda\Sigma_s$	$A_{1g}{}^* = \psi_s - \lambda\Sigma_s$
B_{1g}	$B_{1g}{}^b = \psi_{dx^2-y^2} + \omega\Sigma_{dx^2-y^2}$	$B_{1g}{}^* = \psi_{dx^2-y^2} - \omega\Sigma_{dx^2-y^2}$
E_u	$E_u{}^b = \psi_{px} + k\Sigma_x$ $= \psi_{py} + k\Sigma_y$	$E_u{}^* = \psi_{px} - k\Sigma_x$ $= \psi_{py} - k\Sigma_y$

8.7.2 π-bonding in square planar complexes

The π-bonding in square planar complexes is similar to the π-bonding in octahedral complexes. Accordingly, we may have complexes wherein ligands are both σ- and π-donor type (π-bases) or those which contain ligands involving, σ, π and vacant π*orbitals (π-acids).

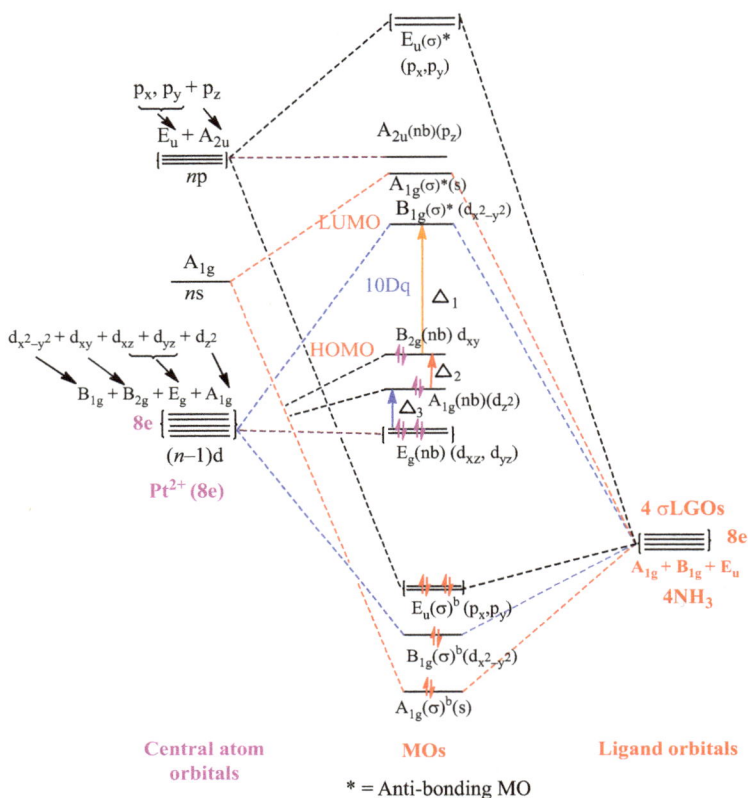

Figure 8.37: MO diagram of AB_4 (square planar) involving σ-bonding with reference to $[Pt(NH_3)_4]^{2+}$.

From the MO diagram of square planar complexes involving only σ-bonding, it is seen that d_{xy} (B_{2g}), d_{xz} and d_{yz} (E_g) and d_{z^2} (A_{1g}) orbitals remain non-bonding. If the ligand atoms have π-orbitals also, they can combine to form MOs of π-symmetry.

In the σ-bonding case, we have utilized one pσ-orbital on each ligand for bonding to the metal. This leaves another two mutually perpendicular p-orbitals on each ligand which is capable of π-**bonding to the metal.** Thus, we can assign **a pair of vectors** to represent these two π-orbitals for each of the four ligands in square planar ML_4 molecule. Of the two vectors on each L atom, one will be **perpendicular** to the plane of the molecule and the other (**parallel**) will be in plane (Figure 8.38).

Taking these vectors as the basis and applying the concept of group theory, one can get the number and nature of symmetries of the resulting LGOs. Accordingly, the symmetries of the LGOs are:

$$\Gamma_\perp(\pi) = A_{2u} + B_{2u} + E_g \text{ and}$$

$$\Gamma_\parallel(\pi) = A_{2g} + B_{2g} + E_u$$

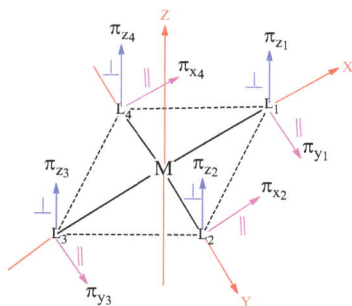

Figure 8.38: Directions of p-orbitals, 4 perpendicular to the plane (parallel to the z-axis, shown by blue arrows) and 4 in plane or parallel to the plane (perpendicular to the z- axis, shown by pink arrow) in square planar molecule.

The central atom orbitals that correspond to these symmetries can be found out from the character table (Table 8.12) of the D_{4h} point group.

Table 8.12: Character table of D_{4h} point group.

D_{4h}	E	$2C_4$	C_2	$2C_2$	$2C_2$	i	$2S_4$	σ_h	$2\sigma_v$	$2\sigma_d$		
A_{1g}	1	1	1	1	1	1	1	1	1	1		$(x^2+y^2+z^2)$
A_{2g}	1	1	1	-1	-1	1	1	1	-1	-1	R_z	
B_{1g}	1	-1	1	1	-1	1	-1	1	1	-1		(x^2-y^2)
B_{2g}	1	-1	1	-1	1	1	-1	1	-1	1		xy
E_g	2	0	-2	0	0	2	0	-2	0	0	(R_x, R_y)	(xz, yz)
A_{1u}	1	1	1	1	1	-1	-1	-1	-1	-1		
A_{2u}	1	1	1	-1	-1	-1	-1	-1	1	1	z	
B_{1u}	1	-1	1	1	-1	-1	1	-1	-1	1		
B_{2u}	1	-1	1	-1	1	-1	1	-1	1	-1		
E_u	2	0	-2	0	0	-2	0	2	0	0	(x, y)	

Accordingly,

$$\underline{\quad T_\perp(\pi) \quad} \qquad \underline{\quad T_{II}(\pi) \quad}$$

A_{2u}	B_{2u}	E_g		A_{2g}	B_{2g}	E_u
p_z	none	d_{xz}, d_{yz}		none	d_{xy}	p_x, p_y

This result shows that all the three p-orbitals, p_x, p_y (of E_u symmetry) and p_z (of A_{2u} symmetry) and three of the d-orbitals, d_{xz}, d_{yz} (of E_g symmetry) and d_{xy} (of B_{2g} symmetry) of the central metal can take part in metal–ligand π-bonding.

Of them, only p_z, d_{xz}, d_{yz} and d_{xy} metal orbitals form exclusively π-bonds, whereas p_x and p_y form both σ and π-bonds. As the p_x and p_y orbitals give much better overlap for σ-bonding than for π-bonding and hence they are predicted to take part mainly in σ-bonding.

Thus, it is noteworthy here that only four π-bonds are supposed to be formed and of them three are perpendicular π-bonds (formed by p_z, d_{xz} and d_{yz}) and one is parallel π-bond formed by d_{xy}. There are no orbitals of A_{2g} and B_{2u} symmetry on the metal valence s-, p- and d-orbitals. It may be possible to find such orbitals from f-orbitals, which do not participate in bonding.

The composition of the composite orbitals or LGOs can be worked out by looking at the way the π-orbitals of the ligands overlap the central atom orbitals. Accordingly, the compositions of the composite orbitals based of the way of the overlaps are given as under (Figure 8.39).

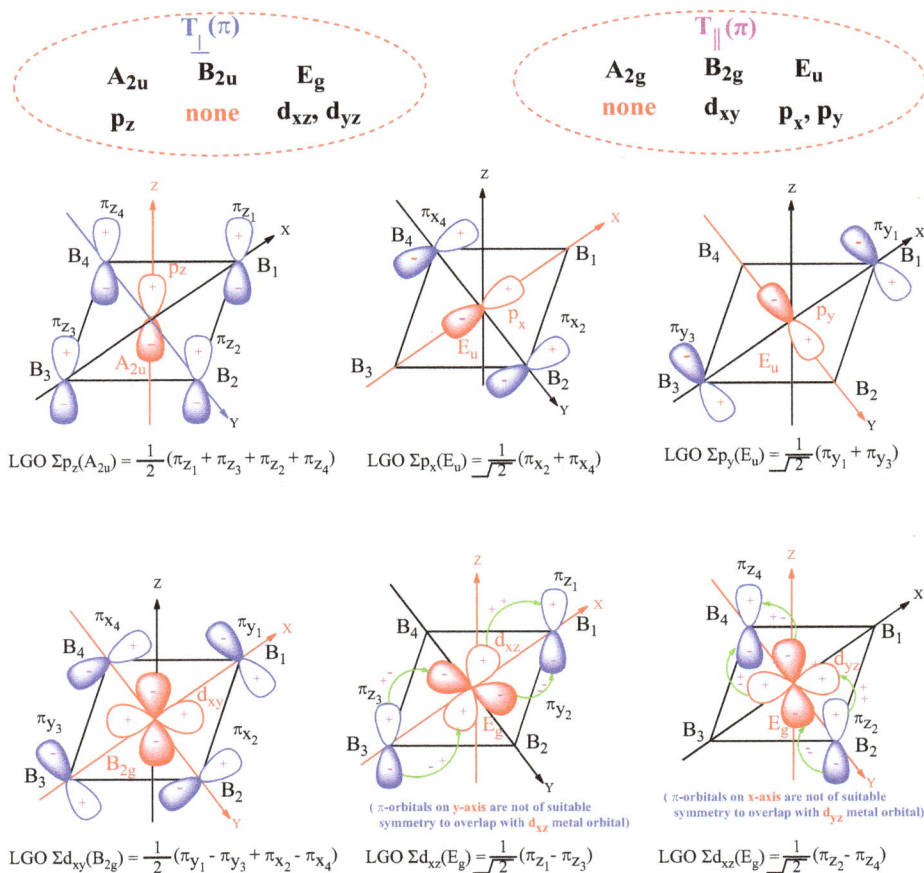

$$\text{LGO } \Sigma p_z(A_{2u}) = \frac{1}{2}(\pi_{z_1} + \pi_{z_3} + \pi_{z_2} + \pi_{z_4})$$

$$\text{LGO } \Sigma p_x(E_u) = \frac{1}{\sqrt{2}}(\pi_{x_2} + \pi_{x_4})$$

$$\text{LGO } \Sigma p_y(E_u) = \frac{1}{\sqrt{2}}(\pi_{y_1} + \pi_{y_3})$$

$$\text{LGO } \Sigma d_{xy}(B_{2g}) = \frac{1}{2}(\pi_{y_1} - \pi_{y_3} + \pi_{x_2} - \pi_{x_4})$$

$$\text{LGO } \Sigma d_{xz}(E_g) = \frac{1}{\sqrt{2}}(\pi_{z_1} - \pi_{z_3})$$

(π-orbitals on y-axis are not of suitable symmetry to overlap with d_{xz} metal orbital)

$$\text{LGO } \Sigma d_{xz}(E_g) = \frac{1}{\sqrt{2}}(\pi_{z_2} - \pi_{z_4})$$

(π-orbitals on x-axis are not of suitable symmetry to overlap with d_{yz} metal orbital)

Figure 8.39: Compositions of the ligand group orbitals (LGOs) suitable for overlapping with p_x, p_y, p_z and d_{xy}, d_{xz} and d_{yz} metal orbitals.

The resulting four LGOs combine with the central atom orbitals of the same symmetry to form bonding and anti-bonding orbitals as given in Table 8.13.

Table 8.13: Formation of bonding and anti-bonding MOs.

Symmetry	Bonding MO	Anti-bonding MO
A_{2u}	$A_{2u}b = \psi p_z + \lambda \Sigma p_z$	$A_{2u}^* = \psi p_z - \lambda \Sigma p_z$
E_u	$E_u b = \psi p_x + \omega \Sigma_x$ $= \psi p_y + \omega \Sigma_y$	$E_u^* = \psi p_x - \omega \Sigma_x$ $= \psi p_y - \omega \Sigma_y$
B_{2g}	$B_{2g}b = \psi d_{xy} + k \Sigma d_{xy}$	$B_{2g}^* = \psi d_{xy} - k \Sigma d_{xy}$
E_g	$E_g b = \psi d_{xz} + \gamma \Sigma d_{xz}$ $= \psi d_{yz} + \gamma \Sigma d_{yz}$	$E_g^* = \psi d_{xz} - \gamma \Sigma d_{xz}$ $= \psi d_{yz} - \gamma \Sigma d_{yz}$

The symbols λ, ω, k and γ represent the ratio of the coefficients of the central atom orbital and the LGO. As p_x and p_y are degenerate and Σ_x and Σ_y are also degenerate, the coefficient ω is same in the two degenerate MOs. Similarly, d_{xz} and d_{yz} are degenerate and also Σd_{xz} and Σd_{yz} are degenerate, the coefficient γ is same in the two degenerate MOs.

Qualitative MO energy level diagrams incorporating the above information can be built up for ligands containing either **π-donor** or **π-acceptor** orbitals separately.

8.7.2.1 MO energy level diagram for AB$_4$ square planar complexes involving both σ- and π-donor ligands

The complexes such as PdX_4^{2-}, PtX_4^{2-} and AuX_4^{2-} (X = halide ion) come under this category because halide ligands are both σ- and π-donors. The MO energy level diagram in such case is shown in Figure 8.40.

Herein, the order of energy of AOs of the central metal atom and the LGOs is: σ-LGOs < π-LGOs < (n-1)d < ns < np. It is notable from the diagram that the energy of the strongly anti-bonding $B_{1g}(\sigma^*)$ MO is higher in energy than the anti-bonding MOs: $E_g(\pi^*)$, $A_{1g}(\sigma^*)$ and $B_{2g}(\pi^*)$ because B_{1g} ($d_{x^2-y^2}$) orbital interacts with all the four ligands (Figure 8.41).

Also, the energy of the $B_{2g}(\pi^*) > E_g(\pi^*)$ since the $B_{2g}(d_{xy})$ orbital interacts with all the four ligands whereas $E_g(d_{xz}, d_{yz})$ orbitals interact with only two ligands in each. Pictorial representation of LGO $B_{2g}(d_{xy})$ and LGO $E_g(d_{xz}, d_{yz})$ are shown in Figure 8.42 for clarity.

It can be easily understood that the complex $PtCl_4^{2-}/PdCl_4^{2-}$ is diamagnetic by accommodating both σ- and π-electrons of the ligands.

The 8e due to σ-donation from the ligands are filled up into the core σ-bonding MOs: $A_{1g}(\sigma)^b$(2e), $B_{1g}(\sigma)^b$ (2e) and $E_u(\sigma)^b$ (4e) and 16e due to π-donation can be accommodated into π-bonding MOs: $B_{2g}(\pi)^b$ (2e), $E_g(\pi)^b$ (4e), $A_{2u}(\pi)^b$ (2e) and $E_u(\pi)^b$ (4e) (bonding) as well as the non-bonding A_{2g}(2e) and B_{2u}(2e).

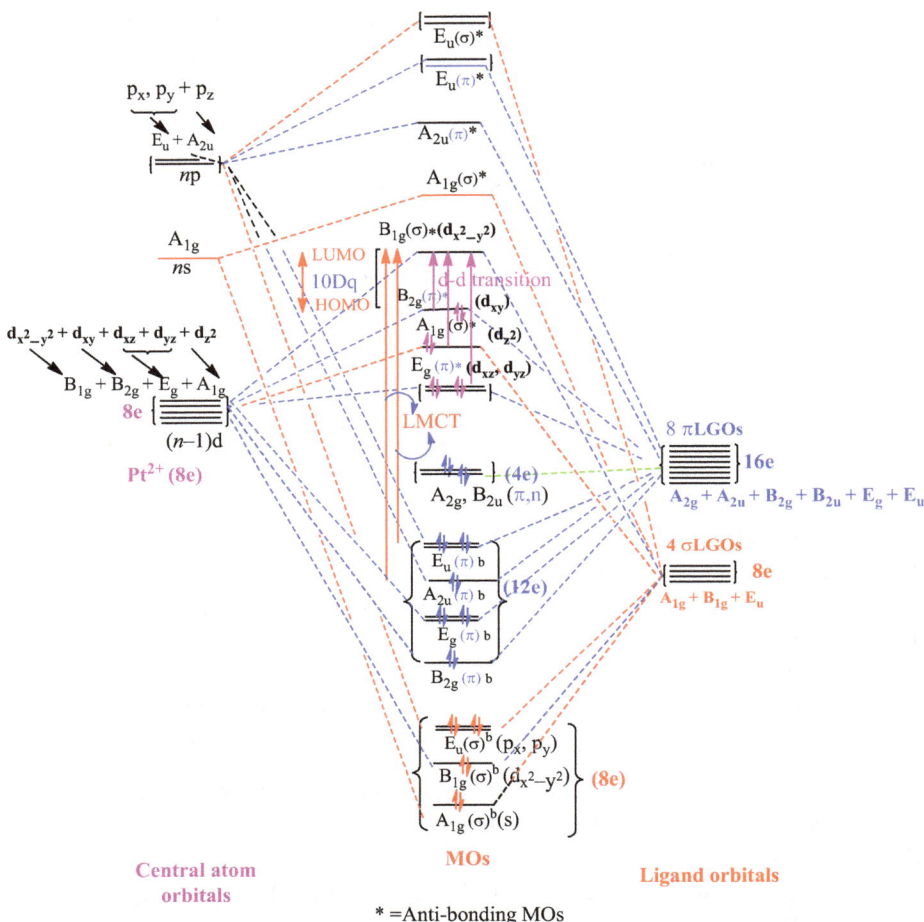

Figure 8.40: MO diagram of AB$_4$ (square planar complex) involving both σ- and π-donors with reference to [PtCl$_4$]^2.

The metal d^8 electrons are now accommodated in the MOs primarily metal based, that is, E$_g$(π*)(4e), A$_{1g}$(σ)* (2e) and B$_{2g}$(π*)(2e). The HOMO in this complex is B$_{2g}$(π*) and LUMO is B$_{1g}$(σ*) and their separation is equal to the ligand field parameter 10Dq (or Δ$_1$).

This explains why low-spin d^8-system, which are Ni(II), Pd(II)and Pt(II) and the d^9-system, such as Cu(II) adopts square planar geometry. There are three possible one-electron spin-allowed d–d transitions (Figure 8.43) and these are:

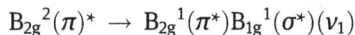

$$B_{2g}^{2}(\pi)^{\star} \rightarrow B_{2g}^{1}(\pi^{\star})B_{1g}^{1}(\sigma^{\star})(\nu_1)$$

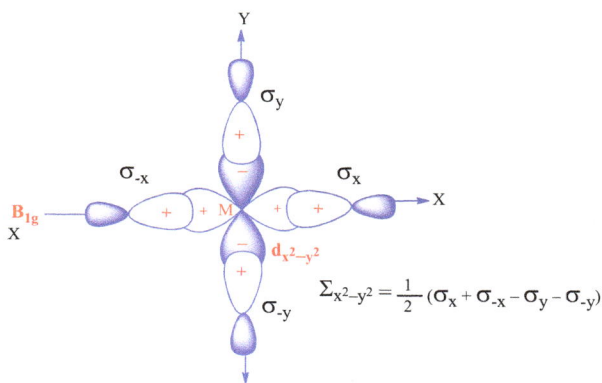

Figure 8.41: Interaction of B_{1g} ($d_{x^2-y^2}$) orbital with all the four ligands.

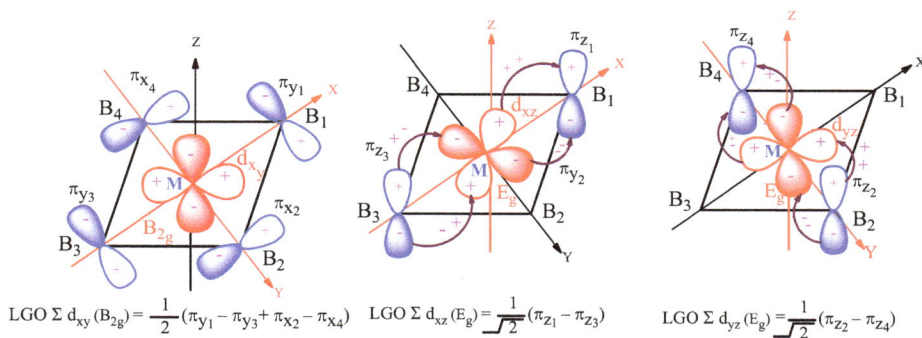

$$\text{LGO } \Sigma\, d_{xy}\, (B_{2g}) = \frac{1}{2}(\pi_{y_1} - \pi_{y_3} + \pi_{x_2} - \pi_{x_4}) \quad \text{LGO } \Sigma\, d_{xz}\, (E_g) = \frac{1}{\sqrt{2}}(\pi_{z_1} - \pi_{z_3}) \quad \text{LGO } \Sigma\, d_{yz}\, (E_g) = \frac{1}{\sqrt{2}}(\pi_{z_2} - \pi_{z_4})$$

Figure 8.42: Pictorial representation of LGO $B_{2g}(d_{xy})$ and LGO E_g (d_{xz}, d_{yz}).

$$A_{1g}{}^2(\sigma)^* \;\rightarrow\; A_{1g}{}^1(\sigma)^* B_{1g}{}^1(\sigma^*)(\nu_2)$$

$$E_g{}^4(\pi^*) \;\rightarrow\; E_g{}^3(\pi^*) B_{1g}{}^1(\sigma^*)(\nu_3)$$

The allowed L→M charge transfer transition may be anticipated corresponding to the one-electron transitions as shown in Figure 8.44.

$$A_{2u}{}^2(\pi)^b \;\rightarrow\; A_{2u}{}^1(\pi)^b B_{1g}{}^1(\sigma^*)$$

$$E_u{}^4(\pi)^b \;\rightarrow\; E_g{}^3(\pi)^b B_{1g}{}^1(\sigma^*)$$

These are the transitions from MOs essentially localized on the ligands to the MOs essentially localized on the metal atom as notable from the transitions shown above.

B_{1g} $\dfrac{(\sigma)^*(d_{x^2-y^2})}{}$ d–d

B_{2g} $(\pi)^*$ (d_{xy})

$A_{1g}(\sigma)^*$ (d_{z^2})

$E_g(\pi)^*$ (d_{xz}, d_{xz})

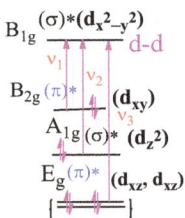

Figure 8.43: Three possible one-electron spin-allowed d–d transitions in square planar complexes of Ni(II), Pd(II) and Pt(II).

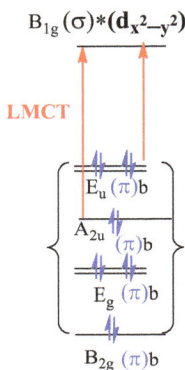

$B_{1g}(\sigma)^*(d_{x^2-y^2})$

LMCT

$E_u (\pi)b$

A_{2u} $(\pi)b$

$E_g (\pi)b$

$B_{2g} (\pi)b$

Figure 8.44: The allowed one-electron L→M charge transfer transition in square planar Ni(II), Pd(II) and Pt(II) complexes.

8.7.2.2 MO energy level diagram for AB₄ square planar complexes involving σ-donor and filled and empty ligand π-orbitals of higher energy π-acceptor (π-acid) ligands

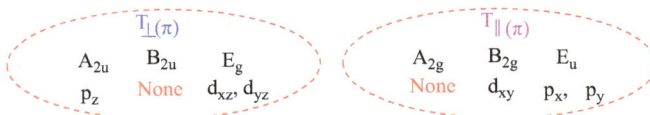

$T_\perp(\pi)$		
A_{2u}	B_{2u}	E_g
p_z	None	d_{xz}, d_{yz}

$T_\parallel(\pi)$		
A_{2g}	B_{2g}	E_u
None	d_{xy}	p_x, p_y

When the ligand in square planar complex is **π-acid** type, for example, as in [Ni(CN)₄]²⁻, then the MO diagram shown in Figure 8.45 is applicable for such ligands. The diagram, though qualitative, accounts for most of the observations of [M(CN)₄]ⁿ⁻ type square planar complexes. The diagram appears a bit complicated, but is not having lack of clarity.

A careful look at the MO energy level diagram reveals that the **π-MOs** resulting from $A_{2u}(p_z)$ and $B_{2g}(d_{xy})$-orbitals are very important. The LGO formed from a proper combination of four $\perp_r$ π-orbitals (p_z) on the ligands actually look like a +ve ring above and a −ve ring below the square plane as shown in Figure 8.46.

The interaction of the LGO [$\Sigma p_z(A_{2u})$] with a centrally located metal p_z-orbital of A_{2u} symmetry results in a strong bonding and anti-bonding $A_{2u}(\pi)^b$ and $A_{2u}(\pi)^*$ MOs. The interaction of the LGO [$\Sigma d_{xy}(B_{2g})$] with d_{xy} orbital of B_{2g} symmetry lowers

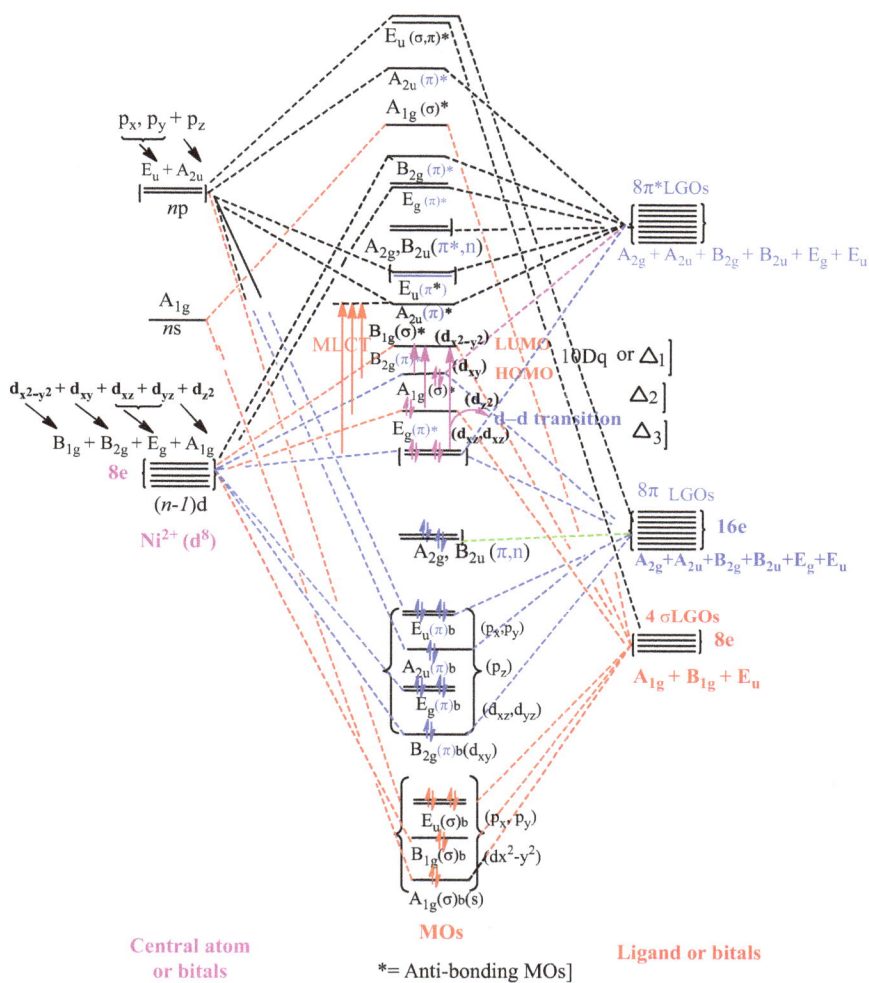

Figure 8.45: MO diagram of AB$_4$ square planar complex involving σ-donor and filled and empty ligand π-orbitals of higher energy π-acceptor (π-acid) ligands with reference to [Ni(CN)$_4$]$^{2-}$.

the energy of B$_{2g}(\pi)$* MO compared to non-π acid ligands. This increases the separation from B$_{1g}(\sigma)$* resulting in a higher value of 10Dq or Δ_1.

As shown in Figure 8.47, the possible one-electron d–d transitions in this case ([Ni(CN)$_4$]$^{2-}$), similar to π-donor case, are:

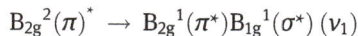

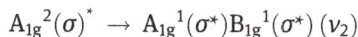

$$B_{2g}^{2}(\pi)^{*} \rightarrow B_{2g}^{1}(\pi^*)B_{1g}^{1}(\sigma^*) \ (\nu_1)$$

$$A_{1g}^{2}(\sigma)^{*} \rightarrow A_{1g}^{1}(\sigma^*)B_{1g}^{1}(\sigma^*) \ (\nu_2)$$

$$\text{LGO } \Sigma_{p_z}(A_{2u}) = \frac{1}{2}(\pi_{z1} + \pi_{z3} + \pi_{z2} + \pi_{z4}) \qquad \text{LGO } \Sigma \, d_{xy}(B_{2g}) = \frac{1}{2}(\pi_{y1} - \pi_{y3} + \pi_{x2} - \pi_{x4})$$

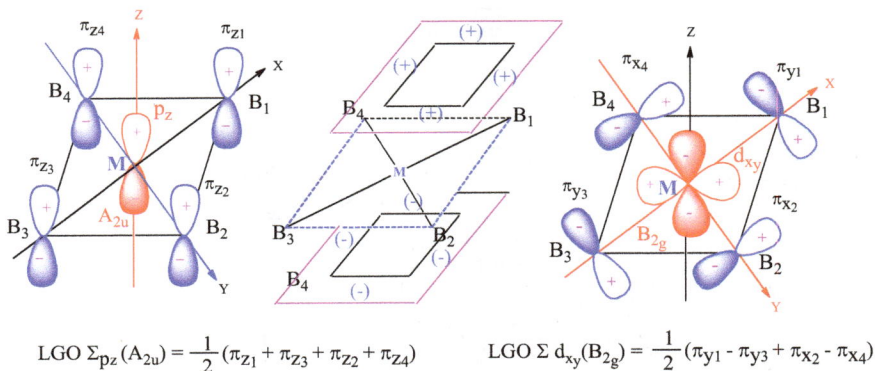

Figure 8.46: π-MOsresulting from $A_{2u}(p_z)$ and $B_{2g}(d_{xy})$-orbitals look like a + ve ring above and a −ve ring below the square plane.

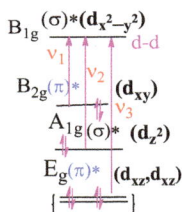

Figure 8.47: Three possible one-electron spin-allowed d–d transitions in square planar $[Ni(CN)_4]^{2-}$ complex.

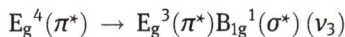

$$E_g{}^4(\pi^*) \;\rightarrow\; E_g{}^3(\pi^*)B_{1g}{}^1(\sigma^*)\,(v_3)$$

The three expected M → L charge transfer transitions (Figure 8.48) may be anticipated corresponding to the one-electron transition:

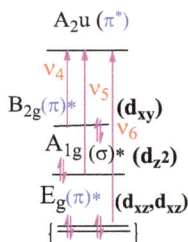

Figure 8.48: Three expected one-electron M→L charge transfer transition in square planar $[Ni(CN)_4]^{2-}$ complex.

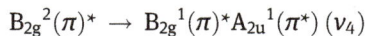

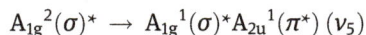

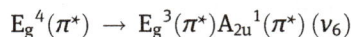

$$B_{2g}{}^2(\pi)^* \;\rightarrow\; B_{2g}{}^1(\pi)^*A_{2u}{}^1(\pi^*)\,(v_4)$$

$$A_{1g}{}^2(\sigma)^* \;\rightarrow\; A_{1g}{}^1(\sigma)^*A_{2u}{}^1(\pi^*)\,(v_5)$$

$$E_g{}^4(\pi^*) \;\rightarrow\; E_g{}^3(\pi^*)A_{2u}{}^1(\pi^*)\,(v_6)$$

These are the transitions from MOs essentially localized on the metal to the MO essentially localized on the ligand as notable from the transitions shown above. It is predicted that the three charge transfer transitions with an intensity order $v_6 > v_5 > v_4$ should be typical of square planar complexes with π-acid ligands.

For $[Ni(CN)_4]^{2-}$ in aqueous solution, there are two shoulders at 22,500 cm^{-1}($\varepsilon = 2$) ((v_1) and 30,500 cm^{-1}($\varepsilon = 250$) (v_2) assignable to d–d transitions. The three bands observed at 32,300 cm^{-1}($\varepsilon = 700$) (v_4), 35,200 cm^{-1}($\varepsilon = 4200$) (v_5) and 37,600 cm^{-1} ($\varepsilon = 10,600$) (v_6) with increasing intensity are assignable to charge transfer transitions.

Exercises

Multiple choice questions/fill in the blanks

1. Crystal field theory (CFT) was first time applied to the ionic-type crystalline substances to explain their colours and magnetic properties in 1929 by:
   (a) L. Pauling and J. L. Slater
   (b) G. Wilkinson
   (c) H. Bethe and J. H. Van Vleck
   (d) A. VanArked and J. DeBoer

2. The name adjusted crystal field theory was first given to the modified form of CFT by:
   (a) F. A. Cotton and G. Wilkinson
   (b) H. Van Vleck
   (c) A. VanArked and J. DeBoer
   (d) None of these

3. According to CFT, the bonds between the metal and ligands are:
   (a) Partly covalent
   (b) Purely ionic
   (c) Ionic and covalent both
   (d) All of these

4. Out of five d-orbitals, suitable orbitals for metal–ligand σ-bonding in octahedral complexes are:
   (a) d_{xy}, d_{yz}, d_{xz} (b) d_{z^2} (c) $3d_{x^2-y^2}, 3d_{z^2}$ (d) All of these

5. Sigma bonding molecular orbitals (σ-MOs) in octahedral complexes are mainly filled by:
   (a) Metal electrons
   (b) Ligand electrons
   (c) Both metal and ligand electrons
   (d) All of these

6. In π-bonding octahedral complexes, suitable d-orbitals that participate in bonding are:
   (a) d_{z^2} (b) d_{xy}, d_{yz}, d_{xz} (c) $3d_{x^2-y^2}, 3d_{z^2}$ (d) All of these

7. Sort out a ligand which has filled pπ-orbital and vacant dπ-orbital along with σ-orbital from the following:
   (a) Cl⁻   (b) CN⁻   (c) PPh₃   (d) CO

8. Which one of the following ligands has empty ligand dπ-orbitals of higher energy than metal t_{2g} orbitals?
   (a) CO   (b) R₂S   (c) CN⁻   (d) o-Phenanthroline

9. Find out a ligand which has filled π-molecular orbital and vacant π*-molecular orbital form the following:
   (a) R₂S   (b) o-Phenanthroline   (c) AsR₃   (d) PR₃

10. In MO diagram of an octahedral [Cr(CO)₆] involving filled π-molecular orbitals and high energy vacant π*MO, has correct sequence of energy order is:
    (a) σ-LGOs > π-LGOs > $(n-1)d$ > ns > π*-LGOs > np
    (b) σ-LGOs < π-LGOs < $(n-1)d$ < ns < π*-LGOs < np
    (c) σ-LGOs < π-LGOs < $(n-1)d$ < ns < π*-LGOs = np
    (d) π-LGOs < σ-LGOs < $(n-1)d$ < ns < π*-LGOs < np

11. In [Cr(o-phen)₃] (o-phen=o-phenanthroline) the number of electrons to be accommodated in the molecular orbitals is:
    (a) 6   (b) 12   (c) 36   (d) 42

12. In case of octahedral π-bonding Mo (0) complex involving vacant ligand dπ-orbital of higher energy (i.e. of PR₃, PX₃ or As R₃ ligand) than metal t_{2g} orbitals, the correct MO electronic configuration is:
    (a) $[(a_{1g})^2(t_{1u})^6(e_g)^4]^\sigma \, [(t_{2g})^6]^\pi \, [e_g(\sigma)*]^0 \, [(t_{1u})^0]^\pi$
    (b) $[(t_{1u})^6[(a_{1g})^2(e_g)^4]^\sigma \, [(t_{2g})^6]^\pi \, [e_g(\sigma)*]^0 \, [(t_{1u})^0]^\pi$
    (c) $[(a_{1g})^2(e_g)^4(t_{1u})^6]^\sigma \, [(t_{2g})^6]^\pi \, [e_g(\sigma)*]^0 \, [(t_{1u})^0]^\pi$
    (d) $[(e_g)^4(a_{1g})^2(t_{1u})^6]^\sigma \, [(t_{2g})^6]^\pi \, [e_g(\sigma)*]^0 \, [(t_{1u})^0]^\pi$

13. In σ-bonding tetrahedral complexes, the correct symmetries of the ligand group orbitals (LGOs) are:
    (a) $\Gamma_\sigma = A_2 + T_2$ (b) $\Gamma_\sigma = A_1 + T_2$ (c) $\Gamma_\sigma = A_2 + T_1$ (d) $\Gamma_\sigma = A_1 + T_1$

14. The correct symmetries of the ligand group orbitals (LGOs) in σ-bonding square planar complexes are:
    (a) $\Gamma\sigma = A_{2g} + B_{2g} + E_u$
    (b) $\Gamma\sigma = A_{2g} + B_{1g} + E_u$
    (c) $\Gamma\sigma = A_{1g} + B_{1g} + E_u$
    (d) None of these

15. The number of one-electron spin-allowed d–d transitions in square planar [Ni $(CN)_4]^{2-}$ complex ion is . . . .

16. There are . . . . number of possible one-electron spin-allowed d–d transitions in $[NiCl_4]^{2-}$ / $[PdCl_4]^{2-}$/$[PtCl_4]^{2-}$.

Short answer type questions

1. Point out the shortcomings of the crystal field theory.
2. Explain what is meant by nephelauxetic effect. How does this effect explain the contribution of covalent bonding in metal–ligand bonds?
3. Just draw the molecular orbital energy diagram of $[Co(NH_3)_6]^{3+}$ complex ion and comment on its magnetic behaviour.
4. Looking over the MOT energy level diagram of $[CoF_6]^{3-}$ involving σ-bonding only justify that this compound is paramagnetic.
5. Develop the suitable ligand group orbitals by linear combination of σ-orbitals together, that can overlap with the matching symmetry of the central metal atomic orbitals, which are 4s, $4p_x$, $4p_y$, $4p_z$, $3d_{z^2}$ and $3d_{x^2-y^2}$ for σ-bonding in octahedral complexes.
6. Give the pictorial representation of ligand group orbitals (LGOs) and matching atomic orbitals (AOs) of the same symmetry to form bonding σ-Mos and σ*-Mos in octahedral complexes.
7. Considering the π-bonding octahedral cases, give the pictorial representation of directions of the atomic orbitals of the ligands L_1, L_2 . . .. . .L_6, which can form π-bonds with atomic orbitals of the same symmetry on the metal atom M. Also give the total number of participation p-orbitals in π-bonding.
8. Classify the types of π-bonds in metal complexes according to the nature of the ligand π-orbitals with their pictorial representations.
9. Just draw the MO diagram of an octahedral $[Cr(CO)_6]$ involving high energy vacant π^*MO. Also comment on its magnetic behaviour.
10. Develop the compositions of ligand group orbitals (LGOs) of A_1 symmetry and T_2 symmetry suitable to overlap with metal s-orbital and p_x, p_y and p_z orbitals, respectively, in σ-bonding tetrahedral complexes. Also give their pictorial representations.
11. Generate the compositions of LGO of A_1 symmetry and T_2 symmetry suitable to overlap with metal s-orbital and d_{xy}, d_{yz}, d_{xz} orbitals, respectively, in σ-bonding tetrahedral complexes. Also give their pictorial representations.
12. Just draw the MO diagram of tetrahedral $[Zn(NH_3)_4]^{2+}$ complex and comment on its magnetic behaviour.
13. Draw the MO diagram of tetrahedral MnO_4^{2-} compound and show that it is paramagnetic.

14. Work out the compositions of σ-LGOs suitable to overlap with metal s, d_{z^2}, $d_{x^2-y^2}$, p_x and p_y orbitals of respective symmetries in σ-bonding square planar complexes.

15. Develop the compositions of π-LGOs suitable for overlapping with metal p_x, p_y, p_z, d_{xy}, d_{xz} and d_{yz} of respective symmetries in π-bonding square planar complexes.

16. Draw the MO diagram of $[Ni(CN)_4]^{2-}$ square planar complex. Is this a diamagnetic species?

Long answer type questions

1. Present a detailed view of molecular orbital theory (MOT) of octahedral complexes involving σ-bonding only when the ligands are strong.

2. Discuss the sigma bonding in transition metal complexes when the ligands are weak with reference to octahedral complexes.

3. Present a detailed view of the formation of ligand group orbitals (LGOs) of t_{2g} symmetry taking part in π-bond formation with metal t_{2g} orbitals. Also give the pictorial representation of the overlap of metal d_{xy}, d_{yz} and d_{xz} AOs with LGOs π_{xy}, π_{yz} and π_{xz}.

4. The π-bonding in octahedral complexes having filled ligand π-orbitals of lower in energy than metal t_{2g} orbitals affects in lowering of magnitude of 10Dq compared to purely σ-bonding complexes. Justify this statement.

5. Explain with the help of MOT why halide ions act as weak field whereas CN^-/CO acts a strong ligand in octahedral complexes.

6. Draw and discuss the MO diagram of complexes involving empty ligand π-orbitals (PR_3, PX_3, AsR_3, R_2S) of higher energy than metal t_{2g} orbitals.

7. Present a detailed view of bonding in complexes having filled and empty ligand π-orbitals of higher energy than metal T_{2g} orbitals.

8. Present a detailed account of metal–ligand σ-bonding in tetrahedral complexes taking a suitable example.

9. Discuss the π-bonding in tetrahedral complexes, such as, $[Zn(CN)_4]^{2-}$ and $[Ni(CO)_4]$.

10. Present a detailed view of MO energy level diagram for square planar complexes involving both σ- and π-donor ligands.

11. Draw and discuss the MO energy level diagram of square planar complexes involving σ-donor and filled and empty ligand π-orbitals of higher energy.

Chapter IX
Bonding in organometallic sandwich compounds: molecular orbital theory approach

9.1 Introduction

Compounds in which a metal atom is 'sandwiched' between two parallel carbocyclic ring systems are known as sandwich compounds; for example, $Fe(C_5H_5)_2$, $Cr(C_6H_6)_2$, etc. The chemistry of this special class of organometallic compounds, called 'sandwich complexes' began in 1950s, when an orange-yellow compound of composition, $Fe(C_5H_5)_2$, now known as ferrocene was made from the reaction of $FeCl_3$ and C_5H_5MgBr (cyclopentadienyl magnesium bromide) in diethyl ether (*Landmarks in Organo-Transition Metal Chemistry: A Personal View* by Helmut Werner (2008)). This new compound eluded many structural propositions and defied all the existing and conventional structural setups and bonding descriptions.

The determination of the crystal structure of Ferrocene in 1956 revealed that it contains an iron atom sandwiched between two identical planar anion $C_5H_5^-$ (cyclopentadienyl anion, abbreviated as Cp^-) species with centre of symmetry. Two structural possibilities were realized with respect to the arrangement of the two planar Cp^- rings: (i) staggered (pentagonal anti-prism) (D_{5d} symmetry in solid state) and (ii) eclipsed (pentagonal prismatic) (D_{5h} symmetry in gaseous phase) (Figure 9.1).

The eclipsed conformation of Ferrocene was further confirmed by X-ray crystallographic and electron diffraction studies in 1979. All the C–C bonds are equal and

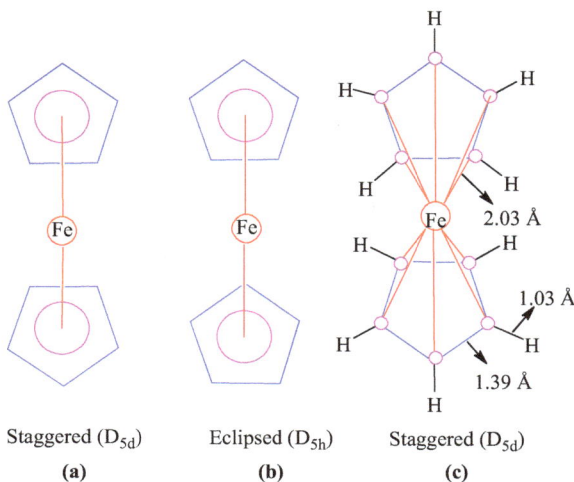

Staggered (D_{5d}) Eclipsed (D_{5h}) Staggered (D_{5d})

(a)   (b)   (c)

Figure 9.1: Structures of two possible rotamers of ferrocene.

https://doi.org/10.1515/9783110727289-009

all the carbons were involved in bonding to the metal; that is, the two planar cyclo-pentadienyl rings are bonded to the metal in pentahapto, η^5 fashion. It may, how-ever, be noted that the energy difference between the staggered and eclipsed rotamers of ferrocene is apparently small (~5 kJ mol^{-1}), and even at low tempera-tures in solution, separate conformation cannot be detected. Thus, the two different conformations in the solid state possibly arise from packing considerations.

The sandwich structure of ferrocene gave the first chemical evidence that the organic ligands could bind to metals via their π-system, particularly, planar cyclic systems (carbocyclic systems) of various ring sizes.

Figures 9.1(a) and (b) represent the common method of depicting the structures of ferrocene. In the figures, the circles within the rings express the aromatic charac-ter of the rings and lines from the iron atom indicate that the metal is bound to the whole ring rather than to the particular carbon atom of the ring. This is very clearly shown in Figure 9.1(c) that all the five C atoms of the ring are bonded to iron with $d(C–C) = 1.39$ Å; $d(C–H) = 1.03$ Å and $d(Fe–C) = 2.03$ Å.

The synthesis of several metallocenes followed soon and today there are large varieties of such sandwiches, half and full, with many polyhapto cyclic systems. Many more metallocenes are now known with staggered and eclipsed conforma-tions. For, examples, ruthenocene [(η^5-C$_5$H$_5$)$_2$Ru] and osmocene [(η^5-C$_5$H$_5$)$_2$Os] have eclipsed and [(η^5-C$_5$H$_5$)$_2$V], [η^5-C$_5$H$_5$)$_2$Cr], [(η^5-C$_5$H$_5$)$_2$Co] and [(η^5-C$_5$H$_5$)$_2$Ni] have stag-gered conformations.

Like C$_5$H$_5$-, neutral arenes like benzene, also form sandwich complexes. Since benzene is a neutral 6-electron planar cyclic π-system, one expects the most stable sandwich compound to be Cr(C$_6$H$_6$)$_2$, based on 18-electrons rule or effective atomic number (EAN) rule. This was what exactly observed and the full structural charac-terization bis(π-benzene)chromium became available. The studies confirmed that it has eclipsed conformation.

Woodward observed that these metallocenes have properties similar to hydro-carbon benzene. It has been shown chemically that the unsaturation in the cyclo-pentadienyl ring is aromatic in nature rather than olefinic. Also, there exists only one band for C–H bonds in the IR spectrum of the ferrocene. These observations clearly suggest that the iron atom is linked equally to all the carbon atoms of the cyclopentadienyl ring. For this to exist, the iron atom must lie along the fivefold axis of symmetry (C$_5$) of the cyclopentadienyl rings. Hence, the two rings are paral-lel and the iron atom is held in between these two rings, resulting into sandwich structure. The ferrocene has a centre of symmetry.

Now, the question arises that what type of bonding is there in Ferrocene/ Metallocenes and how to treat such molecules? The treatment which we have used for AB$_n$ type of molecules in Chapter VIII cannot be applied here because the carbon atoms of cyclopentadienyl ring interact strongly with each other as well as with the central atom and the valence bond (VB) description requires several res-onating structures for Ferrocene/Metallocenes. Therefore, the best way of treating such compounds by molecular orbital theory (MOT) is to work out the local MOs and their energies and then considering the interactions of the π-MOs with the valence orbitals of the metal.

The present chapter will focus on the MO treatment of sandwich compounds with special reference to ferrocene and bis(π-benzene)chromium using the concept of group theory.

9.2 Bonding in ferrocene

9.2.1 Valance bond approach

The valance bond representation of the cyclopentadienide anion is a resonance hybrid of the following energetically equivalent canonical structures (Figure 9.2):

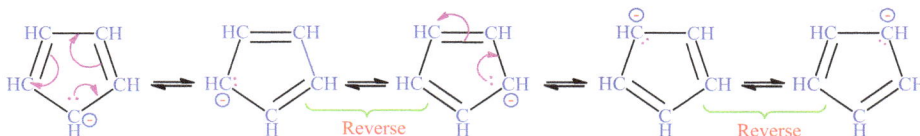

Figure 9.2: Canonical structures of cyclopentadienide anion.

The simplest approach to demonstrate the nature of bonding in ferrocene, $[(\eta^5\text{-}C_5H_5)_2Fe]$, is the VB approach. This approach is helpful to predict the molecular geometry of a complex in terms of the number and types of metal orbitals available for bonding. Based on this approach, a German worker Fritz Fischer in 1959 suggested that the bonding in ferrocene is similar to that in a low-spin octahedral complex of Fe(II), namely $[Fe(CN)_6]^{4-}$. Accordingly, the bonding in ferrocene can be best understood by considering its formation from Fe^{2+} and two cyclopentadienyl anions, C_5H_5-. Herein, each C_5H_5- donates its three pairs of π-electrons to the vacant d^2sp^3-hybrid orbitals of Fe^{2+} (Figure 9.3).

Valence bond approach has thus proved to be quite useful for predicting the octahedral geometry and magnetic properties of such η^5-complexes. However, the aromaticity of ferrocene implies a ready availability of 6π-electrons [(4n + 2), i.e. 6] in each ring, and this is inconsistent with the involvement of 6π-electrons in bonding with the metal centre.

Any suitable theory advanced for explaining the bonding in mtallocenes must be capable of explaining the following observed facts related to ferrocene:
(i)   Magnetic behaviour.
(ii) All the carbons of the two cyclopentadienyl rings are involved in bonding [pentahapto nature (η^5) of the each ring] to the metal; that is, metal is bound to the whole ring rather than to the particular carbon atom of the ring.

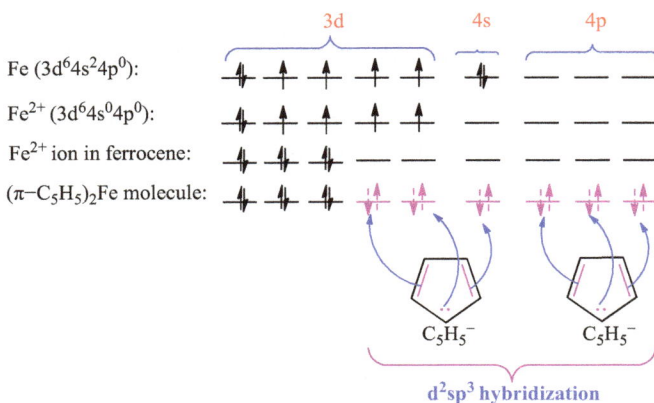

Figure 9.3: d^2sp^3 Hybridization scheme in Ferrocene.

(iii) Aromatic character of the cyclopentadienyl ring in ferrocene. For example, it undergoes Friedel–Crafts acetylation reactions.
(iv) Extra stability of the compound.

As discussed above, VB approach failed to explain the characteristics (ii) to (iv) of ferrocene/metallocenes.

9.2.2 Molecular orbital approach

The MO approach, however, gives satisfactory explanation of all the characteristics of metallocenes, the details of the various steps of this approach are as follows:

(i) Bonding in ferrocene can be best understood if we consider its formation from Fe^{2+} and the two cyclopentadienyl anion, C_5H_5- as this anion contains six π-electrons which obeys the $(4n + 2)$ rule of aromaticity and not the radical $[C_5H_5-]$. The carbon atoms in C_5H_5- ring are sp^2 hybridized, and sp^2 hybridized orbitals are completely utilized in formation of C–C (with two neighbouring carbons) and C–H σ-bonds (Figure 9.4). Each carbon atom of the ring now contains an unused p_z-orbital (supposing the rings lie in xy plane and the line passing through the metal atom and the centres of the two parallel rings is the z-axis).

(ii) The next step in working out the MO treatment of ferrocene is to develop first the possible delocalized π-molecular orbitals (π_D MOs) for C_5H_5- ligand(s), that is, the ligand group orbitals (LGOs), and then select the metal orbitals which have the correct symmetry and energies to overlap with LGOs for effective bonding.

 C_5H_5- ion contains delocalized π-structure. All the 5C and 5H are co-planar, with the same C–C bond distance of 1.39 Å which is intermediate between

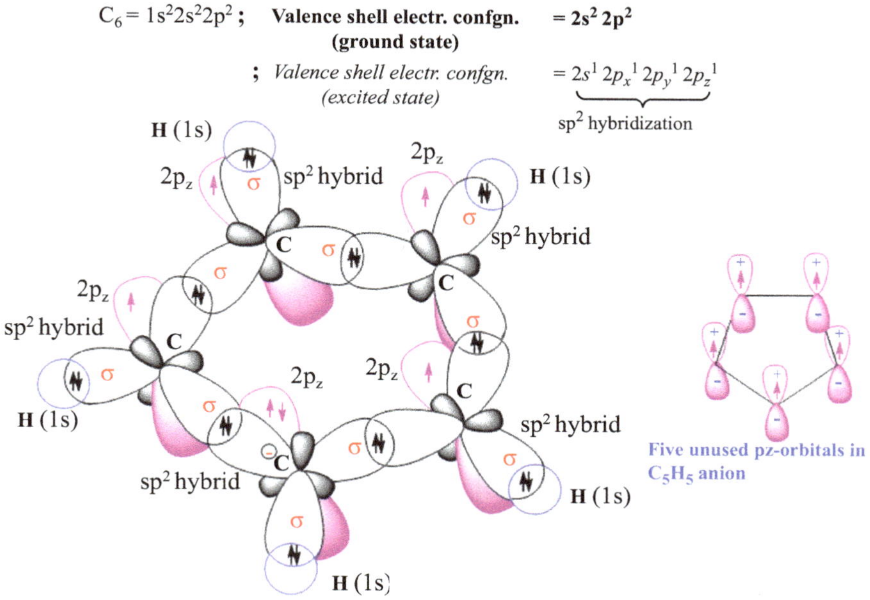

$C_6 = 1s^2 2s^2 2p^2$;   **Valence shell electr. confgn.** $= 2s^2\, 2p^2$
(ground state)

;   *Valence shell electr. confgn.* $= 2s^1\, 2p_x{}^1\, 2p_y{}^1\, 2p_z{}^1$
(excited state)

sp² hybridization

Figure 9.4: sp² hybrid orbitals overlap and formation of 5(C–C) and 5(C–H) σ-bonds in C_5H_5.

ethylene (1.33 Å) and ethane (1.54 Å), and is similar to that in benzene. The $5p_z$ orbitals (containing in all 6e) of each ring, extending above and below the plane of the ring combine linearly in five ways (Figure 9.5) to give five delocalized π- molecular orbitals (π_D MOs), three bonding and two anti-bonding.

(iii) The anion $C_5H_5{}^-$ belongs to D_{5h} symmetry point group with symmetry operations: E, $2C_5$, $2C_5{}^2$, $5C_2$, σ_h, $2S^5$, $2S_5{}^3$ and $5\sigma_v$. Considering the five ligand $p_z\pi$-orbitals as the basis, the symmetry operations of the D_{5h} point group are performed on the $C_5H_5{}^-$ $p_z\pi$-orbitals, and now using the concept of group theory, one can get the number and nature of symmetries of the resulting π_DMOs shown above. Accordingly, the symmetries of the resulting π_DMOs are:

$$\Gamma\pi = A_2{}'' + E_1{}'' + E_2{}''$$

Thus, $5\pi_D$MOs/5π-LGOs are to be formed: one singly degenerate of A_2'' symmetry, and the rest two sets doubly degenerate of E_1'' and E_2'' symmetry.

(iv) The formulation of $5\pi_D$MOs/5π-LGOs can be done out using the concept applied in carbocyclic system $(CH)_n$. Accordingly, for a $(CH)_n$ carbocyclic system, all the essential symmetry properties of the πLGOs are determined by the operations of the uni-axial rotational sub-group, C_n. Hence, in the present case one can formulate the πLGOs by using the character table (Table 9.1) of rotational point group C_5 given below.

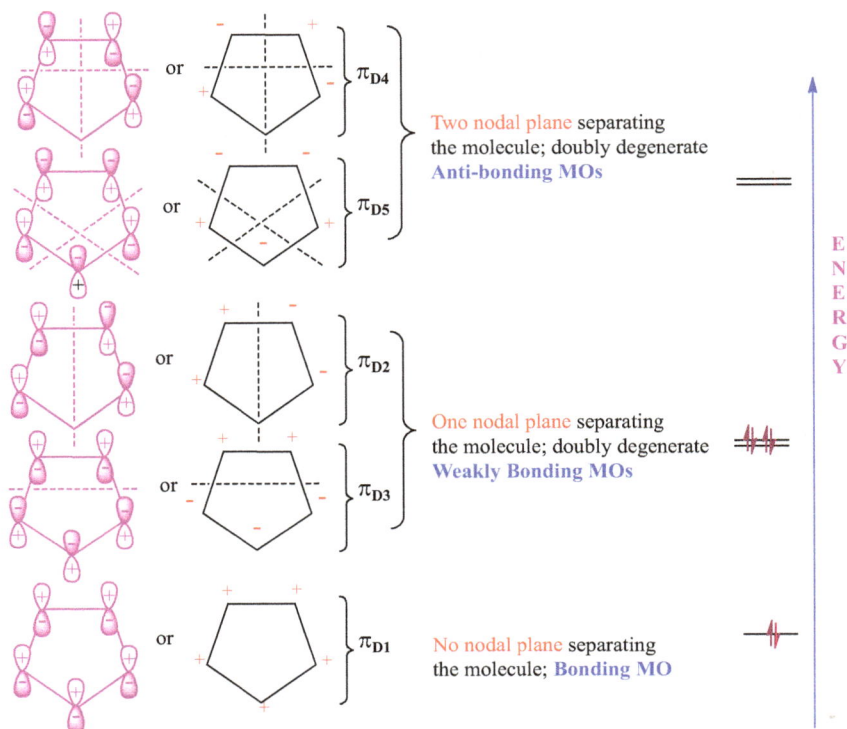

Two nodal plane separating the molecule; doubly degenerate **Anti-bonding MOs**

π_{D4}

π_{D5}

One nodal plane separating the molecule; doubly degenerate **Weakly Bonding MOs**

π_{D2}

π_{D3}

No nodal plane separating the molecule; **Bonding MO**

π_{D1}

E N E R G Y

Figure 9.5: Formation of five delocalized π MOs in C_5H_5- anion, three of bonding and two of antibonding in nature.

Table 9.1: Character table of C_5 point symmetry.

C_5	E	C_5	C_5^2	C_5^3	C_5^4
A	1	1	1	1	1
E_1	1	ε	ε^2	ε^{2*}	ε^*
	1	ε^*	ε^{2*}	ε^2	ε
E_2	1	ε^2	ε^*	ε	ε^{2*}
	1	ε^{2*}	ε	ε^*	ε^2

The formulation of 5πDMOs/5LGOs of various symmetries A_2'', E_1'' and E_2'' [or simply of symmetry A, E_1 and E_2 on account of rotational subgroup C_5 of D_{5h} point group] can be done just by inspecting the C_5 character table. Let ϕ_1, ϕ_2, ϕ_3, ϕ_4 and ϕ_5 be the wave functions associated with the p_z-orbitals on the carbon atoms. Then the formulations of 5πDMOs/5LGOs of the said symmetries in terms of wave functions are as follows:

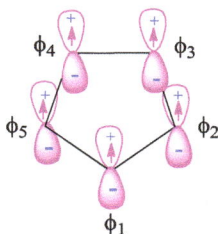

$$\psi A \quad = (\phi_1 + \phi_2 + \phi_3 + \phi_4 + \phi_5)$$

$$\psi E_1(1) = (\phi_1 + \varepsilon\phi_2 + \varepsilon^2\phi_3 + \varepsilon^2{}^*\phi_4 + \varepsilon^*\phi_5)$$

$$\psi E_1(2) = (\phi_1 + \varepsilon^*\phi_2 + \varepsilon^2{}^*\phi_3 + \varepsilon^2\phi_4 + \varepsilon\phi_5)$$

$$\psi E_2(1) = (\phi_1 + \varepsilon^2\phi_2 + \varepsilon^*\phi_3 + \varepsilon\phi_4 + \varepsilon^2{}^*\phi_5)$$

$$\psi E_2(2) = (\phi_1 + \varepsilon^2{}^*\phi_2 + \varepsilon\phi_3 + \varepsilon^*\phi_4 + \varepsilon^2\phi_5)$$

Where $\varepsilon = e^{2\pi i/5}$ and $\varepsilon^* = e^{-2\pi i/5}$

The next step is elimination of imaginary terms and normalization.

The LGO, A can be easily normalized to get

$$\psi A = 1/\sqrt{5}\,(\phi_1 + \phi_2 + \phi_3 + \phi_4 + \phi_5)$$

The $E_1(1)$ and $E_1(2)$ set of functions have imaginary coefficients. In order to have real, rather than complex coefficients, **the two components are added term by term**. To eliminate the complex number using **De Moivre's theorem**, we write

$$\varepsilon = \cos(2\pi/5) + i\sin(2\pi/5) \text{ and}$$

$$\varepsilon^* = \cos(2\pi/5) - i\sin(2\pi/5)$$

Then,

$$\varepsilon + \varepsilon^* = 2\cos(2\pi/5)$$

$$\varepsilon - \varepsilon^* = 2i\sin(2\pi/5)$$

$$\varepsilon^2 = \cos(4\pi/5) + i\sin(4\pi/5)$$

$$\varepsilon^{2*} = \cos(4\pi/5) - i\sin(4\pi/5)$$

$$\varepsilon^2 + \varepsilon^{2*} = 2\cos(4\pi/5)$$

$$\varepsilon^2 - \varepsilon^{2*} = 2i\sin(4\pi/5)$$

$$E_1(1) + E_1(2) = 2\phi_1 + (\varepsilon + \varepsilon^\star)\phi_2 + (\varepsilon^2 + \varepsilon^{2\star})\phi_3 + (\varepsilon^{2\star} + \varepsilon^2)\phi_4 + (\varepsilon^\star + \varepsilon)\phi_5$$

$$= 2\phi_1 + \phi_2 2\cos(2\pi/5) + \phi_3 2\cos(4\pi/5) + \phi_4 2\cos(4\pi/5) + \phi_5 2\cos(2\pi/5)$$

or $\psi E_1(a)$ $= 2[\phi_1 + \phi_2\cos(2\pi/5) + \phi_3\cos(4\pi/5) + \phi_4\cos(4\pi/5) + \phi_5\cos(2\pi/5)]$

The normalized wave function now becomes

$$\psi E_1(a) = \sqrt{2/5}[\phi_1 + \phi_2\cos(2\pi/5) + \phi_3\cos(4\pi/5) + \phi_4\cos(4\pi/5) + \phi_5\cos(2\pi/5)]$$

Now $E_1(1)$ and $E_1(2)$ are subtracted,

$$E_1(1) - E_1(2) = 0\phi_1 + (\varepsilon - \varepsilon^\star)\phi_2 + (\varepsilon^2 - \varepsilon^{2\star})\phi_3 + (\varepsilon^{2\star} - \varepsilon^2)\phi_4 + (\varepsilon^\star - \varepsilon)\phi_5$$

or $\quad E_1(1) - E_1(2) = 0\phi_1 + \phi_2 2i\sin(2\pi/5) + \phi_3 2i\sin(4\pi/5) - \phi_4 2i\sin(4\pi/5)$

$$- \phi_5 2i\sin(2\pi/5)$$

Dividing both sides by i, we get

$$E_1(1) - E_1(2)/i = 0\phi_1 + \phi_2 2\sin(2\pi/5) + \phi_3 2\sin(4\pi/5) - \phi_4 2\sin(4\pi/5)$$

$$- \phi_5 2\sin(2\pi/5)$$

$$= 2[0\phi_1 + \phi_2\sin(2\pi/5) + \phi_3\sin(4\pi/5) - \phi_4\sin(4\pi/5) - \phi_5\sin(2\pi/5)]$$

The normalized wave function now becomes

$$\psi E_1(b) = \sqrt{2/5}\,[0\phi_1 + \phi_2\sin(2\pi/5) + \phi_3\sin(4\pi/5) - \phi_4\sin(4\pi/5) - \phi_5\sin(2\pi/5)]$$

Similarly,

$$E_2(1) + E_2(2) = 2\phi_1 + \phi_2(\varepsilon^2 + \varepsilon^{2\star}) + \phi_3(\varepsilon^\star + \varepsilon) + \phi_4(\varepsilon + \varepsilon^\star) + \phi_5(\varepsilon^{2\star} + \varepsilon^2)$$

or $E_2(1) + E_2(2) = 2[\phi_1 + \phi_2\cos(4\pi/5) + \phi_3\cos(2\pi/5) + \phi_4\cos(2\pi/5) + \phi_5\cos(4\pi/5)]$

The normalized wave function now becomes

$$\psi E_2(a) = \sqrt{2/5}\,[\phi_1 + \phi_2\cos(4\pi/5) + \phi_3\cos(2\pi/5) + \phi_4\cos(2\pi/5) + \phi_5\cos(4\pi/5)]$$

Now $E_2(1)$ and $E_2(2)$ are subtracted,

$$E_2(1) - E_2(2) = 0\phi_1 + \phi_2(\varepsilon^2 - \varepsilon^{2\star}) + \phi_3(\varepsilon^\star - \varepsilon) + \phi_4(\varepsilon - \varepsilon^\star) + \phi_5(\varepsilon^{2\star} - \varepsilon^2)$$

$$= 0\phi_1 + \phi_2 2i\sin(4\pi/5) - \phi_3 2i\sin(2\pi/5) + \phi_4 2i\sin(2\pi/5)$$

$$- \phi_5 2i\sin(4\pi/5)$$

Dividing both sides by i, we get

$$E_2(1) - E_2(2)/i = 0\phi_1 + \phi_2 2\sin(4\pi/5) - \phi_3 2\sin(2\pi/5) + \phi_4 2\sin(2\pi/5)$$

$$- \phi_5 2\sin(4\pi/5)$$

$$= 2[0\phi_1 + \phi_2\sin(4\pi/5) - \phi_3\sin(2\pi/5) + \phi_4\sin(2\pi/5) - \phi_5\sin(4\pi/5)]$$

The normalized wave function now becomes

$$\psi E_2(b) = \sqrt{2}/5[0\phi_1 + \phi_2 2\sin(4\pi/5) - \phi_3 2\sin(2\pi/5) + \phi_4 2\sin(2\pi/5) - \phi_5 2\sin(4\pi/5)]$$

Taking $\omega = 2\pi/5 = 72°$, all the 5πDMOs/5LGOs in normalized form will be

$$\psi A \quad = 1/\sqrt{5}\,(\phi_1 + \phi_2 + \phi_3 + \phi_4 + \phi_5)$$

$$\psi E_1(a) = \sqrt{2}/5[\phi_1 + \phi_2\cos\omega + \phi_3\cos2\omega + \phi_4\cos2\omega + \phi_5\cos\omega]$$

$$\psi E_1(b) = \sqrt{2}/5[0\phi_1 + \phi_2\sin\omega + \phi_3\sin2\omega - \phi_4\sin2\omega - \phi_5\sin\omega]$$

$$\psi E_2(a) = \sqrt{2}/5[\phi_1 + \phi_2\cos2\omega + \phi_3\cos\omega + \phi_4\cos\omega + \phi_5\cos2\omega]$$

$$\psi E_2(b) = \sqrt{2}/5[0\phi_1 + \phi_2\sin2\omega - \phi_3\sin\omega + \phi_4\sin\omega - \phi_5\sin2\omega]$$

9.2.2.1 Eclipsed form of ferrocene

Let us consider the eclipsed form of ferrocene (Figure 9.6) belonging to D_{5h} point group with symmetry operations – E, $2C_5$, $2C_5^2$, $5C_2$, σ_h, $2S^5$, $2S_5^3$ and $5\sigma_v$. In this molecule, two C_5H_5- rings are present (Fig.). Hence, ten $p\pi$-orbitals are involved in bonding.

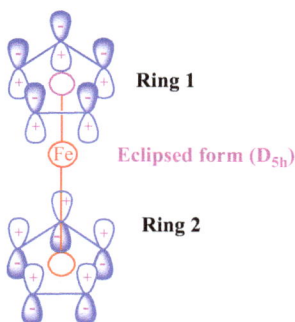

Ring 1

Eclipsed form (D_{5h})

Ring 2

Figure 9.6: Eclipsed form of Ferrocene.

In order to determine the number and nature of symmetries of the LGOs formed by the combination of 10 pπ-orbitals, these 10 pπ-orbitals are subjected to the symmetry operations of the D_{5h} point group. Without going to detailed group theoretical calculations, symmetries of the LGOs are presented below:

$$\Gamma_\pi = A_1' + E_1' + E_2' + A_2'' + E_1'' + E_2''$$

$$= A_1' + A_2'' + E_1' + E_1'' + E_2' + E_2''$$

This gives the symmetry characteristics of the 10 LGOs formed by the combination of 10 pπ-orbitals. Of these, two are non-degenerate LGOs of A_1' and A_2'' (A orbitals), two sets of doubly degenerate LGOs of E_1' and E_1'' (E_1 orbitals) and two sets of doubly degenerate LGOs of E_2' and E_2'' (E_2 orbitals) symmetries.

The next step is to work out the wave functions cum composition of 10 LGOs resulting from $2C_5H_{5-}$ rings of the aforesaid symmetries. These are worked out as follows:

An observation of the character table of D_{5h} point group (Table 9.2) given below shows that, A_1', E_1', E_2' are symmetric and A_2'', E_1'' and E_2'' are anti-symmetric with respect to horizontal plane (σ_h). Hence,

Table 9.2: Character table of D_{5h} point group.

D_{5h}	E	$2C_5$	$2C_5{}^2$	$5C_2$	σ_h	$2S_5$	$2S_5{}^3$	$5\sigma_v$		
A_1'	1	1	1	1	1	1	1	1		(x^2+y^2, z^2)
A_2'	1	1	1	−1	1	1	1	−1	R_z	
E_1'	2	$2\cos 72°$	$2\cos 144°$	0	2	$2\cos 72°$	$2\cos 144°$	0	(x, y)	
E_2'	2	$2\cos 144°$	$2\cos 72°$	0	2	$2\cos 144°$	$2\cos 72°$	0		(x^2-y^2, xy)
A_1''	1	1	1	1	−1	−1	−1	−1		
A_2''	1	1	1	−1	−1	−1	−1	1	z	
E_1''	2	$2\cos 72°$	$2\cos 144°$	0	−2	$-2\cos 72°$	$-2\cos 144°$	0	(R_x, R_y)	(xz, yz)
E_2''	2	$2\cos 144°$	$2\cos 72°$	0	−2	$-2\cos 144°$	$-2\cos 72°$	0		

(i) The LGOs of the two individual rings of symmetry 'A' combined linearly to give wave functions of LGO of A_1' and A_2'' symmetries of the ligand ($2C_5H_{5-}$ rings) in eclipsed form of Ferrocene. This is pictorially shown in Figure 9.7.

(ii) The LGOs of the two individual rings of symmetry 'E_1' combined linearly to give wave functions of LGO of E_1' and E_1'' symmetries of the ligand ($2C_5H_{5-}$ or 2CP) rings in ferrocene (Figure 9.8).

(iii) The LGOs of the two individual rings of symmetry 'E_2' combined linearly as follows to give wave functions of LGO of E_2' and E_2'' symmetries of the ligand ($2Cp^-$ rings) in ferrocene (Figure 9.9).

$$\psi(A_1') = \frac{1}{\sqrt{2}} [\psi_1(A) + \psi_2(A)] \ (1) \qquad \psi(A_2'') = \frac{1}{\sqrt{2}} [\psi_1(A) - \psi_2(A)] \ (2$$

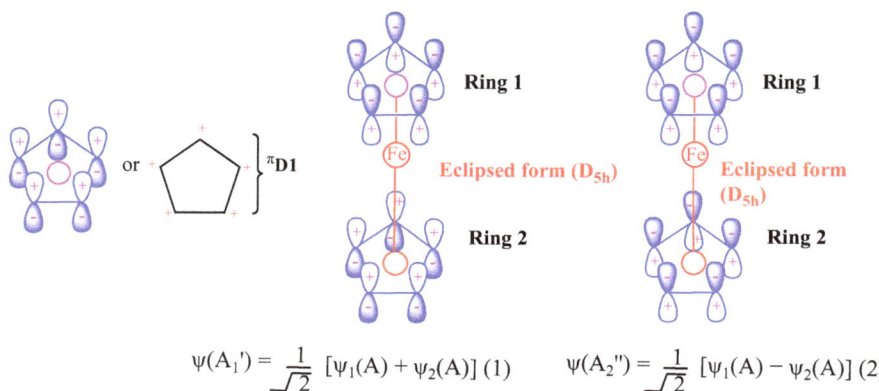

Figure 9.7: Wave functions of ligand group orbitals (LGO) of A_1' and A_2'' symmetries of the ligand in eclipsed form of Ferrocene.

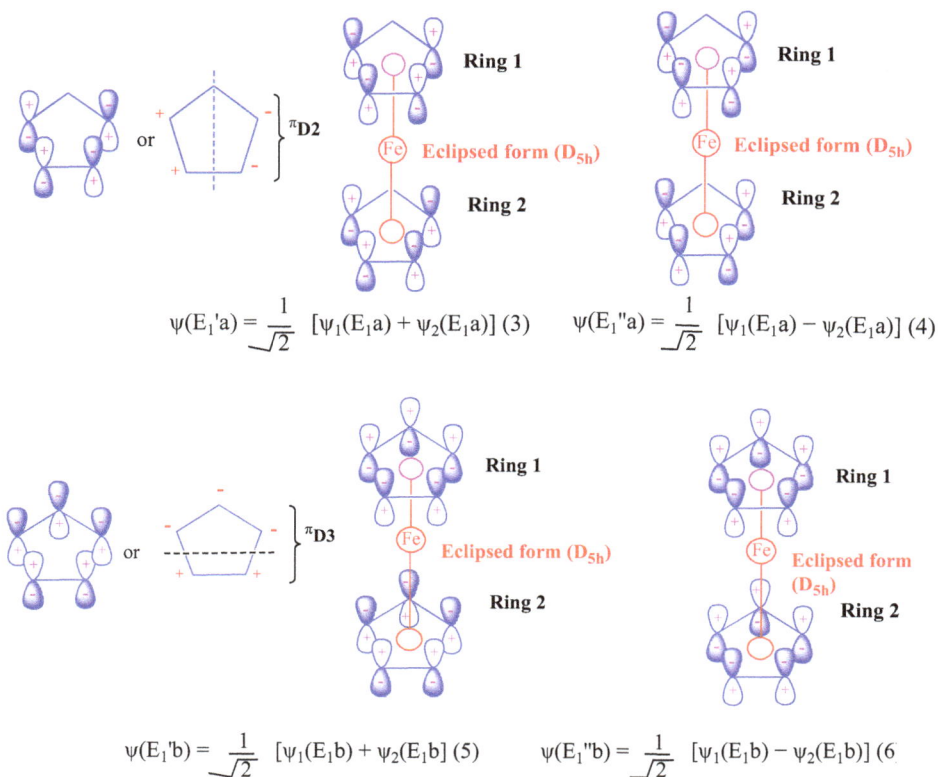

$$\psi(E_1'a) = \frac{1}{\sqrt{2}} [\psi_1(E_1a) + \psi_2(E_1a)] \ (3) \qquad \psi(E_1''a) = \frac{1}{\sqrt{2}} [\psi_1(E_1a) - \psi_2(E_1a)] \ (4)$$

$$\psi(E_1'b) = \frac{1}{\sqrt{2}} [\psi_1(E_1b) + \psi_2(E_1b)] \ (5) \qquad \psi(E_1''b) = \frac{1}{\sqrt{2}} [\psi_1(E_1b) - \psi_2(E_1b)] \ (6$$

Figure 9.8: Wave functions of LGO of E_1' and E_1'' symmetries of the ligand in eclipsed form of Ferrocene.

$$\psi(E_2\text{'a}) = \frac{1}{\sqrt{2}} \ [\psi_1(E_2a) + \psi_2(E_2a)] \ (7) \qquad \psi(E_2\text{"a}) = \frac{1}{\sqrt{2}} \ [\psi_1(E_2a) - \psi_2(E_2a)] \ (8)$$

$$\psi(E_2\text{'b}) = \frac{1}{\sqrt{2}} \ [\psi_1(E_2b + \psi_2(E_2b)] \ (9) \qquad \psi(E_2\text{"b}) = \frac{1}{\sqrt{2}} \ [\psi_1(E_2b) - \psi_2(E_2b)] \ (10)$$

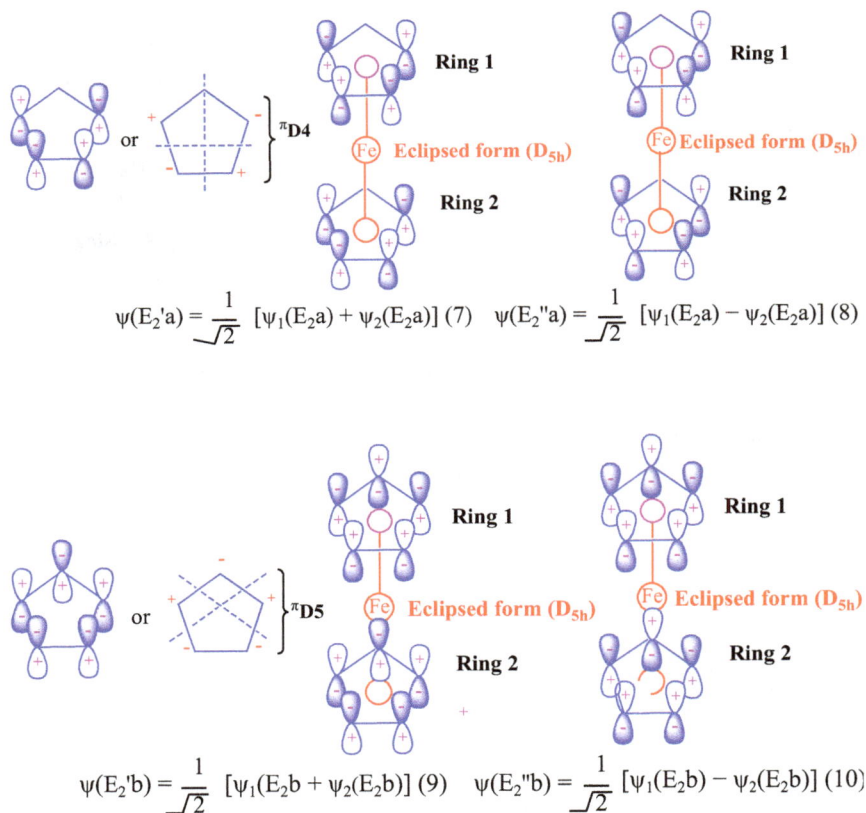

Figure 9.9: Wave functions of LGO of E_2' and E_2'' symmetries of the ligand in eclipsed form of Ferrocene.

(iv) The next step is to work out the matching symmetry orbitals of the metal atom for overlapping with 10 LGOs to form respective MOs. The orbitals available on the metal atom are five 3d-obitals, one s-orbital and three 4p-orbitals. From the character table for D_{5h} point grou of eclipsed form of ferrocene, it is easy to find out the symmetry species to which they belong.

Character table of D_{5h} point group

D_{5h}	E	$2C_5$	$2C_5^2$	$5C_2$	σ_h	$2S_5$	$2S_5^3$	$5\sigma_v$		
A_1'	1	1	1	1	1	1	1	1		$(x^2 + y^2, z^2)$
A_2'	1	1	1	−1	1	1	1	−1	R_z	
E_1'	2	$2\cos 72°$	$2\cos 144°$	0	2	$2\cos 72°$	$2\cos 144°$	0	(x, y)	
E_2'	2	$2\cos 144°$	$2\cos 72°$	0	2	$2\cos 144°$	$2\cos 72°$	0		$(x^2 - y^2, xy)$

A_1''	1	1	1	1	-1	-1		-1	-1	
A_2''	1	1	1	-1	-1	-1		-1	1	z
E_1''	2	$2\cos 72°$	$2\cos 144°$	0	-2	$-2\cos 72°$	$-2\cos 144°$	0	(R_x, R_y)	(xz, yz)
E_2''	2	$2\cos 144°$	$2\cos 72°$	0	-2	$-2\cos 144°$	$-2\cos 72°$	0		

These are:

$$A_1': 4s, 3d_z{}^2;\ E_1': p_x, p_y;\ E_1'': d_{xz}, d_{yz};\ E_2': d_{x^2-y^2}, d_{xy};\ E_2'': Nil;\ A_2'': p_z$$

Thus, the matching symmetry orbitals are available for all the LGOs on the metal atom except for E_2'', and hence they will remain as no-bonding.

The central metal AOs and the matching LGOs now combine in phase (added) giving bonding MOs, and out of phase (with opposite sign) to give anti-bonding MOs. These are tabulated in Table 9.3.

Table 9.3: Formation of bonding and anti-bonding MOs in eclipsed for of Ferrocene.

Symmetry	Bonding MO	Non-bonding MO	Anti-bonding MO
A_1'	$\psi_s + C_1\psi(A_1')$ $\psi_{dz}{}^2 + C_1\psi(A_1')$		$\psi_s - C_1\psi(A_1')$ $\psi_{dz}{}^2 - C_1\psi(A_1')$
A_2''	$\psi_{pz} + C_2\psi(A_2'')$		$\psi_{pz} - C_2\psi(A_2'')$
E_1'	$\psi_{px} + C_3\psi(E_1')$ $\psi_{py} + C_3\psi(E_1')$		$\psi_{px} - C_3\psi(E_1')$ $\psi_{py} - C_3\psi(E_1')$
E_1''	$\psi_{dxz} + C_4\psi(E_1'')$ $\psi_{dyz} + C_4\psi(E_1'')$		$\psi_{dxz} - C_4\psi(E_1'')$ $\psi_{dyz} - C_4\psi(E_1'')$
E_2'	$\psi_{dx^2-z^2} + C_5\psi(E_2')$ $\psi_{dxy} + C_5\psi(E_2')$		$\psi_{dx^2-z^2} - C_5\psi(E_2')$ $\psi_{dxy} - C_5\psi(E_2')$
E_2''		$\psi(E_2''a)$ $\psi(E_2''b)$	

Here the notations C_1, C_2, C_3, C_4 and C_5 represent the ratio of the coefficients of the central atom and the respective LGO.

Based on the above results, the simplified molecular orbital (MO) energy level diagram of eclipsed form of the Ferrocene is shown in Figure 9.10 that gives full information of bonding.

A careful observation of MO energy level diagrams of eclipsed ferrocene reveals that the ligand group orbitals (LGOs) can be categorized in three sets: (i) a filled pair of A_1' and A_2'' (ii) a higher energy filled doubly degenerate set E_1' and E_1'' and

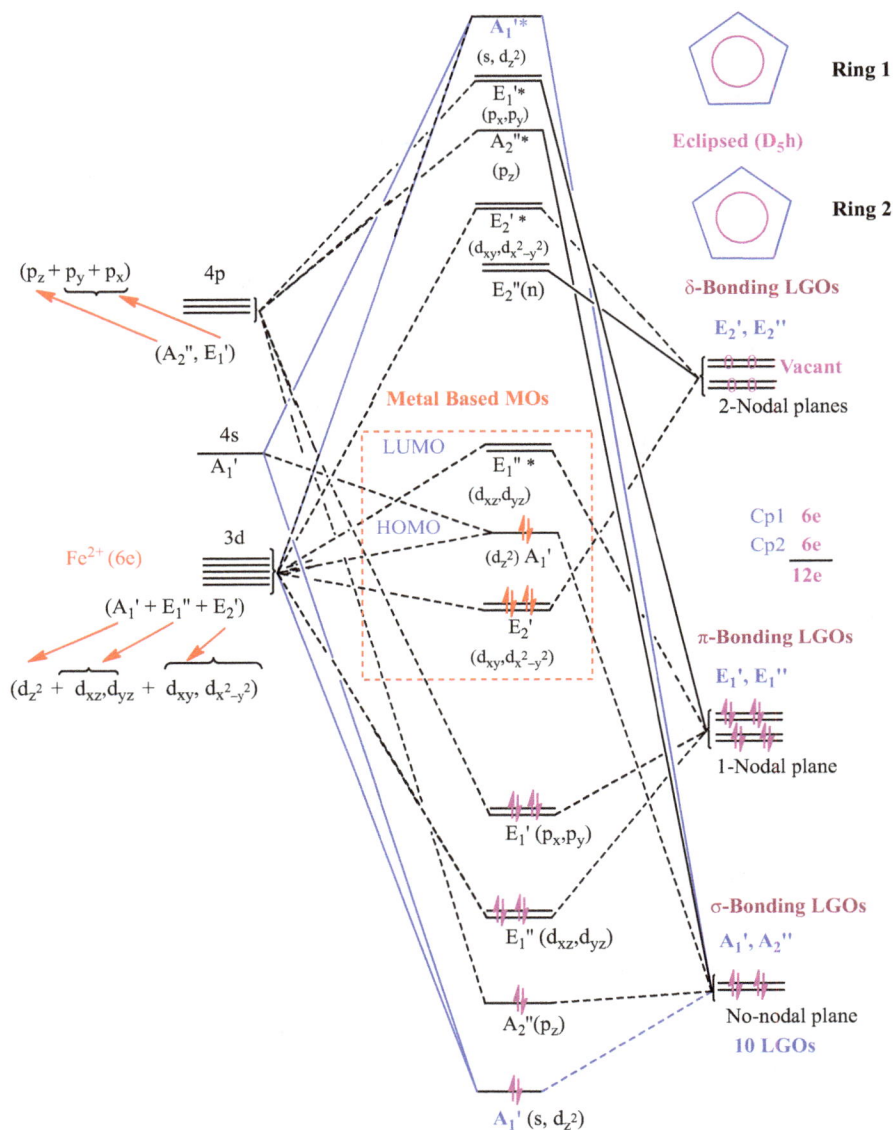

Figure 9.10: MO diagram of eclipsed conformation of ferrocene (D_{5h}).

(iii) still higher energy vacant doubly degenerate set of E_2' and E_2''. The relative energies of these three sets of LGOs are based on the presence of nodal planes in them.

The σ-bonding A_1' LGO of Cp_2 rings in eclipsed form is of lower energy relative to the atomic orbitals, 4s and $3d_{z^2}$ of iron of the same symmetry (A_1'), and hence it interacts less effectively producing weakly σ-bonding A_1' (4s, $3d_{z^2}$) MO essentially localized on the Cp_2 rings. Contrary to this, δ bonding E_2' LGOs of the rings are so high in energy relative to the same symmetry metal atomic orbitals, d_{xy} and $d_{x^2-y^2}$ that they overlap very little with these atomic orbitals leads to slightly δ-bonding MOs essentially localized on the metal d-orbitals. This δ-type of bonding represents certain amount of back donation of electrons from the filled metal orbitals to the vacant ring orbitals of the Cp_2 rings. Similarly, 4p-orbitals ($E_{1u} + A_{2u}$ symmetry) on the iron are at a very high energy, and so the E_1' and A_2'' MOs don't contribute much to the bonding and essentially localized on rings.

The only orbitals that are well matched are the π-bonding E_1'' ligand group orbitals (LGO) and the corresponding 3d-orbitals (d_{xz}, d_{yz}) which form two strong π-bonds. One such π-bond formation is pictorially shown in Figure 9.11. These π-bonds are supposed to supply most of the stabilization that holds the ferrocene molecule together.

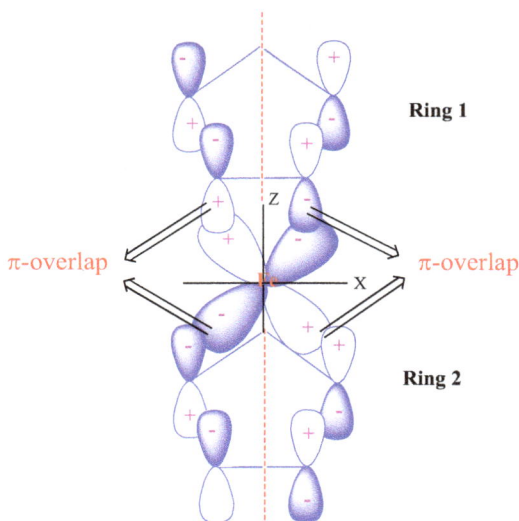

$$\psi(E_1''a) = \frac{1}{\sqrt{2}} [\psi_1(E_1a) - \psi_2(E_1a)] \ (4)$$

Figure 9.11: Formation of two strong π-bonds stabilizing the ferrocene molecule.

As can be seen in the MO diagram, $A_1'(d_z^2)$ MO is non-bonding in nature because energy of this MO is raised relative to metal d_z^2 orbital.

If electrons are supplied to fill all the bonding MOs but none of the anti-bonding molecular orbitals, 18 electrons will be required. Thus, we see that 18 electron rule is a reflection of filling strongly stabilized MOs. The ferrocene [an 18e system; $Fe^{2+}(6e) + 2Cp^-(12e) = 18e$] in which all the nine lowest energy bonding MOs are completely filled is thus the most stable metallocene.

The stability of $[Fe(\eta^5\text{-}C_5H_5)_2]$ with 18 electrons compared to $[Co(\eta^5\text{-}C_5H_5)_2]$ with 19 electrons and the twenty electron system $[Ni(\eta^5\text{-}C_5H_5)_2]$ is readily interpreted on the basis of this bonding scheme since these two latter species have 1 and 2 easily oxidizable electron(s) in anti-bonding $E_1'''^*$ MOs, whereas ferrocene has no electron in $E_1'''^*$ MOs. Similarly, $[Cr(\eta^5\text{-}C_5H_5)_2]$ with 16 electrons and vanadocene with 15 electrons have incompletely filled bonding MOs, and are highly reactive. So, these systems can easily get reduced by accepting 2 and 3 electrons, respectively.

It is notable here that the MOs are *delocalized* over the iron and both the rings in ferrocene. Thus, the linking of iron with both the rings is not *localized*. This accounts for the single υ(C-H) in IR spectrum and the aromatic character of the molecule.

Lastly, HOMO and LUMO in this diagram have shown to have essentially d-character. It is notable that HOMO in eclipsed form of ferrocene is $A_1'(d_z^2)$ and is doubly occupied while the LUMO is $E_1'''^*$ (d_{xz}, d_{yz}) and empty.

It is remarkable to note that the attachment of additional groups or ligands destroys the D_{5d} symmetry of simple metallocene and this modifies the MO diagram. This also happens when ferrocene is protonated to give $[Fe(\eta^5\text{-}C_5H_5)_2H]^+$, and when the isoelectronic neutral molecules, namely $[Re((\eta^5\text{-}C_5H_5)_2H]$ and $[Mo((\eta^5\text{-}C_5H_5)_2H_2]$ are considered.

9.2.2.2 Staggered form of ferrocene
The point group of staggered form of ferrocene (Figure 9.12) is D_{5d} with symmetry operations: E, $2C_5$, $2C_5^2$, $5C_2$, i, $2S_{10}$, $2S_{10}^3$, $5\sigma_d$. In the staggered form the pπ-orbitals are oriented as shown in the Fig. The nature of the MOs formed by the 10 pπ-orbitals will be different from that formed in the eclipsed form.

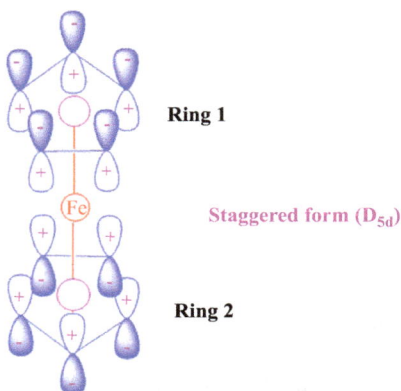

Ring 1

Staggered form (D_{5d})

Ring 2

Figure 9.12: Staggered form of Ferrocene.

For determining the number and nature of symmetries of the LGOs formed by the combination of 10 pπ-orbitals, these pπ-orbitals are subjected to the symmetry operations of the D_{5d} point group. Without going to detailed group theoretical calculations, symmetries of the LGOs are given here:

$$\Gamma\pi = A_{1g} + A_{2u} + E_{1g} + E_{1u} + E_{2g} + E_{2u}$$

This gives the symmetry characteristics of the 10 LGOs formed by the combination of 10 pπ-orbitals. Of these, two are non-degenerate LGOs of A_{1g} and A_{2u} (A orbitals), two sets of doubly degenerate LGOs of E_{1g} and E_{1u} (E_1 orbitals) and two sets of doubly degenerate LGOs of E_{2g} and E_{2u} (E_2 orbitals) symmetries.

The next step is to work out the wave functions cum composition of 10 LGOs resulting from 2Cp⁻ rings of the aforesaid symmetries. These are worked out as follows:

As evidenced from the character table (Table 9.4) of D_{5d} point group given below, A_{1g}, E_{1g}, E_{2g} are symmetric and A_{2u}, E_{1u} and E_{2u} are anti-symmetric with respect to inversion centre.

Table 9.4: Character table of D_{5d} point group.

D_{5d}	E	$2C_5$	$2C_5^2$	$5C_2$	i	$2S_{10}^3$	$2S_{10}$	$5\sigma_d$		
A_{1g}	1	1	1	1	1	1	1	1		(x^2+y^2, z^2)
A_{2g}	1	1	1	−1	1	1	1	−1	R_z	
E_{1g}	2	2 cos 72°	2 cos 144°	0	2	2 cos 72°	2 cos 144°	0	(R_x, R_y)	(xz, yz)
E_{2g}	2	2 cos 144°	2 cos 72°	0	2	2 cos 144°	2 cos 72°	0		(x^2-y^2, xy)
A_{1u}	1	1	1	1	−1	−1	−1	−1		
A_{2u}	1	1	1	−1	−1	−1	−1	1	z	
E_{1u}	2	2 cos 72°	2 cos 144°	0	−2	−2 cos 72°	−2 cos 144°	0	(x, y)	
E_{2u}	2	2 cos 144°	2 cos 72°	0	−2	−2 cos 144°	−2 cos 72°	0		

Hence,
(i) The LGOs of the two individual rings of symmetry 'A' combined linearly as follows to give wave functions of LGO of A_{1g} and A_{2u} symmetries of the ligand (2Cp⁻ rings) in ferrocene (Figure 9.13).

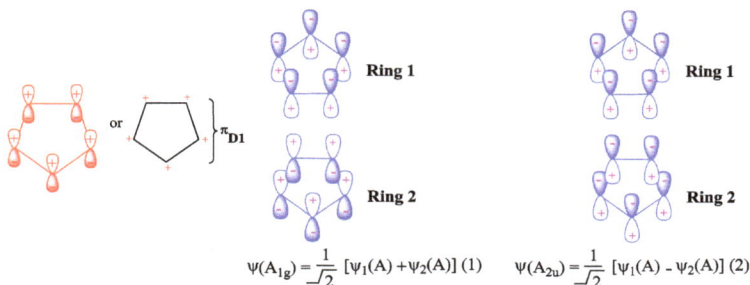

$$\psi(A_{1g}) = \frac{1}{\sqrt{2}}[\psi_1(A) + \psi_2(A)] \ (1) \qquad \psi(A_{2u}) = \frac{1}{\sqrt{2}}[\psi_1(A) - \psi_2(A)] \ (2)$$

Figure 9.13: Wave functions of LGO of A_{1g} and A_{2u} symmetries of the ligand in staggered form of Ferrocene.

(ii) The LGOs of the two individual rings of symmetry 'E$_1$' combined linearly as follows to give wave functions of LGO of E$_{1g}$ and E$_{1u}$ symmetries of the ligand (2Cp rings) in ferrocene (Figure 9.14).

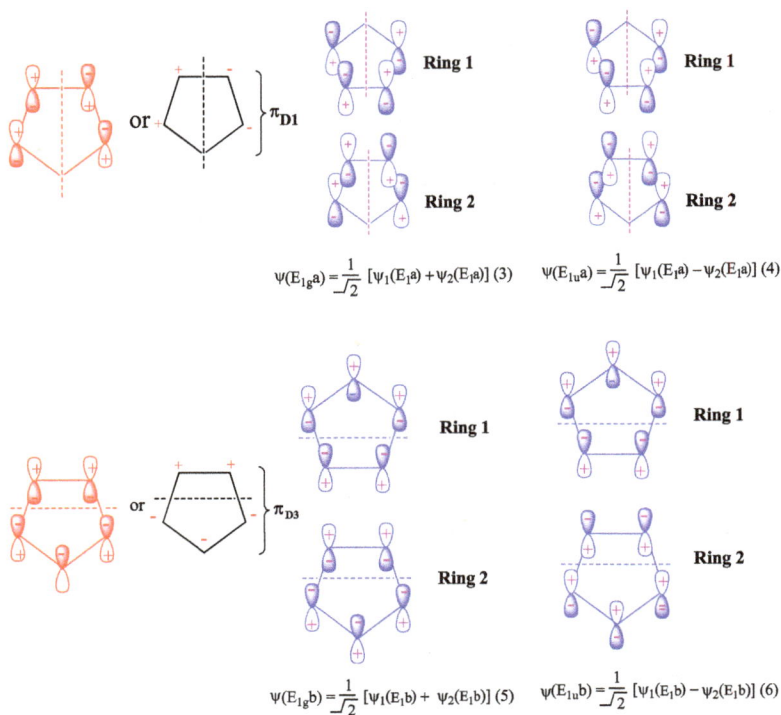

$$\psi(E_{1g}a) = \frac{1}{\sqrt{2}} [\psi_1(E_1a) + \psi_2(E_1a)] \ (3) \qquad \psi(E_{1u}a) = \frac{1}{\sqrt{2}} [\psi_1(E_1a) - \psi_2(E_1a)] \ (4)$$

$$\psi(E_{1g}b) = \frac{1}{\sqrt{2}} [\psi_1(E_1b) + \psi_2(E_1b)] \ (5) \qquad \psi(E_{1u}b) = \frac{1}{\sqrt{2}} [\psi_1(E_1b) - \psi_2(E_1b)] \ (6)$$

Figure 9.14: Wave functions of LGO of E$_{1g}$ and E$_{1u}$ symmetries of the ligand in staggered form of Ferrocene.

(iii) The LGOs of the two individual rings of symmetry 'E$_2$' combined linearly as follows to give wave functions of LGO of E$_{2g}$ and E$_{2u}$ symmetries of the ligand (2Cp rings) in ferrocene (Figure 9.15).

$$\psi(E_{2g}a) = \frac{1}{\sqrt{2}} [\psi_1(E_2a) + \psi_2(E_2a)] \ (7) \quad \psi(E_{2u}a) = \frac{1}{\sqrt{2}} [\psi_1(E_2a) - \psi_2(E_2a)] \ (8)$$

$$\psi(E_{2g}b) = \frac{1}{\sqrt{2}} [\psi_1(E_2b) + \psi_2(E_2b)] \ (9) \quad \psi(E_{2u}b) = \frac{1}{\sqrt{2}} [\psi_1(E_2b) - \psi_2(E_2b)] \ (10)$$

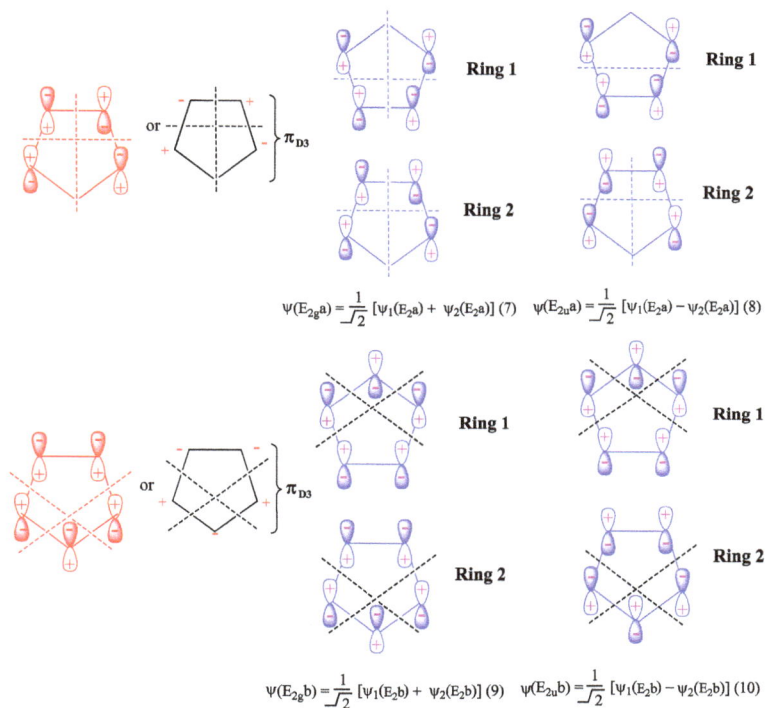

Figure 9.15: Wave functions of LGO of E_{2g} and E_{2u} symmetries of the ligand in staggered form of Ferrocene.

(iv) The next step is to work out the matching symmetry orbitals of the metal atom for overlapping with 10 LGOs of A_{1g}, A_{2u}, E_{1g}, E_{1u}, E_{2g} and E_{2u} symmetries to form respective MOs. Again, the orbitals available on the metal atom are five 3d-obitals, one s-orbital and three 4p-orbitals. From the character table for D_{5d} point group (given below) of staggered form of ferrocene, one can find out the symmetry species to which they belong.

Character table of D_{5d} point group

D_{5d}	E	$2C_5$	$2C_5^2$	$5C_2$	i	$2S_{10}^3$	$2S_{10}$	$5\sigma_d$		
A_{1g}	1	1	1	1	1	1	1	1		(x^2+y^2, z^2)
A_{2g}	1	1	1	−1	1	1	1	−1	R_z	
E_{1g}	2	$2\cos 72°$	$2\cos 144°$	0	2	$2\cos 72°$	$2\cos 144°$	0	(R_x, R_y)	(xz, yz)
E_{2g}	2	$2\cos 144°$	$2\cos 72°$	0	2	$2\cos 144°$	$2\cos 72°$	0		(x^2-y^2, xy)
A_{1u}	1	1	1	1	−1	−1	−1	−1		
A_{2u}	1	1	1	−1	−1	−1	−1	1	z	
E_{1u}	2	$2\cos 72°$	$2\cos 144°$	0	−2	$-2\cos 72°$	$-2\cos 144°$	0	(x, y)	
E_{2u}	2	$2\cos 144°$	$2\cos 72°$	0	−2	$-2\cos 144°$	$-2\cos 72°$	0		

These are

$$A_{1g}: 4s, 3d_z^2; A_{2u}: 4p_z; E_{1g}: 3d_{xz}, 3d_{yz}; E_{1u}: 4p_x, 4p_y; E_{2g}: d_{x^2-y^2}, d_{xy}; E_{2u}: Nil$$

Thus, matching orbitals are available for all the LGOs except E_{2u}. Hence, E_{2u} remain non-bonding MOs.

The central metal AOs and the matching LGOs now combine in phase (added) giving bonding MOs, and out of phase (with opposite sign) to give anti-bonding MOs. These are tabulated in Table 9.5.

Table 9.5: Formation of bonding and anti-bonding MOs.

Symmetry	Bonding MO	Non-bonding MO	Anti-bonding MO
A_{1g}	$\psi_s + C_1\psi(A_{1g})$ $\psi_{dz^2} + C_1\psi(A_{1g})$		$\psi_s - C_1\psi(A_{1g})$ $\psi_{dz^2} - C_1\psi(A_{1g})$
A_{2u}	$\psi_{pz} + C_2\psi(A_{2u})$		$\psi_{pz} - C_2\psi(A_{2u})$
E_{1g}	$\psi_{dxz} + C_3\psi(E_{1g}a)$ $\psi_{dyz} + C_3\psi(E_{1g}b)$		$\psi_{dxz} - C_3\psi(E_{1g}a)$ $\psi_{dyz} - C_3\psi(E_{1g}b)$
E_{1u}	$\psi_{px} + C_4\psi(E_{1u}a)$ $\psi_{py} + C_4\psi(E_{1u}b)$		$\psi_{px} - C_4\psi(E_{1u}a)$ $\psi_{py} - C_4\psi(E_{1u}b)$
E_{2g}	$\psi_{dx^2-z^2} + C_5\psi(E_{2g}a)$ $\psi_{dxy} + C_5\psi(E_{2g}b)$		$\psi_{dx^2-z^2} - C_5\psi(E_{2g}a)$ $\psi_{dxy} - C_5\psi(E_{2g}b)$
E_{2u}		$\psi(E_{2u}a)$ $\psi(E_{2u}b)$	

Here the notations C_1, C_2, C_3, C_4, and C_5 represent the ratio of the coefficients of the central atom and the respective LGO.

A simplified MO energy level diagram of staggered form of the ferrocene that results from the above considerations is given in Figure 9.16.

A perusal of MO energy level diagrams of staggered ferrocene reveals that the ligand group orbitals (LGOs) can be grouped in three sets: (i) a filled pair of A_{1g} and A_{2u} (ii) a higher energy filled doubly degenerate set E_{1g} and E_{1u} and (iii) still higher energy vacant doubly degenerate set of E_{2g} and E_{2u}. The relative energies of these three sets of LGOs are based on the presence of nodal planes in them.

The σ-bonding A_{1g} LGO of Cp_2 rings in staggered form is of lower energy relative to the atomic orbitals, 4s and $3d_z^2$ of iron of the same symmetry (A_{1g}), and hence it interacts less effectively producing weakly σ-bonding $A_{1g}(4s, 3d_{z^2})$ molecular orbital essentially localized on the Cp_2 rings. Contrary to this, δ bonding E_{2g} LGOs of the rings are so high in energy relative to the same symmetry metal atomic orbitals, d_{xy} and $d_{x^2-y^2}$ that they overlap very little with these atomic orbitals leads to slightly δ-bonding

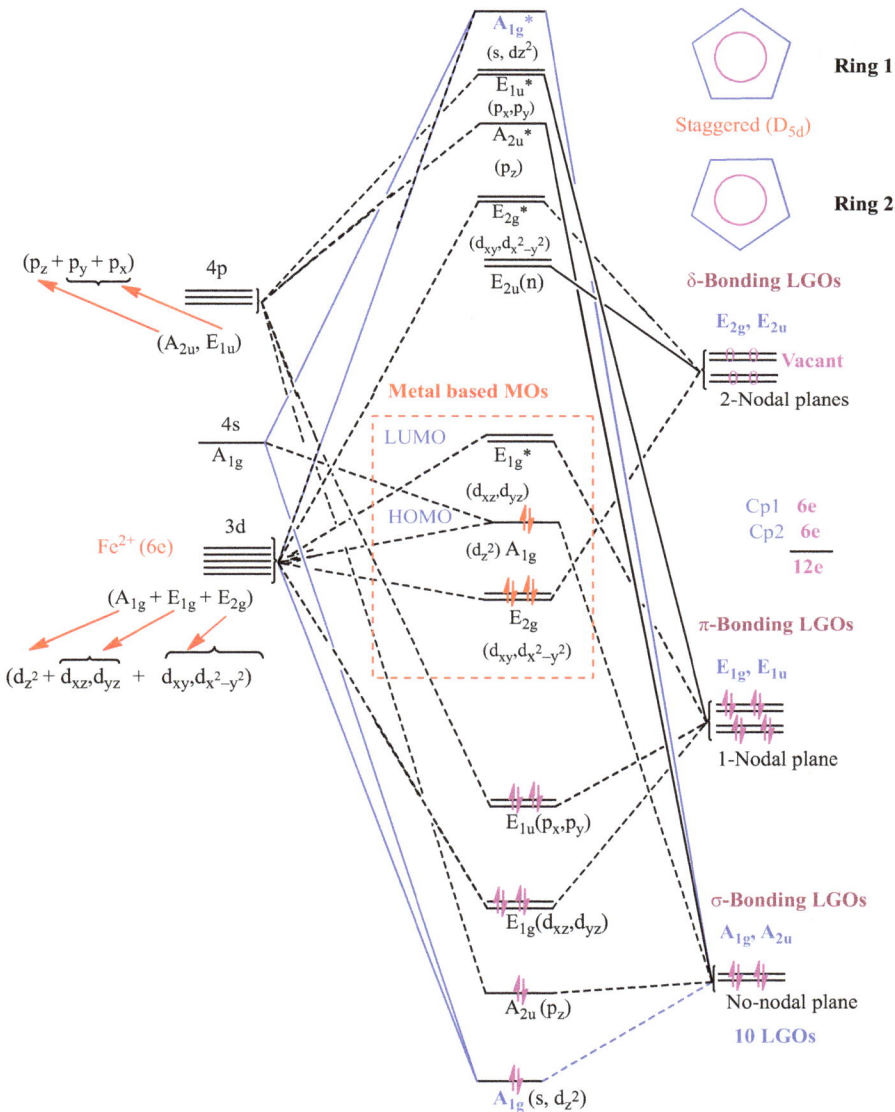

Figure 9.16: MO diagram of staggered form of Ferrocene (D_{5d}).

molecular orbitals essentially localized on the metal d-orbitals. This δ-type of bonding signifies certain amount of back donation of electrons from the filled metal orbitals to the vacant ring orbitals of the Cp_2 rings. Similarly, 4p-orbitals ($E_{1u} + A_{2u}$ symmetry) on the metal ion (Fe^{2+}) are at a very high energy, and so the E_{1u} and A_{2u} molecular orbitals (MOs) don't contribute much to the bonding and essentially localized on rings.

The only orbitals that are well matched are the π-bonding E_{1g} ligand group orbitals (LGO) and the corresponding 3d-orbitals (d_{xz}, d_{yz}) which form two strong π-bonds. One such π-bond formation is pictorially shown in Figure 9.17. These π-bonds are supposed to supply most of the stabilization that holds the ferrocene molecule together.

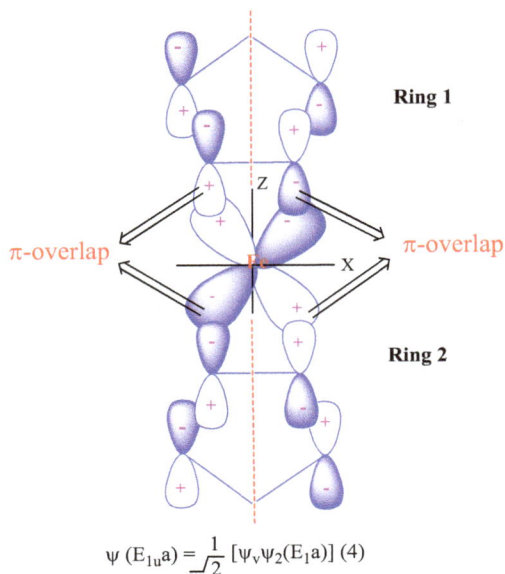

$$\psi\,(E_{1u}a) = \frac{1}{\sqrt{2}}\,[\psi_v\psi_2(E_1a)]\;(4)$$

Figure 9.17: Formation of two strong π-bonds stabilizing the ferrocene molecule.

It can be seen in the MO diagram, $A_{1g}(d_{z^2})$ MO is non-bonding in nature because energy of this MO is raised relative to metal d_{z^2} orbital.

In order to fill all the bonding MOs but none of the anti-bonding molecular orbitals, 18 electrons will be required. Thus, we see that 18 electron rule is a reflection of filling strongly stabilized MOs. The ferrocene [an 18e system; $Fe^{2+}(6e) + 2Cp^-(12e) = 18e$] in which all the nine lowest energy MOs are completely filled is thus the most stable metallocene.

The stability of $[Fe(\eta^5\text{-}C_5H_5)_2]$ with 18 electrons compared to $[Co(\eta^5\text{-}C_5H_5)_2]$ with 19 electrons and the twenty electron system $[Ni(\eta^5\text{-}C_5H_5)_2]$ is readily interpreted on the basis of this bonding scheme since these two latter species have 1 and 2 easily oxidizable electron(s) in anti-bonding E_{1g}^{*} MOs, whereas ferrocene has no electron in E_{1g}^{*} MOs. Similarly, $[Cr(\eta^5\text{-}C_5H_5)_2]$ with 16 electrons and vanadocene with 15 electrons have incompletely filled bonding MOs, and are highly reactive. So, these systems can easily get reduced by accepting 2 and 3 electrons, respectively.

It is notable here that the MOs are *delocalized* over the iron and both the rings in ferrocene. Thus, the linking of iron with both the rings is not localized. This accounts for the single υ(C-H) in IR spectrum and the aromatic character of the molecule.

Lastly, HOMO and LUMO in this diagram have shown to have essentially d-character. It is notable that HOMO in staggered form of ferrocene is A_{1g} (d_{z^2}) and is doubly occupied while the LUMO is E_{1g}^* and empty.

It is remarkable to note that the attachment of additional groups or ligands destroys the D_{5d} symmetry of simple metallocene and this modifies the MO diagram. This also happens when ferrocene is protonated to give $[Fe(\eta^5\text{-}C_5H_5)_2H]^+$, and when the isoelectronic neutral molecules, namely $[Re((\eta^5\text{-}C_5H_5)_2H]$ and $[Mo((\eta^5\text{-}C_5H_5)_2H_2]$ are considered.

9.2.2.3 Magnetic properties of metallocenes: M.O. diagram

The M.O. diagram given in Figure 10.10 also allows rationalizing magnetic properties of metallocenes. Although, it is probable that there may be some change in relative energies of molecular orbitals in going from Ti to Co: self-consistent results can be obtained with the qualitative ordering. It appears that the A_{1g} (d_{z^2}) and E_{2g} (d_{xy}, $d_{x^2-y^2}$) energy levels may converge (come together) and may even cross as one proceeds back from Fe towards Ti.

The observed magnetic susceptibility indicates the number of unpaired spins/ electrons in a complex. Further, the difference between the observed magnetic moments and the calculated 'spin only' values provides more detailed information. Thus, when an unpaired electron is in an orbital where the magnetic quantum number, $m = 0$ (i.e. 4s, $4p_z$ and $3d_{z^2}$ orbitals), the orbital component of the magnetic moment is zero. However, an unpaired electron in either of the orbitals $d_{x^2-y^2}$ or d_{xy} (where $m = \pm 2$) has an orbital component which is non-zero and the magnetic moment is consequently greater than is given by a 'spin only value'. Similarly, the presence of an unpaired electron in either of the d_{xz} and d_{yz} orbitals, orbital component which is non-zero, will contribute to the magnetic moment. Consequently, the magnetic moment is greater than the spin only value (vide the case of $[Ni^{III}(\pi\text{-}C_5H_5)_2]^+$ given in Table 9.6). Finally, in the cases where there is one unpaired electron in each of the $d_{x^2-y^2}$ and d_{xy} orbitals, the contribution of the orbital component to the magnetic moment is again zero. Based on these arguments, magnetic data and electron assignment for some 3d-transition metal bis(π-cyclopentadienyl) complexes are given in Table 9.6.

Table 9.6: Magnetic data and electron assignment for $[M(\eta^5\text{-}C_5H_5)_2]^{n+}$.

Compound	Electron assignment $(A_{1g})^2(A_{2u})^2(E_{1g})^4$- $(E_{1u})^4$ . . . (Total electrons)	No. of unpaired spins/electron(s)	Spin only value	Magnetic moment expected	Magnetic moment observed
$[Ti^{III}(\pi\text{-}C_5H_5)_2]^+$ $-3e$ $(d^2s^2){\to}(d^1s^0)$	. . . .$(E_{2g})^1$ (13e)	1	1.73	>1.73	2.29 ± 0.05
$[V^{IV}(\pi\text{-}C_5H_5)_2]^{2+}$ $-4e$ $(d^3s^2){\to}(d^1s^0)$	. . . .$(E_{2g})^1$ (13e)	1	1.73	>1.73	1.90 ± 0.05
$[V^{III}(\pi\text{-}C_5H_5)_2]^+$ $-3e$ $(d^3s^2){\to}(d^2s^0)$	. . . .$(E_{2g})^2$ (14e)	2	2.83	~2.83	2.86 ± 0.06
$[V^{II}(\pi\text{-}C_5H_5)_2]$ $-2e$ $(d^3s^2){\to}(d^3s^0)$	. . . .$(E_{2g})^2 (A_{1g})^1$ (both energy level converged) (15e)	3	3.87	~3.87	3.84 ± 0.04
$[Cr^{III}(\pi\text{-}C_5H_5)_2]^+$ $-3e$ $(d^5s^1){\to}(d^3s^0)$	. . . .$(E_{2g})^2 (A_{1g})^1$ (both energy level converged)(15e)	3	3.87	~3.87	3.73 ± 0.08
$[Cr^{II}(\pi\text{-}C_5H_5)_2]$ $-2e$ $(d^5s^1){\to}(d^4s^0)$	. . . .$(E_{2g})^3 (A_{1g})^1$ (both energy level converged)(16e)	2	2.83	>2.83	3.20 ± 0.16
$[Fe^{III}(\pi\text{-}C_5H_5)_2]^+$ $-3e$ $(d^6s^2){\to}(d^5s^0)$	. . . . $(A_{1g})^2(E_{2g})^3$ (both energy level converged)(17e)	1	1.73	>1.73	2.34 ± 0.12
$[Fe^{II}(\pi\text{-}C_5H_5)_2]$ $-2e$ $(d^6s^2){\to}(d^6s^0)$	. . . . $(A_{1g})^2(E_{2g})^4$ (both energy level converged)(18e)	0	0	0	0
$[Co^{III}(\pi\text{-}C_5H_5)_2]^+$ $-3e$ $(d^7s^2){\to}(d^6s^0)$	. . . . $(A_{1g})^2(E_{2g})^4$ (both energy level converged)(18e)	0	0	0	0
$[Co^{II}(\pi\text{-}C_5H_5)_2]$ $-2e$ $(d^7s^2){\to}(d^7s^0)$	.. $(E_{2g})^4(A_{1g})^2(E_{1g}^*)^1$ (d_{xz}, d_{yz}) (19e)	1	1.73	~1.73	1.76 ± 0.07

Table 9.6 (continued)

Compound	Electron assignment $(A_{1g})^2(A_{2u})^2(E_{1g})^4$-$(E_{1u})^4$ ... (Total electrons)	No. of unpaired spins/electron(s)	Spin only value	Magnetic moment expected	Magnetic moment observed
$[Ni^{III}(\pi\text{-}C_5H_5)_2]^+$ $-3e$ $(d^8s^2)\rightarrow(d^7s^0)$	.. $(E_{2g})^4(A_{1g})^2(E_{1g}^*)^1$ (d_{xz}, d_{yz}) (19e)	1	1.73	>1.73	1.82 ± 0.09
$[Ni^{II}(\pi\text{-}C_5H_5)_2]$ $-2e$ $(d^8s^2)\rightarrow(d^8s^0)$	.. $(E_{2g})^4(A_{1g})^2(E_{1g}^*)^2$ (20e)	2	2.83	~2.83	2.86 ± 0.11

9.3 Bonding in bis(arene) complexes with special reference to bis(π-benzene)-chromium

9.3.1 Introduction

The simplest and the most common **six π-electron donor** arene ligand is benzene, but other arene moieties can also function in the same way. A wide range of η^6-arene complexes of transition metals, both neutral and charged are known by now. Although the first η^6-arene complexes were prepared by Hein in 1919, yet the structural nature of these compounds could be known only after the synthesis of bis(π-benzene)chromium in 1955 by E. O. Fischer and his co-workers. Since, then the chemistry of arene metal complexes has undergone remarkable transformations leading to the synthesis of η^6-arene complexes of almost all transition metals.

Electron diffraction studies (1965) on bis(π-benzene)chromium in the vapour state show it to have sandwich structure with D_{6h} point symmetry. The C–C bond distances are equal at 1.423 ± 0.002 Å, C–H bonds are 1.090 ± 0.005 Å and the Cr–C distances are 2.150 ± 0.002 Å. X-ray studies of the crystal structure of this compounds have been carried out by several workers, and their results favour a symmetrical D_{6h} molecule (**eclipsed**) with all C–C, C–H and Cr–C distances the same the same within experimental errors.

Figure 9.18(a) given below represents the common method of describing the structure of bis(π-benzene)chromium. In Figure 9.18(b), the circles within the rings express the aromatic character of the rings and lines from the chromium atom indicate that the metal is bound to the whole ring rather than to the particular carbon atom of the ring.

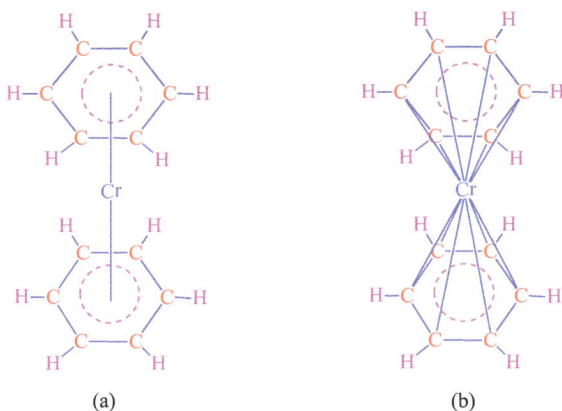

Figure 9.18: Sandwich structure of bis(π-benzene)chromium.

9.4 Bonding in bis(π-benzene)chromium

9.4.1 Valance Bond (VB) approach

The simplest approach to demonstrate the nature of bonding in bis(π-benzene)chromium is the VB approach. This approach is helpful to predict the molecular geometry of a complex in terms of the number and types of metal orbitals available for bonding. Accordingly, the bonding in this compound can be best understood by considering its formation from Cr and two C_6H_6 molecules. Herein, each C_6H_6 molecule donates its three pairs of π-electrons to the vacant d^2sp^3-hybrid orbitals of Cr as shown in the hybridization scheme (Figure 9.19).

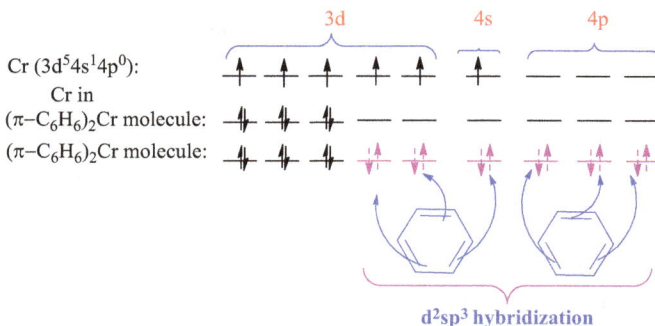

Figure 9.19: d^2sp^3 Hybridization scheme in bis(π-benzene)chromium.

Valence bond approach has thus proved to be quite useful for predicting the octahedral geometry and magnetic property of $(\pi\text{-}C_6H_6)_2Cr$. However, the aromatic character of this compound implies a ready availability of 6π-electrons [(4n + 2), i.e. 6] in each ring, and this is inconsistent with the involvement of 6π-electrons in bonding with the metal centre.

Therefor, a suitable theory advanced for explaining the bonding in $(\pi\text{-}C_6H_6)_2Cr$ must be capable of explaining the following observed facts related to this compound:
(i)   Magnetic behaviour.
(ii) All the carbons of the two benzene rings are involved in bonding [hexahapto nature (η^6) of the each ring] to the metal; that is, metal is bound to the whole ring rather than to the particular carbon atom of the ring.
(iii) Aromatic character of the benzene ring in the compound.
(iv) Extra stability of the compound.

As discussed above, VB approach failed to explain the characteristics (ii) to (iv) of $(\pi\text{-}C_6H_6)_2Cr$.

9.4.2 Molecular orbital (MO) approach

The MO approach, however, gives satisfactory explanation of all the characteristics of $(\pi\text{-}C_6H_6)_2Cr$, the details of the various steps of this approach are as follows:
(i)   Bonding in $(\pi\text{-}C_6H_6)_2Cr$ can be best understood if we consider its formation from neutral Cr and the two benzene molecules. The carbon atoms in C_6H_6 ring are sp^2 hybridized, the sp^2 hybridized orbitals being completely utilized in forming C–C (with two neighbouring carbons) and C–H σ-bonds (Figure 9.20). Each carbon atom of the ring now contains an unused p_z-orbital (assuming the rings lie in xy plane and the line passing through the metal atom and the centres of the two parallel rings is the z-axis).
(ii) The next step in working out the M.O. treatment of $(\pi\text{-}C_6H_6)_2Cr$ is to develop first the possible delocalized π-molecular orbitals for C_6H_6 ligand(s); that is, the ligand group orbitals (LGOs), and then select the metal orbitals which have the correct symmetry and energies to overlap with LGOs.

   C_6H_6 contains delocalized π-structure (i.e. non-localized multiple bonds). All the twelve (6C and 6H) atoms are co-planar, with the same C–C bond distance of 1.39 Å which is intermediate between ethylene (1.33 Å) and ethane (1.54 Å).
(iii) The $6p_z$ orbitals (containing in all 6e) of each ring, extending above and below the plane of the ring combine linearly in six ways (Figure 9.21) to give six delocalized π-MOs, three bonding and three anti-bonding MOs.

$C_6 = 1s^2 2s^2 2p^2$; **Valence shell electr. confgn.** $= 2s^2 2p^2$
(ground state)

; *Valence shell electr. confgn.* $= 2s^1 2p_x{}^1 2p_y{}^1 2p_z{}^1$
(excited state)

$\underbrace{}$

sp² hybridization

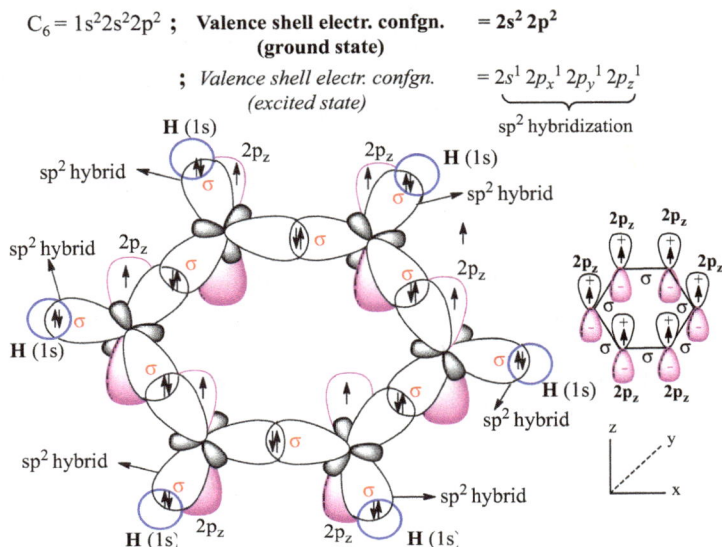

Figure 9.20: sp² Hybrid orbital overlaps and formation of six (C–C)σ and six (C–H)σ bonds.

(iv) The C_6H_6 molecule belongs to D_{6h} point group with symmetry operations: E, $2C_6$, $2C_3$, C_2, $3C_2'$, $3C_2''$, i, $2S_3$, $2S_6$, σ_h, $3\sigma_v$, $3\sigma_d$ (**24 symmetry operations**). Considering the ligand $p_z\pi$-orbitals as the basis, the symmetry operations of the D_{6h} point group are performed on the C_6H_6 $p_z\pi$-orbitals, and now using the concept of group theory, one can get the number and nature of symmetries of the resulting π_DMOs shown above. Accordingly, the symmetries of the resulting π_DMOs are:

$$T\pi = A_{2u} + B_{2g} + E_{1g} + E_{2u}$$

Thus, two non-degenerates and two sets of doubly degenerate π_DMOs or LGOs are to be formed. This can be easily achieved using the concept applied in carbocyclic system $(CH)_n$. Accordingly, for a $(CH)_n$ carbocyclic system, all the essential symmetry properties of the π-SALCs (symmetry adopted linear combinations) (or π_DMOs or LGOs here, that are formed by linear combination of p_z-orbitals) are determined by the operations of the uni-axial rotational sub-group, C_n. Hence, in the present case one can formulate the π_DMOs by using the character table of rotational point group C_6.

No nodal plane separating the molecule; π_{D1} is **bonding MO**

One nodal plane separating the molecule; π_{D2} & π_{D3} are degenerate **weakly bonding MOs**

Two nodal planes separating the molecule; π_{D4} & π_{D5} are degenerate **anti-bonding MOs**

Three nodal planes separating the molecule; π_{D6} is **anti-bonding MO**

E N E R G Y

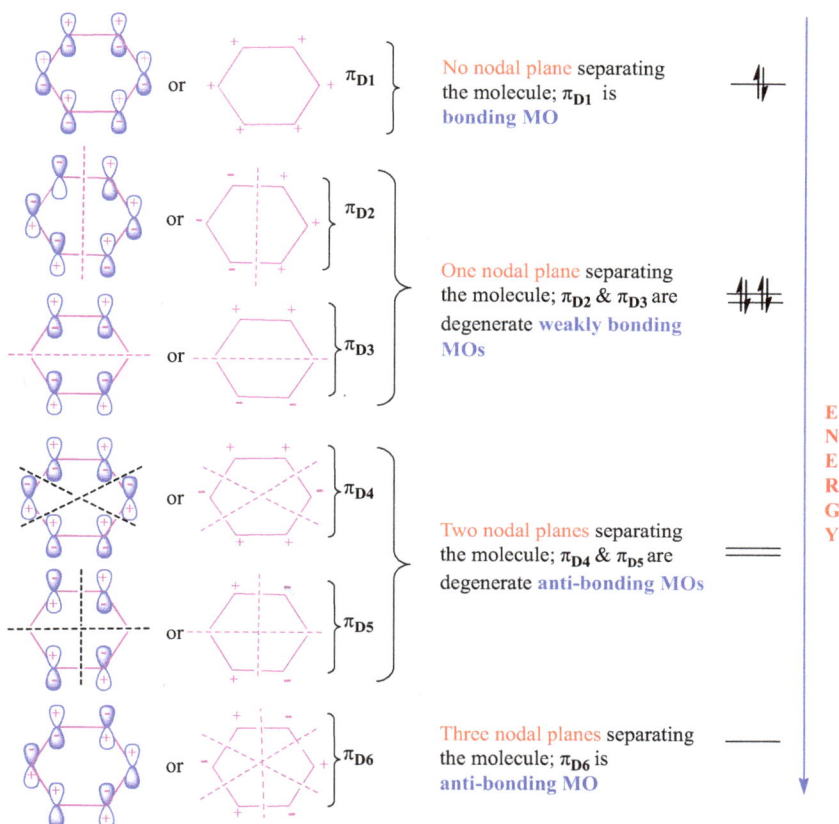

Figure 9.21: Formation of six delocalized π MOs in C_6H_6 molecule, three of bonding and three of anti-bonding in nature.

Character table of C_6 point group

C_6	E	C_6	C_3	C_2	C_3^2	C_6^5
A	1	1	1	1	1	1
B	1	−1	1	−1	1	−1
E_1	1	ε	$-\varepsilon^*$	−1	$-\varepsilon$	ε^*
	1	ε^*	$-\varepsilon$	−1	$-\varepsilon^*$	ε
E_2	1	$-\varepsilon^*$	$-\varepsilon$	1	$-\varepsilon^*$	$-\varepsilon$
	1	$-\varepsilon$	$-\varepsilon^*$	1	$-\varepsilon$	$-\varepsilon^*$

The formulations of π_DMOs/LGOs of various symmetries A_{2u}, B_{2g}, E_{1g} and E_{2g} [or simply of symmetry A, B, E_1 and E_2 on account of rotational subgroup C_6 of D_{6h} point group] can be done just by inspecting the C_6 character table. Let ϕ_1, ϕ_2, ϕ_3, ϕ_4, ϕ_5, and ϕ_6 be the wave functions associated with the p_z-orbitals on the carbon atoms (Figure 9.22). Then, the formulations of πDMOs/LGOs of the said symmetries in terms of wave functions are as follows:

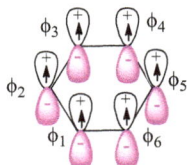

Figure 9.22: Wave functions associated with p_z-orbitals on the carbon atoms in benzene.

$$\psi A \quad = (\phi_1 + \phi_2 + \phi_3 + \phi_4 + \phi_5 + \phi_6)$$

$$\psi B \quad = (\phi_1 - \phi_2 + \phi_3 + \phi_4 + \phi_5 - \phi_6)$$

$$\psi E_1(1) = (\phi_1 + \varepsilon\phi_2 - \varepsilon^*\phi_3 - \phi_4 - \varepsilon\phi_5 + \varepsilon^*\phi_6)$$

$$\psi E_1(2) = (\phi_1 + \varepsilon^*\phi_2 - \varepsilon\phi_3 - \phi_4 - \varepsilon^*\phi_5 + \varepsilon\phi_6)$$

$$\psi E_2(1) = (\phi_1 + \varepsilon^*\phi_2 - \varepsilon\phi_3 + \phi_4 - \varepsilon^*\phi_5 - \varepsilon\phi_6)$$

$$\psi E_2(2) = (\phi_1 - \varepsilon\phi_2 - \varepsilon^*\phi_3 + \phi_4 + \varepsilon\phi_5 + \varepsilon^*\phi_5)$$

Where $\varepsilon = e^{2\pi i/6}$ and $\varepsilon^* = e^{-2\pi i/6}$

The next step is elimination of imaginary terms and normalization. The LGO, A and B can be easily normalized to get

$$\psi A = 1/\sqrt{6}(\phi_1 + \phi_2 + \phi_3 + \phi_4 + \phi_5 + \phi_6) \tag{9.1}$$

$$\psi B = 1/\sqrt{6}(\phi_1 + \phi_2 + \phi_3 + \phi_4 + \phi_5 + \phi_6) \tag{9.2}$$

The $E_1(1)$ and $E_1(2)$ set of functions have imaginary coefficients. In order to have real, rather than complex coefficients, **the two components are added term by term.** To eliminate the complex number using ***De Moivre's theorem***, we write:

$$\varepsilon = \cos(2\pi/6) + i\,\sin(2\pi/6) \text{ and } \varepsilon^* = \cos(2\pi/6) - i\,\sin(2\pi/6)$$

Then,

$$\varepsilon + \varepsilon^* = 2\cos(2\pi/6)$$

$$\varepsilon - \varepsilon^* = 2i\,\sin(2\pi/6)$$

$$\varepsilon^2 \quad = \cos(4\pi/6) + i\,\sin(4\pi/6)$$

$$\varepsilon^{2*} \quad = \cos(4\pi/6) - i\,\sin(4\pi/6)$$

$$\varepsilon^2 + \varepsilon^{2*} \quad = 2\cos(4\pi/6)$$

$$\varepsilon^2 - \varepsilon^{2*} \quad = 2i\,\sin(4\pi/6)$$

$$E_1(1) + E_1(2) = 2\phi_1 + (\varepsilon + \varepsilon^\star)\phi_2 - (\varepsilon + \varepsilon^\star)\phi_3 - 2\phi_4 + (\varepsilon^\star + \varepsilon)\phi_5 + ((\varepsilon^\star + \varepsilon)\phi_6$$

$$= 2\phi_1 + \phi_2 2\cos(2\pi/6) + \phi_3 2\cos(2\pi/6) - 2\phi_4 - \phi_5 2\cos(2\pi/6)$$

$$+ \phi_6 2\cos(2\pi/6)$$

$$E_1(a) \quad = 2\phi_1 + \phi_2 2.1/2 - \phi_3 2.1/2 - 2\phi_4 - \phi_5 2.1/2 + \phi_6 2.1/2 + [\cos 2\pi/6$$

$$= \cos 360°/6 = \cos 60° = 1/2]$$

$$= 2\phi_1 + \phi_2 - \phi_3 - 2\phi_4 - \phi_5 + \phi_6$$

The normalized wave function now becomes

$$\psi E_1(a) = 1/\sqrt{12} \, (2\phi_1 + \phi_2 - \phi_3 - 2\phi_4 - \phi_5 + \phi_6) \tag{9.3}$$

Now $E_1(1)$ and $E_1(2)$ are combined subtractively,

$$E_1(1) - E_1(2) = (\varepsilon - \varepsilon^\star)\phi_2 + (\varepsilon - \varepsilon^\star)\phi_3 + (\varepsilon^\star - \varepsilon)\phi_5 + (\varepsilon^\star - \varepsilon)\phi_6$$

$$\text{or } E_1(1) - E_1(2) = 0\phi_1 + \phi_2 2i\sin(2\pi/6) + \phi_3 2i\sin(2\pi/6) - \phi_5 2i\sin(2\pi/6)$$

$$- \phi_6 2i\sin(2\pi/6)$$

Dividing both sides by i, we get

$$E_1(1) - E_1(2)/i = [\phi_2 2i\sin(2\pi/6) + \phi_3 2i\sin(2\pi/6) - \phi_5 2i\sin(2\pi/6) - \phi_6 2i\sin(2\pi/6)]/i$$

$$= [\phi_2 2\sin(2\pi/6) + \phi_3 2\sin(2\pi/6) - \phi_5 2\sin(2\pi/6) - \phi_6 2\sin(2\pi/6)]$$

$$= (\phi_2 2.\sqrt{3}/2 + \phi_3 2.\sqrt{3}/2 - \phi_5 2.\sqrt{3}/2 - \phi_6 2.\sqrt{3}/2)$$

$$= [\sin 2\pi/6 \, \sin 60° = \sqrt{3}/2]$$

$$\psi E_1(b) = \sqrt{3}(\phi_2 + \phi_3 - \phi_5 - \phi_6) \approx (\phi_2 + \phi_3 - \phi_5 - \phi_6)$$

The normalized wave function now becomes

$$\psi E_1(b) = 1/2(\phi_2 + \phi_3 - \phi_5 - \phi_6) \tag{9.4}$$

We already have

$$\psi E_2(1) = (\phi_1 - \varepsilon^\star\phi_2 - \varepsilon\phi_3 + \phi_4 - \varepsilon^\star\phi_5 - \varepsilon\phi_6)$$

$$\psi E_2(2) = (\phi_1 - \varepsilon\phi_2 - \varepsilon^\star\phi_3 + \phi_4 - \varepsilon\phi_5 - \varepsilon^\star\phi_6)$$

On additive combination we have

$$E_2(1) + E_2(2) = 2\phi_1 - (\varepsilon + \varepsilon^*)\phi_2 - (\varepsilon + \varepsilon^*)\phi_3 + 2\phi_4 - (\varepsilon + \varepsilon^*)\phi_5 - (\varepsilon + \varepsilon^*)\phi_6$$

$$= 2\phi_1 - 2\cos(2\pi/6)\phi_2 - 2\cos(2\pi/6)\phi_3 + 2\phi_4 - 2\cos(2\pi/6)\phi_5$$
$$- 2\cos(2\pi/6)\phi_6$$

$$= 2\phi_1 - 2.1/2\phi_2 - 2.1/2\phi_3 + 2\phi_4 - 2.1/2\phi_5 - 2.1/2\phi_6$$

$$\psi E_2(a) = (2\phi_1 - \phi_2 - \phi_3 + 2\phi_4 - \phi_5 - \phi_6)$$

The normalized wave function now becomes

$$\psi E_2(a) = 1/\sqrt{12}\ (2\phi_1 - \phi_2 - \phi_3 + 2\phi_4 - \phi_5 - \phi_6) \tag{9.5}$$

On subtractive combination of $E_2(1)$ and $E_2(2)$, we have

$$E_2(1) - E_2(2) = (\varepsilon - \varepsilon^*)\phi_2 - (\varepsilon - \varepsilon^*)\phi_3 + (\varepsilon - \varepsilon^*)\phi_5 - (\varepsilon - \varepsilon^*)\phi_6$$

$$= 2i\sin(2\pi/6)\phi_2 - 2i\sin(2\pi/6)\phi_3 + 2i\sin(2\pi/6)\phi_5 - 2i\sin(2\pi/6)\phi_6$$

Dividing both sides by i, we get

$$E_2(1) - E_2(2)/i = [\phi_2 2i\sin(2\pi/6) - \phi_3 2i\sin(2\pi/6) + \phi_5 2i\sin(2\pi/6) - \phi_6 2i\sin(2\pi/6)]/i$$

$$= \phi_2 2\sin(2\pi/6) - \phi_3 2\sin(2\pi/6) + \phi_5 2\sin(2\pi/6) - \phi_6 2\sin(2\pi/6)$$

$$= \phi_2 2.\sqrt{3}/2 - \phi_3 2.\sqrt{3}/2 + \phi_5 2.\sqrt{3}/2 - \phi_6 2.\sqrt{3}/2$$

$$\psi E_2(b) = \sqrt{3}(\phi_2 - \phi_3 + \phi_5 - \phi_6) \approx (\phi_2 - \phi_3 + \phi_5 - \phi_6)$$

The normalized wave function now becomes

$$1/\sqrt{2}\ (\phi_2 - \phi_3 + \phi_5 - \phi_6) \tag{9.6}$$

Thus, the compositions of six πDMOs/LGOs of various symmetries of benzene molecule in normalized form are

$$\psi A = 1/\sqrt{6}(\phi_1 + \phi_2 + \phi_3 + \phi_4 + \phi_5 + \phi_6) \tag{9.1}$$

$$\psi B = 1/\sqrt{6}(\phi_1 + \phi_2 + \phi_3 + \phi_4 + \phi_5 + \phi_6) \tag{9.2}$$

$$\psi E1(a) = 1/\sqrt{12}(2\phi_1 + \phi_2 + \phi_3 + \phi_4 + \phi_5 + \phi_6) \tag{9.3}$$

$$\psi E_1(b) = 1/2(\phi_2 + \phi_3 - \phi_5 - \phi_6) \tag{9.4}$$

$$\psi E_2(a) = 1/\sqrt{12}(2\phi_1 - \phi_2 - \phi_3 + 2\phi_4 - \phi_5 - \phi_6) \tag{9.5}$$

$$\psi E_2(b) = 1/2(\phi_2 - \phi_3 + \phi_5 - \phi_6) \tag{9.6}$$

9.4.3 Eclipsed form of bis(π-C₆H₆)₂Cr

X-ray crystal structure studies along with infrared and Raman spectral data have shown that the structure of bis(π-C$_6$H$_6$)$_2$Cr has eclipsed conformation with molecular point group D$_{6h}$. In the formation of eclipsed conformation of bis(π-C$_6$H$_6$)$_2$Cr (wherein the orientation of the π-orbitals of the two benzene rings with their +ve lobes lies towards the metal ion as shown in Figure 9.23) 12$p\pi$-orbitals of two benzene rings are involved in bonding.

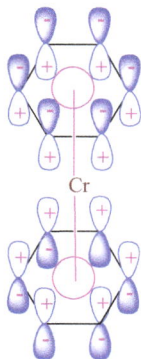

Figure 9.23: Eclipsed conformation of bis(π-C$_6$H$_6$)$_2$Cr.

In order to determine the number and nature of symmetries of the LGOs formed by the combination of 12 $p\pi$-orbitals, these 12 $p\pi$-orbitals are subjected to the symmetry operations of the D$_{6h}$ point group. Without going to detailed group theoretical calculations, symmetries of the LGOs are presented below:

$$\Gamma\pi = A_{1g} + B_{2g} + E_{1g} + E_{2g} + A_{2u} + B_{1u} + E_{1u} + E_{2u}$$
$$= A_{1g} + A_{2u} + B_{2g} + B_{1u} + E_{1g} + E_{1u} + E_{2g} + E_{2u}$$

This gives the symmetry characteristics of the 12LGOs formed by the combination of 12 $p\pi$-orbitals. The twelve LGOs group themselves into four sets: two A orbitals, two B orbitals, two sets of doubly degenerate E$_1$ orbitals and two doubly degenerate sets of E$_2$ orbitals.

The next step is to work out the **wave functions cum composition** of twelve LGOs resulting from two benzene rings of the aforesaid symmetries. These are worked out as follows:

An observation of the character table (Table 9.7) of D$_{6h}$ point group (given below) shows that A$_{1g}$, B$_{2g}$, E$_{1g}$ and E$_{2g}$ are symmetric and A$_{2u}$, B$_{1u}$, E$_{1u}$, and E$_{2u}$ are antisymmetric with respect to inversion centre (i).

Table 9.7: Character table of D_{6h} point group.

D_{6h}	E	$2C_6$	$2C_3$	C_2	$3C_2'$	$3C_2''$	i	$2S_3$	$2S_6$	σh	$3\sigma_d$	$3\sigma_v$		
A_{1g}	1	1	1	1	1	1	1	1	1	1	1	1		x^2+y^2, z^2
A_{2g}	1	1	1	1	-1	-1	1	1	1	1	-1	-1	R_z	
B_{1g}	1	-1	1	-1	1	-1	1	-1	1	-1	1	-1		
B_{2g}	1	-1	1	-1	-1	1	1	-1	1	-1	-1	1		
E_{1g}	2	1	-1	-2	0	0	2	1	-1	-2	0	0	(R_x, R_y)	(xz, yz)
E_{2g}	2	-1	-1	2	0	0	2	-1	-1	2	0	0		(x^2-y^2, xy)
A_{1u}	1	1	1	1	1	1	-1	-1	-1	-1	-1	-1		
A_{2u}	1	1	1	1	-1	-1	-1	-1	-1	-1	1	1	z	
B_{1u}	1	-1	1	-1	1	-1	-1	1	-1	1	-1	1		
B_{2u}	1	-1	1	-1	-1	1	-1	1	-1	1	1	-1		
E_{1u}	2	1	-1	-2	0	0	-2	-1	1	2	0	0	(x, y)	
E_{2u}	2	-1	-1	2	0	0	-2	1	1	-2	0	0		

Hence,

(i) The LGOs of the two individual rings of symmetry 'A' combined linearly as follows to give wave functions of LGO of A_{1g} and A_{2u} symmetries of the ligand (2 benzene rings) in bis(π-benzene)chromium (Figure 9.24).

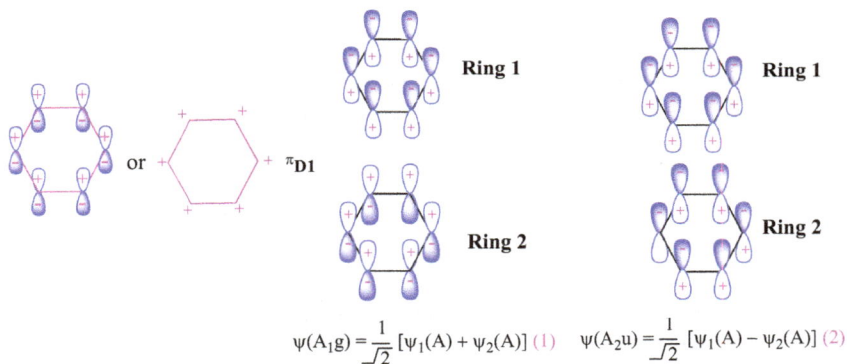

$$\psi(A_1g) = \frac{1}{\sqrt{2}}[\psi_1(A) + \psi_2(A)] \,(1) \qquad \psi(A_2u) = \frac{1}{\sqrt{2}}[\psi_1(A) - \psi_2(A)] \,(2)$$

Figure 9.24: Wave functions of LGO of A_{1g} and A_{2u} symmetries of the ligand (2 benzene rings) in bis(π-benzene)chromium.

(ii) The LGOs of the two individual rings of symmetry 'B' combined linearly as follows to give wave function of LGO of B_{2g} and B_{1u} symmetries (Figure 9.25).

(iii) The LGOs of the two individual rings of symmetry 'E_1' combined linearly as follows to give wave function of LGO of E_{1g} and E_{1u} symmetries (Figure 9.26).

(iv) The LGOs of the two individual rings of symmetry 'E_2' combined linearly as follows to give wave function of LGO of E_{2g} and E_{2u} symmetries (Figure 9.27).

$$\psi(B_{2g}) = \frac{1}{\sqrt{2}}\,[\psi_1(B) + \psi_2(B)] \ (3) \qquad \psi(B_1u) = \frac{1}{\sqrt{2}}\,[\psi_1(B) - \psi_2(B)] \ (4)$$

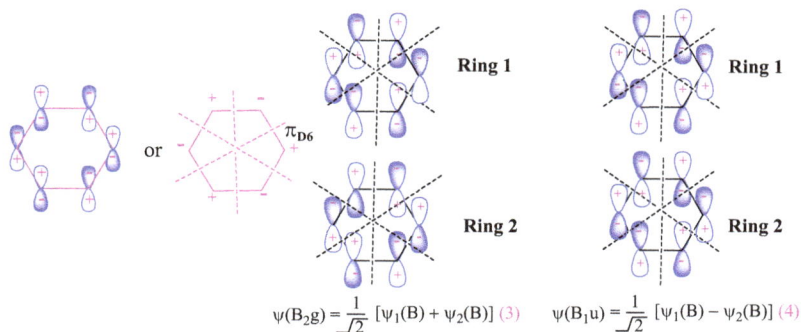

Figure 9.25: Wave functions of LGO of B_{2g} and B_{1u} symmetries of the ligand (2 benzene rings) in bis (π-benzene)chromium.

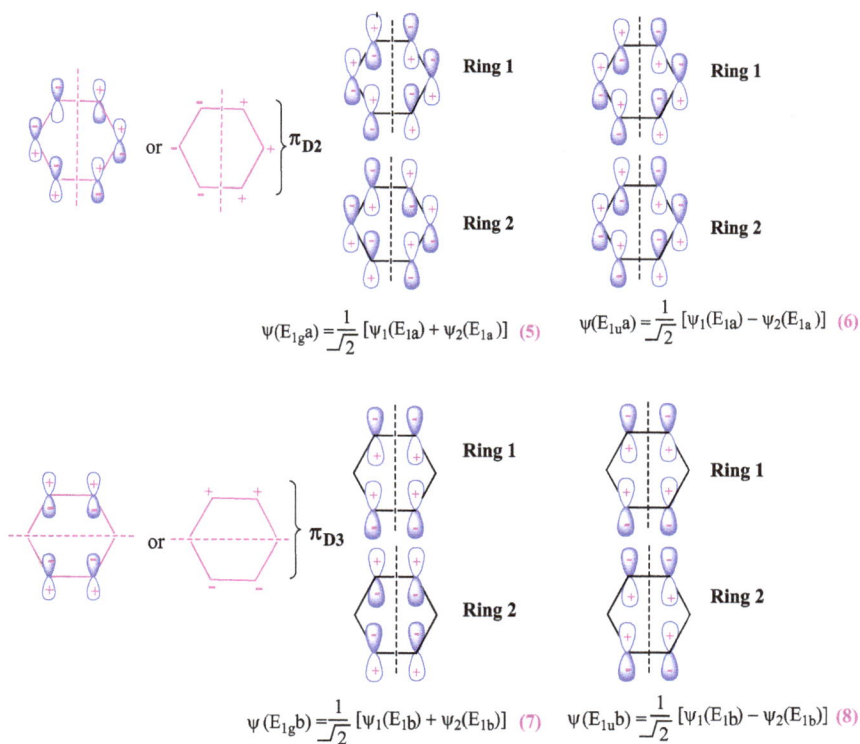

$$\psi(E_{1g}a) = \frac{1}{\sqrt{2}}\,[\psi_1(E_{1a}) + \psi_2(E_{1a})] \ (5) \qquad \psi(E_{1u}a) = \frac{1}{\sqrt{2}}\,[\psi_1(E_{1a}) - \psi_2(E_{1a})] \ (6)$$

$$\psi(E_{1g}b) = \frac{1}{\sqrt{2}}\,[\psi_1(E_{1b}) + \psi_2(E_{1b})] \ (7) \qquad \psi(E_{1u}b) = \frac{1}{\sqrt{2}}\,[\psi_1(E_{1b}) - \psi_2(E_{1b})] \ (8)$$

Figure 9.26: Wave functions of LGO of E_{1g} and E_{1u} symmetries of the ligand (2 benzene rings) in bis(π-benzene)chromium.

$$\psi(E_{2g}a) = \frac{1}{\sqrt{2}}[\Psi_1(E_{2a}) + \Psi_2(E_{2a})] \quad (9) \qquad \psi(E_{2u}a) = \frac{1}{\sqrt{2}}[\Psi_1(E_{2a}) - \Psi_2(E_{2a})] \quad (10)$$

$$\psi(E_{2g}b) = \frac{1}{\sqrt{2}}[\Psi_1(E_{2b}) + \Psi_2(E_{2b})] \quad (11) \qquad \psi(E_{2u}b) = \frac{1}{\sqrt{2}}[\Psi_1(E_{2b}) - \Psi_2(E_{2b})] \quad (12)$$

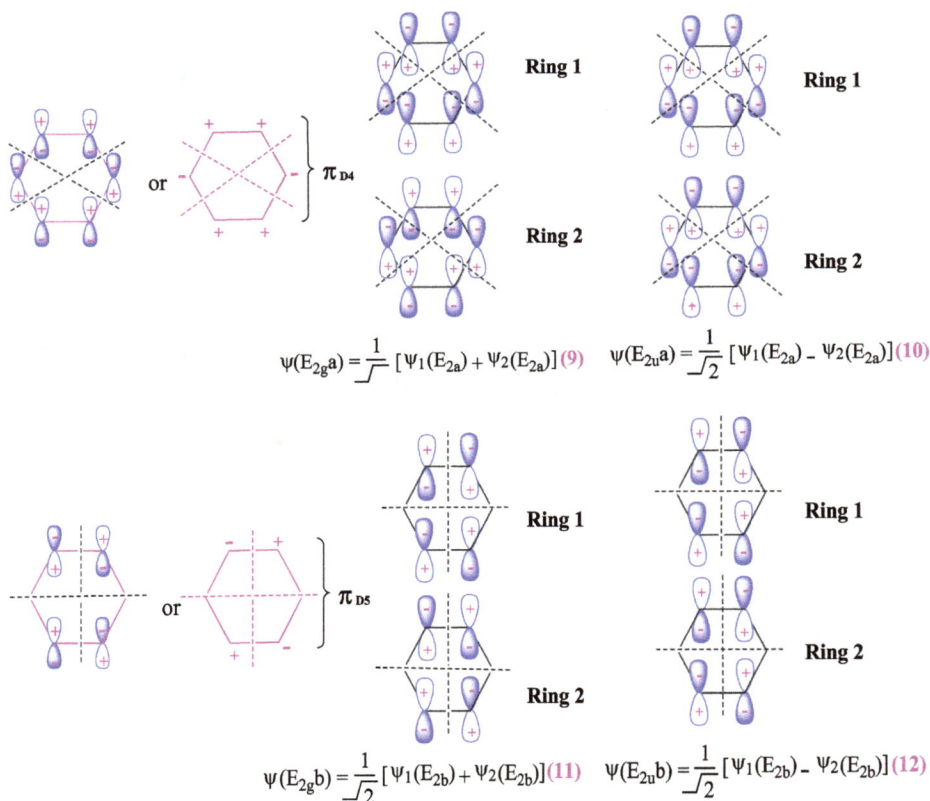

Figure 9.27: Wave functions of LGO of E_{2g} and E_{2u} symmetries of the ligand (2 benzene rings) in bis (π-benzene)chromium.

The next step is to work out the matching symmetry orbitals of the metal atom to overlap with 12 LGOs of A_{1g}, A_{2u}, B_{2g}, B_{1u}, E_{1g}, E_{1u}, E_{2g} and E_{2u} symmetries to form respective MOs. Again, the orbitals available on the metal atom are five 3d-obitals, one s-orbital and three 4p-orbitals. From the character table for D_{6h} point group, one can find out the symmetry species to which they belong.

Character table of D_{6h} point symmetry

D_{6h}	E	$2C_6$	$2C_3$	C_2	$3C_2'$	$3C_2''$	i	$2S_3$	$2S_6$	σh	$3\sigma_d$	$3\sigma_v$		
A_{1g}	1	1	1	1	1	1	1	1	1	1	1	1		$x^2 + y^2$, z^2
A_{2g}	1	1	1	1	-1	-1	1	1	1	1	-1	-1	R_z	
B_{1g}	1	-1	1	-1	1	-1	1	-1	1	-1	1	-1		
B_{2g}	1	-1	1	-1	-1	1	1	-1	1	-1	-1	1		
E_{1g}	2	1	-1	-2	0	0	2	1	-1	-2	0	0	(R_x, R_y)	(xz, yz)
E_{2g}	2	-1	-1	2	0	0	2	-1	-1	2	0	0		$(x^2 - y^2, xy)$

A_{1u}	1	1	1	1	1	1	-1	-1	-1	-1	-1	-1	
A_{2u}	1	1	1	1	-1	-1	-1	-1	-1	-1	1	1	z
B_{1u}	1	-1	1	-1	1	-1	-1	1	-1	1	-1	1	
B_{2u}	1	-1	1	-1	-1	1	-1	1	-1	1	1	-1	
E_{1u}	2	1	-1	-2	0	0	-2	-1	1	2	0	0	(x, y)
E_{2u}	2	-1	-1	2	0	0	-2	1	1	-2	0	0	

These are

$$A_{1g}: 4s, 3d_z^2; \quad A_{2u}: 4p_z; \quad B_{2g}: \text{none}; \quad B_{1u}: \text{none}; \quad E_{1g}: 3d_{xz}, 3d_{yz}$$
$$E_{1u}: 4p_x, 4p_y; \quad E_{2g}: d_{x^2-y^2}, d_{xy}; \quad E_{2u}: \text{none}$$

Thus, matching orbitals are available for all the LGOs except B_{2g}, B_{1u} and E_{2u}. Hence, these LGOs remain non-bonding MOs.

The central metal AOs and the matching LGOs now combine in phase (added) giving bonding MOs, and out of phase (with opposite sign) to give anti-bonding MOs. These are tabulated in Table 9.8.

Table 9.8: Formation of bonding and anti-bonding MOs.

Symmetry	Bonding MO	Non-bonding MO	Anti-bonding MO
A_{1g}	$\psi_s + C_1\psi(A_{1g})$ $\psi_{dz^2} + C_1\psi(A_{1g})$		$\psi_s - C_1\psi(A_{1g})$ $\psi_{dz^2} - C_1\psi(A_{1g})$
A_{2u}	$\psi_{pz} + C_2\psi(A_{2u})$		$\psi_{pz} - C_2\psi(A_{2u})$
E_{1g}	$\psi_{dxz} + C_3\psi(E_{1g}a)$ $\psi_{dyz} + C_3\psi(E_{1g}b)$		$\psi_{dxz} - C_3\psi(E_{1g}a)$ $\psi_{dyz} - C_3\psi(E_{1g}b)$
E_{1u}	$\psi_{px} + C_4\psi(E_{1u}a)$ $\psi_{py} + C_4\psi(E_{1u}b)$		$\psi_{px} - C_4\psi(E_{1u}a)$ $\psi_{py} - C_4\psi(E_{1u}b)$
E_{2g}	$\psi_{dx^2-y^2} + C_5\psi(E_{2g}a)$ $\psi_{dxy} + C_5\psi(E_{2g}b)$		$\psi_{dx^2-y^2} - C_5\psi(E_{2g}a)$ $\psi_{dxy} - C_5\psi(E_{2g}b)$
B_{2g}		$\psi(B_{2g})$	
E_{2u}		$\psi(E_{2u}a)$ $\psi(E_{2u}b)$	

The simplified MO energy level diagram of eclipsed form of bis(π-C_6H_6)$_2$Cr that results from the above considerations is shown in Figure 9.28.

The molecular orbital energy diagram of bis(π-benzene) chromium is similar to that of ferrocene except that there are two additional higher energy non-bonding MOs of the symmetry B_{2g} and B_{1u}.

If electrons are supplied to fill all the bonding and non-bonding orbitals but none of the anti-bonding orbitals, 18 electrons will be required. Thus, we see that

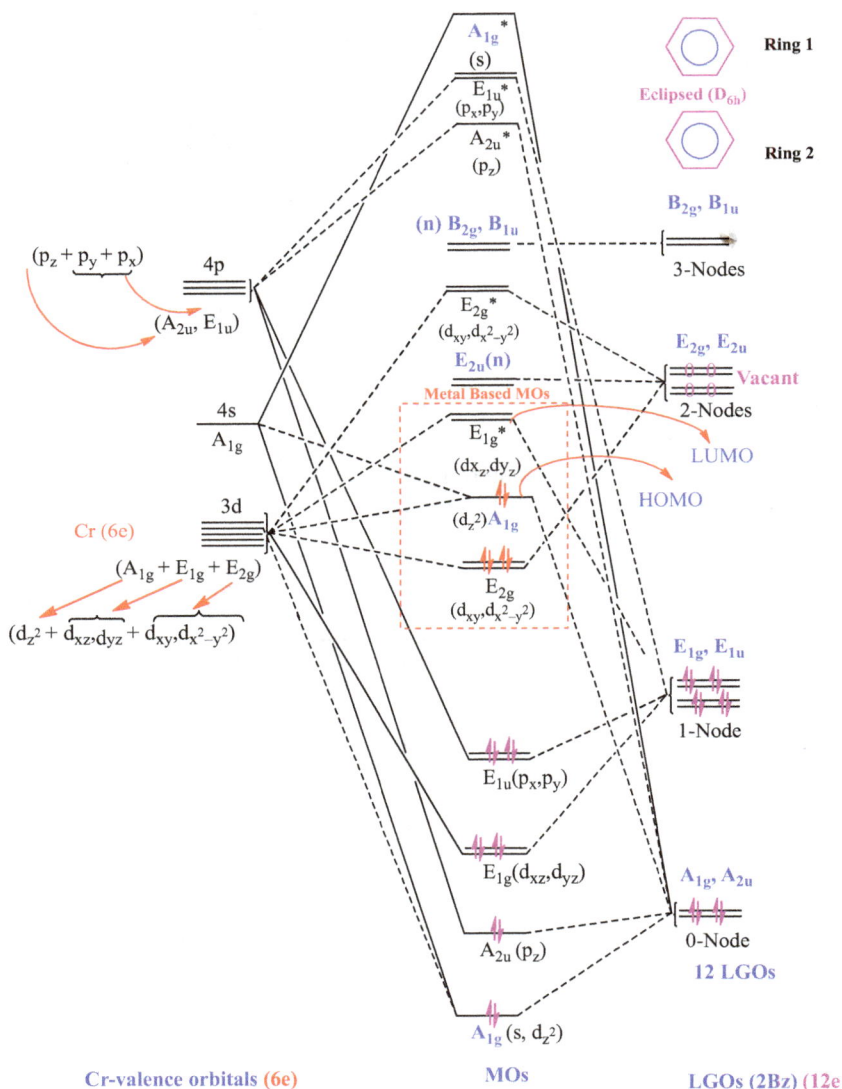

Figure 9.28: M.O. diagram of eclipsed conformation of $[\pi\text{-}(C_6H_6)_2Cr]$ (D_{6h}).

18 electron rule is a reflection of filling strongly stabilized Mos. The bis(π-benzene) chromium [an 18e system; Cr(6e) + 2Bz (12e) = 18e] in which all the nine lowest energy MOs are completely filled, is thus the most stable compound.

It is notable here that the MOs are delocalized over the chromium and both the rings. Thus, the linking of iron with both the rings is not localized. In other words the MO treatment favours a completely delocalized ring structure for the molecule.

This accounts for the single ν(C–H) in IR spectrum and the aromatic character of the molecule.

Lastly, HOMOs and LUMOs in this diagram have shown to have essentially d-character. It is notable that HOMO in the compound is $A_{1g}(dz^2)$ and is doubly occupied while the LUMO is $E_{1g}^\star$.

Exercises

Multiple choice questions/fill in the blanks

1. The point group of staggered form of ferrocene is:
    (a) D_5 (b) D_{5h} (c) C_{5h} (d) D_{5d}

2. Compounds in which metal atom is placed between two parallel carbocyclic ring systems are known as:
    (a) Bridge compounds
    (b) Semibridge Compounds
    (c) Sandwich compounds
    (d) Any one of these

3. The bis(π-benzene)chromium is a system having electrons in number:
    (a) 19   (b) 18   (c) 17   (d) 16

4. Synthesis of bis(π-benzene)-chromium in 1955 was first time reported by:
    (a) F.A. Cotton
    (b) G. Wilkinson
    (c) J. E. Huheey
    (d) E. O. Fischer et al.

5. The point group of eclipsed form of bis(π-benzene)chromium is:
    (a) C_{6h} (b) D_{6h} (c) D_6 (d) D_{6d}

6. IR spectrum of bis(π-benzene)chromium shows ν(C–H) in number:
    (a) 6   (b) 4   (c) 2   (d) 1

7. The point group of the eclipsed form of ferrocene is:
    (a) D_5 (b) D_{5h} (c) C_{5h} (d) D_{5d}

8. The carbon- carbon bond distance in ferrocene is:
    (a) 1.54 Å   (b) 1.39 Å   (c) 1.33 Å   (d) None of these

9. The hapticity of cyclopentadienyl ring in ferrocene is:
    (a) 4   (b) 5   (c) 6   (d) 7

10. The aromaticity of ferrocene implies a ready availability of π-electrons in number:
    (a) 2   (b) 6   (c) 10   (d) 14

11. The proper axis of symmetry of cyclopentadienyl rings passing through Fe in $Fe(C_5H_5)_2$ is:
    (a) C_4 (b) C_5 (c) C_6 (d) None of these

12. The proper axis of symmetry of benzene rings passing through Cr in $Cr(C_6H_6)_2$ is:
    (a) C_4 (b) C_5 (c) C_6 (d) C_7

13. In order to eliminate the complex number in certain functions one uses . . . .
    theorem.

Short answer type questions

1. How valance bond theory is helpful to explain the geometry and magnetic behaviour of ferrocene?
2. Highlight the bonding in bis(π-benzene)chromium in terms of valence bond theory.
3. Point out the shortcomings of valance bond theory with regard to the bonding in ferrocene and bis(π-benzene)chromium, which are responsible for coming up of the group theoretical approach of molecular orbital theory to explain their bonding fully.
4. What are the observed facts related to $(\pi\text{-}C_6H_6)_2Cr$ and $Fe(C_5H_5)_2$ which are expected to be explained by molecular orbital theory?
5. Explain the formation of five delocalized π-molecular orbitals ($\pi_D MOs$) in C_5H_5 anion. Give the number of bonding and anti-bonding $\pi_D MOs$.
6. Present the compositions of of $SALCs/\pi_D MOs/LGOs$ of different symmetries in cyclopentadienyl anion in their normalized form.
7. Give the compositions of of $SALCs/\pi_D MOs/LGOs$ of different symmetries in benzene in their normalized form.
8. Briefly explain the formation of six delocalized π-molecular orbitals ($\pi_D MOs$) in benzene. Give the number of bonding and anti-bonding $\pi_D MOs$.
9. Explain why cobaltocene and nickelocene are easily oxidizable compared to ferrocene while chromocene and vanadocene easily reducible.
10. Just draw the MO diagram of eclipsed conformation of ferrocene and comment on its magnetic behaviour.
11. Present a brief note on magnetic properties of metallocenes.
12. Draw the M.O. diagram of eclipsed conformation of $[\pi\text{-}(C_6H_6)_2Cr]$ and highlight its salient features.
13. Workout the compositions of six $\pi DMOs$ of various symmetries of benzene molecule in normalized form.

Long answer type questions

1. Draw and discuss the molecular orbital diagram of bis(π-benzene)chromium and justify that it is a diamagnetic molecule.
2. Derive the SLACs for π-molecular orbitals of cyclopentadienyl anion. Also develop the π-molecular orbitals diagram and label all the orbitals in eclipsed form of ferrocene.
3. Present the detained view of bonding in staggered form of ferrocene.
4. Describe the bonding in eclipsed form of (π-C_6H_6)$_2$Cr and comment on its magnetic behaviour.
5. Discuss in detail the bonding in one of the well-known sandwich compounds with full description of molecular orbital diagram.
6. Give the pictorial presentation of ten ligand group orbitals (LGOs) of A_1', A_2'', E_1', E_1'', E_2' and E_2'' symmetries of the ligand (2Cp$^-$ rings) in eclipsed form of ferrocene.
7. Develop the 12 LGOs of A_{1g}, A_{2u}, B_{2g}, B_{1u}, E_{1g}, E_{1u}, E_{2g} and E_{2u} symmetries of the ligand (2C_6H_6 rings) in eclipsed form of bis(π-benzene)chromium.

Chapter X
Some aspects of safe and economical inorganic experiments at UG and PG levels

10.1 Introduction

We are very well aware that the prices of chemicals are increasing day by day. In this situation, the inorganic experiments such as mixture analysis and synthesis of coordination compounds at undergraduate and postgraduate levels are becoming more and more expensive. Taking this serious problem into consideration, particularly, in the colleges where the budgets are meagre, it has now become essential to design some innovative techniques and experiments with the following objectives:

(i)   To reduce the consumption of chemicals with a considerable saving in the laboratory budget
(ii)   The greater speed of analyses with smaller quantities of materials, and thus saving the times in carrying out the tests/experiments
(iii) To make the experiments/tests safe
(iv)   To increase the visibility of the tests
(v)   To save space on the reagent shelve

10.2 Inorganic qualitative analysis

The branch of chemical analysis, which aims to find out the constituents of a mixture or compounds, is known as qualitative analysis. In this analysis, it is to learn the methods used for identification of radicals present in the given salt or the mixture of two or more salts.

Each salt consists of two parts, usually called radicals, known as the acid and basic radicals. For example, $NaNO_3$ is formed by the interaction of $NaOH$ and HNO_3. In $NaNO_3$, Na^+ radical has come from $NaOH$ (a base), hence it is called basic radical. The basic radical is also called the cation or the positive radical because it always bears a positive charge. On the other hand, the NO_3^- radical has come from HNO_3 (an acid), hence it is known as acid radical. The detection or identification of acid and basic radicals in a salt or mixture is known as inorganic qualitative analysis.

For the sake of convenience and also to fulfil the various objectives, the present section is divided into the following parts:

(i)   Design of ignition tube apparatus (ITA)
(ii)   Test of radicals using ignition tube apparatus (ITA) technique
(iii) Test of acid radicals using reagent filter papers

https://doi.org/10.1515/9783110727289-010

(iv) Test of acid radicals using reagent solutions
(v) Test of basic radicals using reagent solutions

10.2.1 Design of ignition tube apparatus

The design of the ignition tube apparatus (ITA) is very simple and too cheap and provides at least two pieces to each student. It requires test tubes, ignition tubes, rubber corks of suitable sizes to be fitted into the mouth of test tubes, a cork's borer, ordinary filter papers and reagent droppers. The assembled ITA is shown in Figure 10.1.

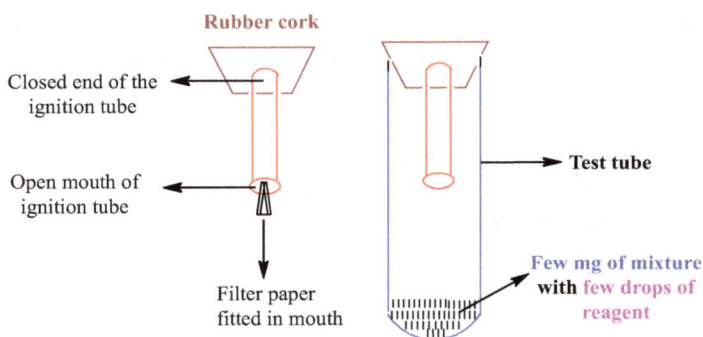

Figure 10.1: Ignition tube apparatus (ITA).

It is notable here that the **closed end of the ignition tube** should be within the rubber cork.

10.2.2 Test of radicals using ignition tube apparatus technique

The simple ITA assembly can be successfully used for the qualitative tests of some radicals as follows:

10.2.2.1 Chloride (Cl⁻)
Take ~50 mg of a salt or a mixture of salts with 3–4 drops of concentrated H_2SO_4 in a dry test tube of the ITA. Now, moisten the open mouth of the ignition tube fitted in the rubber cork with $AgNO_3$ solution (shake well to remove excess of the $AgNO_3$ solution), and now put it into the mouth of the test tube already containing the mixture. On heating the salt/mixture, a curdy white deposit of AgCl occurs around the

mouth of the ignition tube due to reaction of the evolved HCl gas with $AgNO_3$ solution. A typical reaction is shown below:

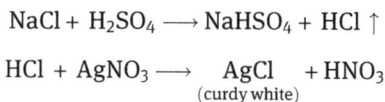

$$NaCl + H_2SO_4 \longrightarrow NaHSO_4 + HCl \uparrow$$

$$HCl + AgNO_3 \longrightarrow \underset{\text{(curdy white)}}{AgCl} + HNO_3$$

10.2.2.2 Bromide (Br⁻)

It can be tested in the same way as given above. In this case, light yellow colour appears around the mouth of the ignition tube due to evolution of HBr gas and its reaction with $AgNO_3$.

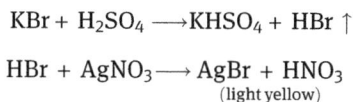

$$KBr + H_2SO_4 \longrightarrow KHSO_4 + HBr \uparrow$$

$$HBr + AgNO_3 \longrightarrow \underset{\text{(light yellow)}}{AgBr} + HNO_3$$

10.2.2.3 Iodide (I⁻)

It can be tested by moistening the ignition tube with starch solution. Intense blue colouration appears due to liberation of I_2 gas when the salt/mixture is heated with concentrated H_2SO_4. This test can also be performed by taking a small piece of the ordinary filter paper moistened with starch solution and inserting it into the mouth of the ignition tube by folding it circularly.

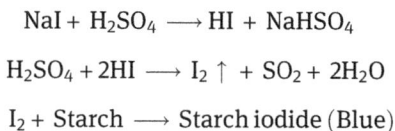

$$NaI + H_2SO_4 \longrightarrow HI + NaHSO_4$$

$$H_2SO_4 + 2HI \longrightarrow I_2 \uparrow + SO_2 + 2H_2O$$

$$I_2 + Starch \longrightarrow Starch\ iodide\ (Blue)$$

10.2.2.4 Sulphide (S²⁻)

(a) For the test of this radical, a small piece of the ordinary filter paper moistened with lead acetate solution is fitted into the mouth of the ignition tube by folding it circularly, and inserting it into the mouth of the test tube containing the mixture. When the mixture is heated with dilute H_2SO_4, the filter paper turns black due to evolution of H_2S gas, and reacts with lead acetate, forming PbS.

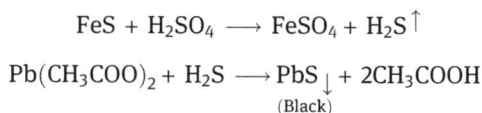

$$FeS + H_2SO_4 \longrightarrow FeSO_4 + H_2S \uparrow$$

$$Pb(CH_3COO)_2 + H_2S \longrightarrow \underset{\text{(Black)}}{PbS \downarrow} + 2CH_3COOH$$

(b) A small piece of the ordinary filter paper moistened with 1% sodium nitroprusside solution, $Na_2[Fe(NO)(CN)_5]$, in distilled water is fitted into the mouth of ignition tube and put into the mouth of the test tube containing the mixture. On heating the mixture with dilute H_2SO_4, no colour change occurs in nitroprusside solution with H_2S. However, filter paper turns purple when put over NH_4OH bottle.

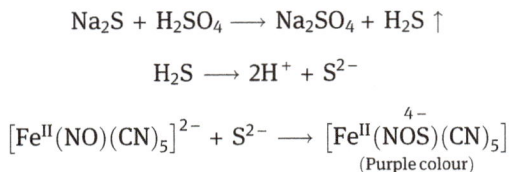

$$Na_2S + H_2SO_4 \longrightarrow Na_2SO_4 + H_2S \uparrow$$

$$H_2S \longrightarrow 2H^+ + S^{2-}$$

$$\left[Fe^{II}(NO)(CN)_5\right]^{2-} + S^{2-} \longrightarrow \left[Fe^{II}(NOS)(CN)_5\right]^{4-}$$
$$\text{(Purple colour)}$$

10.2.2.5 Sulphite (SO_3^-)

In this case, the filter paper moistened with acidified (H_2SO_4) potassium dichromate solution turns **green** due to liberation of SO_2 on heating the mixture with dilute H_2SO_4 forming chromic sulphate $[Cr_2(SO_4)_3]$.

$$Na_2SO_3 + H_2SO_4 \longrightarrow Na_2SO_4 + H_2O + SO_2 \uparrow$$

$$K_2Cr_2O_7 + H_2SO_4 + 3SO_2 \longrightarrow K_2SO_4 + Cr_2(SO_4)_3 + H_2O$$
$$\text{(\textbf{Green})}$$

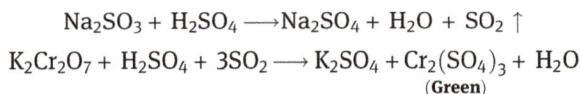

Acidified potassium dichromate solution may be prepared by putting a pinch of $K_2Cr_2O_7$ grains on a watch glass with few drops of dilute H_2SO_4 (Figure 10.2) and triturating it with the closed end of a test tube.

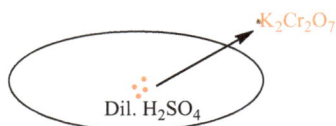

Figure 10.2: Watch glass with pinch of $K_2Cr_2O_7$ grains and few drops of dilute H_2SO_4.

10.2.2.6 Carbonate (CO_3^{2-})

A small piece of the ordinary filter paper first moistened with aqueous solution of 0.05 M Na_2CO_3 and then with 0.5% phenolphthalein solution in alcohol (resulting Pink colour) is fitted into the mouth of the ignition tube. On heating the mixture containing carbonate with dilute H_2SO_4, the pink filter paper is decolourized either immediately or after a short time depending on the quantity of CO_2 formed.

This test depends upon the fact that phenolphthalein is turned pink by soluble carbonate and colourless by soluble bicarbonate.

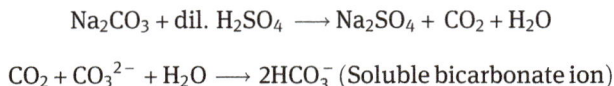

$$Na_2CO_3 + \text{dil. } H_2SO_4 \longrightarrow Na_2SO_4 + CO_2 + H_2O$$

$$CO_2 + CO_3^{2-} + H_2O \longrightarrow 2HCO_3^- \text{ (Soluble bicarbonate ion)}$$

10.2.2.7 Ammonium (NH$_4^+$)

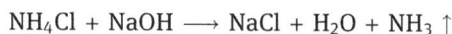

$$NH_4Cl + NaOH \longrightarrow NaCl + H_2O + NH_3 \uparrow$$

The only basic radical that can be successfully tested with the help of the ITA is the NH$_4^+$ radical. For this test, a filter paper moistened with phenolphthalein solution is fitted into the mouth of the ignition tube (fitted in the rubber cork). This is now tightly put over the mouth of test tube already containing few mg of the mixture with a few drops of NaOH solution.

It is important to note here that the phenolphthalein paper of the ignition tube should not touch the wall of the test tube; otherwise, the filter paper will turn pink even in absence of NH$_4^+$.

When the mixture is heated, the phenolphthalein paper turns pink due to liberation of ammonia. This test is based on the fact that phenolphthalein gives pink colour with dilute ammonia in the pH range 8.3–10.

10.2.3 Test of radicals with reagent filter paper

10.2.3.1 Zirconium-alizarin-S filter paper test for fluoride (F$^-$)

Preparation of test filter paper
Immerse ordinary filter papers in a 10% solution of zirconium nitrate Zr(NO$_3$)$_4$.5H$_2$O in 2M HCl; drain and place them in a 2% aqueous solution of alizarin-S. Wash the papers with water until the washings are nearly colourless. Dry the red-violet-coloured filter papers in air and preserve in a stoppered wide-mouth bottle.

Test procedure
Take ~50 mg of the salt or mixture of salts with 3–4 drops of concentrated H$_2$SO$_4$ in a dry test tube. Put a small piece of red-violet zirco-alizarin-S filter paper over the mouth of the test tube containing few mg of a salt or a mixture with 3–4 drops of concentrated H$_2$SO$_4$. On heating the test tube, the filter paper changes to pale yellow because of the formation of hexafluorozirconate(IV) ion, [Zr^{IV}F$_6$]$^{2-}$.

$$2NaF + H_2SO_4 \longrightarrow Na_2SO_4 + HF \uparrow$$

This test can be performed using ITA apparatus by taking a small piece of the zirco-alizarin-S filter paper and fitting it into the mouth of the ignition tube by folding it circularly.

10.2.3.2 Turmeric paper test for borate (BO₃³⁻)

Turmeric papers: How to prepare?
Immerse ordinary filter papers in an aqueous solution of turmeric powder. Dry the yellow filter papers in air and thereafter preserve them in a stoppered wide-mouth bottle.

Test procedure
Dip a small piece of the turmeric paper in sodium carbonate extract acidified with dilute HCl containing borate. If a piece of turmeric paper is dipped into this solution and dried at 100 °C, it becomes reddish brown. The drying of the turmeric paper may be carried out by winding it on the outside near the rim of the test tube (Figure 10.3) containing water and boiling it for 2–3 min. On moistening the paper with 1–2 drops of dilute NaOH solution, it becomes bluish black or greenish black.

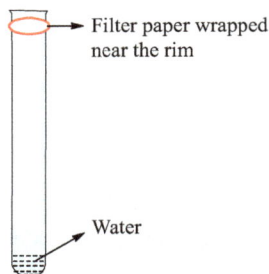

Filter paper wrapped near the rim

Water

Figure 10.3: Turmeric paper dipped in acidified sodium carbonate extract and wrapped near the rim of a test tube.

10.2.3.3 2,2′-Dipyridyl filter paper test for Fe(II)

2,2′-Dipryridyl reagent solution: How to prepare?
Dissolve 0.01 g (10 mg) of 2,2′-dipryridyl in 0.5 mL alcohol or in 0.5 mL of 0.1 M hydrochloric acid.

Test procedure
Moisten the ordinary filter paper in the 2,2′-dipryridyl solution and thereafter put 2–3 drops of the slightly acidified (dilute HCl) test solution containing Fe(II); a red colouration is obtained because of the formation of the Fe(II) dipyridyl complex, $[Fe(C_5H_4N_2)_2]^{2+}$ (Figure 10.4).

Fe(III) does not react with 2,2′-dipryridyl, and hence no red colouration is observed in the presence of Fe(III) salt in the mixture. Other metallic ions react with the reagent in acidic solution but the intensity of the resulting colours are so feeble that they don't interfere with the test of iron(II).

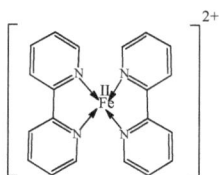

Figure 10.4: Chemical structure of Fe(II) dipyridyl complex.

10.2.3.4 o-Phenanthroline filter paper test for Fe(II)

o-Phenanthroline reagent solution: How to prepare?
The reagent consists of 0.1% solution of o-phenanthroline in water.

Test procedure
Moisten the ordinary filter paper in the reagent solution and thereafter put a drop of the slightly acidified (dilute HCl) test solution containing Fe(II); a red colouration is obtained because of the formation of the Fe(II) o-phenanthroline complex (Figure 10.5).

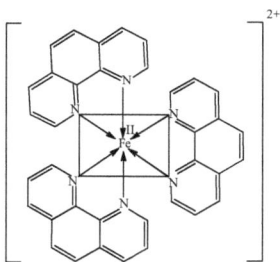

Figure 10.5: Chemical structure of Fe(II) o-phenanthroline complex.

Fe(III) has no effect and must be reduced to the bivalent state with $NH_2OH.HCl$ if the reagent is to be used in testing for iron(III).

10.2.4 Test of acid radicals using reagent solution

10.2.4.1 Test for nitrite (NO_2^-): Griess reagent test

Preparation of the Griess reagent
This requires sulphanilic acid, α-naphthylamine, glacial acetic acid and distilled water. This reagent is prepared as follows:

Solution (a): 1 g of sulphanilic acid is dissolved in 30 mL glacial acetic acid and 70 mL of distilled water by gentle heating. Filter if necessary.

Solution (b): 0.3 g (300 mg) of α-naphthylamine is dissolved in 30 mL glacial acetic acid and 70 mL of distilled water by boiling. Filter if necessary.

Mixing of the two solutions (a) and (b) gives Griess reagent which can be collected in a stoppered bottle for the test of **NO_2^-**. The nitrite reaction was discovered in 1879 by P. Griess, a German chemist (Griess 1879).

Test procedure
Sodium carbonate extract is first acidified with dilute acetic acid and then 1–2 drops of the Griess reagent is added. A pink-red colour is formed in the presence of nitrite ion present in the mixture.

This test depends upon the diazotization of sulphanilic acid by nitrous acid followed by coupling with α-naphthylamine to form a pink colour azo dye (Figure 10.6).

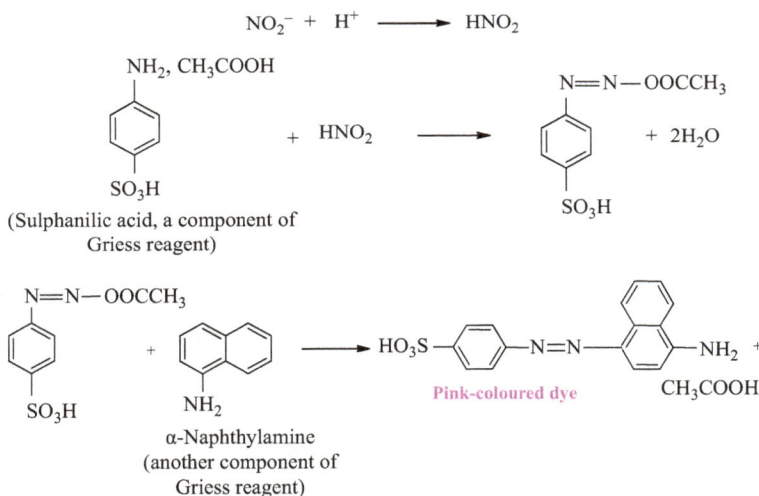

Figure 10.6: Griess reaction generating pink-coloured dye.

10.2.4.2 Test for nitrate (NO_3^-): Griess reagent rest
Sodium carbonate extract is first acidified with dilute acetic acid in little excess and then a small piece of zinc is added which reduces nitrate to nitrite due to evolution of nascent hydrogen.

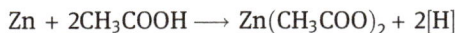

$$Zn + 2CH_3COOH \longrightarrow Zn(CH_3COO)_2 + 2[H]$$

$$NO_3^- + 2[H] \longrightarrow NO_2^- + H_2O$$

Now, addition of 1–2 drops of Griess reagent generates pink colour dye similar to nitrite. Initially, pink colour is generated around revolving zinc piece where nitrate reduces to nitrite, and after sometime the whole solution becomes pinkish.

10.2.4.3 Nitrite (NO_2^-) in presence of nitrate (NO_3^-): Griess reagent test

(a) Nitrite (NO_2^-) can be tested by taking sodium carbonate extract by Griess reagent as mentioned above. But for the test of nitrate, nitrite is to be removed from the sodium carbonate extract as follows:

   When a solution of nitrite (acidified sodium carbonate extract with dilute CH_3COOH containing nitrite) is treated with sulphamic acid, it is completely decomposed.

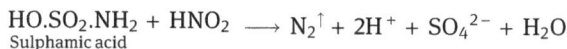

$$\underset{\text{Sulphamic acid}}{HO.SO_2.NH_2} + HNO_2 \longrightarrow N_2\uparrow + 2H^+ + SO_4^{2-} + H_2O$$

   No nitrate is formed in this reaction. It is, therefore, an excellent method for the complete removal of nitrite.

(b) After removal of nitrite, nitrate (**NO_3^-**) can be tested by Griess reagent after reducing it to nitrite by a zinc piece as mentioned above.

It is remarkable to note that the preparation of the Griess reagent is cheap, and 1–2 drops of the reagent is required in a test. It replaces costly KI for confirmatory test of NO_2^-. This test can be carried out in combinations, such as nitrate and bromide, nitrate and iodide, nitrite and sulphide and so on.

10.2.4.4 Chemical testing of coordinated NO^+ in nitrosyl complexes: Griess reagent test

The principle involved in this method is the use of the electrophillic behaviour of the coordinated NO group in the metal nitrosyls. The linearly bonded M-N-O group may-be assumed to contain NO^+ which can react with strong alkali (OH^-, nucleophile) to produce NO^-.

$$NO^+ + 2OH^- \longrightarrow NO_2^- + H_2O$$

The resulting NO_2^- gives pink colour with the Griess reagent. The probable reaction scheme (Figure 10.7) for the said reaction is as follows:

10.2.4.5 Iodine-azide test for sulphide (S^{2-}), thiosulphate ($S_2O_3^{2-}$) and thiocyanate (SCN^-)

The sodium azide-iodine reagent is prepared by dissolving 3 g of sodium azide (NaN_3) in 100 mL of 0.05 M iodine. The test is rendered more sensitive by employing

$$Na_2[Fe(NO)(CN)_5] \xrightarrow{\text{KOH, H}_2\text{O}} NO_2^- + \text{Metal oxide, etc.}$$
(A metal nitrosyl)

$$NO_2^- + H^+ \longrightarrow HNO_2$$

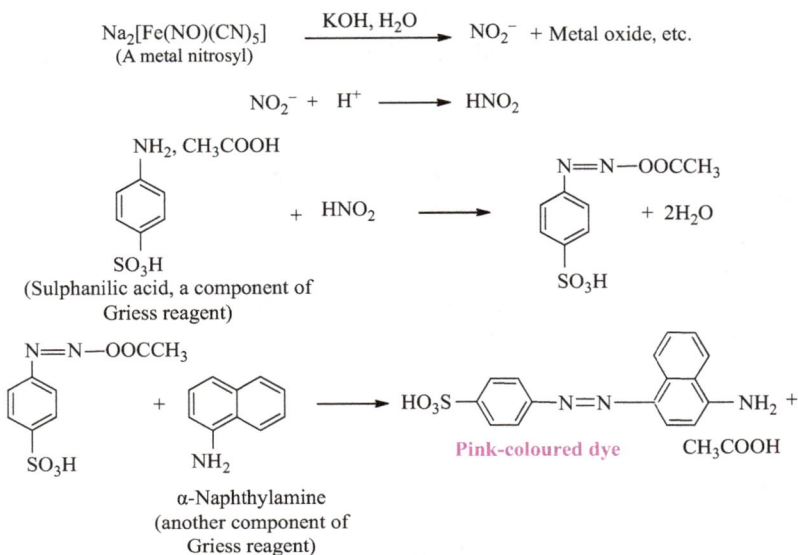

Figure 10.7: Griess reaction generating pink-coloured dye.

a more concentrated reagent composed of 1 g of NaN_3 and a few crystals of iodine in 3 mL of distilled water.

Test procedure

Mix a few drops (2–3 drops) of the reagent to 3–4 drops of sodium carbonate extract containing the acid radical to be tested. A vigorous evolution of nitrogen bubbles starts.

The solutions, NaN_3 and iodine (as I_3^-), do not react, but on the addition of a trace of $S^{2-}/ S_2O_3^{2-}/ SCN^-$, which acts as a catalyst, there is an immediate evolution of nitrogen. This test should be performed on a watch glass for clear vision.

$$2N_3^- + I_3^- \longrightarrow 3I^- + 3N_2 \uparrow$$

10.2.4.6 Zinc nitroprusside test for sulphite (SO_3^{2-})

Preparation of zinc nitroprusside, $Zn[Fe(NO)(CN)_5]$

The pinkish-white precipitate of zinc nitroprusside can be prepared by mixing a few drops of aqueous sodium nitroprusside, $Na_2[Fe(NO)(CN)_5]$, with few drops of aqueous $ZnSO_4$ solution.

Test procedure using ITA
Dip the open mouth of the ignition tube in the freshly prepared precipitate of $Zn[Fe(NO)(CN)_5]$ (pinkish white) and fit it into the test tube already containing the mixture, with few drops of dilute H_2SO_4. On heating, SO_2 evolves, which reacts with $Zn[Fe(NO)(CN)_5]$ to give a red compound of unknown composition. The test is seen more sensitive when the reaction product is held over a bottle of 1:1 NH_3 solution, which decolourizes the unused zinc nitroprusside.

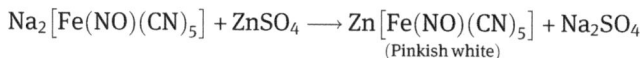

$$Na_2\big[Fe(NO)(CN)_5\big] + ZnSO_4 \longrightarrow Zn\big[Fe(NO)(CN)_5\big] + Na_2SO_4$$
$$\text{(Pinkish white)}$$

$$Na_2SO_3 + H_2SO_4 \longrightarrow Na_2SO_4 + H_2O + SO_2\uparrow$$

10.2.5 Test of basic radicals using reagent solution

10.2.5.1 Cinchonine-potassium iodide reagent for Bi^{3+}

Preparation of the reagent
1% aqueous solution of cinchonine plus 1% aqueous solution of potassium iodide makes the reagent.

Test procedure
Moisten a piece of filter paper with the cinchonine-potassium iodide reagent solution, and place a drop of slightly acidic (HCl) test solution containing Bi^{3+} upon it. An orange-red spot is obtained.

Lead(II), copper(II) and mercury(II) salts interfere in this test if present with bismuth because they react with iodide. However, bismuth may be detected in the presence of salts of these metals as they diffuse at different rates through the capillaries of the paper and are fixed in different zones. When a drop of the test solution containing Bi^{3+}, Pb^{2+}, Cu^{2+} and Hg^{2+} is placed upon the absorbent filter paper impregnated with the cinchonine-potassium iodide reagent, four zones can be observed: (i) a white central ring containing mercury, (ii) an orange-red ring due to bismuth, (iii) a yellow ring of lead iodide and (iv) a brown ring of iodine liberated by the reaction of copper.

10.2.5.2 4-Nitronaphthalene-diazoamino-azo-benzene (cadion 2B) reagent for Cd^{2+}

Reagent preparation
The reagent is prepared by dissolving 20 mg of **cadion 2B** in 100 mL ethanol to which 1 mL 2 M KOH solution is added. The solution must not be warmed.

Test procedure

Place a few drops of the reagent upon a piece of filter paper; add one drop of the test solution containing Cd^{2+} (which should be slightly acidified with 2 M CH_3COOH containing a little sodium potassium tartrate) and then one drop of 2 M KOH solution. A bright pink spot surrounded by blue circle (Figure 10.8) is produced.

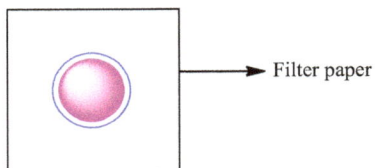

→ Filter paper

Figure 10.8: A bright pink spot surrounded by blue circle if Cd^{2+} is present in the mixture.

The chemical structure of the reagent along with sodium potassium tartrate is shown in Figure 10.9.

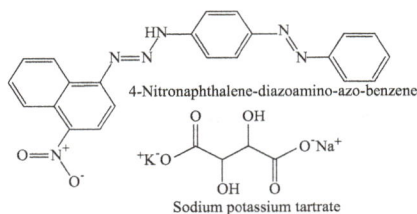

4-Nitronaphthalene-diazoamino-azo-benzene

Sodium potassium tartrate

Figure 10.9: Chemical structure of the reagent cadion 2B along with sodium potassium tartrate.

It is notable here that sodium potassium tartrate is added in the test solution to prevent the interference of Cu, Ni, Co, Cr, Fe and Mg. Only Ag and Hg then interfere. Ag is removed as AgI by the addition of a little KI solution. Mercury is best removed as mercury metal by adding a little sodium potassium tartrate, a few crystals of $NH_2OH \cdot HCl$ followed by NaOH solution until alkaline. The mercury is precipitated as metal.

The chemistry of this test is that $Cd(OH)_2$ forms a pink-coloured lake with the reagent, which contrasts with the blue tint of the latter (reagent).

10.2.5.3 Rodamine-B (or tetraethylrhodamine) reagent for Sb^{3+}

Reagent preparation

The reagent (Figure 10.10) is prepared by dissolving 0.01 g (10 mg) of rhodamine-B in 100 mL of water. The reagent so prepared is bright red in colour.

Test procedure

The test solution containing Sb^{3+} is strongly acidified with HCl first, and then add a little of solid $NaNO_2/KNO_2$ to convert Sb^{3+} to Sb^{5+}. Place 3–5 drops of the reagent on

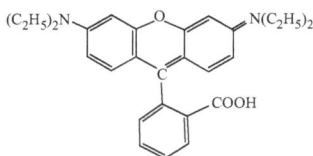

Figure 10.10: Chemical structure of tetraethylrhodamine/rodamine-B.

a piece of the filter paper and one drop of test solution. The bright red colour of the reagent changes to blue. Trivalent antimony does not respond to this test, hence it must be oxidized with $NaNO_2/KNO_2$ in the presence of strong HCl.

10.2.5.4 Cacothelin reagent for Sn^{2+}

Reagent preparation
The reagent consists of 0.25% aqueous solution of cacotheline (Figure 10.11). The reagent so prepared is yellow in colour.

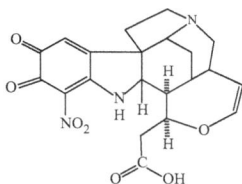

Figure 10.11: Chemical structure of cacotheline.

Test procedure
The test solution should be acidic (2M HCl), and if tin is in +4 state, it should be reduced previously with Al and Mg turnings. Impregnate a piece of filter with the reagent, and before the paper is quite dry, add a drop of the test solution. A violet spot surrounded by a less coloured zone appears on the yellow paper.

Alternatively, treat a little of the test solution in a test tube with a few drops of the reagent, a violet (purple) colouration is produced.

10.2.5.5 7-Ido-8-hydroxyquinoline-5-sulphonic acid (or Ferron reagent) for Fe(III)

Reagent preparation
The reagent (Figure 10.12) consists of 0.2% aqueous solution of 7-ido-8-hydroxyquinoline-5-sulphonic acid.

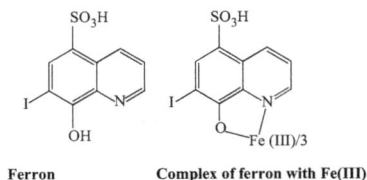

Ferron   Complex of ferron with Fe(III)

Figure 10.12: Chemical structure of ferron and its complexes with Fe(III).

Test procedure

Place a few drops of the slightly acidic (HCl or HNO_3) solution on a piece of filter paper, and add one drop of the reagent. A green colouration appears when Fe(III) is present in the test solution. Fe(II) does not react with the reagent.

10.2.5.6 Alizarin reagent for Al^{3+}

Reagent preparation

A saturated solution of alizarin in alcohol makes the reagent. The chemical structure of alizarin is given in Figure 10.13.

Test procedure

Soak some filter papers in a saturated solution of alizarin and alcohol and dry it. Place a drop of the acidic (HCl) test solution on the alizarin filter paper and hold it over ammonia fumes until a violet colour (due to ammonium alizarinate) appears on the filter paper. In the presence of large amounts of aluminium, the colour is visible almost immediately. It is best to dry the paper at 100 °C (as said in the turmeric paper test for borate) when the violate colour due to ammonium alizarinate disappears owing to its conversion into ammonia and alizarin; the **red colour of the alizarin lake is then clearly visible.**

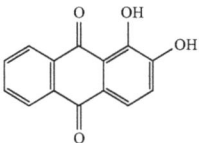

Figure 10.13: Chemical structure of alizarin.

Iron, chromium, uranium, thorium, titanium and manganese interfere in this test but this may be obviated/prevented using filter paper previously treated with aqueous K_4 $[Fe(CN)_6]$ solution. The interfering ions are thus fixed on the paper as insoluble hexacyanoferrates(II), and the aluminium solution diffuses beyond as a damping. Upon adding a drop of a saturated alcoholic solution of alizarin, exposing to ammonia vapour and drying at 100 °C, **a red ring of the alizarin-aluminium lake forms around the precipitate.** Uranium hexacyano-ferrate(II), owing to its slimy/slippery nature, has a tendency to spread outwards from the spot and thus obscure the aluminium lake. This problem is surmounted by dipping the paper after alizarin treatment in $(NH_4)_2CO_3$ solution which dissolves the uranium hexacyanoferrate(II).

10.2.5.7 Diphenylcarbazide reagent for Cr(III)

Reagent preparation
The reagent consists of 1% solution of diphenylcarbazide in alcohol.

Test procedure
Mix 2–3 drops of acidified (HCl) test solution containing Cr^{3+} in a small test tube with 2 drops of 0.1 M potassium peroxysulphate ($K_2S_2O_8$) and 1 drop of $AgNO_3$ solution (catalyst), and allow standing for 2–3 min. Add a drop of diphenylcarbazide reagent. A violet-red colour is formed owing to the following reactions (Figure 10.14).

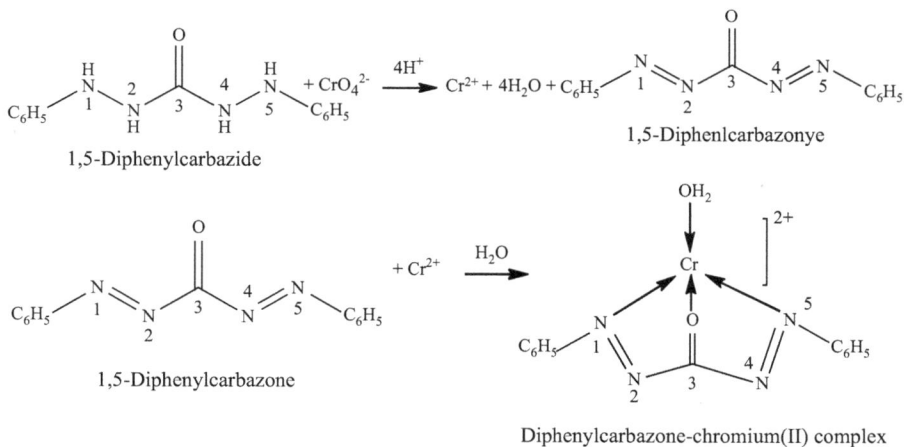

$$2Cr^{3+} + 3S_2O_8^{2-} + 8H_2O \longrightarrow 2CrO_4^{2-} + 16H^+ + 6SO_4^{2-}$$

Figure 10.14: Formation of diphenylcarbazone-chromium(II) complex involving various steps.

10.2.5.8 Sodium 1-nitroso-2-hydroxy-naphthalene-3,6-disulpohonate (nitroso-R salt) reagent for Co^{2+}

Reagent preparation
The reagent consists of a 1% solution of nitroso-R salt in water.

Test procedure
The test is applicable in the presence of nickel. Tin and iron interfere and should be removed (this test may be performed with the filtrate of the IIIrd group). The colouration produced by iron is prevented by the addition of NaF (FeF_6^{3-} complex).

Place a few drops of the neutral test solution (buffered with sodium acetate) on a piece of filter paper and add 2–3 drops of the reagent. A red colouration is obtained owing to the formation of Co(II) complex with the reagent (Figure 10.15).

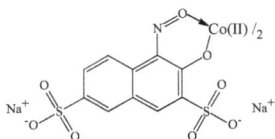

Figure 10.15: Structure of sodium 1-nitroso-2-hydroxynaphthalene-3,6-disulpohonate-Co(II) complex.

10.2.5.9 Dimethydlyoxime (DMG) reagent for Ni^{2+}

Reagent preparation
The reagent is prepared by dissolving 1 g of DMG in 100 mL alcohol.

Test procedure
Place a few drops of the slightly ammoniacal test solution containing Ni^{2+} on a piece of filter paper and add a few drops of the reagent. A red precipitate or colouration is developed on the filter paper due to formation of $[Ni(DMG)_2]$ complex (Figure 10.16).

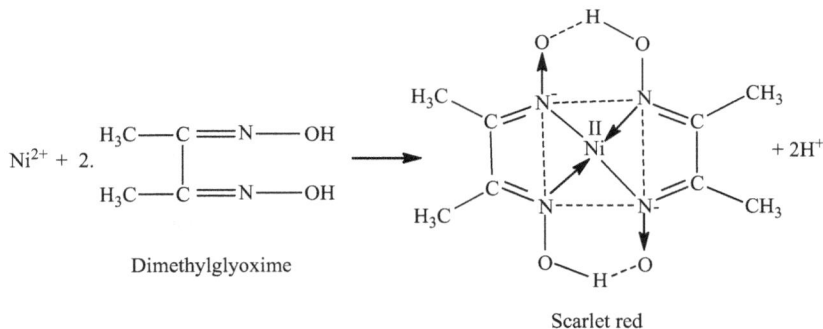

Scarlet red

Figure 10.16: Formation of bis(dimethylglyoximato)nickel(II) complex.

Fe^{2+} also gives red colouration with the reagent, so Fe^{2+} must be oxidized to Fe^{3+} by H_2O_2. When large quantities of cobalt salts are present, they react with the DMG. So, when Ni^{2+} is present with Co^{2+}, the solution containing Co^{2+} and Ni^{2+} is treated with excess of KCN solution followed by 30% H_2O_2, whereby complex cyanides $[Co^{III}(CN)_6]^{3-}$ and $[Ni(CN)_4]^{2-}$, respectively, are formed. Upon adding 40% formaldehyde (HCHO) solution, $[Co^{III}(CN)_6]^{3-}$ is unaffected (and hence remains inactive to DMG), whereas $[Ni(CN)_4]^{2-}$ decomposes with the formation of $Ni(CN)_2$ which reacts immediately with DMG.

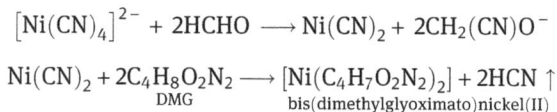

$$[Ni(CN)_4]^{2-} + 2HCHO \longrightarrow Ni(CN)_2 + 2CH_2(CN)O^-$$

$$Ni(CN)_2 + \underset{\text{DMG}}{2C_4H_8O_2N_2} \longrightarrow \underset{\text{bis(dimethylglyoximato)nickel(II)}}{[Ni(C_4H_7O_2N_2)_2]} + 2HCN \uparrow$$

10.2.5.10 Ammonium peroxodisulphate reagent for Mn^{2+}

Place 2–3 drops of the test solution (in dilute HNO_3) free of chloride in a test tube and add 1 drop of 0.1 M $AgNO_3$ solution (acts as a catalyst) and stir. Introduce a few mg of solid $(NH_4)_2S_2O_8$ and heat gently. A reddish-violet solution is formed owing to the formation of permanganate. If the test solution is prepared in HCl, it decomposes permanganate.

$$2Mn^{2+} + 5S_2O_8{}^{2-} + 8H_2O \longrightarrow 2MnO_4{}^- + 10SO_4{}^{2-} + 16H^+$$

The catalytic action of silver is due to transitional formation of silver(II), Ag^{2+} and /or silver(III), Ag^{3+}, which as powerful oxidants, oxidize manganese(II) to permanganate.

10.2.5.11 Dithizone (diphenylthiocarbazone) reagent for Zn^{2+}

Reagent preparation
The reagent is freshly prepared by dissolving 1 g of dithizone in 100 mL CCl_4.

Test procedure
Take a few drops of the test solution in a test tube and acidify it with acetic acid. Now, add a few drops of the reagent. The organic phase (CCl_4) turns red in presence of zinc. Cu^{2+}, Hg^{2+}, $Hg_2{}^{2+}$ and Ag^+ ions interfere. Upon adding NaOH solution to the mixture and shaking it, the aqueous phase also turns red. The reaction is characteristic of zinc ion only. In this reaction, Zn^{2+} forms a red complex with dithizone (Figure 10.17).

10.2.5.12 Sodium rhodizonate reagent for Ba^{2+} and Sr^{2+}

Reagent preparation
The reagent (Figure 10.18) consists of 0.5% aqueous solution of sodium rhodizonate.

This reagent forms a reddish-brown precipitate of the barium salt of rhodizonic acid in neutral solution. Calcium and magnesium salts don't interfere. Strontium

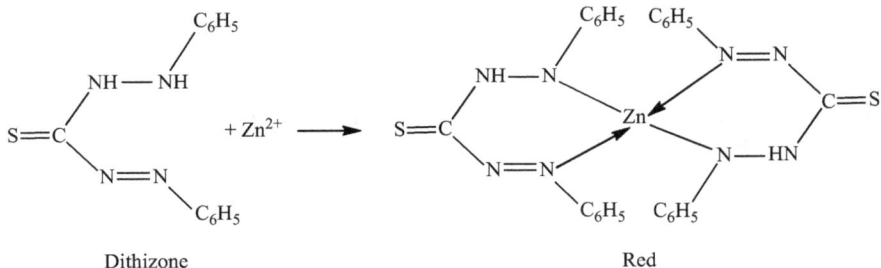

Figure 10.17: Formation of dithizone-zinc(II) complex.

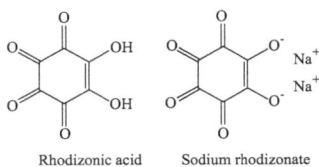

Rhodizonic acid   Sodium rhodizonate   **Figure 10.18:** Chemical structure of sodium rhodizonate.

salts react like those of barium salts giving reddish-brown precipitate but is completely soluble in dilute HCl. Other elements, for example those precipitated by H_2S, should be absent.

Test procedure
Place a drop of the neutral or slightly acidic test solution on a piece of filter paper, and add a drop of the reagent. Appearance of a brown or reddish-brown spot indicates the possibility of both Ba^{2+} and Sr^{2+}. The reddish-brown spot is now treated with a few drops of 0.5 M HCl; strontium rhodizonate dissolves while the barium derivative is converted into the brilliant red acid salt.

Test of Sr^{2+} in presence of Ba^{2+} with sodium rhodizonate reagent
Impregnate a filter paper with saturated solution of K_2CrO_4 and dry it. Place a drop of the test solution containing Ba^{2+} and Sr^{2+} on this filter paper. After a few minutes, place a drop of sodium rhodizonate reagent solution on the moistened spot. A brown-red spot or a ring at this time indicates the presence of Sr^{2+} only. In fact, barium forms insoluble $BaCrO_4$ which does not react with sodium rhodizonate reagent.

10.2.5.13 Sodium dihydroxytartrate osazone reagent for Ca^{2+}

Reagent preparation
This reagent (Figure 10.19) is used as solid.

C$_6$H$_5$.NH—N$=$C—COONa

C$_6$H$_5$.NH—N$=$C—COONa

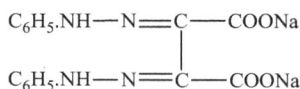

Figure 10.19: Chemical structure of sodium dihydroxytartrate osazone.

It forms a yellow sparingly soluble precipitate of the calcium salt. All other metals, with the exception of alkali and ammonium salts, must be absent. Magnesium does not interfere provided its concentration does not exceed 10 times that of calcium.

Test procedure
Place a few drops of the neutral test solution on a watch glass and add a small fragment of the solid reagent. If calcium is absent, the reagent dissolves completely. In the presence of calcium, a white film forms over the surface which ultimately separates as a dense yellow precipitate.

10.2.5.14 Titan yellow reagent for reagent Mg^{2+}

Reagent preparation
The reagent (Figure 10.20) consists of 0.1% aqueous solution of titan yellow.

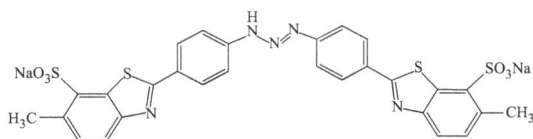

Figure 10.20: Chemical structure of titan yellow.

It is adsorbed by Mg(OH)$_2$ producing deep red colour or precipitate. Barium and calcium do not react but intensify the red colour. All elements of group I and III should be removed before applying the test.

Test procedure
Place 2–3 drops of the neutral or slightly acidic test solution on a watch glass and add a drop of the reagent and a drop of 2 M NaOH solution. A red colour or precipitation is produced due to adsorption of yellow dye by Mg(OH)$_2$.

10.3 Synthesis of inorganic compounds and metal complexes

In this section, illustrative examples of synthesis of some inorganic compounds and metal complexes will be presented taking into consideration the various objectives mentioned in the beginning of this chapter.

10.3.1 Synthesis of mercuric thiocyanate, Hg(SCN)$_2$ from mercuric chloride

Reagents required
HgCl$_2$ (0.2716 g), KSCN (0.1940 g) and ethanol (10 mL)

Reaction

$$HgCl_2 + 2KSCN \xrightarrow{C_2H_5OH} Hg(SCN)_2 + 2KCl$$

271.59	2 × 97	316.59
0.2716	0.1940	0.3166

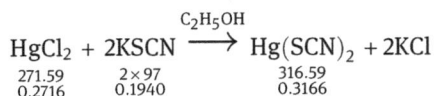

Synthetic procedure
(i) Dissolve 0.2716 g of mercuric chloride and 0.1940 g of potassium thiocyanate separately in two test tubes taking 5 mL of ethanol in each by gentle heating.
(ii) Mix the two solutions whereby KCl (being an electrolyte) is precipitated out in ethanol leaving the desired product, Hg(SCN)$_2$, in the form of a solution.
(iii) The precipitated KCl is filtered off, and the filtrate so obtained on concentration by gentle heating gives Hg(SCN)$_2$.

Colour: White
Decomposition Temperature: Report the decomposition temperature in °C
Yield: y g; % yield = y × 100/0.3166

Structure: The structure of Hg(SCN)$_2$ is given in Figure 10.21

NCS-Hg-SCN   **Figure 10.21:** Linear structure of Hg(SCN)$_2$.

Qualitative test for SCN$^-$
(a) Take 0.1 g of the compound in ~2 mL of distilled water in a test tube, and decompose it with 2–3 pellets of KOH with rigorous heating over a burner.
(b) The residue, if any, is filtered off and the extract thus obtained is acidified with dilute HCl and then added with a freshly prepared solution of FeCl$_3$. The appearance of blood red colour indicates the presence of SCN$^-$ in the compound.

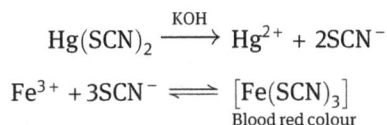

$$Hg(SCN)_2 \xrightarrow{KOH} Hg^{2+} + 2SCN^-$$

$$Fe^{3+} + 3SCN^- \rightleftharpoons \left[Fe(SCN)_3\right]$$

Blood red colour

(c) Instead of FeCl$_3$ solution, if freshly prepared CoCl$_2$ solution is added in acidified extract, blue colouration is obtained due to formation of [Co(SCN)$_4$]$^{2-}$.

$$Co^{2+} + 4SCN^- \longrightarrow \left[Co(SCN)_4\right]^{2-}$$
Blue colour

10.3.2 Synthesis of zinc thiocyanate Zn(SCN)$_2$ from zinc chloride

Reagents required
ZnCl$_2$ (0.1363 g), KSCN (0.1940 g) and ethanol (10 mL)

Reaction

$$ZnCl_2 + 2KSCN \xrightarrow{C_2H_5OH} Zn(SCN)_2 + 2KCl$$

136.38	2×97	181.38	
0.1364	0.1940	0.1814	

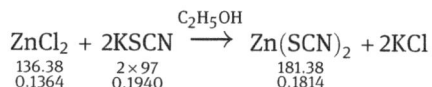

Synthetic procedure
(i)   Dissolve 0.1364 g of zinc chloride and 0.1940 g of potassium thiocyanate separately in two test tubes, taking 5 mL of ethanol in each by gentle heating.
(ii) Mix the two solutions whereby KCl (being an electrolyte) is precipitated out leaving the desired product, Zn(SCN)$_2$, in the form of a solution.
(iii) The precipitated KCl is filtered off and the filtrate so obtained on concentration by gentle heating gives Zn(SCN)$_2$.

Colour: White
Decomposition Temperature: .... °C
Yield: y g; % yield = y × 100/0.1814

Structure: The structure of Zn(SCN)$_2$ is given in Figure 10.22

NCS-Zn-SCN   **Figure 10.22:** Linear structure of Zn(SCN)$_2$.

Qualitative test for SCN$^-$
(a) Take 0.1 g of the compound in ~2 mL of distilled water in a test tube, and decompose it with 2–3 pellets of KOH with rigorous heating over a burner.

(b) The residue, if any, is filtered off and the extract thus obtained is acidified with dilute HCl and then added to a freshly prepared solution of $FeCl_3$. The appearance of blood red colour indicates the presence of SCN^- in the compound.

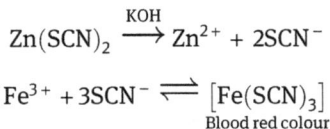

$$Zn(SCN)_2 \xrightarrow{KOH} Zn^{2+} + 2SCN^-$$

$$Fe^{3+} + 3SCN^- \rightleftharpoons [Fe(SCN)_3]$$
Blood red colour

(c) Instead of $FeCl_3$ solution, if freshly prepared $CoCl_2$ solution is added in acidified extract, blue colouration is obtained due to formation of $[Co(SCN)_4]^{2-}$.

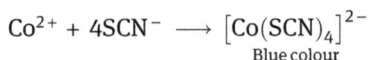

$$Co^{2+} + 4SCN^- \longrightarrow [Co(SCN)_4]^{2-}$$
Blue colour

10.3.3 Synthesis of cadmium thiocyanate Cd(SCN)₂ from cadmium chloride

Reagents required
$CdCl_2 \cdot 5/2\, H_2O$ (0.2284 g), KSCN (0.1940 g) and ethanol (10 mL)

Reaction

$$CdCl_2 \cdot 5/2H_2O + 2KSCN \xrightarrow{C_2H_5OH} Cd(SCN)_2 + 2KCl + 5/2H_2O$$

| 228.41 | 2×97 | 228.41 |
| 0.2284 | 0.1940 | 0.2284 |

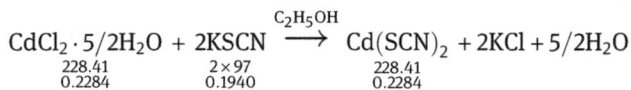

Synthetic procedure
(i)   Dissolve 0.2284 g of cadmium chloride and 0.1940 g of potassium thiocyanate separately in two test tubes, taking 5 mL of ethanol in each by gentle heating.
(ii) Mix the two solutions whereby KCl (being an electrolyte) is precipitated out leaving the desired product, $Cd(SCN)_2$, in the form of a solution.
(iii) The precipitated KCl is filtered off and the filtrate so obtained on concentration by gentle heating gives $Cd(SCN)_2$.

Colour: White
Decomposition Temperature: .... °C
Yield: y g; % yield = y × 100/0.2284

Structure: The structure of $Cd(SCN)_2$ is given in Figure 10.23

NCS-Cd-SCN **Figure 10.23:** Linear structure of $Cd(SCN)_2$.

Qualitative test for SCN⁻
(a) Take 0.1 g of the compound in ~2 mL of distilled water in a test tube, and decompose it with 2–3 pellets of KOH with rigorous heating over a burner.
(b) The residue, if any, is filtered off and the extract thus obtained is acidified with
dilute HCl and then added with a freshly prepared solution of $FeCl_3$. The appearance of blood red colour indicates the presence of SCN⁻ in the compound.

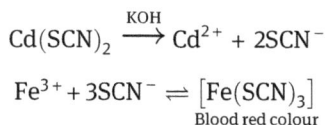

$$Cd(SCN)_2 \xrightarrow{KOH} Cd^{2+} + 2SCN^-$$

$$Fe^{3+} + 3SCN^- \rightleftharpoons \underset{\text{Blood red colour}}{[Fe(SCN)_3]}$$

(c) Instead of $FeCl_3$ solution, if freshly prepared $CoCl_2$ solution is added in acidified
extract, blue colouration is obtained due to formation of $[Co(SCN)_4]^{2-}$.

$$Co^{2+} + 4SCN^- \longrightarrow [Co(SCN)_4]^{2-}$$

10.3.4 Synthesis of copper(I) sulphate, Cu_2SO_4

Reagents required
Copper turnings, concentrated H_2SO_4, methanol and alcohol

Reaction

$$\underset{127.1}{2Cu} + \underset{196.2}{2H_2SO_4} \xrightarrow{CH_3OH} \underset{223.1}{Cu_2SO_4} + 2H_2O + SO_2 \uparrow$$

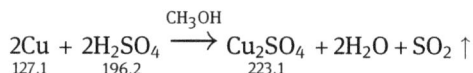

Synthetic procedure
(i) Take 5 g of copper turnings in a beaker and then add ~10 mL concentrated
H_2SO_4 in it.
(ii) The resulting mixture is heated at ~200 °C for few minutes until a green solution is obtained.
(iii) The green solution so obtained is added dropwise through a filter paper in another beaker already containing 15 mL methanol, causing Cu_2SO_4 to precipitate
in the form of almost white crystals.

(iv) Filter the precipitate through a G-3 sintered glass crucible. Wash the precipitate with alcohol, and dry it in vacuum.

Colour: Nearly white crystals or grayish powder
Yield: Report the yield in g

10.3.5 Synthesis of potassium chlorochromate (VI), K[CrO₃Cl]

Reagents required
$K_2Cr_2O_7$ (2.5 g) and concentrated HCl (3.25 mL)

Reaction

$$K_2Cr_2O_7 + 2HCl \xrightarrow{H_2O} 2K\left[Cr^{VI}O_3Cl\right] + H_2O$$
$$\underset{294.18}{\quad} \underset{72.9}{\quad} \underset{349.1}{\quad}$$

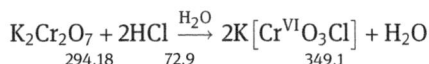

Synthetic procedure
(i) Dissolve 2.5 g of fine potassium dichromate in a mixture of 3.25 mL of concentrated HCl and 3 mL of distilled water by heating it to ~70 °C. The solution is filtered while hot.
(ii) The resulting filtrate is kept overnight on a watch glass whereby the needle-shaped orange crystals of the desired compound appears.

Colour: Orange **(needle shaped)**
Decomposition Temperature: .... °C
Yield: .... g
Structure: Tetrahedral (Figure 10.24)

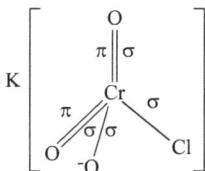

Figure 10.24: Tetrahedral structure of K[CrO₃Cl].

Qualitative test for Cl⁻
(i) Take 0.1 g of the compound in ~2 mL of distilled water in a test tube, and decompose it with 2–3 pellets of KOH with rigorous heating over a burner.
(ii) The residue, if any, is filtered off and the extract thus obtained is acidified with dilute HNO_3 and then added with a few drops of $AgNO_3$ solution. The appearance

of curdy white precipitate of AgCl, soluble in dilute NH_3 solution, indicates the presence of Cl^- in the compound.

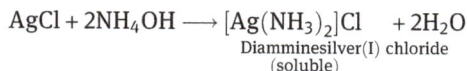

$$Cl^- + Ag^+ \longrightarrow AgCl$$

$$AgCl + 2NH_4OH \longrightarrow \underset{\substack{\text{Diamminesilver(I) chloride} \\ \text{(soluble)}}}{[Ag(NH_3)_2]Cl} + 2H_2O$$

Qualitative test of chromium

(i) Take 0.1 g of the compound in ~2 mL of distilled water in a test tube, and decompose it with 2–3 pellets of KOH with rigorous heating over a burner followed by dissolving it in dilute HNO_3.

(ii) The residue, if any, is filtered off, and the resulting solution is added with aqueous ammonia (1:1) solution whereby grey-green or grey-blue gelatinous precipitate of $Cr(OH)_3$ is formed.

$$Cr(NO_3)_3 + 3NH_4OH \longrightarrow Cr(OH)_3 + 3NH_4NO_3$$

Hybridization in $KCrO_3Cl$

The tetrahedral geometry and the diamagnetic behaviour of the molecule are well explained by the d^3s hybridization scheme shown below. Unlike other tetrahedral molecules, $KCrO_3Cl$ does not undergo sp^3 hybridization (Figure 10.25).

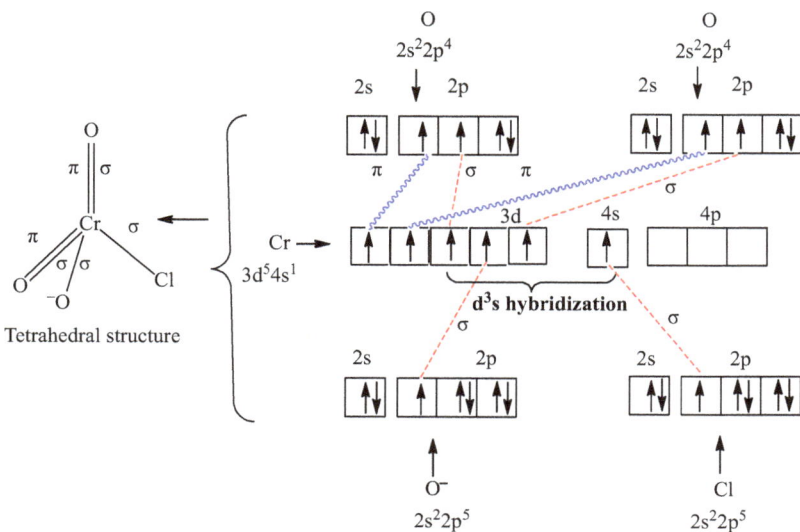

Figure 10.25: d^3s Hybridization scheme in $K[CrO_3Cl]$.

10.3.6 Synthesis of mercury(II) tetrathiocyanatocobaltate(II) complex, $Hg^{II}[Co(SCN)_4]$

Reagents required

$HgCl_2$ (0.540 g), $CoCl_2.6H_2O$ (0.475 g) and NH_4SCN (0.600 g)

Reaction

$$HgCl_2 + CoCl_2.6H_2O + 4NH_4SCN \longrightarrow Hg[Co(SCN)_4] + 4NH_4Cl$$

| 271.50 | 237.93 | $4 \times 76 = 304.00$ | 491.52 |
| 543.00 | 475.86 | 608.00 | 983.04 |

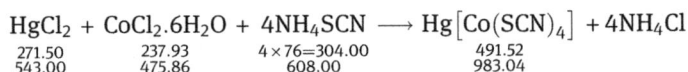

Synthetic procedure

(i)   Take 0.475 g of the $CoCl_2.6H_2O$ and 0.600 g of NH_4SCN with ~5 mLof distilled water in a 100 mL beaker and heat it to dissolve (**solution 1**).

(ii) Dissolve 0.540 g of mercuric chloride in ~2 mL of distilled water in a clean test tube (**solution 2**).

(iii) Mix the **solutions 1 and 2**, and heat again for a few minutes and allow it to cool in order to get the desired compound precipitated.

(iv) The resulting precipitate is suction filtered in a sintered glass crucible and washed 3–4 times with distilled water to remove unreacted materials.

Colour: Prussian blue

Decomposition Temperature .... °C

Yield: .... g

Structure of the complex: Tetrahedral (Figure 10.26)

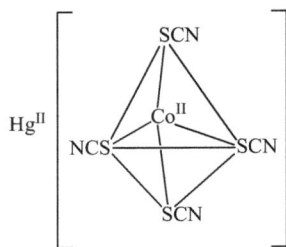

Figure 10.26: Tetrahedral structure of Hg[Co(SCN)₄].

Qualitative test for SCN⁻

(a) Take 0.1 g of the compound in ~2 mL of distilled water in a test tube, and decompose it with 2–3 pellets of KOH with rigorous heating over a burner.

(b) The residue, if any, is filtered off and the extract thus obtained is acidified with dilute HCl and then added with a freshly prepared solution of $FeCl_3$. The appearance of blood red colour indicates the presence of SCN⁻ in the compound.

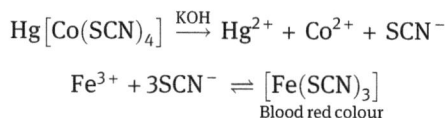

$$Hg[Co(SCN)_4] \xrightarrow{KOH} Hg^{2+} + Co^{2+} + SCN^-$$

$$Fe^{3+} + 3SCN^- \rightleftharpoons \underset{\text{Blood red colour}}{[Fe(SCN)_3]}$$

(c) The blood red colour bleaches on addition of 1–2 mLs of sodium fluoride solution because of the formation of a more stable hexafluoroferrate(III), $[FeF_6]^{3-}$.

$$[Fe(SCN)_3] + 6F^- \longrightarrow [FeF_6]^{3-} + 3SCN^-$$

Qualitative test for Hg²⁺ and Co²⁺

(i) Take 0.1 g of the compound in ~2 mL of distilled water in a test tube, and decompose it with 2–3 pellets of KOH with rigorous heating over a burner followed by dissolving it in dilute HNO_3.

(ii) The residue, if any, is filtered off, and the resulting solution is slowly added with KI solution when red precipitate/colouration of HgI_2 is obtained. The colour/precipitate so obtained dissolves in excess of KI solution owing to the formation of tetraiodomercurate(II), $[HgI_4]^{2-}$.

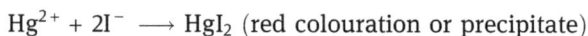

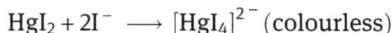

$$Hg^{2+} + 2I^- \longrightarrow HgI_2 \text{ (red colouration or precipitate)}$$

$$HgI_2 + 2I^- \longrightarrow [HgI_4]^{2-} \text{ (colourless)}$$

(iii) The solution obtained in step (i) containing Co²⁺ when added to a few crystals of NH_4SCN results in a blue colouration which appears owing to the formation of tetrathiocyanato-cobltate(II), $[Co(SCN)_4]^{2-}$.

$$Co^{2+} + 4SCN^- \longrightarrow [Co(SCN)_4]^{2-}$$

10.3.7 Synthesis of diaquabis(methyl acetoacetato)nickel(II) complex

Reagents required

$NiCl_2.6H_2O$ (1 g), methyl acetoacetate (1 mL), 1:1 ammonia solution and ethyl alcohol (4 mL)

Reaction

$$CH_3-\overset{\overset{O}{\|}}{C}-CH_2-\overset{\overset{O}{\|}}{C}-OCH_3 \quad \rightleftharpoons \quad CH_3-\overset{\overset{O}{\|}}{C}-CH=\overset{\overset{OH}{|}}{C}-OCH_3$$

Methyl acetoacetate   Enol form
Keto form

$$NiCl_2.6H_2O \; + \; 2(CH_3-\overset{\overset{O}{\|}}{C}-CH=\overset{\overset{OH}{|}}{C}-OCH_3) \; \xrightarrow[\text{Water}]{\text{Ethanol}}$$

237.69   2x116 = 232

$$[Ni(CH_3-\overset{\overset{O}{\|}}{C}-CH=\overset{\overset{O^-}{|}}{C}-OCH_3)_2(H_2O)_2] + 2HCl + 6H_2O$$

Diaquabis(methyl acetoacetato)nickel(II)

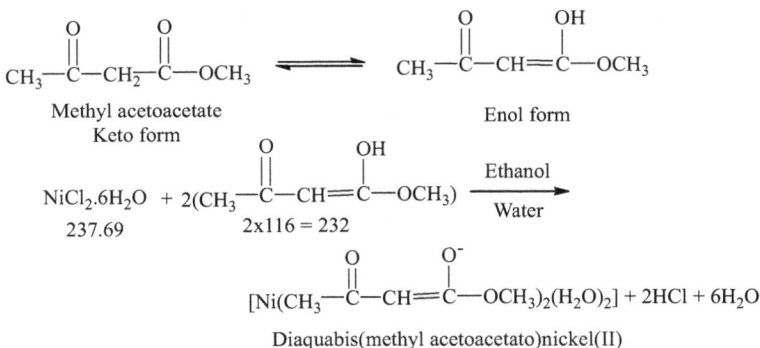

Synthetic procedure

(i) Dissolve 1 g of the nickel chloride hexahydrate in 1:1 ethanol–water (2 mL water and 2 mL ethanol) in a 100 mL beaker. Gentle heating may be done for dissolution if required.

(ii) Mix up 1 mL of methyl acetoacetate in 2 mL of ethanol in a test tube and shake well.

(iii) Mix the methyl acetoacetate solution into the nickel chloride solution prepared in step (i) and shake well.

(iv) The resulting solution is made alkaline by dropwise addition of 1:1 ammonia solution when the desired compound is precipitated as a greenish-blue solid.

(v) The precipitate is filtered off in a sintered glass crucible and washed 2–3 times with distilled water and finally with 1:10 ethanol–water 2–3 times to remove the unreacted metal salt and the ligand.

Colour: Greenish blue
Decomposition Temperature: .... °C
Yield: .... g
Structure of the complex: Octahedral (Figure 10.27)

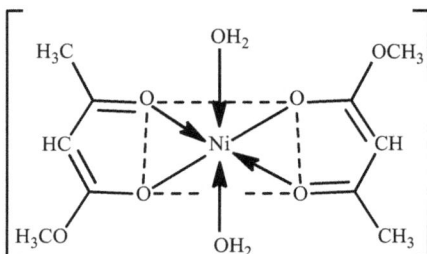

Figure 10.27: Octahedral structure of the diaquabis(methyl acetoacetato)nickel(II).

Determination of nickel as nickel oxide (NiO)
(i) A weighed amount (~200 mg) of the complex in a silica crucible is first decomposed by heating with concentrated HNO_3 (~1 mL) and then strongly heating the residue over 500 °C for ~60 min in a muffle furnace until constant weight is obtained. After heating it in the furnace, the final residue changes to NiO.
(ii) The residue is weighed as nickel oxide (NiO). From the weight of NiO, nickel content of the complex in question is calculated using the following formula:

$$\% \text{ Ni} = \frac{\text{Weight of NiO}}{\text{Weight of nickel complex}} \times 0.7858^* \times 100$$

$$* = \frac{\text{Atomic weight of nickel } (58.69)}{\text{Molecular weight of NiO } (74.69)}$$

$$= 0.7858$$

Qualitative test for Ni^{2+}
(a) Take 0.1 g of the compound in ~2 mL of distilled water in a test tube, and decompose it with 2–3 pellets of KOH with rigorous heating over a burner.
(b) The residue, if any, is filtered off and the extract thus obtained is acidified with dilute HCl and then added with a 1% ethanolic dimethylglyoxime solution. On making the solution just alkaline by addition of 1:1 aqueous ammonia solution, a scarlet red precipitate or solution is obtained due to the formation of nickel dimethylglyoxime (Figure 10.28).

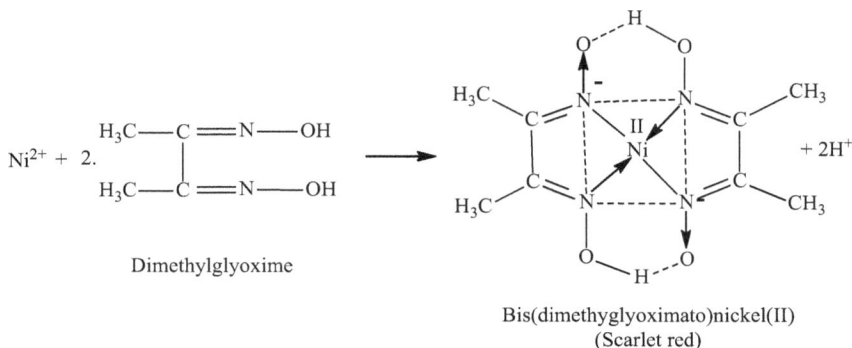

Figure 10.28: Formation of nickel dimethylglyoxime complex.

10.3.8 Synthesis of diaquabis(ethyl acetoacetato)cobalt(II) complex

Reagents required

$CoCl_2.6H_2O$ (1 g), ethyl acetoacetate (1 mL), 1:1 ammonia solution and ethyl alcohol (4 mL)

Reaction

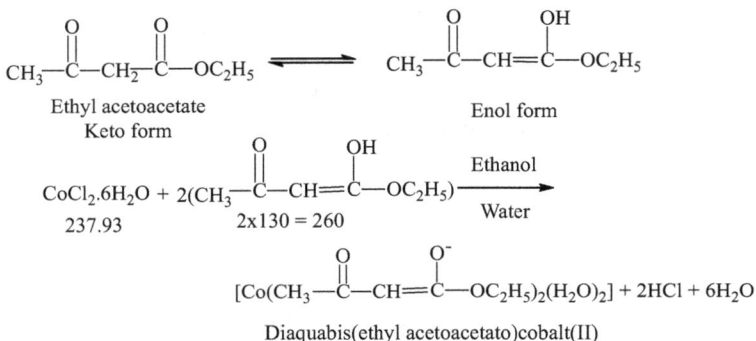

$$CoCl_2.6H_2O + 2(CH_3-\overset{O}{\overset{\|}{C}}-CH=\overset{OH}{\overset{|}{C}}-OC_2H_5) \xrightarrow[\text{Water}]{\text{Ethanol}}$$

$$237.93 \qquad 2\times130 = 260$$

$$[Co(CH_3-\overset{O}{\overset{\|}{C}}-CH=\overset{O^-}{\overset{|}{C}}-OC_2H_5)_2(H_2O)_2] + 2HCl + 6H_2O$$

Diaquabis(ethyl acetoacetato)cobalt(II)

Synthetic procedure

(i)   Dissolve 1 g of the $CoCl_2.6H_2O$ in 1:1 ethanol–water (2 mL water and 2 mL ethanol) in a 100 mL beaker. Gentle heating may be done for dissolution if required.

(ii) Mix up 1 mL of ethyl acetoacetate and 2 mL of ethanol in a test tube and shake well.

(iii) Mix the ethyl acetoacetate solution into the cobalt chloride solution prepared in step (i) and shake well.

(iv) The resulting solution is made alkaline by dropwise addition of 1:1 ammonia solution when the desired compound is precipitated as a light pink solid.

(v)   The precipitate is filtered off in a sintered glass crucible and washed 2–3 times with distilled water and finally with 1:10 ethanol–water 2–3 times to remove the unreacted metal salt and the ligand.

Colour: Light pink
Decomposition Temperature: .... °C
Yield: ....g

Structure of the complex: Octahedral (Figure 10.29)

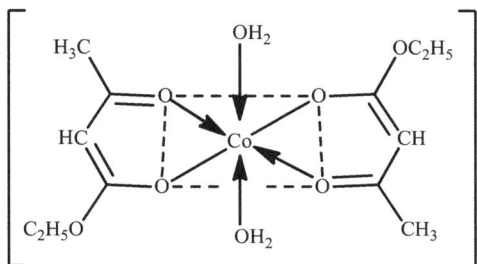

Figure 10.29: Octahedral structure of the diaquabis(ethyl acetoacetato)cobalt(II) complex.

Determination of cobalt as cobalt oxide (CoO)

(i) A weighed amount (~200 mg) of the complex in a silica crucible is first decomposed by heating with concentrated HNO_3 (~1 mL) and then strongly heating the residue over 500 °C for ~60 min in a muffle furnace until constant weight is obtained. After heating it in the furnace, the final residue changes to CoO.

(ii) The residue is weighed as cobalt oxide (CoO). From the weight of CoO, cobalt content of the complex in question is calculated using the following formula:

$$\% \text{ Co} = \frac{\text{Weight of CoO}}{\text{Weight of cobalt complex}} \times 0.7865^* \times 100$$

$$^* = \frac{\text{Atomic weight of cobalt (58.93)}}{\text{Molecular weight of CoO (74.93)}}$$

$$= 0.7865$$

Qualitative test for Co^{2+}

(i) Take 0.1 g of the compound in ~2 mL of distilled water in a test tube, and decompose it with 2–3 pellets of KOH with rigorous heating over a burner followed by dissolving in dilute HCl.

(ii) The residue, if any, is filtered off, and the resulting solution containing Co^{2+} when added with few crystals of NH_4SCN produces a blue colouration which appears owing to the formation of tetrathiocyanatocobltate(II), $[Co(SCN)_4]^{2-}$.

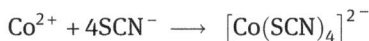

$$Co^{2+} + 4SCN^- \longrightarrow [Co(SCN)_4]^{2-}$$

10.3.9 Synthesis of bis(methyl acetoacetato)copper(II) monohydrate complex

Reagents required

$CuCl_2.2H_2O$ (0.750 g), methyl acetoacetate (1 mL), 1:1 ammonia solution and ethyl alcohol (4 mL)

Reaction

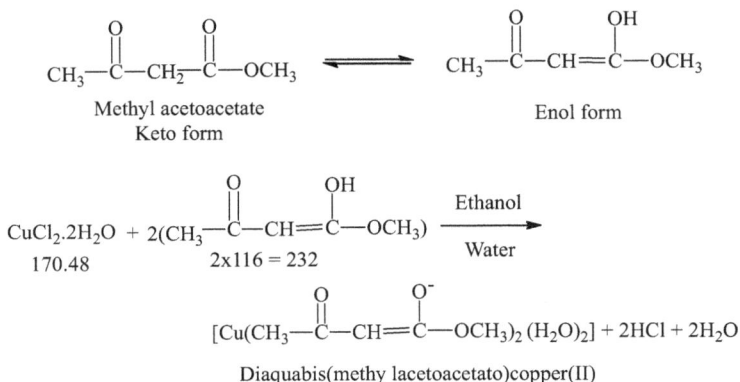

Methyl acetoacetate
Keto form

Enol form

$$CuCl_2.2H_2O + 2(CH_3-C(=O)-CH=C(OH)-OCH_3) \xrightarrow[\text{Water}]{\text{Ethanol}}$$

170.48 $\qquad$ 2x116 = 232

$$[Cu(CH_3-C(=O)-CH=C(O^-)-OCH_3)_2 (H_2O)_2] + 2HCl + 2H_2O$$

Diaquabis(methy lacetoacetato)copper(II)

Synthetic procedure

(i) Dissolve 0.750 g of the $CuCl_2.2H_2O$ in 1:1 ethanol–water (2 mL water and 2 mL ethanol) in a 100 mL beaker. Gentle heating may be done for dissolution if required.

(ii) Mix up 1 mL of methyl acetoacetate in 2 mL of ethanol in a test tube and shake well.

(iii) Mix the methyl acetoacetate solution into the nickel chloride solution prepared in step (i) and shake well.

(iv) The resulting solution is made alkaline by dropwise addition of 1:1 ammonia solution when the desired compound is precipitated as a coloured solid.

(v) The precipitate is filtered off in a sintered glass crucible and washed 2–3 times with distilled water and finally with 1:10 ethanol–water 2–3 times to remove the unreacted metal salt and the ligand.

Colour: Bluish green
Decomposition Temperature: .... °C
Yield: ....g
Structure of the complex: Square planar (Figure 10.30)

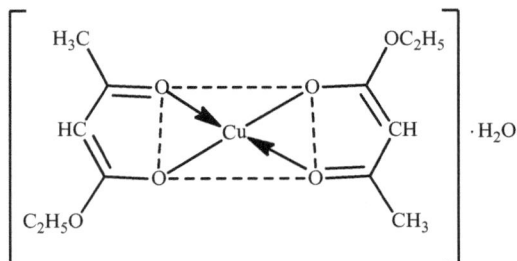

$\cdot H_2O$

Figure 10.30: Square planar structure of bis(methyl acetoacetato)copper(II) monohydrate.

Determination of copper as cupric oxide (CuO)

(i) A weighed amount (~200 mg) of the complex in a silica crucible is first decomposed by heating with concentrated HNO_3 (~1 mL) and then strongly heating the residue over 500 °C for about 60 min in a muffle furnace until constant weight is obtained. After heating it in the furnace, the final residue changes to CuO.

(ii) The residue is weighed as cupric oxide (CuO). From the weight of CuO, copper content of the complex in question was calculated using the following formula:

$$\% \text{ Cu} = \frac{\text{Weight of CuO}}{\text{Weight of copper complex}} \times 0.7988^* \times 100$$

$$^* = \frac{\text{Atomic weight of copper } (63.54)}{\text{Molecular weight of CuO } (79.54)}$$

$$= 0.7988$$

Qualitative test for Cu^{2+}

(i) Take 0.1 g of the compound in ~2 mL of distilled water in a test tube, and decompose it with 2–3 pellets of KOH with rigorous heating over a burner followed by dissolving it in dilute H_2SO_4.

(ii) The residue, if any, is filtered off, and the resulting solution containing Cu^{2+} and SO_4^{2-} ions when added with ammonia solution scarcely forms a blue precipitate of a basic copper sulphate.

$$2Cu^{2+} + SO_4^{2-} + 2NH_3 + 2H_2O \longrightarrow Cu(OH)_2 \cdot CuSO_4 \downarrow + 2NH_4^+$$

This blue precipitate is soluble in excess ammonia solution giving it a deep blue colouration owing to the formation of tetraamminecopper(II) complex ion.

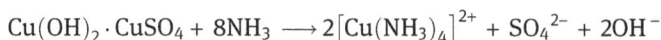

$$Cu(OH)_2 \cdot CuSO_4 + 8NH_3 \longrightarrow 2[Cu(NH_3)_4]^{2+} + SO_4^{2-} + 2OH^-$$

10.3.10 Synthesis of dichlorobis(dimethyl sulphoxide)copper(II), [CuCl₂(DMSO)₂] (DMSO = dimethyl sulphoxide)

Reagents required

Anhydrous $CuCl_2$ (0.6 g), absolute alcohol (4 mL) and DMSO (1 mL).

Reaction with other details of DMSO

Dimethyl sulphoxide is structurally similar to acetone, with sulphur replacing the carbonyl carbon.

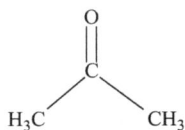

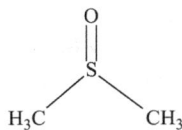

Acetone   Dimethyl sulphoxide

The $v(S=O)$ occurs at 1,650 cm^{-1}. This is lower than $v(C=O)$ (1,700 cm^{-1}) since the SO bond has a larger reduced mass than the CO bond resulting in the frequency shift. Metal can bond to DMSO either through its oxygen or its sulphur. If the bonding is through sulphur, the metal donates electron from its π-orbitals (t_{2g}) into an empty π-orbital on the DMSO ligand, thereby increasing the S-O bond order. Thus, if the metal is bonded through sulphur, the $v(S=O)$ increases. If the bonding is through oxygen, the metal forms a bond through one of the lone pairs on the oxygen, and thereby withdraws electron density from the oxygen. The net effect is that the S=O bond order reduces and the $v(S=O)$ appears at lower frequency.

The copper(II) chloride forms a metal complex in 1:2 stoichiometry as given below:

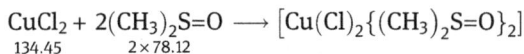

$$CuCl_2 + 2(CH_3)_2S=O \longrightarrow [Cu(Cl)_2\{(CH_3)_2S=O\}_2]$$
$$\underset{134.45}{} \qquad \underset{2 \times 78.12}{}$$

Synthetic procedure
(i)   Place 0.6 g of anhydrous copper(II) chloride in a 100 mL beaker. Add 4 mL of absolute alcohol and stir it until the metal salt is dissolved.
(ii) 1 mL of DMSO is added drop by drop with stirring in the above solution. The immediate exothermic reaction yields a coloured precipitate. The mixture is stirred further for ~10 min for complete precipitation.
(iii) The desired compound is isolated by suction filtration in a sintered glass crucible; wash it several times with cold absolute ethanol and dry it in a desiccator over anhydrous $CaCl_2$.

Colour: Light green
Decomposition Temperature: .... °C
Yield: ....g
Structure of the complex: Square planar (Figure 10.31)

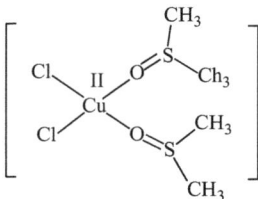

Figure 10.31: Square planar structure of dichlorobis(dimethyl sulphoxide)copper(II) complex.

Qualitative test for Cl⁻

(i) Take 0.1 g of the compound in ~2 mL of distilled water in a test tube, and de-compose it with 2–3 pellets of KOH with rigorous heating over a burner.

(ii) The residue, if any, is filtered off and the extract thus obtained is acidified with dilute HNO_3 and then added with a few drops of $AgNO_3$ solution. The appearance of curdy white precipitate soluble in dilute NH_3 solution which re-precipitates by addition of concentrated HNO_3 indicates the presence of Cl⁻ in the compound.

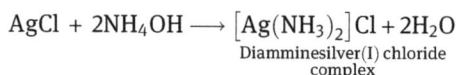

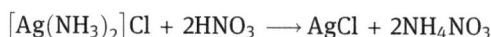

$$Cl^- + Ag^+ \longrightarrow AgCl \, (\text{curdy white})$$

$$AgCl + 2NH_4OH \longrightarrow \underset{\substack{\text{Diamminesilver(I) chloride} \\ \text{complex}}}{\left[Ag(NH_3)_2\right]Cl} + 2H_2O$$

$$\left[Ag(NH_3)_2\right]Cl + 2HNO_3 \longrightarrow AgCl + 2NH_4NO_3$$

Determination of copper as cupric oxide (CuO)

(i) A weighed amount (~200 mg) of the complex in a silica crucible is first decom-posed by heating it in concentrated HNO_3 (~1 mL) and then strongly heating the residue over 500 °C for about 60 min in a muffle furnace until constant weight is obtained. After heating it in the furnace, the final residue changes to CuO.

(ii) The residue is weighed as cupric oxide (CuO). From the weight of CuO, copper content of the complex in question was calculated using the following formula:

$$\% \, Cu = \frac{\text{Weight of CuO}}{\text{Weight of copper complex}} \times 0.7988^* \times 100$$

$$^* = \frac{\text{Atomic weight of copper } (63.54)}{\text{Molecular weight of CuO } (79.54)}$$

$$= 0.7988$$

10.3.11 Synthesis of dimeric tris(thiourea)copper(I) sulphate

Reagents required

Thiourea (NH_2CSNH_2) and copper(II) sulphate pentahydrate ($CuSO_4 \cdot 5H_2O$)

Principle and reaction

The thiourea acts as a reducing agent converting Cu(II) to Cu(I), and then as a com-plexing agent in this synthesis. When an element can exist in more than one oxidation state in aqueous solution, each oxidation state will have a different thermodynamic stability. In this experiment, the stability of Cu(I), which is generally less stable in com-parison to Cu(II) in aqueous solution, will be achieved by complex formation using thiourea ligand that can function both as σ-donor and π- acceptor. Thiourea can coordinate to the metal via the sulphur atom.

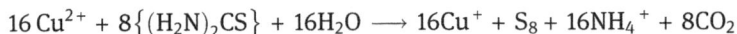

$$16\,Cu^{2+} + 8\{(H_2N)_2CS\} + 16H_2O \longrightarrow 16Cu^+ + S_8 + 16NH_4^+ + 8CO_2$$

$$Cu^+ + 3(H_2N)_2CS \longrightarrow \left[Cu\{(H_2N)_2CS\}_3\right]^+$$

Synthetic procedure
(i) Dissolve 2.5 g of thiourea in 15 mL of distilled water in a 100 mL beaker at room temperature (**solution I**).
(ii) In another beaker, dissolve 2.5 g of $CuSO_4.5H_2O$ in 15 mL of distilled water. Cool this solution in an ice bath (**solution II**).
(iii) Gradually and slowly add the solution **II** to solution **I** while stirring continuously. Allow the mixture to stand in the ice bath until white crystals or oily drops are formed.
(iv) Prepare another solution of thiourea (1 g in 10 mL of water), and add this solution to the cool reaction mixture while stirring vigorously.
(v) Allow the final mixture to stand for a while whereby the desired compound is precipitated as white crystals.
(vi) The resulting precipitate is filtered off in a sintered glass crucible, and washed, first 2–3 times with water and finally with alcohol.

Recrystallization of the crude compound
Make a solution of NH_2SNH_2 by dissolving ~0.15 g of it in 30 mL of water containing a few drops of dilute (1 M) sulphuric acid. Now, dissolve the crude product of the complex using this solution while heating it at around 75 °C. The solution so obtained is then filtered through a folded/pleated filter paper. The filtrate may also be heated until any precipitated crystals have dissolved. As a little of colloidal sulphur passes through the filter, so the solution will not become completely clear. The solution is cooled down afterward to room temperature by standing for 2–3 h. The crystals are collected by filtration with suction, washed with 5 mL portion of cold water followed by 5 mL portion of alcohol and dried in air overnight.

Colour: White
Decomposition Temperature: .... °C
Yield: y g

Structure of the complex: The structure of the complex compound dimeric tris(thiourea)copper(I) sulphate is shown in Figure 10.32

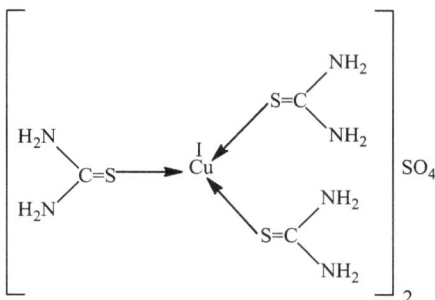

Figure 10.32: Dimeric triangular planarstructure of tris(thiourea)copper(I) sulphate.

Determination of copper as cupric oxide (CuO)

(i) A weighed amount (~200 mg) of the complex is first decomposed by heating it with concentrated HNO_3 (~1 mL) and then strongly heating the residue over 500 °C for ~60 min in a muffle furnace until constant weight is obtained. After heating it in the furnace, the final residue transforms to the stable CuO.

(ii) The residue is weighed as cupric oxide (CuO). The copper content of the complex in question was calculated from the weight of CuO using the formula given below:

$$\% \, Cu = \frac{\text{Weight of CuO}}{\text{Weight of copper complex}} \times 0.7988^* \times 100$$

$$* \quad = \frac{\text{Atomic weight of copper (63.54)}}{\text{Molecular weight of CuO (79.54)}}$$

$$= 0.7988$$

Qualitative test for copper

(i) Take 0.1 g of the compound in ~2 mL of distilled water in a test tube, and decompose it with 2–3 pellets of KOH with rigorous heating over a burner followed by dissolving it in dilute HNO_3. During decomposition, aerial oxidation of Cu(I) to Cu (II) takes place, and in nitric acid it will be in the form of cupric nitrate $Cu(NO_3)_2$.

(ii) The residue, if any, is filtered off, and the resulting solution containing Cu^{2+} ions when added with aqueous solution of potassium ferrocyanide [potassium hexacyanoferrate(II)] forms a reddish-brown precipitate of copper hexacyanoferrate(II).

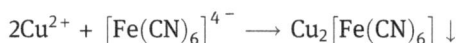

$$2Cu^{2+} + \left[Fe(CN)_6\right]^{4-} \longrightarrow Cu_2\left[Fe(CN)_6\right] \downarrow$$

The precipitate so obtained is soluble in 1:1 aqueous ammonia solution when dark blue tetraamminecopper(II) complex ions are formed.

$$Cu_2[Fe(CN)_6] + 8NH_3 \longrightarrow 2[Cu(NH_3)_4]^{2+} + [Fe(CN)_6]^{4-}$$

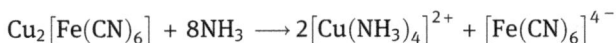

10.3.12 Synthesis of bis(acetylacetonato)dioxomolybdenum(VI), [MoO₂(acac)₂] (acacH = acetylacetone)

Reagents required
$(NH_4)_2MoO_4$ (1.5 g), acetylacetone (2 mL), nitric acid (10%), distilled water (6 mL) and ether (2 mL)

Principle and reaction
Ammonium molybdate in acidic medium in situ provides $[MoO_2]^{2+}$ species:

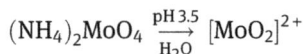

$$(NH_4)_2MoO_4 \xrightarrow[H_2O]{pH\,3.5} [MoO_2]^{2+}$$

Acetylacetone present in the reaction vessel, being a strong chelating molecule, forms a complex with $[MoO_2]^{2+}$ as shown below:

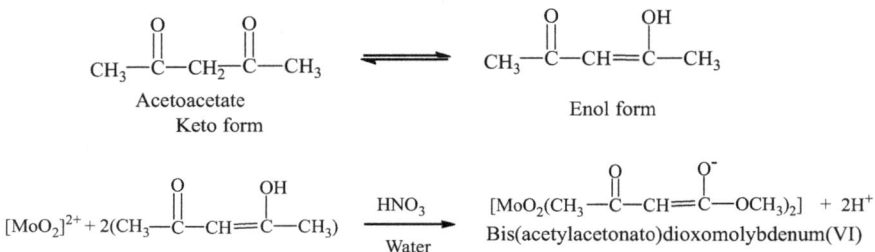

Acetoacetate
Keto form

Enol form

$[MoO_2]^{2+} + 2(CH_3-C-CH=C-CH_3) \xrightarrow[\text{Water}]{HNO_3} [MoO_2(CH_3-C-CH=C-OCH_3)_2] + 2H^+$
Bis(acetylacetonato)dioxomolybdenum(VI)

Synthetic procedure
(i) Dissolve 1.5 g of ammonium molybdate in 6 mL of distilled water in a 100 mL beaker.
(ii) Add 2 mL of acetylacetone in the ammonium molybdate solution (prepared above) drop by drop, and stir it in 1–2 min.
(iii) The pH of the resulting solution is now adjusted to 3.5 by dropwise addition of 10% nitric acid solution with stirring. The mixture is ice cooled while stirring. After ~1 h, a coloured precipitate of the desired compound is obtained.
(iv) The desired compound is isolated by suction filtration, washed several times with water followed by ether and dried in a desiccator over anhydrous $CaCl_2$.

Colour: Greenish yellow
Decomposition Temperature: .... °C
Yield: y g; %

Structure of the complex: The structure of the complex [MoO$_2$(acac)$_2$] is given in Figure 10.33

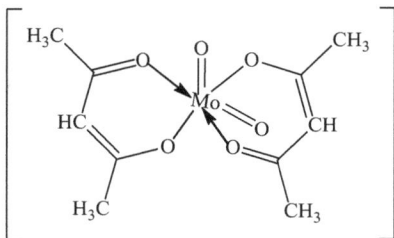

Figure 10.33: Structure of bis(acetylacetonato)-dioxomolybdenum(VI).

Determination of molybdenum as molybdenum trioxide (MoO$_3$)

(i) A weighed amount (~200 mg) of the complex in a silica crucible is first decomposed by heating it with concentrated HNO$_3$ and then strongly heating the residue over 500 °C for 40 min until constant weight was obtained. After heating it in the furnace, the final residue changes to MoO$_3$.

(ii) The residue is weighed as MoO$_3$. The molybdenum content of the complex in question is calculated from the weight of MoO$_3$ using the following formula:

$$\% \text{ Mo} = \frac{\text{Weight of MoO}_3}{\text{Weight of molybdenum complex}} \times 0.6665^* \times 100$$

$$^* = \frac{\text{Atomic weight of molybdenum (95.94)}}{\text{Molecular weight of MoO}_3 \text{ (143.94)}}$$

$$= 0.6665$$

Qualitative test for molybdenum

(i) Take 0.1 g of the compound in ~2 mL of distilled water in a test tube, and decompose it with 2–3 pellets of KOH with rigorous heating over a burner followed by acidifying/ dissolving it in dilute HCl wherein molybdenum will be present in the form of MoO$_4^{2-}$ as yellow colouration.

(ii) The residue, if any, is filtered off, and the resulting yellow solution when added with aqueous ammonium thiocyanato and a zinc piece, produces a blood-red colouration which appears owing to the formation of hexathiocyanatomolybdenum(III) [Mo(SCN)$_6$]$^-$.

$$\text{Zn} + 2\text{HCl} \longrightarrow \text{ZnCl}_2 + \text{H}_2$$

$$\text{MoO}_4^{2-} + 4\text{H}_2 \longrightarrow \text{Mo}^{3+} + 4\text{H}_2\text{O}$$

$$\text{Mo}^{3+} + 6\text{SCN}^- \longrightarrow \left[\text{Mo(SCN)}_6\right]^{3-}$$

10.3.13 Synthesis of aquabis(acetylacetonato)nitrosylchromium(I), [Cr^I(NO)(acac)$_2$(H$_2$O)] (acacH = acetylacetone)

Reagents required

Chromium(VI)oxide/chromic anhydride (CrO$_3$) (1.75 g), sodium carbonate (2.50 g + 1.25 g = 3.75 g), thiourea (NH$_2$CSNH$_2$) (1.25 g), acetylacetone (3.75 mL), NH$_2$OH·HCl (2.25 g) and distilled water (5 mL)

Principle

The synthetic strategy for low-spin nitrosyl complexes using NH$_2$OH.HCl as a reductive nitrosylating source requires basic medium along with the respective metal salt in high oxidation states. The utility of Na$_2$CO$_3$ for providing basic medium and Cr^{VI}O$_3$ as the source of chromium metal for the desired chromium(I) nitrosyl may be understood as follows:

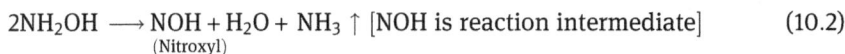

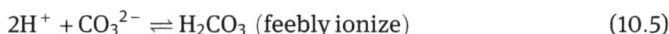

$$NH_2OH \cdot HCl \xrightarrow{H_2O} NH_2OH + HCl \tag{10.1}$$

$$2NH_2OH \longrightarrow \underset{(Nitroxyl)}{NOH} + H_2O + NH_3 \uparrow \text{[NOH is reaction intermediate]} \tag{10.2}$$

$$NOH \rightleftharpoons NO^- + H^+ \tag{10.3}$$

$$Na_2CO_3 \rightleftharpoons 2Na^+ + CO_3^{2-} \tag{10.4}$$

$$2H^+ + CO_3^{2-} \rightleftharpoons H_2CO_3 \text{ (feebly ionize)} \tag{10.5}$$

In presence of base Na$_2$CO$_3$, H$^+$ ion is being removed from the right hand side of the reaction (10.3) by the CO$_3^{2-}$ ion to form feebly ionized H$_2$CO$_3$. So, according to Le Châtelier's principle, reaction (10.3) proceeds in the forward direction to generate more and more NO$^-$. The resulting NO$^-$ species in situ reduce Cr^{6+} to Cr$^+$, and the NO$^+$ so generated and the co-ligand acetylacetone present in the reaction vessel form the desired chromium(I) nitrosyl complex.

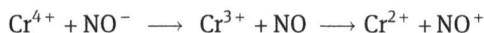

$$2CrO_3 + 2H_2O \rightleftharpoons 2H_2CrO_4 \rightleftharpoons 2Cr^{6+} + 2H_2O + 3O_2$$

$$Cr^{6+} + NO^- \longrightarrow Cr^{5+} + NO \longrightarrow Cr^{4+} + NO^+$$

$$Cr^{4+} + NO^- \longrightarrow Cr^{3+} + NO \longrightarrow Cr^{2+} + NO^+$$

$$Cr^{2+} + NO^- \longrightarrow Cr^+ + NO$$

The thiourea, a well-established reducing agent, also assists in reducing Cr(VI) to Cr(I) and probably converts itself into its well-known oxidized form formamidine disulphide.

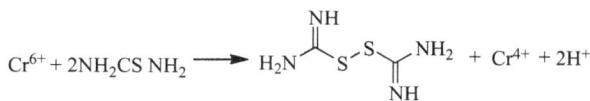

$$Cr^{6+} + 2NH_2CS\,NH_2 \longrightarrow \underset{\text{Formamidine disulphide}}{H_2N\overset{\overset{\displaystyle NH}{\|}}{}\!\!\!\!\!\overset{}{}S\!-\!\!S\overset{\overset{\displaystyle NH_2}{}}{\underset{\underset{\displaystyle NH}{\|}}{}}} + Cr^{4+} + 2H^+$$

Synthetic procedure

(i) 1.75 g of CrO_3 is dissolved in 5 mL of previously cooled distilled water, and then 2.5 g of solid Na_2CO_3 is added into it.

(ii) A solution of thiourea (1.25 g) and Na_2CO_3 (1.25 g) dissolved together in 6 mL of distilled water is added to the above solution [prepared in step (i)], and subsequently 3.75 mL of acetylacetone is added dropwise to it.

(iii) The resulting solution is then heated to room temperature and a solution containing 2.25 g of $NH_2OH.HCl$ dissolved in 2 mL of water is added to it in parts by shaking. Approximately after 30–40 min, the desired compound is precipitated as a fine brown mass.

(iv) The precipitate so obtained is filtered off and washed 3–4 times with distilled water to remove any untreated matter.

Colour: Brown
Decomposition Temperature: .... °C
Yield: .... g

Structure of the complex: The structure of the complex compound $[CrI(NO)(acac)_2(H_2O)]$ is shown in Figure 10.34.

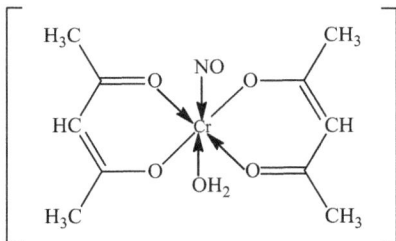

Figure 10.34: Structure of aquabis-(acetylacetonato)nitrosylchromium(I).

Chemical identification of NO group

Principle

The basic principle involved in this method is the use of electrophilic behaviour of the coordinated NO moiety in metal nitrosyls. The linearly bonded M–N–O group may be assumed to contain NO^+, which can react with strong alkali to produce NO_2^-. It is notable here that NO is present as NO^+ in Cr(I) nitrosyl compound. This is because nitrosyl complexes synthesized using alkaline hydroxyl amine as a nitrosylating agent usually contains nitrosyl grouping as NO^+.

$$NO^+ + 2OH^- \longrightarrow NO_2^- + H_2O$$

Procedure

(i) Dissolve ~100 mg of $[Cr(NO)(acac)_2(H_2O)]$ in ~1 mL of distilled water in a test tube, and decompose it with 2–3 pellets of KOH with rigorous heating over a burner.

(ii) The residue, if any, is filtered off and the extract thus obtained is acidified with dilute acetic acid and then added to 1–2 drops of the Griess reagent. The appearance of pink colour indicates the presence of nitrosyl grouping as NO^+ in the complex.

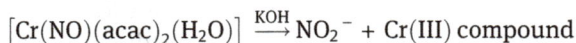

$$[Cr(NO)(acac)_2(H_2O)] \xrightarrow{\text{KOH}} NO_2^- + Cr(III) \text{ compound}$$

The reaction scheme (Figure 10.35) for the Griess reaction as suggested by Maurya et al. (1994) is given below:

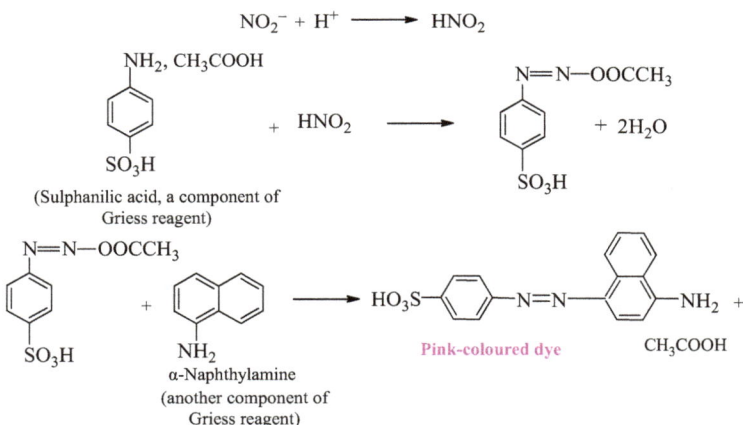

Figure 10.35: Griess reaction generating pink-coloured dye.

Determination of chromium as chromic oxide (Cr₂O₃)

(i) An aqueous suspension of a weighed amount (~200 mg) of the complex taken in an 100 mL beaker is decomposed by vigorous heating with 3–4 pellets of KOH followed by dissolving it in dilute nitric acid.

(ii) Chromium is precipitated as chromium hydroxide, $Cr(OH)_3$ by adding 1:1 aqueous NH_3 solution in the solution prepared in step (i).

$$Cr(NO_3)_3 + 3NH_4OH \longrightarrow 3NH_4NO_3 + Cr(OH)_3$$

(iii) The precipitate so obtained is filtered off in a weighed G-4 sintered glass crucible and washed several times with distilled water.
(iv) The crucible having $Cr(OH)_3$ is kept in an electric oven at around 120 °C for ~1 h
whereby the final residue left is Cr_2O_3. It is better to keep the crucible in a muffle
furnace for ~30 min at about 500 °C.

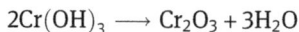

$$2Cr(OH)_3 \longrightarrow Cr_2O_3 + 3H_2O$$

(v) From the weight of Cr_2O_3, chromium content of the complex in question is calculated using the following formula:

$$\% \, Cr = \frac{\text{Weight of } Cr_2O_3}{\text{Weight of chromium complex}} \times 0.6841^* \times 100$$

$$^* \qquad = 2 \times \frac{\text{Atomic weight of chromium (51.99)}}{\text{Molecular weight of } Cr_2O_3 \text{ (151.98)}}$$

$$= 0.6841$$

10.3.14 Synthesis of sodium nitrosylpentacyanoferrate(II) dihydrate/sodium nitroprusside, $Na_2[Fe(NO)(CN)_5] \cdot 2H_2O$

Significance
In recent years, interest in metal nitrosyl complexes has been renewed following
the successful use of sodium nitroprusside as a NO donor drug during hyperten-
sive episodes. Sodium nitroprusside, $Na_2[Fe(NO)(CN)_5]\cdot 2H_2O$ (SNP, Nipride^{TM}),
which releases NO both thermally and photochemically, is used widely to induce
hypotension to reduce bleeding during surgery. Except when used briefly or at
low-infusion rates, nitroprusside in vivo releases a large quantity of cyanide, cre-
ating cyanide toxicity to whom medical treatment is given. Hence, the synthesis
of nitroprusside and qualitative tests of both coordinated nitric oxide and cyanide
are of much significance.

Reagent required
Potassium hexacyanoferrate(II), $K_4[Fe(CN)_6]$ (4 g), concentrated nitric acid (6 mL)
and Na_2CO_3.

Reaction

$$K_4[Fe(CN)_6] \cdot 3H_2O + 6HNO_3 \longrightarrow H_2[Fe(NO)(CN)_5] + 4KNO_3 + NH_4NO_3 + CO_2$$
422.4

$$H_2[Fe(NO)(CN)_5] + Na_2CO_3 \longrightarrow Na_2[Fe(NO)(CN)_5] \cdot 2H_2O + H_2O + CO_2$$
298.0

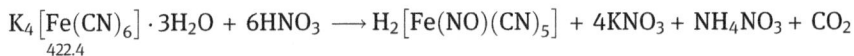

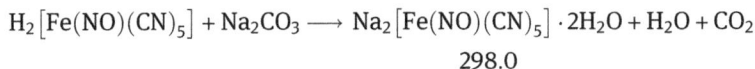

Synthetic procedure

(i)   Dissolve 4 g of $K_4[Fe(CN)_6] \cdot 3H_2O$ in 6 mL of distilled water in a 100 mL beaker with slight heating. Then add 6.5 mL of concentrated HNO_3 (density: 1.24) in parts with stirring.

(ii) The resulting mixture is now digested on a water bath at moderate temperature until a test drop of the brown solution reacts with $FeSO_4$ solution to give a dark green (rather than blue) precipitate.

(iii) After standing for 1–2 days, the digested mixture is neutralized with Na_2CO_3 (excess must be avoided). The neutralized solution is heated to a boil and concentrated to a small volume. After cooling, an equal volume of ethanol is added to precipitate most of the KNO_3.

(iv) The precipitate of KNO_3 so obtained is filtered off, and the filtrate is reconcentrated to remove the ethanol. The dark red solution yields crystals on standing for sometime.

(v) The crystals so obtained are filtered off and washed 3–4 times with cold distilled water. Further crystalline material is obtained by repeating the evaporation of the mother liquors.

Colour: Ruby red crystals
Decomposition Temperature: .... °C
Yield: .... g
Structure of the complex: Octahedral (Figure 10.36)

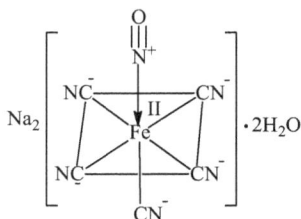

Figure 10.36: Octahedral structure of sodium nitroprusside dihydrate.

(i) Chemical identification of coordinated NO

Principle

The basic principle involved in this method is the use of electrophilic behaviour of the coordinated NO moiety in metal nitrosyls. The linearly bonded M–N–O group may be assumed to contain NO^+, which can react with strong alkali to produce NO_2^-. It is notable here that NO is present as NO^+ in nitroprusside ion.

$$NO^+ + 2OH^- \longrightarrow NO_2^- + H_2O$$

Procedure

(i) Dissolve ~100 mg of $Na_2[Fe(NO)(CN)_5]\cdot 2H_2O$ in ~1 mL of distilled water in a test tube, and decompose it with 2–3 pellets of KOH with rigorous heating over a burner.

$$Na_2\left[Fe(NO)(CN)_5\right]\cdot 2H_2O \xrightarrow{KOH} NO_2^- + \text{iron compound}$$

(ii) The residue, if any, is filtered off and the extract thus obtained containing NO_2^- is acidified with dilute acetic acid and then added with 1–2 drops of the Griess reagent. The appearance of pink colour indicates the presence of nitrosyl grouping as NO^+ in the complex. The reaction scheme is similar to that as given in case of $[Cr^I(NO)(acac)_2(H_2O)]$ (acacH = acetylacetone).

(ii) Chemical identification of coordinated cyanide group

Principle

When an aqueous solution of sodium nitroprusside is decomposed by rigorous heating with an alkali, it decomposes to Fe^{2+} and CN^-, forming ultimately $[Fe(CN)_6]^{4-}$. The filtrate containing $[Fe(CN)_6]^{4-}$ when acidified with dilute HCl and added with a pinch of solid $FeCl_3$ or a few drops of aqueous $FeCl_3$ solution gives a deep greenish-blue precipitate or colour (Prussian blue).

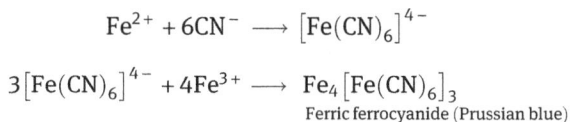

$$Fe^{2+} + 6CN^- \longrightarrow \left[Fe(CN)_6\right]^{4-}$$

$$3\left[Fe(CN)_6\right]^{4-} + 4Fe^{3+} \longrightarrow Fe_4\left[Fe(CN)_6\right]_3$$
$$\text{Ferric ferrocyanide (Prussian blue)}$$

Procedure for testing

(a) Dissolve 0.1 g of sodium nitroprusside in ~2 mL of distilled water in a test tube, and decomposed it with 2–3 pellets of KOH with rigorous heating over a burner.

(b) The residue, if any, is filtered off and the extract thus obtained is acidified with dilute HCl, and a pinch of solid $FeCl_3$ is then added. The development of greenish-blue precipitate or colouration indicates the presence of cyano grouping in nitroprusside.

10.3.15 Synthesis of nitrosylbis(diethyldithiocarbamato)iron(I), $[Fe(NO)(S_2CNEt_2)_2]$

Reagents required
Ferrous sulphate heptahydrate ($FeSO_4 \cdot 7H_2O$; FW 278.01), sodium nitrite ($NaNO_2$; FW 69.00) and sodium diethyldithiocarbamate trihydrate $[(C_2H_5)_2NCSSNa \cdot 3H_2O$; FW 225.31]

Reaction

$$2NaNO_2 + 2FeSO_4 + 3H_2SO_4 \longrightarrow Fe_2(SO_4)_3 + 2NaHSO_4 + 2H_2O + 2NO$$

or

$$2NaNO_2 + 2FeSO_4 + 2H_2SO_4 \longrightarrow Fe_2(SO_4)_3 + Na_2SO_4 + 2NO + 2H_2O$$

As NO is an odd electron molecule, it gives up an unpaired electron to Fe^{3+} to reduce it from +3 to +2 and ultimately +1, and NO becomes NO^+. NO^+ along with co-ligand diethyldithiocarbamate binds Fe^+ to form a stable complex $[Fe(NO)(S_2CNEt_2)_2]$.

$$Fe^{3+} + NO^{\cdot} \longrightarrow Fe^{2+} + NO^+$$

$$Fe^{2+} + NO^{\cdot} \longrightarrow Fe^+ + NO^+$$

Synthetic procedure
This reaction should be carried out in a fume hood/cupboard
(i) Dissolve ferrous sulphate heptahydrate (5.6 g) and sodium nitrite (1.5 g) in 25 mL of dilute (0.5 M) H_2SO_4.
(ii) Dissolve 10 g of sodium diethyldithiocarbamate trihydrate in ~4 mL of distilled water, and then add it immediately to the solution obtained in step (i).
(iii) The dark reaction mixture is stirred vigorously for 5 min and extracts the complex with small portions of chloroform (3 × 20 mL).
(iv) Separate the chloroform layer in a 100 mL separating funnel, and dry it over anhydrous $CaCl_2$.
(v) Filer the solution, and allow the filtrate to evaporate in a bowl (basin) to obtain **dark green crystals** of the complex. [The volume of the solution may be

reduced using the rotary evaporator if available. Petroleum ether is added to the solution to precipitate the **dark green solid**].

(vi) Finally collect the crystals by filtration and wash them with petroleum ether and then dry in vacuum.

Colour: Dark green crystals
Decomposition Temperature: .... °C
Yield: .... g
Structure of the complex: Square pyramidal (Figure 10.37)

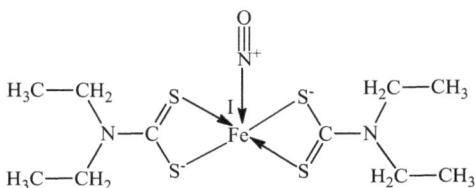

Figure 10.37: Square pyramidal structure of [Fe^I(NO)(S$_2$CNEt$_2$)$_2$].

Hybridization Scheme: This is shown in Figure 10.38

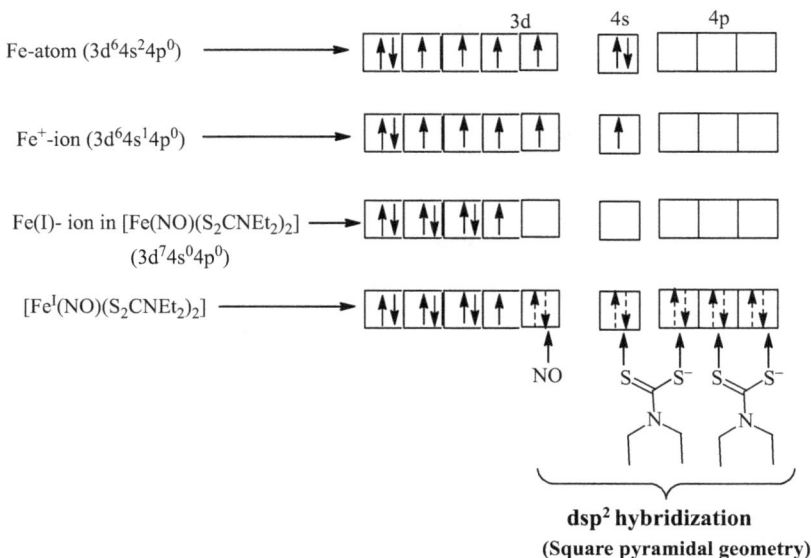

dsp^2 hybridization
(Square pyramidal geometry)

Figure 10.38: dsp^2 Hybridization scheme showing the formation of [Fe^I(NO)(S$_2$CNEt$_2$)$_2$].

10.3.16 Synthesis of dichlorobis(triphenylphosphine)nickel(II), [NiCl$_2$(PPh$_3$)$_2$]

Significance
The four-coordinated dichloropbis(phosphine)nickel(II), NiCl$_2$(PR$_3$)$_2$, compounds have been demonstrated as efficient catalysts for Suzuki cross-coupling reactions (Thananatthanachon and Lecklider 2017). These colourful nickel compounds are not only attractive towards these cross-coupling reactions due to their high reactivity, but they are also air-stable in the solid form. Furthermore, they can be easily prepared from the inexpensive NiCl$_2$·6H$_2$O precursor and the appropriate phosphine.

$$NiCl_2 \cdot 6H_2O + 2PR_3 \longrightarrow NiCl_2(PR_3)_2$$
$$R = \text{alkyl, aryl}$$

Hence, the synthesis of dichlorobis(triphenylphosphine)nickel(II)is of much significance.

Reagents required
Nickel chloride hexahydrate (NiCl$_2$·6H$_2$O), triphenylphosphine (PPh$_3$), ethanol and nitromethane (CH$_3$NO$_2$)

Reaction

$$\underset{237.69}{NiCl_2 \cdot 6H_2O} + \underset{\underset{524.58}{2 \times 262.29}}{2PPh_2} \overset{\text{Ethanol}}{\underset{N_2\text{gas}}{\longrightarrow}} [NiCl_2(PPh_3)_2] + 6H_2O$$

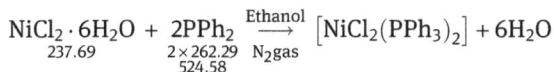

Synthetic procedure
(i)   Add 2.38 g of NiCl$_2$·6H$_2$O and 5.30 g of triphenylphosphine (PPh$_3$) in a reaction flask which is purged with a constant flow of N$_2$ gas to ensure that all of the air is removed.
(ii) 20 mL of ethanol is then quickly added to the reaction flask and the reaction mixture is heated at reflux.
(iii) Since the desired compound is colourful and insoluble in ethanol, the progress of the reaction is monitored by the colour change and formation of precipitate.
(iv) The desired compound is air-stable in the solid form and is conveniently collected by filtration using a Buchner funnel or sintered glass crucible, and immediately recrystallized from nitromethane. The crystals should be stored in a desiccator over anhydrous CaCl$_2$.

Colour: Dark blue crystals
Decomposition Temperature: .... °C
Yield: .... g

Structure of the complex: Tetrahedral (Figure 10.39)

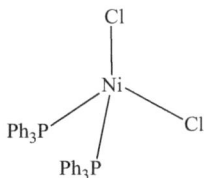

Figure 10.39: Tetrahedral structure of dichlorobis-(triphenylphosphine)nickel(II), [NiCl$_2$(PPh$_3$)$_2$].

Hybridization Scheme: It is shown in Figure 10.40.

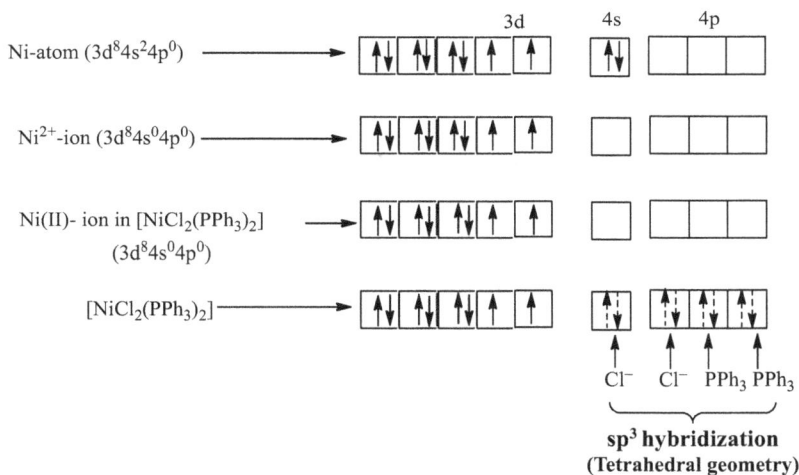

Figure 10.40: Sp^3 Hybridization scheme showing the formation of [NiCl$_2$(PPh$_3$)$_2$].

10.3.17 Synthesis of nitrosylmonobromobis(triphenylphosphine)nickel(0), [NiBr(NO)(PPh$_3$)$_2$]

Reagents required

Dibromobis(triphenylphosphine)nickel(II) [[NiBr$_2$(PPh$_3$)$_2$]; FW 743.07], sodium nitrite (NaNO$_2$; FW 69.00) and triphenylphosphine (PPh$_3$; FW 262.29)

Reaction

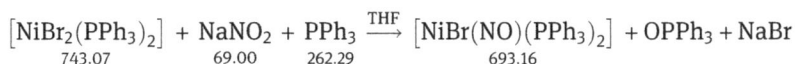

$$[\text{NiBr}_2(\text{PPh}_3)_2] + \text{NaNO}_2 + \text{PPh}_3 \xrightarrow{\text{THF}} [\text{NiBr}(\text{NO})(\text{PPh}_3)_2] + \text{OPPh}_3 + \text{NaBr}$$

743.07   69.00   262.29   693.16

Synthetic procedure

(i) Dissolve 6.73 mmoles (5 g) of [NiBr$_2$(PPh$_3$)$_2$] (Aldrich chemical) and 6.75 mmoles (1.8 g) of triphenylphosphine (PPh$_3$) in 50 mL of tetrahydrofuran (THF).

(ii) Add 115 mmoles (8 g) of freshly powdered dry NaNO$_2$ in the solution prepared in step (i) and then reflux for 35 min on a water bath.

(iii) Cool the resulting mixture, thereafter filter and the filtrate is evaporated to ~30 mL.

(iv) Addition of 25 mL of n-hexane in the concentrated filtrate gives blue oil.

(v) Separation of the blue oil and treatment with 50 mL of cold methanol give the desired mixed nitrosyl compound, [NiBr(NO)(PPh$_3$)$_2$]. The compound is purified by recrystallization from benzene-hexane mixtures.

(vi) Experiments involving other nickel halide (chloride or iodide) complexes may be carried out in an analogous manner to give respective nitrosyl compound.

Colour: Blue

Decomposition Temperature: .... °C

Yield: .... g

Structure of the complex: Tetrahedral (Figure 10.41)

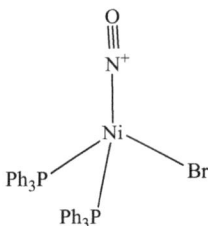

Figure 10.41: Tetrahedral structure of nitrosylmonobromobis(triphenylphosphine)nickel(0), [NiBr(NO)(PPh$_3$)$_2$].

Hybridization scheme: It is shown below (Figure 10.42)

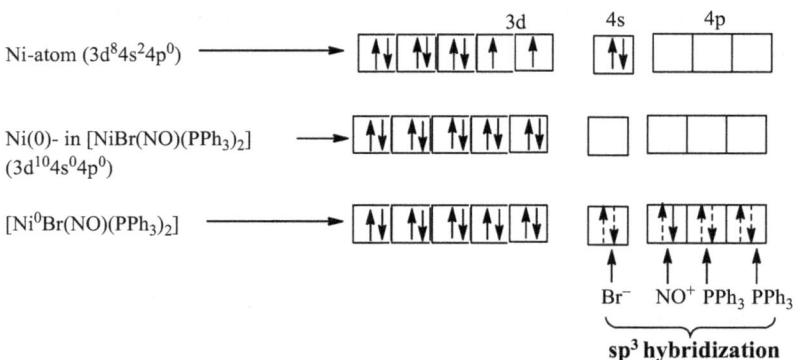

Figure 10.42: Sp^3 Hybridization scheme showing the formation of [NiBr(NO)(PPh$_3$)$_2$].

10.3.18 Synthesis of potassium hexathiocyanatochromate(III), $K_3[Cr(SCN)_6]$

Reagents required
Potassium chromium(III) sulphate (chrome alum) $[KCr(SO_4)_2]$, potassium thiocyanare (KSCN) and absolute alcohol

Reaction

$$\underset{499.4}{KCr(SO_4)_2 \times 12\ H_2O} + \underset{\substack{6 \times 97.18 \\ =583.0}}{6KSCN} \xrightarrow{H_2O} \underset{589.8}{K_3\left[Cr(SCN)_6\right] \times 4H_2O} + 2K_2SO_4$$

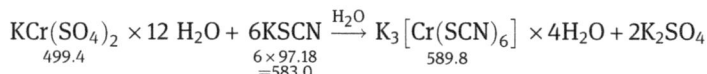

Synthetic procedure
(i)   Dissolve 5 parts (12.5 g) of chrome alum and 6 parts (3 g) of KSCN in 25 mL of water, and heat it for 2 h on a water bath.
(ii) The resulting mixture is concentrated in a beaker until the cooled residual liquid solidifies to a mass of red crystals.
(iii) The resulting solid mass is extracted with absolute alcohol, in which the desired compound dissolves very readily while K_2SO_4 remains as a residue.
(iv) After evaporation of the filtered alcoholic extract, the salt is recrystallized once more from alcohol.

Colour: Red
Decomposition Temperature: .... °C
Yield: .... g
Structure of the complex: Octahedral (Figure 10.43)

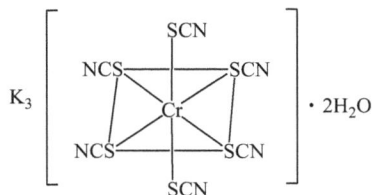

Figure 10.43: Octahedral structure of potassium hexathiocyanatochromate(III), $K_3[Cr(SCN)_6]$.

10.3.19 Synthesis of potassium trioxalatochromate(III) trihydrate, $K_3[Cr(C_2O_4)_3] \cdot 3H_2O$

Reaction

$$\underset{294.18}{\underset{Cr=+6}{K_2Cr_2O_7}} + \underset{126.07}{\underset{C=+3}{7H_2C_2O_4}} \times 2H_2O + \xrightarrow{H_2O} \underbrace{\underset{Cr=+3}{K_2C_2O_4} + Cr_2(C_2O_4)_3}_{\text{Solution}} + 21H_2O + \underset{C=+4}{6CO_2 \downarrow} \qquad (i)$$

$$\underbrace{K_2C_2O_4 + Cr_2(C_2O_4)_3}_{\text{Solution}} + \underset{184.23}{2K_2C_2O_4 \times H_2O}\ 2K_3\left[Cr(C_2O_4)_3\right] \qquad (ii)$$

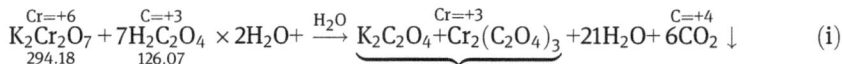

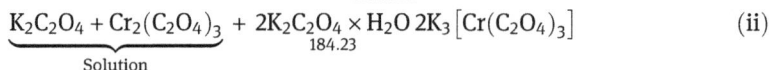

In reaction (10.6), $K_2Cr_2O_7$ is reduced to $Cr_2(C_2O_4)_3$, and $H_2C_2O_4$ is oxidized to CO_2. The solution obtained in reaction (10.6) contains $K_2C_2O_4$ and $Cr_2(C_2O_4)_3$. When this solution is treated with $K_2C_2O_4$, desired compound $K_3[Cr(C_2O_4)_3]$ is formed as shown in equation (10.7).

Synthetic procedure

(i) A concentrated aqueous solution of 6.45 mmole (1.9 g) of $K_2Cr_2O_7$ is added drop-wise with stirring to a solution containing 43.6 mmole (5.5 g) of oxalic acid dihy-drate and 12.5 mmole (2.3 g) of neutral potassium oxalate monohydrate.

(ii) Concentrate the resulting solution to a small volume. On cooling, deep green shiny crystals of $K_3[Cr(C_2O_4)_3] \cdot 3H_2O$ is formed.

(iii) Filter the solid in a sintered glass crucible and keep in a desciccator over anhydrous $CaCl_2$.

Colour: Black green
Decomposition Temperature: .... °C
Yield: .... g
Structure of the complex: Octahedral (Figure 10.44)

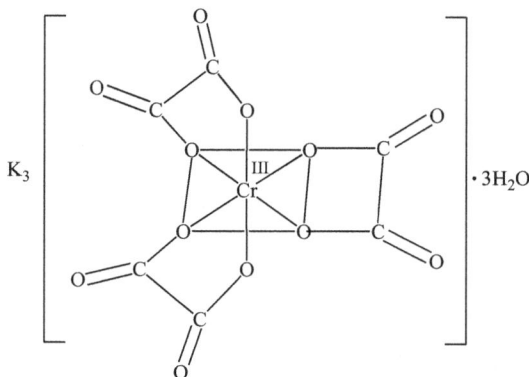

Figure 10.44: Octahedral structure of potassium trioxalatochromate(III) trihydrate, $K_3[Cr(C_2O_4)_3] \cdot 3H_2O$.

10.3.20 Synthesis of hexaamminenickel(II) chloride, [Ni(NH₃)₆]Cl₂

Reagents required
Nickel chloride hexahydrate, $NiCl_2 \cdot 6H_2O$ and concentrated ammonia (density: 0.90 g/cm³)

Reaction

$$NiCl_2 \times 6H_2O + 6NH_3 \longrightarrow \left[Ni(NH_3)_6\right]Cl_2$$

$$\underset{237.69}{} \qquad \underset{102.2}{} \qquad \underset{231.8}{}$$

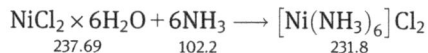

Synthetic procedure
(i) Dissolve 3 g of $NiCl_2 \cdot 6H_2O$ in 5 mL of warm water, and filter it to remove any insoluble impurities.
(ii) Add ~6 mL concentrated aqueous ammonia solution slowly to the rapidly stirred solution of nickel chloride prepared in step (i) until the green precipitate of $Ni(OH)_2$ has dissolved.
(iii) Allow the mixture to stand at room temperature for 30 min, and then remove the crystals of desired compound by filtration on a Buchner funnel.
(iv) Wash the precipitate with concentrated ammonia and then wash it with acetone/alcohol and ether, and allow the product to dry in air at room temperature.

Colour: Blue violet
Decomposition Temperature: .... °C
Yield: .... g
Structure of the complex: Octahedral (Figure 10.45)

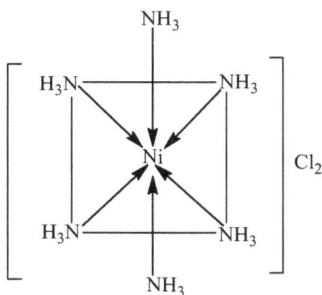

Figure 10.45: Octahedral structure of hexaamminenickel(II) chloride, [Ni(NH₃)₆]Cl₂.

Determination of nickel as bis(dimethylglyoximato)nickel(II (Figure 10.46))
Accurately weigh 0.3 g of the complex into a 500 mL beaker provided with a watch glass and stirring rod. Add 5 mL of dilute HCl (6 M) and dilute the solution to 200 mL. Heat the solution to 75 °C, and add 35 mL of 1% alcoholic solution of dimethylglyoxime. Immediately add 2M ammonia solution dropwise with stirring until precipitation

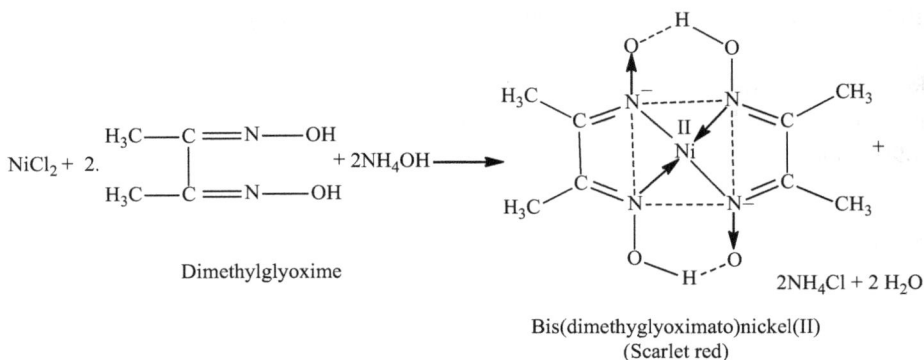

Figure 10.46: Formation of bis(dimethylglyoximato)nickel(II) complex.

occurs, and then add 10–15 drops of excess ammonia. Heat the mixture on a water bath for ~30 min. Test the solution for complete precipitation, when the red precipitate has settled, by adding few drops of dimethylglyoxime solution. Allow the precipitate to stand at room temperature for 1 h and collect the precipitate by filtration through a weighed sintered glass crucible (G-4). Wash the precipitate with cold water until the washings are free from chloride ion. Dry the precipitate at 110–120 °C for 60 min. Allow to cool in a desciccator and weigh. The process of heating and cooling is repeated until constant weight is obtained.

Factor for calculation

$$\underset{288.93}{Ni(DMG)_2} \equiv \underset{58.69}{Ni}$$

Hence, weight of nickel = 0.2031 × weight of nickel dimethylglyoxime complex.

10.3.21 Synthesis of tetraamminecopper(II) sulphate monohydrate, $[Cu(NH_3)_4]SO_4 \cdot H_2O$

Reagents required
Cupric sulphate pentahydrate ($CuSO_4 \cdot 5H_2O$), concentrated aqueous ammonia (density: 0.90 g/ cm³) and ethanol

Reaction

$$\underset{249.69}{CuSO_4 \cdot 5H_2O} + \underset{17.03 \times 4\, =\, 68.12}{4NH_3} \longrightarrow \underset{245.8}{[Cu(NH_3)_4]SO_4 \cdot H_2O}$$

Synthetic procedure

(i)   Dissolve 5 g of finely divided $CuSO_4 \cdot 5H_2O$ in 15 mL of concentrated aqueous ammonia solution and 10 mL of distilled water.

(ii) Stir the resulting solution for 5 min, and filter it to remove in any insoluble matter.

(iii) The filtrate so obtained is added with 15 mL of alcohol; allow the mixture to remain at room temperature for 30 min. Cool it in an ice-water bath, and collect the crystals that are deposited by filtration on a Büchner funnel.

(iv) Wash the product with 1:1 concentrated ammonia and ethanol followed by pure ethanol and then ether to remove alcohol, and dry it by suction.

Colour: Deep blue

Decomposition Temperature: .... °C

Yield: .... g

Structure of the complex: Square planar (Figure 10.47)

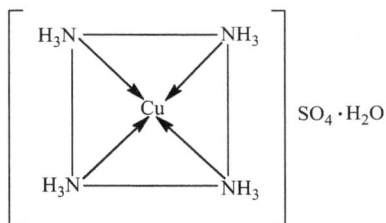

Figure 10.47: Square planar structure of $[Cu(NH_3)_4]SO_4 \cdot H_2O$.

Determination of sulphate as BaSO₄ gravimetrically

Dissolve 0.4 g of the accurately weighed complex in 150 mL of distilled water acidified with 2 mL of concentrated HCl. Add to this solution, with stirring, a solution of 10 mL of 5% $BaCl_2$, cover with a watch glass and leave the solution on a water bath for 1 h. Remove the solution from water bath, allow the precipitate ($BaSO_4$) to settle and test the clear supernatant liquid with a few drops of $BaCl_2$ solution for complete precipitation. If no precipitation is obtained, collect the precipitate by suction filtration in a sintered glass crucible (G-4).

$$CuSO_4 + BaCl_2 \longrightarrow BaSO_4 \downarrow + CuCl_2$$

Wash the precipitate with warm water until the filtrate gives no precipitate of AgCl with $AgNO_3$ solution. Dry the crucible and precipitate in an electric oven at 110 °C for 1 h. Allow to cool in a descriccator and weigh. The process of heating and cooling is repeated until constant weight is obtained.

Factor for calculation

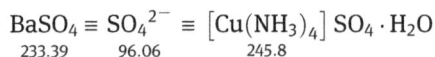

$$BaSO_4 \equiv \underset{96.06}{SO_4{}^{2-}} \equiv \underset{245.8}{[Cu(NH_3)_4]\,SO_4 \cdot H_2O}$$
$$\underset{233.39}{}$$

Weight of $SO_4{}^{2-} = 0.4115 \times$ weight of $BaSO_4$

Weight of $[Cu(NH_3)_4]SO_4 \cdot H_2O = (245.8/233.39) \times$ weight of $BaSO_4$

Determination of copper iodometrically

Dissolve 0.5 g of accurately weighed complex in 60 mL of 2M HCl, and add concentrated aqueous ammonia solution drop by drop until the solution is deep blue. Add 2M acetic acid from a measuring cylinder to give a pale blue-green solution, and then add the same volume of acetic acid solution. To this solution add 2.5 g of potassium iodide, and titrate the liberated iodine (I_2) with the standard 0.1 M sodium thiosulphate (hypo) solution using starch as the indicator. As soon as all the liberated iodine has been reduced to iodide (NaI), the blue colour of iodine-starch complex will disappear, and the colour of the precipitate in conical flask will turn white due to formation of cuprous white (Cu_2I_2).

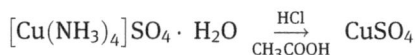

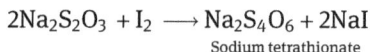

$$[Cu(NH_3)_4]SO_4 \cdot H_2O \xrightarrow[CH_3COOH]{HCl} CuSO_4$$

$$CuSO_4 + 2KI \longrightarrow CuI_2 + K_2SO_4$$

$$2CuI_2 \longrightarrow \underset{\text{Cuprous iodide (white)}}{Cu_2I_2 + I_2}$$

$$I_2 + Starch \longrightarrow I_2 - Starch\,complex\,(Blue\,colour)$$

$$2Na_2S_2O_3 + I_2 \longrightarrow \underset{\text{Sodium tetrathionate}}{Na_2S_4O_6 + 2NaI}$$

Hence,

$$2CuSO_4 \equiv I_2 \equiv 2Na_2S_2O_3$$

10.3.22 Synthesis of hexaamminecobalt(III) chloride, [Co(NH₃)₆]Cl₃

The formation of this compound exemplifies the stabilization of Co(III) in coordination compound.

Reagents required

Cobaltous chloride hexahydrate ($CoCl_2 \cdot 6H_2O$), ammonium chloride (NH_4Cl), 20 Vol. hydrogen peroxide (H_2O_2), activated charcoal and concentrated aqueous ammonia (density: 0.90 g/ cm^3)

Reaction

$$2CoCl_2 \cdot 6H_2O + 2NH_4Cl + 10NH_3 + H_2O_2 \xrightarrow{\text{Charcoal}} 2[Co(NH_3)_6]Cl_3 + 14H_2O$$

Synthetic procedure
(i) Dissolve 6 g of NH_4Cl and 9 g of $CoCl_2.6H_2O$ in 15 mL of boiling water in a 250 mL beaker, and carefully add 1 g of activated charcoal.
(ii) Cool the mixture to 0 °C in an ice bath, and add 20 mL concentrated aqueous ammonia, keeping the temperature below 10 °C. Then add 18 mL 20 Vol. H_2O_2 from a burette while the solution is stirred rapidly and the temperature is kept below 20 °C. When all the H_2O_2 has been added, heat the mixture to 60 °C until the pink colour disappears.
(iii) Now, cool the mixture in ice and collect the precipitate by filtering it on a Buchner funnel. Dissolve the precipitate in a boiling mixture of 80 mL of water and 2.5 mL of concentrated HCl, and remove the charcoal by filtration. Add 10 mL of concentrated HCl to the filtrate, and then cool the mixture in ice when crystals of [Co(NH_3)_6]Cl_3 are deposited. Collect the orange crystalline product by suction filtration, wash with acetone and dry the crystals in a desciccator over anhydrous $CaCl_2$.

Colour: Orange
Decomposition Temperature: .... °C
Yield: .... g
Structure of the complex: Octahedral (Figure 10.48)

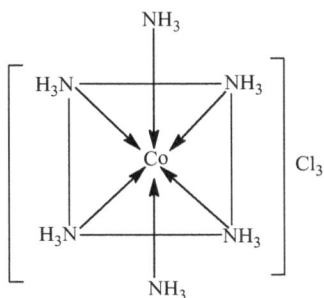

Figure 10.48: Octahedral structure of hexaamminecobalt(III) chloride.

Orange colour of this compound is due to its diamagnetic behaviour. The octahedral structure and diamagnetic nature of this compound are well explained by d^2sp^3 hybridization scheme (Figure 10.49).

Determination of cobalt iodometrically
Take 0.1 g of accurately weighed $[Co(NH_3)_6]Cl_3$ complex into a beaker, add 20 mL of 2M NaOH and boil off the ammonia. To the cooled residual alkaline suspension of

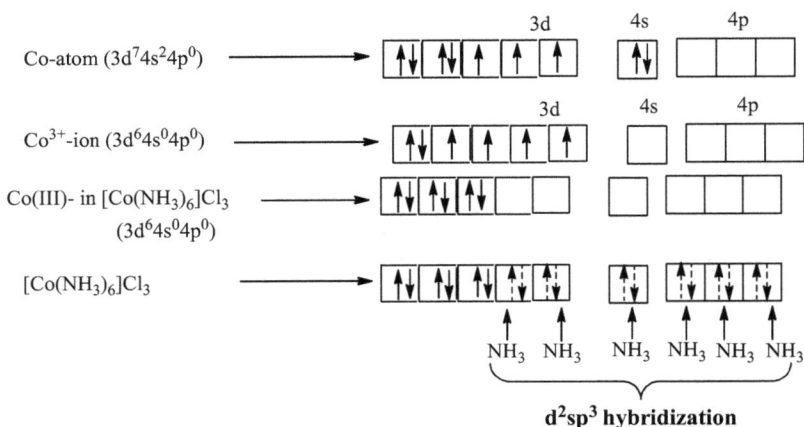

Figure 10.49: d^{2}sp^3 Hybridization scheme showing the formation of [Co(NH$_3$)$_6$]Cl$_3$.

cobaltic oxide (Co$_2$O$_3$), add 1 g of KI and acidify the liquid with 2M HCl. When the solution is acidic, add a further 40 mL of 2M HCl. When all the cobaltic oxide has dissolved, titrate the liberated iodine with 0.1 M sodium thiosulphate (hypo) solution using starch as the indicator. At the end point the blue colour disappears, and a pink solution due to cobaltous ion (Co^{2+}) remains.

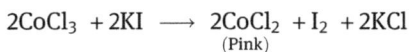

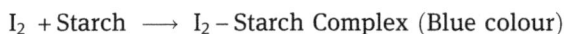

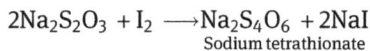

$$2[Co(NH_3)_6]Cl_3 + 6NaOH \longrightarrow Co_2O_3 + 12NH_3 \uparrow + 6NaCl + 3H_2O$$

$$Co_2O_3 + 6HCl \longrightarrow 2CoCl_3 + 3H_2O$$

$$2CoCl_3 + 2KI \longrightarrow \underset{\text{(Pink)}}{2CoCl_2} + I_2 + 2KCl$$

$$I_2 + Starch \longrightarrow I_2 - Starch\ Complex\ (Blue\ colour)$$

$$2Na_2S_2O_3 + I_2 \longrightarrow \underset{\text{Sodium tetrathionate}}{Na_2S_4O_6} + 2NaI$$

Hence,

$$2[Co(NH_3)_6]Cl_3 \equiv Co_2O_3 = 2CoCl_3 \equiv I_2 \equiv 2Na_2S_2O_3$$

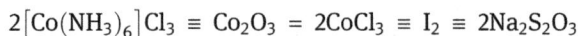

Determination of chloride

Take 0.3 g of accurately weighed [Co(NH$_3$)$_6$]Cl$_3$ complex into a beaker, add 20 mL of 2M NaOH and boil off the ammonia. Remove the cobaltic oxide (Co$_2$O$_3$) by filtration, wash it several times with hot water and combine the washings with filtrate. Acidify this solution with 2M nitric acid, and to the cold solution add 0.1 M AgNO$_3$ dropwise with stirring. Only a slight excess of AgNO$_3$ should be added. When precipitation of AgCl is completed, heat the suspension nearly to a boil until the precipitation coagulates (~5 min). Collect the precipitate by suction filtration in a G-4 sintered glass crucible. Wash the precipitate with 0.01 M nitric acid until the washings give no test of Ag^+ with HCl. Dry the precipitate in an oven at 150 °C. Now, allow to cool in a desiccator and weigh. Repeat the drying and cooling until the sintered glass crucible and the contents are a constant weight.

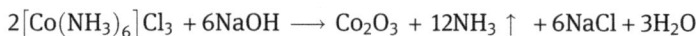

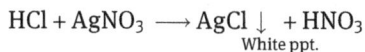

$$2[Co(NH_3)_6]Cl_3 + 6NaOH \longrightarrow Co_2O_3 + 12NH_3 \uparrow + 6NaCl + 3H_2O$$

$$NaCl + HNO_3 \longrightarrow NaNO_3 + HCl$$

$$HCl + AgNO_3 \longrightarrow AgCl \downarrow + HNO_3$$
$$\text{White ppt.}$$

Factors for calculation:

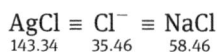

$$\underset{143.34}{AgCl} \equiv \underset{35.46}{Cl^-} \equiv \underset{58.46}{NaCl}$$

Hence,

$$\text{weight of } Cl^- = 0.2474 \times \text{weight of AgCl}$$

10.3.23 Preparation of monochloropentaamminecobalt(III) chloride, [Co(NH₃)₅Cl]Cl₂

Reagents required
Cobaltous chloride hexahydrate ($CoCl_2.6H_2O$), ammonium chloride (NH_4Cl), 30% hydrogen peroxide (H_2O_2) and concentrated aqueous ammonia (density: 0.90 g/cm³).

Reaction

$$Co^{2+} + NH_4^+ + 4NH_3 + 1/2H_2O_2 \longrightarrow [Co(NH_3)_5(H_2O)]^{3+}$$

$$[Co(NH_3)_5(H_2O)]^{3+} + 3HCl \longrightarrow [Co(NH_3)_5Cl]Cl_2 + H_2O + 3H^+$$

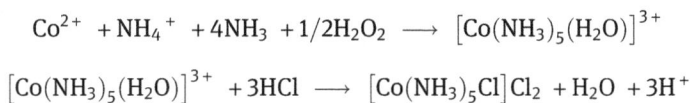

Synthetic procedure

Procedure I
This reaction should be carried out in a fume hood/cupboard.
(i) In a fume hood, add 5 g of NH_4Cl to 30 mL concentrated aqueous ammonia in a 250 mL flask (the combination of NH_4Cl and aqueous NH_3 guarantees a large excess of the NH_3 ligand).
(ii) Stir the NH_4Cl solution vigorously on a magnetic stirrer while adding 10 g of finely divided $CoCl_2 \cdot 6H_2O$ in small portions. Now add 8 mL of 30% H_2O_2 to the brown Co slurry (mixture) using a burette at a rate of ~2 drops/sec. As the reaction is exothermic, care should be taken to avoid excessive effervescence. In case of excessive effervescence, turn off the magnetic stirrer.

(iii) When the effervescence has ceased, add 30 mL of concentrated HCl slowly with continuous stirring. At this time, the reaction vessel may be removed from the hood. Now, heat the solution to 60 °C with occasional stirring by holding the temperature between 55 and 65 °C for 15 min.

(iv) Now, add 25 mL of distilled water, and allow the solution to cool to room temperature. Collect the purple product by filtration through a Buchner funnel, wash it three times with cold distilled water and finally two times with ice-cold ethanol. The washing solutions must be cold to prevent undue loss of product by redissolving.

(v) Transfer the product to a crystallizing dish, loosely cover with aluminum foil, and allow drying.

Procedure II

(i) 1.7 g of NH_4Cl is dissolved in   10 mL of concentrated ammonia NH_3 in a 250 mL beaker. 3.3 g of $CoCl_2 \cdot 6H_2O$ is now added to this solution gradually with stirring.

(ii) With continued stirring of the resulting brown slurry over a hot plate cum magnetic stirrer, 2.7 mL of 30% H_2O_2 is added slowly, and after ceasing of effervescence, 10 mL of concentrated HCl is add slowly. With continued stirring, the mixture is heated at 85 °C for 20 min.

(iii) The resulting mixture so obtained is then cooled in an ice bath whereby the desired compound appears as crystals. Collect the shiny purple crystalline product by filtration through a Buchner funnel, and wash it three times with cold distilled water and finally two times with ice-cold ethanol.

Colour: Purple
Decomposition Temperature: .... °C
Yield: .... g
Structure of the complex: Octahedral (Figure 10.50)

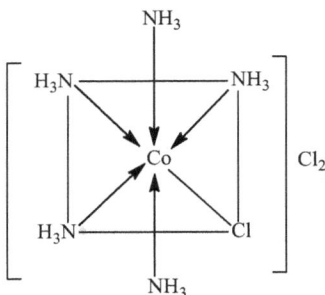

Figure 10.50: Octahedral structure of monochloropentaamminecobalt(III) chloride.

10.3.24 Synthesis of potassium tri(oxalato)ferrate(III) trihydrate, $K_3[Fe(C_2O_4)_3]\cdot 3H_2O$

This compound can be prepared by the metathetical reaction of ferric chloride and potassium oxalate. Metathetical reactions are those in which two compounds react to form two new compounds with exchange of their ions.

Reagents required
Ferric chloride hexahydrate ($FeCl_3\cdot 6H_2O$), potassium oxalate monohydrate ($K_2C_2O_4\cdot H_2O$) and distilled water

Reaction

$$\underset{\substack{270.30 \\ 5.4g}}{FeCl_3\times 6H_2O} + \underset{\substack{3\times 184.23=552.69 \\ 5.53g}}{3K_2C_2O_4\times H_2O} \longrightarrow K_3\left[Fe(C_2O_4)_3\right] + 3KCl + 7H_2O$$

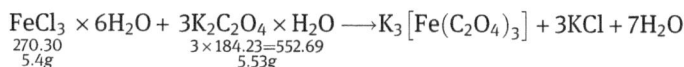

Synthetic procedure
(i) Dissolve 5.4 g of ferric chloride hexahydrate in 10 mL distilled water and 5.53 g of potassium oxalate monohydrate in 10 mL distilled water in two 100 mL beakers separately.

(ii) Mix the ferric chloride solution slowly by constantly stirring it into the potassium oxalate solution at room temperature.

(iii) The resulting mixture is cooled in an ice bath (0 °C) whereby the desired compound starts crystallizing. After crystallizing, remove the green solid by suction filtration, wash it with 1:1 mixture of ethanol and water and finally with acetone.

(iv) The complex is photosensitive, but it is quite stable if it is stored in the dark. Upon exposure to light of <450 nm, the anion $[Fe(C_2O_4)_3]^{3-}$ undergoes an intramolecular redox reaction in which the Fe(III) anion is reduced to Fe(II) while one of the oxalate groups is oxidized to CO_2.

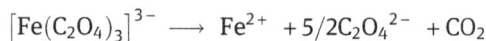

$$\left[Fe(C_2O_4)_3\right]^{3-} \longrightarrow Fe^{2+} + 5/2C_2O_4{}^{2-} + CO_2$$

The Fe^{2+} so produced can readily be detected by adding a solution of potassium ferricyanide, $K_3[Fe(CN)_6]$. A deep blue-coloured ferro ferricyanide complex is formed.

$$3Fe^{2+} + 2\left[Fe(CN)_6\right]^{3-} \longrightarrow \underset{\text{Ferro ferricyanide (Deep blue)}}{Fe_3\left[Fe(CN)_6\right]_2}$$

Colour: It is a green crystalline salt, which is soluble in hot water but rather insoluble when cold.

Decomposition Temperature: .... °C

Yield: .... g

Structure of the complex: Octahedral (Figure 10.51)

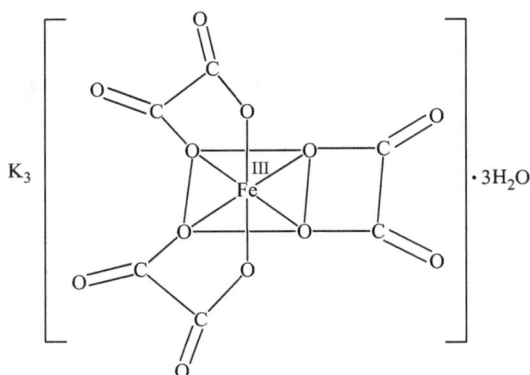

Figure 10.51: Octahedral structure of potassium tri(oxalato)ferrate(III) trihydrate.

The green colour of this compound is due to its paramagnetic behaviour (high spin complex; $\mu = 5.96$ B.M.). The octahedral structure and paramagnetic nature of this compound are well explained by sp^3d^2 hybridization scheme (Figure 10.52) shown below:

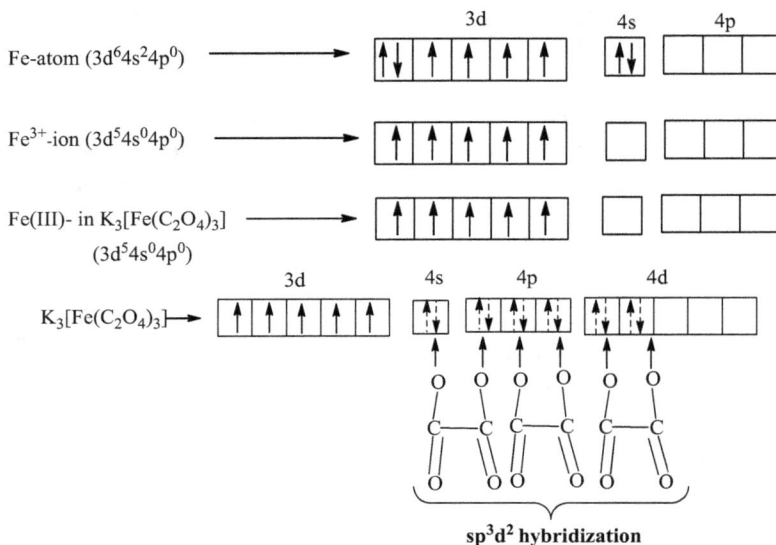

sp^3d^2 hybridization

Figure 10.52: Sp^3d^2 Hybridization scheme showing the formation of $K_3[Fe(C_2O_4)_3] \cdot 3H_2O$.

Determination of oxalate in K$_3$[Fe(C$_2$O$_4$)$_3$] · 3H$_2$O
Dissolve 0.2 g of the accurately weighed complex in 25 mL of 2M sulphuric acid. Heat the solution to 60 °C, and the free oxalic acid is titrated against standard 0.02 molar KMnO$_4$ solution until the warm solution retains light pink colour after standing for about 30 s.

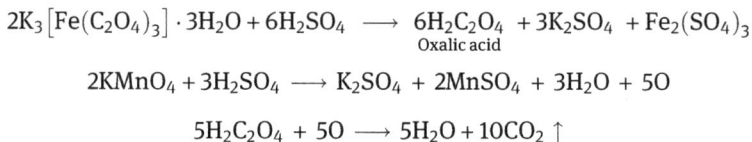

$$2K_3\left[Fe(C_2O_4)_3\right] \cdot 3H_2O + 6H_2SO_4 \longrightarrow \underset{\text{Oxalic acid}}{6H_2C_2O_4} + 3K_2SO_4 + Fe_2(SO_4)_3$$

$$2KMnO_4 + 3H_2SO_4 \longrightarrow K_2SO_4 + 2MnSO_4 + 3H_2O + 5O$$

$$5H_2C_2O_4 + 5O \longrightarrow 5H_2O + 10CO_2 \uparrow$$

Determination of iron in K$_3$[Fe(C$_2$O$_4$)$_3$] · 3H$_2$O
Dissolve 0.2 g of the accurately weighed complex in 25 mL of 2M sulphuric acid. Heat the solution to 60 °C, and the resulting ferric sulphate is reduced to ferrous sulphate by adding 2.0 g of zinc dust to the solution, and boil it for few minutes. Filter the resulting solution through glass wool and wash the residual zinc with 2M H$_2$SO$_4$. The filtrate so obtained is titrated against standard 0.02 molar KMnO$_4$ solution. In the beginning, KMnO$_4$ should be added in drops with constant shaking. At the end point, when all the ferrous salts have been oxidized, the slight excess of KMnO$_4$ will make the solution light pink.

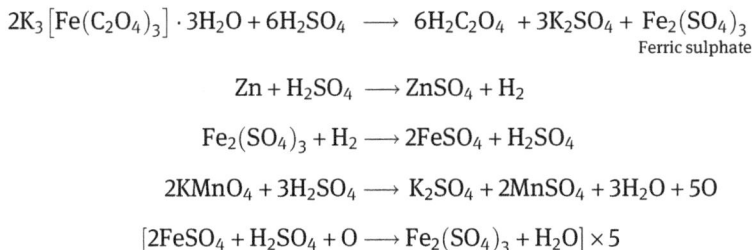

$$2K_3\left[Fe(C_2O_4)_3\right] \cdot 3H_2O + 6H_2SO_4 \longrightarrow 6H_2C_2O_4 + 3K_2SO_4 + \underset{\text{Ferric sulphate}}{Fe_2(SO_4)_3}$$

$$Zn + H_2SO_4 \longrightarrow ZnSO_4 + H_2$$

$$Fe_2(SO_4)_3 + H_2 \longrightarrow 2FeSO_4 + H_2SO_4$$

$$2KMnO_4 + 3H_2SO_4 \longrightarrow K_2SO_4 + 2MnSO_4 + 3H_2O + 5O$$

$$[2FeSO_4 + H_2SO_4 + O \longrightarrow Fe_2(SO_4)_3 + H_2O] \times 5$$

....

$$2KMnO_4 + 8H_2SO_4 + 10FeSO_4 \longrightarrow K_2SO_4 + 2MnSO_4 + 5Fe_2(SO_4)_3 + 8H_2O$$

From the above two set of titrations, determination of oxalate and iron in the complex can be carried out.

10.3.25 Synthesis of bis(acetylacetonato)oxovanadium(IV), [VO(acac)₂] (acacH = acetylacetone)

(Fedorova et al. 2005)

Reagents required
Vanadium pentoxide (V_2O_5), ethanol, concentrated H_2SO_4, acetylacetone and sodium carbonate

Basic principle reactions
(i) Vanadyl sulphate ($VOSO_4$) is prepared by the reduction of a solution of vanadium pentoxide (V_2O_5) in concentrated sulphuric acid by ethanol (C_2H_5OH). It is notable here that ethanol reduces V_2O_5 (vanadium in +V oxidation state) to $VOSO_4$ (vanadium in +IV oxidation state) and itself oxidized to acetaldehyde (CH_3CHO) [Reaction (10.8)].
(ii) $VOSO_4$ reacts with acetylacetone and gives bis(acetylacetonato)oxovanadium(IV), [VO(acac)₂] [Reaction (10.9)].

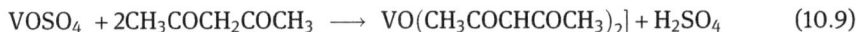

$$V_2O_5 + 2H_2SO_4 + C_2H_5OH \longrightarrow 2VOSO_4 + CH_3CHO + 3H_2O \qquad (10.8)$$

$$VOSO_4 + 2CH_3COCH_2COCH_3 \longrightarrow VO(CH_3COCHCOCH_3)_2] + H_2SO_4 \qquad (10.9)$$

Synthetic procedure
(i) Cautiously add 10 mL of concentrated H_2SO_4 to an equal volume of distilled water. Then add 5 g of vanadium pentoxide and 25 mL of ethanol to it.
(ii) Boil the resulting mixture with stirring for ~1 h until the colour changes from green to blue.
(iii) Collect the residue, if any, by filtration and add 13 mL of acetylacetone to the filtrate. Then a 10% solution of Na_2CO_3 (about 200 mL) is carefully added with stirring to neutralize the reaction mixture whereby a blue precipitate of the desired compound appears.
(iv) Collect the blue product by filtration on a Buchner funnel, wash with water and allow it to dry in air.
(v) The compound may be recrystallized from chloroform ($CHCl_3$).

Colour: Blue crystals
Decomposition Temperature: .... °C
Yield: .... g
Structure of the complex: Square pyramidal (Figure 10.53)

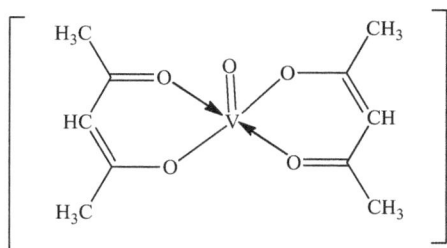

Figure 10.53: Square pyramidal structure of [VO(acac)$_2$].

The blue colour of this compound is due to its paramagnetic behaviour. The square pyramidal structure and paramagnetic nature of this compound are well explained by dsp^3 hybridization scheme (Figure 10.54) shown below:

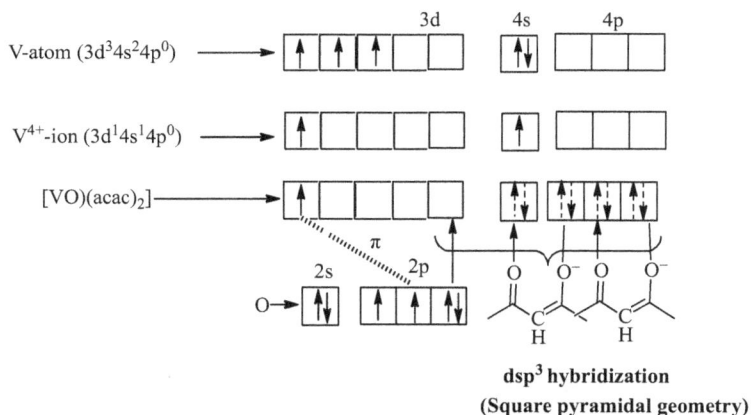

dsp^3 **hybridization**
(Square pyramidal geometry)

Figure 10.54: dsp^3 Hybridization scheme showing the formation of [VO(acac)$_2$].

Because of the π-bond formation of the oxo group with the central vanadium metal, V=O bond is about 50 pm shorter than the four equatorial V–O bonds.

10.3.26 Synthesis of tris(acetylacetonato)chromium(III), [Cr(acac)$_3$] (acacH = acetylacetone)

Reagents required
Chromium(III) chloride hexahydrate (CrCl$_3$ · 6H$_2$O), acetylacetone (CH$_3$COCH$_2$COCH$_3$), urea (NH$_2$CONH$_2$) and distilled water

Principle and reactions

In the presence of a base, acetylacetone, acacH readily loses a proton to form the acetylacetonate anion, acac⁻, as shown below:

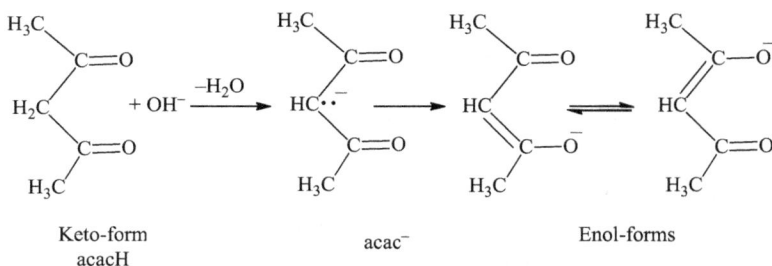

| Keto-form acacH | acac⁻ | Enol-forms |

Hydrogen atoms on α-carbon atom that are adjacent to carbonyl, $C = O$, groups are relatively acidic. The three different representations of the acetylacetonate anion are called resonance forms. They differ only in the location of the electrons. In this experiment, the basic solution needed to remove the proton from the acetylacetone is provided by generating ammonia, NH_3, via the hydrolysis of urea:

$$(NH_2)_2C=O + H_2O \longrightarrow 2NH_3 + CO_2$$

In water, ammonia acts as a base:

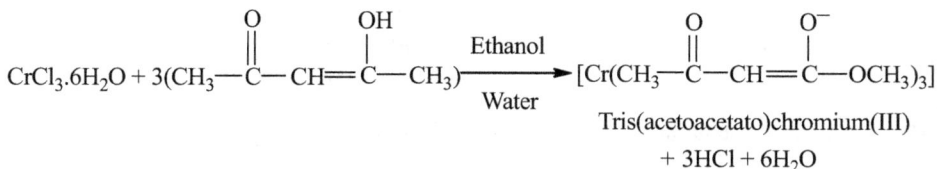

$$NH_3 + H_2O \longrightarrow NH_4^+ + OH^-$$

Tris(acetoacetato)chromium(III)
$+ 3HCl + 6H_2O$

Synthetic procedure

(i)   Dissolve 130 mg (0.49 mmol) of $CrCl_3 \cdot 6H_2O$ in 2 mL of distilled water in a 100 mL beaker. Add 500 mg (8.3 mmol) of urea and 0.4 mL (3.8 mmol) of acetylacetone. A large excess of acetylacetone is used as it helps the reaction go to completion.

(ii) Heat the mixture on a magnetic stirrer to just below boiling point with stirring for 1 h. As the urea releases ammonia and the solution becomes basic, deep maroon crystals will begin to form.

(iii) After 1 h, cool the reaction vessel to room temperature. Collect the crystals by suction filtration. Wash the crystals several times with distilled water. Dry the crystals in a desiccator over anhydrous $CaCl_2$.

Colour: Maroon
Decomposition Temperature: .... °C
Yield: .... g
Structure of the complex: Octahedral (Figure 10.55)

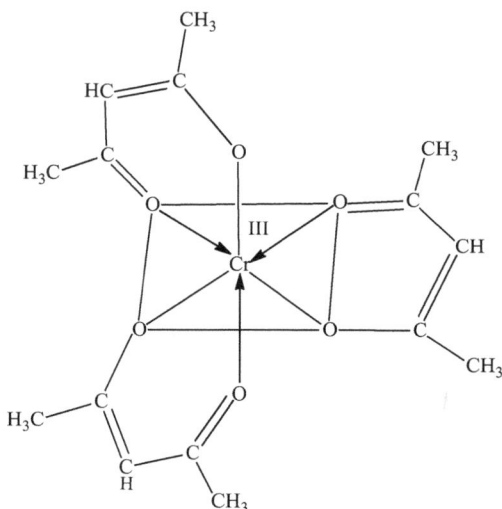

Figure 10.55: Octahedral structure of [Cr(acac)$_3$].

The paramagnetic behaviour and octahedral structure of this compound are well clear by the d^{2}sp^3 hybridization scheme (Figure 10.56) given below:

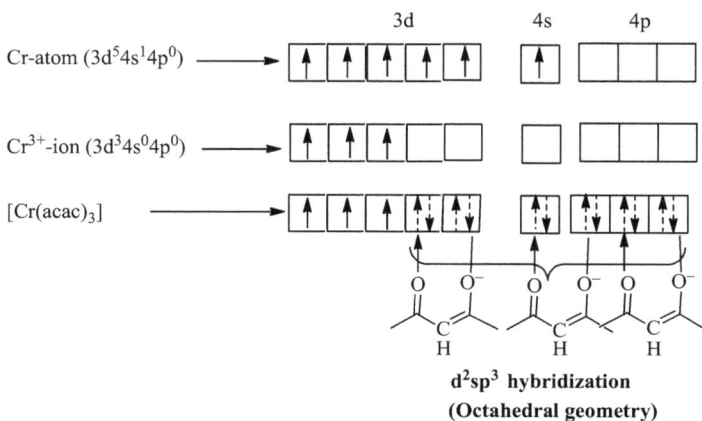

d^{2}sp^3 hybridization
(Octahedral geometry)

Figure 10.56: d^{2}sp^3 Hybridization scheme showing the formation of [Cr(acac)$_3$].

Qualitative test of chromium

(i) Take 0.1 g of the compound in ~2 mL of distilled water in a test tube, and decompose it with 2–3 pellets of KOH with rigorous heating over a burner followed by dissolving it in dilute HNO_3.

(ii) The residue, if any, is filtered off, and the resulting solution is added with aqueous ammonia (1:1) solution whereby grey-green or grey-blue gelatinous precipitate of $Cr(OH)_3$ is formed.

$$Cr(NO_3)_3 + 3NH_4OH \longrightarrow Cr(OH)_3 + 3NH_4NO_3$$

10.3.27 Synthesis of potassium trioxalatomanganate(III) trihydrate, $K_3[Mn(C_2O_4)_3] \cdot 3H_2O$

The compound is a 'classic' within inorganic photo chemistry.

Reagents required

Potassium permanganate ($KMnO_4$), oxalic acid dihydrate ($H_2C_2O_4 \cdot 2H_2O$), potassium ocarbonate (K_2CO_3) and ethanol

Principle and reaction

The synthesis comprises a reduction of Mn(VII) to Mn(II) with oxalic acid at 70–75 °C in aqueous medium. After the addition of the sufficient amount of potassium ions in the form of potassium carbonate, Mn(III) is formed due the oxidation of Mn(II) by addition of Mn(VII) at a temperature below 2 °C. The resulting Mn(III) coordinates with oxalate ions to form the desired compound.

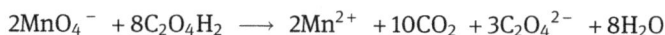

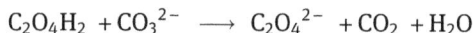

$$2MnO_4^- + 8C_2O_4H_2 \longrightarrow 2Mn^{2+} + 10CO_2 + 3C_2O_4^{2-} + 8H_2O$$

$$C_2O_4H_2 + CO_3^{2-} \longrightarrow C_2O_4^{2-} + CO_2 + H_2O$$

$$4Mn^{2+} + MnO_4^- + 11C_2O_4^{2-} + 4C_2O_4H_2 \longrightarrow 5[Mn(C_2O_4)_3]^{3-} + 4H_2O$$

The overall reaction may be written as

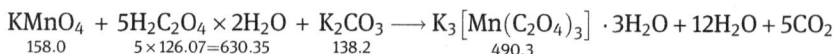

$$KMnO_4 + 5H_2C_2O_4 \times 2H_2O + K_2CO_3 \longrightarrow K_3[Mn(C_2O_4)_3] \cdot 3H_2O + 12H_2O + 5CO_2$$
$$\underset{158.0}{} \quad \underset{5 \times 126.07=630.35}{} \quad \underset{138.2}{} \qquad \underset{490.3}{}$$

Synthetic procedure

(i) A solution of 31.5 g (0.25 mol) of oxalic acid in 200 mL of water is heated in a 500 mL beaker to 70–75 °C. Then add 6.32 g (0.04 mol) of $KMnO_4$ powder little by little with constant stirring, and as soon as the solution turns colourless, 6.9 g (0.05 mol) of K_2CO_3 is added little by little with constant stirring.

(ii) The resulting mixture is cooled to 4–5 °C by keeping it in an ice bath with fre-
quent stirring and diluting with 150 mL of 0–1 °C water.
(iii) In all the following operations, light must be excluded as much as possible.
(iv) The oxidation of Mn^{2+} to Mn^{3+} is affected through the gradual addition of 1.58 g
(0.01 mol) of $KMnO_4$ powder. The solution is then stirred for about 10 min at
0–2 °C. This gives an intense cherry red solution/liquid.
(v) The resulting red solution is suction filtered through a sintered glass crucible
precooled to 0 °C, and is collected in a similarly cooled beaker.
(vi) The filtrate so obtained is added with half of its volume of ice-cold ethanol
and left to crystallize for 2 h in an ice-salt mixture.
(vii) The precipitate is collected in a precooled sintered glass crucible, washed
several times with 1:1 alcohol, then with 95 % absolute alcohol and finally
2–3 times with ether. It is notable here that all the wash liquids must be ice-
cold.
(viii) Dry the deep reddish-purple crystals in air by spreading these in a thin layer,
and protect them from light for at least 1 h. The crystals are stored in a brown
bottle.

Colour: Reddish purple
Decomposition Temperature: .... °C
Yield: .... g
Structure of the complex: Octahedral (Figure 10.57)

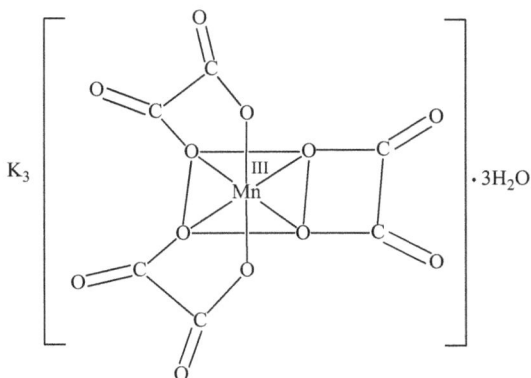

Figure 10.57: Octahedral structure of $K_3[Mn(C_2O_4)_3] \cdot 3H_2O$.

The paramagnetic behaviour and octahedral structure of this compound are well clear by the sp^3d^2 hybridization scheme (Figure 10.58) given below:

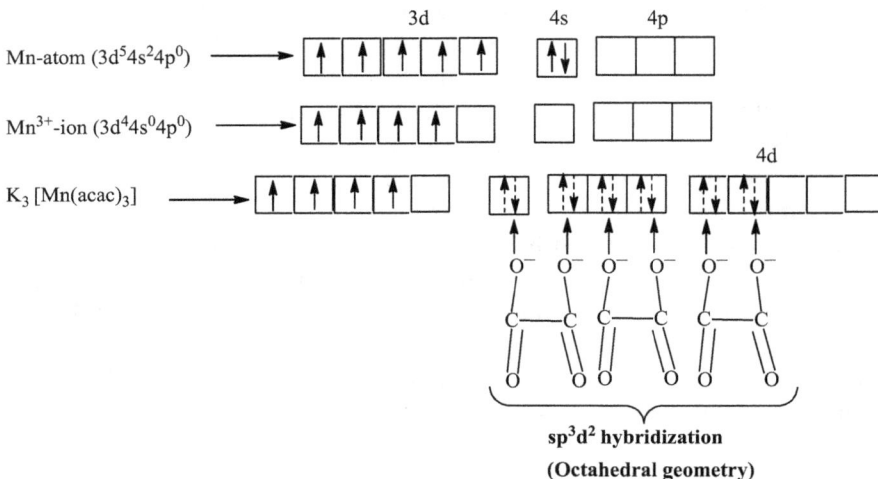

Figure 10.58: sp^3d^2 Hybridization scheme showing the formation of $K_3[Mn(C_2O_4)_3]$ along with its paramagnetic nature.

Determination of manganese in $K_3[Mn(C_2O_4)_3] \cdot 3H_2O$

Manganese(III) is reduced to manganese(II) by iodide ions and the triiodide ions formed are then titrated with sodium thiosulphate.

$$2Mn^{III}(aq) + 3I^-(aq) \longrightarrow 2Mn^{II}(aq) + I_3^-(aq)$$

$$I_3^-(aq) + 2S_2O_3^{2-}(aq) \longrightarrow 3I^-(aq) + S_4O_6^{2-}(aq)$$

In a 250 mL conical flask, dissolve 1.0 g of KI in 25 mL of distilled water, and add 10 mL of 4 M HCl. Thereafter, an accurately preweighed amount (approx. 200 mg) of the complex is transferred (as much as possible is poured directly into the KI solution in small portions before the residue is washed down) quantitatively with distilled water into the flask. Titrate the I_3^- formed with the standard 0.025 M $Na_2S_2O_3$ solution. When the brown colour has faded to light yellow, add 2 mL of starch indicator solution and continue the titration until the colour changes from blue to colourless.

Determination of oxalate in $K_3[Mn(C_2O_4)_3] \cdot 3H_2O$

Manganese(III) is reduced to manganese(II) by the oxalate ligands in acidic medium, and excess oxalate is titrated with permanganate.

$$2\left[Mn(C_2O_4)_3\right]^{3-}(aq) + 10H^+(aq) \longrightarrow 2Mn^{2+}(aq) + 2CO_2(g) + 5C_2O_4H_2(aq)$$

$$5C_2O_4H_2(aq) + 2MnO_4^-(aq) + 6H^+(aq) \longrightarrow 10CO_2(g) + 2Mn^{2+}(aq) + 8H_2O$$

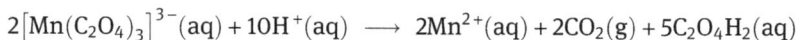

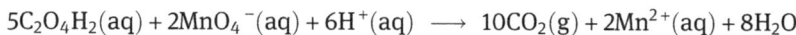

Transfer an accurately preweighed sample (approx. 200 mg) of the complex in question quantitatively with distilled water to a 250 mL conical flask. Add 25 mL of 2M sulphuric acid, and heat the solution to 75–80 °C. Without further heating, titrate with the standard 0.025 M $KMnO_4$ solution. Near the end of the titration add the titrant slowly until one drop gives the solution a pink/rose colour which does not fade on standing for 30 s.

10.3.28 Synthesis of potassium trioxalatoaluminate(III) trihydrate, $K_3[Al(C_2O_4)_3] \cdot 3H_2O$

Reagents required
Aluminium sulphate hexahydrate $[Al_2(SO_4)_3 \cdot 6H_2O]$, sodium hydroxide (NaOH), oxalic acid dihydrate $(H_2C_2O_4 \cdot 2H_2O)$, potassium carbonate (K_2CO_3) and distilled water

Principle and reaction
Aluminium is a 3rd group element with characteristic of the p-block having oxidation state of +3. Because of its small size and high charge, Al^{3+} ion has a high polarizing power, thus, like many of the transition metals, it forms octahedral complex ions. The formation of the complex in question is shown through the equations given below:

$$Al_2(SO_4)_3 \cdot 6H_2O + 6NaOH \longrightarrow 2Al(OH)_3 + 3Na_2SO_4 + 6H_2O$$

$$2Al(OH)_3 + 3H_2C_2O_4 + 3K_2C_2O_4 \longrightarrow 2K_3[Al(C_2O_4)_3] \cdot 3H_2O$$

Synthetic procedure
(i) Dissolve 3.5 g of $Al_2(SO_4)_3 \cdot 6H_2O$ in 25 mL of distilled water in a 100 mL beaker. A solution containing 1.5 g of NaOH in 10 mL of water is added to it. This precipitates $Al(OH)_3$.

(ii) The precipitate of $Al(OH)_3$ so obtained is filtered off in a sintered glass crucible and washed several times with distilled water.

(iii) In another beaker, the precipitate of $Al(OH)_3$ is boiled with aqueous solution of oxalic acid dihydrate (2.5 g) and potassium oxalate (3.0 g), formed by dissolving them in 50 m L of distilled water.

(iv) The unreacted $Al(OH)_3$ is filtered off and the filtrate is evaporated to saturation to effect crystallization. The product is then filtered off in a crucible, washed with small quantities of cold water and dried in an oven at 70 °C.

(v) The product is weighed after cooling in a desiccator over anhydrous $CaCl_2$ and the weight is used to calculate the percentage yield.

Colour: Colourless crystal
Decomposition Temperature: .... °C
Yield: .... g
Structure of the complex: Octahedral (Figure 10.59)

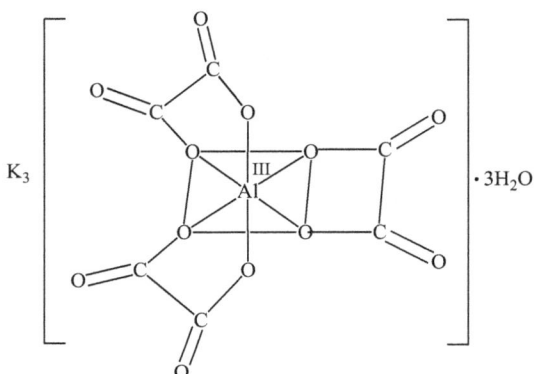

Figure 10.59: Octahedral structure of $K_3[Al(C_2O_4)_3] \cdot 3H_2O$.

The octahedral structure of this compound are well clear by the sp^3d^2 hybridization scheme (Figure 10.60) given below:

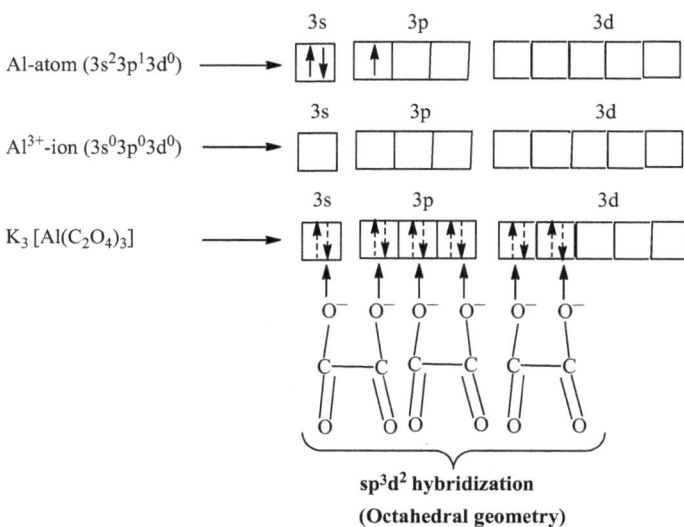

sp^3d^2 **hybridization**
(Octahedral geometry)

Figure 10.60: sp^3d^2 Hybridization scheme showing the formation of $K_3[Al(C_2O_4)_3]$ along with its diamagnetic nature.

10.3.29 Synthesis of sodium hexanitritocobaltate(III), Na₃[Co(NO₂)₆]

Reagent required

Cobalt nitrate hexahydrate [Co(NO₃)₂·6H₂O], sodium nitrite (NaNO₂), glacial acetic acid and alcohol

Reaction

$$2Co(NO_3)_2 \cdot 6H_2O + 12NaNO_2 + 2CH_3COOH + \tfrac{1}{2}O_2 \longrightarrow 2Na_3\left[Co(NO_2)_6\right]$$
$$\underset{291.03 \times 2 = 582.06}{} \qquad \underset{69 \times 12 = 828.0}{} \qquad \underset{60.05 \times 2 = 120.1}{}$$
$$+ 4NaNO_3 + 2CH_3COONa + 13H_2O$$

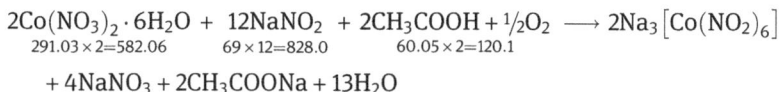

Synthetic procedure

(i)   Take 15 g of NaNO₂ in a 250 mL beaker, dissolve it in 15 mL of distilled water by heating and then cool it to 50–60 °C. Then 5 g of Co(NO₃)₂·6H₂O is added to this solution followed by 5 mL of 50% acetic acid in small portions with shaking.

(ii)   In the resulting mixture, a fast stream of air is bubbled through for 30 min. After standing for 2 h, the brown precipitate is filtered off. The filtrate must be clear at this point, and preserve it in a beaker.

(iii) The precipitate is stirred with 5 mL of water at 70–80 °C. The solution is separated from undissolved Na₃[Co(NO₂)₆] on a filter paper and mixed with the above mentioned filtrate.

(iv)   The combined solution is treated with 25 mL of absolute alcohol, and the resulting precipitate is allowed to settle for at least 1 h.

(v)   The precipitate is collected in a sintered glass crucible, washed several times with alcohol, then twice with ether and dried in air.

Colour: Yellow crystalline powder
Decomposition Temperature: .... °C
Yield: .... g
Structure of the complex: Octahedral (Figure 10.61)

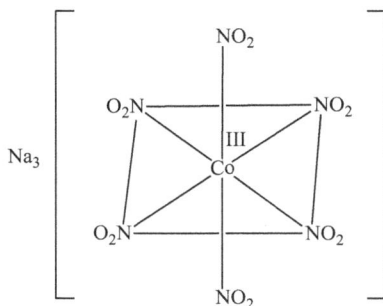

Figure 10.61: Octahedral structure of Na₃[Co(NO₂)₆].

The paramagnetic behaviour and octahedral structure of this compound are well clear by the sp^{3}d^2 hybridization scheme (Figure 10.62) given below:

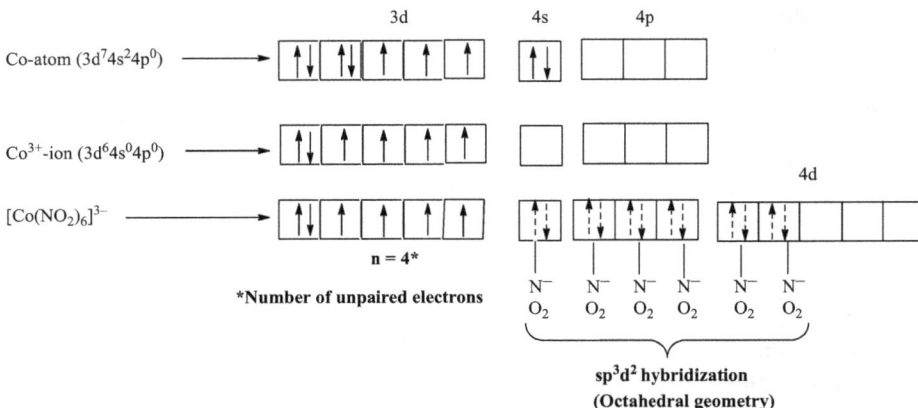

Figure 10.62: sp^{3}d^2 Hybridization scheme showing the formation of Na$_3$[Co(NO$_2$)$_6$] along with its paramagnetic behaviour.

10.3.30 Synthesis of tetraamminezinc(II) tetrafluoroborate, [Zn(NH$_3$)$_4$](BF$_4$)$_2$

Reagent required
Zinc sulphate heptahydrate (ZnSO$_4 \cdot$ 7H$_2$O), concentrated aqueous ammonia (density: 0.90 g/cm^3), 2M Ammonia solution and ammonium tetrafluoroborate (NH$_4$BF$_4$)

Reaction

$$\underset{287.56}{ZnSO_4 .7H_2O} + \underset{\substack{17.03 \times 4 \\ =68.12}}{4NH_3} + \underset{\substack{104.84 \times 2 \\ =209.68}}{2NH_4BF_4} \longrightarrow \left[Zn(NH_3)_4\right](BF_4)_2 + (NH_4)_2SO_4 + 7H_2O$$

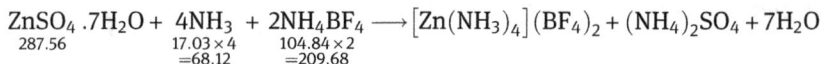

Synthetic procedure
(i)   Dissolve 4 g of ZnSO$_4 \cdot$7H$_2$O in the minimum quantity (~8 mL) of water, and add concentrated aqueous ammonia solution until the precipitate of Zn(OH)$_2$ that has initially formed just redissolves.

(ii) Dissolve 6 g of ammonium tetrafluoroborate in minimum quantity of hot 2M ammonia solution, and add the cooled solution slowly to zinc sulphate solution prepared in step (i).

(iii) Allow the resulting mixture to stand at room temperature for about 30 min and then collect the crystalline product by filtration in a sintered glass crucible. Wash it with acetone, and dry it in a desiccator over anhydrous CaCl$_2$.

Colour: Yellowish white
Decomposition Temperature: .... °C
Yield: .... g
Structure of the complex: Tetrahedral (Figure 10.63)

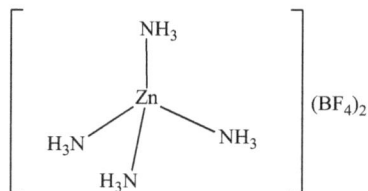

Figure 10.63: Tetrahedral structure of $[Zn(NH_3)_4](BF_4)_2$.

The diamagnetic behaviour and tetrahedral structure of this compound are well clear by the sp^3 hybridization scheme (Figure 10.64) shown below:

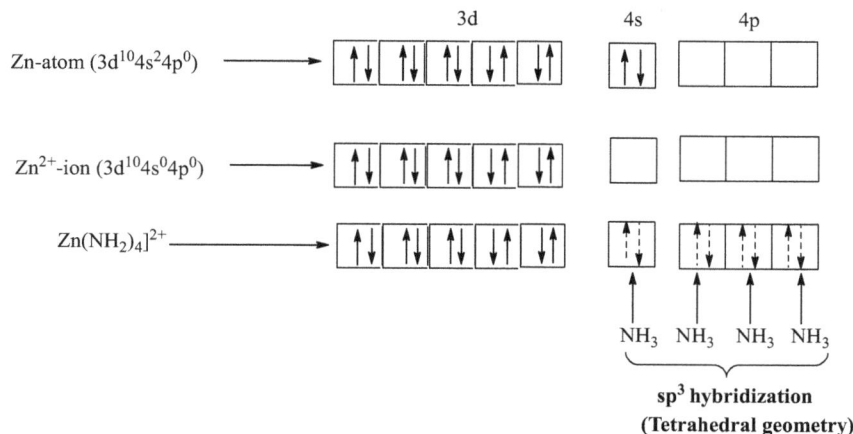

Figure 10.64: sp^3 Hybridization scheme showing the formation of $[Zn(NH_3)_4](BF_4)_2$ along with its diamagnetic behaviour.

10.3.31 Synthesis of ammonium tetrathiocyanatodiamminechromate(III) mono-hydrate, $NH_4[Cr(SCN)_4(NH_3)_2] \cdot H_2O$ (Reinecke salt)

This compound was prepared for the first time by A. Reinecke in 1983, hence it is called Reinecke salt (Reinecke 1863).

Reagents required
Ammonium thiocyanato (NH_4SCN), ammonium dichromate $[(NH_4)_2Cr_2O_7]$ and alcohol

Reaction

$$NH_4SCN \xrightarrow[\substack{(NH_4)_2Cr_2O_7,\ H_2O}]{140-150\ °C} NH_4\left[Cr(SCN)_4(NH_3)_2\right]\cdot H_2O + NH_3 \uparrow$$

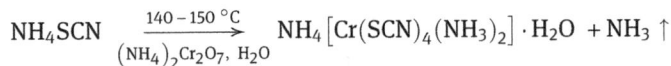

Synthetic procedure

(i) Put 8 g of finely powdered NH_4SCN in a big silica dish, and heat it until it begins to melt. Now add 3 g of finely powdered ammonium dichromate in small portions in the molten ammonium thiocyanate. A violent evolution of NH_3 takes place.

(ii) Cool and powder this mass, and dissolve out any excess of ammonium thiocyanate in hot alcohol and any dichromate by washing with cold water.

(iii) The remaining purified mass is dissolved in hot water and set aside for crystallization. After a short time, glistering red plates of Reinecke's salt are obtained.

Colour: Glistering red
Decomposition Temperature: .... °C
Yield: ....g
Structure of the complex: Octahedral (Figure 10.65)

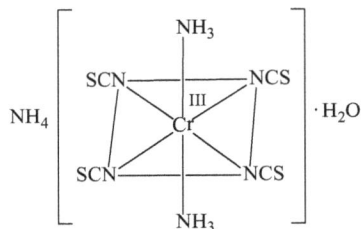

Figure 10.65: Octahedral structure of $NH_4[Cr(SCN)_4(NH_3)_2]\cdot H_2O$.

The paramagnetic behaviour and octahedral structure of this compound are well clear by the d^2sp^3 hybridization scheme (Figure 10.66) shown below:

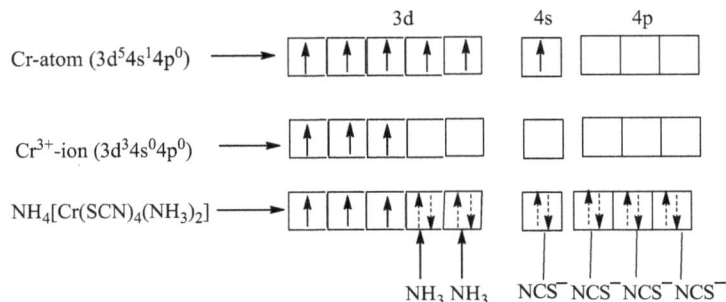

Figure 10.66: d^2sp^3 Hybridization scheme showing the formation of $NH_4[Cr(SCN)_4(NH_3)_2]\cdot H_2O$ along with its paramagnetic behaviour.

10.3.32 Synthesis of potassium dichromate, $K_2Cr_2O_7$

Reagent required
Chromic acetate $[Cr(CH_3COO)_3]$, potassium hydroxide (KOH), hydrogen peroxide (H_2O_2), acetic acid (CH_3COOH) and ethanol (C_2H_5OH).

Principle and reaction
When chromic acetate solution is treated with excess of KOH solution, chromic hydroxide first forms and redissolves due to formation of potassium tetrahydrochromate(III), $K[Cr(OH)_4]$. On adding H_2O_2 to **the alkaline solution of $K[Cr(OH)_4]$** and boiling, a yellow solution is obtained owing to the oxidation of chromium(III) to chromate(VI), that is, K_2CrO_4. On acidifying it with concentrated CH_3COOH, potassium dichromate is formed.

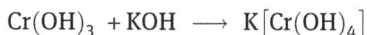

$$Cr(CH_3COO)_3 + 3KOH \longrightarrow Cr(OH)_3 + 3CH_3COOK$$

$$Cr(OH)_3 + KOH \longrightarrow K[Cr(OH)_4]$$

$$2K[Cr(OH)_4] + 3H_2O_2 + 2KOH \longrightarrow 2K_2CrO_4 + 8H_2O$$

$$2K_2CrO_4 + 2CH_3COOH \longrightarrow K_2Cr_2O_7 + 2CH_3COOK + H_2O$$

Synthetic procedure
(i) Dissolve 2.5 g of $Cr(CH_3COO)_3$ in 20 mL of distilled water in a beaker. The solution may be heated if required.

(ii) Now, dissolve 9 g of KOH in 25 mL of distilled water, and add this solution to the solution of chromic acetate slowly till $Cr(OH)_3$ first formed is redissolved.

(iii) The resulting solution is added with 5 mL of 20 Vol. hydrogen peroxide and boiled. If the solution does not become yellow in colour after boiling, add more H_2O_2. If it has turned yellow, boil it gently until volume has been reduced to 10 mL.

(iv) The reduced solution is then cooled and acidified with concentrated CH_3COOH. On concentrating the solution to a small bulk, cool it on an ice bath whereby a fine crop of orange yellow $K_2Cr_2O_7$ crystals is obtained.

Colour: Orange yellow
Decomposition Temperature: .... °C
Yield: .... g

Structure of the compound: It is shown in Figure 10.67

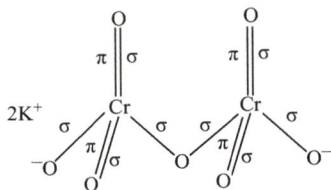

Figure 10.67: Structure of potassium dichromate involving two CrO_4^{2-} tetrahedra linked through O-atom.

Hybridization scheme in $K_2Cr_2O_7$

The diamagnetic behaviour of potassium dichromate and its structure involving two CrO_4^{2-} tetrahedra linked through O-atom are well clear by the d^3s hybridization scheme (Figure 10.68) in each CrO_4^{2-} unit shown below:

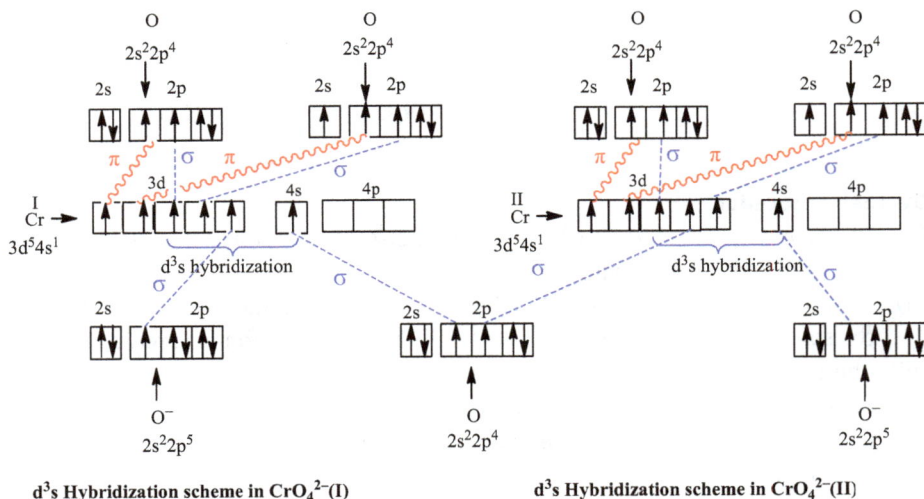

d^3s Hybridization scheme in CrO_4^{2-}(I) d^3s Hybridization scheme in CrO_4^{2-}(II)

Figure 10.68: d^3s Hybridization scheme in two CrO_4^{2-} tetrahedra linked through O-atom showing the formation of $K_2Cr_2O_7$.

10.3.33 Synthesis of potassium manganate (VI), K_2MnO_4

Reagents required

Potassium permanganate ($KMnO_4$), potassium hydroxide (KOH) and ice

Reaction

$$2KMnO_4 + 2KOH \longrightarrow 2K_2MnO_4 + H_2O + 1/2O_2$$
$$\underset{316.1}{} \quad \underset{112.22}{} \quad \underset{394.2}{}$$

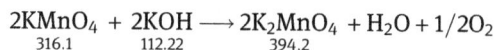

Synthetic procedure

(i) Shake 5 g of KOH with 5 mL of distilled water in a beaker, and add 5 g of $KMnO_4$ to it.

(ii) Maintain the temperature of the mixture at about 130 °C on a hot plate, and stir it with a glass rod until it is green in colour and potassium manganate begins to separate as a granular black solid. As the mixture may spurt, carry out the heating in the fume hood, wear safety spectacles and protect the reaction vessel with a safety screen.

(iii) Allow the beaker and its contents to cool, and add a solution of 5 g KOH in 5 mL distilled water.

(iv) Cool the beaker in ice with stirring, and then collect the crystalline product in a G-4 sintered glass crucible by suction filtration. Transfer the material without delay to the vacuum desiccator containing solid KOH, and evacuate it for several hours. The dried product contains an appreciable amount of alkali as impurity (~10%) which is difficult to remove.

Colour: Black green
Decomposition Temperature: .... °C
Yield: .... g
Structure of the complex: Tetrahedral (Figure 10.69)

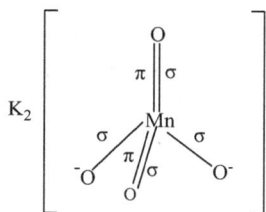

Figure 10.69: Tetrahedral structure of K_2MnO_4.

The paramagnetic behaviour and tetrahedral structure (with four σ- and two π-bonds) of this compound are well clear by the d^3s hybridization scheme (Figure 10.70) shown below:

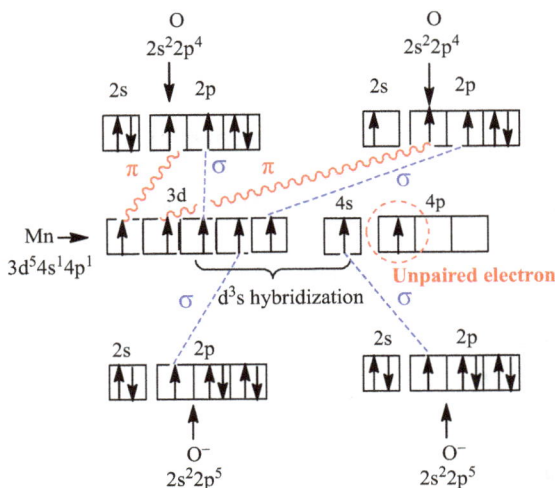

Figure 10.70: d^3s Hybridization scheme with four σ- and two π-bonds leading to the formation of K_2MnO_4.

Determination of manganese

Manganese is determined in the potassium manganate by reducing it to manganese(II) with potassium iodide.

Accurately weigh 0.15 g of the compound and dissolve it in 20 mL 2M sulphuric acid. Add 1 g of KI and titrate the free Iodine with the standard 0.1 M sodium thiosulphate ($Na_2S_2O_3$) solution using starch as indicator. As soon as all the liberated iodine has been reduced to iodide (NaI), the blue colour of iodine–starch complex will disappear and the colour of the precipitate in conical flask will turn brown due to formation of MnO_2.

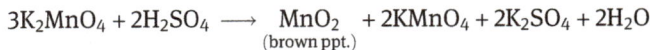

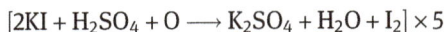

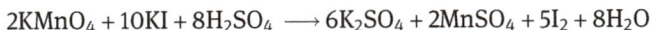

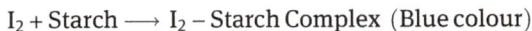

$$3K_2MnO_4 + 2H_2SO_4 \longrightarrow \underset{\text{(brown ppt.)}}{MnO_2} + 2KMnO_4 + 2K_2SO_4 + 2H_2O$$

$$2KMnO_4 + 3H_2SO_4 \longrightarrow K_2SO_4 + 2MnSO_4 + 3H_2O + 5O$$

$$[2KI + H_2SO_4 + O \longrightarrow K_2SO_4 + H_2O + I_2] \times 5$$

....

$$2KMnO_4 + 10KI + 8H_2SO_4 \longrightarrow 6K_2SO_4 + 2MnSO_4 + 5I_2 + 8H_2O$$

$$I_2 + Starch \longrightarrow I_2 - Starch\ Complex\ (Blue\ colour)$$

$$2Na_2S_2O_3 + I_2 \longrightarrow \underset{\text{Sodium tetrathionate}}{Na_2S_4O_6} + 2NaI$$

Hence,

$$3K_2MnO_4 \equiv 2KMnO_4 \equiv 5I_2 \equiv 10Na_2S_2O_3$$

10.3.34 Synthesis of potassium dioxalatodiaquachromate(III) dihydrate, $K[Cr(C_2O_4)_2(H_2O)_2] \cdot 2H_2O$

Reagents required
Potassium dichromate ($K_2Cr_2O_7$), oxalic acid dihydrate ($H_2C_2O_4 \cdot 2H_2O$) and ethanol

Reaction
The complex is prepared by the reaction of oxalic acid with crystalline potassium dichromate.

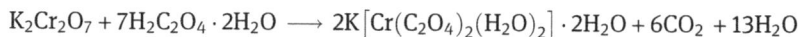

$$K_2Cr_2O_7 + 7H_2C_2O_4 \cdot 2H_2O \longrightarrow 2K\left[Cr(C_2O_4)_2(H_2O)_2\right] \cdot 2H_2O + 6CO_2 + 13H_2O$$

This compound exists as two geometrical isomers: (i) *cis*-$K[Cr(C_2O_4)_2(H_2O)_2] \cdot 2H_2O$ and (ii) *trans*-$K[Cr(C_2O_4)_2(H_2O)_2] \cdot 2H_2O$.

Synthetic procedure

The *cis*-isomer
(i)   2 g of potassium dichromate and 6 g of oxalic acid dihydrate are first powdered by grinding them separately.
(ii) Mix the two powders and place the mixture carefully in a single heap in an evaporating basin to which 6–8 drops of water has already been added.
(iii) Place the basin covered with a watch glass on a hot plate. After a short induction period, a vigorous exothermic reaction begins, and the reaction mixture becomes a dark viscous liquid.
(iv) After the reaction is subsided, add 15 mL of ethanol to the hot liquid. Stir the product with a spatula until it solidifies and crush the large lumps, if any. If the product remains partly as oil, discard the supernatant ethanol and a fresh portion (10 mL) of ethanol to get the solid product.
(v)   Collect the product by filtration in a sintered glass crucible, wash it with ethanol two to three times and dry it in the vacuum desciccator.

The *trans*-isomer
(i)   Take 6 g of $H_2C_2O_4 \cdot 2H_2O$ in a 250 mL beaker and add boiling water until the solid just dissolves. Also prepare a solution of 2 g of $K_2Cr_2O_7$ in a separate beaker by addition of minimum amount of boiling water.

(ii)   Cover the oxalic acid solution with a watch glass and add the dichromate solution in a small portion. Keep the reaction mixture covered between additions of dichromate solution.

(iii) When the dichromate has been added, cool the beaker to room temperature. Allow the solvent to evaporate for 2–3 days until the volume is reduced by two third.

(iv)   Collect the pink crystalline product by filtration on a Büchner funnel. Wash it with water and then ethanol. Dry it in a vacuum desiccator.

Note

(i)   If the solution is reduced too much by evaporation, the *cis*-isomer will also crystallize. In fact, in solution, the two isomers are in equilibrium, and the *trans*- isomer crystallizes first as it is least soluble.

(ii)   Differences in the nature of *cis* and *trans*- geometrical isomers of $K[Cr(C_2O_4)_2$ $(H_2O)_2] \cdot 2H_2O$ is the reaction with aqueous ammonia. *Trans* isomer is not soluble in aqueous ammonia and produces light brown precipitate, whereas the *cis* isomer is soluble in aqueous ammonia to produce a dark green solution. The possible reactions of these two isomers with ammonia forming hydroxo complex are shown below:

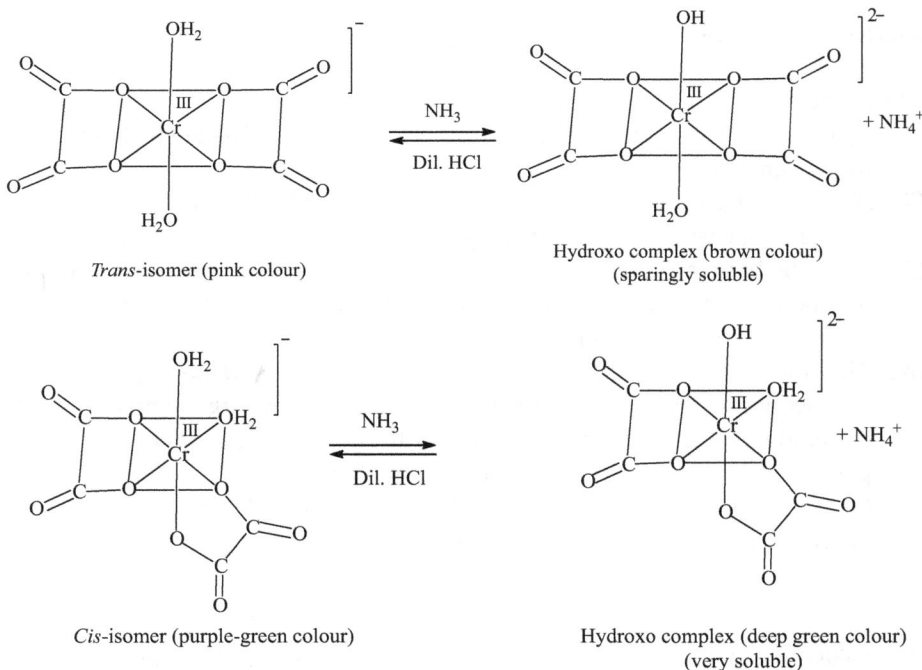

Trans-isomer (pink colour)

Hydroxo complex (brown colour)
(sparingly soluble)

Cis-isomer (purple-green colour)

Hydroxo complex (deep green colour)
(very soluble)

Colour: *Cis*-isomer: Purple-green; *Trans*-isomer: Pink
Decomposition Temperature: .... °C
Yield: .... g
Structure of the complex: Octahedral (Figure 10.71)

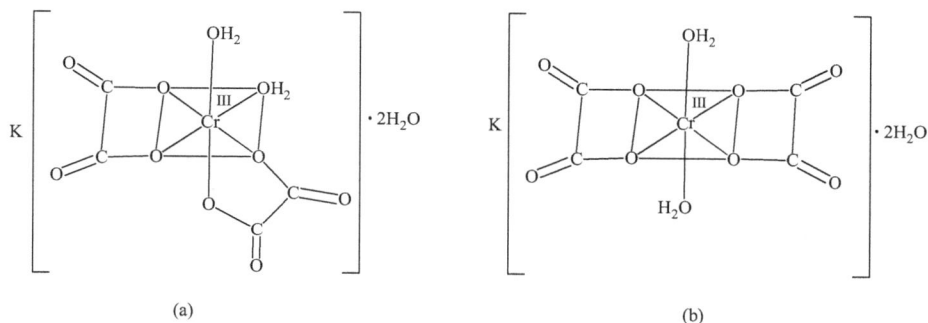

(a)   (b)

Figure 10.71: (a) *cis*- and (b) *trans*-octahedral structure of $K[Cr(C_2O_4)_2(NH_3)_2] \cdot 2H_2O$.

The paramagnetic behaviour and octahedral structure of these two isomers are well clear by the sp^3d^2 hybridization scheme (Figure 10.72) shown below:

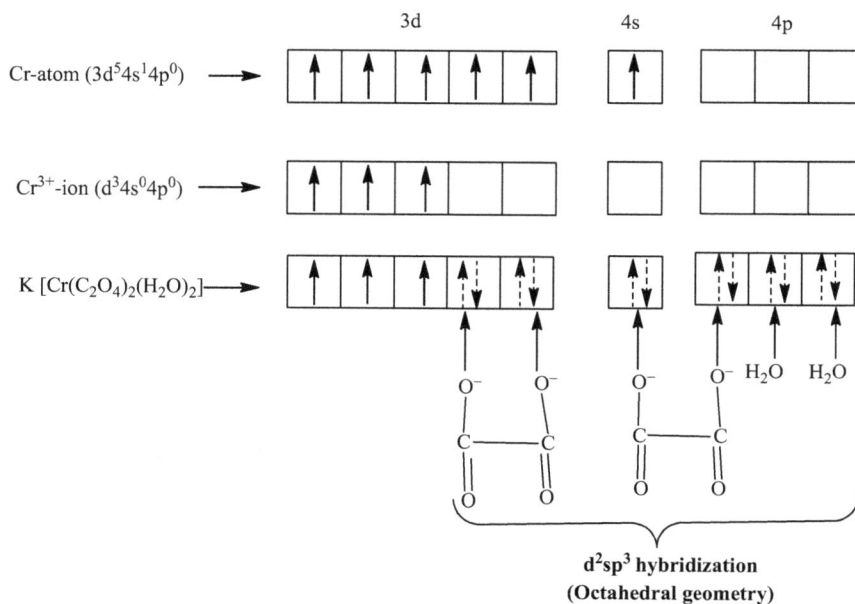

d²sp³ hybridization
(Octahedral geometry)

Figure 10.72: d^2sp^3 Hybridization scheme leading to the formation of $K[Cr(C_2O_4)_2(NH_3)_2] \cdot 2H_2O$ with three unpaired electrons.

10.3.35 Synthesis of potassium dioxalatocuprate(II) dihydrate, $K_2[Cu(C_2O_4)_2] \cdot 2H_2O$

Reagents required

Copper sulphate pentahydrate ($CuSO_4 \cdot 5H_2O$), potassium oxalate monohydrate ($K_2SO_4 \cdot H_2O$), absolute alcohol and acetone

Reaction

$$CuSO_4 \cdot 5H_2O + 2K_2SO_4 \cdot H_2O \longrightarrow K_2[Cu(C_2O_4)_2] \cdot 2H_2O + K_2SO_4 + 5H_2O$$

Synthetic procedure

(i)   Take 6.2 g of $CuSO_4 \cdot 5H_2O$ in a beaker with 12 mL of distilled water, and shake it well to dissolve. Heat this solution to 90 °C, and add it rapidly with vigorous stirring to a hot (90 °C) solution of 10 g of $K_2SO_4 \cdot H_2O$ in 50 mL of water contained in a 100 mL beaker.

(ii) Cool the resulting mixture by keeping the beaker in an ice bath for ~30 min. This crystallizes the desired compound as blue crystals.

(iii) Collect the blue crystalline product by suction filtration in a sintered glass crucible. Wash the crystals successively with about 10 mL of cold water, then 10 mL of absolute ethanol and finally 10 mL of acetone. Dry it in a vacuum desiccator.

Colour: Blue
Decomposition Temperature: .... °C
Yield: .... g
Structure of the complex: Square planar (Figure 10.73)

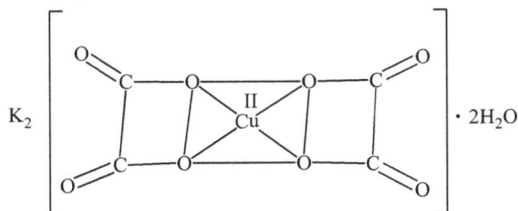

Figure 10.73: Square planar structure of $K_2[Cu(C_2O_4)_2] \cdot 2H_2O$.

The paramagnetic behaviour and square planar structure of complex in question are well explained by the sp^{2}d hybridization scheme (Figure 10.74) shown below:

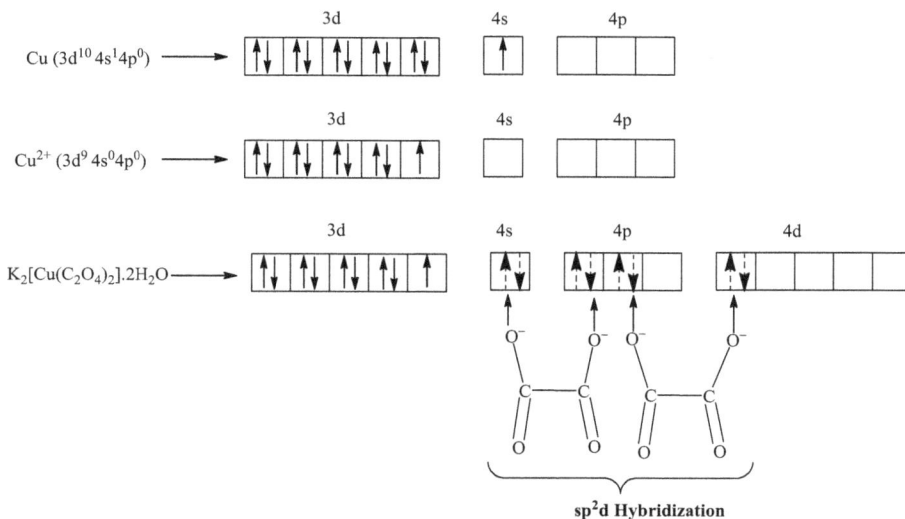

Figure 10.74: sp^{2}d Hybridization scheme leading to the formation of K[Cu(C$_2$O$_4$)$_2$]·2H$_2$O with one unpaired electron.

10.3.36 Synthesis of tetraaqua-bis(*o*-sulphobenzimide)copper(II), [Cu(Sac)$_2$(H$_2$O)$_4$]·2H$_2$O

Introduction

Accidentally discovered by Fahlberg in 1878, saccharin (*o*-sulphobenzimide; 1,2-benzothiazole-3(2H)-one 1,1-dioxide; Figure 10.75) is one of the best known and most widely used artificial sweetening agents. The imino hydrogen is acidic and, thus, the molecule can be easily converted into the corresponding nitranion. It has been shown that the coordination chemistry of this anion is very interesting and versatile, taking

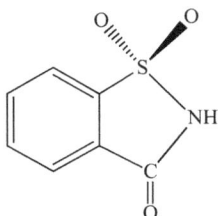

Figure 10.75: Chemical structure of saccharin (*o*-sulphobenzimide).

into account that it offers different coordination sites to metallic centres, that is, one N, one O (carbonylic) and two O (sulphonic) atoms. Using these donor atoms, the anion can generate either N- or O-monodentate or bidentate (N, O) coordination, and also more complex polymeric species with the participation of all possible donor atoms (Baran and Yilmaz 2006).

Reagent required
Sodium salt of saccharin monohydrate ($NaSac \cdot H_2O$), copper sulphate pentahydrate ($CuSO_4 \cdot 5H_2O$) and distilled water

Reaction
The general reaction for the synthesis of metal complexes of saccharin taking copper sulphate as an example is ($Sac = NSO_3H_4C_7$):

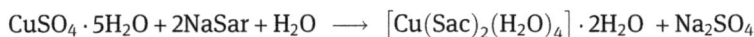

$$CuSO_4 \cdot 5H_2O + 2NaSar + H_2O \longrightarrow \left[Cu(Sac)_2(H_2O)_4\right] \cdot 2H_2O + Na_2SO_4$$

Synthetic procedure
(i) In a 20 mL beaker containing magnetic stirrer bar, place 52 mg (0.2 mmol) of $CuSO_4 \cdot 5H_2O$, 100 mg (0.49 mmol) sodium salt of saccharin monohydrate and 6 mL of distilled water.

(ii) The mixture is stirred until dissolution occurs. Slight warming on a magnetic stirring hot plate hastens this operation.

(iii) Place the light blue solution on a hot sand bath (~140 °C), and with stirring concentrate the solution to a volume of ~3 mL.

(iv) Remove the beaker and content from the sand bath and allow them to cool slowly to room temperature. Light green crystals start forming during this time.

(v) Cool the beaker further in an ice bath (~30 min) and collect the resulting crystals by suction filtration using sintered glass crucible. Wash the crystals with two 1 mL portions of ice-cold water, and dry them over silica gel in a desiccator.

Colour/Nature: Robin's egg blue crystals
Decomposition Temperature: .... °C
Yield: .... g

Structure of the complex: Octahedral (Figure 10.76)

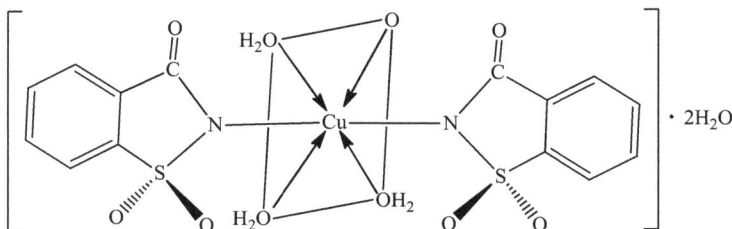

Figure 10.76: Octahedral structure of $[Cu(Sac)_2(H_2O)_4] \cdot 2H_2O$.

10.3.37 Synthesis of tetraaqua-bis(o-sulphobenzimide)cobalt(II)

$$CoCl_2 \cdot 6H_2O + 2NaSar \longrightarrow [Co(Sac)_2(H_2O)_4] \cdot 2H_2O + 2NaCl$$

Synthetic procedure

This compound is prepared using the same procedure given above for tetraaqua-bis(o-sulphobenzimide)copper(II). In the present case, use 48 mg (0.2 mmol) of $CoCl_2 \cdot 6H_2O$, 100 mg (0.49 mmol) sodium salt of saccharin monohydrate and 6 mL of distilled water.

Colour/Nature: Rose red
Decomposition Temperature: .... °C
Yield: .... g
Structure of the complex: Octahedral (Figure 10.77)

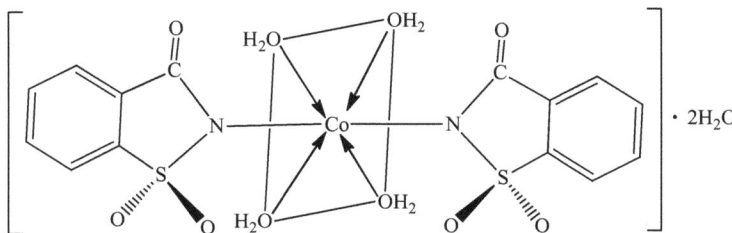

Figure 10.77: Octahedral structure of $[Co(Sac)_2(H_2O)_4] \cdot 2H_2O$.

10.3.38 Synthesis of tetraaqua-bis(*o*-sulphobenzimide)nickel(II)

$$Ni(NO_3)_2 \cdot 6H_2O + 2NaSar \longrightarrow [Ni(Sac)_2(H_2O)_4] \cdot 2H_2O + 2NaNO_3$$

Synthetic procedure

Nickel nitrate hexahydrate (0.98 g, 0.0034 mol) and 1.6 g of the sodium salt of saccharin (0.0078 mol) are dissolved in 100 mL of water. The solution is heated gently on a water bath until the volume is reduced to 50 mL. It is then allowed to stand overnight, whereupon **green crystals** separate. The mixture is cooled in an ice bath, and the crystals **are** collected on a filter paper, washed with two 10 mL portions of cold water and dried over silica gel.

Colour/Nature: Green
Decomposition Temperature: .... °C
Yield: .... g
Structure of the complex: Octahedral (Figure 10.78)

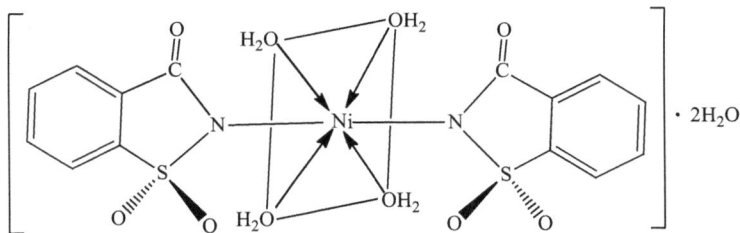

Figure 10.78: Octahedral structure of [Ni(Sac)$_2$(H$_2$O)$_4$] · 2H$_2$O.

10.3.39 Synthesis of tetraaqua-bis(*o*-sulphobenzimide)iron(II)

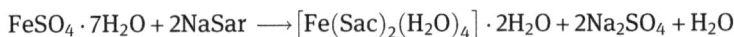

$$FeSO_4 \cdot 7H_2O + 2NaSar \longrightarrow [Fe(Sac)_2(H_2O)_4] \cdot 2H_2O + 2Na_2SO_4 + H_2O$$

Iron(II) sulphate heptahydrate: 0.95 g (0.0034 mol); sodium saccharinate: 1.6 g (0.0078 mol). This compound is prepared using the same procedure given above for tetraaqua-bis(*o*-sulphobenzimide)copper(II) (Figure 10.79). The yield of **greenish-yellow** crystalline product is 1.08 g (60.0%).

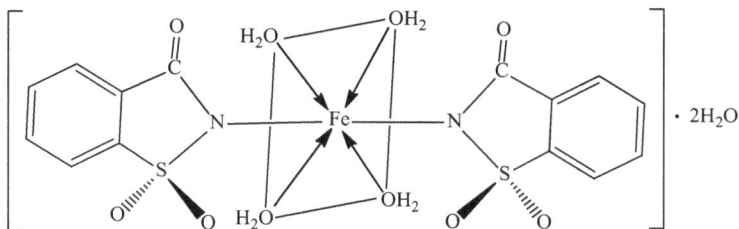

Figure 10.79: Octahedral structure of [Fe(Sac)$_2$(H$_2$O)$_4$]·2H$_2$O.

10.3.40 Synthesis of tetraaqua-bis(*o*-sulphobenzimide)Zn(II)

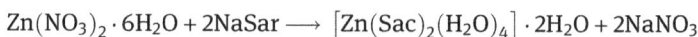

$$Zn(NO_3)_2 \cdot 6H_2O + 2NaSar \longrightarrow [Zn(Sac)_2(H_2O)_4] \cdot 2H_2O + 2NaNO_3$$

Zinc nitrate hexahydrate: 1.00 g (0.0034 mol); sodium saccharinate: 1.60 g (0.0078 mol).

This compound is prepared using the same procedure given above for tetraaqua-bis(*o*-sulphobenzimide)copper(II) (Figure 10.80). The yield of **white, crystalline product** is 0.95 g (51.6%).

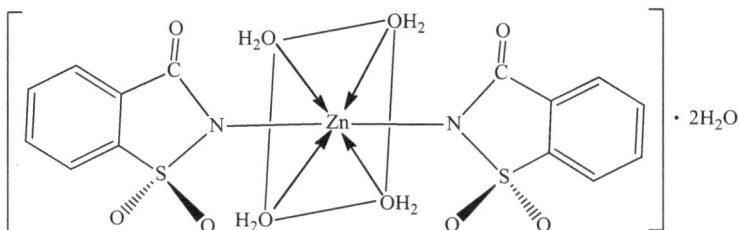

Figure 10.80: Octahedral structure of [Zn(Sac)$_2$(H$_2$O)$_4$]·2H$_2$O.

10.3.41 Synthesis of *cis-*/*trans*-dichlorobis(ethylenediamine)cobalt(III) chloride, *trans*-[Co(en)$_2$Cl$_2$]Cl

Introduction
Co(III) exhibits a particular tendency to coordinate with ligands containing nitrogen. A majority of these complexes have ammonia-, amines- or nitrogen- bonded NCS$^-$ groups. Several of these complexes have *cis* and *trans* isomerism, and one of them isdichlorobis(ethylenediamine)cobalt(III) chloride, [Co(en)$_2$Cl$_2$]Cl.

Reagents required
Cobalt(II) chloride hexahydrate (CoCl$_2$·6H$_2$O), ethylenediamine (NH$_2$CH$_2$CH$_2$NH$_2$) [en] and distilled water

Principle and reaction
The *trans*-dichlorobis(ethylenediamine)cobalt(III) chloride is synthesized by the aerial oxidation of an aqueous solution of $CoCl_2.6H_2O$ and ethylenediamine followed by
addition of concentrated HCl. The synthesis uses a Co^{2+} species rather than a Co^{3+}
salt because cobaltic ion reacts with water and is, therefore, unstable in the presence
of moisture.

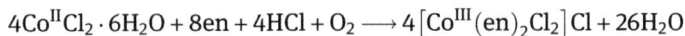

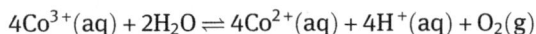

$$4Co^{II}Cl_2 \cdot 6H_2O + 8en + 4HCl + O_2 \longrightarrow 4\left[Co^{III}(en)_2Cl_2\right]Cl + 26H_2O$$

$$4Co^{3+}(aq) + 2H_2O \rightleftharpoons 4Co^{2+}(aq) + 4H^+(aq) + O_2(g)$$

(a) Synthetic procedure for *trans*-complex
(i)   Take 300 mg (1.26 mmol) of cobalt(II) chloride hexahydrate, 2 mL of water
        and 1.0 mL of 10% ethylenediamine in a side arm test tube equipped with an
        air inlet (Figure 10.81).
(ii)   Clamp the reaction tube on a hot water bath (90–95 °C), and connect the side
        arm to a water aspirator through a water trap. Turn on the aspirator so as to
        draw air through the solution at a slow but steady rate.
(iii)   The reaction mixture, which is purple in colour, is maintained under such
        conditions for a period of 1 h. From time to time, add additional water to the
        reaction flask (down the air inlet tube) to maintain the water volume.
(iv)   After 1 h of heating time, disconnect the aspirator and remove the reaction
        tube from the water bath. Now, allow the reaction tube to cool to ~50–60 °C.

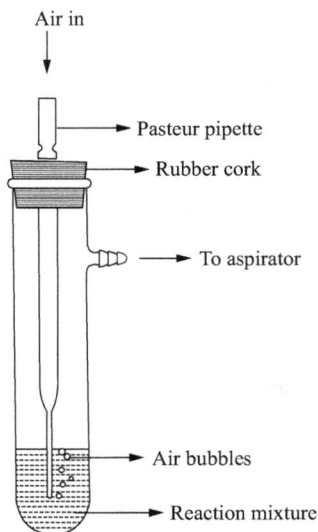

Figure 10.81: Side arm test tube apparatus.

(v) Using the pasture pipette, slowly add 600 μL (0.6 mL) of concentrated HCl down the inlet tube. Swirl the reaction assembly by hand for several minutes.

(vi) Now, place the reaction assembly back into the water bath. Reconnect the aspirator and adjust it so that a steady flow of air is once again pulled through the solution. This process is continued until the volume of the solution is decreased to the point that crystal of the product is evident in the tube.

(vii) Disconnect the tube from the aspirator, remove it from the water bath and place it in an ice bath to cool. Scrap the resulting solid from the tube, and collect it by suction filtration in a sintered glass crucible. Wash the crystals with two 2 mL portions of cold methanol which is added to the reaction tube to assist in the removal of additional crystalline product. This is followed by washing with two 2 mL portions of cold diethyl ether. The green crystals that form are actually the hydrochloride salts of the desired product.

(viii) In order to obtain the *trans*-dichlorobis(ethylenediamine)cobalt(III) chloride, place the crystals on a small watch glass and heat them in an oven at 110 °C for 1.5 h.

Structure: Octahedral (Figure 10.82)

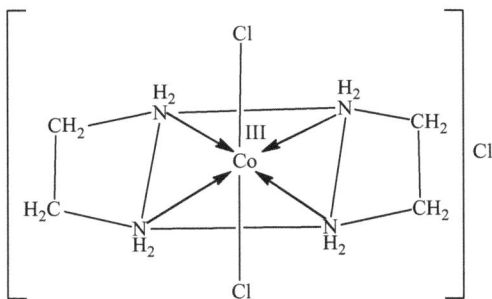

Figure 10.82: Octahedral structure of *trans*-dichlorobis(ethylenediamine)cobalt(III) chloride.

(b) Synthetic procedure for *cis*-complex

Synthetic procedure

(i) Place 10 mg (0.06 mmol) of the green *trans*-dichlorobis(ethylenediamine)cobalt (III) chloride on a 5 cm watch glass. Dissolve this sold material in 300 μL (0.3 mL) of water, and allow the solution to stand for about 10 min at room temperature.

(ii) Place the watch glass on a hot water bath prepared from a 10 mL beaker containing a magnetic stirring bar filled with boiling water. The beaker is previously positioned in a sand bath on a magnetic stirring hot plate. Heat the green solution and concentrate it to dryness. A deep violet glassy material is formed on the watch glass.

(iii) The product so obtained is the desired *cis*-complex.

Structure: Octahedral (Figure 10.83)

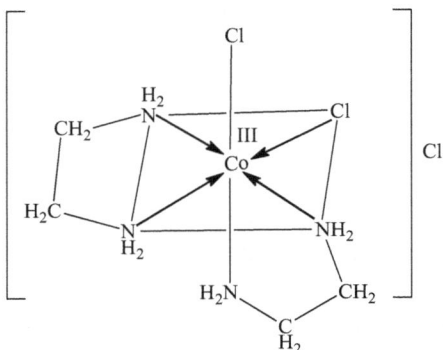

Figure 10.83: Octahedral structure of *cis*-dichlorobis(ethylenediamine)cobalt(III) chloride.

10.3.42 Synthesis of *cis*/*trans*-bis(glycinato)copper(II) monohydrate

Introduction

Like the more familiar methyl acetoacetate ($CH_3COCH_2COOCH_3$) (vide supra), the amino acid glycine (glyH) (Figure 10.84) dissociates to form (*gly⁻*) (and H^+). The anion *gly* can coordinate to a wide variety of metal complexes.

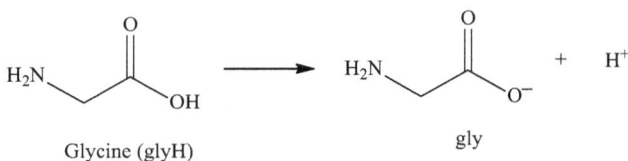

Glycine (glyH)

gly

Figure 10.84: Dissociation of glycine (glyH) to form (*gly⁻*) (and H^+).

One major difference is that the glycine anion is not symmetric and structural isomers can arise depending on the relative orientation of the ligands (Figure 10.85).

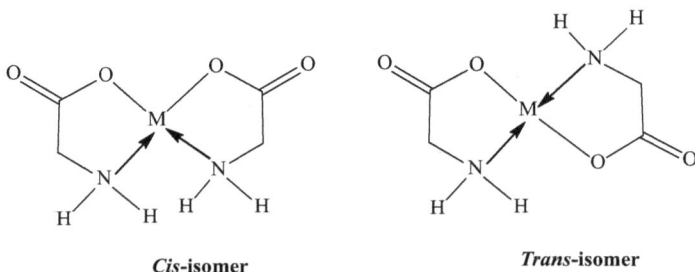

Cis-isomer

Trans-isomer

Figure 10.85: Structural isomers of bis(glycinato)copper(II)monohydrate.

Principle and reaction

The direct reaction of copper(II) acetate monohydrate [$(CH_3COO)_2Cu \cdot H_2O$] and glycine (NH_2CH_2COOH) (glyH) results in an equilibrium mixture of the two isomers.

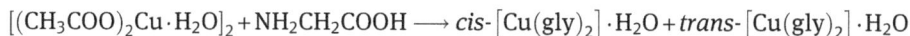

$$[(CH_3COO)_2Cu \cdot H_2O]_2 + NH_2CH_2COOH \longrightarrow cis\text{-}[Cu(gly)_2] \cdot H_2O + trans\text{-}[Cu(gly)_2] \cdot H_2O$$

The *cis* isomer precipitates much more quickly than the *trans*, leading to a shift in the equilibrium away from the other, producing only the one isomer. The *cis* isomer (kinetically favoured product) is converted to the other (thermodynamically favoured product) simply by heating.

(a) Synthetic procedure for *cis*-[Cu(gly)₂] · H₂O

$$(CH_3COO)_2Cu \cdot H_2O + 2NH_2CH_2COOH \longrightarrow cis\text{-}[(NH_2CH_2COO)_2 Cu] \cdot H_2O$$
$$+ 2CH_3COOH$$

(i)   Copper(II) acetate monohydrate ($CH_3COO)_2Cu \cdot H_2O$ (2.0 g) is dissolved in 25 mL of hot distilled water. Then 25 mL of ethyl alcohol is added to this solution and the temperature is maintained at 70 °C.

(ii) Dissolve 1.5 g of glycine in 25 mL of hot water. Then, mix this solution to the above solution [prepared in step (i)] while hot (ca. 70 °C).

(iii) On cooling the resulting solution over ice for 10 min, a needle-like precipitate of the *cis*-isomer is obtained.

(iv) Collect the resulting product by suction filtration using sintered glass crucible. Wash the product with ice-cold ethanol, and dry it in a vacuum desiccator.

Colour/Nature: Blue crystals
Decomposition Temperature: .... °C
Yield: .... g
Structure of the complex: Square planar (Figure 10.86)

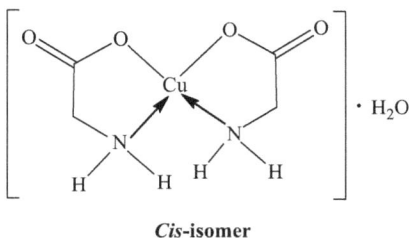

Cis-isomer

Figure 10.86: Square planar structure of *cis*-[(NH₂CH₂COO)₂Cu] · H₂O.

(b) Synthetic procedure for *trans*-[Cu(gly)$_2$] · H$_2$O

(i) Place a weighed amount of *cis*-[Cu(gly)$_2$] · H$_2$O compound [synthesized in part (a)] in a muffle furnace at 220 °C for 30 min.

(ii) Remove the crucible from the furnace, and allow it to cool to room temperature. The *cis* product is now changed to *trans* product.

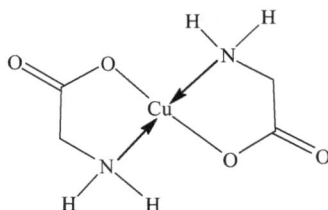

Trans-isomer

Figure 10.87: Square planar structure of *trans*-[(NH$_2$CH$_2$COO)$_2$Cu] · H$_2$O.

10.3.43 Synthesis of Chlorotris(triphenylphosphine)rhodum(I), [RhCl(PPh$_3$)$_3$]: Wilkinson's catalyst

Significance

This compound was synthesized for the first time in the laboratory of Sir Geoffrey Wilkinson (Figure 10.88) in 1965. The importance of this compound arises from the observation that the first homogeneous hydrogenation of unsatureated hydrocarbons

Figure 10.88: Sir Geoffrey Wilkinson, Nobel laureate in chemistry, 1973.

was achieved using this compound as a catalyst. This is why it is also known as Wilkinsons's catalyst.

Reagents required
Rhodium(III) chloride hydrate ($RhCl_3 \cdot 3H_2O$), triphenylphosphine (PPh_3) and ethanol

Basic principle and reaction
This compound is prepared from the reaction of hydrated rhodium trichloride and excess triphenylphosphine in boiling ethanol. During the synthesis of Wilkinson's catalyst, one equivalent of triphenylphosphine reduces rhodium(III) to rhodium(I) while three other equivalents bind themselves to the metal as ligands in the final product.

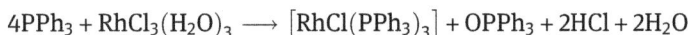

$$4PPh_3 + RhCl_3(H_2O)_3 \longrightarrow \left[RhCl(PPh_3)_3\right] + OPPh_3 + 2HCl + 2H_2O$$

Synthetic procedure
(i) Place 20 mL of absolute ethanol in a 25 mL round-bottom flask equipped with a magnetic stirring bar. Attach a water condenser and place the apparatus in a sand bath upon a magnetic stirring hot plate.

(ii) Heat the ethanol just below the boiling point. Now, remove the condenser momentarily, and add 600 mg (2.29 mmol, a large excess) of triphenylphosphine to the hot ethanol, and stir until dissolution is achieved.

(iii) Remove the condenser for a moment once again, and add 100 mg (0.48 mmol) of $RhCl_3 \cdot 3H_2O$ to the solution and continue stirring.

(iv) Now, heat the solution to a gentle reflux. At first, a deep red-brown solution is obtained, which during the course of heating under reflux will slowly form yellow crystals. After 20–30 min of reflux, the yellow crystals are converted into shining burgundy red crystals.

(v) Collect the resulting product by suction filtration using sintered glass crucible. Wash the product with three 1 mL portions of ether, and dry it in a vacuum desiccator.

Colour/Nature: Burgundy red crystals
Decomposition Temperature: .... °C
Yield: .... g

Structure of the complex: Square planar (Figure 10.89)

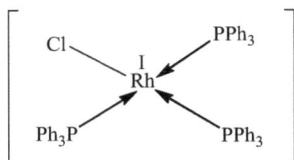

Figure 10.89: Square planar structure of [RhCl(PPh₃)₃].

Burgundy red colour of this compound is due to its diamagnetic behaviour. The square planar structure and diamagnetic nature of this compound are well explained by dsp^2 hybridization scheme (Figure 10.90) shown below:

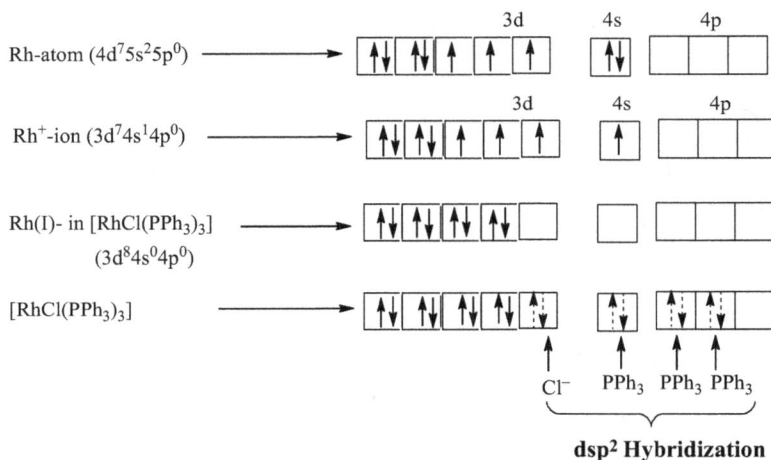

Figure 10.90: dsp^2 Hybridization scheme leading to the formation of diamagnetic square planar [RhCl(PPh₃)₃].

10.3.44 Synthesis of bromotris(triphenylphosphine)rhodum(I), [RhBr(PPh₃)₃]

Reagents required
Rhodium(III) chloride hydrate ($RhCl_3 \cdot 3H_2O$), triphenylphosphine (PPh_3), lithium bromide (LiBr) and ethanol

Reaction

$$RhCl_3 \cdot 3H_2O + 4PPh_3 + LiBr \longrightarrow \left[RhBr(PPh_3)_3\right] + OPPh_3 + 2HCl + LiCl + 2H_2O$$

Synthetic procedure
(i)   The same experimental setup is utilized as in the synthesis of [RhCl(PPh$_3$)$_3$].
(ii) Add 25 mg (0.12 mmol) of RhCl$_3 \cdot$3H$_2$O to a hot solution of 125 mg (0.57 mmol) of PPh$_3$ in 5 mL of ethanol.
(iii) The solution so obtained is heated under reflux for about 5 min, whereupon colour of the solution becomes lighter. At this moment, add 100 mg (1.15 mmol) of LiBr dissolved in 0.5 mL of ethanol. Heat the mixture, with stirring, under reflux for an additional 1 h whereby orange crystalline product is formed.
(iv) The resulting orange product is collected by suction filtration using a sintered glass crucible; wash it with few drops of ether and dry it in a vacuum desiccator.

Colour/Nature: Orange crystals
Decomposition Temperature: .... °C
Yield: .... g
Structure of the complex: Square planar (Figure 10.91)

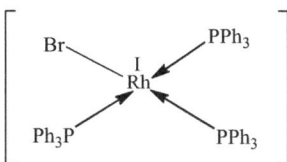

Figure 10.91: Square planar structure of [RhBr(PPh$_3$)$_3$].

10.3.45 Synthesis of *trans*-chlorocarbonylbis(triphenylphosphine)rhodium(I), *trans*-[RhCl(CO)(PPh$_3$)$_2$]

Introduction
Compounds in which a metal atom is directly bonded to carbon are known as organometallic compounds. Such compounds are widely used in the area of organic synthesis and in indudtrial chemistry. Metals in organometallic compounds are generally found in low oxidation states, with the most common carbon ligand being CO (called a carbonyl ligand).

Metal carbonyls are most often prepared by the direct reaction of a metal with CO gas. Synthesis using CO gas is quite dangerous as CO binds irreversibly with hemoglobin, and is, therefore, extremely toxic. In the synthesis of the compound in question, that is, [RhCl(CO)(PPh$_3$)$_2$], CO is generated in situ (within the reaction system) employing the much safer dimethylformamide (DMF) as the source of the CO group.

Reagents required
Rhodium(III) chloride trihydrate (RhCl$_3 \cdot$3H$_2$O), triphenylphosphine (PPh$_3$) and dimethylformamide [HCON(CH$_3$)$_2$]

Reaction

$$RhCl_3 \cdot 3H_2O + 2PPh_3 + HCON(CH_3)_2 \longrightarrow [RhCl(CO)(PPh_3)_2]$$

Synthetic procedure
(i)   In a 10 mL round-bottom flask equipped with a magnetic stirring bar, place 2 mL of DMF followed by 25 mg of $RhCl_3 \cdot 3H_2O$.
(ii)   Now, attach a water condenser to the flask, place the apparatus on a sand bath set upon a magnetic stirring hot plate, and stir the solution for 5 min.
(iii) The resulting solution is then heated at reflux for ~20 min until the colour changes from dark brown to lemon yellow.
(iv)   Cool the solution to room temperature, and then remove any solid that may still be present by suction filtration. Wash the crucible with a few drops of DMF to ensure that no rhodium(I) carbonyl adheres to it.
(v)   Return the filtrate to the reaction flask, and place it in the **fume hood.** Now, add 100 mg of PPh_3 in small portions to the filtrate until no further evolution of gas (CO) is evident. By the end of PPh_3 addition, shiny yellow crystals of the desired compound should precipitate from the solution.
(vi)   In order to complete the precipitation, add few drops of ethanol, and cool the solution in a salt-ice bath for 30 min.
(vii) The resulting crystalline product is collected by suction filtration using a sintered glass crucible. Wash the product with a 0.5 mL portion of absolute ethanol, then with ether and finally dry it in a desiccator over anhydrous $CaCl_2$.

Colour/Nature: Shiny yellow crystals
Decomposition Temperature: .... °C
Yield: .... g
Structure of the complex: Square planar (Figure 10.92)

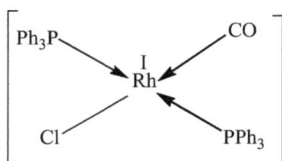

Figure 10.92: Square planar structure of [RhCl(CO)(PPh₃)₂].

10.3.46 Synthesis of *mer*-carbonyltrichlorobis(triphenylphosphine)rhodium(III), *mer*-[Rh^{III}Cl₃(CO)(PPh₃)₂]

Reagents required
Rhodium(III) chloride trihydrate ($RhCl_3 \cdot 3H_2O$), triphenylphosphine (PPh_3), N,N'-dimethylformamide [$HCON(CH_3)_2$], chloroform ($CHCl_3$) and carbon tetrachloride (CCl_4)

Reaction
Addition of cholrine (Cl_2) to the square planar complex, $[RhCl(CO)(PPh_3)_2]$ results in a six coordinate rhodium(III) species, *mer*-$[RhCl_3(CO)(PPh_3)_2]$

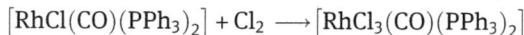

$$[RhCl(CO)(PPh_3)_2] + Cl_2 \longrightarrow [RhCl_3(CO)(PPh_3)_2]$$

Synthetic procedure
(i)   The square planar complex, $[RhCl(CO)(PPh_3)_2]$ is first synthesized taking $RhCl_3 \cdot 3H_2O$, triphenylphosphine and DMF as described in experiment (**10.3.45**).
(ii) In a 5 mL conical flask equipped with a magnetic stirring bar, place 25 mg of $[RhCl(CO)(PPh_3)_2]$ and 1.5 mL of chloroform. When the solid has completely dissolved to from a yellow solution, add 1 mL of **chlorine saturated CCl$_4$**. The solution immediately turns red-brown.
(iii) Allow the mixture to stand for 10 min, whereupon the precipitation of the product should begin. In order to achieve a complete precipitation of the de-sired compound, a gentle stream of nitrogen gas is passed over the solution until solvent is evaporated. Alternatively, conical flask may be attached to a rotatory evaporator to strip off the solvent.
(iv) Disperse the precipitated solid in 1 mL ethanol, suction filter the resulting mix-ture in a sintered glass crucible and wash it with 0.5 mL portions each of etha-nol and ether. Dry the solid in desiccator over anhydrous $CaCl_2$.

Preparation of chlorine saturated CCl$_4$
The chlorine saturated CCl_4 solution may be easily prepared by bubbling Cl_2 gas through the CCl_4 for 30 s. Cl_2 can be generated by dropwise addition of concentrated HCl to the solid $KMnO_4$ taken in a measuring flask (Figure 10.93).

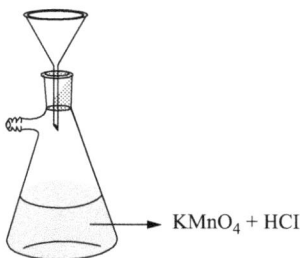

$$2KMnO_4 + 16HCl \longrightarrow 2MnCl_2 + 2KCl + 8H_2O + 5Cl_2$$

KMnO$_4$ + HCl

Figure 10.93: Measuring flask containing KMnO$_4$.

Colour/Nature: Red-brown crystals
Decomposition Temperature: .... °C
Yield: .... g
Structure of the complex: Octahedral (Figure 10.94)

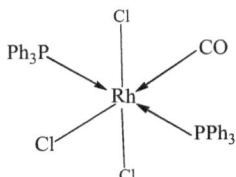

Figure 10.94: Octahedral structure of [RhCl$_3$(CO)(PPh$_3$)$_2$].

Red-brown colour of this compound is due to its diamagnetic behaviour. The octahedral structure and diamagnetic nature of this compound are well explained by d^{2}sp^3 hybridization scheme (Figure 10.95) shown below.

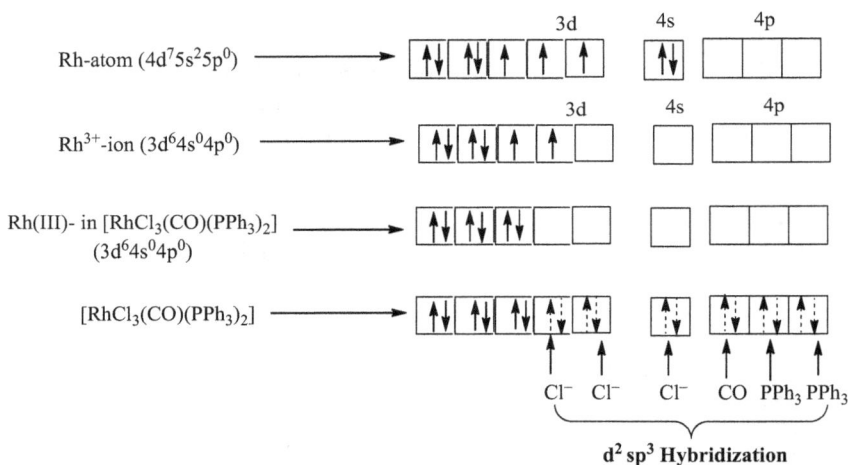

Figure 10.95: dsp^2 Hybridization scheme leading to the formation of diamagnetic square planar [RhCl$_3$(CO)(PPh$_3$)$_2$].

10.4 Synthesis of metal oxide nanoparticles

The nanotechnology is a branch of science and engineering that deals with materials having dimensions in the range of 100 nm scale or less. The idea of nanotechnology was first recognized by American Scientist and Nobel Prize winner, Richard P. Feynman (Figure 10.96), who gave a fanciful speech in 1959. The history of nanotechnology dates very long back to 1959, when Richard Feynman, a physicist at California Institute of Technology, in one of his famous lectures said 'There is plenty of room at bottom' and

Figure 10.96: Richard P. Feynman, Nobel Prize winner in physics (1965).

suggested that scaling down to nanolevel and starting from the bottom would be a key to future technology.

Nanoparticles have wide applications due to the unique size-dependent properties. Magnetic nanoparticles have been receiving considerable attention because of their wide range of applications, such as the immobilization of the proteins and enzymes, bioseparation, immunoassays, drug delivery and biosensors. Nanoparticles possess high surface-to-volume ratio due to their small size which give very distinctive features to nanoparticles.

Transition metal oxide nanoparticles (NPs) have many applications such as catalysts, sensors, superconductors and adsorbents. Metal oxides constitute an important class of materials that are involved in environmental science, electrochemistry, biology, chemical sensors, magnetism and other fields.

In view of various applications of nanoparticles, in general, and transition metal oxide, in particular, synthesis methods of some metal oxide nanoparticles are being given in this section.

10.4.1 Synthesis of copper oxide (CuO) nanoparticles

Reagents required
Copper nitrate, potassium hydroxide and polyvinylpyrrolidone

Principle
In the synthesis of metal oxide nanoparticles, polymers are used to stabilize the aggregation of metal atoms. Polyvinylpyrrolidone (PVP) is the most commonly used

polymer in the synthesis of metal oxides. PVP consists of repeating chains of polar groups, which help in dissolving metal salts and facilitate transport. Without PVP, the metal nanoparticles will be unstable.

(i) Take 2.9 g of copper nitrate in a 250 mL beaker and mix it with 1.2 g of polyvinylpyrrolidone (PVP) and 100 mL of distilled water. Stir this mixture using magnetic stirrer and heat the solution till it reaches 600 °C.
(ii) Take 1 M of sodium hydroxide solution. Once the desired temperature (600 °C) is reached, add sodium hydroxide solution to the above solution drop by drop. Heat and stir for 1 h. Black precipitate of CuO will be formed.
(iii) The resulting precipitate is suction filtered in a sintered glass crucible, and dry it in a muffle furnace at 500 °C for 2 h to get copper oxide nanopowder.

$$Cu(NO_3)_2 + 2NaOH \longrightarrow Cu(OH)_2 + 2NaNO_3$$

$$Cu(OH)_2 \longrightarrow CuO + H_2O$$

The pH of the solution plays an important role in the synthesis of copper oxide nanoparticles. For 0.2 M, 0.4 M, 0.6 M and 0.8 M concentrations of sodium hydroxide, blue solution is obtained and the pH is of the order of 4–8, but for 1 M solution of sodium hydroxide black precipitate is formed indicating the formation of copper oxide nanoparticles. The pH of the solution is 10.5.

The actual size of nanoparticles is estimated from scanning electron microscopy (SEM) micrograph. Most of the nanoparticles have size around less than 50 nm.

10.4.2 Synthesis of manganese dioxide nanoparticles

Manganese dioxide nanoparticles are synthesized by co-precipitation method.

Method I

Reagents required
Manganese(II) sulphate monohydrate, manganese(II) oxalate and NaOH

Synthetic procedure
(i) Mix 0.2 M aqueous solution of manganese(II) sulphate monohydrate and 0.2 M aqueous solution of manganese(II) oxalate with continuous stirring at a constant temperature of 60 °C.
(ii) During stirring, an aqueous solution NaOH is added to the above solution till the pH of the solution becomes 12. Stirring is continued for 1 h at a constant temperature of 60 °C. This gives brown precipitate of MnO_2.

(iii) The precipitate so obtained is suction filtered in a sintered glass crucible and washed several times with ethanol to remove unreacted matters and dried overnight in an electric oven at 100 °C. Then the precipitate is kept in a muffle furnace at 500 °C for 4 h.

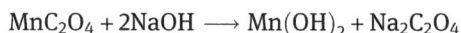

$$MnSO_4 + 2NaOH \longrightarrow Mn(OH)_2 + Na_2SO_4$$

$$MnC_2O_4 + 2NaOH \longrightarrow Mn(OH)_2 + Na_2C_2O_4$$

$$Mn^{II}(OH)_2 + 1/2O_2 \longrightarrow Mn^{IV}O_2 + H_2O \text{ (Aerial oxidation)}$$

XRD studies predict the average size MnO_2 particles to be 25–30 nm.

Method II

Reagents required
Manganese(II) sulphate monohydrate and NaOH

Synthetic procedure
(i) Freshly prepared 100 mL aqueous solution of 2M NaOH is added drop by drop to 100 mL aqueous solution of 1 M $MnSO_4 \cdot H_2O$. The solution is stirred continuously at 60 °C for 2 h to precipitate the MnO_2 nanoparticles.
(ii) The precipitate is then separated from the reaction mixture through suction filtration in a sintered glass crucible and washed several times with deionized water and dried in a hot air oven at 100 °C for 12 h.

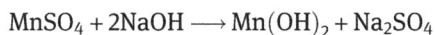

$$MnSO_4 + 2NaOH \longrightarrow Mn(OH)_2 + Na_2SO_4$$

$$Mn^{II}(OH)_2 + 1/2O_2 \longrightarrow Mn^{IV}O_2 + H_2O(\text{Areal oxidation})$$

The size of the synthesized manganese dioxide nanoparticles from SEM micrograph is found to be in the range of 40.5–70 nm.

Method III

Reagents required
Manganese(II) sulphate monohydrate, potassium permanganate and NaOH

Synthetic procedure
(i) Dissolve 1.69 g (10 mmol) of $MnSO_4.H_2O$ in 3 mL of distilled water in a test tube. In this solution, 0.5 g of cetyltrimethyl ammonium bromide (CTAB),

$CH_3(CH_2)_{15}NBr(CH_3)_3$, is added as a surfactant in order to narrow the range of obtained nanomaterial.

(ii) In another test tube take 2.4 mL aqueous solution of 2.5 M NaOH.

(iii) Add simultaneously the above two solutions into a hot solution of $KMnO_4$ obtained by dissolving 1.9 g (1 mmol) in 15 mL of distilled water. Stir the solution for 1 h, and isolate the brown precipitate of MnO_2 by suction filtration in a sintered glass crucible.

(iv) Dry the precipitate in a desiccator under vacuum and grind finely.

In this synthesis, $KMnO_4$ in alkaline medium plays a role of oxidizing manganese(II) to manganese(IV) while in methods I and II aerial oxidation is responsible for oxidation of manganese(II) to manganese(IV).

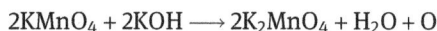

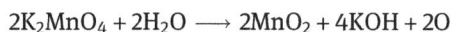

$$2KMnO_4 + 2KOH \longrightarrow 2K_2MnO_4 + H_2O + O$$

$$2K_2MnO_4 + 2H_2O \longrightarrow 2MnO_2 + 4KOH + 2O$$

$$MnSO_4 + 2NaOH \longrightarrow Mn(OH)_2 + Na_2SO_4$$

$$Mn^{II}(OH)_2 + O \longrightarrow Mn^{IV}O_2 + H_2O$$

10.4.3 Synthesis of chromium oxide (Cr_2O_3) nanoparticles

Reagents required

Chromium (III) sulphate hydrate [$Cr_2(SO_4)_3 \cdot xH_2O$] and aqueous ammonia solution

Synthetic procedure

(i) 500 mL of 0.1 M solution of $Cr_2(SO_4)3 \cdot xH_2O$ in deionized water was taken and concentrated aqueous ammonia was added into it dropwise with constant stirring until the pH of the solution reached 10.

(ii) The precipitate thus obtained is filtered on Buckner funnel and washed several times with distilled water. The precipitate is dried in an oven at 70 °C for 24 h and then calcined at 600 °C in a muffle furnace for 5 h to get chromium oxide (Cr_2O_3) nanoparticles.

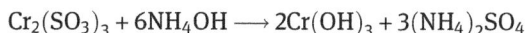

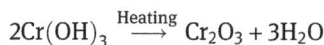

$$Cr_2(SO_3)_3 + 6NH_4OH \longrightarrow 2Cr(OH)_3 + 3(NH_4)_2SO_4$$

$$2Cr(OH)_3 \xrightarrow{\text{Heating}} Cr_2O_3 + 3H_2O$$

TEM image shows that Cr_2O_3 nanoparticles are having particle size in the range of 20–70 nm.

10.4.4 Synthesis of Co_3O_4 nanoparticles

Reagents required
Cobalt(II) chloride hexahydrate, glycerol, aqueous ammonia solution and distilled water

Reaction
Co_3O_4 can be prepared by calcinations of cobalt hydroxide at 450 °C in air.

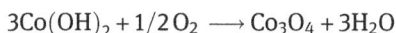

$$CoCl_2 + 2NH_4OH \longrightarrow Co(OH)_2 + 2NH_4Cl$$
$$3Co(OH)_2 + 1/2O_2 \longrightarrow Co_3O_4 + 3H_2O$$

Synthetic procedure
(i) The 0.01 M cobalt chloride was dissolved in 500 mL distilled water and 10% of glycerol. This suspension was stirred for 20 min at 50 °C over magnetic stirrer. After 20 min, 50 mL of aqueous ammonium solution was slowly added to control the agglomeration.
(ii) The obtained cobalt hydroxide was calcined in air for 3 h at 450 °C, which resulted in the Co_3O_4 nanoparticles.

The SEM of Co_3O_4 nanoparticles shows the spherical agglomerated particles.

10.4.5 Synthesis of ZnO nanoparticles

Reagents required
Zinc nitrate hexahydrate [$Zn(NO_3)_2 \cdot 6H_2O$] and sodium hydroxide
(i) Dissolve 2.28 g of zinc nitrate hexahydrate in 75 mL of deionized water in a beaker, and then a solution of 0.6 g of NaOH dissolved in 150 mL of deionized water is added dropwise under magnetic stirring. After the addition was completed, the stirring is continued for 30 min and then cooled with cold water.
(ii) The precipitate so obtained is suction filtered and washed several times with distilled water. Then the precipitate is dried at 60 °C for 24 h and finally calcined at 200 °C for 2 h.

Reaction
The probable reaction scheme showing the formation of ZnO nanoparticles is shown below:

$$Zn(NO_3)_2 + 2NaOH \longrightarrow Zn(OH)_2$$

$$Zn(OH)_2 \longrightarrow ZnO + H_2O$$

SEM studies show that zinc oxide nanoparticles are spherical in shape and their sizes are about 30–63 nm.

10.4.6 Synthesis of NiO nanoparticles

Reagents required
Anhydrous nickel(II) chloride, aqueous ammonia solution and deionized water
(i)   20 mL ammonia was gradually added (approximately 30 min) to 2 g of anhydrous $NiCl_2$ powder under vigorous stirring (550–700 rpm) to produce the solution containing Ni^{2+}. Because of the exothermic reaction of $NiCl_2$ with ammonia, it should be slowly added to the precursor.
(ii) The solution temperature is set at 75 °C and the pH is also set in the range of 7.5 to 9. Distilled water was subsequently added as a precipitant agent while vigorous stirring of solution continued until a green precipitate was formed.
(iii) The resulting precipitate was filtered in a sintered glass crucible and washed twice with deionized water and ethanol. To produce black NiO nanoparticles, green precipitates were dried at 105 °C for 90 min in an oven and then heated at 410 °C for 1 h in a muffle furnace.

$$NiCl_2 + 2NH_4OH \longrightarrow Ni(OH)_2$$

$$Ni(OH)_2 \longrightarrow NiO + H_2O$$

The SEM micrograph of nanoparticles reveals that the particles have walls.

10.4.7 Synthesis of MgO nanoparticles

Reagents required
Magnesium nitrate hexahydrate [$Mg(NO_3)_2 \cdot 6H_2O$], Sodium hydrogen carbonate/ sodium bicarbonate ($NaHCO_3$) and poly vinyl alcohol (PVA) as surfactant

Synthetic procedure
(i)   Take 50 mL of 0.5 M $Mg(NO_3)_2$ and PVA solutions (both solutions prepared in distilled water) in a 250 mL three-necked round-bottom flask and keep over the magnetic stirrer and allow to stir for 5–10 min at room temperature.
(ii) Now, 50 mL of 1 M solution of $NaHCO_3$ is slowly added to the above solution using addition funnel drop by drop under constant stirring condition. Then

50 mL of 1 M NaOH solution was slowly added into the above resulting solution under stirring.

(iii) After addition of surfactant and precipitating agent, the constituent mixture is allowed to be stirred for 3 h without altering any parameters. After completion of this whole reaction process, very finely powdered white precipitate is settled at the bottom of RB flask.

(iv) The resulting precipitate is separated carefully using Buckner funnel; wash thoroughly with the help of doubly distilled water to make the precipitate free from traces of foreign elements. The resulting substrate [$Mg(OH)_2$] is kept in an air oven for 1 h at 80 °C for complete drying. Then the MgO nanoparticles are obtained via controlled calcination process using muffle furnace for 5hrs at 350 °C.

$$Mg(NO_3)_2 + 2NaOH \longrightarrow Mg(OH)_2 + 2NaNO_3$$

$$Mg(OH)_2 \longrightarrow MgO + H_2O$$

SEM image shows that MgO nanoparticles are having particle size in the range of 60–100 nm.

Exercises

Multiple choice questions

1. Design of ignition tube apparatus (ITS) requires:
   (a) Test tube   (b) Ignition tube   (c) Rubber cork   (d) All of these

2. Qualitative tests of acid and basic radicals are expected to be:
   (a) Safe   (b) Less time consuming   (c) Economic   (d) All of these

3. When a salt/mixture containing iodide is heated with concentrated H_2SO_4, the appearance of intense blue colouration with starch solution is due to liberation of:
   (a) HI   (b) I_2 (c) HI and I_2 (d) None of these

4. When a salt/mixture containing chloride is heated with concentrated H_2SO_4, the appearance of curdy white deposit with $AgNO_3$ solution around the mouth of the ignition tube is due to liberation of:
   (a) HCl   (b) Cl_2 (c) HCl and Cl_2 (d) None of these

5. In the test of sulphide (S^{2-}) radical using ignition tube apparatus, when taking filter paper moistened with 1% aqueous $Na_2[Fe(NO)(CN)_5]$ solution, the filter paper that reacted with H_2S (generated from salt/mixture and H_2SO_4) turns purple when put over the bottle of:
   (a) HCl   (b) NaOH   (c) NH_4OH (d) All of these

6. In the test of CO_3^{2-} radical using ignition tube apparatus (ITA), ignition tube fitted with filter paper first moistened with aqueous solution of 0.05 M Na_2CO_3 and then with 0.5% phenolphthalein solution in alcohol (now pink colour) is decolourized when it comes in contact with CO_2 generated by heating a mixture having carbonate radical with dilute H_2SO_4 due to the formation of:
    (a) Soluble carbonate ion   (b) Soluble bicarbonate ion
    (c) Both (a) and (b)   (d) None of these

7. In the zirconium-alizarin-S filter paper test of fluoride (F^-), the red-violet zirco-alizarin-S filter paper when put over the mouth of the test tube containing few mg of a salt or a mixture containing fluoride with 3–4 drops of concentrated H_2SO_4 on heating the test tube changes to pale yellow because of the formation of a complex of zirconium with fluoride of composition:
    (a) $[ZrF_6]^{3-}$ (b) $[ZrF_6]^{2-}$ (c) $[ZrF_6]^-$ (d) None of these

8. The nitrite reaction was discovered by P. Griess, a German Chemist, in year:
    (a) 1869   (b) 1879   (c) 1889   (d) None of these

9. The nitrate ion can be tested by Griess reagent by its reduction to nitrite using:
    (a) Zn, CH_3COOH (b) Zn, HCl   (c) Zn, H_2SO_4 (d) All of these

10. When the mixture contains both nitrite and nitrate, nitrate can be tested by Griess reagent for which nitrite is removed by treating the acidic mixture solution with:
    (a) Thiourea   (b) Sulphamic acid
    (c) Ammonium bromide   (d) None of these

11. Coordinated NO^+ in nitrosyl complexes can be chemically tested by Griess Reagent because of the . . . . behaviour of coordinated NO group in metal nitrosyls in which M–N–O bond is linear.
    (a) Nucleophilic behaviour   (b) Electrophilic behaviour
    (c) Electromeric behaviour   (d) None of these

12. The dimethylglyoxime reagent in the test of Ni^{2+} gives scarlet red precipitate or colouration in a medium which should be:
    (a) Slightly acidic   (b) Slightly ammoniacal
    (c) Neutral   (d) All of these

13. In the test of Mn^{2+} basic radical using ammonium peroxodisulphate, $(NH_4)_2S_2O_8$ reagent, the oxidation state of Mn^{2+} changes to:
    (a) Mn^{5+} (b) Mn^{6+} (c) Mn^{7+} (d) None of these

14. In the synthesis of mercuric thiocyanate, $Hg(SCN)_2$, when ethanolic solutions of mercuric chloride and potassium thiocyanate are mixed together, KCl is precipitated because KCl is an:
    (a) Electrovalent compound   (b) Covalent compound
    (c) Mixture of (a) and (b) both (d) None of these

15. In tetrahedral potassium chlorochromate, $K[CrO_3Cl]$, the hybridization is:
(a) sp^2 (b) sp^3 (c) d^3s (d) None of these

16. Which of the following compound is widely used to induce hypotension to reduce bleeding during surgery?
(a) $K_3[Cr(NO)(CN)_5]$ (b) $Na_2[Fe(NO)(CN)_5]\cdot 2H_2O$
(c) $[Fe(NO)(S_2CNEt_2)_2]$ (d) All of these

17. In the synthesis of $[Cr^I(NO)(acac)_2(H_2O)]$ (acacH = acetylacetone) starting from chromic anhydride, $Cr^{VI}O_3$ is a source of Cr metal using $NH_2OH.HCl$ in basic medium, which one of the following generated nitric oxide species in situ reduce Cr^{6+} to Cr^+ ?
(a) NO   (b) NO^+ (c) NO^- (d) All of these

18. In the synthesis of nitrosylbis(diethyldithiocarbamato)iron(I), $[Fe^I(NO)(S_2CNEt_2)_2]$ taking $FeSO_4\cdot 7H_2O$, $NaNO_2$ and diethyldithiocarbamate trihydrate $[(C_2H_5)_2NCSSNa \cdot 3H_2O]$ as starting materials, which one of the following generated nitric oxide species in situ reduce Fe^{3+} to Fe^+ ?
(a) NO   (b) NO^+ (c) NO^- (d) All of these

19. Which one of the following statement is correct with regard to structure and magnetic behaviour of dichlorobis(triphenylphosphine)nickel(II), $[NiCl_2(PPh_3)_2]$?
(a) Compound is tetrahedral and paramagnetic with respect to 2 unpaired electrons
(b) Compound is square planar and diamagnetic
(c) Compound is tetrahedral and diamagnetic
(d) Compound is square planar and paramagnetic with respected to 2 unpaired electrons

20. In $[NiBr(NO)(PPh_3)_2]$ (nitrosyl compound), the oxidation state of nickel is:
(a) 0   (b) +I   (c) +II   (d) None of these

21. In potassium trioxalatochromate trihydrates using $K_2Cr_2O_7$ and $H_2C_2O_4\cdot 2H_2O$, the oxidation state of chromium changes from +6 to:
(a) +2   (b) + 3   (c) +4   (d) None of these

22. Tetraamminecopper(II) sulphate monohydrate $[Cu(NH_3)_4]SO_4\cdot H_2O$ is insoluble at low temperature in:
(a) Water   (b) Ethanol
(c) N, N-Dimethylformamide   (d) None of these

23. In the synthesis of monochloropentaamminecobalt(III) chloride, $[Co(NH_3)_5Cl]Cl_2$, using cobaltous chloride hexahydrate $(CoCl_2.6H_2O)$, a suitable oxidizing agent for the oxidation of Co(II) to Co(III) in this synthesis is:
(a) Aerial oxygen   (b) H_2O_2 (c) HNO_3 (d) All of these

24. Vanadyl sulphate ($VOSO_4$) is prepared by the reduction of a solution of vanadium pentoxide (V_2O_5) in concentrated sulphuric acid by:
(a) Zn and HCl   (b) Zn and H_2SO_4 (c) C_2H_5OH (d) None of these

25. The structure of bis(acetylacetonato)oxovanadium(IV), [VO(acac)$_2$] (acacH = acety-lacetone) is:
(a) Trigonal bipyramidal   (b) Square pyramidal
(c) Distorted octahedral   (d) None of these

26. In the octahedral structure of $K_3[Al(C_2O_4)_3]\cdot 3H_2O$, the number of members in the chelate ring formed by oxalate ion including Al is:
(a) 4   (b) 5   (c) 6   (d) None of these

27. In the structure of potassium dichromate ($K_2Cr_2O_7$) involving two $CrO_4{}^{2-}$ tetrahedra linked through O-atom involves a hybridization known as:
(a) sp^3 (b) dsp^3 (c) d^3s (d) None of these

28. [$NiCl_2(PPh_3)_2$] is insoluble in:
(a) Nitromethane   (b) Ethanol   (c) Chloform   (d) None of these

29. In the synthesis of [$Ni(NH_3)_6$]Cl_2 taking $NiCl_2 \cdot 6H_2O$ and concentrated ammo-nia, precipitation of the complex takes place in:
(a) Aqueous medium   (b) Ethanolic medium
(c) Acidic medium   (d) None of these

30. In the synthesis of [$Cu(NH_3)_4$]$SO_4\cdot H_2O$ from $CuSO_4\cdot 5H_2O$ and concentrated am-monia, precipitation of the complex takes place in:
(a) Aqueous medium   (b) Aqueous ethanolic medium
(c) Acidic medium   (d) None of these

31. In the synthesis of sodium hexanitritocobaltate(III), $Na_3[Co(NO_2)_6]$ using Co $(NO_3)_2\cdot 6H_2O$ and sodium nitrite, $NaNO_2$, a suitable oxidizing agent for the oxidation of Co(II) to Co(III) in this synthesis is:
(a) Aerial oxygen   (b) H_2O_2 (c) HNO_3 (d) All of these

32. In tetrahedral potassium manganate (VI), K_2MnO_4, the hybridization is:
(a) sp^2 (b) sp^3 (c) d^3s (d) None of these

33. Saccharin was discovered by Fahlberg in year:
(a) 1879   (b) 1878   (c) 1880   (d) None of these

34. The colour of tetraaqua-bis(o-sulphobenzimide)nickel(II) dihydrate is:
(a) White   (b) Red   (c) Green   (d) None of these

35. In the synthesis of *trans*-dichlorobis(ethylenediamine)cobalt(III) chloride using aqueous solution of $CoCl_2.6H_2O$ and ethylenediamine followed by addition of concentrated HCl, a suitable oxidizing agent for the oxidation of Co(II) to Co(III) in this synthesis is:
   (a) Aerial oxygen   (b) H_2O_2 (c) HNO_3 (d) None of these

36. Glycine anion forms *cis*- and *trans*-isomer of bis(glycinato)copper(II)monohydrate because this anion is . . . . in nature and structural isomers can arise depending on the relative orientation of the ligands.
   (a) Symmetric   (b) Asymmetric   (c) Cannot be said   (d) None of these

37. The colour of Wilkinsons's catalyst, Chlorotris(triphenylphosphine)rhodum(I), $[RhCl(PPh_3)_3]$, is:
   (a) White   (b) Burgundy red   (c) Yellow   (d) None of these

38. In the synthesis of *trans*-chlorocarbonylbis(triphenylphosphine)rhodium(I), *trans*-$[RhCl(CO)(PPh_3)_2]$, safer source of CO group is:
   (a) CO gas   (b) Dimethylformamide   (c) Acetone   (d) None of these

39. The colour of *mer*-carbonyltrichlorobis(triphenylphosphine)rhodium(III), $[Rh^{III}Cl_3(CO)(PPh_3)_2]$, is:
   (a) White   (b) Red-brown   (c) Yellow   (d) None of these

40. The idea of nanotechnology was first recognized by Richard P. Feynman, a Nobel Prize winner in year 1965 in the subject:
   (a) Chemistry   (b) Physics   (c) Mathematics   (d) None of these

41. MgO nanoparticles synthesized by taking $Mg(NO_3)_2 \cdot 6H_2O$, sodium hydrogen carbonate/sodium bicarbonate ($NaHCO_3$) and poly vinyl alcohol (PVA) as surfactant have particles in the range of:
   (a) 80–100 nm   (b) 60–100 nm   (c) 25–40 nm   (d) 40–60 nm

Appendix I
Some aspects of modern periodic table

(a) Modern periodic table

The periodic table (Table I.1) has undergone extensive changes in the time since it was originally developed by Mendeleev and Moseley. Many new elements have been discovered, while others have been artificially synthesized. Each fits properly into a **group** of elements with similar properties. The periodic table is an arrangement of the elements in order of their atomic numbers so that elements with similar properties appear in the same vertical column or group. The table thus consists of 18 columns or groups.

Table I.1 shows the most commonly used form of the periodic table. Each square shows the chemical symbol of the element along with its atomic number. The name of each element is given in a separate table along with other details because of the compact size of the square in a single page.

A **period** is a horizontal row of the periodic table. There are seven periods in the periodic table, with each one beginning at the far left. When we talk about the periods of a modern periodic table, one should keep in mind that the number of shells present in an atom determines its period number. The elements of period will have only one shell, elements of period 2 will have two shells and so on. In other words, a new period begins when a new principal energy level begins filling with electrons. Period 1 has only two elements (hydrogen and helium), whilst periods 2 and 3 have 8 elements. Periods 4 and 5 have 18 elements. Periods 6 and 7 have 32 elements because the two bottom rows that are separated from the rest of the table belong to those periods. They are pulled out in order to make the table itself fit more easily onto a single page.

A **group** is a vertical column of the periodic table, based on the organization of the outer shell electrons. There are a total of 18 groups. There are two different numbering systems that are commonly used to designate groups and one should be familiar with both. The traditional system used in the United States involves the use of the letters A and B. The first two groups are 1A and 2A, whilst the last six groups are 3A through 8A. The middle groups use B in their titles. Unfortunately, there was a slightly different system in place in Europe. To eliminate confusion the International Union of Pure and Applied Chemistry (IUPAC) decided that the official system for numbering groups would be a simple **1** through **18** from left to right. Many periodic tables show both systems simultaneously.

The elements of group 1, 2, 13, 14, 15, 16 and 17 are known as the main group elements or normal elements. The elements of groups 3, 4, 5, 6, 7, 8, 9, 11 and 12 are known as the **transition elements.** Group 18 is called **noble gases** or **inert gases**. Their outermost shell is completely filled. Due to this stable electronic configuration, they generally do not react with the other elements.

https://doi.org/10.1515/9783110727289-011

Table I.1: Modern periodic table.

Period \ Group	1	2	3	4	5	6	7	8	9	10	11	12	13	14	15	16	17	18
1	1 H																	2 He
2	3 Li	4 Be											5 B	6 C	7 N	8 O	9 F	10 Ne
3	11 Na	12 Mg											13 Al	14 Si	15 P	16 S	17 Cl	18 Ar
4	19 K	20 Ca	21 Sc	22 Ti	23 V	24 Cr	25 Mn	26 Fe	27 Co	28 Ni	29 Cu	30 Zn	31 Ga	32 Ge	33 As	34 Se	35 Br	36 Kr
5	37 Rb	38 Sr	39 Y	40 Zr	41 Nb	42 Mo	43 Tc	44 Ru	45 Rh	46 Pd	47 Ag	48 Cd	49 In	50 Sn	51 Sb	52 Te	53 I	54 Xe
6	55 Cs	56 Ba	57–71	72 Hf	73 Ta	74 W	75 Re	76 Os	77 Ir	78 Pt	79 Au	80 Hg	81 Tl	82 Pb	83 Bi	85 Po	85 At	86 Rn
7	55 Fr	56 Ra	89–103	104 Rf	105 Db	106 Sg	107 Bh	108 Hs	109 Mt	110 Ds	111 Rg	112 Cn	113 Nh	114 Fl	115 Mc	116 Lv	117 Ts	118 Og

(Group)

Period

Lanthanide series	57 La	58 Ce	59 Pr	60 Nd	61 Pm	62 Sm	63 Eu	64 Gd	65 Tb	66 Dy	67 Ho	68 Er	69 Tm	70 Yb	71 Lu
Actinide series	89 Ac	90 Th	91 Pa	92 U	93 Np	94 Pu	95 Am	96 Cm	97 Bk	98 Cf	99 Es	100 Fm	101 Md	102 No	103 Lr

(b) Features of modern periodic table

The features of modern periodic table are summarized in Table I.2.

Table I.2: Features of modern periodic table.

Group number	Group name	Property
Group 1 or IA.	Alkali metals	They form strong alkalis with water
Group 2 or IIA	Alkaline earth metals	They also form alkalis but are weaker than group 1 elements.
Group 13 or IIIA	Boron family	Boron is the first member of this family.
Group 14 or IVA	Carbon family	Carbon is the first member of this property.
Group 15 or VA	Nitrogen family	This group has non-metals and metalloids.
Group 16 or VIA	Oxygen family	They are also known as **chalcogens**.
Group 17 or VIIA	Halogen family	The elements of this group form salts.
Group 18	Zero group	They are noble gases and under normal conditions they are inert.

(c) List of chemical elements

Table I.3 consists of 118 elements of the periodic table, sorted by atomic number, atomic weight, symbols, discovery year and the group.

Table I.3: Elements of periodic table sorted out by atomic weight, symbols, discovery year and the group.

Atomic number	Atomic weight	Symbol	Name of the element	Group	Discovery year
1	1.008	H	Hydrogen	1	1776
2	4.003	He	Helium	18	1895
3	6.941	Li	Lithium	1	1817
4	9.012	Be	Beryllium	2	1797
5	10.811	B	Boron	13	1808
6	12.011	C	Carbon	14	Ancient
7	14.007	N	Nitrogen	15	1772
8	15.999	O	Oxygen	16	1774

Table I.3 (continued)

Atomic number	Atomic weight	Symbol	Name of the element	Group	Discovery year
9	18.998	F	Fluorine	17	1886
10	20.180	Ne	Neon	18	1898
11	22.990	Na	Sodium	1	1807
12	24.305	Mg	Magnesium	2	1755
13	26.982	Al	Aluminium	13	1825
14	28.086	Si	Silicon	14	1824
15	30.974	P	Phosphorous	15	1669
16	32.065	S	Sulphur	16	Ancient
17	35.453	Cl	Chlorine	17	1774
18	39.948	Ar	Argon	18	1894
19	39.098	K	Potassium	1	1807
20	40.078	Ca	Calcium	2	1808
21	44.956	Sc	Scandium	3	1879
22	47.867	Ti	Titanium	4	1791
23	50.942	V	Vanadium	5	1830
24	51.996	Cr	Chromium	6	1797
25	54.938	Mn	Manganese	7	1774
26	55.845	Fe	Iron	8	Ancient
27	58.933	Co	Cobalt	9	1735
28	58.693	Ni	Nickel	10	1751
29	63.546	Cu	Copper	11	Ancient
30	65.390	Zn	Zinc	12	Ancient
31	69.723	Ga	Gallium	13	1875
32	72.640	Ge	Germanium	14	1886
33	74.922	As	Arsenic	15	Ancient
34	78.960	Se	Selenium	16	1817
35	79.904	Br	Bromine	17	1826

Table I.3 (continued)

Atomic number	Atomic weight	Symbol	Name of the element	Group	Discovery year
36	83.800	Kr	Krypton	18	1898
37	85.468	Rb	Rubidium	1	1861
38	87.620	Sr	Strontium	2	1790
39	88.906	Y	Yttrium	3	1794
40	91.224	Zr	Zirconium	4	1789
41	92.906	Nb	Niobium	5	1801
42	95.940	Mo	Molybdenum	6	1781
43	98.000	Tc	Technetium	7	1937
44	101.070	Ru	Ruthenium	8	1844
45	102.906	Rh	Rhodium	9	1803
46	106.420	Pd	Palladium	10	1803
47	107.868	Ag	Silver	11	Ancient
48	112.411	Cd	Cadmium	12	1817
49	114.818	In	Indium	13	1863
50	118.710	Sn	Tin	14	Ancient
51	121.760	Sb	Antimony	15	Ancient
52	127.600	Te	Tellurium	16	1783
53	126.905	I	Iodine	17	1811
54	131.293	Xe	Xenon	18	1898
55	132.906	Cs	Cesium	1	1860
56	137.327	Ba	Barium	2	1808
57	138.906	La	Lanthanum	3	1839
58	140.116	Ce	Cerium	101	1803
59	140.908	Pr	Praseodymium	101	1885
60	144.240	Nd	Neodymium	101	1885
61	145.000	Pm	Promethium	101	1945
62	150.360	Sm	Samarium	101	1879
63	151.964	Eu	Europium	101	1901

Table I.3 (continued)

Atomic number	Atomic weight	Symbol	Name of the element	Group	Discovery year
64	157.250	Gd	Gadolinium	101	1880
65	158.925	Tb	Terbium	101	1843
66	162.500	Dy	Dysprosium	101	1886
67	164.930	Ho	Holmium	101	1867
68	167.259	Er	Erbium	101	1842
69	168.934	Tm	Thulium	101	1879
70	173.040	Yb	Ytterbium	101	1878
71	174.967	Lu	Lutetium	101	1907
72	178.490	Hf	Hafnium	4	1923
73	180.940	Ta	Tantalum	5	1802
74	183.840	W	Tungsten	6	1783
75	186.207	Re	Rhenium	7	1925
76	190.230	Os	Osmium	8	1803
77	192.217	Ir	Iridium	11	Ancient
78	195.078	Pt	Platinum	9	1803
79	196.967	Au	Gold	10	1735
80	200.590	Hg	Mercury	12	Ancient
81	204.383	Tl	Thallium	13	1861
82	207.200	Pb	Lead	14	Ancient
83	208.980	Bi	Bismuth	15	Ancient
84	209.000	Po	Polonium	16	1898
85	210.000	At	Astatine	17	1940
86	222.000	Rn	Radon	18	1900
87	223.000	Fr	Francium	1	1939
88	226.000	Ra	Radium	2	1898
89	227.000	Ac	Actinium	3	1899
90	232.038	Th	Thorium	102	1829
91	231.036	Pa	Protactinium	102	1913

Table I.3 (continued)

Atomic number	Atomic weight	Symbol	Name of the element	Group	Discovery year
92	238.029	U	Uranium	102	1789
93	237.000	Np	Neptunium	102	1940
94	239.050	Pu	Plutonium	102	1940
95	241.060	Am	Americium	102	1944
96	244.060	Cm	Curium	102	1944
97	249.080	Bk	Berkelium	102	1949
98	252.080	Cf	Californium	102	1950
99	252.080	Es	Einsteinium	102	1952
100	257.100	Fm	Fermium	102	1952
101	258.100	Md	Mendelevium	102	1955
102	259.100	No	Nobelium	102	1958
103	262.110	Lr	Lawrencium	102	1961
104	261.000	Rf	Rutherfordium	4	1964
105	262.000	Db	Dubnium	5	1967
106	266.000	Sg	Seaborgium	6	1974
107	264.000	Bh	Bohrium	7	1981
108	277.000	Hs	Hassium	8	1984
109	268.000	Mt	Meitnerium	9	1982
110	281.000	Ds	Darmstadtium	10	1994
111	272.000	Rg	Roentgenium	11	1994
112	285.000	Cn	Copernicium	12	1996
113	286.000	Nh	Nihonium	13	2003
114	289.000	Fl	Flerovium	14	1998
115	288.000	Mc	Moscovium	15	2003
116	292.00	Lv	Livermorium	16	2000
117	294.000	Ts	Tennessine	17	2010
118	294.00	Og	Oganesson	18	2006

Note: Since 2016, the periodic table has 118 confirmed elements, from element 1 (hydrogen) to 118 (oganesson). Elements 113, 115, 117 and 118, the most recent discoveries, were officially confirmed by the IUPAC in December 2015. Their proposed names, nihonium (Nh), moscovium (Mc), tennessine (Ts) and oganesson (Og), respectively, were made official in November 2016 by IUPAC.

The first 94 elements occur naturally; the remaining 24, americium to oganesson (95–118), occur only when synthesized in laboratories. Of the 94 naturally occurring elements, 83 are primordial and 11 occur only in decay chains of primordial elements. No element heavier than einsteinium (element 99) has ever been observed in macroscopic quantities in its pure form, nor has astatine (element 85); francium (element 87) has been only photographed in the form of light emitted from microscopic quantities (300,000 atoms).

(d) Melting point, boiling point, density, earth crust %, ionization energy and electronic configuration of elements

The above information are given in Table I.4.

Table I.4: Different physical properties of elements along with electronic configurations.

At. No.	Symbol	Melting point (°C)	Boiling point (°C)	Density (g/cm³)	Earthcrust %	Ionization energy (eV)	Electronic configuration
1	H	−259	−253	0.09	0.14	13.60	$1s^1$
2	He	−272	−269	0.18		24.60	$1s^2$
3	Li	180	1,330	0.53		5.39	$[He]2s^1$
4	Be	1,287	2,469	1.85		9.32	$[He]2s^2$
5	B	2,076	3,927	2.08		8.30	$[He]2s^2 2p^1$
6	C	3,550	4,827	2.26	0.09	11.26	$[He]2s^2 2p^2$
7	N	−210	−196	1.25		14.53	$[He]2s^2 2p^3$
8	O	−219	−183	1.43	46.71	13.62	$[He]2s^2 2p^4$
9	F	−220	−188	1.70	0.03	17.42	$[He]2s^2 2p^5$
10	Ne	−249	−246	0.90		21.56	$[He]2s^2 2p^6$
11	Na	98	883	0.97	2.75	5.14	$[Ne]3s^1$
12	Mg	650	1,091	1.74	2.08	7.65	$[Ne]3s^2$
13	Al	660	2,470	2.70	8.07	5.99	$[Ne]3s^2 3p^1$
14	Si	1,414	3,265	2.33	27.69	8.15	$[Ne]3s^2 3p^2$
15	P	44	280	1.82	0.13	10.49	$[Ne]3s^2 3p^3$

Table I.4 (continued)

At. No.	Symbol	Melting point (°C)	Boiling point (°C)	Density (g/cm³)	Earthcrust %	Ionization energy (eV)	Electronic configuration
16	S	115	445	2.07	0.05	10.36	$[Ne]3s^23p^4$
17	Cl	−101	−35	3.21	0.05	12.97	$[Ne]3s^23p^5$
18	Ar	−189	−186	1.78		15.76	$[Ne]3s^23p^6$
19	K	64	1,032	0.89	2.58	4.34	$[Ar]4s^1$
20	Ca	842	1,484	1.55	3.65	6.11	$[Ar]4s^2$
21	Sc	1,541	2,836	2.99		6.56	$[Ar]3d^14s^2$
22	Ti	1,668	3,287	4.51	0.62	6.83	$[Ar]3d^24s^2$
23	V	1,910	3,407	6.11		6.75	$[Ar]3d^34s^2$
24	Cr	1,907	2,671	7.19	0.04	6.77	$[Ar]3d^54s^1$
25	Mn	1,246	2,061	7.43	0.09	7.43	$[Ar]3d^54s^2$
26	Fe	1,538	2,862	7.87	5.05	7.90	$[Ar]3d^64s^2$
27	Co	1,495	2,927	8.86		7.88	$[Ar]3d^74s^2$
28	Ni	1,455	2,730	8.90	0.02	7.64	$[Ar]3d^84s^2$
29	Cu	1,085	2,562	8.96		7.73	$[Ar]3d^{10}4s^1$
30	Zn	420	907	7.14		9.39	$[Ar]3d^{10}4s^2$
31	Ga	30	2,400	5.90		6.00	$[Ar]3d^{10}4s^24p^1$
32	Ge	938	2,833	5.32		7.90	$[Ar]3d^{10}4s^24p^2$
33	As	614	817	5.77		9.79	$[Ar]3d^{10}4s^24p^3$
34	Se	221	685	4.81		9.75	$[Ar]3d^{10}4s^24p^4$
35	Br	−7.2	59	3.10		11.81	$[Ar]3d^{10}4s^24p^5$
36	Kr	−157	−153	3.75		14.00	$[Ar]3d^{10}4s^24p^6$
37	Rb	39	688	1.53		4.18	$[Kr]5s^1$
38	Sr	777	1,377	2.64		5.69	$[Kr]5s^2$
39	Y	1,526	2,930	4.47		6.22	$[Kr]4d^15s^2$
40	Zr	1,855	4,377	6.52	0.03	6.63	$[Kr]4d^25s^2$
41	Nb	2,477	4,741	8.57		6.76	$[Kr]4d^35s^2$
42	Mo	2,623	4,639	10.28		7.09	$[Kr]4d^45s^2$

Table I.4 (continued)

At. No.	Symbol	Melting point (°C)	Boiling point (°C)	Density (g/cm³)	Earthcrust %	Ionization energy (eV)	Electronic configuration
43	Tc	2,157	4,265	11.50		7.28	$[Kr]4d^55s^2$
44	Ru	2,310	3,900	12.20		7.36	$[Kr]4d^65s^2$
45	Rh	1,964	3,695	12.41		7.46	$[Kr]4d^75s^2$
46	Pd	1,555	2,963	12.02		8.34	$[Kr]4d^85s^2$
47	Ag	961	2,177	9.34		7.58	$[Kr]4d^{10}5s^1$
48	Cd	321	765	8.65		8.99	$[Kr]4d^{10}5s^2$
49	In	157	2,072	7.31		5.79	$[Kr]4d^{10}5s^25p^1$
50	Sn	232	2602	7.26		7.34	$[Kr]4d^{10}5s^25p^2$
51	Sb	631	1,635	6.69		8.61	$[Kr]4d^{10}5s^25p^3$
52	Te	449	988	6.24		9.01	$[Kr]4d^{10}5s^25p^4$
53	I	114	184	4.93		10.45	$[Kr]4d^{10}5s^25p^5$
54	Xe	−112	−108	0.006		12.13	$[Kr]4d^{10}5s^25p^6$
55	Cs	29	671	1.87		3.89	$[Xe]6s^1$
56	Ba	727	1,845	3.51	0.05	5.21	$[Xe]6s^2$
57	La	920	3,464	6.16		5.58	$[Xe]5d^16s^2$
58	Ce	795	3,443	6.77		5.54	$[Xe]4f^15d^16s^2$
59	Pr	935	3,520	6.77		5.47	$[Xe]4f^36s^2$
60	Nd	1,024	3,074	6.89		5.53	$[Xe]4f^46s^2$
61	Pm	1,042	3,000	6.48		5.58	$[Xe]4f^56s^2$
62	Sm	1,072	1,900	7.52		5.64	$[Xe]4f^66s^2$
63	Eu	826	1,489	5.24		5.67	$[Xe]4f^76s^2$
64	Gd	1,312	3,000	7.90		6.15	$[Xe]f^75d^16s^2$
65	Tb	1,360	3,041	8.27		5.86	$[Xe]f^96s^2$
66	Dy	1,407	2,562	8.54		5.94	$[Xe]f^{10}6s^2$
67	Ho	1,470	2,720	8.54		6.02	$[Xe]f^{11}6s^2$
68	Er	1,529	2,868	9.07		6.11	$[Xe]f^{12}6s^2$
69	Tm	1,545	1,950	9.32		6.18	$[Xe]f^{13}6s^2$

Table I.4 (continued)

At. No.	Symbol	Melting point (°C)	Boiling point (°C)	Density (g/cm³)	Earthcrust %	Ionization energy (eV)	Electronic configuration
70	Yb	824	1,196	6.90		6.25	$[Xe]f^{14}6s^2$
71	Lu	1,656	3,315	9.85		5.43	$[Xe]f^{14}5d^16s^2$
72	Hf	2,233	4,603	13.31		6.83	$[Xe]f^{14}5d^26s^2$
73	Ta	3,017	5,458	16.69		7.55	$[Xe]f^{14}5d^36s^2$
74	W	3,422	5,930	19.30		7.86	$[Xe]f^{14}5d^46s^2$
75	Re	3,186	5,630	21.02		7.83	$[Xe]f^{14}5d^56s^2$
76	Os	3,033	5,012	22.59		8.44	$[Xe]f^{14}5d^66s^2$
77	Ir	2,446	4,130	22.56		8.97	$[Xe]f^{14}5d^76s^2$
78	Pt	1,774	3,827	21.45		8.96	$[Xe]f^{14}5d^96s^1$
79	Au	1,064	2,970	19.30		9.23	$[Xe]f^{14}5d^{10}6s^1$
80	Hg	−39	357	13.53		10.44	$[Xe]4f^{14}5d^{10}\,6s^2$
81	Tl	304	1,473	11.85		6.11	$[Xe]4f^{14}5d^{10}\,6s^26p^1$
82	Pb	327	1,749	11.34		7.42	$[Xe]4f^{14}5d^{10}\,6s^26p^2$
83	Bi	272	1,564	9.78		7.29	$[Xe]4f^{14}5d^{10}\,6s^26p^3$
84	Po	254	962	9.20		8.42	$[Xe]4f^{14}5d^{10}\,6s^26p^4$
85	At	302	337	–		9.30	$[Xe]4f^{14}5d^{10}\,6s^26p^5$
86	Rn	−71	−62	9.73		10.75	$[Xe]4f^{14}5d^{10}\,6s^26p^6$
87	Fr	8	620	2.48		4.07	$[Rn]7s^1$
88	Ra	700	1,737	5.50		5.28	$[Rn]7s^2$
89	Ac	1,227	3,200	10.00		5.17	$[Rn]6d^17s^2$
90	Th	1,800	4,500	11.70		6.31	$[Rn]6d^27s^2$
91	Pa	1,568	4,027	15.37		5.89	$[Rn]5f^26d^17s^2$
92	U	1,132	4,131	19.10		6.19	$[Rn]5f^36d^17s^2$
93	Np	640	4,174	20.45		6.27	$[Rn]5f^46d^17s^2$
94	Pu	639	3,228	19.82		6.03	$[Rn]5f^67s^2$
95	Am	1,176	2,607	12.00		5.97	$[Rn]5f^77s^2$
96	Cm	1,340	3,110	13.51		5.99	$[Rn]5f^76d^17s^2$
97	Bk	986	2,627	14.78		6.20	$[Rn]5f^97s^2$

Table I.4 (continued)

At. No.	Symbol	Melting point (°C)	Boiling point (°C)	Density (g/cm³)	Earthcrust %	Ionization energy (eV)	Electronic configuration
98	Cf	900	1,470	15.10		6.28	$[Rn]5f^{10}7s^2$
99	Es	860	996	8.84		6.42	$[Rn]5f^{11}7s^2$
100	Fm	1,527	–	–		6.50	$[Rn]5f^{12}7s^2$
101	Md	827	–	–		6.58	$[Rn]5f^{13}7s^2$
102	No	827	–	–		6.65	$[Rn]5f^{14}7s^2$
103	Lr	1,627	–	–		4.90	$[Rn]5f^{14}6d^17s^2$
104	Rf	–	–	–		6.02	$[Rn]5f^{14}6d^27s^2$
105	Db	–	–	–		6.89	$[Rn]5f^{14}6d^37s^2$
106	Sg	–	–	–		7.84	$[Rn]5f^{14}6d^47s^2$
107	Bh	–	–	–		7.69	$[Rn]5f^{14}6d^57s^2$
108	Hs	–	–	–		7.59	$[Rn]5f^{14}6d^67s^2$
109	Mt	–	–	–		8.30	$[Rn]5f^{14}6d^77s^2$
110	Ds	–	–	–		9.89	$[Rn]5f^{14}6d^87s^2$
111	Rg	–	–	–		10.59	$[Rn]5f^{14}6d^97s^2$
112	Cn	–	–	–		11.96	$[Rn]5f^{14}6d^{10}7s^2$
113	Nh	–	–	–		7.30	$[Rn]5f^{14}6d^{10}7s^27p^1$
114	Fl	–	–	–		8.54	$[Rn]5f^{14}6d^{10}7s^27p^2$
115	Mc	–	–	–		5.57	$[Rn]5f^{14}6d^{10}7s^27p^3$
116	Lv	–	–	–		7.50	$[Rn]5f^{14}6d^{10}7s^27p^4$
117	Ts	–	–	–		7.69	$[Rn]5f^{14}6d^{10}7s^27p^5$
118	Og	–	–	–		8.69	$[Rn]5f^{14}6d^{10}7s^27p^6$

Note: (–) means respective value is not known.

Atomic number: The number of protons in an atom. Each element is uniquely defined by its atomic number.

Atomic mass: The mass of an atom is primarily determined by the number of protons and neutrons in its nucleus. Atomic mass is measured in atomic mass units (amu) which are scaled relative to carbon, ^{12}C, that is, taken as a standard element with an atomic mass of 12. This isotope of carbon has six protons and six neutrons. Thus, each proton and neutron has a mass of about 1 amu.

Isotope: Atoms of the same element with the same atomic number, but different number of neutrons. Isotope of an element is defined by the sum of the number of protons and neutrons in its nucleus. Elements have more than one isotope with varying numbers of neutrons. For example, there are two common isotopes of carbon, ^{12}C and ^{13}C, which have six and seven neutrons, respectively. The abundances of different isotopes of elements vary in nature depending on the source of materials.

Atomic weight: Atomic weight values represent weighted average of the masses of all naturally occurring isotopes of an element. The values shown here are based on the IUPAC Commission determinations (2001). For these elements, the weight value shown represents the mass number of the longest-lived isotope of the element.

Electron configuration: The distribution of electrons according to the energy sublevels (subshells) in uncharged atoms. The noble gas shown in square brackets (e.g. [He]), marks that all the subshells associated with that element are fully occupied by electrons.

Ionization energy: The energy required to remove the outermost electron from an atom or a positive ion in its ground level. The table lists only the first ionization energy (IE) in eV units. To convert to kJ/mol multiply by 96.4869. IE decreases going down a column of the periodic table, and increases from left to right in a row. Thus, alkali metals have the lowest IE in a period and rare gases have the highest.

Appendix II
Units, fundamental physical constants and conversions

(a) Base units of the SI

The International System of Units (SI [Système International d'Unitès]) is the inter-nationally agreed basis for expressing measurements at all levels and in all areas of science and technology. There are two classes of units in the SI: (i) base or basic units and (ii) derived units. The seven base units of the SI and their base quantities provide the reference used to define all the measurement units of the International System. The seven base units are given in the following table.

(i) Base SI units and physical quantities

Physical Quantity	SI unit	Symbol for SI unit
Length	Metre	m
Mass	Kilogram	kg
Time	Second	s
Electric current	Ampere	A
Thermodynamic temperature	Kelvin	K
Luminous intensity	Candela	cd
Amount of substance	Mole	mol

(ii) Prefixes used for SI units

Factor	Prefix	Symbol
10^{21}	Zetta	Z
10^{18}	Exa	E
10^{15}	Peta	P
10^{12}	Tera	T
10^{9}	Giga	G
10^{6}	Mega	M
10^{3}	Kilo	K

https://doi.org/10.1515/9783110727289-012

(continued)

Factor	Prefix	Symbol
10^2	Hecto	h
10	Deca	da
10^{-1}	Deci	d
10^{-2}	Centi	c
10^{-3}	Milli	m
10^{-6}	Micro	μ
10^{-9}	Nano	n
10^{-12}	Pico	p
10^{-15}	Femto	f
10^{-18}	Atto	a
10^{-21}	Zepto	z

(iii) Derived SI units with special names and symbols

Physical quantity	SI unit		Expression in terms of base or derived SI units
	Name	Symbol	
Frequency	Hertz	Hz	$1 \text{ Hz} = 1 \text{ s}^{-1}$
Force	Newton	N	$1 \text{ N} = 1 \text{ kg m s}^{-2}$
Pressure; stress	Pascal	Pa	$1 \text{ Pa} = 1 \text{ Nm}^{-2}$
Energy; work; quantity of heat	Joule	J	$1 \text{ J} = 1 \text{ N m}$
Electric charge; quantity of electricity	Coulomb	C	$1 \text{ C} = 1 \text{ A s}$
Electric potential; potential difference; electromotive force	Volt	V	$1 \text{ V} = 1 \text{ J C}^{-1}$
Electric capacitance	Farad	F	$1 \text{ F} = 1 \text{ C V}^{-1}$
Electric resistance	Ohm	Ω	$1 \text{ Ω} = 1 \text{ V A}^{-1}$
Electric conductance	Siemens	S	$1 \text{ S} = 1 \text{ Ω}^{-1}$

(continued)

Physical quantity	SI unit		Expression in terms of base or derived SI units
	Name	Symbol	
Magnetic flux; flux of magnetic induction	Weber	Wb	1 Wb = 1 V s
Magnetic flux density; magnetic induction	Tesla	T	$1\ T = 1\ Wb\ m^{-2}$
Inductance	Henry	H	$1\ H = 1\ Wb\ A^{-1}$
Temperature	Degree Celsius	°C	$t[°C] = T[K]\text{-}273.15$
Luminous flux	Lumen	lm	1 lm = 1 cd sr
Illuminance	Lux	lx	$1\ lx = 1\ lm\ m^{-2}$
Activity (of a radionuclide)	Becquerel	Bq	$1\ Bq = 1\ s^{-1}$

(b) Fundamental physical constants

The fundamental physical constants such as the speed of light, the Planck constant and the mass of the electron provide a system of natural units. The constants provide the link between the SI units and theory and also between scientific fields. The values of the fundamental physical constants are taken from the recommended values of the constants which are produced by the CODATA Task Group on Fundamental Constants (http://www.codata.org/about/index.html) based on a review of all the available data. The latest review is available at the CODATA fundamental physical constants webpage (http://physics.nist.gov/cuu/-Constants/index.html). Selected fundamental physical constants and their values are given in the table below which is taken from the latest (2006) values of the constants. The figures in parentheses in the 'Value' column represent the best estimates of the standard deviation uncertainties in the last two digits quoted. Where no uncertainty is given, the value is exact through definition.

Constant	Symbol	Value	SI unit
Speed of light in vacuum	c	$2.997\ 924\ 58 \times 10^8$	$m\ s^{-1}$
Electron charge (elementary charge)	e	$1.602\ 176\ 487(40) \times 10^{-19}$	C
Planck constant (in eV s)	h	$6.626\ 068\ 96\ (33) \times 10^{-34}$	J s
The Avogadro constant	N_A	$6.022\ 141\ 79\ (30) \times 10^{23}$	mol^{-1}
Atomic mass constant (unified atomic mass unit) $m_u = 1/12\ m(^{12}C) = 1\ u$	m_u	$1.660\ 538\ 782(83) \times 10^{-27}$	kg

(continued)

Constant	Symbol	Value	SI unit
Electron mass	m_e	$9.109\ 382\ 15(45) \times 10^{-31}$	kg
(in u)		$5.485\ 799\ 0943(23) \times 10^{-4}$	u
Proton mass	m_p	$1.672\ 621\ 637(83) \times 10^{-27}$	kg
(in u)		$1.007\ 276\ 466\ 77(10)$	u
Neutron mass	m_n	$1.674\ 927\ 211(84) \times 10^{-27}$	kg
(in u)		$1.008\ 664\ 915\ 97(43)$	u
Molar gas constant	R	$8.314\ 472(15)$	$J\ mol^{-1}\ K^{-1}$
The Boltzmann constant	k	$1.380\ 6504(24) \times 10^{-23}$	$J\ K^{-1}$
The Rydberg constant	R_∞	$1.097\ 372 \times 10^7$	m^{-1}
The Bohr radius	a_0	$0.529\ 177 \times 10^{-10}$	m
Electron magnetic moment	μ_e	$9.284\ 83 \times 10^{-24}$	$J\ T^{-1}$
Proton magnetic moment	μ_p	$1.410\ 617 \times 10^{-26}$	$J\ T^{-1}$
The Bohr magneton	μ_B	$9.274\ 08 \times 10^{-24}$	$J\ T^{-1}$
Nuclear magneton	μ_N	$5.050\ 82 \times 10^{-27}$	$J\ T^{-1}$
Molar volume of ideal gas (stp)	V_m	$0.\ 0.022\ 413\ 8$	$M^3\ mol^{-1}$
Acceleration of free fall	g	$9.806\ 65$	$m\ s^{-2}$
(Acceleration due to gravity)			
Gravitational constant	G	$6.672\ 59 \times 10^{-11}$	$m^2\ s^{-2}\ kg$
Debye (electric dipole moment of	D	1.0×10^{-18}	esu cm
molecules)		3.33×10^{-32}	C cm

(c) Conversion factors

1 eV	=	1.602×10^{-12} erg	=	1.602×10^{-19} J		
1 eV/molecule	≈	23.063 kcal/mol	=	96.496 kJ/mol	≈	$8065.46\ cm^{-1}$
1 kcal mol^{-1}	=	$349.76\ cm^{-1}$				
1 kJ mol^{-1}	=	$83.54\ cm^{-1}$				
10,000 cm^{-1}	=	1000 nanometers (nm)	≈ 1.24 eV/molecule			
20,000 cm^{-1}		500 nm				
1 wavenumber (cm^{-1})	=	2.8591×10^{-3} kcal mol^{-1}				

(continued)

1 nanometer (nm)	=	10 A	=	10^{-9} m	= 1 millimicron (mµ)
1 picometer (pm)	=	10^{-2} A			
1 cm	=	10^8 A	=	10^7 nm	
1 erg	=	2.390×10^{-11} kcal			
1 J		10^7 ergs			
1 L	=	1 dm^3	=	10^3 cm^3	
1 dyne	=	1 g cm/sec^2	=	10^{-5} N	
1 cal	=	4.1840 J			
1 atm		760 torr	=	1.101325×10^5 Pascal (Pa)	
1 esu	=	1 dyn$^{1/2}$ cm			

(d) Chemical numerical prefixes

1	Mono
2	Di (bis)
3	Tri (tris)
4	Tetra (tetrakis)
5	Penta (pentakis)
6	Hexa (hexakis)
7	Hepta (heptakis)
8	Octa (octakis)
9	Nona (nonakis)
10	Deka (dekakis)
11	Undeca
12	Dodeca
15	Pentadeca
16	Hexadeca
20	Icosa
30	Triaconta
50	Pentaconta
100	Hecta

Appendix III
Nomenclature of inorganic compounds: the rules

Introduction

The standards of nomenclature in chemistry are the rules published by the International Union of Pure and Applied Chemistry (IUPAC) from time to time. Based on these rules, the universal adoption of an agreed chemical nomenclature is a key tool for communication in the chemical sciences, computer-based searching in databases and regulatory purposes, such as those associated with health and safety or commercial activity. The IUPAC provides recommendations on the nature and use of chemical nomenclature which are freely available at http://www.iupac.org-/publi cations/pac/ and http://www.chem.qmul.ac.uk/iupac/. An overall summary of chemical nomenclature can be found in *Principles of Chemical Nomenclature* (Leigh 2011). Further details can be found in *Nomenclature of Inorganic Chemistry*, colloquially known as *The Red Book* (Connelly et al. 2005).

It should be noted that many compounds may have non-systematic or semi-systematic names (some of which are not accepted by IUPAC for several reasons) and IUPAC rules allow more than one systematic name in many cases. IUPAC is working towards identification of single names which are to be preferred for regulatory purposes.

In this appendix, the nomenclature types described are applicable to compounds, molecules and ions that do not contain carbon, and to many structures that do contain carbon.

(i) Atomic symbols, mass, atomic number and so on of elements

The approved symbols and names of the elements are given in Appendix I. The atomic number, mass number, number of atoms and atomic charges are to be represented as follows:

Mass number: Left upper index
Atomic number: Left lower index
Number of atoms: Right lower index
Ionic charge: Right upper index

For example, $^{200}_{80}Hg_2^{2+}$ represents the doubly charged ion containing two mercury atoms each of which has mass number 200 and atomic number 80. The charge is to be written as Hg_2^{2+}, not Hg_2^{+2}.

https://doi.org/10.1515/9783110727289-013

The periodic A and B subgroups have become hopelessly confused by disparate use in different parts of the world (the traditional system used in the United States involves the use of the letters A and B. The first two groups are 1A and 2A, while the last six groups are 3A through 8A. The middle groups use B in their titles). The IUPAC recommends numbering the groups 1–18 as shown in the table given in Appendix I. Designation of families by names such as the alkali metals, the halogens or the carbon family is unambiguous.

(ii) Formulae and names of compounds in general

The molecular formula is used for compounds that exist as discrete molecules. For example S_2Cl_2, not SCl; $Co_2(CO)_8$, not $Co(CO)_4$ and so on. The electropositive constituent (cation) is placed first in the formula, for example KCl, $MgCO_3$, $CaSO_4$ and so on.

In the case of binary compounds between non-metals, in accordance with the established practice that constituent should be placed first which appears earlier in the sequence:

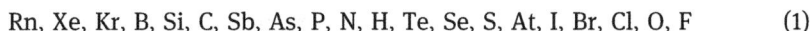

$$\text{Rn, Xe, Kr, B, Si, C, Sb, As, P, N, H, Te, Se, S, At, I, Br, Cl, O, F} \qquad (1)$$

For example: XeF_2, NH_3, H_2S, S_2Cl_2, Cl_2O, OF_2 and so on.

The name of the electropositive constituent is not modified and is placed first. The name of the electronegative constituent (or the element later in sequence (1) for compounds of non-metals) is modified to end in -*ide* if it is monoatomic or homopolyatomic (e.g. O_2^{2-} or N_3^-). For example:

NaCl: Sodium chloride; **MgS:** magnesium sulphide; **Li₃N:** lithium nitride; **NiAs:** nickel arsenide; **SiC:** silicon carbide; **SF₆:** sulphur hexafluoride; **OF₂:** oxygen difluoride; **KI₃:** potassium triiodide and so on.

If the compound contains more than one kind of electropositive constituent, the names should be spaced and cited in alphabetical order for the name of the element. Hydrogen is cited last among the electropositive constituents and is separated from the anion name(s) by a space unless it is known to be bound to an anion. For example, the compound **NaNH₄HPO₄** having three electropositive constituents is named as *ammonium sodium hydrogen phosphate*.

If the electronegative constituent is heteronuclear, it should be named by termination -*ate*. Exceptions include OH^-: hydroxide ion, NH^{2-}: imide ion, NH_2^-: amide ion and CN^-: cyanide ion. In case of two or more electronegative constituents, their sequence should be alphabetical order. This order might differ in names and formulae. For examples,

$Na_2[SO_4]$:   Sodium tetraoxosulphate(VI) or disodium tetraoxosulphate
$Na_2[SO_3]$:   Sodium trioxosulphate(IV) or disodium trioxosulphate
$Na_2[S_2O_3]$:   Disodium trioxothiosulphate
$Na[SO_3F]$:   Sodium trioxofluorosulphate(VI)

Complex anions can be named using the name of the characteristic or central atom modified to end in -*ate*. Ligands attached to the central atom are indicated by the termination -*O*. The oxidation number of the central atom should be indicated by a Roman numeral. For example,

$Na[ICl_4]$ Sodium tetrachloroiodate(III)
$Na[ICl_6]$ Sodium hexachlorophosphate(V)

(iii) Stoichiometric or compositional names

A stoichiometric or compositional name provides information only on the composition of an ion, molecule or compound, and may be related to either the empirical or molecular formula for that entity. It does not provide any structural information.

For homoatomic entities, where only one element is present, the name is formed (Table III.1) by combining the element name with the appropriate multiplicative prefix (multiplicative prefixes are given in a table at the end of Appendix II). Ions are named by adding charge numbers in parentheses, for example (1+), (3+), (2−) and for (most) homoatomic anion names 'ide' is added in place of the 'en', 'ese', 'ic', 'ine', 'ium', 'ogen', 'on', 'orus', 'um','ur', 'y' or 'ygen' endings of element names.

Table III.1: Naming of homoatomic entities.

Formula	Name	Formula	Name
O_2	Dioxygen	Cl^-	Chloride(1−) or chloride
S_8	Octasulphur	I_3^-	Triiodide(1−)
Na^+	Sodium (1+)	O_2^{2-}	Dioxide(2−) or peroxide
Fe^{2+}	Iron (2+)	N_3^-	Trinitride(1−) or azide

Binary compounds are named stoichiometrically by combining the element names and treating, by convention, the element reached first when following the arrow in the element sequence (Figure III.1) as if it were an anion. Thus, the name of this formally 'electronegative' element is given an 'ide' ending and is placed after the name of the formally 'electropositive' element followed by a space (Table III.2). Again, multiplicative prefixes are applied as needed, and certain acceptable alternative names may be used. Stoichiometry may be implied in some cases by the use of oxidation numbers, but is often omitted for common cases, such as in calcium fluoride. Heteropolyatomic entities in general can be named similarly using compositional nomenclature.

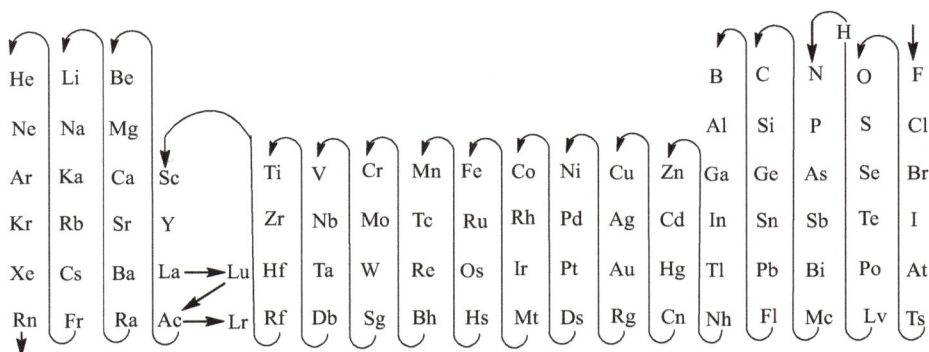

Figure III.1: Arrow in the element sequence.

Table III.2: Naming of binary compounds.

Formula	Name	Formula	Name
GaAs	Gallium arsenide	$FeCl_2$	Iron dichloride or iron(II) chloride
CO_2	Carbon dioxide	$FeCl_3$	Iron trichloride or iron(III) chloride
CaF_2	Calcium difluoride or calcium fluoride	H_2O_2	Dihydrogen dioxide or hydrogen peroxide

Certain ions have traditional short names, which are commonly used and are still acceptable. For example: Ammonium: NH_4^+, Hydroxide: OH^-, Nitrite: NO_2^-, Phosphate: PO_4^{3-} and diphosphate: $P_2O_7^{4-}$.

Inorganic compounds in general can be combinations of cations, anions and neutral entities. By convention, the name of a compound is made up of the names of its component entities: cations before anions and neutral components last. Names of some of the inorganic compounds are given in Table III.3.

The number of each entity present has to be specified in order to reflect the composition of the compound. For this purpose, multiplicative prefixes are added to the name of each entity. The multiplicative prefixes 'di', 'tri', 'tetra' and so on, are used with names for simple entities, or 'bis()', 'tris()', 'tetrakis()' and so on for names for most entities which themselves contain multiplicative prefixes or locants. Care must also be taken in situations when use of a simple multiplicative prefix may be misinterpreted, for example tris(iodide) must be used for $3I^-$ rather than triiodide (which is used for I_3^-), and bis(phosphate) rather than diphosphate (which is used for $P_2O_7^{4-}$). Examples are shown in Table III.3.

Table III.3: Use of multiplicative prefixes in compositional names of some inorganic compounds.

Formula	Name of the compound
$Ca_3(PO_4)_2$	Tricalcium bis(phosphate)
$Ca_2P_2O_7$	Dicalcium diphosphate
BaO_2	Barium(2+) dioxide(2−) or barium peroxide
$MgSO_4 \cdot 7H_2O$	Magnesium sulphate heptahydrate
$CdSO_4 \cdot 6NH_3$	Cadmium sulphate—ammonia (1/6)
$AlK(SO_4)_2 \cdot 12H_2O$	Aluminum potassium bis(sulphate)—water (1/12) or Aluminum potassium bis(sulphate) dodecahydrate
$Al_2(SO_4)_3K_2SO_4 \cdot 24H_2O$	Dialuminum tris(sulphate) —dipotassium sulphate—water (1/1/24)

Names of neutral components are separated by 'em' dashes without spaces. Inorganic compounds may themselves be components in (formal) **addition compounds** (last four examples in Table III.3). The ratios of component compounds can be indicated, in general, using a stoichiometric descriptor in parentheses after the name (see the last three examples in Table III.3). In the special case of hydrates, multiplicative prefixes can be used with the term 'hydrate'.

(iv) Name for ions and radicals

(a) Cations

Names of monoatomic cations are the same as the name of the elements. Oxidation numbers are designated by the use of the stock system. **Stock nomenclature** for inorganic compounds is a widely used **system** of chemical **nomenclature** developed by the German chemist **Alfred Stock** and first published in 1919. In the 'Stock system', the oxidation states of some or all of the elements in a compound are indicated in parentheses by Roman numerals. For example:

Cu^+: Copper (I) ion, Cu^{2+}: Copper(II) ion, I^+: Iodine(I) cation

Polyatomic cations formed from radicals who have special names use those names without change. For example:

NO^+: Nitrosyl cation, NO_2^+: Nitryl cation, UO_2^{2+}: Uranyl(VI) ion or Uranyl(2+) ion

Polyatomic cations derived from the addition of protons to monoatomic anions are named by adding the ending -*onium* to the root name of anions.

Examples: Phosphonium, arsonium, stibonium, oxonium, sulphonium, selenonium, telluronium and iodonium ions

Exceptions: NH_4^+: ammonium (or azanium) ion, $OHNH_3^+$: hydroxylammonium ion (or Diazanium) ion, $C_6H_5NH_2^+$ anilinium ion, $C_5H_5NH^+$: pyridinium, $HOOCCH_2NH_2+$: glycinium and so on.

(b) Anions

The names of the monoatomic anions shall consist the name of the elements with the termination *-ide*. For example:

H^-: Hydride ion, F^-: Fluoride ion, N_3^-: Nitride ion, O^{2-}: Oxide ion

Certain polyatomic anions have names ending in *-ide*. Their systematic names are given in brackets (Table III.4).

Table III.4: Name of the polyatomic anions with their systematic names.

Anion	Name of the anion	Anion	Name of the anion
OH^-	Hydroxide ion	CN^-	Cyanide ion
N_3^-	Azide ion [trinitride(1–)]	O_2^{2-}	Peroxide ion [dioxide(2–)]
C_2^{2-}	Acetylide ion [dicarbide(2–)]	NH^{2-}	Imide ion
NH_2^-	Amide	$NHOH^-$	Hydroxylamine ion
O_2^-	Hyperoxide ion or superoxide ion [dioxide(1–)]	$N_2H_3^-$	Hydrazide ion
HF_2^-	Hydrogendifluoride ion	S_2^{2-}	Disulphide ion [disulphide(2–)]
HS^-	Hydrogen sulphide ion	HO_2^-	Hydrogen peroxide

(c) Oxoacid anions

The systematic names for oxoanions can always be used. Certain anions have names using prefixes (hypo, per, etc.) that are well established. These are in accordance with the names of corresponding acids. The termination *-ite* has been used to denote a lower oxidation state and may be retained in trivial names in the following cases (Table III.5)

Table III.5: Name of the polyatomic anions with their systematic names.

Oxoanion	Name of oxoanion	Oxoanion	Name of oxoanion	Oxoanion	Name of oxoanion
NO_2^-	Nitrite	SO_3^{2-}	Sulphite	ClO_2	Chlorite
$N_2O_2^{2-}$	Hyponitrite	$S_2O_6^{2-}$	Disulphite	ClO	Hypochlorite
NOO_2^-	Peroxonitrite	$S_2O_4^{2-}$	Dithionite	BrO	Hypobromite
AsO_3^{3-}	Arsenite	SeO_3^{2-}	Selenite	IO	Hypoiodite

(d) Radicals

A radical is regarded as a group of atoms which occurs repeatedly in a number of different compounds. Certain neutral and cationic radicals containing oxygen and other chalcogens have, irrespective of charge, special names ending in -*yl*. The Commission approves the names of the following radicals (Table III.6).

Table III.6: Approved naming of some radicals.

Radical	Name	Radical	Name
HO	Hydroxyl	SO	Sulphinyl (Thionyl)
CO	Carbonyl	SO_2	Sulphonyl (Sulphuryl)
NO	Nitrosyl	S_2O_5	Disulphuryl
NO_2	Nitryl	SeO	Seleninyl
PO	Phosphoryl	SeO_2	Selenonyl
ClO	Chlorosyl	CrO_2	Chromyl
ClO_2	Chloryl	UO_2	Uranyl
ClO_3	Perchloryl	NpO_2	Neptunyl
Similarly for other halogens		Similarly for other actinides	

Radicals similar to the above having other chalcogens (S, Se, Te, Po or Lv) in place of oxygen are named adding the prefixes thio-, seleno- and so on.

These polyatomic radicals are always treated as forming the positive part of the compounds. For example:

$COCl_2$ Carbonyl chloride
$PSCl_2$ Thiophosphoryl chloride
SO_2NH Sulphonyl (sulphuryl) imide

(v) Acids

(a) Acid giving rise to the -*ide* anions [vide Section (IV)b] are named as binary and pseudo binary compounds of hydrogen. For example, HCl (hydrogen chloride), HCN (hydrogen cyanide) and HN_3 (hydrogen azide) are binary compounds of hydrogen.

Other acids may be named as pseudo binary compounds of hydrogen. For example, H_2SO_4 (hydrogen sulfate) and $H_2[Fe(CN)_6]$ {hydrogen hexacyanferrate(II)}.

(b) Oxoacids

Most of the common acids are oxoacids. These are commonly named using -*ic* and -*ous* endings in place of the anion's endings -*ate* and -*ite*, respectively. The acids using the -*ous* ending should be restricted to those listed above for the -*ite* anions [vide Section (IV)c].

The prefix *hypo-* is used to denote a lower oxidation state and the prefix *per-* is used to denote a higher oxidation state. These prefixes should be limited to the following cases (Table III.7):

Table III.7: Naming of some oxoacids with prefix *hypo-*.

Oxoacid	Name	Oxoacid	Name
$H_2N_2O_2$	Hyponitrous acid	HOCl	Hypochlorous acid
$H_4P_2O_6$	Hypophosphoric acid	HOBr	Hypobromous acid
		HOI	Hypoiodous acid

The prefix *per-* has been used to designate a higher oxidation state and is retained for $HClO_4$, perchloric acid and the corresponding acids of the other elements of Group 16 such as $HBrO_4$ (perbromic acid) and HIO_4 (periodic acid). This use of the prefix *per-* should not be extended to elements of other groups.

The prefix *ortho-* and *meta-* have been used to distinguish acids differing in the 'content of water'. The following names are approved (Table III.8):

Table III.8: Approved name of some oxoacids with prefix *ortho-* and *meta-*.

Oxoacid	Name	Oxoacid	Name
H_3BO_3	Orthoboric acid or boric acid	$(HBO_3)_n$	Metaboric acid
H_4SiO_4	Orthosilicic acid	$(H_2SiO_3)_n$	Metasilicic acid
H_3PO_4	Orthophosphoric acid	$(HPO_3)_n$	Metaphosphoric acid
H_5IO_6	Orthoperiodic acid		
H_6TeO_6	Orthotelluric acid		

(c) Peroxoacids

The prefix *peroxo-* specifies the replacement of —O— by —O—O— in oxoacids. The names of some peroxoacids are given below (Table III.9):

Table III.9: Naming of some peroxoacids.

Peroxoacid	Name
HNO_4	Dioxoperoxonitric acid or peroxonitric acid
H_3PO_5	Trioxoperoxophosphoric acid
$H_4P_2O_8$	μ-Peroxo-bis-trioxophosphoric acid
H_2SO_5	Trioxoperoxosulphuric acid or peroxosulphuric acid
$H_2S_2O_8$	μ-Peroxo-bis-trioxosulphuric acid or peroxodisulphuric acid

(d) Thioacids

The prefix *thio-* means replacement of one oxygen atom by the sulphur atom in oxoacids. The prefixes *seleno-* and *telluro-* may be used in a similar manner. Examples are (Table III.10):

Table III.10: Naming of some thiocids.

Thiocids	Name
$H_2S_2O_2$	Thiosulphurous acid
$H_2S_2O_3$	Thiosulphuric acid
H_3PO_3S	Thiophosphoric acid
$H_3PO_2S_2$	Dithiophosphoric acid
H_2CS_3	Thiocarbonic acid
HSCN	Thiocyanic acid

Acids containing ligands other than O and S are normally named as complexes. Examples are Table III.11:

Table III.11: Naming of the ligand containing F and Cl ligands.

Acid	Name
$H[PF_6]$	Hydrogen hexafluoroposphate(1–)
$H[AuCl_4]$	Hydrogen tetrachloroaurate(1–)

(vi) Salts and salt-like compounds

(a) Simple salts

Simple salts come under the broad definition of binary compounds given in Section (ii) and their names are formed from those of constituent ions [given in Section (iv)] in the manner set out in Section (ii). Examples are (Table III.12):

Table III.12: Naming of some simple salts.

Salt	Name
Na_2CO_3	Sodium trioxocarbonate(IV)
$Na_2SO_4 \cdot 10H_2O$	Sodium tetraoxosulphate(VI) decahydrate
$FeCl_3$	Iron(III) chloride
$SnCl_4$	Tin(IV) chloride
MnO_2	Manganese(IV) oxide
BaO_2	Barium(II) peroxide

(b) Salt containing acidic hydrogen

Salts that contain acidic hydrogen are named by adding the word 'hydrogen' with numerical prefix where necessary to denote the replaceable hydrogen in the salt. The word hydrogen shall be placed immediately in front of the name of the anion. Examples are (Table III.13):

Table III.13: Naming of some salt containing acidic hydrogen.

Salt with acidic hydrogen	Name
$NaHCO_3$	Sodium hydrogencarbonate
LiH_2PO_4	Lithium dihydrogenphosphate
KHS	Potassium hydrogensulphide

(c) Double salt

(i) Cations

Cations other than hydrogen are cited in alphabetical order, which may be different in formula and name. Examples are (Table III.14):

Table III.14: Naming of the cations in double salts.

Formula	Name
$KMgF_3$	Magnesium potassium fluoride
$NaTl(NO_3)_2$	Sodium thallium(I) nitrate
$KNaCO_3$	Potassium sodium carbonate
$MgNH_4PO_4 \cdot 6H_2O$	Ammonium magnesium phosphate hexahydrate [or water(1/6)]
$NaNH_4HPO_4 \cdot 4H_2O$	Ammonium sodium hydrogenphosphate tetrahydrate
$AlK(SO_4)_2 \cdot 12H_2O$	Aluminium potassium sulphate water(1/12)

(ii) Anions

Anions are cited in alphabetical order, which may be different in formulae and names. Examples are (Table III.15):

Table III.15: Naming of the anions in double salts.

Formula	Name
$Na_6ClF(SO_4)_2$ or $NaCl \cdot NaF \cdot 2Na_2SO_4$	Hexasodium chloride fluoride bis(sulphate)
$Ca_5F(PO_4)_3$	Pentacalcium fluoride tris(phosphate)

(iii) Basic salts

Basic salts should be treated as double salts, not as oxo or hydroxo salts. Naming of some basic salts is given in Table III.16:

Table III.16: Naming of the basic salts.

Basic salt	Name
$MgCl(OH)$	Magnesium chloride hydroxide
$BiClO$	Bismuth chloride oxide (not bismuth oxychloride)

Table III.16 (continued)

Basic salt	Name
$ZrCl_2O \cdot 8H_2O$	Zirconium dichloride oxide octahydrate
$Cu_2Cl(OH)_3$	Dicopper chloride trihydroxide

(vii) Coordination compounds

(a) Overall approach

Additive nomenclature was developed in order to describe the structures of coordination entities, or complexes, but this method is readily extended to other molecular entities as well. Mononuclear complexes are considered to consist a central atom, often a metal ion, which is bonded to surrounding small molecules or ions, which are referred to as ligands. The names of complexes are constructed by adding the names of the ligands *before* those of the central atoms, using appropriate multiplicative prefixes. Formulae are constructed by adding the symbols or abbreviations of the ligands *after* the symbols of the central atoms.

In the development of formula of coordination compounds, the symbol of the central atom is placed first followed by anionic ligands in alphabetical order of the symbols and then neutral ligands in alphabetical order. The formula of the complex molecule or ion is enclosed in square brackets []. In names, the central atom is placed after the ligands. The ligands are listed in alphabetical order regardless of charge and regardless of number of each. The diammine is listed under 'a' whereas diethylamine is listed under 'd'.

The oxidation number of the central atom is shown by the Stock notation (Roman numbers). Instead, proportion of the constituents may be given by means of stoichiometric prefixes or the charge on the entire ion can be designated the Ewens–Bassett number (Arabic numbers). It is notable here that the usages of Ewens–Bassett number are no longer recommended by IUPAC. Formulae and names of the complexes may be supplemented by italicized prefixes *cis*, *trans*, *fac* (*facial*), *mer* (*meridional*) and so on. Name of the complex anions end in *-ate*. Complex cations and neutral molecules are given no distinguishing ending.

(b) Naming of ligands

Names of anionic ligands that end in 'ide', 'ate' or 'ite' are modified within the full additive name for the complex to end in 'ido', 'ato' or 'ito', respectively. Note that the 'ido' ending is now used for halide and oxide ligands as well. By convention, a single coordinated hydrogen atom is always considered anionic and it is represented in the name by the term 'hydrido', whereas coordinated dihydrogen is usually treated as a neutral two-electron donor entity (Crabtree 2016). *Enclosing marks* are required for inorganic anionic ligands containing numerical prefixes, as (trisphosphato) and for thio, seleno and telluro analogues of oxo anions containing more than one atom, as (thiosulphato). Examples of organic anionic ligands named in this fashion are:

$$CH_3COO^-: acetate, (CH_3)_2N^-: dimethylamido.$$

Examples:

Complex compound	Name of the ligands in complex
Li[AlH$_4$]	Lithium tetrahydridoaluminate(III) or lithium tetrahydridoaluminate(1−)
K$_2$[OsCl$_5$N]	Potassium pentachloronitridoosmate(VI) or potassium pentachloronitridoosmate(2−)
Na$_3$[Ag(S$_2$O$_3$)$_2$]	Sodium bis(thiosuphato)argentate(I) or (3−)
[Ni(C$_4$H$_7$O$_2$N$_2$)$_2$]	Bis(2,3-butanedione dioximato)nickel(II)
[Cu(C$_5$H$_7$O$_2$)$_2$]	Bis(2,4-pentanedionato)copper(II)
	Bis(8-quinolinato)copper(II)
K$_2$[Cr(CN)$_2$O$_2$(O$_2$)NH$_3$]	Potassium amminecyano**dioxo**peroxo-chromate(VI) or (2−)

The names of some common anionic ligands are:

Anionic ligand	Name	Anionic ligand	Name
H$^-$	Hydrido	HS$^-$	Mercapto (hydrogensulphido)
F$^-$	Fluoro/fluorido	S^{2-}	Thio (sulphide)
Cl$^-$	Chloro/chlorido	CN$^-$	Cyano
Br$^-$	Bromo/bromido	CH$_3$O$^-$	Methoxo
I$^-$	Iodo/iodido	C$_6$H$_5^-$	Phenyl
OH$^-$	Hydroxo (hydroxide)	C$_5$H$_5^-$	Cyclopentadienyl
O$_2^{2-}$	Peroxo [dioxido(2−)]	Other hydrocarbons are also given radical names without the -*o* ending.	

Neutral and cationic ligands have no special ending. H_2O and NH_3 are called aqua and ammine, respectively, in complexes. Groups such as NO and CO are named as radicals and treated as neutral ligands.

Examples:

Complex	Name of the ligand in the complex
$K[CrF_4O]$	Potassium tetraflurooxochrmate(V) or (1–)
$Na[BH(OCH_3)_3]$	Sodium hydroxotrimethoxoborate(III) or (1–)
$[CuCl_2(CH_3NH_2)_2]$	Dichlorobis(methylamine)copper(II) or **omit (II)***
$[Pt(py)_4][PtCl_4]$	Tetrakis(pyridine)platinum(II) tetrachloroplatinate(II) or tetrakis(pyridine)platinum(2+) tetrachloroplatinate(2–)
$[Co(en)_3]_2(SO_4)_3$	Tris(ethylenediamine)cobalt(III) sulphate or tris(ethylenediamine)cobalt(3+) sulphate
$K[PtCl_3(C_2H_4)]$	Potassium trischloro(ethylene)platinate(II) or (1–)**
$K[SbCl_5C_6H_5]$	Potassium pentachloro(phenyl)animonate(V) or (1–)**
$[Al(OH)(H_2O)_5]^{2+}$	Pentaaquahydroxoaluminium(III) or (2+)
$Na_2[Fe(CN)_5NO]$	Sodium pentacyanonitrosylfrrate(III) or (2–)***
$K_3[Fe(CN)_5CO]$	Potassium corbonylpentacyanoferrate(II) or (3–)***
$[Ru(NH_3)_5N_2]Cl_2$	Pentaammine(dinitrogen)ruthenium(II) chloride
$[Fe(C_5H_5)_2]$	Bis(cyclopentadienyl)iron(II)
$Cr(C_6H_6)_2]$	Bis(benzene)chromium(0)

Note:

(i) *For many compounds, the oxidation number of central atom and/or charge of the ion are so well known that there is no need to use either a Stock number (Roman number) or an Ewens–Bassett number (Arabic number). However, it is not wrong to use such numbers and they are included here.

(ii) **Normally, ethylene/phenyl would not be placed within enclosing marks. They are used here to avoid confusion with a chloroethylene/chlorophenyl.

(iii) ***The groups NO and CO, when linked directly to a metal atom, are called 'nitrosyl' and 'carbonyl', respectively. In computing the oxidation number, these ligands are treated as number. However, NO grouping in $Na_2[Fe(CN)_5NO]$ is found to be NO^+ [v(NO) = 1,944 cm^{-1}] (Cotton et al. 1959), and accordingly, the oxidation state of Fe in this compound is (II).

(c) Alternative modes of bonding of some ligands

A ligand may be attached to a metal centre through different atoms and that may be denoted by adding the italicized symbol(s) for the atom or atoms through which attachment occurs at the end of the name of the ligand. For example:

(i) Dithiooxalato anion possibly may be attached through S or O, and that are distinguished as dithiooxalato-*S S'* and dithiooxalato-*O O'*, respectively.

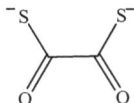

(ii) In some cases, different names are already in use for alternative modes of attachment as, for example, thiocyanato (–SCN) and isothiocyanato (–NCS), nitro (–NO$_2$) and nitrito (–ONO).

Example:

Complex	Name of the ligand in the complex
[Co(NO$_2$)$_3$(NH$_3$)$_3$]	Triamminetri(nitro-*N*)cobalt(III)
Na$_3$[Co(NO$_2$)$_6$]	Sodium hexa(nitro-*N*)cobaltate(III)
[Co(ONO)(NH$_3$)$_5$]SO$_4$	Pentaammine(nitro-*O*)cobalt(III) sulphate or (2+)
[Co(NCS)(NH$_3$)$_5$]Cl$_2$	Pentaammine(thiocyanato-*N*)cobalt(III) chloride
K$_2$ [Ni(...)$_2$]	Potassium bis(dithiooxalato-*S,S'*)nickelate(II) or (2−)

(d) The Kappa convention

The coordinating atom of a ligand can also be designated by the Greek kappa, κ, preceding italicized element symbol as thiocyanato-κ*N* for SCN bonded through N and thiocyanato-κ*S* for SCN bonded through S. Similarly, nitrite, which can be bound through either the nitrogen atom (M–NO$_2$, nitrito-κ*N*), or an oxygen atom (M–ONO, nitrito-κ*O*) are designated in terms of kappa, κ. This convention is particularly helpful for polydentate ligands with not all donor atoms ligating and where the ligating atoms are not obvious. A right superscript to κ indicates the number of identically bound ligating atom of a ligand. **Examples are** (Figure III.2):

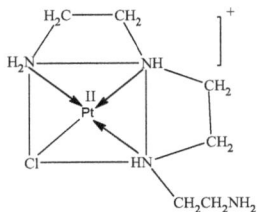

[N-(2-amino-κ*N*)-ethyl)-N'-(2-aminoethyl)-
1,2-ethanediamine-κ²*N*, *N*')]chloroplatinum(II)

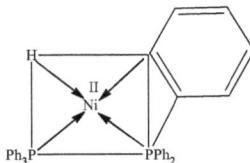

[2-(Diphenylphosphino)-κ*P*)-phenyl-
κ*C*¹]hydrido(triphenylphosphine-κ*P*)-
nickel(II)

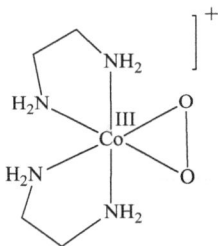

Bis(ethane-1,2-diamine-κ²*N*)-(η²-peroxido)-
cobalt(III)

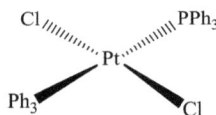

Dichloridobis(triphenylphosphine-κ*P*)-
platinum(II)

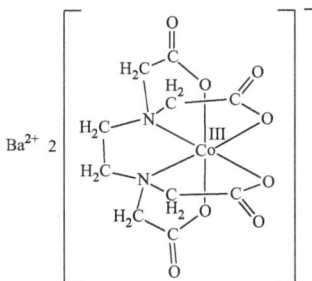

Barium [2, 2', 2'', 2'''-(ethane-1,2-diyldinitrilo-κ²*N*)-
tetraacetato-κ⁴*O*]cobaltate(III)

Figure III.2: Structure of some ligands forming complexes.

(e) Compounds with bridging atoms or groups

(i) A bridging group is indicated by adding the Greek letter μ (mu) immediately before the name and separating the name from rest of the complex by hyphens.

(ii) Two or more bridging groups of the same kind are indicated by the di-μ- (or bis-μ-).

(iii) If the bridging group bridges more than two atoms, use μ₃, μ₄ and so on.

(iv) The bridging groups are listed with other groups in alphabetical order unless the symmetry of the molecule permits a simpler name by the use of multiplicative prefixes.

(v) If the same ligand is present in bridging and non-bridging roles, it is cited first as a bridging ligand.

Examples:

Compound	Name of bridging ligand along with others
$[(NH_3)_5Cr-OH-Cr(NH_3)_5]Cl_5$	μ-Hydroxo-bis[pentaamminechromium(III)] chloride μ-hydroxo-bis(pentaamminechromium)(5+) chloride
$[(CO)_3Fe(CO)_3 Fe(CO)_3]$ (Figure III.3)	Tris-μ-carbonyl-bis[tricarbonyliron(0)] tris-μ-carbonyl-bis(tricarbonyliron)
$[Br_2Pt(SMe_2)_2 Br_2Pt]$	Bis(μ-dimethylsulphide)-bis[dibromoplatinum(II)]
$[(CH_3Hg)_4S]^{2+}$	$μ_4$-Thio-tetrakis[methylmercurt(II)] ion or (2+)
$[Be_4O(CH_3COO)_6]$ (Figure III.3)	Hexa-μ-acetato(O,O')-$μ_4$-oxo-tetraberyllium(II) or (2+)

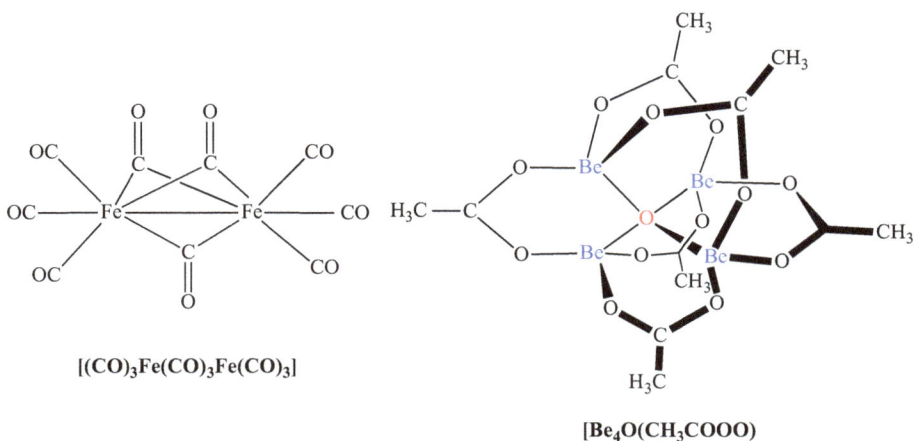

$[(CO)_3Fe(CO)_3Fe(CO)_3]$

$[Be_4O(CH_3COOO)$

Figure III.3: Structure of complexes $[(CO)_3Fe(CO)_3 Fe(CO)_3]$ and $[Be_4O(CH_3COO)_6]$.

(f) Dinuclear compounds without bridging groups: direct linking between metal–metal centres

There are a number of compounds containing metal–metal bonds. Such compounds, when symmetrical, are named by the use of multiplicative prefixes. In such compounds, metal–metal bonding may be designated by italicized atomic symbols of the appropriate metal atoms, separating by a long dash and enclosed in parentheses. Examples are:

Dinuclear compound	Name of compound
$[(CO)_5Mn-Mn(CO)_5]$	Bis(pentacarbonyl)manganese(*Mn—Mn*) or decacarbonyldimanganese(*Mn—Mn*)
$[Br_4Re-ReBr_4]^{2-}$	Bis(tetrabromorhenate)(*Re—Re*)(2−)
$Cs_3[ReCl_4]_3$	Cesium *cyclo*-tris(tetrachlororhenate)(3*Re—Re*)(3−)
$[(\eta-C_5H_5)(CO)_3Mo-Mo(CO)_3(\eta-C_5H_5)$	Bis(tricarbonyl-η-cyclopentadienyl)molybdenum- (*Mo—Mo*)

(g) Unsymmetrical dinuclear species

Such dinuclear entities may have different central atoms and/or different ligands or different number of ligands on the central atoms. They are named by citing the ligands and bridging groups in a single alphabetical ordered series, followed by the names of central atoms in alphabetical order. Moreover, the central atoms are numbered 1 and 2 to provide numerical locants for the ligands and the points of attachment of bridging groups.

In case the central atoms are different, the central atom 1 is the more electropositive element in the sequence going down group 17, then groups 16, 15, 14 and so on. It is approximate sequence of electronegativity by groups. If the central atoms are identical, the central atom 1 is designated with priority (Block 1990) decreasing from criteria (i) to (iv).

(i)   Atom with higher oxidation number

(ii) Atom with higher coordination number

(iii) Atom with greater variety of bonding atoms

(iv) Atom to which is attached the side group with the name first in alphabetical order.

Each criterion mentioned previously is applied only if the previous criterion fails to make a distinction. The kappa convention is used to indicate the ligating atom(s) and its (their) role. A right superscript to κ designates the number of equivalent ligating atoms bonded to a central atom. The numerical locant (1or 2) of the central atom is placed before κ.

Examples:

Unsymmetrical compound	Name of unsymmetrical compound
$(CO)_4Co - Re(CO)_5$ $\quad\;\; 2 \quad\quad 1$	Nonacarbonyl-$1\kappa^5C$, $2\kappa^4C$-2-cobalt-1-rhenium$(Co-Re)$
$[(Ph_3As)Au - Mn(CO)_5$ $\quad\quad\;\; 2 \quad\quad 1$	Pentacarbonyl-$1\kappa^5C$-(triphenylarsine-$2\kappa As$)-2-gold-1-manganese$(Au-Mn)$
$[(H_3N)_5Cr(OH)Cr(NH_3)_4(NH_2Me)]Cl_5$ $\quad\quad 1 \quad\quad\;\; 2$	Nonaammine-$1\kappa^5N,2\kappa^4N$-μ-hydroxo-(methylamine)-$2\kappa N$-dichromium(5+) chloride
$(Et_2PhP)_3(Cl)_2Re-N-Pt(PEt_3)(Cl)_2$ $\quad\quad\quad\quad 1 \quad\quad\; 2$	Tetrachloro-1 κ^2, $2\kappa^2$-tris(diethylphenyl-phosphine-$1\kappa P$)-μ-nitrido-(triethylphosphine-$2\kappa P$)-2-platinum-1-rhenium$(Pt-Re)$

(h) Metal cluster compounds

There are several instances of a finite group of metal atoms with bonds directly between the metal atoms but also with some non-metal atoms or groups (ligands) intimately associated with the clusters. The geometrical shape of such aggregates is designated by *triangulo, quarto, tetrahedro, octahedro* and so on, and nature of the bonds to the ligands by the conventions for bridging bonds and simple bonds. Numbers are used as locant designators as they are for homoatomic chains.

Cluster compound	Name of cluster compound
$Os_3(CO)_{12}$	Dodecacarbonyl-*triangulo*-triosmium$(3Os-Os)$
$Cs_3[Re_3Cl_{12}]$	Cesium dodecacarbonyl-*triangulo*-trirhenate$(3Re-Re)$
B_4Cl_4	Tetrachloro-*tetrahedro*-teraboron$(4B-B)$
$[Nb_6Cl_{12}]^{2+}$	Dodeca-μ-chloro-*octahedro*-hexaniobium(2+) ion
$[Mo_6Cl_8]^{4+}$	Octa-μ-chloro-*octahedro*-hexamolybdeum(4+) ion
$Cu_4I_4(PEt_3)_4$	Tetra-$μ_3$-iodo-tetrakis(triethylphosphine)-*tetrahedro*-tetracopper

Organometallic compounds

Organometallic compounds contain at least one bond between a metal atom and a carbon atom. They are named as coordination compounds, using the additive nomenclature system.

The name for an organic ligand *binding through one carbon atom* may be derived either by treating the ligand as an anion or as a neutral substituent group.

The compound [Ti(CH$_2$CH$_2$CH$_3$)Cl$_3$] is thus named as trichlorido(propan-1-ido)tita-nium or as trichlorido(propyl)titanium with *removal of the terminal 'e' of the name of the parent hydrocarbon*. Similarly, 'methanido' or 'methyl' may be used for the ligand –CH$_3$.

When an organic ligand forms *two or three metal–carbon single bonds* (to one or more metal centres), the ligand may be treated as a di- or tri-anion, with the endings 'diido' or 'triido' being used, with *no removal of the terminal 'e' of the name of the parent hydrocarbon*. Again, names derived with regard to such ligands as substituent groups and using the suffixes 'diyl' and 'triyl' are still commonly encountered. Thus, the bidentate ligand –CH$_2$CH$_2$CH$_2$– would be named propane-1,3-diido (or propane-1,3-diyl) when chelating a metal centre, and µ-propane-1,3-diido (or µ-propane-1,3-diyl) when bridging two metal atoms.

Organometallic compounds containing a **metal–carbon multiple bond** are given substituent prefix names derived from the parent hydrides which end with the suffix 'ylidene' for a metal–carbon double bond and with 'ylidyne' for a triple bond. These suffixes either replace the ending 'ane' of the parent hydride, or, more generally, are added to the name of the parent hydride with the insertion of a locant and the elision/omission of the terminal 'e', if present. Thus, the entity CH$_3$CH$_2$CH= as a ligand is named propylidene and (CH$_3$)$_2$C= is called propan-2-ylidene. The 'diido'/'triido' approach, outlined above, can also be used in this situation. The terms 'carbene' and 'carbyne' are not used in systematic nomenclature. The name of a ruthenium compound (figure given below containing **metal–carbon double bond** (Figure III.4) is:

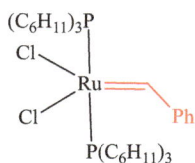

Figure III.4: Dichlorido(phenylmethylidene)bis(tricyclohexylphosphane-κ*P*)ruthenium, dichlorido(phenylmethanediido)bis(tricyclohexylphosphane-κ*P*)ruthenium or (benzylidene)dichloridobis(tricyclohexylphosphane-κ*P*) ruthenium.

The special nature of the bonding to metals of unsaturated hydrocarbons in a 'side-on' fashion *via* their π-electrons requires the **eta (η) convention**. In this 'hapto' nomenclature, the number of *contiguous/adjoining* atoms in the ligand coordinated to the metal (the hapticity of the ligand) is indicated by a right superscript on the eta symbol, for example η^3 ('eta three' or 'trihapto') (Figure III.5). The η-term is added as a prefix to the ligand name, or to that portion of the ligand name most appropriate to indicate the connectivity, with locants if necessary. For example

The ubiquitous ligand η^5-C$_5$H$_5$, strictly η^5-cyclopenta-2,4-dien-1-ido, is also acceptably named η^5-cyclopentadienido or η^5-cyclopentadienyl (Figure III.6). When cyclopenta-2,4-dien-1-ido coordinates through one carbon atom *via* a σ bond, a κ-term is added for explicit indication of that bonding. The symbol η^1 should not be used, as the *eta convention* applies only to the bonding of contiguous atoms in a ligand.

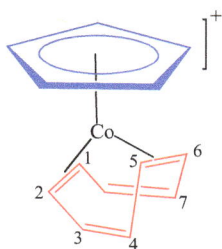

(η^6-Benzene)[(1, 2, 5, 6-η)-cycloocta-1, 3, 5, 7-tetraene]cobalt(1+)

Tris(η^3-prop-2-en-1-ido)chromium,
Tris(η-^{3}prop-2-en-1-yl)chromium,
or Tris(η^3-allyl)chromium

Figure III.5: Organometallic compounds containing hexahapto and trihapto unsaturated hydrocarbon, respectively.

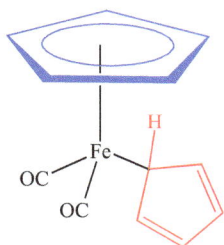

Figure III.6: Dicarbonyl(η^5-cyclopentadienido)(cyclopenta-2,4-dien-1-ido-κC^1)iron or Dicarbonyl(η^5-cyclopentadienyl)(cyclopenta-2,4-dien-1-yl-κC^1) iron.

Discrete molecules containing two *parallel* η^5-cyclopentadienido ligands in a 'sandwich' structure around a transition metal, as in bis(η^5-cyclopentadienido)iron, [Fe(η^5-C$_5$H$_5$)$_2$], are generically called **metallocenes** and may be given 'ocene' names, in this case ferrocene. These 'ocene' names may be used in the same way as parent hydride names are used in substitutive nomenclature, with substituent group names taking the forms 'ocenyl', 'ocenediyl', 'ocenetriyl' (with insertion of appropriate locants) (Figure III.7).

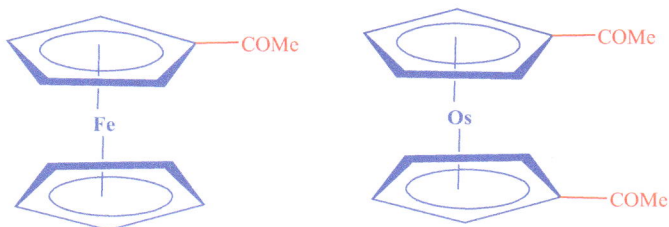

Figure III.7: 1-ferrocenylethan-1-one; 1,1'-(osmocene-1,1'-diyl)di(ethan-1-one).

By convention, 'organoelement' compounds of the **main group elements** are named by substitutive nomenclature if derived from the elements of Groups 13–16, but by additive nomenclature if derived from the elements of Groups 1 and 2. In some cases, compositional nomenclature is used if less structural information is to be conveyed.

Appendix IV
Symmetry operations and point groups in molecules

IV.1 Introduction

The collection of all symmetry elements/operation possessed by a molecule constitute a group. Moreover, this group of symmetry elements is also a mathematical
group as it also satisfies certain conditions for a mathematical group.

A molecular group is called a **symmetry point group** or simply a **point group**
because all the symmetry elements present in the molecule pass through or inter-
sect at a common point (centre of symmetry or in its absence centre of gravity of the
molecule) and this point remains unchanged under all the symmetry operations of
the molecules (Figure IV.1).

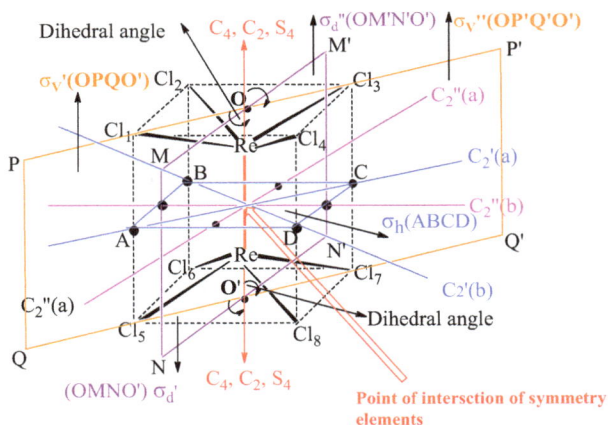

Figure IV.1: Various symmetry elements in $Re_2Cl_8^{2-}$ intersecting at a common point.

Therefore, a set of all symmetry operations constituting a group is known as **point
group and symmetry of a molecule is called point symmetry.**

IV.2 Notation of point groups

Each point group has a descriptive symbol signifying the presence of some definite
combination of symmetry elements. There are two types of notations in this regard.

https://doi.org/10.1515/9783110727289-014

(a) Schöenflies notation

Point groups are labelled by Schöenflies notations which is named after their inventor.

(i)   In this notation, the main symbol (alphabet) used refers to the axis of highest symmetry in the molecule. Thus,

   C stands for highest-fold proper axis,
   S stands for highest-fold improper axis,
   D stands for highest-fold proper axis in combination with nC_2 axes perpendicular to it,
   T, O and **I** are specially chosen to represent the highly symmetric tetrahedral, octahedral, and icosahedral groups.

(ii) The numerical subscripts indicate the order of the highest-order rotational axis.

(iii) Further labelling with alphabetical subscript indicates the presence of certain type of plane of symmetry. Thus, v is used for the presence of vertical planes (σ_v), d is used for the presence of dihedral planes (σ_d) and h is used for a plane perpendicular to the principal axis.

(iv) The subscript i alone (without numerical subscripts) is used when the molecules contain only i element (C_i).

(v)   The subscript s alone is used when the molecules contain a plane of symmetry (σ), and the point group is C_s.

(vi) The linear molecules are given the symbols $C_{\infty v}$ or $D_{\infty h}$ depending on the absence or presence of i, the centre of inversion in linear molecules.

(b) Hermann–Mauguin notation

A brief account of this notation is given here because some details of crystallography are needed for complete understanding of this notation.

(i)   The symbol in this notation contains generally three parts each defining a different direction.

(ii) The orders of proper axes are denoted by number n and those of improper axes by n^-.

(iii) The presence of mirror plane is indicated by m, the position of which indicates the direction perpendicular to the plane. For example, if a molecule contains a C_2 axis along Z-axis and the two vertical planes perpendicular to each other, then the one σ is said to be perpendicular to X-axis (σ_{yz}) and the other to Y-axis (σ_{xz}). The point group symbol may then contain the order $\sigma_{yz}\ \sigma_{xz}\ C_{2(z)}$, which is the same order as their normals, that is, XYZ. The Hermann–Mauguin symbol will be mm2. The corresponding Schoenflies symbol is C_{2v}.

The horizontal plane (σ_h) will be similarly assigned by n/m, where n is the order of the axis perpendicular to the plane. For example, $C_{2h} = 2/m$ and $C_{4h} = 4/m$. If the point group contains other planes in addition to σ_h planes, the symbol will be n/mmm or (n/m) mm. Thus, $D_{3h} = 3/mmm$, $D_{4h} = 4/mmm$ but $D_{2h} = mmm$ since all the planes are mutually perpendicular.

(iv) For molecule with less symmetry, their point group symbols need not contain three parts. For example, the molecules containing only Cn axes are designated by a single number. Thus, $C_2 = 2$, $C_3 = 3$. Also, $S_4 = 4^-$, $S_6 = 6^-$, $S8 = 8^-$. If the molecules contain only rotational axes and neither planes nor inversion centre (i), then they are given the symbols based on order of their principal and secondary axes. For example, $D_2 = 222$ and $D_3 = 32$. The other groups such as C_1, C_s and C_i are given the symbols, 1, m or 2 and 1, respectively.

(v) Since this notation is used primarily for crystalline solids, the ordering of the symbol is dependent on the basic symmetry of the crystal and the further discussion is beyond the scope of this book.

IV.3 Classification of point groups

Several molecules may have the same set of symmetry operations, and hence belong to the same point group. Molecules can thus be classified according to their characteristic set of symmetry. Based on symmetry characteristic, the molecules are classified into the following three broad types.

IV.3.1 Type I: non-axial point group or molecules with low symmetry

This type includes molecules having either no symmetry elements (except the identity E) or one characteristic element: a rotation (C_n) or an inversion centre (i) or a mirror plane σ.

(a) C_1 point group

This point group has no symmetry element, except identity. We may, therefore, consider that it has a onefold proper axis since $C_1 = E$. The order of this group (h) is 1. The symmetry element E means doing nothing or rotates the molecule by 360°. Thus, $E = C_1^1$. Though, it is possible to theoretically predict the existence of such molecules, it is very difficult to practically synthesize them. This point group includes all molecules having one asymmetric atom (C, N, P, Si, etc.) (Figure IV.2).

The $TeCl_2Br_2$ with its structure (Figure IV.3) in gas phase belongs to C_1 point group. Though Nb forms seven bonds with only F atoms, the arrangement of fluorides in NbF_7

around Nb (Figure IV.4) is such that the whole molecule lacks in any kind of symmetry elements. Hence, NbF$_7$ molecule belongs to C$_1$ point group

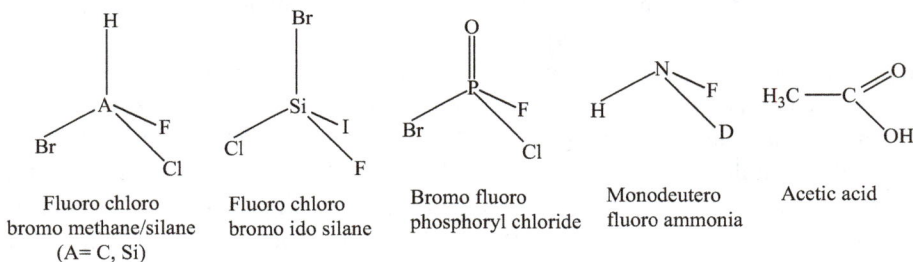

| Fluoro chloro bromo methane/silane (A= C, Si) | Fluoro chloro bromo ido silane | Bromo fluoro phosphoryl chloride | Monodeutero fluoro ammonia | Acetic acid |

Figure IV.2: Molecules having one asymmetric atom.

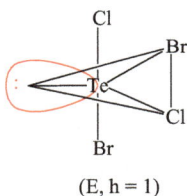

(E, h = 1)   **Figure IV.3:** Tecl$_2$Br$_2$ having C$_1$ point group.

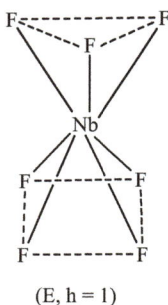

(E, h = 1)   **Figure IV.4:** NbF$_7$ having C$_1$ point group.

(b) C$_n$ point group

These are the groups in which all operations are due to the presence of a proper axis as the only symmetry element. The general symbol for such a group and the operations in it are: C_n: C_n^1, C_n^2, C_n^3, C_n^4 . . . . . C_n^{n-1}, $C_n^n = E$. Examples of molecules exhibiting C_n point group symmetry are (Figure IV.5):

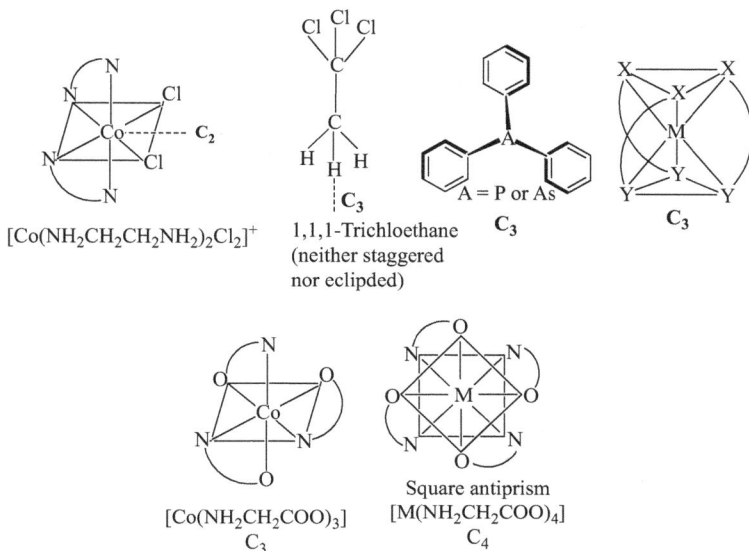

[Co(NH$_2$CH$_2$CH$_2$NH$_2$)$_2$Cl$_2$]$^+$

1,1,1-Trichloethane
(neither staggered
nor eclipded)

A = P or As

C_3

[Co(NH$_2$CH$_2$COO)$_3$]
C_3

Square antiprism
[M(NH$_2$CH$_2$COO)$_4$]
C_4

Figure IV.5: Molecules exhibiting C_n point group symmetry.

(c) C_s point group

If the molecule has only a plane of symmetry, the notation used for the point symmetry is C_s. Thus, this point group has only two symmetry elements, E and σ (h = 2). Examples of molecules belonging to this class are:
(a) formyl chloride (b) thionyl halides (SOCl$_2$, SOF$_2$) and sulphoxides [(CH$_3$)$_2$SO, (C$_6$H$_5$)$_2$SO] (c) BClBrF (d) NOCl (planar) (e) HOD (f) CH$_2$BrCl (g) POBrCl$_2$ (h) chloroethylene (i) 1,2,3-trichlorocyclopropane (j) PF$_3$Cl$_2$ (k) secondary amines R$_2$NH and NHF$_2$ (l) non-symmetrical aromatic compounds (Figure IV.6).
1,2-Benzpyrene: This molecule contains five benzene rings fused together which occurs in **cigarette smoke** and chimney soot. **It is one of the cancer causing substances.**

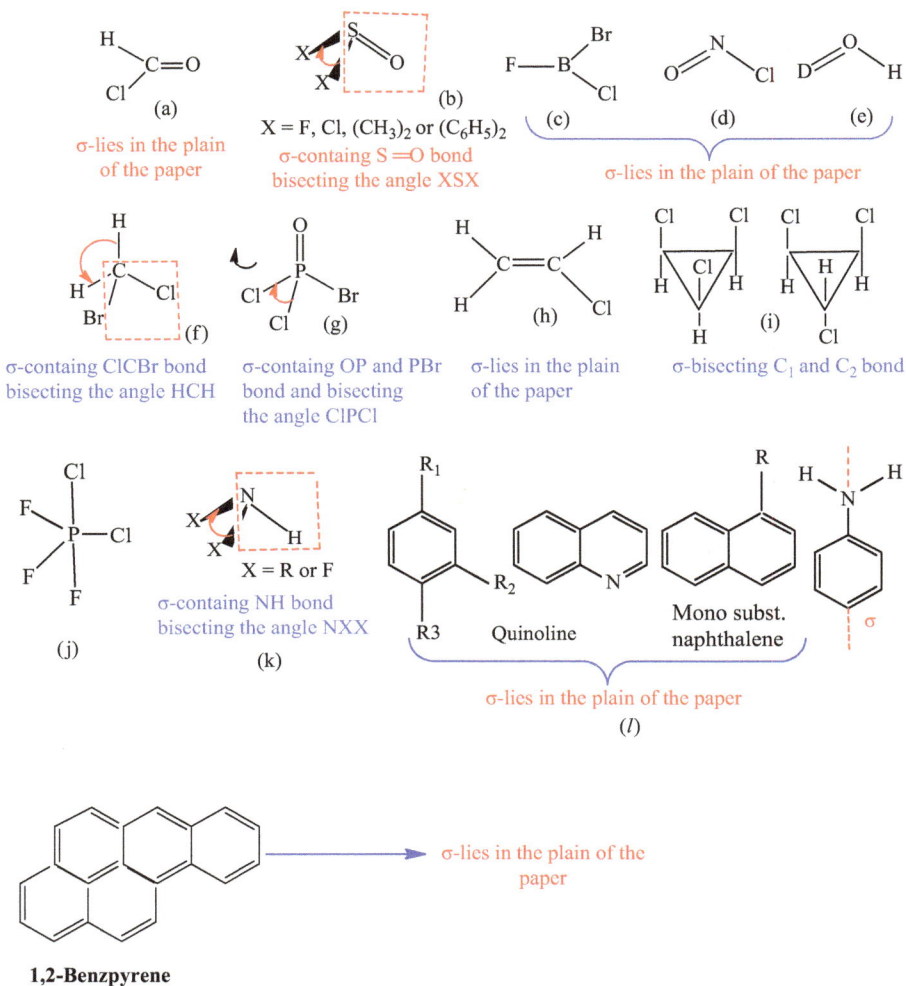

$X = F, Cl, (CH_3)_2$ or $(C_6H_5)_2$

(a) σ-lies in the plain of the paper

(b) σ-containg S=O bond bisecting the angle XSX

(c) (d) (e) σ-lies in the plain of the paper

(f) σ-containg ClCBr bond bisecting the angle HCH

(g) σ-containg OP and PBr bond and bisecting the angle ClPCl

(h) σ-lies in the plain of the paper

(i) σ-bisecting C_1 and C_2 bond

(j)

(k) X = R or F σ-containg NH bond bisecting the angle NXX

R_1

R_2

$R3$ Quinoline

R Mono subst. naphthalene

σ σ-lies in the plain of the paper

(l) σ-lies in the plain of the paper

1,2-Benzpyrene

σ-lies in the plain of the paper

Figure IV.6: Molecules belonging to C_s point group.

(d) C_i point group

Molecules belonging to this group have centre of inversion (i) only as the symmetry element. The number of elements in this point group are E and i, so the order of the group in 2 (h = 2).

Examples belonging to this group are: (a) P_2F_4 (b) staggered form of 1,2-dichloro-1, 2-difluoroethane (c) *meso*-tartaric acid (d) 1,3-dichloro-2,4-dibromocyclobutane (e) 3,

6-dimethylpiperazine-2,5-dione. The structures of these molecules with location of inversion centre (i) in them are given below (Figure IV.7):

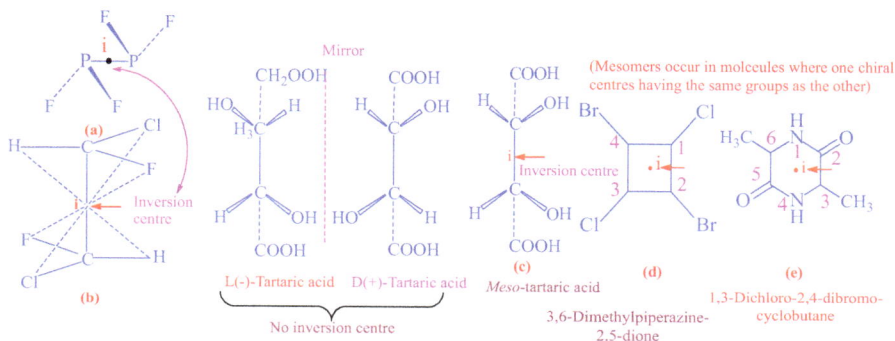

Figure IV.7: Molecules belonging to C_i point group.

IV.3.2 Type II: molecules with intermediate symmetry

Characteristic for this type of point symmetry groups is the presence of a rotation axis, C_n, associated with other symmetry elements ($S_n{}^1$, C_2, i). The lowest point symmetry groups are the S_n groups (n even), and the highest ones are D_{nd} and D_{nh}. In between these two, two more point groups are C_{nv} and C_{nh}.

(a) S_n point group: When an improper axis is present, we must consider whether it is even or odd. When the axis, S_n, is of even order, the group of operations it generates is called S_n, and consists of n elements: E, $S_n{}^1$, $C_{n/2}$, $S_n{}^3$, $S_n{}^4$, ... $S_n{}^{n-1}$. This can be shown by taking examples of S_4 and S_6. Hence,

$$S_4 = S_4{}^1, S_4{}^2, S_4{}^3, S_4{}^4 \; ; \; S_6 = S_6{}^1, S_6{}^2, S_6{}^3, S_6{}^4, S_6{}^5, S_6{}^6$$

In terms of C_n and σ_h operations, these two sets of operations may be written as:

$$C_4{}^1\sigma_h{}^1, C_4{}^2\sigma_h{}^2, C_4{}^3\sigma_h{}^3, C_4{}^4\sigma_h{}^4; \; C_6{}^1\sigma_h{}^1, C_6{}^2\sigma_h{}^2, C_6{}^3\sigma_h{}^3, C_6{}^4\sigma_h{}^4, C_6{}^5\sigma_h{}^5, C_6{}^6\sigma_h{}^6$$

This gives operations:

$$S_4{}^1, C_2{}^1, S_4{}^3, E \, (\textbf{4 No.}); S_6{}^1, C_3{}^1, i \, (C_2{}^1\sigma_h = S_2 = i), C_3{}^2, S_6{}^5, E \, (\textbf{6 No.})$$

The group S_2 is a special case because it is equivalent to i. which is already discussed separately.

The group of operations generated by S_n axis with n odd, consists of *2n* elements, including σ_h and the operations generated by C_n. By convention such groups are assigned point symmetry C_{nh}. This can be shown by taking examples of S_3.

Let us consider the operations generated by S_3 axes:

$$S_3{}^1 \quad S_3{}^2 \quad S_3{}^3 \quad S_3{}^4 \quad S_3{}^5 \quad S_3{}^6$$

In terms of C_n and σ_h operations, we write

$$C_3{}^1\sigma_h{}^1, \; C_3{}^2\sigma_h{}^2, \; C_3{}^3\sigma_h{}^3, \; C_3{}^4\sigma_h{}^4, \; C_3{}^5\sigma_h{}^5, \; C_3{}^6\sigma_h{}^6$$

That is, $S_3{}^1$, $C_3{}^2$, σ_h, $C_3{}^1$, $S_3{}^5$, E = **6** distinct operations, that is $2n$ operation.

The operations, $C_3{}^1$, $C_3{}^2$, E (or $C_3{}^3$) are just the operations generated by C_3 axis. This shows that the element S_3 requires that C_3 and a σ_h must exist independently.

(b) C_{nv} point groups: Molecules which possess C_n (a proper axis of n-fold) and n vertical planes (σ_v) of symmetry, they assigned a point group C_{nv} of order $2n$. They do not possess a centre of inversion, i, and σ_h.

(i) C_{2v}: (Contains C_2 and two vertical planes (σ_v); Symmetry operations: E, C_2, $2\sigma_v$ (**4** No., that is, of order $2n$).

Examples (Figure IV.8):

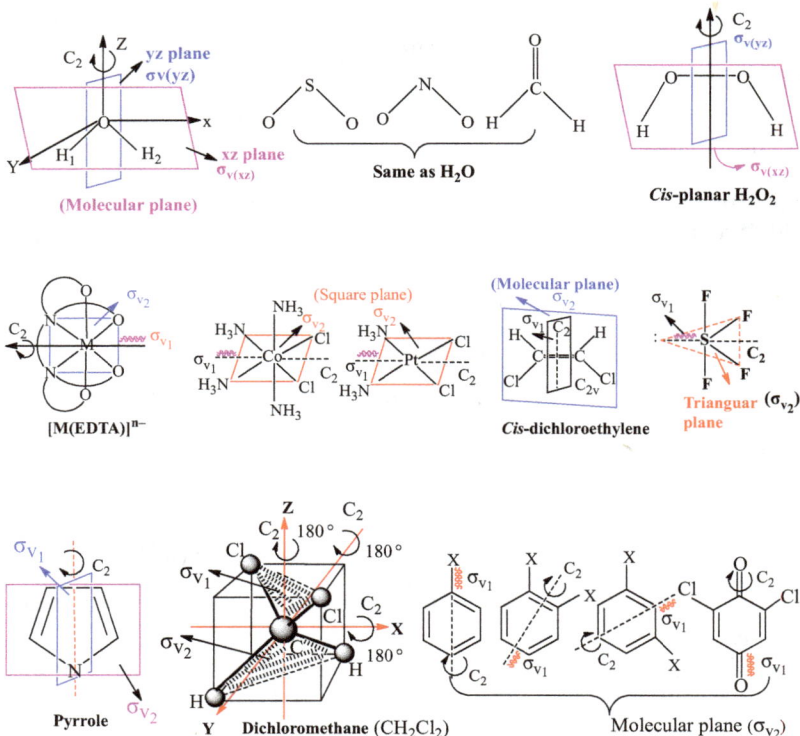

Figure IV.8: Molecules belonging to C_{2v} point group.

(ii) C$_{3v}$: (Contains C$_3$ and three vertical planes (σ_v); Symmetry operations: E, 2C$_3$, 3σ_v (**6 No.**, that is, of order 2n).

Examples (Figure IV.9):

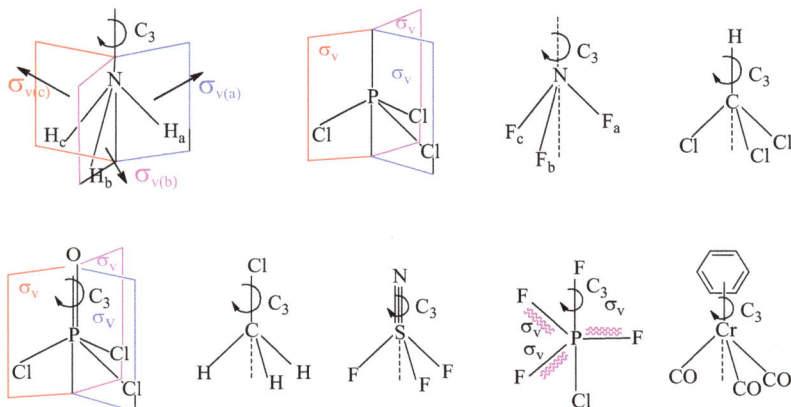

Figure IV.9: Molecules belonging to C$_{3v}$ point group.

(iii) C$_{4v}$: Contains C$_4$ and four vertical planes (σ_v); Symmetry operations: E, 3C$_4$, 4σ_v (**8 No.**, that is, of order 2n).

Examples (Figure IV.10):

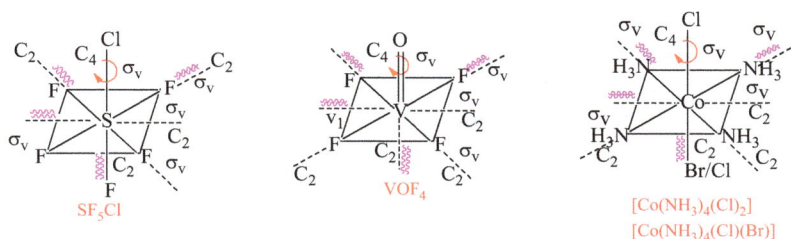

Figure IV.10: Molecules of C$_{4v}$ point group.

(iv) C$_{5v}$: Contains C$_5$ and five vertical planes (σ_v); Symmetry operations: E, 4C$_5$, 5σ_v (**10 No.**, that is, of order 2n).

Examples (Figure IV.11):

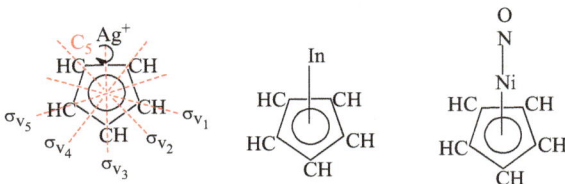

Figure IV.11: Molecules of C_{5v} point group.

(v) C_{6v}: (Contains C_6 and six vertical planes (σ_v); Symmetry operations: E, $5C_6$, $6\sigma_v$ (**12 No.**, that is, of order $2n$).

Examples (Figure IV.12):

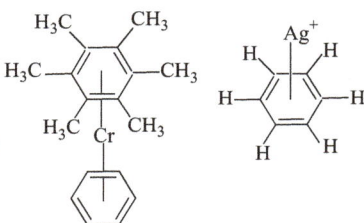

Figure IV.12: Molecules of C_{6v} point group.

(c) C_{nh} point groups: In addition to a proper axis (C_n) of order n, there is also a horizontal plane of symmetry (σ_h), we have a point group C_{nh} of **order $2n$**. It is notable here that when n is even, C_{nh} group is identical to S_n.

The $2n$ operations include S_n^m operations that are products of C_n^m and σ_h for **n odd**, to make a total of $2n$ operations. Thus, for C_{3h}, the operations are:

$$C_3^1, C_3^2, C_3^3 = E \qquad (3)$$

$$\sigma_h \qquad (1)$$

$$C_3^1.\sigma_h = S_3^1 \qquad (1)$$

$$C_3^2.\sigma_h = C_3^3.C_3^2.\sigma_h = C_3^5.\sigma_h = S_3^5 \qquad (1) \; [C_n^m.\sigma_h = S_n^m]$$

Hence, total operations = $3 + 1 + 1 + 1 = 6$, that is, $2n$ ($n = 3$).

Similarly, for C_{4h} (n even), the operations are:

$$C_4^1, C_4^2, C_4^3, C_4^4 = E \quad (4)$$

$$\sigma_h \qquad (1)$$

$$C_4{}^1.\sigma_h = S_4{}^1 \qquad (1)$$

$$C_4{}^2.\sigma_h = C_2{}^1.\sigma_h = i \qquad (1)$$

$$C_4{}^3.\sigma_h = S_4{}^3 \qquad (1)$$

Hence, total operations $= 4 + 1 + 1 + 1 + 1 = 8$, that is, $2n$ $(n = 4)$. **It is notable here that in case of n even, there is inversion centre, i.**

Here, one must **exclude** molecules with other mirror planes (σ_v and σ_d) than the horizontal plane. For example, both boric acid and methenium ion ($CH_3{}^+$) have a C_3 perpendicular to the molecular plane (σ_h), but the $CH_3{}^+$ also contains three vertical planes (σ_v). Hence, boric acid having C_3 and σ_h only belong to C_{3h}, while for $CH_3{}^+$ we must find a higher symmetry group.

Molecules with C_{nh} symmetry with $n > 2$ are relatively rare. Examples with $n = 2$, 3 and 4 are shown in Figure IV.13.

(d) Dihedral point groups: The point groups, D_n, D_{nh} and D_{nd} come under this division.

D_n point groups

The occurrence of a principal axis C_n and nC_2 axes perpendicular to it classify a molecule as belonging to the point group D_n. There is no plane of symmetry in this point group. The total symmetry operations for this group will thus be $2n$ (n due to C_n, plus n due to C_2).

The point group D_1 means C_1 and $1C_2$, that is, elements E ($C_1 = E$) and C_2 (2 operations, order $= 2n$), and is therefore identical with C_2 point group.

The point group D_2 requires the presence three mutually perpendicular C_2 axes, and thus having E, $C_{2(x)}$, $C_{2(y)}$, $C_{2(z)}$ (4 operations, order $= 2n$). D_3 (with C_3 and $3C_2$) contains E, two threefold axes, $C_3{}^1$, $C_3{}^2$ and three twofold axes, $C_2{}^1$ (6 operations, order $= 2n$). It is chemically important group since many trischelates belong to this point group.

C₂ₕ:

Centre of symmetry

(E, C₂, i, σₕ)

Centre of symmetry

Trans-planar H₂O₂
C₂ₕ

Trans-1,2-dichloroethylene
C₂ₕ

C₂ₕ

C₃ₕ

(1E,5E,9E)-Cyclododeca-1,5,9-triene

C₃ₕ

H₃BO₃

Not C₃ₕ

C₃ₕ

C₃ₕ:

C₃ₕ

C₃ₕ

Tri-o-thymotide

Note: Molecular plane is σₕ in each case

C₄ₕ:

C₄ₕ

C₄ₕ

C₄ₕ

C₆ₕ:

1,2,3,4,5,6-Hexa(prop-1-en-2-yl)benzene

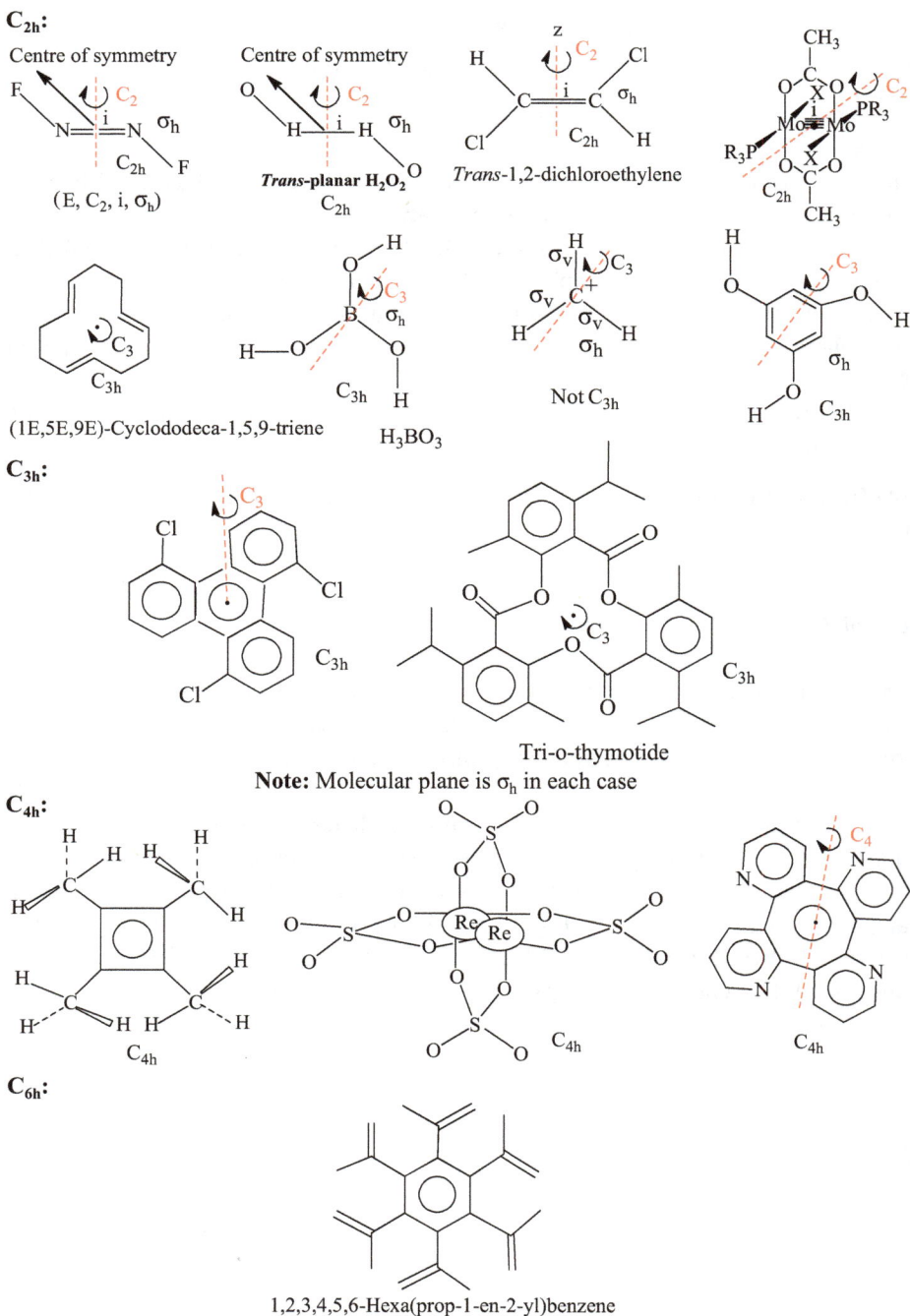

Figure IV.13: Molecules with Cₙₕ point symmetry.

Examples (Figure IV.14):

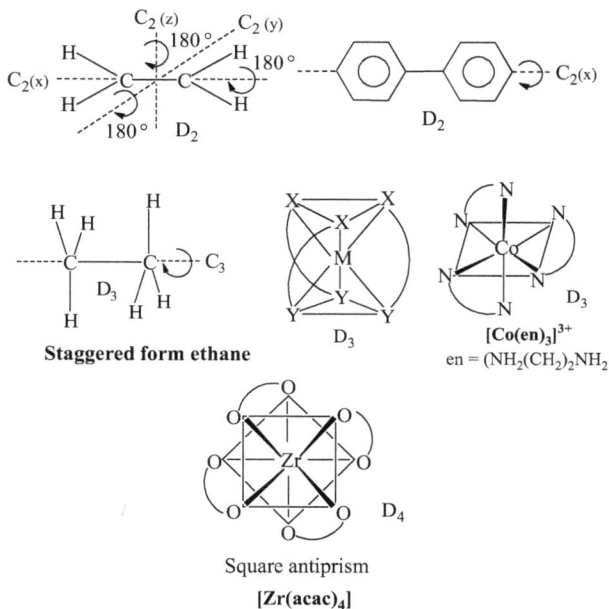

Figure IV.14: Molecules belonging to D_n point group.

D_{nh} point groups

Molecules belonging to the D_n point group and also containing a horizontal mirror plane (σ_h) belong to the point group D_{nh}. In other words, it can be said that the molecules belonging to D_{nh} point group contain C_n as the principal axis and n number of C_2 axes $\perp r$ to this C_n axis. Further, there will be one σ plane $\perp r$ to the C_n axis (σ_h) and n planes all containing the C_n axis, that is, σ_v planes.

D_{nh} point group generates $4n$ symmetry operations, $2n$ operations of the D_n group and $2n$ more that of the products of these operations with σ_h.

Examples (Figure IV.15):

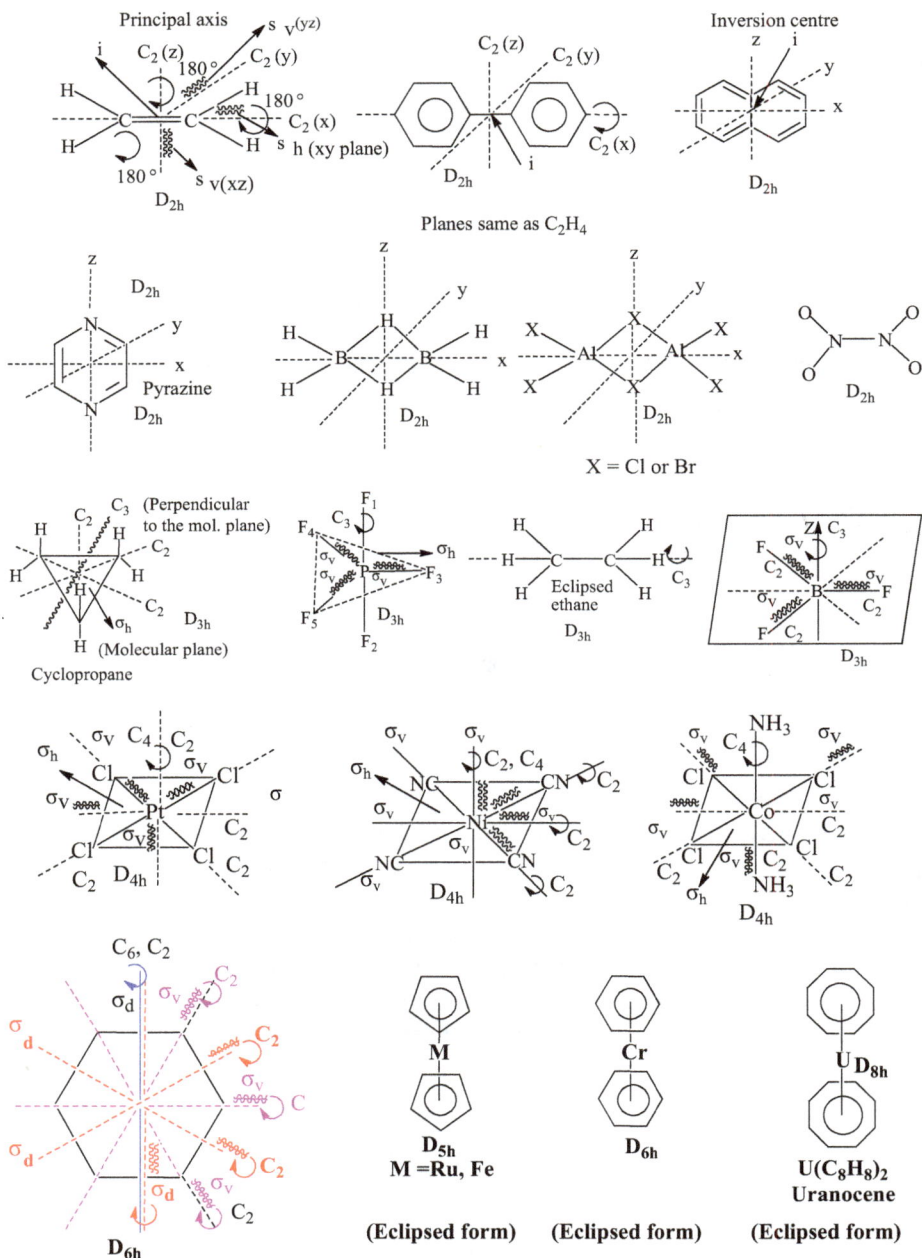

Figure IV.15: Molecules belonging to the D_{nh} point group.

D_{nd} point group

Molecules containing the symmetry elements characterizing D_n point group and possessing in addition the σ_d planes, belong to the point group D_{nd}. In other words, this point group contains one C_n axis and there are n numbers of C_2 axes $\perp r$ to the C_n axis. There are n symmetry planes containing the C_n axis, that is, the vertical planes. These vertical planes are σ_d planes bisecting the angle between the successive C_2 axes.

Let us consider the consequences of adding a set of dihedral planes, σ_ds to C_n and nC_2s. These are vertical planes that bisect the angles between adjacent pairs of C_2 axis. The products of a σ_d with the various $C_n{}^m$ operations are all other σ_d operations. However, among the various products of the type $\sigma_d C_2$ there is a set of n new operations generated by an S_{2n} axis collinear with C_n.

The presence of the proper axis of symmetry of even order generates the S_{2n} axis, and when n assumes odd value the S_{2n} and inversion centre (because of the independent existence of a σ_h plane) must be present in molecules with D_{nd} point group symmetry.

Thus, the symmetry operations in a D_{nd} group include E, $(n\text{-}1)$ proper rotations about C_n, n rotations about C_2 axes, $n\sigma_d$ operations and n operations generated by an S_{2n} axis collinear with C_n. There are $4n$ symmetry operations, $2n$ operations for D_n and $2n$ more which are the products of these operations with σ_d.

Examples:
(a) The point group D_{1d} is identical with C_{2h}.
(b) D_{2d}: A very good example of this point group is allene (Figure IV.16). This molecule has three C_2 axes, one passing through the molecular axis, the other two C_2' and C_2'' are $\perp r$ to it. The first plane passes through the molecular axis and H_1H_2, that is, PQRS, and the second passes through the molecular axis and H_3H_4, that is, ABCD. As these planes are passing in between the two subsidiary axes (the plane PQRS passes in between C_2 and C_2'; plane ABCD passes in between C_2' and C_2''), these are dihedral planes (σ_ds). Hence, the point group is D_{2d}.

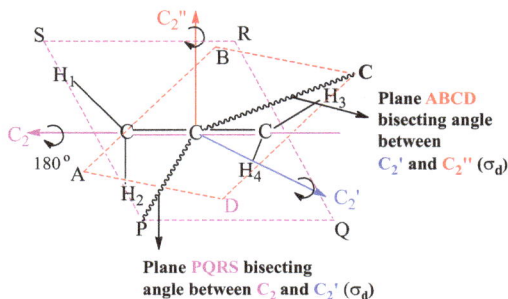

Figure IV.16: Allene molecule of D_{2d} point group.

Other examples (Figure IV.17):

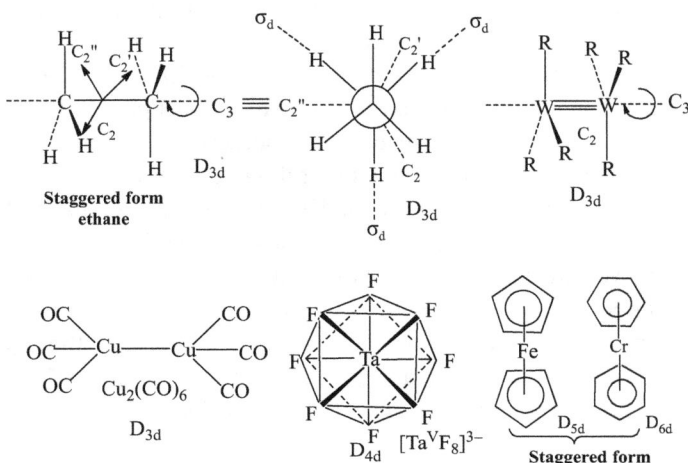

Figure IV.17: Molecules of D_{nd} point group.

IV.3.3 Type III: molecules of higher symmetry or higher symmetry point groups

The point groups considered so far were characterized by a single principal axis of the order equal to or greater than three. There are some points groups in which more than one rotation axis which is at least threefold, are present. They include among them the most important point groups, namely the tetrahedral and octahedral point groups. Since they are derived from a cube, they are also referred to as cubic point groups. In these point groups, the axes of rotation and reflection planes are present, which are not all $\perp r$ to the principal axis in contrast to the other point groups described earlier.

(a) Tetrahedral point group (T_d)

A regular tetrahedron can be inscribed in a cube by taking the alternate vertices. If each of the four vertices of a tetrahedron is occupied by an atom surrounding an atom at the centre, it gives a tetrahedral molecule. CH_4, CCl_4, $NiCl_4^{2-}$, $Ni(CO)_4$, $CoCl_4^{2-}$, $Zn(CN)_4^{2-}$ belong to this group. All tetrahedral molecules do not have Td point symmetry. There should be the following 24 symmetry operations possible on the molecules.

Let us take the general case of AB_4 inscribed in a cube (Figure IV.18). Inspection of this molecule reveals the following symmetry elements and symmetry operations.

(i) There are four C_3 axes, one along each A–B bond and centre of the triangle formed by the other three H atoms (i.e. centre of the opposite trigonal face). Over each C_3 axis, two operations C_3^1 and C_3^2 are possible. Hence a total of $8C_3$ operations.

(ii) There are three C_2 axes passing through centres of two opposite edges as shown in the (Figure IV.18). In fact, each of the Cartesian axes, x, y and z bisecting two BAB bond angles, is the C_2 axis. More specifically, following the notations of four B atoms, one passes through B_1B_3 and B_2B_4, second through B_1B_4 and B_2B_3 and third between B_2B_4 and B_3B_4. Over each C_2 axis, only one possible operation is C_2^1. Hence, a total of $3C_2$ operations.

Each of the C_2 is also S_4, that is, there are three S_4 axes. However, $S_4^2 = C_2^1$, and hence new operations are S_4^1 and S_4^3. Hence, the total new improper operations are $6S_4$.

(iii) There are six planes of symmetry each passing through one edge and centre of the opposite edge. Using the numbering system (Figure IV.18), we may specify these symmetry planes by stating the atoms they contain: AB_1B_2, AB_1B_3, AB_1B_4, AB_2B_3, AB_2B_4 and AB_3B_4. These planes are dihedral planes ($\sigma_d s$) as they pass in between two C_2's. As over each σ plane, only one operation is possible, the total No. operations will be 6σ.

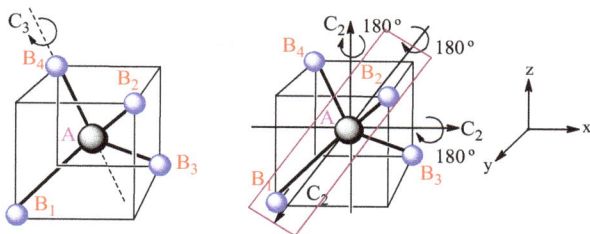

Figure IV.18: A tetrahedral AB_4 molecule inscribed in a cube having three C_2 axes passing through centres of two opposite edges.

(iv) Thus, the total number of symmetry operations possible is 24 as listed below:

$$1E,\ 8C_3,\ 3C_2,\ 6S_4,\ 6\sigma_d$$

Irregular tetrahedral molecules, such as, $CHCl_3$ and CH_3Cl, which don't possess 24 symmetry operations don't belong to this symmetry group.

(b) Octahedral point group (O_h)

Formation of an octahedral complex in a cube

A cube has six faces and eight vertices. The figure obtained by joining the centre points of the faces of a cube is an octahedron. If each of the centres of the faces is occupied by ligands surrounding a metal atom at the centre of the cube, it generates an octahedral molecule as shown in Figure IV.19. An octahedral has eight trigonal equilateral faces, six corners and 12 edges. The most common molecular geometry belonging to this point symmetry is a regular octahedron, such as, $[Co(NH_3)_6]^{3+}$, $[Fe(CN)_6]^{4-}$, $[Pt(Cl)_6]^{2-}$, $[Ti(F)_6]^{3-}$, SF_6, etc. The octahedron (say ML_6) has the following symmetry elements and operations:

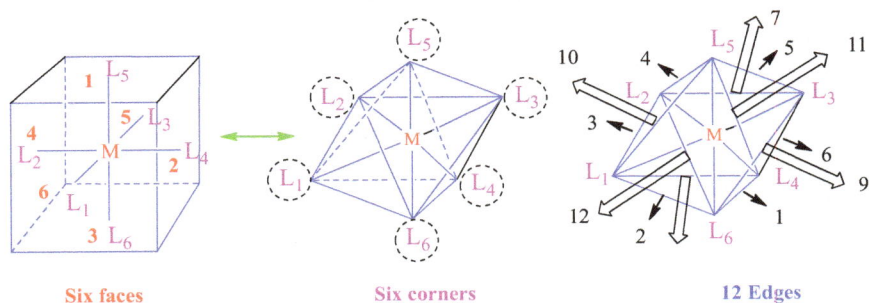

Six faces   Six corners   12 Edges

Figure IV.19: Octahedral complex having six faces, eight trigonal equilateral faces, six corners and 12 edges.

(i)   There are three axes of fourfold symmetry (i.e. $3C_4$) passing through 2Ls at the opposite corners. Over each C_4, three operations C_4^1, C_4^2 and C_4^3 are possible. Hence a total of $9C_4$ rotation operations.

(ii) The $3C_4$ axes are coincident with $3S_4$ axes. Only two operations, S_4^1 and S_4^3 are separate operations on each axis because $S_4^2 = C_4^2$ which has already been counted above. Hence a total of $2 \times 3 = 6S_4$ improper operations.

(iii) The $3C_4$ axes are also coincident with three axes of twofold symmetry, that is $3C_2$. Over C_2, one possible operation is C_2^1, which is equivalent to C_4^2. Thus, there are $1 \times 3 = 3C_2$. In the light of $3C_2 \equiv 3C_4^2$, the total of C_4 operations discussed in point (i) above will be $9C_4 - 3C_4 = 6C_4$.

(iv) There are six C_2, axes, which bisect six pairs of 12 opposite edges. On each axis C_2^1, operation is possible. Thus, a total of $6C_2$, are there.

(v)   There are four axes of threefold symmetry (i.e. $4C_3$) each passing through the centres of two opposite trigonal faces of the octahedron (Figure IV.20a). As there are 8 such trigonal faces, there are $4C_3$. On each axis, the operations possible are C_3^1 and C_3^2. Thus, there are $2 \times 4 = 8C_3$.

(vi) The $4C_3$ axes are also $4S_6$ axes. On each S_6 axis five operations should be possible, but $S_6^2 \equiv C_3^1$, $S_6^4 \equiv C_3^2$ and $S_6^3 \equiv i$. Hence, the new operations are S_6^1, S_6^5. Thus there are $2 \times 4 = 8S_6$.

(vii) The molecule has three σ_h each of which passes through four of the six corners including the central metal, and generate an operation σ_h, and thus a total of $3\sigma_h$. These three planes include the set of atoms: $ML_1L_2L_3L_4$, $ML_1L_5L_3L_6$ and $ML_2L_5L_4L_6$ (Figure IV.20b). As these planes are in the plane of the molecules, and are $\perp$r to the principal axis C_4, they are σh.

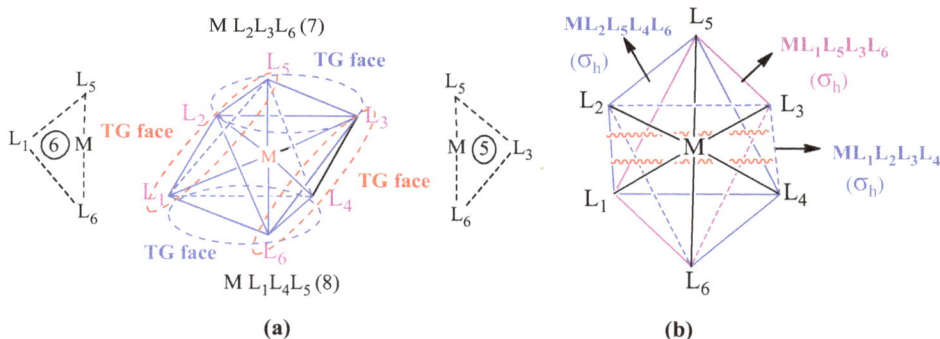

Figure IV.20: An octahedral complex having $4C_3$ and $3\sigma_h$.

(viii) The molecule also contains six σ_v (σ_d in fact). There are three of the one type including atoms, which are ML_5L_6, ML_1L_3 and ML_2L_4. The rest three planes are those which pass through the centres of the opposite edges including 'M'. Three such planes are shown in Figure IV.21. Thus, the six dihedral planes will generate $6\sigma_d$ reflection operations.

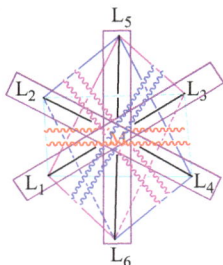

Figure IV.21: An octahedal complex having $6\sigma_d$.

(ix) As mentioned in point (vi), the molecule also contains an inversion centre i. The entire set of operations thus consists of the following 48, grouped by classes:

$$E,\ 6C_4,\ 6S_4,\ 3C_2,\ 6C_2',\ 8C_3,\ 8S_6,\ 3\sigma_h,\ 6\sigma_d,\ i$$

IV.3.4 Type III Type IV: linear molecules with infinite point groups

The linear molecules belong to either of the two infinite groups, namely $C_{\infty v}$ and $D_{\infty h}$.

$C_{\infty v}$

The point group $C_{\infty v}$ is characterized by the presence of a C_∞ axis which is the molecular axis itself, and since the molecule is linear, σ_v planes containing the C_∞ axes are infinite (∞) in number. Examples of molecules belonging to this group are the heteronuclear diatomic molecules and unsymmetrical linear tri or tetratomic molecules: HX (X = F, Cl, Br, I), SCN$^-$, HCN, N_2O (NNO), HC = CX and so on (Figure IV.22).

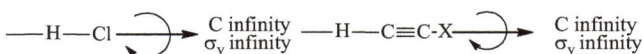

—H—Cl ⟶ C infinity, σ_v infinity ⠀⠀—H—C≡C-X ⟶ C infinity, σ_v infinity

Figure IV.22: Heteronuclear diatomic molecules and unsymmetrical linear tetra-atomic molecules belonging to $C_{\infty v}$ point group.

$D_{\infty h}$

Molecules belonging to this point group have identical ends. Thus, in addition to infinite rotation axis C_∞ coinciding with the molecular axis, an infinite number of vertical planes (σ_vs) containing C_∞ axes and a centre of inversion.

They also have a horizontal plane of symmetry and an infinite number of C_2 axis perpendiculars to C_∞. Therefore, all centrosymmetric molecules belong to this point group. Homonuclear diatomic molecules, such as N_2, H_2, F_2, Cl_2, O_2 are the best examples of this point group. Hg_2^{2+} (mercurous ion) belongs to this point group. Among the linear polyatomics, pseudo halogens, CH=CH, CO_2, $HgCl_2$, N_3^- are some of the examples (Figure IV.23)

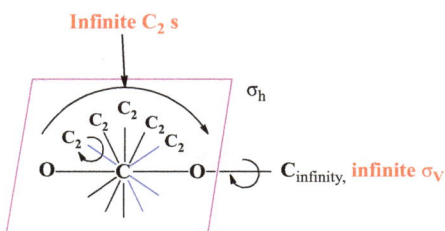

Infinite C_2 s

σ_h

O—C—O ⟶ $C_{infinity}$, infinite σ_v

Figure IV.23: CO_2 molecule belonging to $D_{\infty h}$ point group.

Appendix V
Prediction of infrared and Raman active modes in molecules belonging to icosahedral (I_h) point group

V.1 Introduction

Spectroscopy is concerned with the use of electromagnetic radiations in the study of structures and shapes of molecules. The infrared (IR) and Raman spectroscopy comprise a major part of the spectroscopic techniques. Molecular vibrations can interact with IR radiations and this interaction is the subject matter of IR or vibrational spectroscopy. The use of visible radiations to induce vibrational–rotational transitions in molecules is the basic idea underlying in Raman spectroscopy. Use of IR and Raman spectroscopy, which are complimentary to one another, in conjunction is an extraordinarily efficient tool in the elucidation of molecular structure.

Even though molecular vibrations can interact with radiation in general, these interactions can occur only if the vibrations satisfy certain conditions. Generally molecular vibrations should be associated with a change in dipole moment of the molecule for it to be IR active. Depending on the symmetry of molecules, some vibrations may be IR active and some not. Moreover, the number of normal or fundamental vibrations possible in a molecule depends on the number of atoms present in it.

Normally, a molecule containing N atoms will have 3N degrees of freedom, which involves translational, rotational and vibrational motions. All the molecules, irrespective of linear or non-linear, will have three degrees of translational motion considering whole molecule as a rigid unit. A linear molecule will have 2 degrees of rotational motion because rotation only about the two axes perpendicular to the bond axis (molecular axis) constitute rotations of the system. Thus, the remaining 3N-(3)-(2) = 3N-5 degrees of freedom will be associated with the vibrational motion. A non-linear molecule can rotate about three mutually perpendicular axes that pass through the centre of gravity of the molecule and thus have 3 degrees of freedom for its rotational motion. The remaining 3N-(3)-(3) = 3N-6 degrees of freedom will be associated with the vibrational motion.

Out of (3N-5 or 3N-6) normal modes of vibrations possible for a linear or non-linear molecule, respectively, only those vibrations which accompany with a change in the dipole moment of the molecule will be IR active. Raman activity is associated with polarization of molecules. So, if a normal mode of vibration brings about a polarization in the molecule, the vibration will be Raman active.

Whether or not a particular vibration will bring about a dipole moment change or polarization in a molecule can be understood easily in the case of simple molecules. But as the complexity of the molecule increases it becomes more and more difficult to understand the change in properties during molecular vibration. Hence,

https://doi.org/10.1515/9783110727289-015

it may not be always easy to find out whether a particular vibration is IR and/or Raman active or not.

Group theory can be successfully applied to predict the IR and Raman activity of various vibrations associated with the molecule, for which the knowledge of various symmetry operations of the point group to which the molecule belongs are required. It also helps us to understand the complementarities (mutual exclusion rule) of Raman and IR spectra and thus we can have better insight into structure and shapes of molecules.

V.2 Regular icosahedron structure and possible symmetry operations involved

(a) Icosahedron structure

The structure of a regular icosahedron can be understood with the help of a cardboard model. This model can be constructed in a few steps as follows:

(i) Draw a regular pentagon (named 1) on a cardboard sheet and draw an equilateral triangle (named T_1) on each side of it (Figure V.1a); these triangles may be called as teeth of pentagon 1.

(ii) Another regular pentagon (named 2) of the same size and its teeth (named T_2) is drawn on another cardboard sheet (Figure V.1b); these will look like a pair of five-vertex stars (Figure V.1).

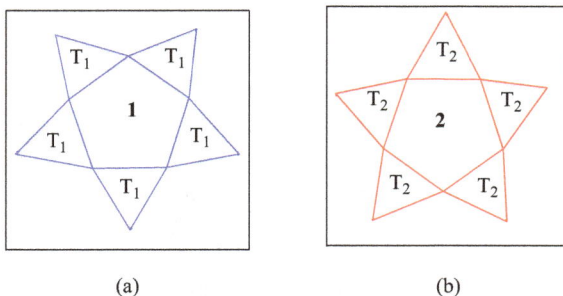

(a)   (b)

Figure V.1: A pair of five-vertex stars.

(iii) Now, cut out the stars and place the pentagon one over the other in a staggered conformation as shown in Figure V.2.

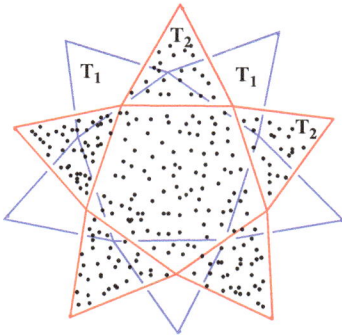

Figure V.2: Two five-vertex stars placed one above the other in a staggered fashion.

(iv) Then folds the teeth such that each T_1 points towards plane of the pentagon 2, and each T_2 points towards plane of pentagon 1. Join them together, and this will give a pentagonal antiprism structure (Figure V.3).

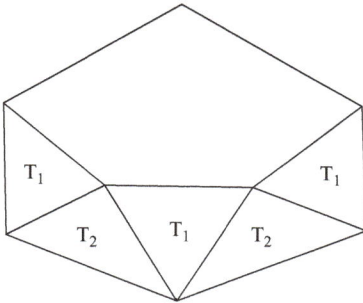

Figure V.3: Formation of pentagonal antiprism structure.

(v) On constructing the pentagonal pyramids above and below the pentagonal antiprism shown above, the icosahedron (I_h) structure is formed (Figure V.4).

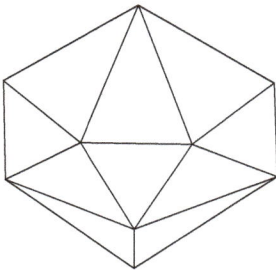

Figure V.4: Front view of icosahedron.

(vi) One can see that icosahedron (I_h) contains 12 equivalent vertices, 20 equilateral triangles as faces and 30 equivalent edges, consistent with Euler's polyhedron formula,

$$V - E + F = 2,$$

where V = No. of vertices, E = No. of edges and F = No. faces in polyhedron. So,

$$12 - 30 + 20 = 2$$

(b) Symmetry operations in icosahedron (I_h)

(i) The line joining the pair of vertices is a C_5 axis. Being 12 equivalent vertices in I_h, there are $12/2 = 6$ such C_5 axes. One such C_5 axis is shown in Figure V.5. Each C_5 generates the symmetry operations: $C_5^1, C_5^2, C_5^3, C_5^4, C_5^5 = E$. Arranged class wise, these are $(C_5^1, C_5^4), (C_5^2, C_5^3)$, E. Thus, there are $4 \times 6 = \mathbf{24C_5}$ rotation operations.

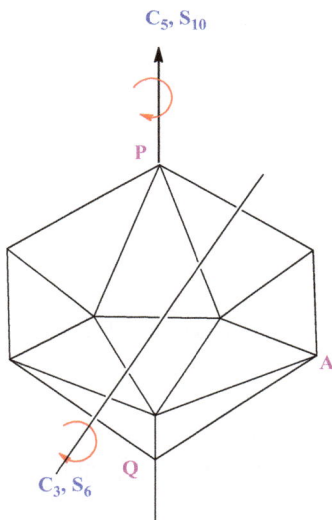

Figure V.5: Symmetry elements in I_h.

(ii) There are five vertical planes (σ_v) each containing a C_5 axis. Hence, there should exist a total of $6 \times 5 = 30$ such planes; among them each plane is counted twice. For example, the σ_v plane containing the C_5 axis PQ (Figure V.5) also contains the one joining the vertex P and its opposite vertex (not shown in Figure V.5). Thus, there are a total of $30/2 = \mathbf{15\sigma_v}$.

(iii) As the oppositely positioned pentagon is staggered, each C_5 axis also acts as an S_{10} axis and every such axis generates the operations: $S_{10}^1, S_{10}^3, S_{10}^5 = S_2 = i$, S_{10}^7 and S_{10}^9 (only odd powers). Even powers of S_{10} results in pure C_5-rotations and E, which have already been considered. Arranged class wise, improper rotations generated by each of the six C axes are: $(S_{10}^1, S_{10}^9), (S_{10}^3, S_{10}^7)$ and $(S_{10}^5 = i)$. Hence a total of $4 \times 6 = \mathbf{24S_{10}}$ improper rotation operations.

(iv) Joining the centroids of the opposite pairs of equilateral triangle one gets a C_3-axis. As the icosahedron contains 20 equilateral triangles, there are $20/2 = 10$ such pairs of equilateral triangles and so $10C_3$ axes (one is shown in Figure V.5). Each C_3 axis generates the operations, (C_3^1, C_3^2) and E. Thus, there are a total of $2 \times 10 = \mathbf{20C_3}$ rotation operations.

(v) As the oppositely positioned triangles are staggered, C_3 axis also acts as S_6 axis. Thus, in all, there are $10S_6$ axes, and each of them generates, S_6^1, $S_6^3 = S_2 = i$, S_6^5 (only odd powers). Even power of S_6 results in pure C_3-rotations and E, which have already been considered. Arranged class-wise, these are $(S_6^1, S_6^5,)$, and (i). Hence a total of $2 \times 10 = \mathbf{20S_6}$ improper rotation operations.

(vi) Joining the centres of oppositely situated edges of I_h, one gets a C_2 axis (Figure V.6). As there are $30/2 = 15$ pairs of opposite edges, hence $15C_2$ axes exist. In a regular icosahedron, one can find three distinct belt regions, each containing five C_2 axes (Figure V.6). Each of the $15C_2$ axes generates a C_2^1 along with $C_2^2 = E$. Hence a total of $1 \times 15 = \mathbf{15C_2}$ rotation operations.

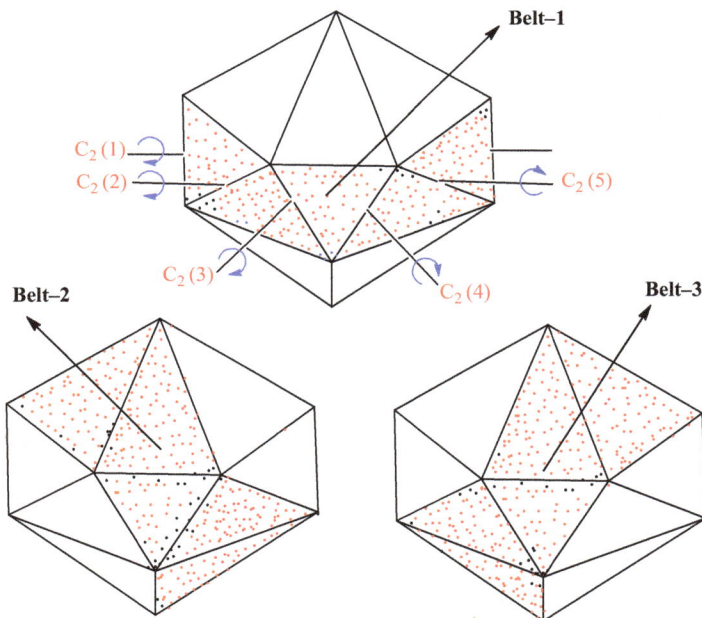

Figure V.6: Three belt regions of an icosahedron. Five twofold symmetry axes are shown for Belt-1.

The following table (Table V.1) gives the complete list of symmetry elements/ symmetry operations of the point group I_h of a regular icosahedron.

Table V.1: Symmetry elements and symmetry operations of a regular icosahedron.

Symmetry elements (No. of such elements)	Symmetry operations	No. of conjugate classes of symmetry operations	No. of symmetry operations
E	E	1	1
i	i	1	1
C_5–axis (6)	$6(C_5^1, C_5^4)$, $6(C_5^2, C_5^3)$	2	$12 + 12 = 24$
S_{10}–axis (6)	$6(S_{10}^1, S_{10}^9)$, $6(S_{10}^3, S_{10}^7)$	2	$12 + 12 = 24$
C_3–axis (10)	$10(C_3^1, C_3^2)$	1	$10 + 10 = 20$
S_6–axis (10)	$10(S_6^1, S_6^5,)$	1	$10 + 10 = 20$
C_2–axis (15)	$15(C_2^1)$	1	15
σ_v plane (15)	$15(\sigma_v)$	1	15
Total No. **8**		Total No. = **10**	Total No. = **120**

Truncation of the icosahedron and the structure of fullerene-60

If the icosahedron is truncated by 12 planes at each of the 12 vertices, 12 pentagonal faces will be generated and they will be separated by 20 hexagons. Two such hexagons are shown in Figure V.7a **(shaded)**. The resulting figure contains 60 vertices. If these 60 sites are occupied by sp^2 hybridized carbon atoms (one at each vertex), we get the structure of Buckminsterfullerene-C_{60} (Figure V.7b). The structure is just like that of a football.

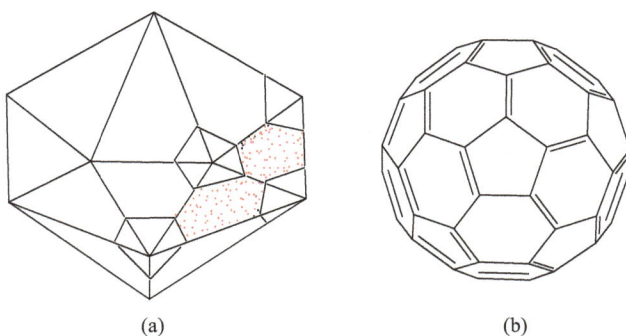

(a)   (b)

Figure V.7: (a) Truncated icosahedron (only three truncations are shown). (b) Buckminsterfullerene-C_{60} [now called (60) fullerene].

Six pairs of pentagons are arranged with opposite orientation at the ends of six diameters of the football, three of these diameters are X, Y and Z axes and the other three are along the meridonial directions. The fullerene-C_{60} structure (Figure V.8), therefore, has the same symmetry operations as those of the original icosahedron. Hence, the point group of fullerene-C_{60} is I_h.

V.3 Prediction of IR and Raman active modes in (60) fullerene of I_h point group

Molecular symmetry and group theory has simplified the problem of finding and predicting IR and Raman active modes in this molecule. Belonging to I_h point group, this molecule contains 12 pentagonal and 20 hexagonal rings. Taking one member from each of the two conjugate classes $(C_5^1, C_5^4)/(C_5^2, C_5^3)$ and $(S_{10}^1, S_{10}^9)/(S_{10}^3, S_{10}^7)$ and other members from single conjugate classes, the symmetry operations in (60) fullerene of I_h point group are: E, $12C_5$, $12C_5^2$, $20C_3$, $15C_2$, i, $12S_{10}$, $12S_{10}^3$, $20S_6$ and 15σ. Thus, the total number of symmetry operations in this point group is 120 and these belong to 10 different classes. Thus, the point group I_h of (60) fullerene is of order 120.

Figure V.8: Structure of (60) fullerene.

The character table (Table V.2) of I_h point group is given below. This shows that characters for various rotational symmetry operations are of the type $\frac{1}{2}(1 + \prod 5)$ and $\frac{1}{2}(1 - \prod 5)$. There are only ten irreducible representations in this point group. These irreducible representations are one-dimensional (A-type), three-dimensional (T-type), four-dimensional (G-type) and five-dimensional (H-type).

If one carefully looks at the structure of C_{60}, it can be seen that no rotation axis pass through any of the atom of the molecule. It means that when rotational symmetry operations are carried out over these axes, then all atoms will be shifted. Hence, contribution by the basis set towards χ of the reducible representation will be zero. Likewise symmetry operation i will shift all the atoms, and so χ will be zero in this case. Each symmetry plane in the molecule contains only four atoms, and so they are not shifted

Table V.2: Character table of I_h point group.

I_h	E (1)	$12C_5$ (2)	$12C_5{}^2$ (3)	$20C_3$ (4)	$15C_2$ (5)	i (6)	$12S_{10}$ (7)
A_g	1	1	1	1	1	1	1
T_{1g}	3	$\frac{1}{2}(1+\sqrt{5})$	$\frac{1}{2}(1-\sqrt{5})$	0	−1	3	$\frac{1}{2}(1-\sqrt{5})$
T_{2g}	3	$\frac{1}{2}(1-\sqrt{5})$	$\frac{1}{2}(1+\sqrt{5})$	0	−1	3	$\frac{1}{2}(1+\sqrt{5})$
G_g	4	−1	−1	1	0	4	−1
H_g	5	0	0	−1	1	5	0
A_u	1	1	1	1	1	−1	−1
T_{1u}	3	$\frac{1}{2}(1+\sqrt{5})$	$\frac{1}{2}(1-\sqrt{5})$	0	−1	−3	$-\frac{1}{2}(1-\sqrt{5})$
T_{2u}	3	$\frac{1}{2}(1-\sqrt{5})$	$\frac{1}{2}(1+\sqrt{5})$	0	−1	−3	$-\frac{1}{2}(1+\sqrt{5})$
G_u	4	−1	−1	1	0	−4	1
H_u	5	0	0	−1	1	−5	0

I_h	$12S_{10}{}^3$ (8)	$20S_6$ (9)	15σ (10)	Linear function, rotation (11)	Quadratic functions (12)
A_g	1	1	1	–	$x^2 + y^2 + z^2$
T_{1g}	$\frac{1}{2}(1+\sqrt{5})$	0	−1	R_x, R_y, R_z	–
T_{2g}	$\frac{1}{2}(1-\sqrt{5})$	0	−1	–	–
G_g	−1	1	0	–	–
H_g	0	−1	1	–	$[2z^2 - x^2 - y^2, x^2 - y^2, xy, xz, yz]$
A_u	−1	−1	−1	–	–
T_{1u}	$-\frac{1}{2}(1+\sqrt{5})$	0	1	(x, y, z)	–
T_{2u}	$-\frac{1}{2}(1-\sqrt{5})$	0	1	–	–
G_u	1	−1	0	–	–
H_u	0	1	−1	–	–

by these planes will contribute to χ. Similar to rotation symmetry operations, in each of the improper symmetry operations, all atoms will be shifted. Hence, character contribution in each case will also be zero.

Cartesian coordinate method or 3N vector method

Looking over the complexity of the molecule, Cartesian coordinate method will be applied here to get fundamental vibrational modes.

The symmetry operations in this case are five proper rotations through $0°(E)$, $72°(C_5)$ $144°(C_5^2)$, $120°(C_3)$ and $180°(C_2)$, and five improper rotations through $180°(i)$, $36°(S_{10})$, $108°(S_{10}^3)$, $60°(S_6)$, and $0°(\sigma)$. The following table (Table V.3) provides the calculations of the total characters.

Table V.3: Determination of total character of C_{60} molecule.

I_h	E	$15C_5$	$15C_5^2$	$20C_3$	$15C_2$	i	$12S_{10}$	$12S_{10}^3$	$20S_6$	15σ
β	0	72	144	120	180	180	36	108	60	0
$\cos\beta$	1	0.3090	−0.8090	−1/2	−1	−1	0.8090	−0.3090	1/2	1
$\pm 1 + 2\cos\beta$	3	1.6180	−0.6180	0	−1	−3	0.6180	−1.6180	0	1
n_R	60	0	0	0	0	0	0	0	0	4
Γ_{3N}	180	0	0	0	0	0	0	0	0	4

The vertical dashed line separates the proper and improper rotations. The corresponding angleβ, $\cos\beta$ and also $(\pm 1 + 2\cos\beta)$ values are tabulated for each operation. The proper rotations have $(+1 + 2\cos\beta)$ and the improper rotations have $(-1 + 2\cos\beta)$ values. The number of atoms retaining position (n_R) for each operation is also given in the table.

The above reducible representation (Γ_{3N}) can be reduced to irreducible representations applying the standard reduction formula (given below) and using the character table of I_h point group as follows:

$$N_i = \frac{1}{h} \sum_R \chi(R) \cdot n \cdot \chi_i(R)$$

where N_i is the number of times the ith irreducible representation occurs in a reducible representation, h is the order of the point group (number of symmetry operations), $\chi(R)$ is the character of a particular operation in the reducible representation, n is the number of operation of that type and $\chi_i(R)$ is the character of the same operation in the irreducible representation.

Going back to the reducible representation in question, the irreducible representation can be worked out by referring to the character table of I_h point group:

$$NA_g = 1/120[(180.1.1) + 0 + 0 + 0 + 0 + 0 + 0 + 0 + 0 + (4.15.1)$$
$$= 1/120[(180 + 60 = 240)] = 2$$

$NT_{1g} = 1/120[(180.1.3) + 0 + 0 + 0 + 0 + 0 + 0 + 0 + 0 + (4.15.-1)$
$\quad = 1/120[(540 - 60 = 480)] = 4$
$NT_{2g} = 1/120[(180.1.3) + 0 + 0 + 0 + 0 + 0 + 0 + 0 + 0 + (4.15.-1)$
$\quad = 1/120[(540 - 60 = 480)] = 4$
$NG_g = 1/120[(180.1.4) + 0 + 0 + 0 + 0 + 0 + 0 + 0 + 0 + (4.15.0)$
$\quad = 1/120[(720 + 0 = 720)] = 6$
$NH_g = 1/120[(180.1.5) + 0 + 0 + 0 + 0 + 0 + 0 + 0 + 0 + (4.15.1)$
$\quad = 1/120[(900 + 60 = 960)] = 8$
$NA_u = 1/120[(180.1.1) + 0 + 0 + 0 + 0 + 0 + 0 + 0 + 0 + (4.15.-1)$
$\quad = 1/120[(180 - 60 = 120)] = 1$
$NT_{1u} = 1/120[(180.1.3) + 0 + 0 + 0 + 0 + 0 + 0 + 0 + 0 + (4.15.1)$
$\quad = 1/120[(540 + 60 = 600)] = 5$
$NT_{2u} = 1/120[(180.1.3) + 0 + 0 + 0 + 0 + 0 + 0 + 0 + 0 + (4.15.1)$
$\quad = 1/120[(540 + 60 = 600)] = 5$
$NG_u = 1/120[(180.1.4) + 0 + 0 + 0 + 0 + 0 + 0 + 0 + 0 + (4.15.0)$
$\quad = 1/120[(720 + 0 = 720)] = 6$
$NH_u = 1/120[(180.1.5) + 0 + 0 + 0 + 0 + 0 + 0 + 0 + 0 + (4.15.-1)$
$\quad = 1/120[(900 - 60 = 840)] = 7$

Thus, the reducible representation Γ_{3N} is reduced to:

$$\Gamma_{3N} = 2A_g + 4T_{1g} + 4T_{2g} + 6G_g + 8H_g + A_u + 5T_{1u} + 5T_{2u} + 6G_u + 7H_u$$

Thus, there are 180 modes $(2 + 12 + 12 + 24 + 40 + 1 + 15 + 15 + 24 + 35 = 180)$ of degrees of freedom with different symmetry. These include the vibrational, rotational and translational degrees of freedom.

The I_h character table point out that the translational modes have the same symmetry as the coordinates x, y and z, that is, T_{1u}. So, the rotational modes have the same symmetry as the rotation axes, R_x, R_y and R_z, that is, T_{1g}. Subtracting these translational and rotational degrees of freedom, we get the vibrational degrees of freedom only of appropriate symmetry

$$\Gamma_{3N} = 2A_g + 4T_{1g} + 4T_{2g} + 6G_g + 8H_g + A_u + 5T_{1u} + 5T_{2u} + 6G_u + 7H_u$$

$$\Gamma_{tr} = \qquad\qquad\qquad\qquad\qquad\qquad\qquad\qquad T_{1u}$$

$$\Gamma_{rot} = \qquad T_{1g}$$

. . . .

$$\Gamma_{vib} = 2A_g + 3T_{1g} + 4T_{2g} + 6G_g + 8H_g + A_u + 4T_{1u} + 5T_{2u} + 6G_u + 7H_u$$

Now, the total dimensionality number is $180 - 3(T_{1u}) - 3(T_{1g}) = 176$.

Identification of IR active vibrations

From the character table of I_h point group, it is notable that x-, y-, z-coordinates belong to T_{1u} irreducible representations. Hence, $4T_{1u}$ vibrational modes are IR active.

An observation of the I_h point group character table indicates that the binary products of the Cartesian coordinates belong to A_g and H_g irreducible representations. Hence $2A_g$ and $8H_g$ vibrational modes will be Raman active.

Bibliography

The following books/reviews/research papers are recommended for further reading:

M. Clyde Day Jr. and J. Selbin, *Theoretical Inorganic Chemistry*, Second Edition, Affiliated East-West Press Pvt. Ltd., New Delhi (India) (1977).

R. J. Gillespie, Fifty years of the VSEPR model, *Coord. Chem. Rev.*, 252 (2008) 1315–1327.

R. J. Gillespie and R. S. Nyholm, Inorganic stereochemistry, *Quart. Rev. Chem. Soc.*, 11 (1957) 339–380.

R. J. Gillespie, The valence-shell electron-pair repulsion (VSEPR) theory of directed valency, *J. Chem. Educ.*, 40 (1963) 295–301.

N. V. Sidgwick and H. E. Powell, Bakerian lecture, stereochemical types and valency groups, *Proc. R. Soc.*, A176 (1940) 153–180.

James E. Huheey, *Inorganic Chemistry: Principles of Structure and Reactivity*, Third Edition, Harper International SI Edition, London (1983).

D. M. P. Mingos, *Essential Trends in Inorganic Chemistry*, Oxford University Press, Oxford, New York, Tokyo (1998).

D. F. Shriver and P. W. Atkins, *Inorganic Chemistry*, Third Edition, Oxford University Press, Oxford (1999).

F. Albert Cotton and G. Wilkinson *Advanced Inorganic Chemistry*, Fifth Edition, John Wiley & Sons, New York (1988).

B. E. Douglas, D. H. McDaniel and J. J. Alexander, *Concepts and Models of Inorganic Chemistry*, Second Edition, John Wiley & Sons, Inc., New York (2001).

J. D. Lee, *Concise Inorganic Chemistry*, Fifth Edition, Chapman and Hall Ltd., London (1996).

O. P. Agrawal, *Chemical Bonding with an Introduction to Mechanism of Reactions of Complex ions*, Fifth Edition, Dhanpat Rai & Co.(P) Ltd., Delhi (India) (2003).

N. N. Greenwood and A. Earnshaw, *Chemistry of the Elements*, First Edition, Pergamon Press, New York (1984).

F. Albert Cotton, G. Wilkinson, C. A. Murillo and M. Bochmann *Advanced Inorganic Chemistry*, Sixth Edition, John Wiley & Sons, New York (1999).

G. Gonzalez-Moraga, *Cluster Chemistry: Introduction to the Chemistry of Transition Metal and Main Group Element Molecular Clusters*, Springer-Verlag, Berlin (1994).

F. A. Cotton and R. A. Walton, *Multiple Bonds between Metal Atoms*, Second Edition, Oxford University press, Oxford (1993).

B. E. Douglas, D. H. McDaniel and J. J. Alexander, *Concepts and Models of Inorganic Chemistry*, Third Edition, John Wiley & Sons, Inc., New York (2001).

J. Inczedy, *Analytical Applications of Complex Equilibria*, John Wiley & Sons, Inc., New York (1976).

F. J. C. Rossotti and H. Rossotti, *The Determination of Stability Constants and Other Equilibrium Constants in Solution*, McGraw-Hill Book Company, Inc., New York, Toronto and London (1961).

K. B. Yatsimirskii and V. P. Vasil'ev, *Instability Constants of Complex Compounds*, Pergamon Press, Oxford, London, New York and Paris (1960).

M. T. Beck, *Chemistry of Complex Equilibria*, Van Nostrand Reinhold Company, London, New York, Toronto and Melbourne (1969).

J. Rose, *Advanced Physico-Chemical Experiments*, Sir Isaac Pitman & Sons Ltd, London (1963).

R. Gopalan and V. Ramalingam, *Concise Coordination Chemistry*, Vikas Publishing House Pvt. Ltd., New Delhi (India) (2001).

F. Basolo and R. G. Pearson, *Mechanisms of Inorganic Reactions: Study of Metal Complexes in Solution*, Second Edition, John Wiley, New York (1967).

B. R. Puri, L. R. Sharma and K. C. Kalia, *Principles of Inorganic Chemistry*, Twenty-Fifth Edition, Vishal Publications, Delhi (India) (2004).

https://doi.org/10.1515/9783110727289-016

R. C. Maurya, *Molecular symmetry and Group Theoretical Approach of Chemical Bonding*, Lap Lambert Academic Publishing, Mauritius (2017).

R. C. Maurya, Some less-time consuming, visible, safe and economic tests in inorganic qualitative analysis, *Res. J. Sci., R. D. Univ., Jabalpur*, 5 (1998) 95–104.

Vogel's Textbook of Macro and Semimicro Qualitative Inorganic Analysis, Fifth Edition, Revised by, G. Svehla, Orient Longman Limited, New Delhi (India) (1982).

Z. Szafran, R. M. Pike and M. M. Singh, *Microscale Inorganic Chemistry: Comprehensive Laboratory Experience*, John Wiley & Sons, Inc., New York (1991).

Handbook of Preparative Inorganic Chemistry, Volume 2, Second Edition, Edited by, G. Brauer, Academic Press, New York (1965).

G. Marr and B. W. Rockett, *Practical Inorganic Chemistry*, Van Nostrand Reinhold, New York (1972).

Z. Vasilyeva, A. Granovskaya and A. Taperova, *Laboratory Manual for general and Inorganic Chemistry*, Mir Publishers, Moscow (1988).

T. Thananatthanachon and M. R. Lecklider, Synthesis of dichlorophosphinenickel(II) compounds and their catalytic activity in suzuki cross-coupling reactions: a simple air-free experiment for inorganic chemistry laboratory, *J. Chem. Edu, Am. Chem. Soc.*, 94 (2017) 786–789; doi: 10.1021-/acs.jchemed.6b00273.

R. D. Feltham, Synthesis of nitrosylmonobromobis(triphenylphosphine)nickel(0), [NiBr(NO)(PPh$_3$)$_2$], *Inorg. Chem.*, 3 (1964) 116–119.

E. J. Baran and V. T. Yilmaz, Metal complexes of Saccharin, *Coord. Chem. Rev.*, 250 (2006) 1980–1999 .

C. Díaz, M. L. Valenzuela, M. A. Laguna-Bercero, A. Orera, D. Bobadilla, S. Abarcaa and O. Peña, Synthesis and magnetic properties of nanostructured metallic Co, Mn and Ni oxide materials obtained from solid-state metal-macromolecular complex precursors, *RSC Adv.*, 2017 (2017) 27729–27736.

R. C. Maurya and J. M. Mir, *Molecular symmetry and Group Theory: Approaches in Spectroscopy and Chemical Reactions*, De Gruyter Publications, Berlin, Germany (2019), 1–480: ISBN 978-3-11-063496-9, e-ISBN (PDF) 978-3-11-063503-4, e-ISBN (EPUB) 978-3-11-063512-6.

IUPAC Commission determinations, *Pure Appl. Chem.* 73 (2001) 667–683.

Principles of Chemical Nomenclature – A Guide to IUPAC Recommendations, 2011 Edition, G. J. Leigh (Ed.), Royal Society of Chemistry, Cambridge, UK, ISBN 978-1-84973-007-5.

Nomenclature of Inorganic Chemistry – IUPAC Recommendations, N. G. Connelly, T. Damhus, R. M. Hartshorn and A. T. Hutton (Eds.), Royal Society of Chemistry, Cambridge, UK (2005), ISBN 0-85404-438-8.

R. H. Crabtree, Dihydrogen complexation, *Chem. Rev.*, 116 (2016) 8750–8769.

F. A. Cotton, R. R. Monchamp, J. M. Henry and R. C. Young, The preparation and properties of a new pentacyanomanganesenitric oxide anion, [Mn(CN)$_5$NO]$^{2-}$, and some observations on other penta-cyanonitrosyl complexes, *J. Inorg. Nucl. Chem.*, 10 (1959) 28–38.

B. P. Block, W. H. Powell and W. C. Fernelius, *Inorganic Chemical Nomenclature*, American Chemical Society, Washington, D. C., (1990) 50–54.

F. A. Cotton and C. B. Harris, The crystal and molecular structure of dipotassium octachlorodirhenate(III) dihydrate, K$_2$[Re$_2$Cl$_8$] · 2H$_2$O, *Inorg. Chem.* 4, (1965) 330–333.

R. M. Adams, Nomenclature of inorganic boron compounds. *Pure Appl. Chem.*, 30, (1972) 683–710.

F. E. Hong, T. J. Coffy, D. A. McCarthy and S. G. Shore, Synthesis and structure of the ruthenium carbonyl "boride" cluster HRu$_6$(CO)$_{17}$B, *Inorg. Chem.*, 28 (1989) 3284–3285.

C. E. Housecroft, D. M. Matthews, A. L. Rheingold and X. Song, The first trigonal prismatic discrete transition-metal boride cluster: preparation and molecular structure of [PPN][Ru$_6$(H)$_2$(CO)$_{18}$B] [PPN = (Ph$_3$P)$_2$N$^+$], *J. Chem. Soc., Chem. Commun.*, (1992) 842–843; Doi.org/10.1039/ C39920000842.

H. C. Longuet-Higgins and R. P. Bell, The structure of the boron hydride, *J. Chem. Soc.*, (1943), 250–255; Doi:10.1039/jr9430000250.

W. N. Lipscomb, *Boron hydrides*, Benjamin, New York, (1963).

K. Wade, Structural and bonding patterns in cluster chemistry, *Adv. Inorg. Chem. Radiochem.*, 18 (1976) 1–66.

M. Calvin and K. W. Wilson, Stability of chelate compounds, *J. Amer. Chem. Soc.*, 67 (1945) 2003.

H. M. Irving and H. S. Rossotti, The calculation of formation curves of metal complexes from pH titration curves in mixed solvents, *J. Chem. Soc.* (1954) 2904; Doi.org/10.1039/JR9540002904.

P. Job, Formation and stability of inorganic complexes in solution, *Ann. Chim. Phys.*, 9, (1928) 113–203.

James J. Lingane, Interpretation of the polarographic waves of complex metal ions, *Chem. Rev.*, 29 (1941) 1–35.

J. P. Griess, Bemerkungen zu der Abhandlung der HH: Wesely und Benedikt Über einige Azoverbindungen, *Ber. Deutsch. Chem. Ges.*, 12 (1879) 426–428

R. C. Maurya, Synthesis and physico-chemical investigation of some mixed-ligand cyanonitrosyl $\{Mo(NO)_2\}^6$ complexes of molybdenum with some aniline derivative *Synth. React. Inorg. Met.–Org. Chem.*, 24 (1994) 53–62.

E. V. Fedorova, V. B. Rybakov, V. M. Senyavin, A. V. Anisimov and L. A. Aslanov, Synthesis and structure of oxovanadium(IV) complexes, [VO(*acac*)$_2$] and [VO(*sal*: L-alanine)(H$_2$O)], *Crystallogr. Rep.*, 50 (2005) 224–229

A. Reinecke, Über Rhodanchromammonium-Verbindungen, Annalen der Chemie und Pharmacie, 126 (1863) 113–118; doi:10.1002/jlac.18631260116.

Henry Taube, Rates and mechanisms of substitution in inorganic complexes in solution, *Chem. Rev.*, 50 (1952) 69–126.

F. J. Garrick, Possible acid-dissociation of metal-ammonia ions, and its bearing on certain reactions. *Nature*, 139 (1937) 507–508.

B. Adell, Über die gegenseitige Umwandlung der chloro- und aquopentamminkobalt (III) – chloride in schwach salzsauren, wäßrigen Lösungen, *Z. anorg.u. allgem. Chem.*, 246 (1941) 303.

T. P. Dasgupta and G. M. Harris, Kinetics and mechanism of aquation of carbonato complexes of cobalt(III). The acid-catalyzed aquation of carbonato-tetraaminecobalt(III) ion, *J. Am. Chem. Soc.*, 91 (1969) 3207–3211.

I. I. Chernyaev, *Ann. Inst. Platine USSR*, 4 (1926) 246–261.

A. D. Hel'man, E. F. Karandashova and L. N. Essen, *Dokl. Akad. Nauk*, 63, (1948) 37.

A. A. Grinberg, *Acta Physiochim.*, USSR, 3 (1935) 573.

J. Chatt, L. A. Duncanson and L. M. Venanzi, Directing effects in inorganic substitution reactions. Part I. A hypothesis to explain the trans-effect, *J. Chem. Soc.*, 4456 (1955).

L. E. Orgel, An electronic interpretation of the trans effect in platinous complexes, *J. Inorg. Nucl. Chem*, 2, (1956) 137–140.

R. C. Maurya, R. Shukla, N. Anandam, D.C. Gupta and M. R. Maurya, Cyanonitrosyl complexes of chromium(I) with aniline and substituted anilines, *Synth. React. Inorg. Met.-Org. Chem.*, 16 (1986) 1059.

L. S. Hollis and S. J. Lippard, Redox properties of cis-diammineplatinum .alpha.-pyridone blue and related complexes. Synthesis, structure, and electrochemical behavior of cis-diammineplatinum(III) dimers with bridging .alpha.-pyridonate ligands, *Inorg. Chem.*, 22 (1983) 2605.

L. S. Hollis, M. M. Roberts and S. J. Lippard, Synthesis and structures of platinum(III) complexes of .alpha.-pyridone, [X(NH$_3$)$_2$Pt(C$_5$H$_4$NO)$_2$Pt(NH$_3$)$_2$X](NO$_3$)$_2$.nH$_2$O, (X$^-$ = Cl$^-$, NO$_2^-$, Br$^-$), *Inorg. Chem.*, 22 (1983) 3637.

G. Wilkinson, R. D. Gillard and J. A. McCleverty (Eds.), *Comprehensive Coordination Chemistry*, Vol.1, Pergamon Press, Oxford, (1987).

M. J. Wovkulich and J. D. Atwood, A trans effect on the rate of ligand dissociation from octahedral organometallic complexes. Dissociation of L' from $Cr(CO)_4LL'$ (LL' = PBu_3, PPh_3, $P(OPh)_3$, $P(OMe)_3$, $AsPh_3$), *Organometallics*, 1, 1316 (1982).

M. S. Lupin and B. L. Shaw, Electronic effects in some ruthenium(II)–carbonyl and –tertiary phosphine or –tertiary arsine complexes, *J. Chem. Soc. A*, 741 (1968).

P. G. Douglas and B. L. Shaw, The relative affinities of tertiary phosphines, tertiary arsines or phosphites (L) for ruthenium in complexes of type $[RuHCl(CO)(PR_2Ph)_2L](R = alkyl)$, *J. Chem. Soc. A*, 1556 (1970).

Index

https://doi.org/10.1515/9783110727289-017

www.ingramcontent.com/pod-product-compliance
Lightning Source LLC
Chambersburg PA